Principles of Electricity and Electronics for the Automotive Technician

Second Edition

Norm Chapman

Certified Master Automotive Technician
Professor, Automotive Technology
South Puget Sound Community College
Olympia, Washington

Australia • Brazil • Japan • Korea • Mexico • Singapore • Spain • United Kingdom • United States

DELMAR
CENGAGE Learning™

Principles of Electricity and Electronics for the Automotive Technician, 2e
Norm Chapman

Vice President, Career and Professional Editorial: Dave Garza

Director of Learning Solutions: Sandy Clark

Executive Editor: Dave Boelio

Managing Editor: Larry Main

Product Manager: Lauren Stone

Editorial Assistant: Lauren Stone

Vice President, Career and Professional Marketing: Jennifer McAvey

Marketing Director: Deborah S. Yarnell

Marketing Manager: Jimmy Stephens

Marketing Coordinator: Mark Pierro

Production Director: Wendy Troeger

Production Manager: Stacy Masucci

Content Project Manager: Barbara LeFleur

Art Director: Benj Gleeksman

Technology Project Manager: Christopher Catalina

Production Technology Analyst: Thomas Stover

Cover Image: Courtesy of Tom Witt

For product information and technology assistance, contact us at
Professional & Career Group Customer Support, 1-800-648-7450

For permission to use material from this text or product,
submit all requests online at **cengage.com/permissions.**
Further permissions questions can be e-mailed to
permissionrequest@cengage.com.

Library of Congress Control Number: 2008905204

ISBN-13: 9781428361218

ISBN-10: 1-4283-6121-9

Delmar
5 Maxwell Drive
Clifton Park, NY 12065-2919
USA

Cengage Learning products are represented in Canada by Nelson Education, Ltd.

For your lifelong learning solutions, visit **delmar.cengage.com**

Visit our corporate website at **cengage.com.**

Notice to the Reader

Publisher does not warrant or guarantee any of the products described herein or perform any independent analysis in connection with any of the product information contained herein. Publisher does not assume, and expressly disclaims, any obligation to obtain and include information other than that provided to it by the manufacturer. The reader is expressly warned to consider and adopt all safety precautions that might be indicated by the activities described herein and to avoid all potential hazards. By following the instructions contained herein, the reader willingly assumes all risks in connection with such instructions. The publisher makes no representations or warranties of any kind, including but not limited to, the warranties of fitness for particular purpose or merchantability, nor are any such representations implied with respect to the material set forth herein, and the publisher takes no responsibility with respect to such material. The publisher shall not be liable for any special, consequential, or exemplary damages resulting, in whole or part, from the readers' use of, or reliance upon, this material.

Printed in United States of America
2 3 4 5 X X 13 12 11

Principles of Electricity & Electronics for the Automotive Technician

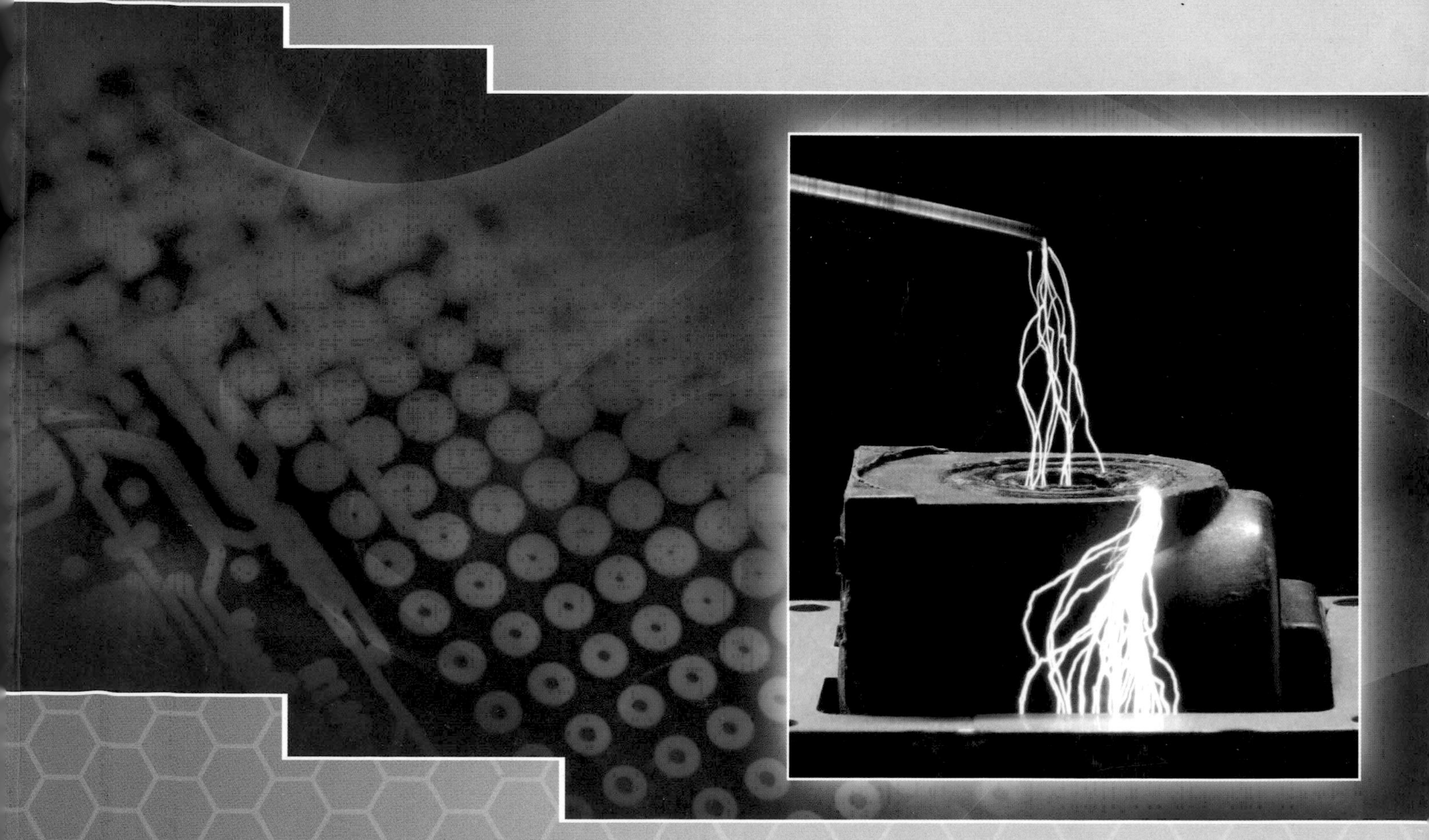

2nd Edition

Norm Chapman

Contents

Chapter 5 Components of an Electrical Circuit

Chapter 6 Meters and Measuring Devices

Chapter 7 Series Circuits

Chapter 8 Parallel Circuits

Chapter 9 Series/Parallel Circuits

Chapter 15 Starting System Theory

Chapter 16 Charging System Theory

Chapter 17 Ignition System Theory

Preface

The second edition of *Principles of Electricity and Electronics for the Automotive Technician* has been written with the student in mind. Learning the basic principles of electricity and electronics should be fun and create a feeling of excitement for the subject, not intimidation. When compared to other textbooks on the same topic, most students find that the relaxed style of writing draws them in. The need to learn the basics of electricity has never been more important than it is now.

The use of electronics in automobiles has grown drastically in the last 20 years. What used to be optional car equipment is now common in most cars, including rear window defoggers, climate control systems, intermittent wipers, air bags, digital instrument clusters, and memory seats. Somebody has to accurately diagnose and fix these vehicle options when they go bad, and it might as well be you!

The automotive technician of the future must be solidly proficient in the basics of electricity. This electrical foundation must be a sixth sense for the technician, who will interpret the information in a wiring diagram and apply electrical concepts to the vehicle in order to accurately diagnose and repair the problem the first time.

This textbook is a comprehensive exercise and resource guide to the basics. It uses a four-step approach for the acquisition of knowledge:

1. *Cognitive theory* To give students plenty of theory-based exercises and to demonstrate *why* the circuit or device acts or reacts as it does, exercises are designed so that repetition brings mastery of the theory.
2. *Case studies* By studying actual electrical case studies, the student can prove that the cognitive exercise is true. This expands the student's ability to see the entire operation of an electrical system before facing live troubleshooting on a vehicle.
3. *Hands-on vehicle tasks* Through the use of custom-designed tasks and the task list created by the National Automotive Technicians Educational Foundation (NATEF), the learning objectives are completed by means of a series of simple hands-on electrical tests common to most vehicles.
4. *ASE-style review questions* The most widely known national standard for the automotive technician is set by the National Institute of Automotive Service Excellence (ASE). This textbook uses primarily the ASE-style multiple-choice review questions, which are helpful to the student in studying for all the ASE tests that have basic electrical questions: Electrical, Engine Repair, Engine Performance, Air Conditioning, and Brakes.

We can all agree that times are changing in the automotive industry. Over the years, the repair industry lost some good technicians because of their inability or unwillingness to upgrade their skills to modern technology. The vehicles of today cannot be repaired with skills from the 1960s or 1970s. Letting the smoke out of a solid-state module because of your improper troubleshooting techniques can be expensive. A technician with a solid background in electrical basics is one of the most sought-after technicians in today's industry. Armed with wiring diagrams, various diagnostic charts, and the ability to use test equipment properly, a technician can efficiently solve the most complex electrical problems.

If this textbook is your first experience with the basics of electricity and electronics, I hope it is a positive and enjoyable one.

Norm Chapman

Acknowledgments

The contributions of the following reviewers of the first and second editions of the text are gratefully acknowledged.

Stephan Baldwin
Sullivan High School
Sullivan, Indiana

George Behrens
Monroe Community College
Rochester, New York

Walter Bertotti
Porter and Chester Institute
Wethersfield, Connecticut

Thomas Broxholm
Skyline College
San Bruno, California

William Caspole
Northwest Career Center
Dublin, Ohio

John Eichelberger
St. Phillip's College
San Antonio, Texas

Earl Friedell
DeKalb Technical Institute
Clarkson, Georgia

John Gahrs
Ferris State University
Big Rapids, Michigan

Drew Goddard
Hennepin Technical College
Brooklyn Park, Minnesota

Robert Klauer
Metro Tech
Phoenix, Arizona

James W. Manning
Santa Ana College
Santa Ana, California

Christopher McNally
Hudson Valley Community College
Troy, New York

Paul Mueller
Brooklyn Center High School
Brooklyn Center, Minnesota

Paul Pate
Community College of Southern Nevada
Boulder City, Nevada

Fred Raadsheer
British Columbia Institute of Technology
General Motors Corp.
Langley, British Columbia, Canada

Charles Statz
Temple Junior College
Temple, Texas

Dr. Robert Wenzlaff
Fullerton College
Fullerton, California

Tom Witt
South Puget Sound Community College
Olympia, Washington

Dedication

This second edition is dedicated to the many people who continue to have patience with my many projects: my dear wife Barbara, and four children Brandon, Brook, Bryan, and Brittney, and those six special grandchildren, Sterling, McKade, Tanner, Hyrum, Molly, and Jax. Without all their positive support, patience, and encouragement over the last several months, this undertaking would not have been possible.

I would also like to acknowledge the numerous staff people at Delmar Cengage Learning and supporting companies, and the many expert reviewers from around the country, whose professionalism, expertise, and insight helped bring this manuscript to print.

Basic Shop Safety

Introduction

Automotive shop safety is *everyone's* responsibility, and a shop is only as safe as the *least* safe person in it. All technicians, students, instructors, and supervisors must include safety as part of their daily routine. This begins with a positive attitude toward ensuring that all precautions are taken to provide a safe working environment. It includes the correct handling and use of hand tools and equipment, proper eye protection, personal attire, hazardous material safety and containment, location and type of fire extinguishers and first aid kits, and preventing electrical shock hazards. This chapter is intended to help you become aware of your working environment and develop the habits for a productive, safe automotive career.

Objectives

When you complete this chapter, you should be able to:

- ❑ Explain basic shop safety rules and protection.
- ❑ Identify hazardous waste products in the shop.
- ❑ Explain the use of material safety data sheets (MSDS).
- ❑ Define the purpose of OSHA and the EPA.
- ❑ Explain the different types of fires and fire extinguishing agents.
- ❑ Explain basic electrical safety.
- ❑ Explain basic hybrid electric vehicle safety procedures.

Personal Safety

Personal Attire

Personal safety begins with the clothes you wear (Figure 1–1). Attire should reflect a professional image and be appropriate for the specific job environment. Loose fitting clothes with the shirt-tail or long sleeves dangling are accidents waiting to happen. Long hair should be tucked under a hat or tied back to prevent being caught in rotating machinery, a fan belt, or drive axles. Rings or other jewelry items should not be worn; they may catch on moving parts and pull fingers or hands into the moving part. Also, most jewelry is made of electrically conductive metal such as gold or silver that heats up instantly when in contact with an electrical current.

FIGURE 1–1 Working clothes should display a professional image.

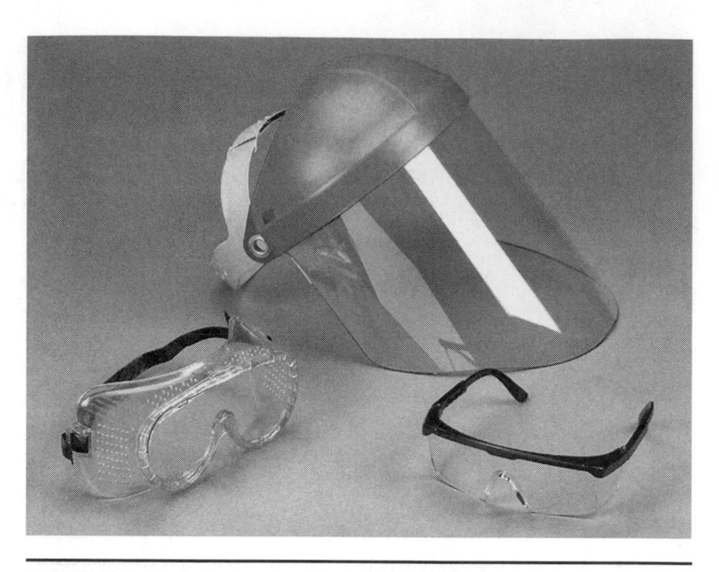

FIGURE 1–2 Typical styles of safety glasses and full-face shield used by technicians in the industry.

(Owen, *Basic Auto Service and Systems* [Clifton Park, NY: Thomson Delmar Learning])

Shoes or boots should be intended for a professional working environment with slip-resistant soles, hardened leather or steel toe, and a grease/oil-resistant material. Sports shoes, street shoes, or sandals are not recommended.

Eye and Ear Protection

Eye Protection Eye protection should be worn at *all times* in the automotive shop environment. Thousands of unnecessary and permanent eye injuries are reported every year by careless workers who do not wear protection or wear it *on top* of their heads. Several different types of eye and face protection are available (Figure 1–2). Good eye protection should meet the following basic requirements:

- Comfortable to wear
- Unobstructed vision
- Safety plastic lens or tempered glass
- Side shields
- Lenses that do not pop out under impact

Regular prescription eye wear does not meet safety impact standards and has no side shield protection. Some eye protection is designed to fit over prescription eye wear. For increased protection from splash or grinding hazards, a full face shield should be worn.

Ear Protection Extended exposure to high noise levels, as measured in **decibels,** in the shop may lead to permanent hearing damage. Loud machinery, air tools, and vehicles running in a closed environment all produce harmful levels of noise. Harmful levels consistently over 90 decibels can cause permanent hearing damage (Figure 1–3). To protect the worker, certified earplugs or earmuffs should be worn (Figure 1–4).

Emergency Eyewash Station An emergency eyewash should be available in the immediate shop environment. When eyes are accidentally contaminated with foreign matter or dangerous liquids, they should be flushed with water (Figure 1–5) and medical attention sought. It is important to know where the eyewash station is located in your shop, that it is in good working order, and that the path is unobstructed.

Hazardous Materials

Dozens of chemicals and materials are considered hazardous to the worker. Some create injury or illness on contact. Other chemicals have long-term side effects that

Representative Sound Levels		
Sound Level (dB)	**Operation or Equipment**	
150	Jet engine test cell	**DANGER ZONE**
145	_____________ Threshold of Pain	
130	Pneumatic press (close range) Pneumatic rock drill Riveting steel tank	
125	Pneumatic chipper Pneumatic riveter	
120	_____________ Threshold of Discomfort Turbine generator	
112	Punch press Sandblasting	
110	Drills, shovels, operating trucks Drop hammer	
105	Circular saw Wire braiders, stranding machine Pin routers Riveting machines	
100	Can-manufacturing plant Portable grinders Ram turret lathes Automatic screw machine	**RISK ZONE**
90	Welding equipment Weaving mill Milling machine Pneumatic diesel compressor Engine lathes Portable sanders	
85	California freeway traffic (overpass)	
Hearing Damage if Continued Exposure Above This Level		
80	Tabulating machines, electric typewriters	
75	Stenographic room	
70	Electronics assembly plant	
65	Department store	**SAFE ZONE**
60	Conversation	
35	Quiet home forced air heating	
10	Whisper	

FIGURE 1–3 Typical decibel levels and zones of common noises.

(a) (b)

FIGURE 1–4 (a) Typical earmuffs and (b) earplugs.
(Courtesy of Dalloz Safety)

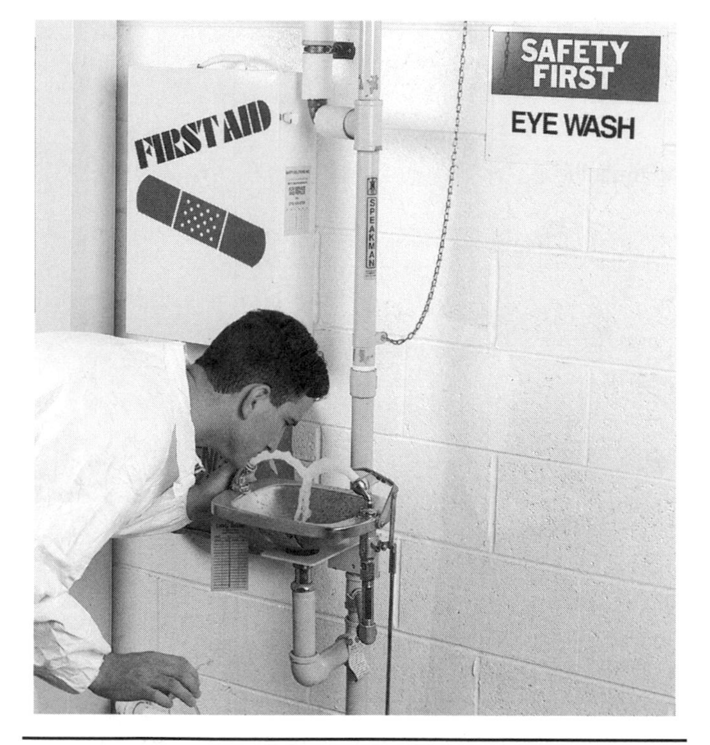

FIGURE 1–5 Typical emergency eyewash fountain.
(Courtesy of Western Emergency Equipment)

create a false sense of safety for the worker. Some of the common hazardous shop chemicals are:

- Antifreeze/coolant (ethylene glycol).
- Air conditioning refrigerants (freon/R-12 and R134A).
- Cleaning solvents and chemicals.
- Battery electrolyte acid.
- Gasoline, diesel, and other fuels.
- Asbestos from brakes and clutches.
- Used motor oils.
- Heavy metals.

The regulation of the use of these chemicals and materials in the United States is by the **Occupational Safety and Health Administration (OSHA).** In 1970, the Federal government created OSHA to ensure consistent safety standards across the country that inform, teach, and train workers and enforce the conditions of a safe working environment.

Material Safety Data Sheet

Information about all chemicals used by the worker are listed on **material safety data sheets (MSDS)** (Figure 1–6). These sheets, which should be available to the worker, detail the chemical composition and health, or safety, hazard precaution information of each product.

Hazardous Waste Management

Hazardous materials normally produce hazardous waste. The disposal of hazardous waste products is regulated by the **Environmental Protection Agency (EPA).** A product is considered a hazardous waste by the EPA if it has been identified specifically or has one or more of the following characteristics:

- *Ignitability* A solid product that will spontaneously ignite, or a liquid with a flash point below 140°F.
- *Corrosivity* Metals or other materials that will dissolve or melt or that will burn the skin on contact.
- *Reactivity* Materials that generate toxic vapors, flammable gases, or any material that reacts violently with water.
- *EP toxicity* The leaching of heavy metals from materials in concentrations greater than 100 times drinking water standards.

```
HEXANE
========================================================
MSDS Safety Information
========================================================
Ingredients
========================================================
Name: HEXANE (N_HEXANE)
% Wt: >97
OSHA PEL: 500 PPM
ACGIH TLV: 50 PPM
EPA Rpt Qty: 1 LB
DOT Rpt Qty: 1 LB
========================================================
Health Hazards Data
========================================================
LD50 LC50 Mixture: LD50:(ORAL,RAT) 28.7 KG/MG
Route Of Entry Inds _ Inhalation: YES
Skin: YES
Ingestion: YES
Carcinogenicity Inds _ NTP: NO
IARC: NO
OSHA: NO
Effects of Exposure: ACUTE:INHALATION AND INGESTION ARE HARMFUL AND MAY BE FATAL.
INHALATION AND INGESTION MAY CAUSE HEADACHE, NAUSEA, VOMITING, DIZZINESS, IRRITATION
OF RESPIRATORY TRACT, GASTROINTESTINAL IRRITATION AND UNCONSCIOUSNESS. CONTACT
W/SKIN AND EYES MAY CAUSE IRRITATION. PROLONGED SKIN MAY RESULT IN DERMATITIS (EFTS
OF OVEREXP)
Signs And Symptions Of Overexposure: HLTH HAZ:CHRONIC:MAY INCLUDE CENTRAL
NERVOUS SYSTEM DEPRESSION.
Medical Cond Aggravated By Exposure: NONE IDENTIFIED.
First Aid: CALL A PHYSICIAN. INGEST:DO NOT INDUCE VOMITING. INHAL:REMOVE TO FRESH AIR. IF
NOT BREATHING, GIVE ARTIFICIAL RESPIRATION. IF BREATHING IS DIFFICULT, GIVE OXYGEN.
EYES:IMMED FLUSH W/PLENTY OF WATER FOR AT LEAST 15 MINS. SKIN:IMMED FLUSH W/PLENTY
OF WATER FOR AT LEAST 15 MINS WHILE REMOVING CONTAMD CLTHG & SHOES. WASH CLOTHING
BEFORE REUSE.
========================================================
Handling and Disposal
========================================================
Spill Release Procedures: WEAR NIOSH/MSHA SCBA & FULL PROT CLTHG. SHUT OFF
IGNIT SOURCES:NO FLAMES, SMKNG/FLAMES IN AREA. STOP LEAK IF YOU CAN DO SO W/OUT
HARM. USE WATER SPRAY TO REDUCE VAPS. TAKE UP W/SAND OR OTHER NON_COMBUST MATL &
PLACE INTO CNTNR FOR LATER (SU PDAT)
Neutralizing Agent: NONE SPECIFIED BY MANUFACTURER.
Waste Disposal Methods: DISPOSE IN ACCORDANCE WITH ALL APPLICABLE FEDERAL, STATE AND
LOCAL ENVIRONMENTAL REGULATIONS. EPA HAZARDOUS WASTE NUMBER:D001 (IGNITABLE
WASTE).
Handling And Storage Precautions: BOND AND GROUND CONTAINERS WHEN TRANSFERRING LIQUID.
KEEP CONTAINER TIGHTLY CLOSED.
Other Precautions: USE GENERAL OR LOCAL EXHAUST VENTILATION TO MEET
TLVREQUIREMENTS. STORAGE COLOR CODE RED (FLAMMABLE).
========================================================
Fire and Explosion Hazard Information
========================================================
Flash Point Method: CC
Flash Point Text: _9F,_23C
Lower Limits: 1.2%
Upper Limits: 77.7%
Extinguishing Media: USE ALCOHOL FOAM, DRY CHEMICAL OR CARBON DIOXIDE. (WATER MAY BE
INEFFECTIVE).
Fire Fighting Procedures: USE NIOSH/MSHA APPROVED SCBA & FULL PROTECTIVE
  EQUIPMENT (FP N).
Unusual Fire/Explosion Hazard: VAP MAY FORM ALONG SURFS TO DIST IGNIT SOURCES & FLASH
BACK. CONT W/STRONG OXIDIZERS MAY CAUSE FIRE. TOX GASES PRDCED MAY INCL:CARBON
MONOXIDE, CARBON DIOXIDE.
========================================================
```

FIGURE 1–6 A typical Material Safety Data Sheet (MSDS).

All businesses that generate hazardous waste must be registered with the EPA and have a hazardous waste policy in force. When handling hazardous waste products, workers must use the proper equipment (Figure 1–7) to contain the waste product.

The storage of hazardous waste products should be in approved containers (Figure 1–8) and cabinets (Figure 1–9).

Fire Extinguishers and Agents

Fire extinguishers (Figure 1–10) are important pieces of shop equipment. Knowledge of the type of fire extinguisher and its location in the shop is extremely important when in a rush to extinguish a fire.

FIGURE 1–7 Typical modern asbestos cleaning method using a low-pressure washer.

FIGURE 1–9 Typical self-closing flammables storage cabinet.

(Courtesy of Justrite Manufacturing Co.)

FIGURE 1–10 Typical portable fire extinguishers.

FIGURE 1–8 Typical combustibles safety containers.

TABLE 1–1 Guide to proper fire extinguisher selection for different classes of fires.

	Class of Fire	Typical Fuel Involved	Type of Extinguisher
Class A Fires (green)	**For Ordinary Combustibles** Put out a class A fire by lowering its temperature or by coating the burning combustibles.	Wood Paper Cloth Rubber Plastics Rubbish Upholstery	Water*[1] Foam* Multipurpose dry chemical[4]
Class B Fires (red)	**For Flammable Liquids** Put out a class B fire by smothering it. Use an extinguisher that gives a blanketing, flame-interrupting effect; cover whole flaming liquid surface.	Gasoline Oil Grease Paint Lighter fluid	Foam* Carbon dixoide[5] Halogenated agent[6] Standard dry chemical[2] Purple K dry chemical[3] Multipurpose dry chemical[4]
Class C Fires (blue)	**For Electrical Equipment** Put out a class C fire by shutting off power as quickly as possible and by always using a nonconducting extinguishing agent to prevent electric shock.	Motors Appliances Wiring Fuse boxes Switchboards	Carbon dioxide[5] Halogenated agent[6] Standard dry chemical[2] Purple K dry chemical[3] Multipurpose dry chemical[4]
Class D Fires (yellow)	**For Combustible Metals** Put out a class D fire of metal chips, turnings, or shavings by smothering or coating with a specially designed extinguishing agent.	Aluminum Magnesium Potassium Sodium Titanium Zirconium	Dry powder extinguishers and agents only

*Cartridge-operated water, foam, and soda-acid types of extinguishers are no longer manufactured. These extinguishers should be removed from service when they become due for their next hydrostatic pressure test.

Notes:

(1) Freezes in low temperatures unless treated with antifreeze solution, usually weighs over 20 pounds, and is heavier than any other extinguisher mentioned.
(2) Also called ordinary or regular dry chemical (sodium bicarbonate).
(3) Has the greatest initial fire-stopping power of the extinguishers mentioned for class B fires. Be sure to clean residue immediately after using the extinguisher so sprayed surfaces will not be damaged (potassium bicarbonate).
(4) The only extinguishers that fight A, B, and C classes of fires. However, they should not be used on fires in liquefied fat or oil of appreciable depth. Be sure to clean residue immediately after using the extinguisher so sprayed surfaces will not be damaged (ammonium phosphates).
(5) Use with caution in unventilated, confined spaces.
(6) May cause injury to the operator if the extinguishing agent (a gas) or the gases produced when the agent is applied to a fire is inhaled.

It is important to select the right fire extinguisher for each class of fire (Table 1–1).

The proper method of extinguishing a fire is partially determined by the type of fire extinguishing agent used (Figure 1–11). Always refer to label instructions printed on the fire extinguisher.

Basic Electrical Safety

You should follow five basic safety precautions when working around shop or vehicle electrical systems:

1. Wear proper attire. Electrical shock can be harmful and in some cases fatal. Insulating your body

Extinguisher	Method
Foam Solution of aluminum sulphate and bicarbonate of soda	Don't play stream into the burning liquid. Allow foam to fall lightly on fire.
Carbon Dioxide Carbon dioxide gas under pressure	Direct discharge as close to fire as possible. First at edge of flames and gradually forward and upward.
Dry Chemical	Direct stream at base of flames. Use rapid left-to-right motion toward flames.
Soda-Acid Bicarbonate of soda solution and sulphuric acid	Direct stream at base of flame.

FIGURE 1–11 Extinguishing methods for each type of fire extinguisher.

through rubber-soled shoes or using approved electrically insulating mats or gloves will help prevent shock hazards (Figure 1–12). Voltage levels as low as 120 volts AC or around 50 volts DC can be dangerous or even fatal when the current level is between 100 and 200 milliamperes (mA) (Figure 1–13).

2. Never attempt to perform repairs on an electrical circuit with which you are unfamiliar.

FIGURE 1–12 Properly rated high-voltage insulating gloves and a pair of leather overprotectors if sharp edges are present.

Current	Effect
0.100–0.200 Amperes	Death: This range generally causes fibrillation of the heart. When the heart is in this condition, it vibrates at a fast rate like a "quiver" and ceases to pump blood to the rest of the body.
0.060–0.100 Amperes	Extreme Difficulty in Breathing
0.040–0.060 Amperes	Breathing Difficulty
0.030–0.040 Amperes	Muscular Paralysis
0.020–0.030 Amperes	Unable to Let Go of the Circuit
0.010–0.020 Amperes	Very Painful
0.009–0.010 Amperes	Moderate Sensation
0.002–0.003 Amperes	Slight Tingling Sensation

FIGURE 1–13 The effect of various electric current thresholds on the human body.

3. Never wear metal objects such as watches, rings, or chains that will conduct electricity. If a metal ring comes into contact with a live electrical current, it can instantaneously glow red hot, causing severe injury to the skin.
4. Treat every electrical circuit like a *live* circuit. Even though the safest electrical circuit to work on is an *inactive* or *dead* circuit, treating the circuit as if it is *active* or *live* makes you more attentive and cautious.
5. Use only a *Type C* fire extinguisher on electrical fires. Table 1–1 shows the proper precautions to fight a class C fire.

Hybrid Electric Vehicle Safety

Hybrid electric vehicles (HEV) are one of the biggest technological changes to arrive in the twenty-first century. Several automotive manufacturers have produced HEVs in the last few years, and these hybrid vehicles have been efficiently rolling along the highways and city streets with few problems. A typical hybrid vehicle has a gasoline engine and an electric motor powered by a high-voltage (HV) battery pack (Figure 1–14). Although most electrical systems in conventional vehicles operate on voltages from 12 to 14 volts, the typical HEV may require operating voltages ranging from 12 to over 600 volts. Voltages over 50 volts are considered *high voltage* and potentially lethal. The high-voltage component connectors and wiring in most HEVs are color coded *orange* for easy identification as special high-voltage circuits. *Always follow the specific safety procedures from the vehicle manufacturer when performing service or repairs on an HEV.* Here are some of the typical safety procedures:

- Identify specific vehicle safety precautions and the approved high-voltage shutdown procedures.
- Remove the key from the ignition. Some newer HEVs are being equipped with a so-called smart key, which can be as far away as 15 feet from the vehicle and still energize the electrical systems.
- Disconnect the vehicle's auxiliary 12-volt battery to ensure that the control circuit for the HV battery pack is disabled (Figure 1–15).
- Wear gloves rated for high-voltage electrical service (Figure 1–12)

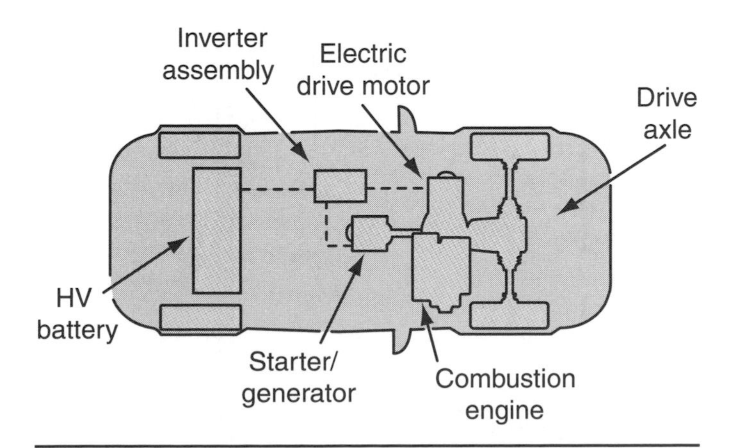

FIGURE 1–14 Example of one type of HEV power combination.

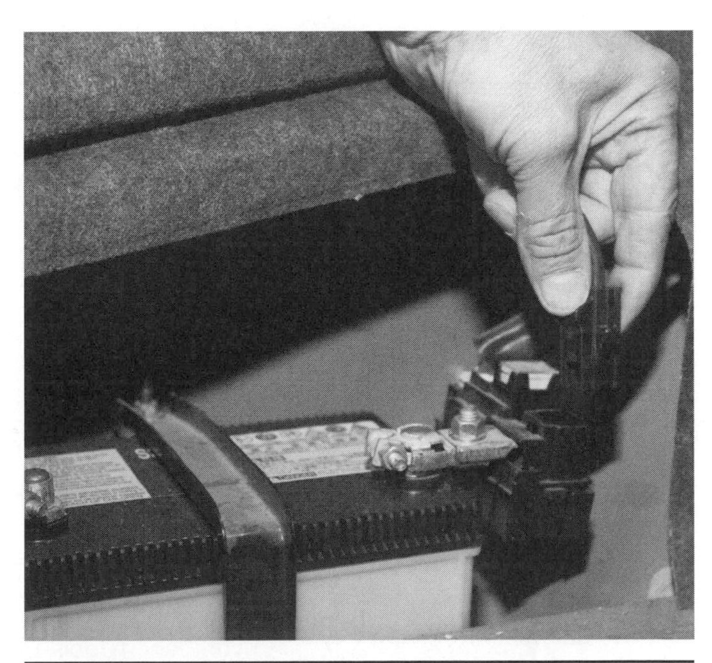

FIGURE 1–15 Disconnecting vehicle auxiliary 12-volt battery.

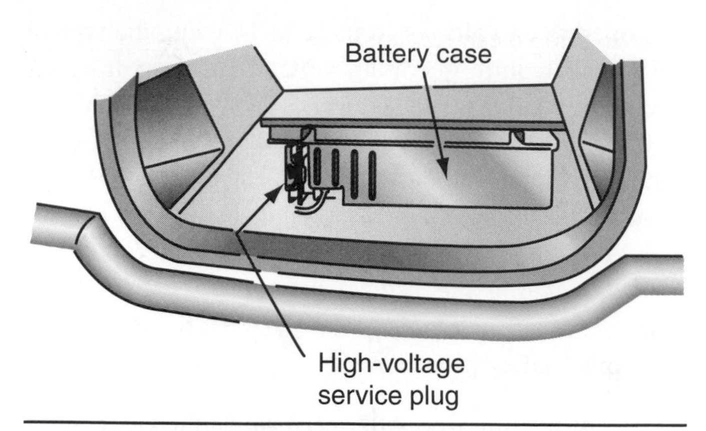

FIGURE 1–16 Typical high-voltage service plug location.

- Disable the vehicles high-voltage service plug, which disconnects the HV battery pack from the system (Figure 1–16).
- Do not attempt to start the vehicle with its high-voltage service plug removed. Damage could occur to vehicle control system components.
- Let the vehicle sit for 5–10 minutes after removing the service plug. This delay allows the system's high-voltage circuits to de-energize internally.
- Before touching any high-voltage components, perform the necessary voltage measurements using a voltage meter suitable for reading high voltages.
- Never probe the orange high-voltage harness through the insulation. A tiny hole in the insulation could prove fatal to someone touching it accidentally.
- Remember that your *safety* is your primary concern. Working on HEVs is safe as long as you follow the appropriate manufacturer's procedures.

Shop Safety Exercise

Each repair shop has a layout and location of necessary safety equipment. Figure 1–17 is the layout of a typical automotive shop and important safety equipment.

❑ On a separate sheet of paper, draw the layout of your automotive shop.

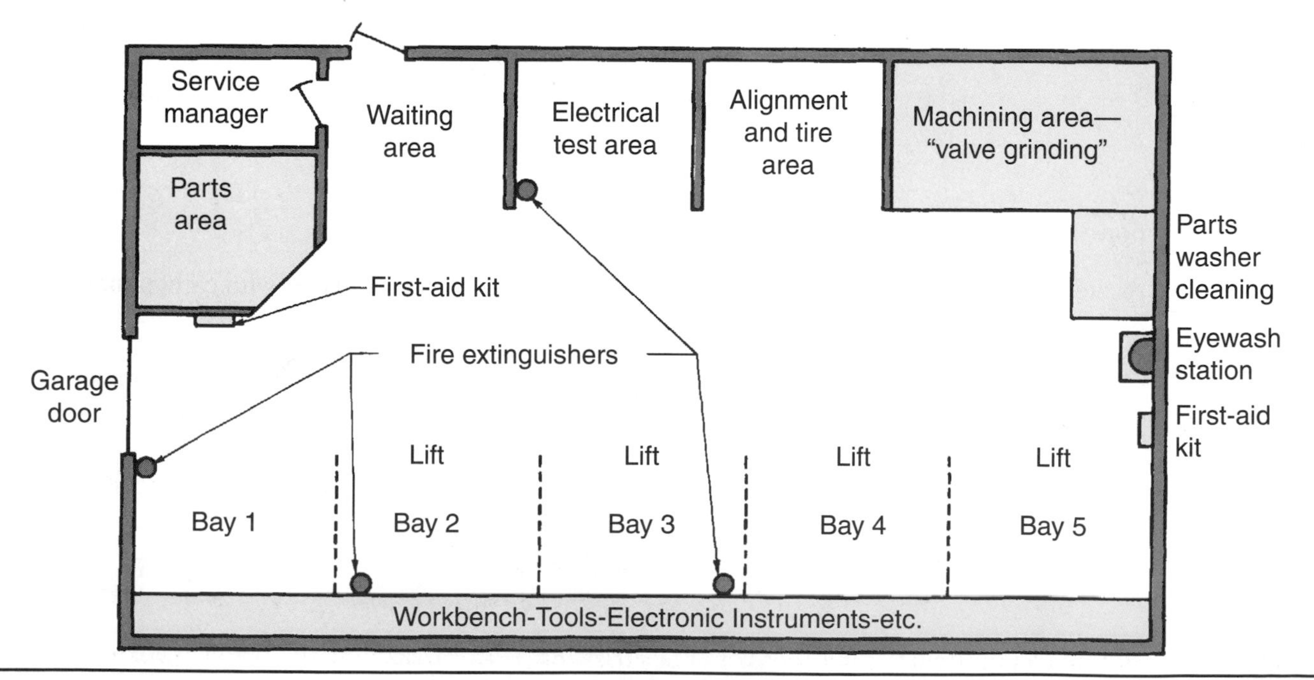

FIGURE 1–17 Typical layout of safety equipment in an automotive shop.

❑ List the location of the following shop safety items.

Equipment	Location
1. First aid kit	______
2. Emergency eyewash station	______
3. Fire extinguishers	______
4. Exit doors	______
5. Hazardous waste containment	______
6. Flammable storage locker	______
7. Electrical circuit panel	______
8. Emergency phone numbers	______
9. Shop cleaning supplies	______
10. MSDS sheets	______
11. Hazardous spill containment supplies	______

❑ List the proper procedure for operating the fire extinguisher(s) in your shop.

______

______

❑ Exercise Summary. After performing the preceding shop safety exercise, state what you found most useful.

Summary

❑ Wear professional clothing that is appropriate for the specific job environment.

❑ Wear eye protection at all times when in the automotive shop environment.

❑ An emergency eyewash station should be available in the immediate shop environment.

❑ The regulation of hazardous chemicals is done through OSHA.

❑ An MSDS should be available for every chemical used in the automotive shop environment.

❑ The disposal of hazardous waste products is regulated by the EPA.

❑ A class A fire extinguisher is for ordinary combustibles, and it extinguishes with water, foam, or dry chemical powder.

❑ A class B fire extinguisher is for flammable liquids, and it extinguishes with foam, dry chemical powder, or carbon dioxide.

❑ A class C fire extinguisher is for electrical equipment, and it extinguishes with carbon dioxide or dry chemical powder.

❑ A class D fire extinguisher is for combustible metals, and it extinguishes with dry chemical powder only.

❑ Voltage levels as low as 120 volts or current flow as low as 0.15 amperes can interrupt the natural electrical rhythm of the heart.

❑ Voltage levels as low as 120 volts AC or around 50 volts DC and current flow as low as 0.15 ampere can interrupt the natural electrical rhythm of the heart and prove fatal.

❑ Hybrid electric vehicles may consist of electrical circuits ranging from 12 volts to over 600 volts.

Key Terms

decibel
Environmental Protection Agency (EPA)
hybrid electric vehicle (HEV)
material safety data sheets (MSDS)
Occupational Safety and Health Administration (OSHA)

Review Questions

Short Answer Essays

1. Discuss some of the safety features of a well designed shop.
2. Describe the four classes of fire and the proper extinguishing agent for each type.
3. Describe the basic safety precautions when working around electrical equipment.
4. Describe some of the typical safety procedures to follow when servicing a hybrid electric vehicle.

Fill in the Blanks

1. A technician with long hair should ________________ or ________________ when working in the automotive shop.
2. Eye protection should be worn ________________ when working in the automotive shop.
3. Sound levels over ________________ decibels are harmful and can cause permanent hearing damage.
4. The regulation of chemicals and hazardous materials is through ________________.
5. Information about all chemicals used by the worker in the automotive shop is listed on the ________________.
6. The high-voltage component connectors and wiring in most HEVs are color coded ________________ for easy identification as special high-voltage circuits.

ASE-Style Review Questions

1. Technician A says safety glasses should be worn at all times while in the shop. Technician B says regular prescription eye wear may be substituted for safety glasses. Who is correct?
   A. A only
   B. B only
   C. Both A and B
   D. Neither A nor B

2. Technician A says if a hazardous substance dissolves metals and other materials, or burns the skin, it is said to have reactivity. Technician B says when a solid hazardous waste product spontaneously ignites, it is considered to have ignitability. Who is correct?
   A. A only
   B. B only
   C. Both A and B
   D. Neither A nor B

3. Technician A ties his long hair behind his head while working in the shop. Technician B covers his long hair with a brimless cap. Who is correct?
   A. A only
   B. B only
   C. Both A and B
   D. Neither A nor B

4. Technician A uses carbon dioxide to extinguish a Class A fire. Technician B uses foam to extinguish a Class D fire. Who is correct?
   A. A only
   B. B only
   C. Both A and B
   D. Neither A nor B

5. Technician A says extended noise levels exceeding 90 decibels can damage the hearing. Technician B says all noises produced in the shop environment are harmful. Who is correct?
   A. A only
   B. B only
   C. Both A and B
   D. Neither A nor B

6. Technician A says good eye protection has side shields. Technician B says good eye protection has a plastic lens or tempered glass. Who is correct?
   A. A only
   B. B only
   C. Both A and B
   D. Neither A nor B

7. Technician A says used motor oil is considered a hazardous waste. Technician B says ethylene glycol coolant is hazardous only when in a used condition. Who is correct?
   A. A only
   B. B only

C. Both A and B

D. Neither A nor B

8. Technician A says OSHA is the federal regulating agency for the disposal of hazardous wastes. Technician B says the EPA is the federal regulating agency for the use of chemicals and hazardous materials. Who is correct?

   A. A only

   B. B only

   C. Both A and B

   D. Neither A nor B

9. Technician A says water is a suitable extinguishing agent on Class C electrical fires. Technician B says electrical shock hazards as low as 0.15 ampere can be fatal. Who is correct?

   A. A only

   B. B only

   C. Both A and B

   D. Neither A nor B

10. Which of the following is/are important when working in the automotive shop?

    A. Using the proper tool for the job.

    B. Avoiding loose fitting clothes.

    C. Wearing steel-tipped shoes.

    D. All of the above.

2 Introduction to Electricity and Electronics

Introduction

As we move toward the end of the first decade of the twenty-first century, the need for the automotive technician to understand the fundamentals of electricity and electronics has never been greater. Complex electrical and electronic systems have replaced many mechanical and hydraulic automobile systems. As a result, repair technicians who gain the requisite education, knowledge, and hands-on skills have found that there are more career opportunities for skilled and certified automotive technicians than ever before.

As consumers of electricity, we sometimes take for granted where it originates. Electricity can best be described as a natural form of energy. We cannot create energy, but we can successfully produce and harness electricity by converting various other forms of natural energy. In this chapter we look at natural forms of energy and the people who have played a part in understanding and harnessing them for our use.

Objectives

When you complete this chapter, you should be able to:

- ❑ Explain the natural forms of energy used to produce electricity.
- ❑ Describe the people who were instrumental in developing modern electrical theory.

Electrical Sources of Power

Six sources of power are discussed in this chapter.

Static Electricity

The first is **static electricity,** which is electricity that is without motion or at rest and that produces an electrical charge whenever two dissimilar nonmetallic materials are rubbed lightly together. Our knowledge of electricity began 2,500 years ago, when the Greeks discovered that, if they rubbed a piece of fossilized resin called amber with other materials, it attracted lint, fibers, feathers, and other lightweight objects. The Greeks looked at this invisible force as a supernatural phenomenon. They didn't realize they discovered static electricity. Nearly everyone has experienced static electricity at some usually unexpected time. Examples are as simple as seeing clothes stick together and crackle when they are removed from a hot dryer, running a comb through your hair and then picking up a piece of paper with the comb, or walking across a carpet on a dry day and then touching a metal object (Figure 2–1).

One way to visualize static electricity is to understand that it is either a surplus or a lack of electrons in

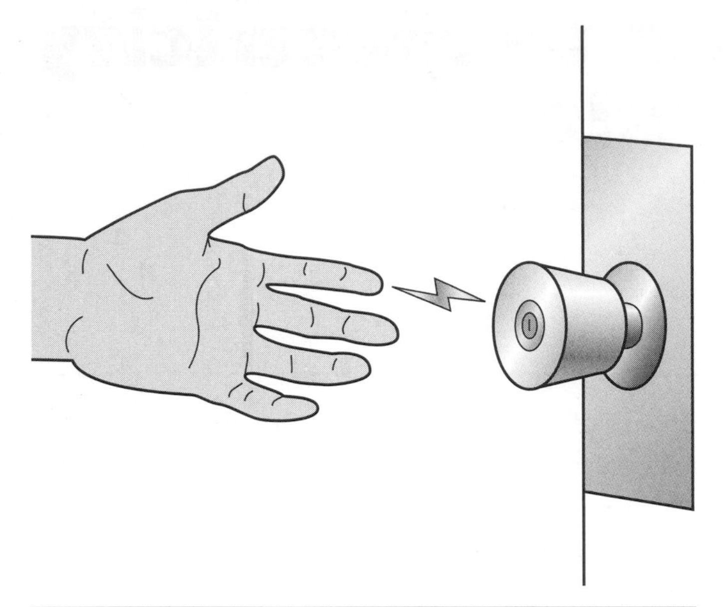

FIGURE 2–1 Example of static electricity.

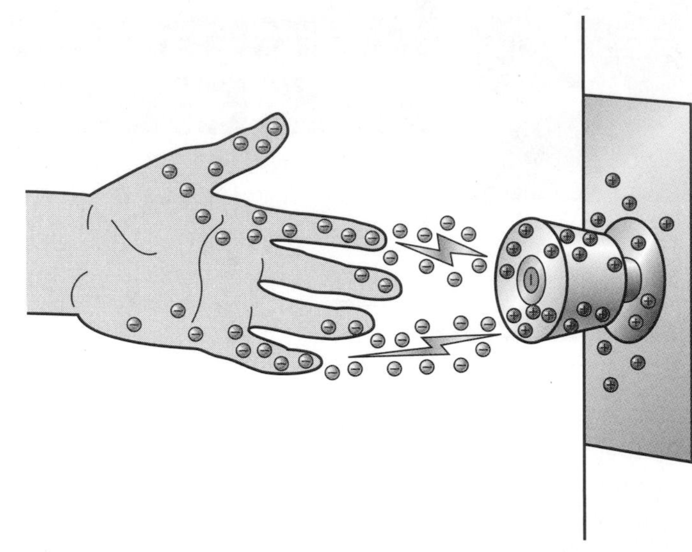

FIGURE 2–3 Example of electron transfer from a hand to a metal object.

FIGURE 2–2 Excess negative electrons are transferred from the atoms in the carpet to the shoes.

an object. Electrons jump from one object to another in order to return an atom to its neutral state of charge. In Figure 2–2, the electrons are transferred from the carpet to the rubber-soled shoe. The lower the humidity is in the air, the greater is the transfer of electrons. These electrons then jump from the surface of your skin to any object that allows the transfer of the electrons (Figure 2–3).

Automotive technicians should know the characteristics of static electricity because of **electrostatic discharge (ESD).** ESD is the process by which electrically charged objects transfer their charge to a nearby object. A common nuisance example of an ESD is a lightning storm. When Benjamin Franklin (inventor and scientist, 1706–1790) flew a kite during a thunderstorm (*not* a safe experiment to try yourself), he was trying to prove that the positive and negative distribution of electrons in the clouds produced the static electricity that causes lightning. A natural electron charge produces lightning when it finds a suitable path to return to its neutral state of charge (Figure 2–4).

For the automotive technician, sliding across the seat of a car to service the solid-state circuitry in the dash panel or to replace the vehicle computer can produce several thousand volts of static electricity, which can damage ESD-sensitive devices. Vehicle components that are ESD sensitive are marked with ESD warning labels. A typical schematic diagram with an ESD warning label is shown in Figure 2–5.

To protect sensitive electronic components from ESD damage, the automotive technician should *always* use a grounding strap (Figure 2–6). These static straps are available from most electronic supply stores.

Solar Light Energy

The second source of power is electricity generated from light energy, commonly known as **solar light energy,** which is light energy from the sun that is captured in a photovoltaic solar cell (Figure 2–7).

Light energy is composed of particles called **photons.** The photon is pure energy that contains no mass.

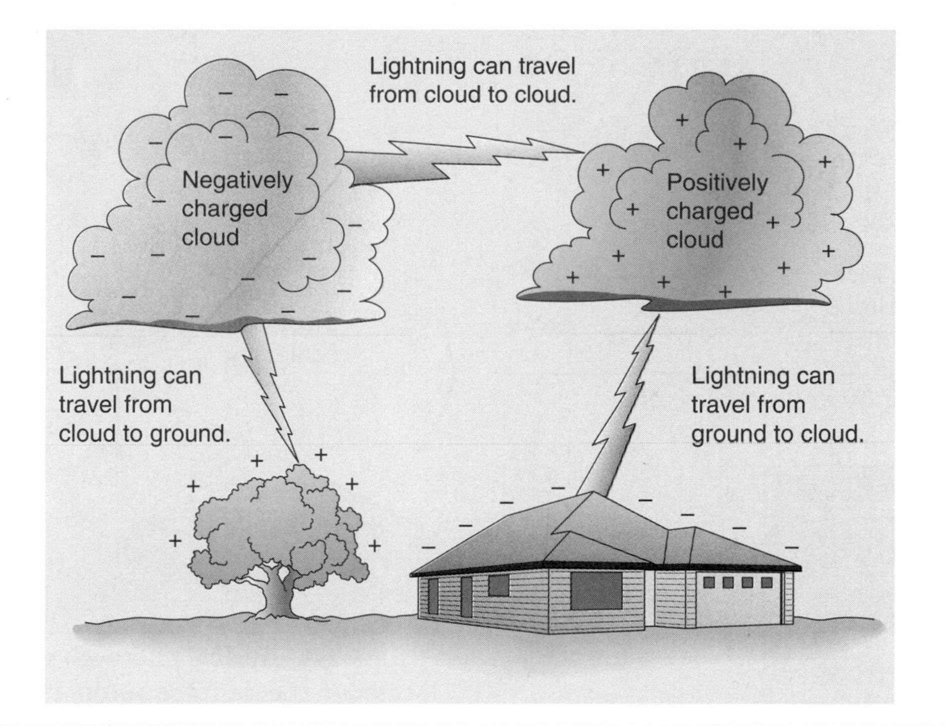

FIGURE 2–4 Electron-induced charges in the earth's atmosphere.

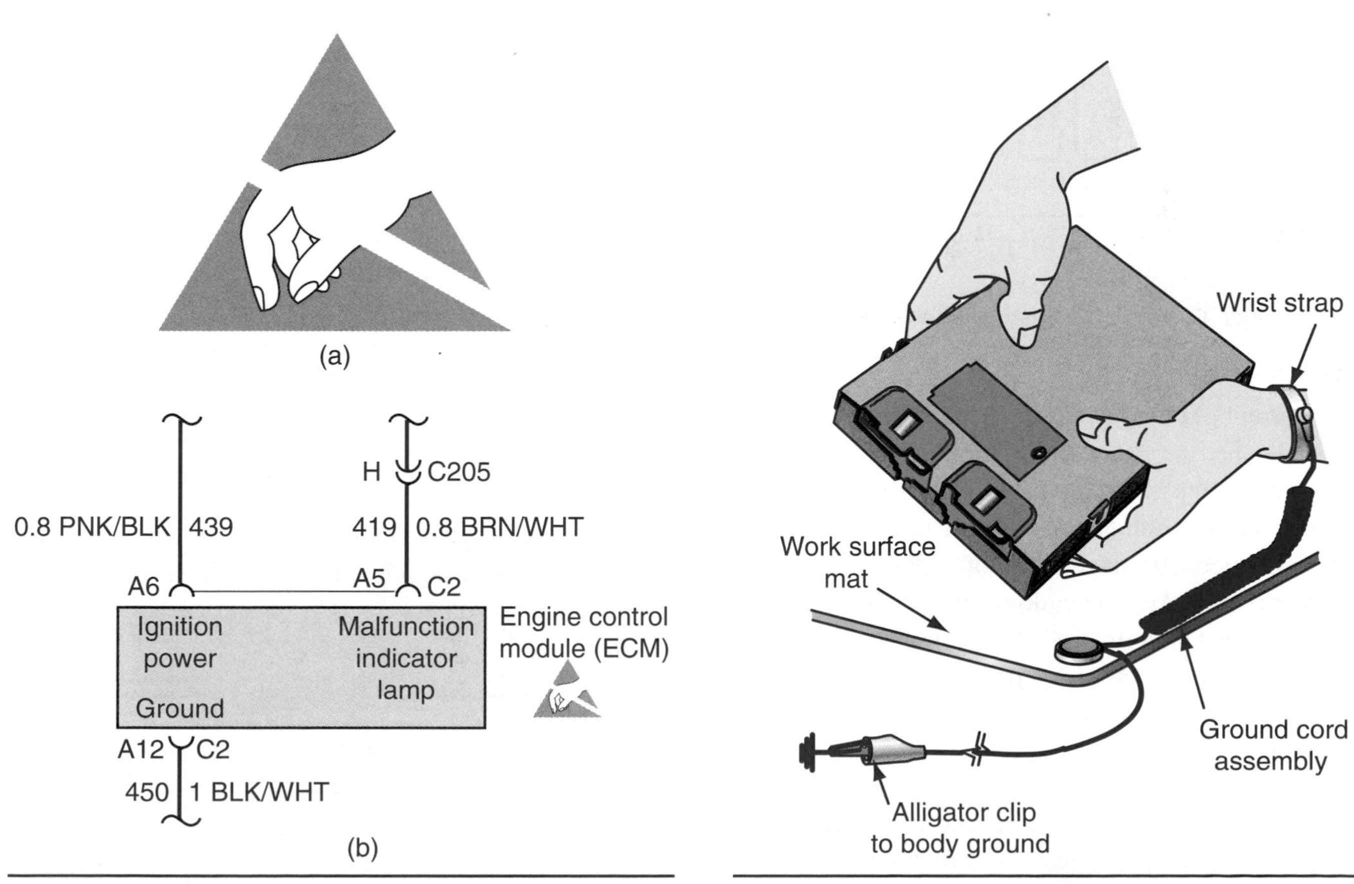

FIGURE 2–5 (a) Electrostatic discharge (ESD) symbol and (b) sample ESD-sensitive wiring circuit.

FIGURE 2–6 Using a grounding strap to protect sensitive electrical components from static discharge.

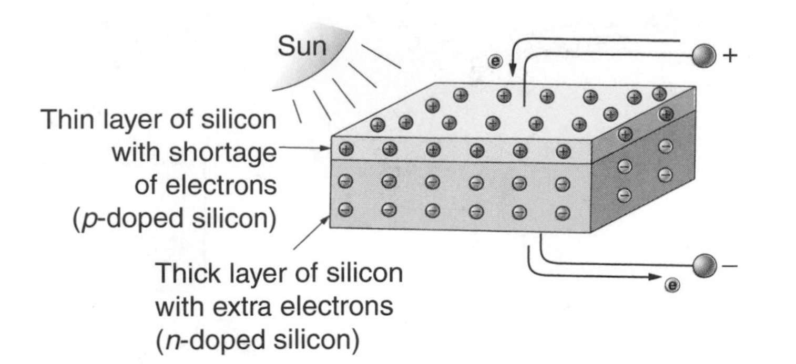

FIGURE 2–7 Simple photovoltaic solar cell.

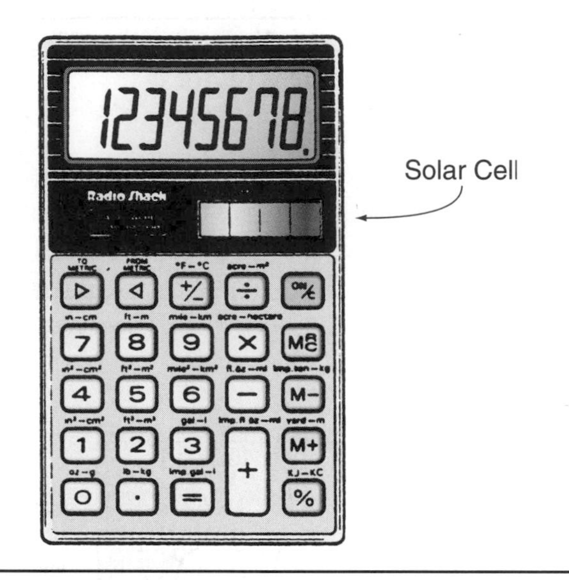

FIGURE 2–8 Pocket calculator powered by a photovoltaic solar cell.

(The EC-352 Metric Conversion Calculator is a trademark of the Radio Shack Division of Tandy Corporation)

When photon light energy strikes the surface of a photovoltaic cell, the energy in the photons is given over to free electrons. This additional energy causes electrons to cross over the semiconductor material in a photovoltaic cell and produce a voltage. One of the most common applications of solar technology is in the common pocket calculator (Figure 2–8). The amount of voltage a photovoltaic cell can produce depends on the type of semiconductor material used, but it is normally around 0.5 volt per cell in sunlight.

In automotive applications, photovoltaic cells are used to control the circuits of several electrical systems. One example is the headlight control circuit that has the automatic on/off feature. In this type of circuit, a photocell is located on top of the vehicle dash to sense the outside light (Figure 2–9). The photocell is like the eye of the complex circuitry that turns the headlight system on and off. When the ambient light level decreases, the resistance of the photocell increases, triggering the headlight control circuitry to command the headlights to go on. Two other popular examples are the automatic adjustment control for a rearview mirror and a photovoltaic crankshaft position sensor (Figure 2–10).

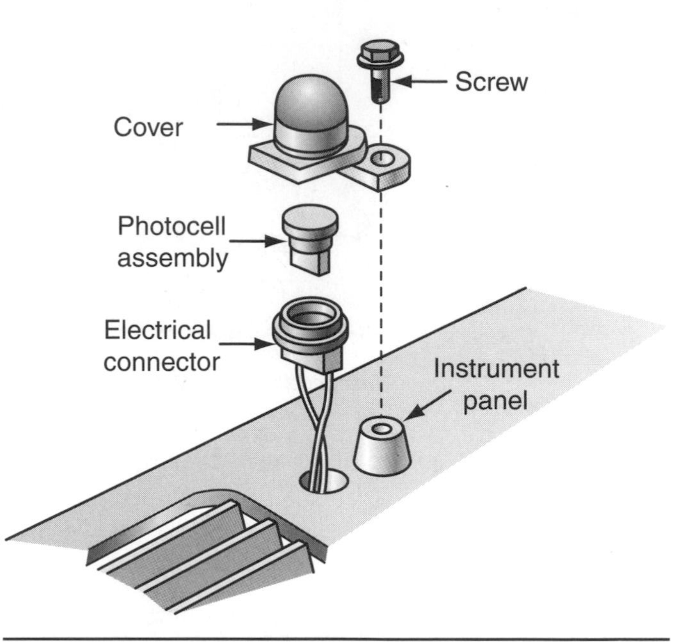

FIGURE 2–9 Typical photocell sensing assembly located in the dash for automatic on/off headlight systems, also known as twilight sentinel headlamps.

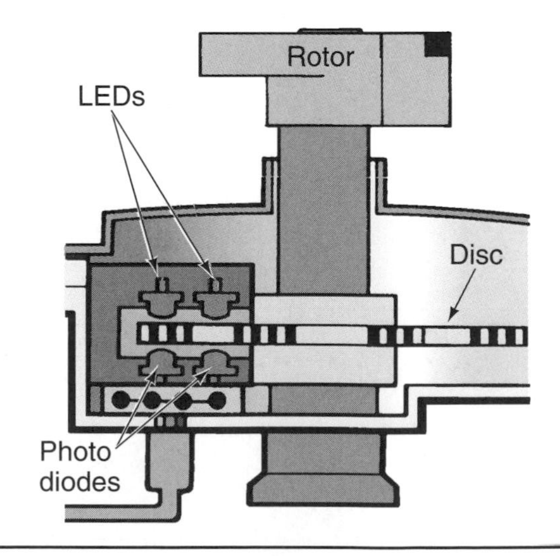

FIGURE 2–10 A photovoltaic crankshaft position sensor.

(Courtesy of DaimlerChrysler Corporation)

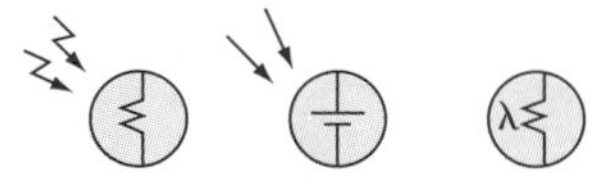

FIGURE 2–11 Common photovoltaic symbols used in wiring diagrams.

The schematic symbol for a photovoltaic cell is typically one of the three shown in Figure 2–11.

Thermoelectricity

The third common source of power is **thermoelectricity,** which is electricity produced when two dissimilar metals are heated to generate an electrical voltage. In 1822, a German scientist named Thomas Johann Seebeck found a direct connection between heating two dissimilar metals joined at one end and a voltage produced at the open ends. This is also known as the **Seebeck effect** (Figure 2–12).

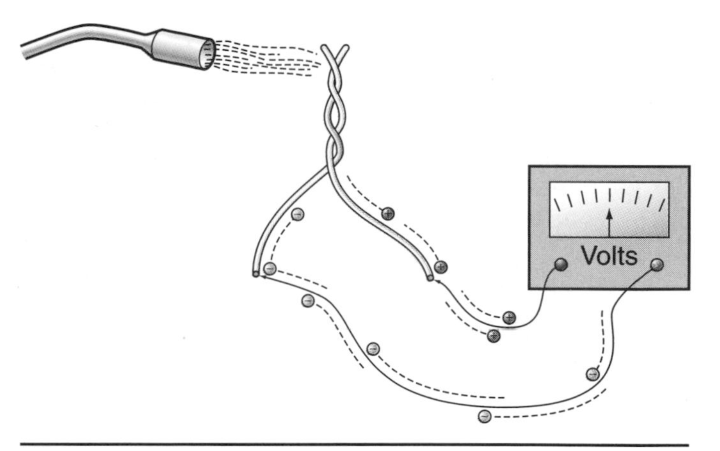

FIGURE 2–12 Thermoelectric production using two dissimilar metals and a heat source.

One application of this source of power is a **thermocouple,** which is a small device that develops a voltage when the two dissimilar metals inside it are heated. These devices are used extensively in the diesel engine field, where they monitor the exhaust temperature of each engine cylinder (Figure 2–13). Because exhaust

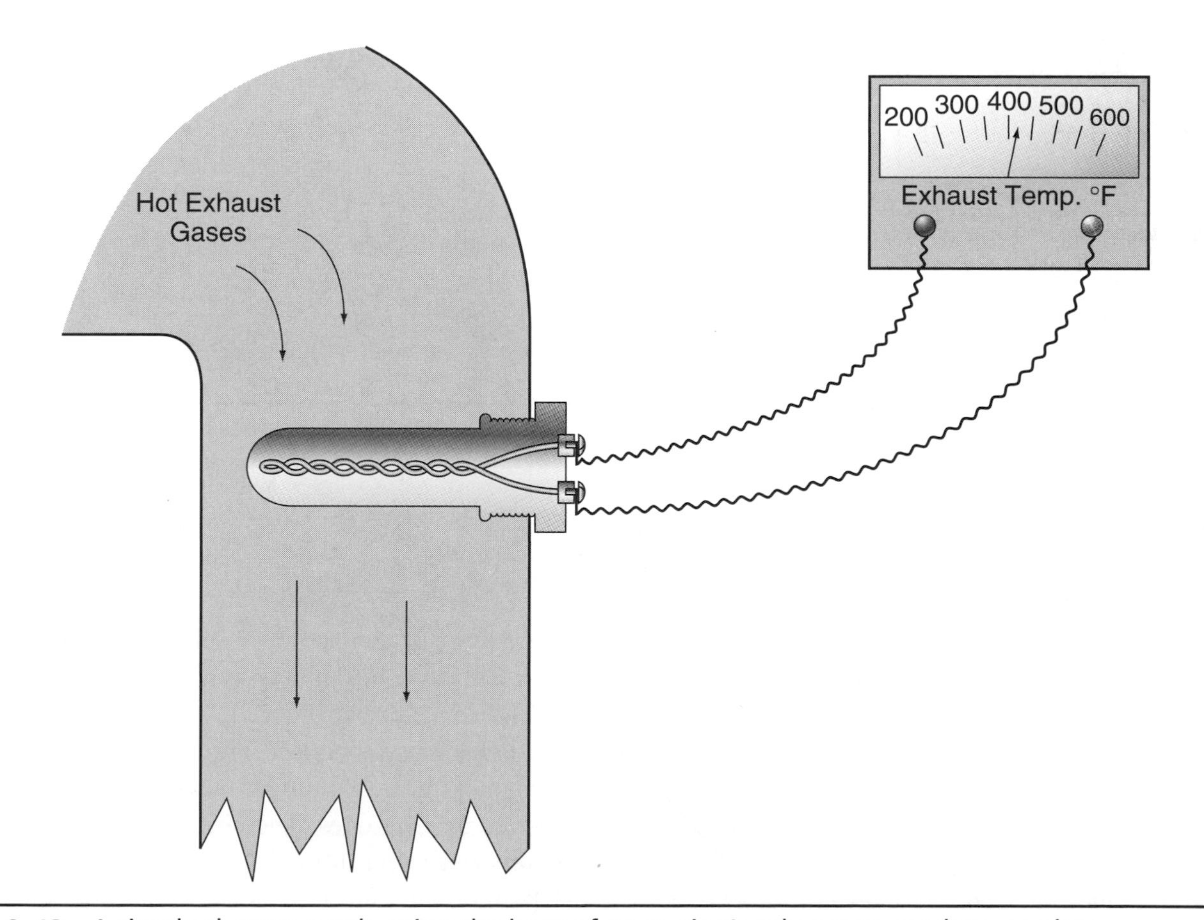

FIGURE 2–13 A simple thermocouple using the heat of an engine's exhaust to produce a voltage.

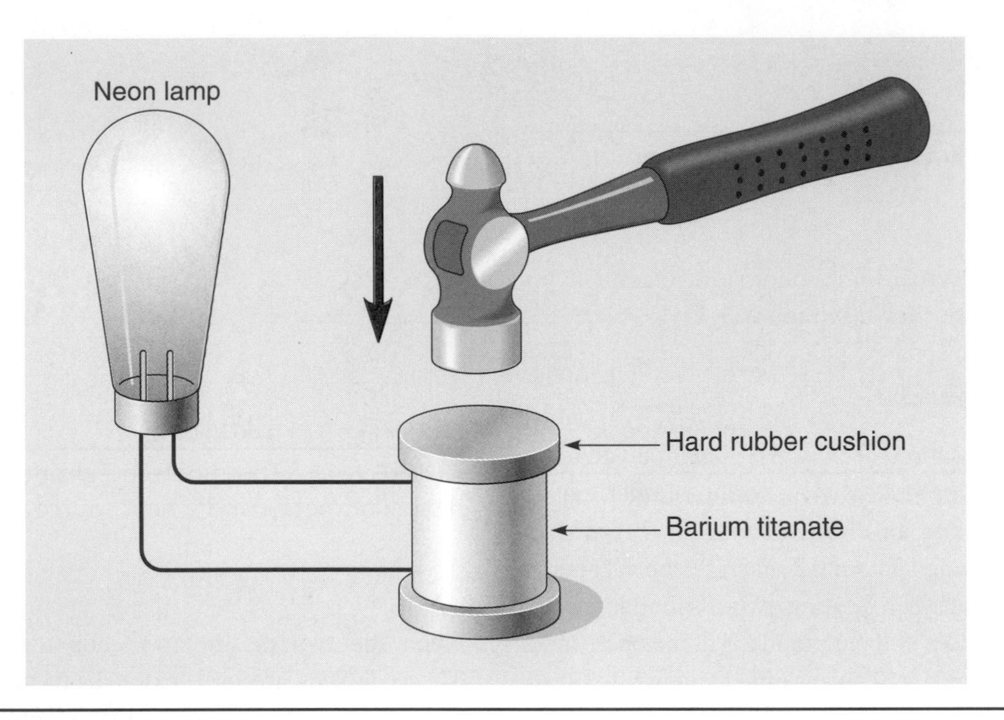

FIGURE 2–14 Producing a piezoelectric voltage.

temperatures usually exceed several hundred degrees, the use of a mercury glass thermometer is impractical.

Piezoelectricity

The fourth source of power is **piezoelectricity,** which is electricity produced from pressure. The word *piezoelectric* comes from the Greek word *piezo,* which means "pressure." When certain materials, such as quartz or barium titanate, undergo physical stress or vibration, a small oscillating voltage is produced (Figure 2–14).

The most common automotive application of a piezoelectric device is the knock sensor, or detonation sensor as it is sometimes called, which is used to control ignition spark knock (Figure 2–15).

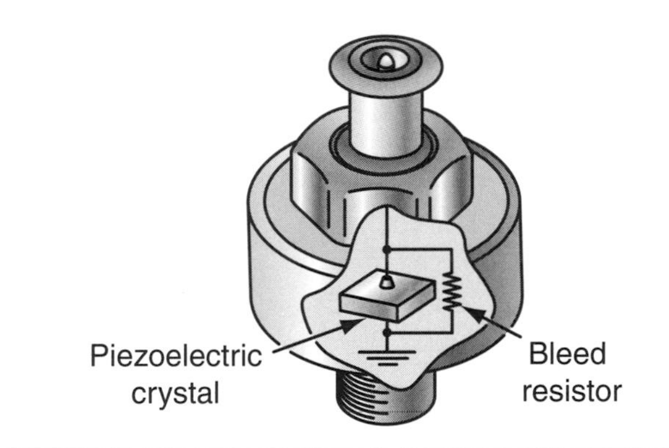

FIGURE 2–15 View of knock sensor with a piezoelectric crystal sensing device.

The typical knock sensor produces about 300 millivolts (mV) and vibrates at a 6,000-hertz (cycles-per-second) frequency, which is about the same frequency of vibration as an ignition spark knock or detonation. When the engine detonates, the fuel burn rate is too fast. One cause could be that the ignition timing advance is out of specifications. The knock sensor picks up this mini explosion and sends the voltage signal to the engine control computer (Figure 2–16). The knock sensor is usually found on the engine block to prevent an erroneous reading from the noise in the upper valve train or from the transmission.

Chemical Energy

The fifth source of power is electricity produced from **chemical energy.** The most common form of this energy is the lead acid battery used in most automotive applications. One of the simplest forms of a chemical energy is a lemon with two dissimilar metals like copper and zinc inserted as electrodes. The citric acid in the lemon acts as the electrolyte needed to produce a small voltage (Figure 2–17). Zinc accepts electrons better than copper, so zinc is considered the negative plate and copper is the positive plate.

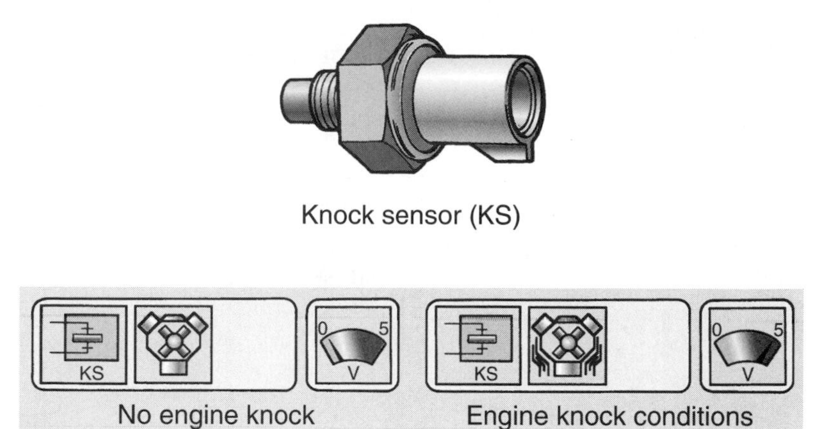

FIGURE 2–16 Knock sensor with related voltage signal.

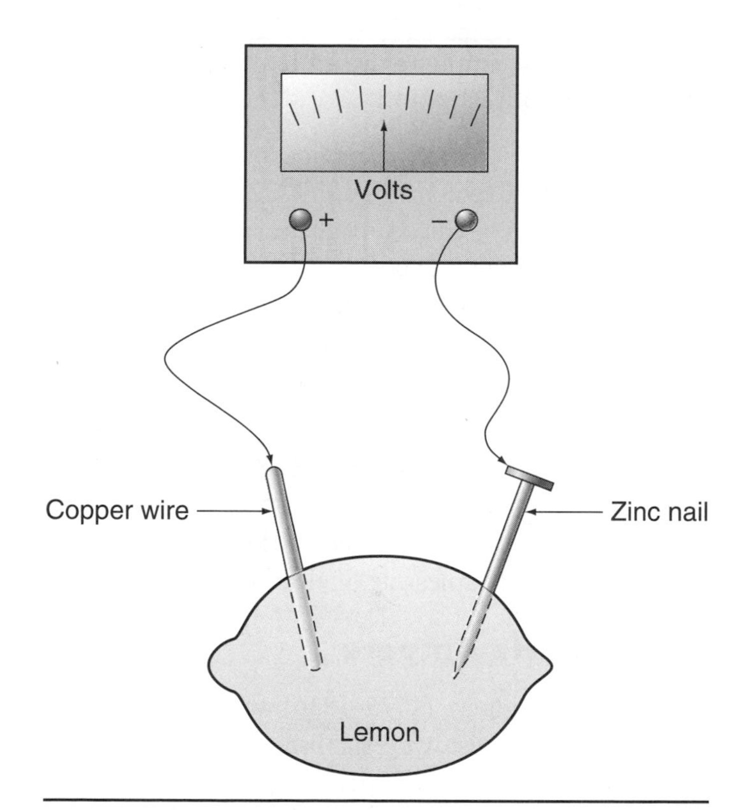

FIGURE 2–17 Chemical energy being produced by a simple lemon.

A simple lead acid battery has a single cell with a positive electrode made from lead peroxide (PbO_2) and a negative electrode made from sponge lead (Pb) (Figure 2–18). The electrolyte is a mixture of sulfuric acid (H_2SO_4) and distilled water. A chemical reaction takes place when the negative electrode releases electrons to the electrolyte and the positive electrode collects these electrons from the electrolyte.

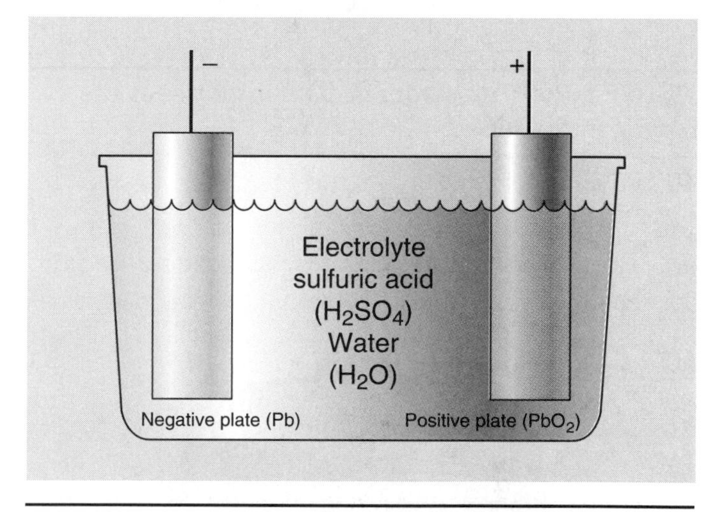

FIGURE 2–18 A simple lead acid battery.

Lead acid battery construction and testing will be discussed in Chapter 14.

Electromagnetic Induction (EMI)

The sixth and last source of electrical power is electricity produced from **electromagnetic induction (EMI).** When a current is carried through a conductor, a magnetic field surrounds the wire and electromagnetism occurs (Figure 2–19).

When more loops are added to the conductor, the strength of the magnetic field increases accordingly (Figure 2–20).

In a large-scale electromagnet application, a lifting magnet, one pole of a magnet is formed in a shell, and the other is formed to the core inside the coil (Figure 2–21).

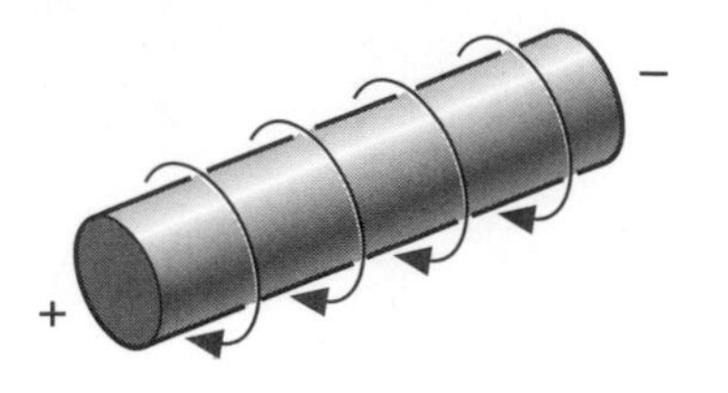

FIGURE 2–19 A magnetic field surrounds a conductor when current flows through it.

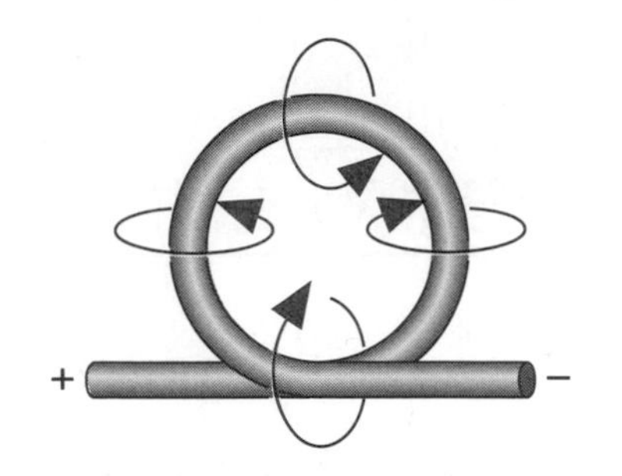

FIGURE 2–20 The magnetic field increases as the conductor is looped.

FIGURE 2–21 A simple lifting magnet.

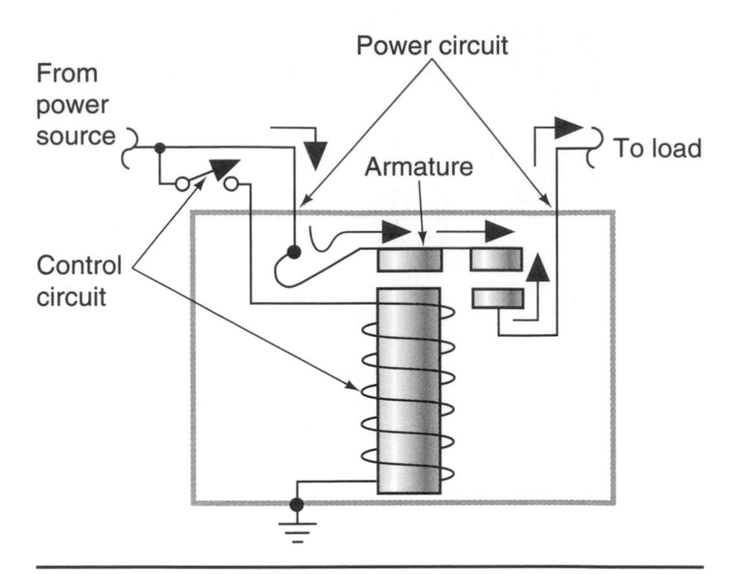

FIGURE 2–22 A simple relay using magnetism to energize the contact points.

A common automotive application of electromagnetism is a relay switch (Figure 2–22). Relays use an electric current to produce a magnetic field that pulls the contact points together to close a switch.

In other chapters of this textbook we will discuss additional applications of producing electricity from electromagnetism, such as generators and alternators.

Historical Perspective

Four people have been instrumental in our modern-day understanding and harnessing of electricity.

André Marie Ampère

André Marie Ampère (1775–1836) was a French mathematician and scientist who established the importance of the relationship between electricity and magnetism. Ampére discovered that wires could behave like magnets by passing current through them and that the polarity of their magnetism depended on the direction of the current. If the current in one wire is opposite that of another wire, the wires repel each other, in much the same way as if you took two magnets of the same polarity and tried to touch them together (Figure 2–23).

Although it was not until sixty years after his death that his knowledge and accomplishments were given the respect they deserved, he did get the unit of electrical current, the ampere, or amp, named after him.

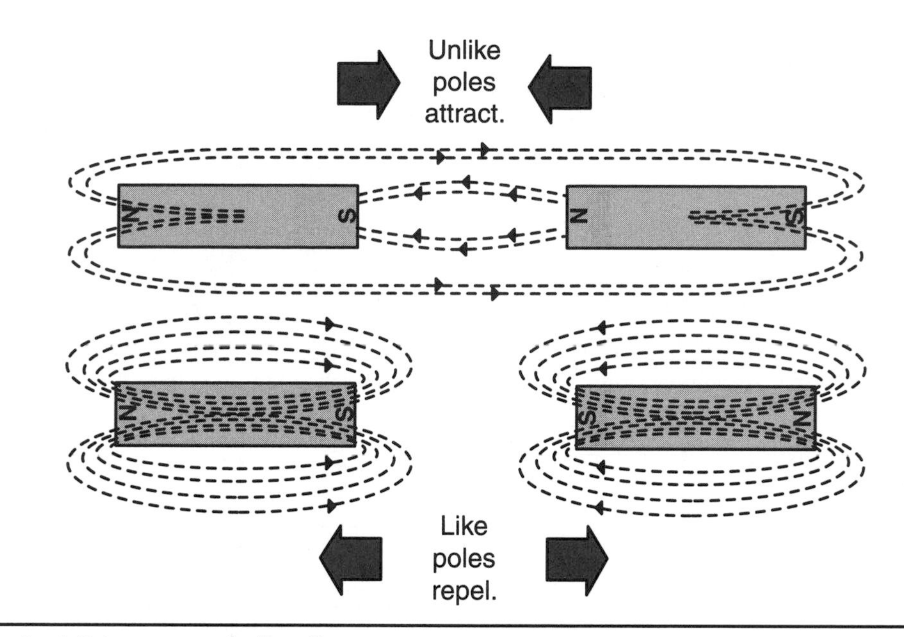

FIGURE 2–23 Invisible magnetic flux lines.

Alassandro Volta

The Italian physicist **Alassandro Volta** (1745–1827) was best-known for his study of static electricity and current. His inventions ranged from the storage of static electricity to the first device for producing electrical current, the battery. In 1775 he invented the electrophorus, a device used for storing an electrical charge. Today we call this device a condenser or capacitor. The invention that made him famous, though, was the battery. He discovered that, if two dissimilar metals were brought in contact with a salt solution, a current is produced. This finding led to what is known as **voltaic piles,** which are a series of small round copper and zinc plates separated by cardboard soaked in saltwater (Figure 2–24).

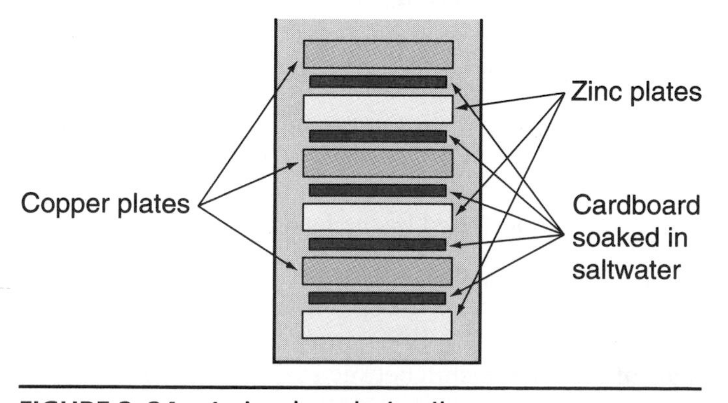

FIGURE 2–24 A simple voltaic pile.

Volta's most enduring reward was to have the unit of electrical potential, the volt, named after him.

George Simon Ohm

In 1827, German physicist **George Simon Ohm** (1789–1854) showed that there was a relationship between resistance, current, and voltage. As a result, the now famous **Ohm's law** states that the current in an electrical circuit is inversely proportional to the resistance of the circuit and directly proportional to the voltage. This law has become the cornerstone in understanding the properties of electricity. With this law, scientists accurately calculate the properties of an electrical circuit. We will be looking at Ohm's law extensively in Chapter 3.

James Watt

The Scottish inventor **James Watt** (1736–1819), of all the inventors who made the Industrial Revolution possible, was probably the best known. In 1764 he realized the steam engine produced a considerable amount of wasted energy. Over the next several years he developed, patented, and refined several of his steam engines. To express the capacity of his engines, he developed a comparison between what his engine could do and the average work capacity of a normal size horse. He found that a horse could produce 550 foot-pounds (ft-lb) of work per second; so he called this unit of work one **horsepower** (Figure 2–25): 1 horsepower (hp) = 550 ft-lb per second. If you consider that there are 60 seconds in a minute, then 1 horsepower = 33,000 ft-lb per minute.

In electricity, the metric unit **joule** is the base unit of energy measurement. This unit is also called a watt, in honor of James Watt, where 1 watt is a rate of 1 joule per second. This relationship is best described in **Watt's law,** which states that 1 watt is the amount of work done in 1 second by 1 volt moving a charge of 1 ampere through a resistance of 1 ohm in a circuit. James Watt found that horsepower could also be expressed as a unit of electrical power, or watt. The simple conversion is

1 horsepower = 746 watts

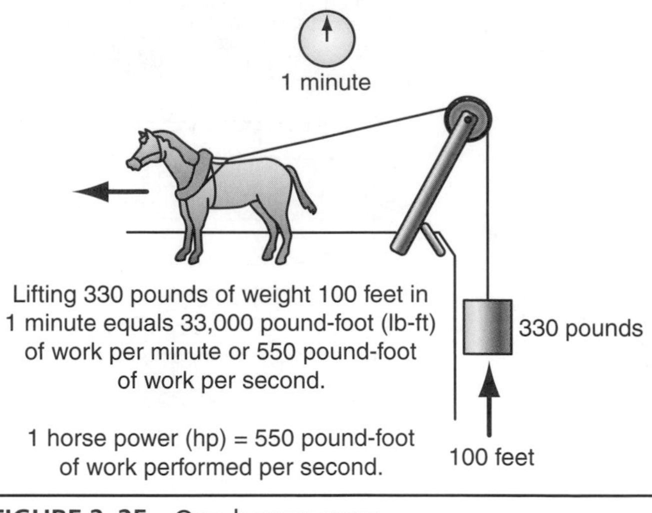

FIGURE 2–25 One horsepower.

All four of these incredible inventors contributed substantial knowledge to the establishment of our current understanding of electricity. As a result, they have become standard household names associated with terms like a *60-watt light bulb*, a *20-amp circuit breaker*, a *120-volt wall outlet*, and a *60-ohm electrical circuit.*

Summary

- Static electricity is electricity without motion or at rest.
- Electrostatic discharge (ESD) is the process by which charged objects transfer their free electron charge to a neighboring object.
- An automotive technician should always use a static grounding strap when working with static-sensitive electronic devices.
- Solar energy is light energy from the sun that is gathered in a photovoltaic solar cell.
- A photon is pure energy that contains no mass.
- Thermoelectricity is electricity produced when two dissimilar metals are heated to generate an electrical voltage.
- A thermocouple is a small device that gives off a low voltage when two dissimilar metals are heated.
- Piezoelectricity is electricity produced when materials, such as quartz or barium titanate, are placed under pressure.
- The production of electricity from chemical energy is demonstrated by the lead acid battery.
- Electromagnetic induction is the production of electricity when a current is carried through a conductor and a magnetic field is produced.
- André Marie Ampère established the importance of the relationship between electricity and magnetism.

- ❑ Alassandro Volta discovered that if two dissimilar metals are brought into contact with a salt solution, a current is produced. This invention is now known as the battery.
- ❑ George Simon Ohm showed a relationship between resistance, current, and voltage in an electrical circuit. He developed what is known as Ohm's law.
- ❑ James Watt developed a method used to express a unit of electrical power known as Watt's law.

Key Terms

André Marie Ampère
chemical energy
electromagnetic induction (EMI)
electrostatic discharge (ESD)
horsepower
joule
George Simon Ohm
Ohm's law
photons
piezoelectricity
Seebeck effect
solar light energy
static electricity
thermocouple
thermoelectricity
Alassandro Volta
voltaic piles
James Watt
Watt's law

Review Questions

Short Answer Essays

1. Describe how static electricity transfers from one object to another.
2. Name two sources of unwanted static electricity in automotive repair work and what is used to prevent its occurrence.
3. Explain the use of photovoltaics in automotive applications.
4. Define the term *thermocouple* and explain its most common use.
5. Describe the four people who have been instrumental in our modern-day understanding and harnessing of electricity.

Fill in the Blanks

1. Electricity that is without ____________________ or at ____________________ will produce an electrical charge whenever two dissimilar nonmetallic materials are rubbed together lightly.
2. To protect sensitive electronic components from ____________________ damage, the automotive technician should always use a ____________________ ____________________.
3. A photovoltaic cell is located on top of the vehicle dash to sense the __________ __________, and trigger the complex circuitry that activates the automatic headlight system.
4. A simple lead acid battery consists of a single cell with a ____________________ electrode made from sponge lead and a ____________________ electrode made from lead peroxide.
5. In Ohm's law, the ____________________ in an electrical circuit is ____________________ ____________________ to the resistance of the circuit, and ____________________ ____________________ to the voltage.

ASE-Style Review Questions

1. Technician A says static electricity can occur whenever the technician slides across a vinyl seat in a car. Technician B says one way to guard against static electricity is the use of a static wrist strap. Who is correct?
   A. A only
   B. B only
   C. Both A and B
   D. Neither A nor B

2. Technician A says static electricity is worse when the air is very humid. Technician B says an electrostatic discharge (ESD) is the process by which a charged object transfers its charge to a neighboring object. Who is correct?
   A. A only
   B. B only
   C. Both A and B
   D. Neither A nor B

3. Technician A says the use of photovoltaic solar cells provides a way to capture energy from the sun. Technician B says a photon is the energy in sunlight that strikes the surface of a photovoltaic cell to create a voltage. Who is correct?
   A. A only
   B. B only
   C. Both A and B
   D. Neither A nor B

4. Technician A says a thermocouple is installed in the exhaust system of a diesel engine to measure the exhaust temperature. Technician B says a mercury glass thermometer can be substituted for a thermocouple in an exhaust system. Who is correct?
   A. A only
   B. B only
   C. Both A and B
   D. Neither A nor B

5. Technician A says a knock sensor produces a small voltage when there is an ignition spark knock. Technician B says the knock sensor is a piezoelectric device that produces a voltage under physical stress or vibration. Who is correct?
   A. A only
   B. B only
   C. Both A and B
   D. Neither A nor B

6. Technician A says a lead acid battery uses positive and negative electrode plates of the same type of material. Technician B says the chemical mixture of electrolyte in a lead acid battery is $PbSO_4$. Who is correct?
   A. A only
   B. B only
   C. Both A and B
   D. Neither A nor B

7. Technician A says the positive plate in a lead acid battery is made of lead peroxide (PbO_2). Technician B says the negative plate is made of a sponge lead material. Who is correct?
   A. A only
   B. B only
   C. Both A and B
   D. Neither A nor B

8. Two technicians are discussing the theory of electromagnetism. Technician A says that any time an electrical current flows through a conductor, a magnetic field is created around the conductor. Technician B says electromagnetism is what causes a relay switch to operate. Who is correct?
   A. A only
   B. B only
   C. Both A and B
   D. Neither A nor B

9. Ohm's law is being discussed: Technician A says the current is directly proportional to the resistance in a circuit and inversely proportional to the voltage. Technician B says the current is inversely proportional to the voltage in a circuit and directly proportional to the resistance. Who is correct?
   A. A only
   B. B only
   C. Both A and B
   D. Neither A nor B

10. Technician A says a joule is a metric unit of energy measurement. Technician B says 1 joule is the same as 100 watts. Who is correct?
    A. A only
    B. B only
    C. Both A and B
    D. Neither A nor B

Basic Electrical Theory

Introduction

During the past several decades, automotive electrical systems have shown a steady growth in complexity. With the introduction of alternators in the 1960s, electronic ignition in the 1970s, computer-controlled systems in the 1980s, advanced onboard diagnostics in the 1990s, and advanced hybrid vehicle technology in the 2000s, automotive technicians need to prepare themselves for these and future advanced systems. By developing an in-depth understanding of electrical and electronic basics you can prepare yourself for the rapidly changing responsibilities that lie ahead. This chapter provides a look at basic electrical theory and some practice using Ohm's law and Watt's law, introduced in Chapter 1.

Objectives

When you complete this chapter you should be able to:

- ❑ Apply electrical laws and theories.
- ❑ Describe atomic structure, atoms, and electrons.
- ❑ Define conductors, semiconductors, and insulators.
- ❑ Define voltage, resistance, and current flow.
- ❑ Define the relationship between voltage, amperage, and resistance in Ohm's law.
- ❑ Use the mathematical relationships associated with Ohm's law to find the value of a missing factor.
- ❑ Define the relationship between watts, voltage, and amperage in Watt's law.
- ❑ Use the mathematical relationships associated with Watt's law to find the value of a missing factor.
- ❑ Identify commonly used electrical symbols.
- ❑ Define magnetism and electromagnetism.

FIGURE 3–1 Three forms of matter depending on temperature and pressure conditions.

Basics of Electricity

To understand electricity we will take a brief look at how atoms are structured, and then find out what makes some atoms better conductors of electricity and others good insulators. **Matter** can be defined as any substance that occupies a space and contains a mass. Using H_2O (water) as an example, mass may be present in three forms: a solid (ice), a liquid (water), or a gas (steam) (Figure 3–1).

This mass is made up of **molecules** which consist of a group of elements. An **element** is a cluster of one type of atom. We are all familiar with the well-known molecule of water and its combination of elements in H_2O (Figure 3–2).

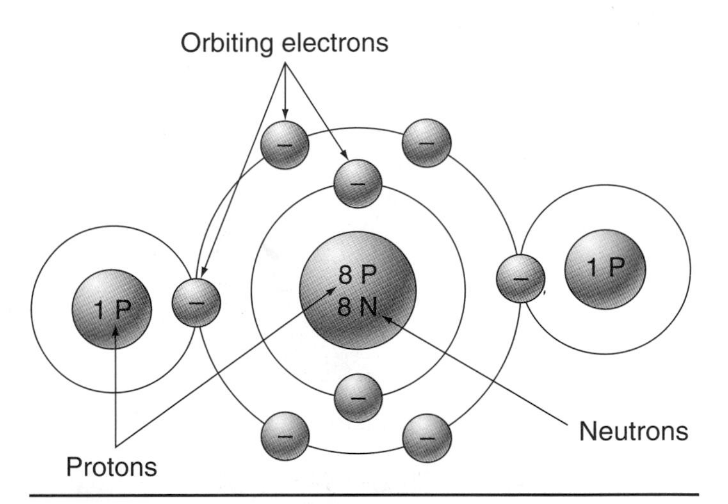

FIGURE 3–2 Two hydrogen atoms and one oxygen atom.

There are over 100 different elements in nature. Table 3–1 lists the elements, both natural and artificial, according to atomic order. This list is also referred to as The Table of Elements.

An **atom** is considered the smallest part of an element. It consists of *positively* charged **protons, neutrons** with *no* charge, and **electrons** that are *negatively* charged. Protons and neutrons combine to form the **nucleus,** or center, of the atom. The simplest atom is a hydrogen atom consisting of 1 proton and 1 orbiting electron, but no neutron (Figure 3–3). All other atoms in the atomic structure have neutrons.

The protons and neutrons form the center of the atom, which is smaller than the electron but over 1,800 times heavier. The number of protons in an atom determines its atomic number. The neutron is a heavy particle and

TABLE 3–1 The Table of Elements

ATOMIC NUMBER	NAME	VALENCE ELECTRONS	SYMBOL	ATOMIC NUMBER	NAME	VALENCE ELECTRONS	SYMBOL	ATOMIC NUMBER	NAME	VALENCE ELECTRONS	SYMBOL
1	Hydrogen	1	H	37	Rubidium	1	Rb	73	Tantalum	2	Ta
2	Helium	2	He	38	Strontium	2	Sr	74	Tungsten	2	W
3	Lithium	1	Li	39	Yttrium	2	Y	75	Rhenium	2	Re
4	Beryllum	2	Be	40	Zirconium	2	Zr	76	Osmium	2	Os
5	Boron	3	B	41	Niobium	1	Nb	77	Iridium	2	Ir
6	Carbon	4	C	42	Molybdenum	1	Mo	78	Platinum	1	Pt
7	Nitrogen	5	N	43	Technetium	2	Tc	79	Gold	1	Au
8	Oxygen	6	O	44	Ruthenium	1	Ru	80	Mercury	2	Hg
9	Fluorine	7	F	45	Rhodium	1	Rh	81	Thallium	3	TI
10	Neon	8	Ne	46	Palladium	–	Pd	82	Lead	4	Pb
11	Sodium	1	Na	47	Silver	1	Ag	83	Bismuth	5	BI
12	Magnesium	2	Ma	48	Cadmium	2	Cd	84	Polonium	6	Po
13	Aluminum	3	Al	49	Indium	3	In	85	Astatine	7	At
14	Silicon	4	Si	50	Tin	4	Sn	86	Radon	8	Rd
15	Phosphorus	5	P	51	Antimony	5	Sb	87	Francium	1	Fr
16	Sulfur	6	S	52	Tellurium	6	Te	88	Radium	2	Ra
17	Chlorine	7	Cl	53	Iodine	7	I	89	Actinium	2	Ac
18	Argon	8	A	54	Xenon	8	Xe	90	Thorium	2	Th
19	Potassium	1	K	55	Cesium	1	Cs	91	Protactinium	2	Pa
20	Calcium	2	Ca	56	Barium	2	Ba	92	Uranium	2	U
21	Scandium	2	Sc	57	Lanthanum	2	La				
22	Titanium	2	Ti	58	Cerium	2	Ce		Artifical Elements		
23	Vanadium	2	V	59	Praseodymium	2	Pr				
24	Chromium	1	Cr	60	Neodymium	2	Nd	93	Neptunium	2	Np
25	Manganese	2	Mn	61	Promethium	2	Pm	94	Plutonium	2	Pu
26	Iron	2	Fe	62	Samarium	2	Sm	95	Americium	2	Am
27	Cobalt	2	Co	63	Europium	2	Eu	96	Curium	2	Cm
28	Nickel	2	Ni	64	Gadolinium	2	Gd	97	Berkelium	2	Bk
29	Copper	1	Cu	65	Terbium	2	Tb	98	Californium	2	Cf
30	Zinc	2	Zn	66	Dysprosium	2	Dy	99	Einsteinium	2	E
31	Gallium	3	Ga	67	Holmium	2	Ho	100	Fermium	2	Fm
32	Germanium	4	Ge	68	Erbium	2	Er	101	Mendelevium	2	Mv
33	Arsenic	5	As	69	Thulium	2	Tm	102	Nobelium	2	No
34	Selenium	6	Se	70	Ytterbium	2	Yb	103	Lawrencium	2	Lw
35	Bromine	7	Br	71	Lutetium	2	Lu				
36	Krypton	8	Kr	72	Hafnium	2	Hf				

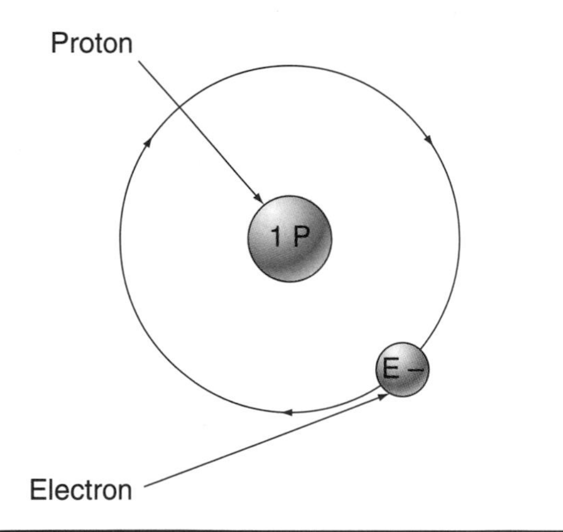

FIGURE 3–3 The hydrogen atom is the simplest atom known.

electrically neutral. The electron rotates in an orbit around the proton, as the earth rotates around the sun. Even though a simple law of physics states that like charges repel each other (Figure 3–4) and unlike charges attract each other (Figure 3–5), the centrifugal force of the rotating, negatively charged electrons keeps them separated from the positively charged protons.

To understand how atomic structure relates to electricity, look at a copper atom that has an atomic number of 29 and contains 29 protons, 34 neutrons, and 29 electrons (Figure 3–6). The number of electrons and protons in the copper atom's structure attempts to remain balanced or equal. If possible, the atom will give or take electrons from neighboring atoms to remain balanced.

An atom has several electron orbits (Figure 3–7). The electrons in the outer most orbit are called **valence electrons** because they are in the valence ring. When valence electrons become dislodged from their orbit by some outside force, they are also called **free electrons.** Thus they are commonly called either valence or free electrons.

The inner orbits have **bound electrons,** which are free to move only within their own orbits at a fixed distance from the nucleus. The free, or valence, electrons are where our study of electron flow begins.

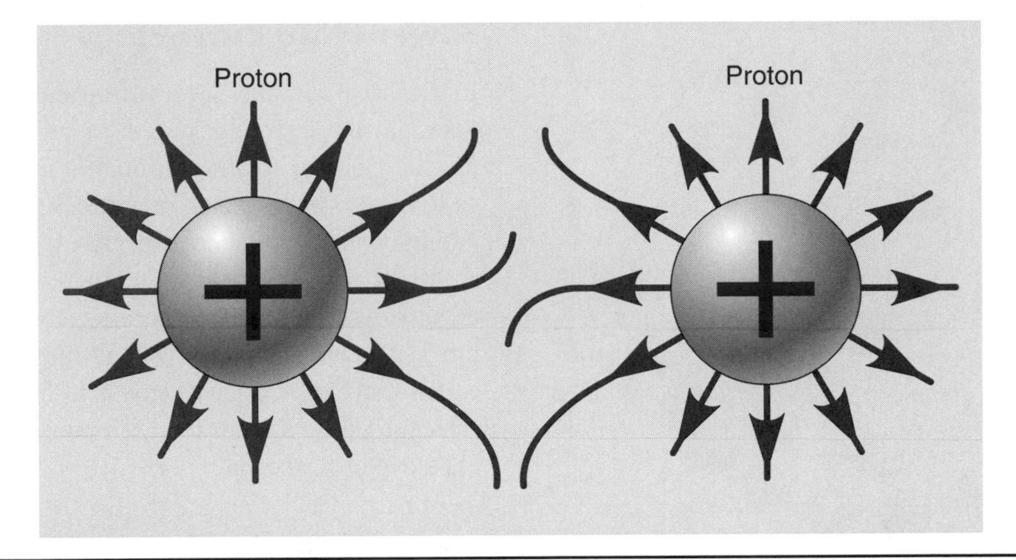

FIGURE 3–4 Like charges repel each other.

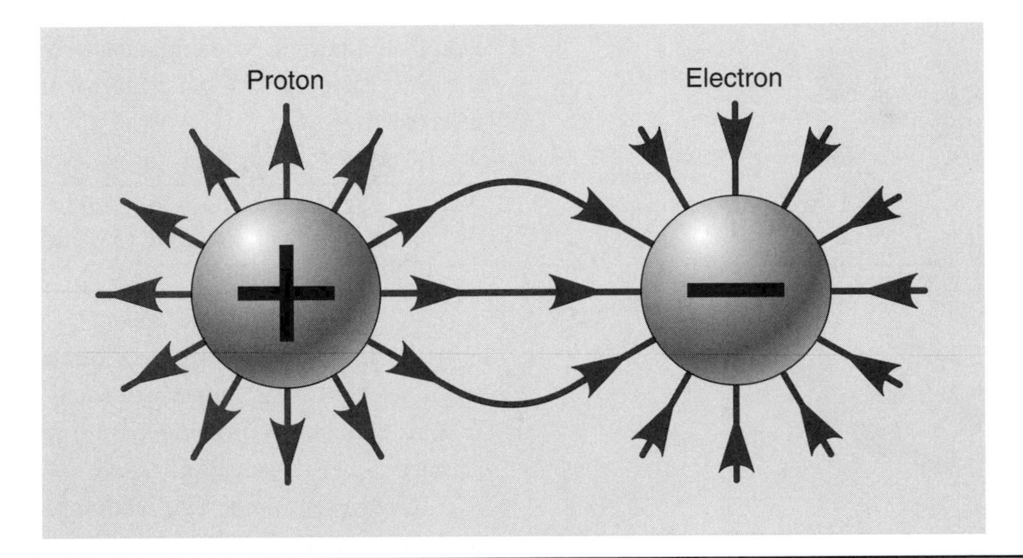

FIGURE 3–5 Unlike charges attract each other.

Note: Electricity is the movement of free electrons from atom to atom in a complete circuit.

Not all materials make good conductors of electricity. Figure 3–8 lists some of the most commonly used atoms in the automotive field. A good conductor must have three or fewer free electrons. The lower the number is of free electrons, the better the conductor is. A semiconductor contains exactly four free electrons and is used extensively in electronic circuits because of the material's electron stability. Insulators are usually combinations of several different molecules that have combined to form a tightly bound molecule with five or more free electrons. The force holding these free electrons is very strong, which makes them good insulators.

Note: Rubber is a good insulator only if it is pure rubber. The more carbon that is impregnated in the rubber, the less insulation properties it has.

We could list several other atoms here, but our main focus is on the conductors and insulators that we normally use in automotive circuits.

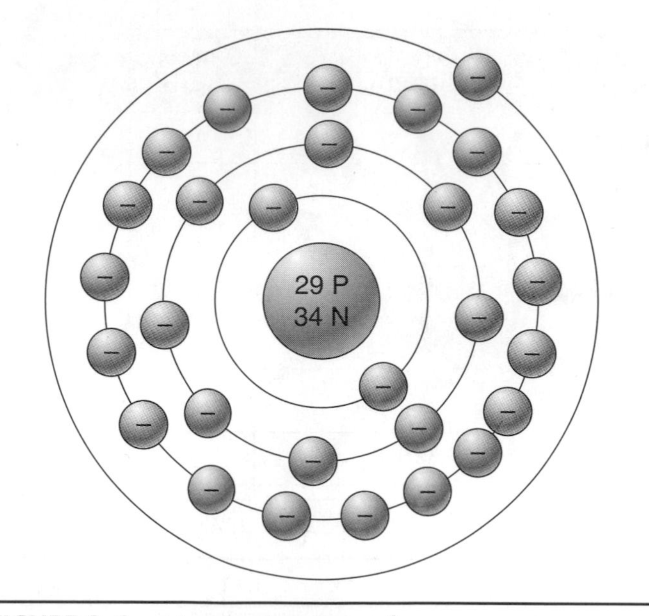

FIGURE 3–6 Atomic structure of a copper atom.

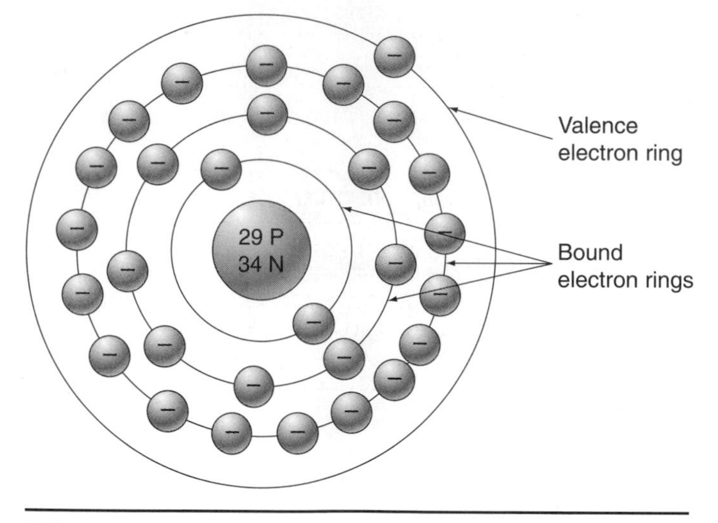

FIGURE 3–7 Bound and valence electron orbits of a copper atom.

Direct Current

Most automotive electrical circuits operate on **direct current (DC)** that is produced by the battery or by the converted direct current from the charging system circuit. Direct current typically is at a specified level of electrical potential that flows in one direction at a specific DC voltage, as indicated on the scope screen (Figure 3–9). Some DC voltages may have a small up-and-down waveform or ripple similar to an AC waveform as the electric component is being operated. However, normally the DC waveform remains on only one side (positive or negative) of the zero baseline.

Alternating Current

In electrical circuits, **alternating current (AC)** is produced in the charging system or any other device that has a conductor rotating through a magnetic field, such as a wheel speed sensor in antilock brakes. With alternating current, the voltage begins at zero and rises to a specified positive voltage, and then reverses to an equal negative value before returning to zero again (Figure 3–9). One complete rise and fall of the AC wave is called a cycle. A good example of an AC circuit is the typical lighting system found in houses of 110/120 volts at 60 cycles per second.

Current Flow

In Figure 3–10, the free electrons move to the left. The unbalanced positive charge on the left pulls a free electron from the right, causing a chain reaction. The flow of these free electrons from negative to positive is called **current flow.**

Note: The current is not used up in the circuit; it just moves from negative to positive in a complete circuit.

It is measured in **amperes** (amps) and flows at the speed of light. One ampere is equal to 6.28 billion billion free electrons flowing past a point in one second (Figure 3–11). That's a lot!

To represent amperage readings, the letter *I* is often used because it stands for intensity of current. The letter *A* is also used, and it stands for amps. Both are used in electrical mathematical formulas. This textbook uses the letter *I* to represent current.

SAFETY TIP *The human body can conduct electricity. Because of this, any time you perform inspection or service on electrical systems, beware of electrical hazards. Observe electrical safety rules at all times, and never wear jewelry while testing or servicing electrical systems.*

Another way to look at current flow is to compare it with the flow of water. Electron flow is measured by an ammeter, and water flow is measured by a meter that reads the flow of water in gallons per minute (gpm) (Figure 3–12).

Good conductors: 3 or fewer free electrons

Atomic Number	Name of Atom	Number of Valence Electrons	Symbol Used
47	Silver	1	Ag
29	Copper	1	Cu
79	Gold	1	Au
13	Aluminum	3	Al
26	Iron	2	Fe

Semiconductors: exactly 4 free electrons

Atomic Number	Name of Atom	Number of Valence Electrons	Symbol Used
32	Germanium	4	Ge
14	Silicon	4	Si

The following nonmetal materials are considered tightly bound molecular combinations of different atoms that share free electrons.

Good insulators: 5 or more free electrons

Material Name	Molecule Combination
Plastics	Carbon, hydrogen, oxygen
Dry wood	Carbon, hydrogen, oxygen, and nitrogen
Glass	Silicon, oxygen
Diamond	Pure carbon
Pure rubber	Carbon and hydrogen. Includes sulfur if vulcanized

FIGURE 3–8 Determining electrical conductivity of an atom.

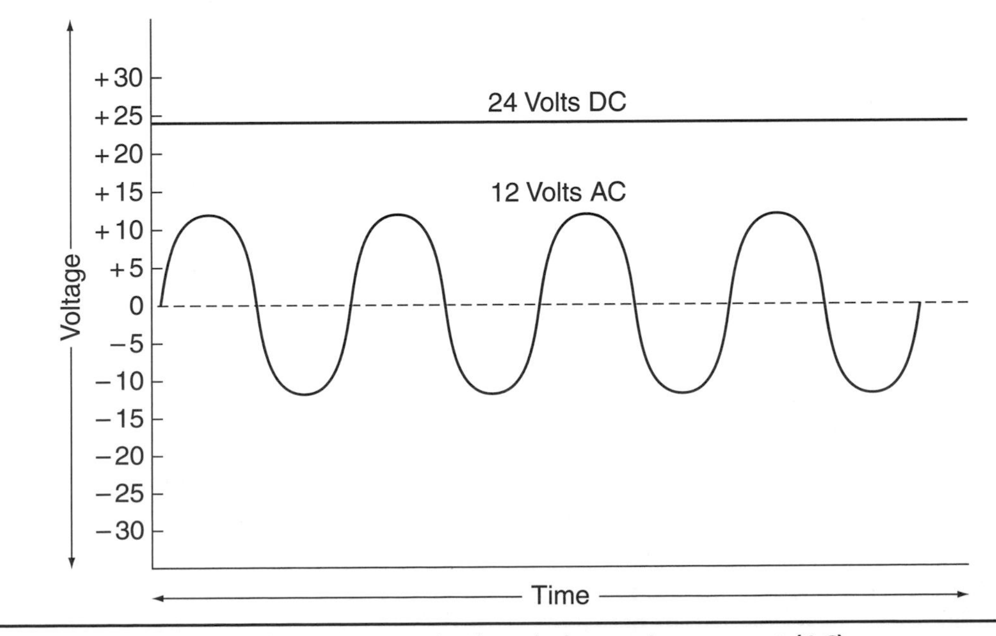

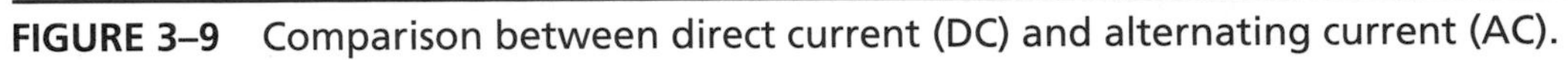

FIGURE 3–9 Comparison between direct current (DC) and alternating current (AC).

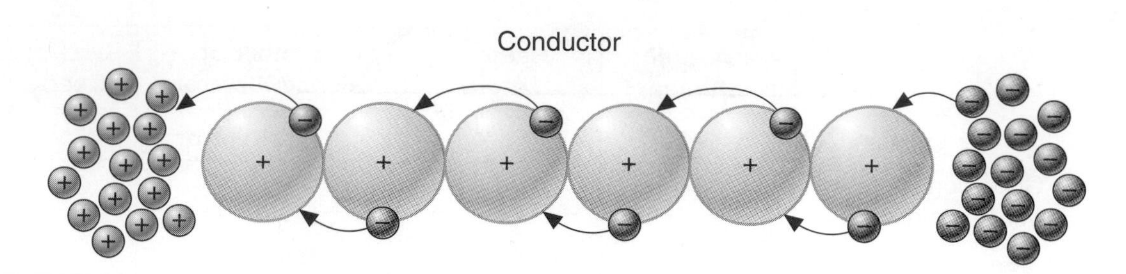

FIGURE 3–10 The flow of free electrons from atom to atom.

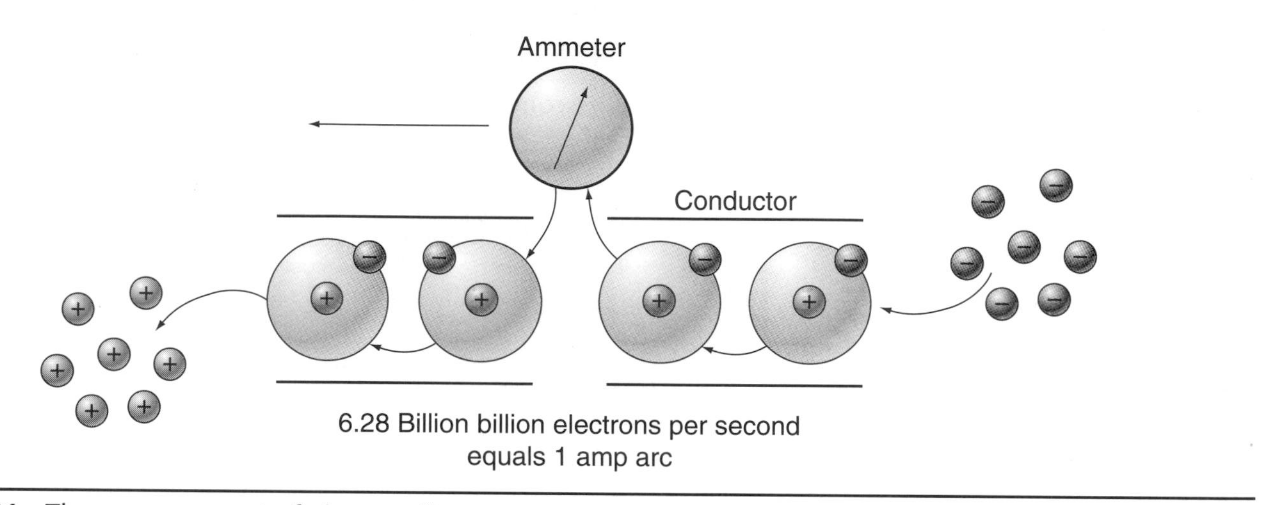

FIGURE 3–11 The measurement of electron flow.

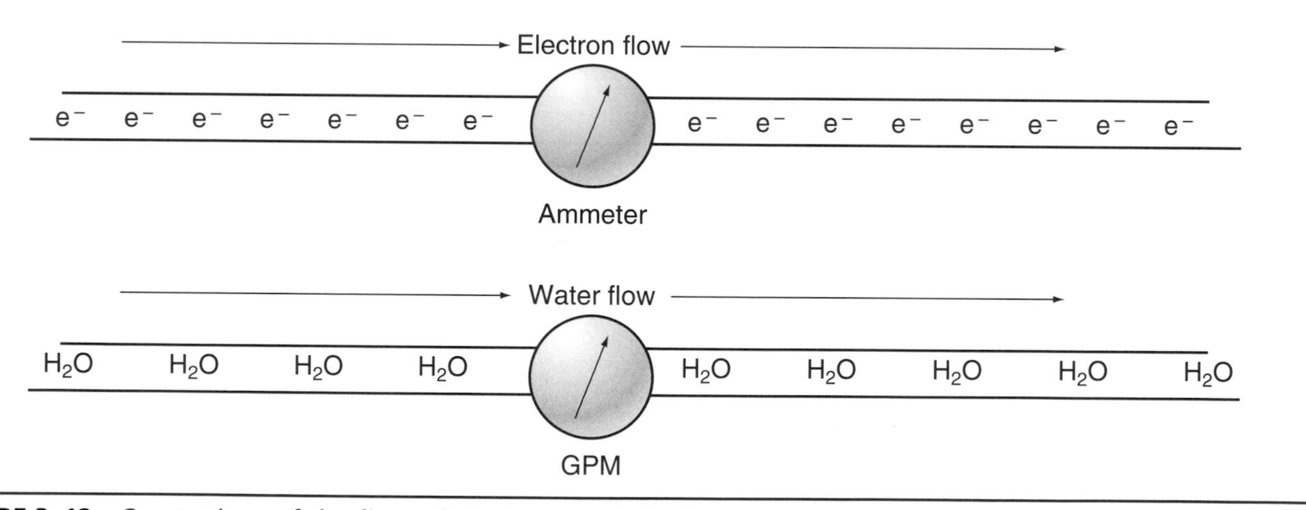

FIGURE 3–12 Comparison of the flow of electrons and the flow of water.

Electron Flow Theory and Conventional Current Flow

Over the years, most automotive repair literature has described the flow of current using **conventional flow theory.** This theory states that current flows from the most positive potential to the least positive, or negative, side of complete circuit. In automotive applications the battery's negative terminal is the ground, and its positive terminal is the hot, or insulated, side. Automotive wiring diagrams generally show the flow from positive to negative (Figure 3–13).

Electron flow theory suggests that, since electrons are negatively charged, free electrons move from the

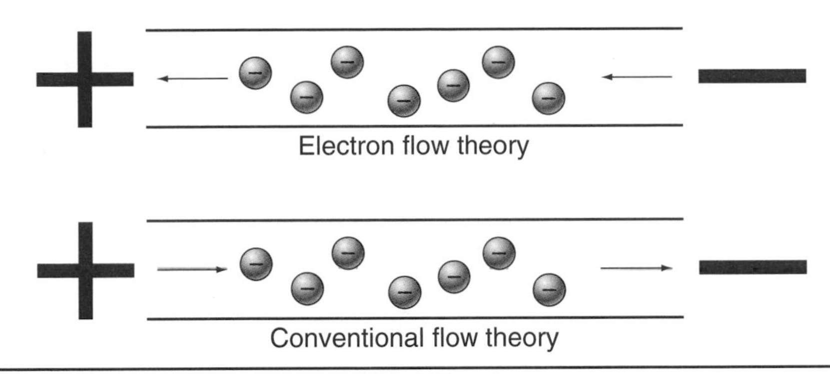

FIGURE 3–13 Comparison of electron flow theory and conventional flow theory.

negative, or surplus, side of the circuit toward an area having fewer free electrons, or the positive side of the circuit. Either theory applies because, to have a current flow from a beginning point to an ending point, there has to be a complete circuit.

Voltage

For the electrons to flow, a force or pressure must push the electrons through the circuit (Figure 3–14). This pressure is called an **electromotive force (EMF).** It is also known as **voltage,** which is a measurement of pressure at a point of excess free electrons in a circuit to a point of a shortage of free electrons. The common symbols used for voltage are *E* for EMF or *V* for voltage. The voltage pressure can be described much like the pressure in a water system (Figure 3–15), which is measured in pounds per square inch (psi). When the pressure is equal on both sides of the circuit, the circuit is considered near zero, or absolute (Figure 3–16).

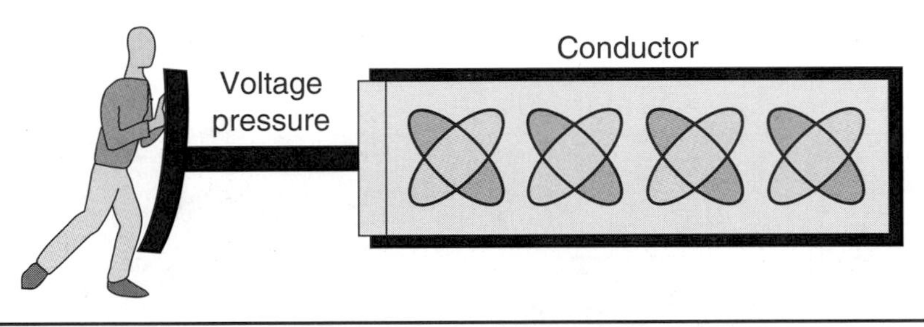

FIGURE 3–14 The pressure that causes electrons to flow is called voltage.

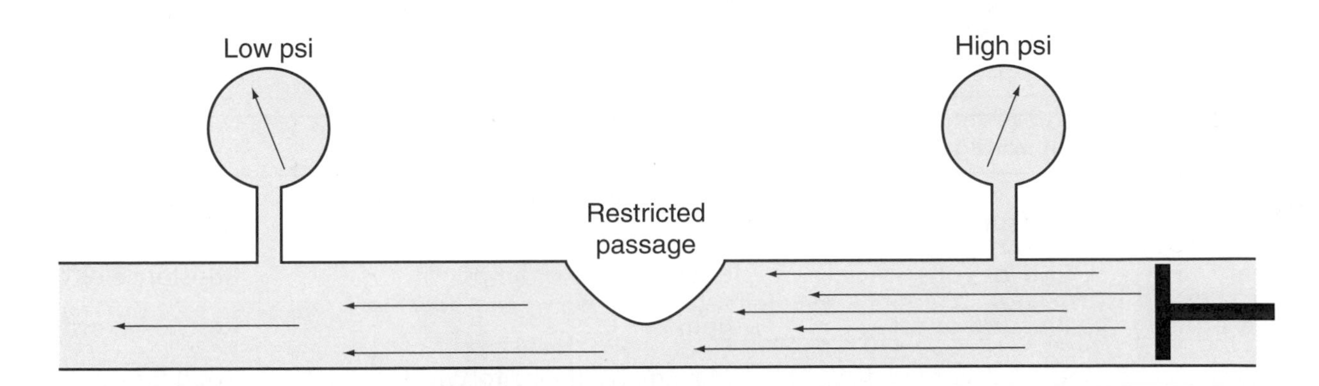

FIGURE 3–15 The psi in a water system operates similarly to voltage in an electrical circuit.

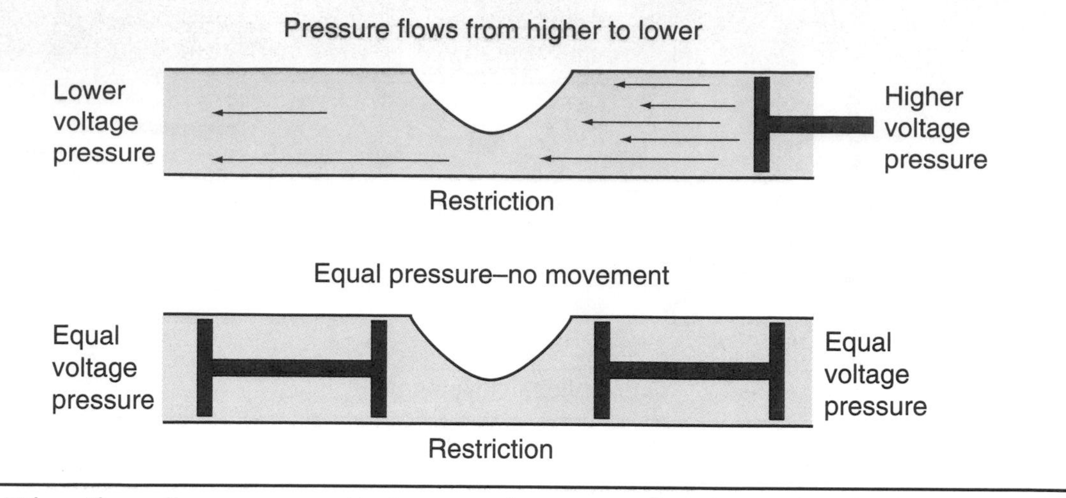

FIGURE 3–16 When the voltage pressure that causes electrons to flow is equal on both sides of a circuit, there is no electron flow.

If the psi at one end is higher than the psi at the other end, the water flows. If the psi is equal on both ends, then there is no water flow.

Resistance

Resistance is any force or substance that restricts or opposes the flow of current in a circuit. The symbol most widely used for resistance is the Greek letter omega (Ω), which is used extensively in this text and in manufacturers' wiring diagrams. The letter *R* is also used in mathematical calculations to represent resistance.

Another way to describe resistance is the use of road blocks in freeway lanes (Figure 3–17). The more lanes that are blocked, the fewer cars will pass. This is similar to the restriction of water flow in a piping circuit. A nozzle on the end, or a small hose, would be a restriction (orifice) that would stop or slow down the flow of water (Figure 3–18).

The analogies we used to describe resistance can be summed up in Table 3–2.

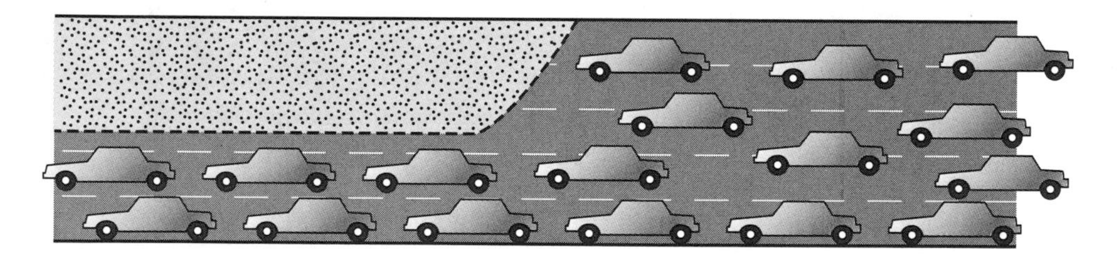

FIGURE 3–17 The more blocked lanes there are, the slower the flow of cars is, and, of course, the more angry drivers!

TABLE 3–2 Summary of voltage analogies.

System	Pressure	Flow	Restriction
Electrical	EMF or volts	Electron flow or amps	Resistance or ohms
Fluids	Pressure or psi	gpm	Restrictive orifice
Freeway		Movement of cars forward	Road blocks

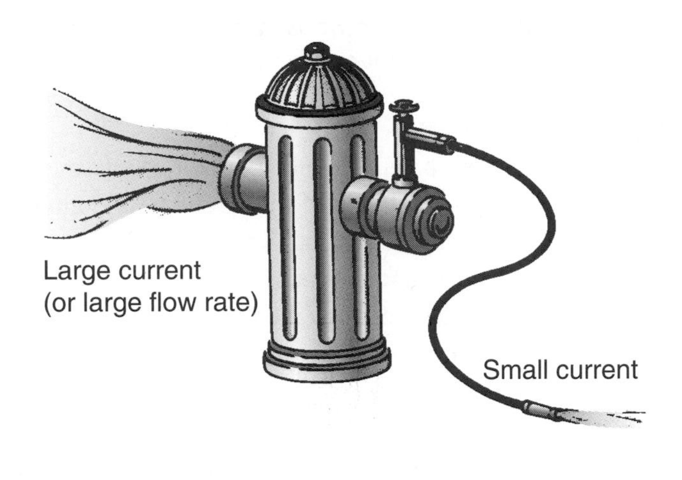

FIGURE 3–18 The higher the restrictive orifice, the smaller the flow.

Resistance is present in all circuits. Without resistance, a circuit would have no way to control current flow. Several factors aid in determining circuit resistance:

1. *Atomic structure* of the resistive material, especially the number of free electrons in the outer shell.

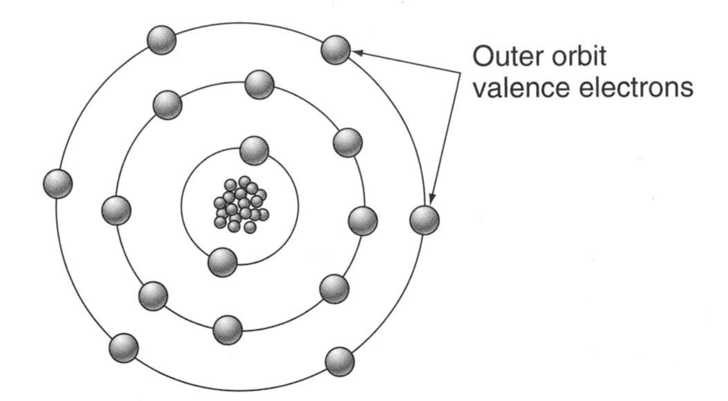

2. *Condition and quality of the circuit wiring,* anything that unintentionally restricts flow because of wire failure or damage.

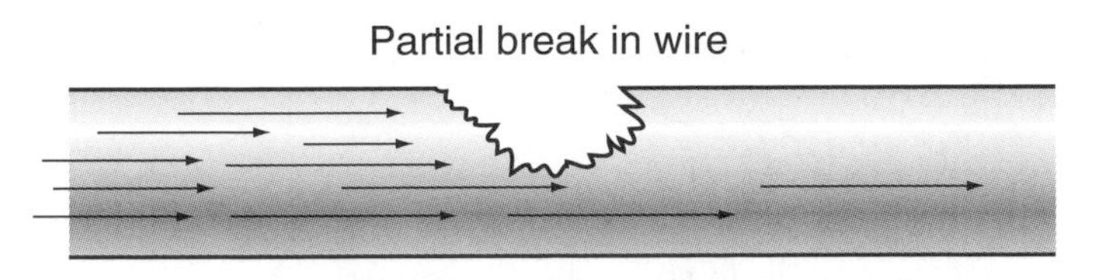

3. *Length of the wire or conductive device,* with resistance going up as the wire gets longer.

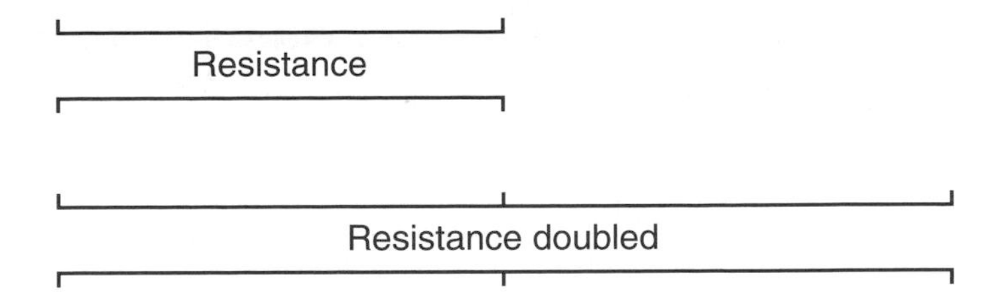

4. *Diameter of wire or conductive device,* with smaller the diameters creating higher circuit resistance.

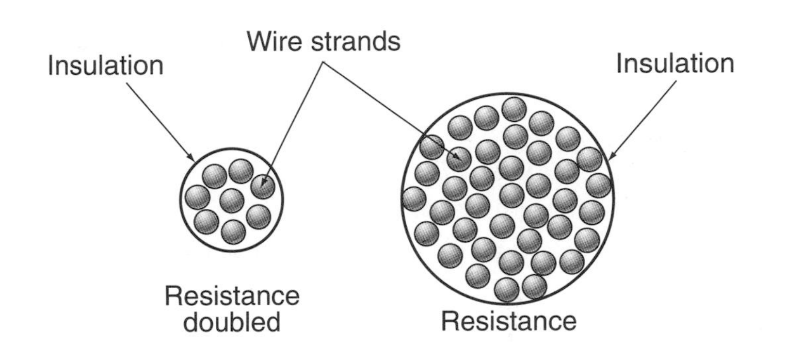

5. *Temperature of the circuit,* with high circuit temperatures creating high resistance.

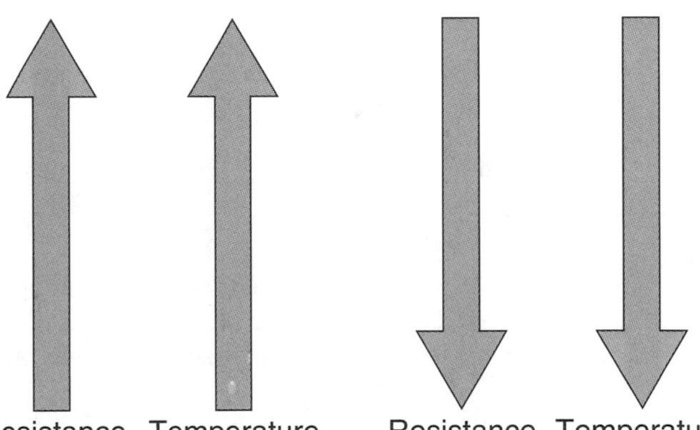

6. *Corrosion or foreign material at electrical connections,* with resistance developing at any electrical connection that develops a loose or corroded connection, such as a battery cable-to-post connection.

Quick Review of Voltage, Amperage, and Resistance

Definition	Symbol	
Voltage	Electromotive force or pressure	*E* or *V*
Amperage	Flow of electrons	*I* or *A*
Resistance	Opposition to electron flow	*R* or Ω

Ohm's Law

The relationship between voltage, current, and resistance in a circuit was proven in 1827 when German physicist George Simon Ohm (1787–1854) demonstrated a clearly defined relationship between voltage, current, and resistance in an electrical circuit. This relationship is known as **Ohm's law,** which, simply stated, is:

In a DC circuit, the current is directly proportional to the voltage and inversely proportional to the resistance.

The amount of current flow in a circuit is determined by the voltage and resistance of the circuit. If the circuit resistance increases, the current flow decreases, or vice versa. When the resistance remains constant, any change in the voltage will also change the current in proportion to the voltage (Figure 3–19).

In automotive circuits, system voltage remains relatively constant at 14.5 volts or at 5 volts for some computer sensor circuits. A change in resistance causes an opposite reaction to the current. A simple way to remember the principle of Ohm's law whenever the voltage in a circuit is constant is to look at a teeter-totter (Figure 3–20).

In any electrical circuit, the voltage, current, and resistance work together as a mathematical trio. One volt of

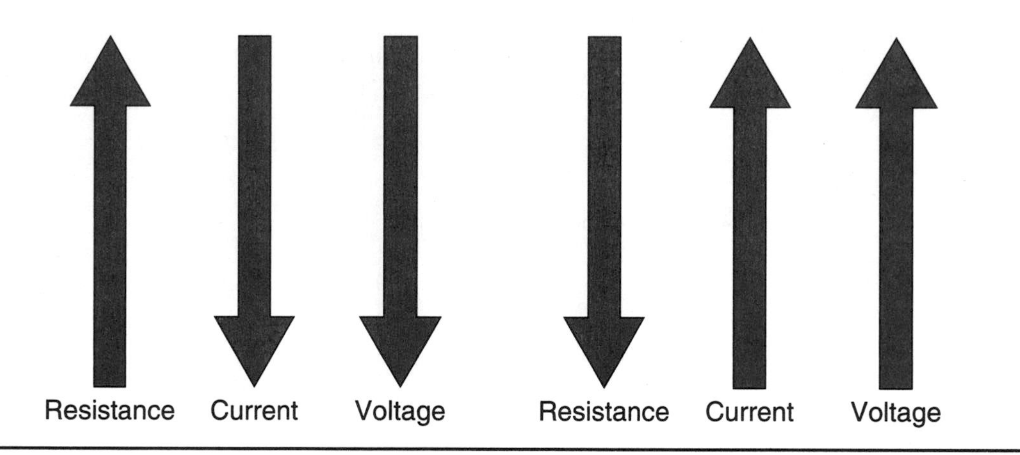

FIGURE 3–19 Relationship of resistance, current, and voltage in Ohm's law.

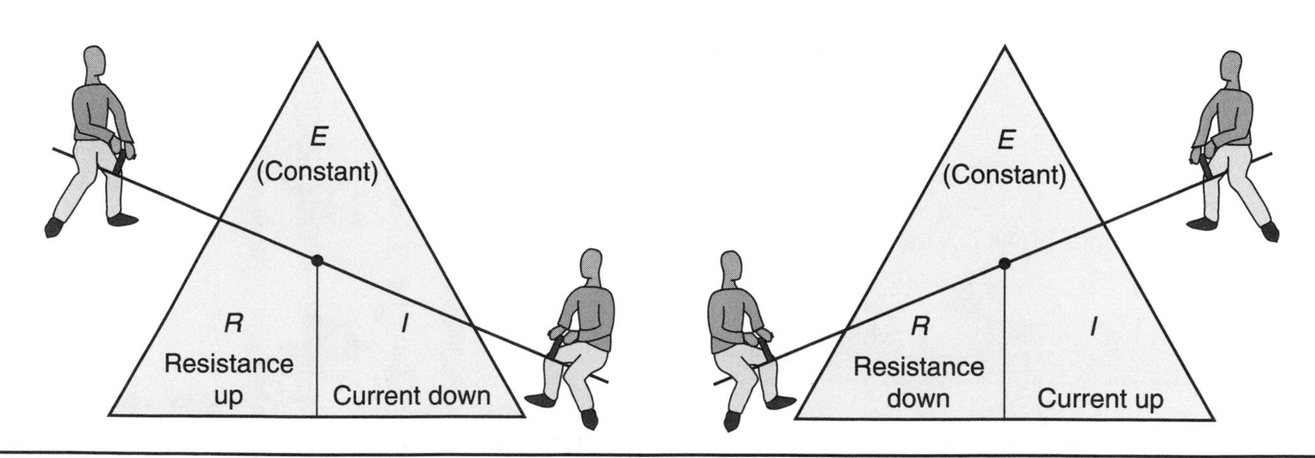

FIGURE 3–20 With constant voltage (*E*), the relationship of current (*I*) to resistance (*R*).

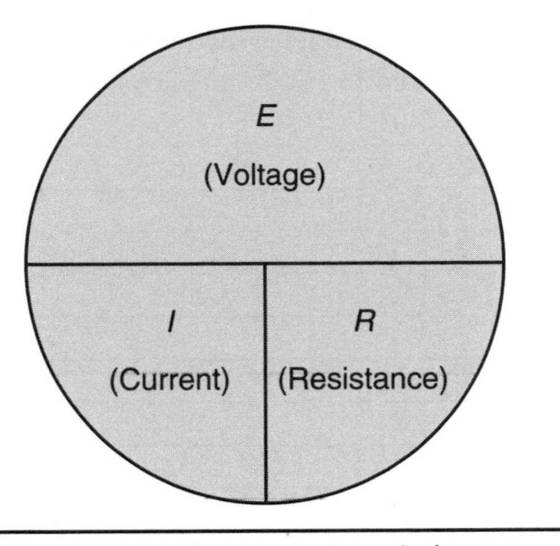

FIGURE 3–21 Ohm's law equation circle.

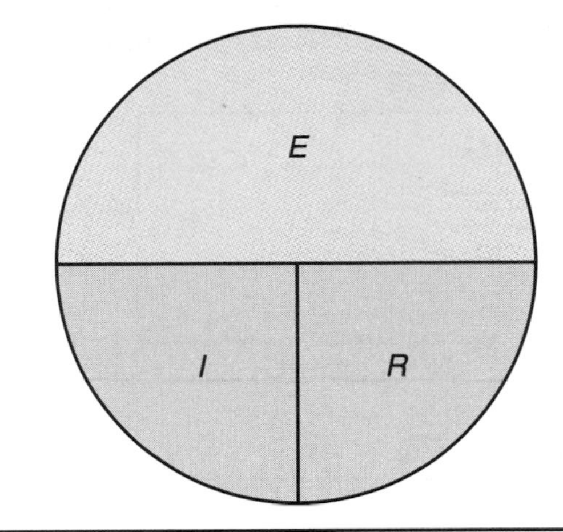

FIGURE 3–22 Use the exposed formula to find the unknown voltage (E).

electromotive force (EMF), pushing 1 ampere of current, is the result of having 1 ohm (Ω) of resistance in the circuit. If you know any two parts of the trio, then you can mathematically figure out the third part. Figure 3–21 is an example of the **Ohm's law circle.** You may see this drawn as a triangle or square, but the equations inside the parameter remain the same.

In the circle, the E represents electromotive force or voltage, the I represents intensity or amperage, and the R represents resistance or ohms. Sometimes you may see a V for voltage, or an A for amperage, or a mixture of these letters. As long as you know the basic Ohm's law circle, the mathematical equations remain the same. When any two of the electrical factors are known, simply cover the unknown value, and calculate as follows.

To find *voltage* in a circuit $E = I \times R$ (Figure 3–22).

To find *amperage* in a circuit $I = E \div R$ (Figure 3–23).

To find *resistance* in a circuit $R = E \div I$ (Figure 3–24).

An example of using Ohm's law in a simple electrical circuit is shown in Figure 3–25. The source voltage is 12 volts, and the bulb resistance is 6 ohms. Using the Ohm's law circle in Figure 3–23, calculate the circuit amperage.

The math equation for this problem looks like the following.

$$I = \frac{E}{R} = \frac{12\text{ V}}{6\ \Omega} = 2\text{ A}$$

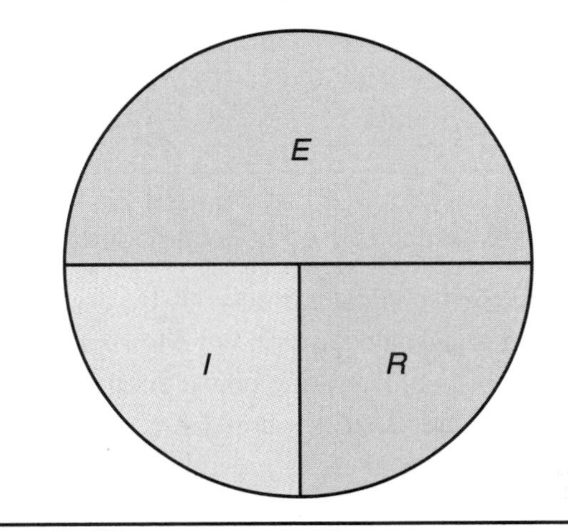

FIGURE 3–23 Use the exposed formula to find the unknown amperage (I).

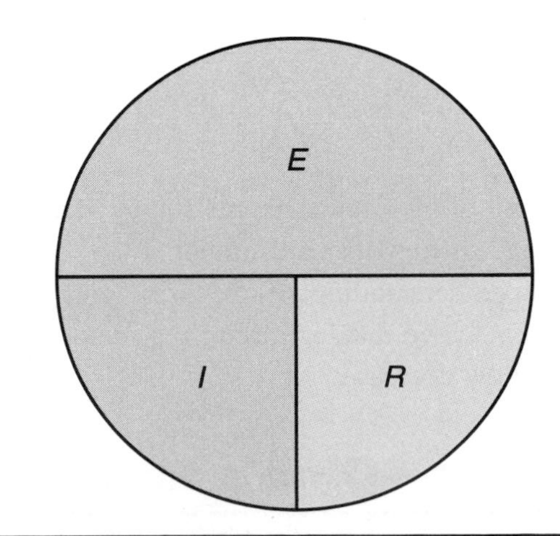

FIGURE 3–24 Use the exposed formula to find the unknown resistance (R).

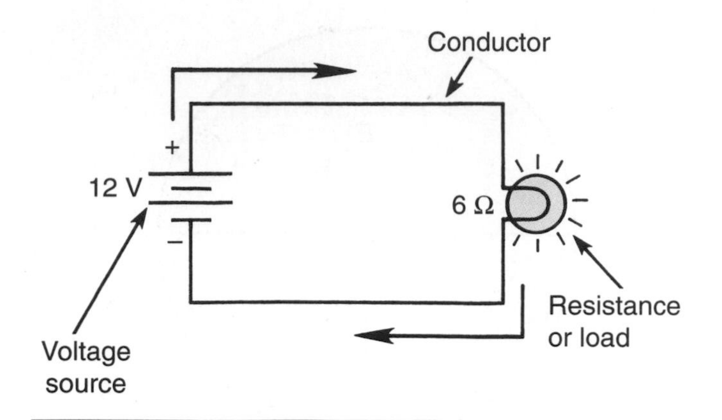

FIGURE 3–25 A simple electrical circuit to show the application of Ohm's law.

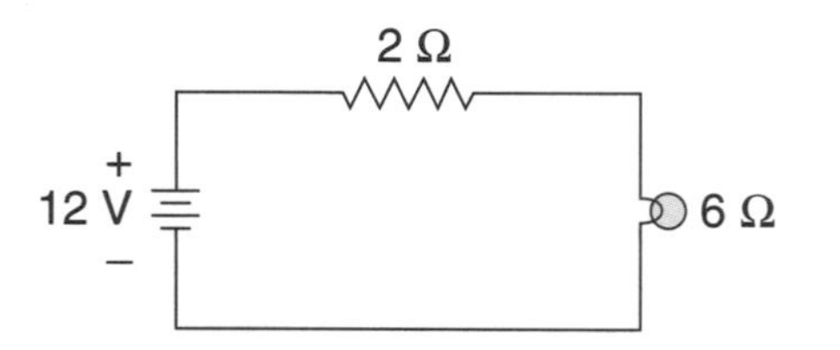

FIGURE 3–26 Same circuit as Figure 3-25, with the additional resistance caused by a corroded wire.

Automotive technicians rarely sit by a vehicle performing Ohm's law calculations, but *knowing* these basic principles helps you determine how a circuit is supposed to operate. Figure 3–26 is the same circuit as Figure 3–25, except that there is an added resistance in the wire or connection between the source battery and the light bulb. The circuit resistance now totals 8 ohms. Using the same math equation as before, but first adding the series resistance for a total of 8 Ω, execute Ohm's law to find the total circuit amperage.

$$I = \frac{E}{R} = \frac{12\text{ V}}{8\ \Omega} = 1.5\text{ A}$$

The result of increasing the resistance to 8 Ω caused the amperage to drop from 2 amperes to 1.5 amperes. Through an understanding of the principles of Ohm's law, it is easy to see that, as circuit resistance increases, the current flow decreases.

Exercises—Ohm's Law

The following exercise is designed to help give you practice using the Ohm's law circle to determine the unknown value of an electrical circuit. Complete your answers to this exercise on a separate sheet of paper.

#	Voltage (*E*)	Resistance (*R*)	Amperage (*I*)
1.	12 V	24 Ω	
2.		0.15 Ω	0.015 A
3.	24,000 V	24,000 Ω	
4.	0.057 V		0.00114 A
5.	36 V	12,000 Ω	
6.	12 V	48 Ω	
7.		48,000 Ω	1 A
8.	24 V		6 A
9.	24 V	0.024 Ω	
10.	12 V		8.0 A
11.	12 V		6 A
12.	12,000 V		6 A
13.	120 V	60 Ω	
14.		4.8 Ω	0.012 A
15.	12 V	30 Ω	
16.		12 Ω	0.0006 A
17.	24,000 V		0.024 A
18.		7.5 Ω	15 A
19.	12 V	8 Ω	
20.	18.0 V	1.8 Ω	

Watt's Law

In the early 1800s, Scottish inventor James Watt was looking for a method to describe electrical activity as a measurement of power. The result is a relationship known as **Watt's law,** which states:

One watt is the amount of work done in 1 second by 1 volt moving a charge of 1 amp in a circuit.

The standard unit of measurement for power is the watt. The conversion factor between watts and horsepower is *1 horsepower equals 746 watts.* In any electrical circuit, power (watts), voltage, and current act as a mathematical trio. If you know any two parts of the trio, you can mathematically figure the third part. The **Watt's law circle** is also known as the power triangle (Figure 3–27).

This circle is similar to the Ohm's law circle. In the circle, *P* represents power or watts of energy, *I* represents

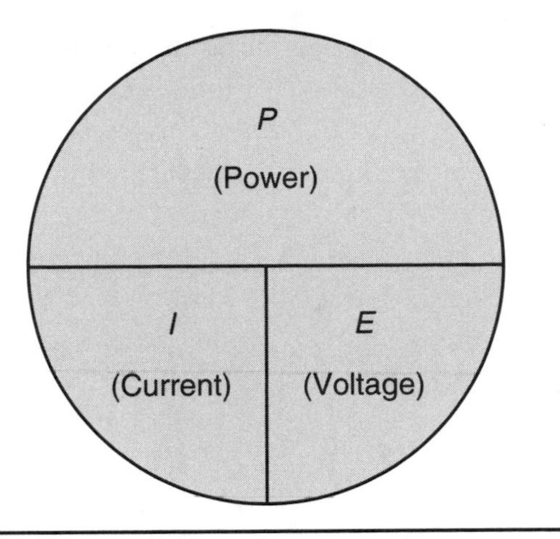

FIGURE 3–27 Watt's law equation circle.

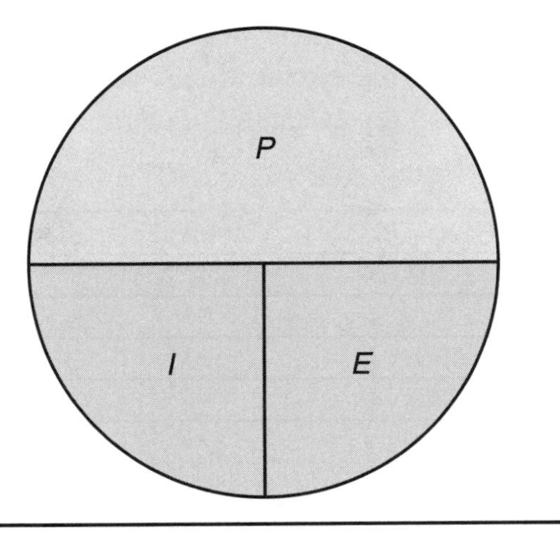

FIGURE 3–28 Use the exposed formula to find the unknown power (*P*).

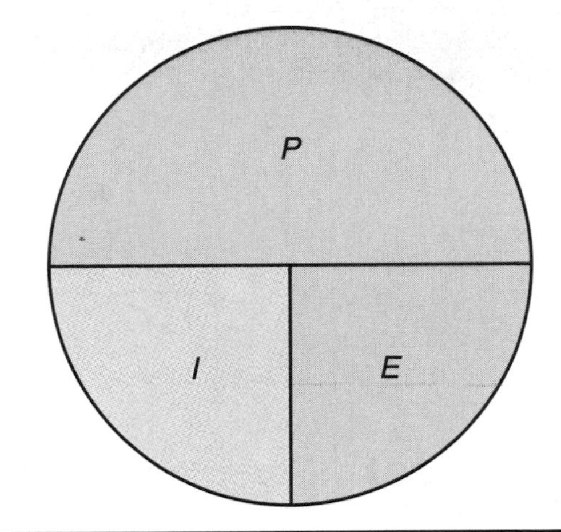

FIGURE 3–29 Use the exposed formula to find the unknown amperage (*I*).

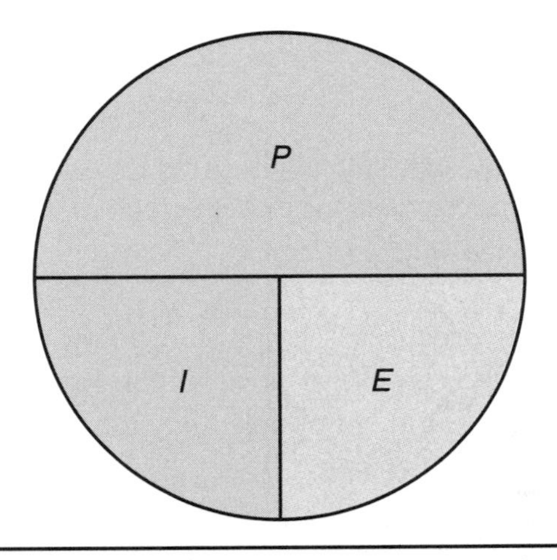

FIGURE 3–30 Use the exposed formula to find the unknown voltage (*E*).

intensity or amperage, and E represents electromotive force or voltage. Sometimes you may see W used to represent power A for amperage, or V for voltage, or a mixture of these. As long as you know the basic Watt's law circle, the mathematical equations remain the same. When any two of the basic factors are known, simply cover the unknown value, and calculate the same way you did when using Ohm's law.

To find *power* in a circuit $P = I \times E$ (Figure 3–28).

To find *amperage* in a circuit $I = P \div E$ (Figure 3–29).

To find *voltage* in a circuit $E = P \div I$ (Figure 3–30).

The most well-known example of power measurement in electrical circuits is the wattage rating on household light bulbs. The higher the wattage rating, the brighter the bulb's intensity or output. Even though automotive technicians do not generally use Watt's law to diagnose an electrical circuit, this law could aid you in a complete understanding of the circuit (Figure 3–31).

In Circuit A in Figure 3–31, the light bulb has 12 volts across its connections and 4 amps of current flow. Using the Watt's law circle from Figure 3–28, calculate the circuit's power output.

$$P = E \times I \qquad P = 12\text{ V} \times 4\text{ A} \qquad P = 48\text{ W}$$

In Circuit B (Figure 3–31), the additional resistance in the circuit lowers the total current flow (remember the relationship between current and resistance in Ohm's law),

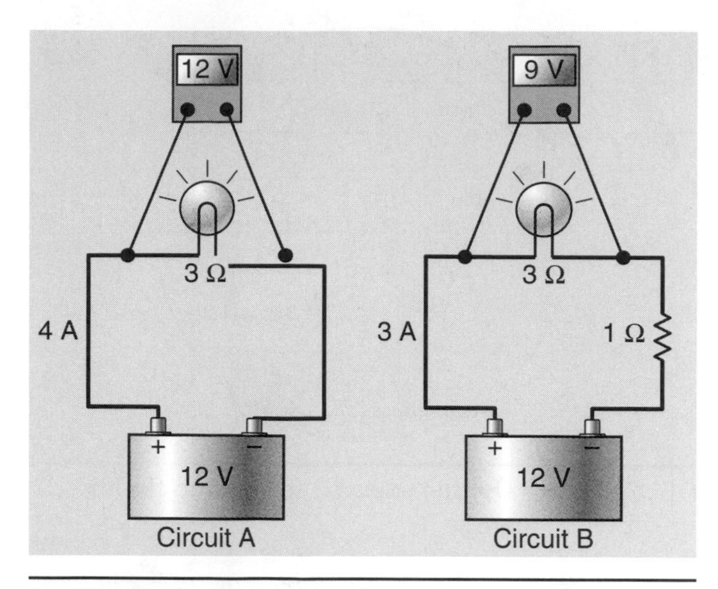

FIGURE 3–31 Simple circuits to demonstrate power measurement output.

and the voltage across the bulb is now 9 volts. Using the same formula, calculate the power output.

$$P = E \times I \qquad P = 9\,\text{V} \times 3\,\text{A} \qquad P = 27\,\text{W}$$

Circuit A produces 21 watts more power than Circuit B; so the bulb is almost twice as bright. Remember that the bulb wattage did not change; only the characteristics of the circuit itself changed.

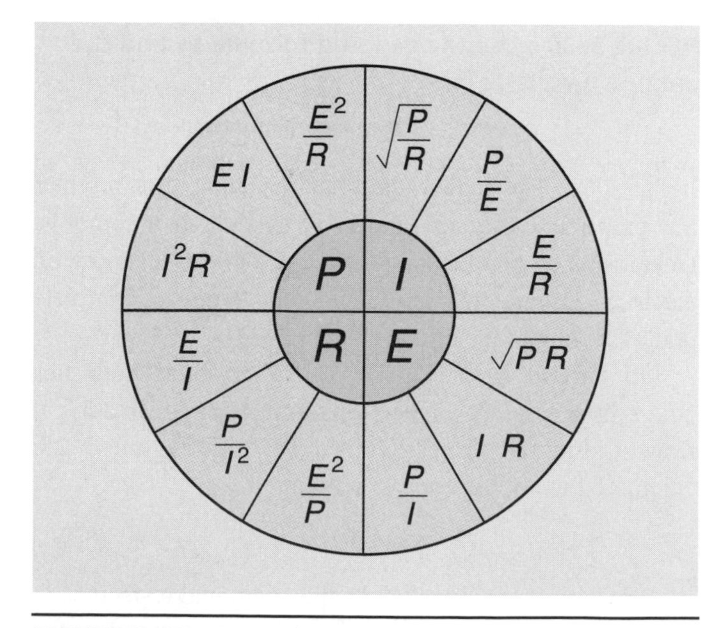

FIGURE 3–32 Formula circle for Ohm's law and Watt's law.

Although not as common to the automotive technician, other formulas in a **formula circle** can be used to *find the missing value* of both Ohm's law and Watt's law, depending on the known values (Figure 3–32). The formula circle has four mathematical sections, with three formulas in each section. This formula circle can be used whenever the basic Ohm's law or Watt's law formulas have more than one of the factors missing, but enough factors to work with if both laws are combined.

Exercise—Watt's Law

The following exercise is designed to give you practice using the Watt's law circle to determine the unknown value of an electrical circuit. Complete your answers to this exercise on a separate sheet of paper.

#	Power (*P*)	Voltage (*E*)	Amperage (*I*)
1.	24 W	24 V	
2.		12.0 V	0.025 A
3.	360 W	12.0 V	
4.	72 W		12 A
5.	18 W	12 V	
6.	12 W	0.036 V	
7.		72 V	12 A
8.	24 W		6 A
9.	24 W	24 V	
10.	36 W		18.0 A
11.	36.0 W		1.8 A
12.	0.036 W		18.0 A
13.	60 W	60 V	
14.		48 V	24 A
15.	24 W	0.006 V	
16.		24 V	0.0003 A
17.	2.4 W		0.024 A
18.		12.5 V	15 A
19.	12 W	3 V	
20.	180 W	5.0 V	

Symbols

In automotive electrical repair we use a variety of **symbols** to abbreviate or describe a device or circuit. Depending on the manufacturer, you may find differences

Symbols Used in Wiring Diagrams			
+	Positive		Temperature switch
—	Negative		Diode
	Ground		Zener diode
	Fuse		Motor
	Circuit breaker	→ ≻ C101	Connector 101
	Condenser	→	Male connector
Ω	Ohms	≻	Female connector
	Fixed value resistor		Splice
	Variable resistor	S101	Splice number
	Series resistors		Thermal element
	Coil		Multiple connectors
	Open contacts	88:88	Digital readout
	Closed contacts		Single filament bulb
	Closed switch		Dual-filament bulb
	Open switch		Light-emitting diode
	Ganged switch (N.O.)		Thermistor
	Single-pole double throw switch		PNP bipolar transistor
	Momentary contact switch		NPN bipolar transistor
	Pressure switch		Gauge

FIGURE 3–33 Some common electrical symbols used by auto manufacturers on wiring diagrams.

in the symbols used for the same device. Figure 3–33 is a sample list of common symbols used by auto manufacturers.

Be sure to use the symbol chart specific to your vehicle. Although there is some commonality of symbols, do not assume all symbols mean the same for each automotive manufacturer. We will be using electrical symbols extensively in this text to study wiring diagrams and circuits.

Magnetism

In the study of electrical production in Chapter 1, we learned that electricity is a form of natural energy that is converted to a usable electrical source. One of the conversions that plays an important role in the production of electrical energy is **magnetism.** Magnetism is an invisible force that can repel or attract through an air space or solid item (Figure 3–34).

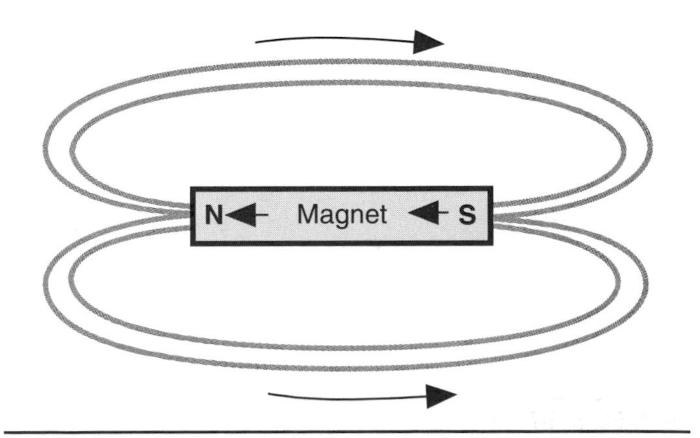

FIGURE 3–34 The invisible forces of magnetism.

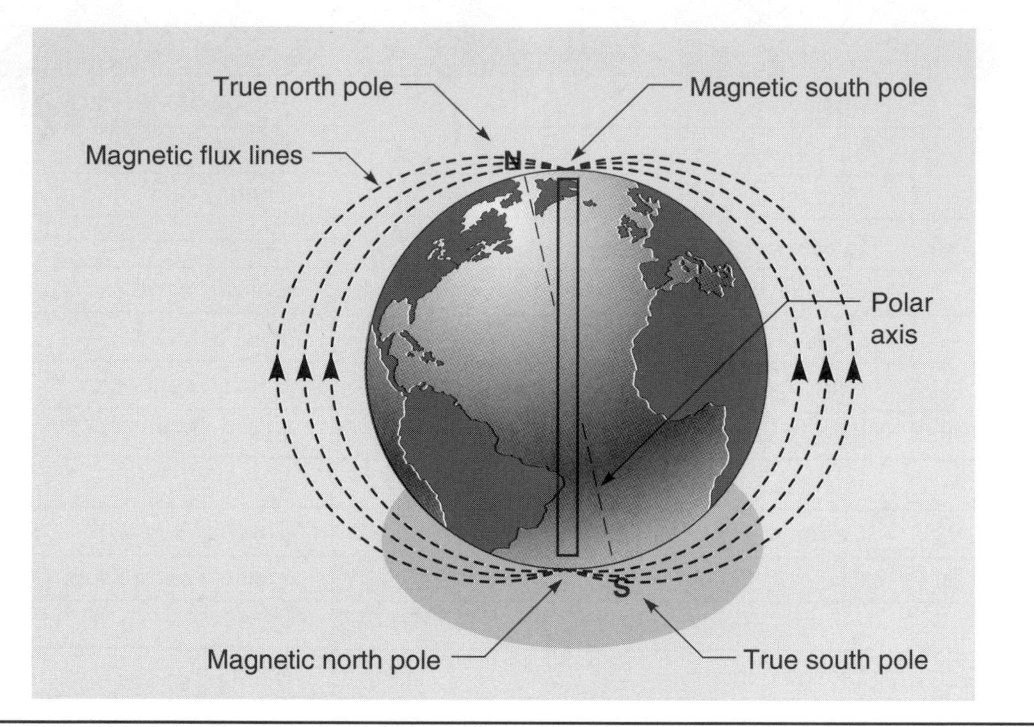

FIGURE 3–35 Earth's magnetic field.

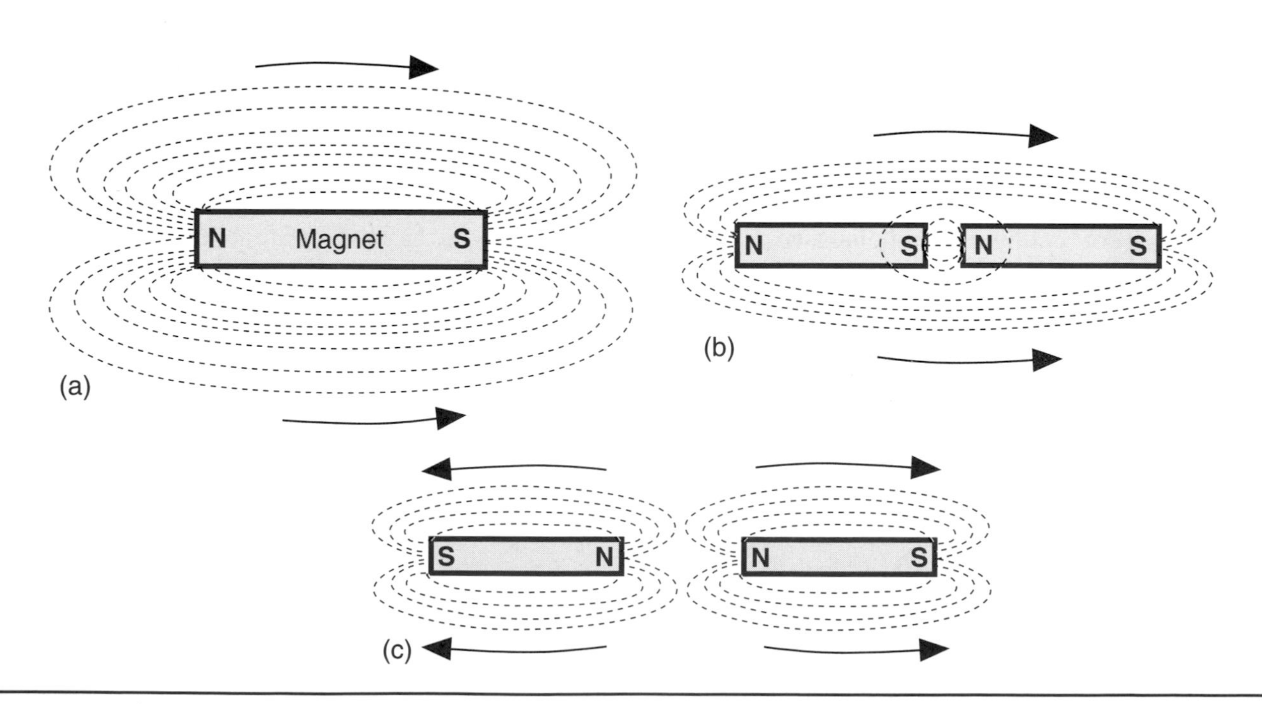

FIGURE 3–36 The principles of magnetism: (a) magnetic north and south poles, (b) unlike poles attract each other, and (c) like poles repel each other.

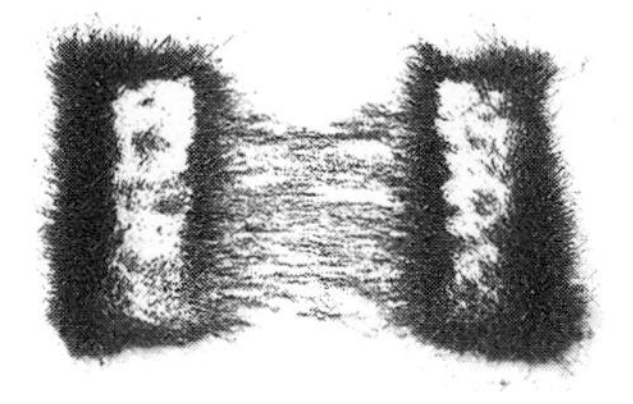

FIGURE 3–37 The invisible magnetic flux lines become visible.

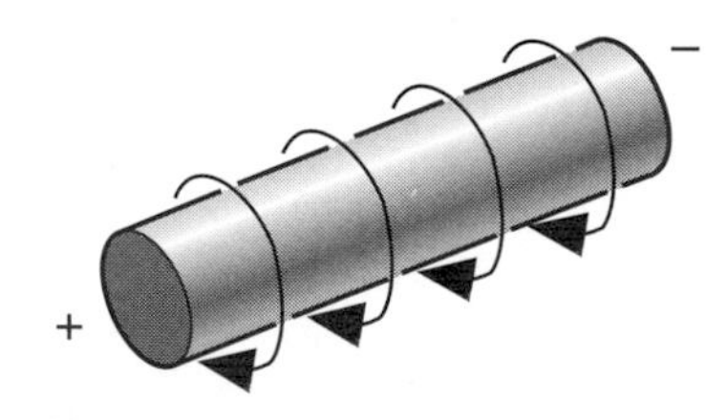

FIGURE 3–39 When current flows through a conductor, a magnetic field surrounds it.

Earth itself is a large magnet with a north pole and a south pole. Magnetic flux lines wrap around the earth like a giant magnetic shell (Figure 3–35).

This phenomenon is based on the fact that a simple magnet has one north pole and one south pole. Magnetic poles that are alike repel each other and poles that are unlike attract each other (Figure 3–36).

Even though the lines of magnetic force are invisible, we can use iron filings surrounding a magnet to indicate where the lines of force are (Figure 3–37).

The closer to the magnet's core, the stronger the magnetic field. This is discussed more in later chapters as we learn how magnetism is harnessed for automotive electrical use.

Electromagnetism

In the early nineteenth century, André Marie Ampère discovered that when a compass is held over a wire carrying current, the needle on the compass points at right angles to the wire (Figure 3–38).

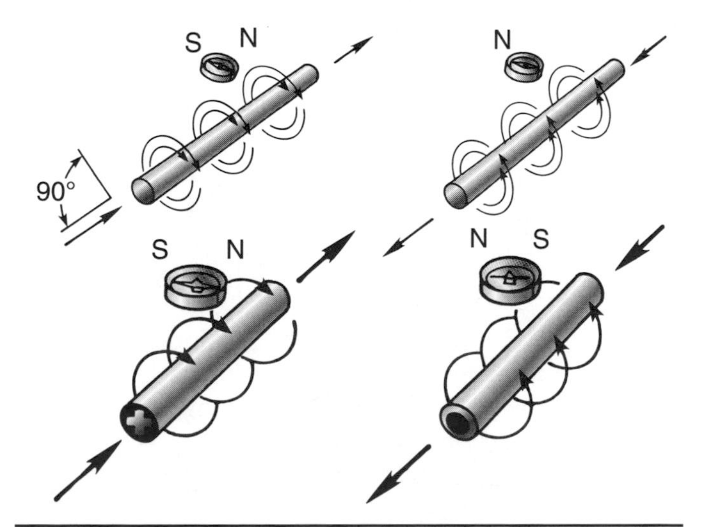

FIGURE 3–38 Magnetic flux lines cause the compass needle to deflect at a right angle to the conductor.

The movement of the compass needle is due to a circular magnetic field that forms around the wire and is perpendicular to the current flow through the wire. If the current increases, so does the strength of the magnetic field. The direction of the circular magnetic flux lines is determined by the **right-hand rule,** which is based on conventional flow theory (positive-to-negative flow) (Figure 3–39).

If you place your right hand on a wire with your thumb pointing in the direction of current flow, your fingers will point in the direction of the magnetic flux lines, and your thumb will point to the negative side of the circuit (Figure 3–40).

If you do not know which side of an electrical circuit is the positive and which is the negative, the following simple compass experiment will assist you (Figure 3–41).

Step 1 Place a compass on a wire that has current flowing through it, and position the west and east compass headings parallel to the conductor, or wire.

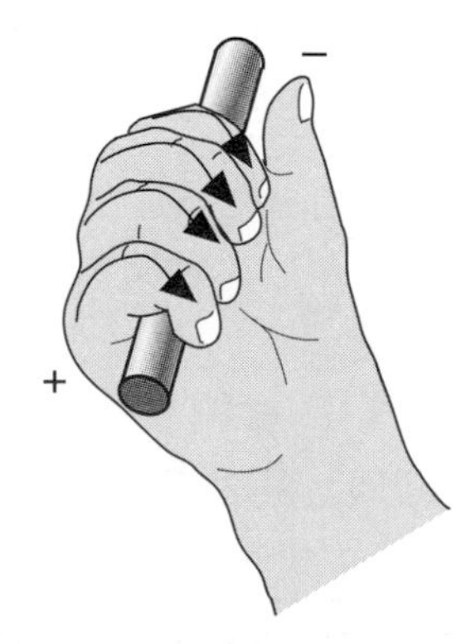

FIGURE 3–40 Right-hand rule of magnetism.

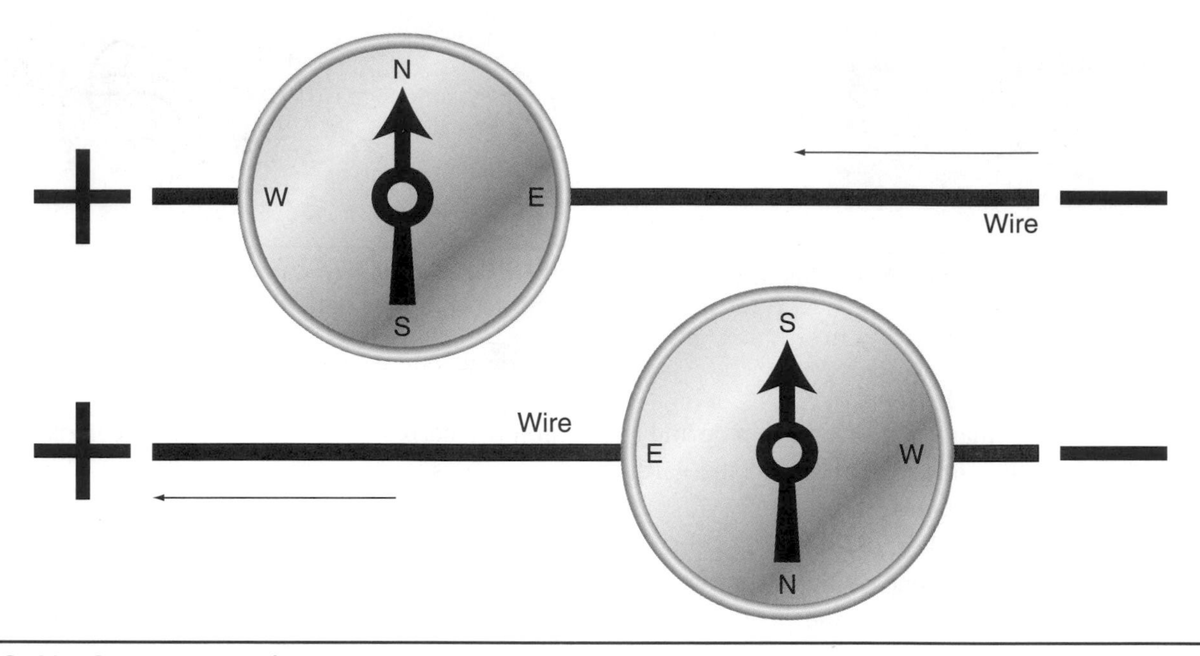

FIGURE 3–41 Compass experiment.

Step 2 Observe the compass needle direction. If the needle pointer points to the north heading, the positive end is to the left of the needle pointer, or west. If the needle pointer points to the south heading, then the positive end is also to the left of needle, or east.

Analysis When you place the west and east compass headings parallel to and over a wire, no matter which way in relation to polarity, the positive polarity is *always* to the left of the compass pointer when the pointer is positioned upward, or pointing away from you.

To increase the magnetic field strength around a wire, another wire can be wound around it in equal size turns. Increasing the number of turns increases the magnetic field (Figure 3–42). When current flows through the wound wire, the magnetic field that normally would surround a straight wire instead combines to form a stronger magnetic field.

If a soft iron core is placed inside a coil of wire and a current is applied to the wire, the magnetic field is further increased, and it becomes an **electromagnet** (Figure 3–43).

Electromagnetic Induction (EMI)

Magnetic fields can also be used to produce electricity. Instead of using current to generate a magnetic field, we can move a conductor within a magnetic field to generate a voltage. Similarly, we can move a magnetic field across a conductor to produce a voltage (Figure 3–44).

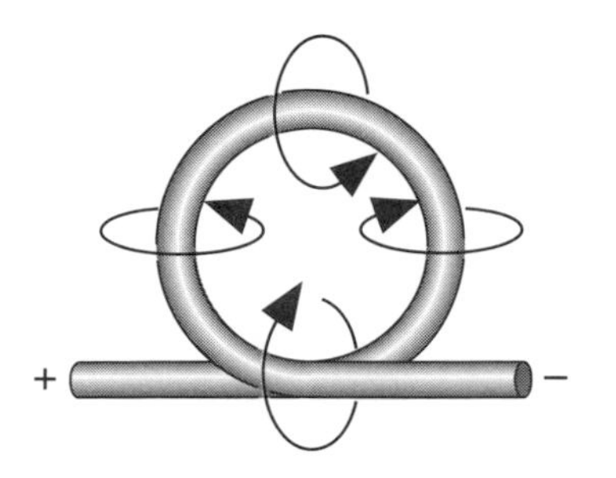

FIGURE 3–42 Magnetic field strengthens by looping the wire conductor into equal size turns.

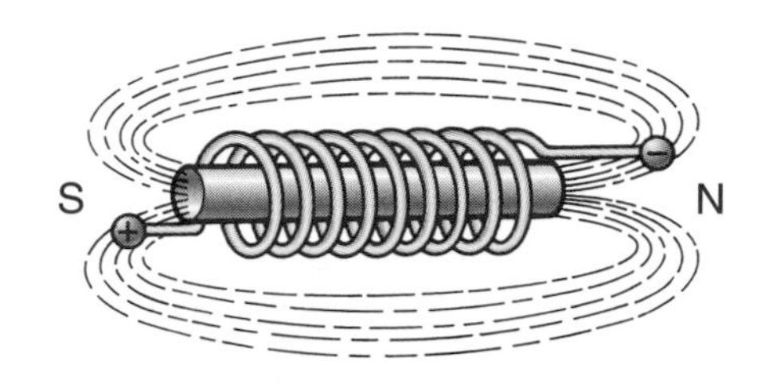

FIGURE 3–43 A simple electromagnet.

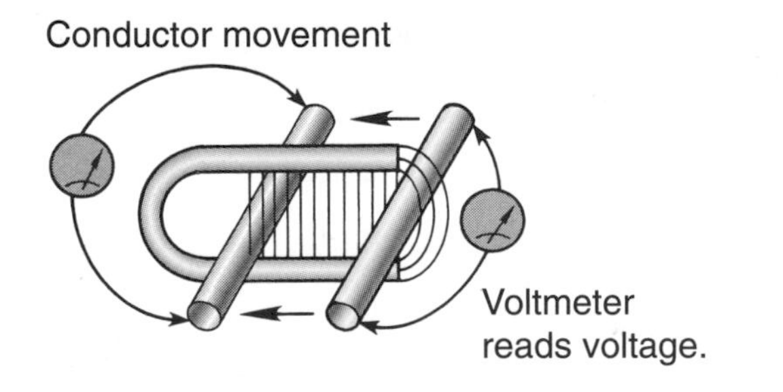

FIGURE 3–44 Moving a conductor through a magnetic field creates a voltage across the conductor.

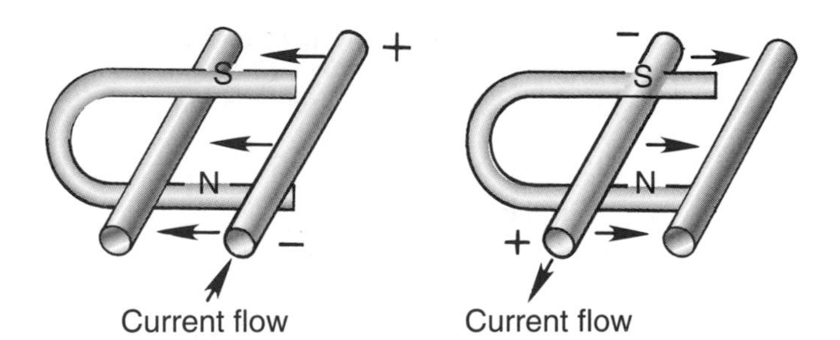

FIGURE 3–45 Voltage polarity is determined by the direction the conductor passes through in the magnetic field.

As the conductor shown in Figure 3–45 moves rapidly through the magnetic field and into the magnet, a small voltage is induced in the conductor. This voltage has a specific polarity with a positive side and a negative side. When the conductor is brought back rapidly through the magnetic field from the inside, the polarity switches (Figure 3–45).

The electromagnetic induction (EMI) induced in a wire or conductor depends on five main factors:

1. The larger the conductor diameter, the less the internal resistance is.
2. The higher the number of conductor turns, the higher the induced voltage is.
3. If the intensity of the magnetic field is increased, then the induced voltage is increased.
4. As the gap between the magnetic field and the conductor decreases, the voltage potential increases.
5. As the speed of the conductors cutting through magnetic flux lines increases, the voltage induced increases.

The significance of learning magnetism and electromagnetism theory will be evident as you study later chapters that cover ignition systems, starting and charging systems, and electrical relay control circuits.

Summary

- ❏ Matter can be any substance that occupies a space and contains a mass.
- ❏ Molecules consist of a group of elements.
- ❏ Elements are known as a cluster of one type of atom.
- ❏ The table of elements contains over 100 elements of nature.
- ❏ An atom is the smallest part of an element.
- ❏ Atoms contain protons, neutrons, and electrons.
- ❏ Protons are positively charged particles.
- ❏ Neutrons have no charge and combine with the protons to form the nucleus of the atom.
- ❏ Electrons are negatively charged particles.

- ❑ The electrons in the outermost rotating orbit of an atom are called valence electrons.
- ❑ Free electrons are valence electrons that become dislodged from their orbit.
- ❑ A bound electron is free to move within its own orbit only, and it remains a fixed distance from the nucleus.
- ❑ Electricity is defined as the movement of free electrons from atom to atom.
- ❑ Good conductors of electricity have three or fewer valance electrons.
- ❑ Semiconductors have exactly four valance electrons.
- ❑ Good insulators of electricity have more then four valance electrons.
- ❑ Current flow is the mass movement of free electrons from atom to atom.
- ❑ One ampere of current flow is equal to 6.28 billion billion free electrons flowing past a point in 1 second.
- ❑ Conventional flow theory says current flows from positive to negative.
- ❑ Electron flow theory says current flows from negative to positive.
- ❑ Amperage is often recorded by using the letter *I* for intensity of current or the letter *A* for amperes.
- ❑ EMF is electromotive force, or voltage pressure in a circuit.
- ❑ Resistance is any force or substance that resists or opposes the flow of current in a circuit.
- ❑ The resistance symbol often used is the letter *R* or the Greek omega symbol Ω.
- ❑ Circuit resistance is determined by five factors: atomic structure, condition and quality of the circuit, length of wire, diameter of wire, temperature of the circuit.
- ❑ Ohm's law is a clearly defined relationship between voltage, current, and resistance.
- ❑ The Ohm's law circle consists of voltage, current, and resistance working together as a mathematical trio.
- ❑ Watt's law is a measurement of power in an electrical circuit.
- ❑ The Watt's law circle consists of power, voltage, and current working together as a mathematical trio.
- ❑ The *formula circle* consists of all the combinations of mathematical formulas used in Ohm's law and Watt's law.
- ❑ Symbols are used by automotive manufacturers to abbreviate or describe a device or circuit.
- ❑ Magnetism is an invisible force causing like poles to repel each other and unlike poles to attract each other.
- ❑ The right-hand rule of magnetism determines which way the magnetic flux lines point.
- ❑ The amount of EMI induced in a wire depends on five factors: diameter of the conductor, number of conductor turns, increasing the magnetic field, lowering the gap between the conductor and magnetic field, and changing the speed of the conductor cutting through the magnetic field.

Key Terms

alternating current (AC)
amperes
atomic number
atom
bound electron
conductors
conventional flow theory
current flow
direct current (DC)
electromagnet
electromotive force (EMF)
electron
electron flow theory
element
formula circle
free electron
insulators
magnetism
matter
molecule
neutron
nucleus
Ohm's law
Ohm's law circle
proton
resistance
right-hand rule
semiconductors
symbols
valence electron
voltage
Watt's law
Watt's law circle

Review Questions

Short Answer Essays

1. What determines whether an atom is a good conductor, semiconductor, or insulator of electricity?
2. Describe current flow in an electrical circuit.
3. Describe voltage in an electrical circuit.
4. Describe resistance in an electrical circuit.
5. Explain the five main factors that determine how much EMI is induced in a wire or conductor.

Fill in the Blanks

1. Magnetism is an ________________________ force that can repel or attract through an ________________________ space or solid item.
2. The current in an electrical circuit is ________________________ proportional to the resistance of the circuit and ________________________ proportional to the voltage.
3. Electron flow theory suggests that free electrons move from the ________________________ side toward the ________________________ side of the circuit.
4. The electrons in the outermost orbit are called ________________________ electrons.
5. A typical atom consists of ________________________ and ________________________ that form the nucleus and ________________________ that rotate in orbits around the nucleus.

ASE-Style Review Questions

1. Technician A says an atom is made up of protons, neutrons, and electrons. Technician B says the center of the atom is made up of protons and electrons only. Who is correct?
   A. A only
   B. B only
   C. Both A and B
   D. Neither A nor B

2. A simple law of physics states that like charges repel each other and unlike charges attract each other. Technician A says this law is what keeps the protons and electrons from being attracted to each other. Technician B says it is the centrifugal force of the rotating electrons in orbit that keeps them from being attracted to the protons. Who is correct?
   A. A only
   B. B only
   C. Both A and B
   D. Neither A nor B

3. A copper atom is an excellent conductor of electricity. Technician A says this is due to having one valance electron. Technician B says the neutrons determine the conductivity of the atom. Who is correct?
   A. A only
   B. B only
   C. Both A and B
   D. Neither A nor B

4. Technician A says the electrons in the outermost orbit of an atom are the valence electrons. Technician B says the free electrons are dislodged from their orbit and jump from atom to atom. Who is correct?
   A. A only
   B. B only
   C. Both A and B
   D. Neither A nor B

5. Technician A says the nucleus of the atom determines whether an atom is a good conductor or insulator of electricity. Technician B says the number of bound electron orbits determines whether an atom is a good conductor or insulator of electricity. Who is correct?
   A. A only
   B. B only
   C. Both A and B
   D. Neither A nor B

6. Technician A says the flow of free electrons in a circuit is called current flow. Technician B says that 6.28 billion billion electrons flowing past a point in 1 second is 1 amp. Who is correct?
   A. A only
   B. B only
   C. Both A and B
   D. Neither A nor B

7. Technician A says that the letter *I* indicates amperage in an electrical circuit. Technician B says the letter *A* indicates amperes in an electrical circuit. Who is correct?
   A. A only
   B. B only
   C. Both A and B
   D. Neither A nor B

8. Technician A says conventional flow theory means current flows from the most positive potential to the least positive or negative. Technician B says conventional flow theory means that the battery's negative terminal is the ground and that the positive terminal is the hot side. Who is correct?
   A. A only
   B. B only
   C. Both A and B
   D. Neither A nor B

9. Technician A says electron flow theory suggests that current flows from negative to positive. Technician B says that in electron flow theory the free electrons move from a surplus side to an area having less free electrons. Who is correct?
   A. A only
   B. B only
   C. Both A and B
   D. Neither A nor B

10. Technician A says the pressure in a circuit is called electromotive force (EMF). Technician B says the pressure in a circuit is called voltage. Who is correct?
    A. A only
    B. B only
    C. Both A and B
    D. Neither A nor B

11. Technician A says current flows when the voltage is equal on both sides of a circuit. Technician B says voltage is a measurement of a pressure difference across a load. Who is correct?
    A. A only
    B. B only

C. Both A and B
D. Neither A nor B

12. Technician A says the greater the circuit resistance is, the greater the current flow is. Technician B says the lower the current flow is, the lower the circuit resistance is. Who is correct?
    A. A only
    B. B only
    C. Both A and B
    D. Neither A nor B

13. Technician A says the resistance of a wire increases as the length increases. Technician B says the resistance of a wire increases as the diameter increases. Who is correct ?
    A. A only
    B. B only
    C. Both A and B
    D. Neither A nor B

14. Technician A says Ohm's law states that current is inversely proportional to the resistance in a circuit. Technician B says that as circuit resistance goes up, the current decreases. Who is correct?
    A. A only
    B. B only
    C. Both A and B
    D. Neither A nor B

15. Technician A says the current in an electrical circuit is directly proportional to the resistance. Technician B says current is inversely proportional to the voltage. Who is correct?
    A. A only
    B. B only
    C. Both A and B
    D. Neither A nor B

16. Technician A says, to find the voltage of a circuit, the amperage is divided by the resistance. Technician B says the letter *E* represents electromotive force. Who is correct?
    A. A only
    B. B only
    C. Both A and B
    D. Neither A nor B

17. Technician A says a simple circuit with 12 volts and 8 ohms should have an amperage total of 1.5 amps. Technician B says if the resistance increases on the same circuit from 8 ohms to 12 ohms, then the amperage total would also increase by the same amount. Who is correct?
    A. A only
    B. B only
    C. Both A and B
    D. Neither A nor B

18. A simple circuit with 24 volts and 8 ohms is being discussed: Technician A says that to find the power of the circuit, you must first find the circuit current. Technician B says this circuit should have 72 watts of power. Who is correct?
    A. A only
    B. B only
    C. Both A and B
    D. Neither A nor B

19. Two technicians are discussing the formula circle that combines Ohm's law and Watt's law into 4 sections and 12 equations. Technician A says this formula circle can be used when only one factor is known for a circuit. Technician B says the known values of both laws can be combined to find a missing factor. Who is correct?
    A. A only
    B. B only
    C. Both A and B
    D. Neither A nor B

20. Technician A says all wiring diagram symbols are standard to all auto manufacturers. Technician B says there may be a difference in symbol usage between auto manufacturers. Who is correct?
    A. A only
    B. B only
    C. Both A and B
    D. Neither A nor B

21. Technician A says the following is the correct symbol for a circuit fuse.

Technician B says the following is the correct symbol for a circuit fuse.

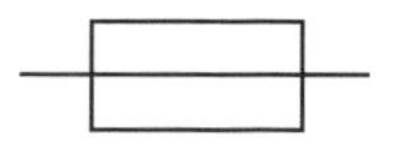

Who is correct?

A. A only

B. B only

C. Both A and B

D. Neither A nor B

22. Technician A says magnetism is an invisible force that can be harnessed for electrical production. Technician B says the magnetic field is strongest at the magnet's core. Who is correct?

    A. A only

    B. B only

    C. Both A and B

    D. Neither A nor B

23. Technician A says the closer you get to the center of the magnetic field lines, the weaker the magnetism is. Technician B says the right-hand rule of magnetism determines the circular direction of the magnetic flux lines. Who is correct?

    A. A only

    B. B only

    C. Both A and B

    D. Neither A nor B

24. Technician A says a magnetic field can be used to produce electricity. Technician B says current flow in a conductor can produce a magnetic field. Who is correct?

    A. A only

    B. B only

    C. Both A and B

    D. Neither A nor B

25. Technician A says if a conductor is moved rapidly through a magnetic field, a voltage is induced in the conductor. Technician B says that when the gap between the magnetic field and conductor increases, the voltage induced in the conductor also increases. Who is correct?

    A. A only

    B. B only

    C. Both A and B

    D. Neither A nor B

4 Measurement System and Calculator Use

Introduction

During the past two hundred years, much of the world has utilized the International System (IS) of metric measurement. In contrast, the United States has used the U.S. Customary System (USC) of measurement, also known as the English System.

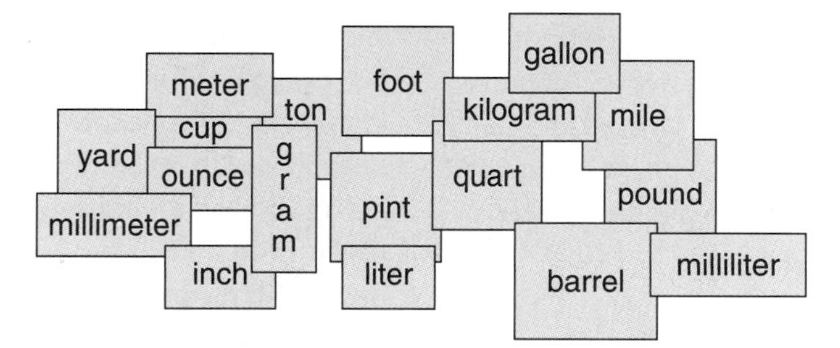

Since the early 1900s, the United States Congress has attempted to get the American public and the industrial community to convert to the metric system. In 1975, when Congress passed the Metric Conversion Act, the scientific community had been using the metric system for decades. The industrial community now started to move toward adopting the metric system, especially the automotive manufacturers who began using more metrics in their products. This chapter provides a brief look at the metric system in preparation for a thorough understanding of the principles of electrical measurement.

Most modern digital multimeters (DMM) display only four or five digits, making it necessary to use measurement abbreviations. We will practice using the scientific notation features of a scientific calculator, which will be helpful in later chapters when we deal with electrical theory and associated mathematical formulas.

Objectives

When you complete this chapter you should be able to:

- ❑ Describe the metric system of measurement.
- ❑ Apply basic metric system conversion principles.
- ❑ Compare metric prefix terms to electrical prefix terms.
- ❑ Demonstrate the basic features of a scientific calculator.
- ❑ Use scientific notation in conversions.

The Metric Measurement System

Today, all over the United States and the world, manufacturers, scientists, businesspeople, government offices, the military, private homes, and even the auto industry use the metric system as the *primary* system of measurement. The metric system is very simple in design—an increase or decrease from the base unit of measurement always operates on the **power of ten** principle. Mastering the metric system will help when you learn electrical theory because you can directly apply its principles to the base unit identifiers used for electrical terms (more on this later). The base units for the metric system are listed in Figure 4–1.

FIGURE 4–1 Basic metric units of measurement.

When each base unit is increased by a power of ten, it becomes a new number that is ten times the previous number. For example, if 1 **gram** (g) is the base unit, increasing it by a power of ten (1×10) makes 10 grams. If you increase this new number by another power of ten, or 10^2 relative to the base unit, you have 10×10, or 100 grams total.

Each power of ten has a name attached to its new unit, which is the **prefix.** Table 4–1 shows the most common units of measurements, prefixes, and symbols for the metric system.

For example, 1,000 grams can be referred to as 1 **kilo**gram (1 kg), or 1 gram $\times 10^3$. Using this type of notation, called **scientific notation,** can help when dealing with awkwardly large or small numbers. When units of measurement are smaller than the base unit, divide by powers of ten to make a smaller value number. For example, 1 gram divided by 10 (10^{-1}) is now 1/10 gram or 0.1 decigram (0.1 dg).

Because you will be working extensively with various units of measurements, you need to convert between the different metric prefix values. An easy way to do this without using math formulas is to use a **linear scale** of metric conversions (Figure 4–2).

Unit values to the left of the base unit (0) are larger, and unit values to the right of the base unit are smaller. For example, if you wanted to change 15,000 **meters** to kilometers (km), you would move the decimal to the left three places to make 15 km (Figure 4–3).

If you had 84 grams and wanted to change the number to **milli**grams (mg), you would move the decimal to the right three places to make 84,000 mg (Figure 4–4).

Just remember the following when you change the unit value bigger or smaller:

Larger unit ←————→ Smaller unit

Let's show a few more examples before giving you several to try yourself. Convert 275.73 **micro**grams (μg) to mg (milligrams). This means the decimal must move 3 places to the left to reach the milli unit of measurement (Figure 4–5).

TABLE 4–1 Metric prefix measurement table.

Prefix	Symbol	Name	Power of 10	Decimal Equivalent
tera	T	one trillion	10^{12}	1,000,000,000,000
giga	G	one billion	10^{9}	1,000,000,000
mega	M	one million	10^{6}	1,000,000
kilo	k	one thousand	10^{3}	1,000
hecto	h	one hundred	10^{2}	100
deka	da	ten	10^{1}	10
base unit		one unit	10^{0}	1
deci	d	one tenth	10^{-1}	0.1
centi	c	one hundredth	10^{-2}	0.01
milli	m	one thousandth	10^{-3}	0.001
micro	μ	one millionth	10^{-6}	0.000001
nano	n	one billionth	10^{-9}	0.000000001
pico	p	one trillionth	10^{-12}	0.000000000001

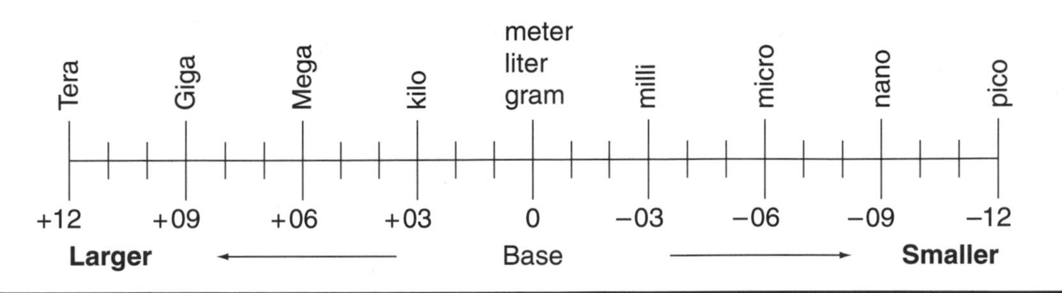

FIGURE 4–2 Linear scale of metric conversion.

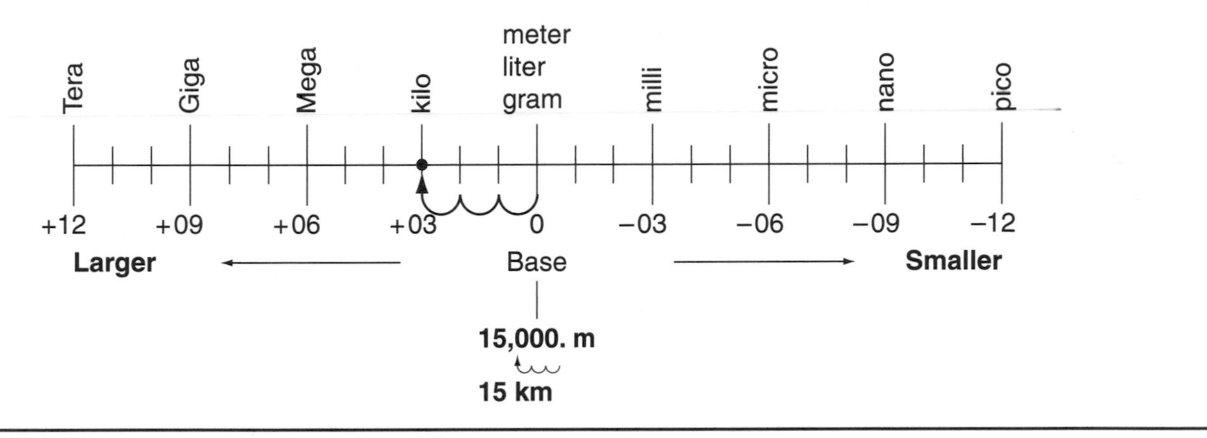

FIGURE 4–3 Conversion to a larger unit of measurement.

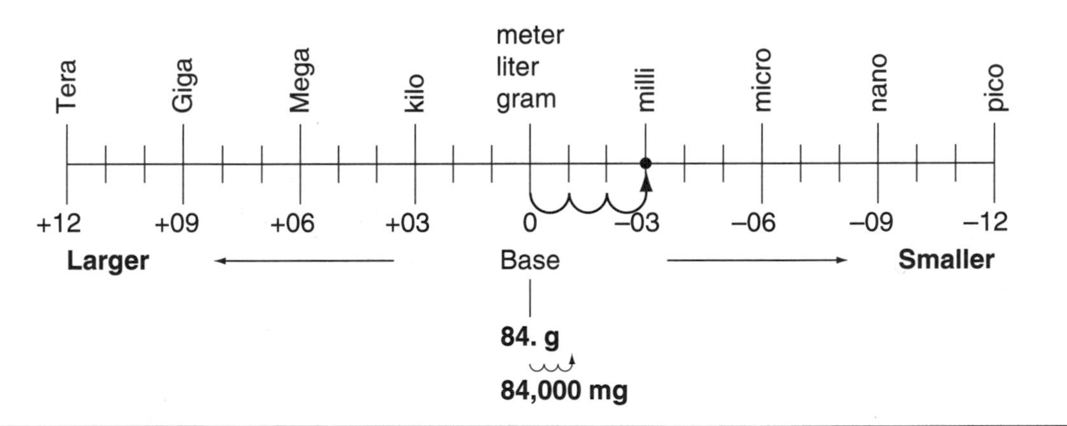

FIGURE 4–4 Conversion to a smaller unit of measurement.

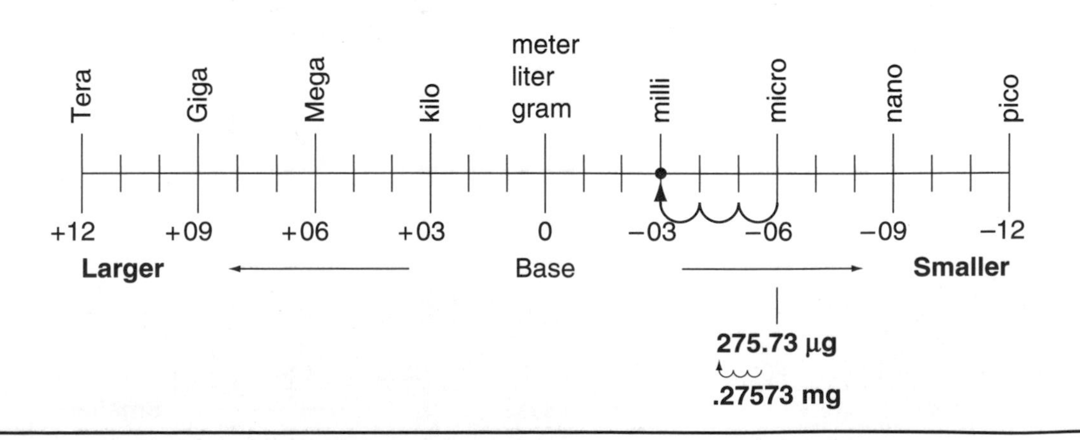

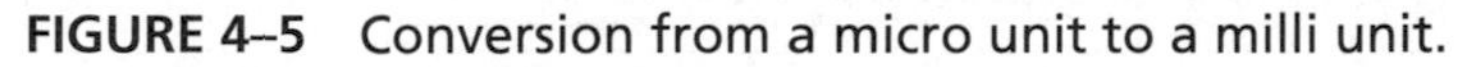
FIGURE 4–5 Conversion from a micro unit to a milli unit.

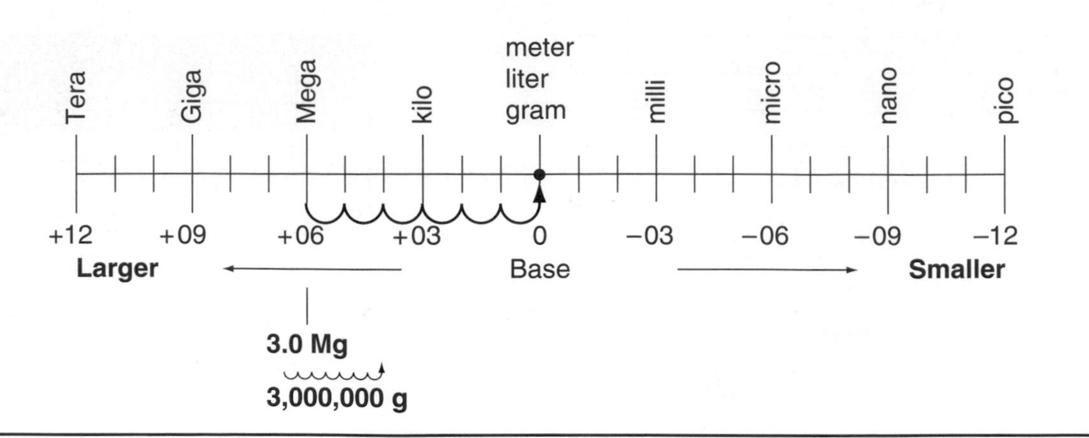

FIGURE 4–6 Conversion from a mega unit to a whole unit.

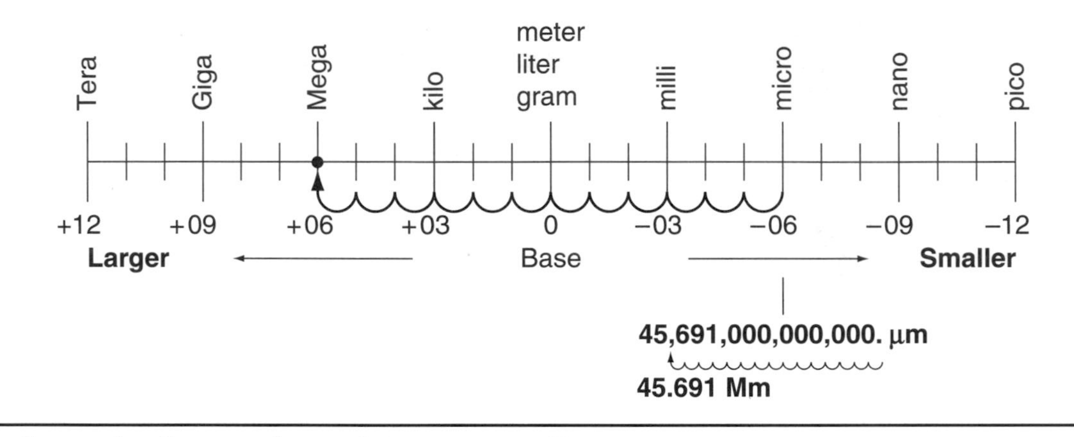

FIGURE 4–7 Conversion from a micro unit to a mega unit.

If you had 3.0 **mega**grams (Mg) and wanted to change to the whole number 3 million, you need to move the decimal to the right 6 places to get to grams (Figure 4–6).

Why not try changing an extra large number to a shorter number to make it easier to mathematically manipulate? If you had 45,691,000,000,000 µm and converted the number to megameters (Mm), you would move the decimal to the left 12 places (Figure 4–7).

Try the practice following conversions using the linear scale conversion method. Write your answers on a *separate* sheet of paper.

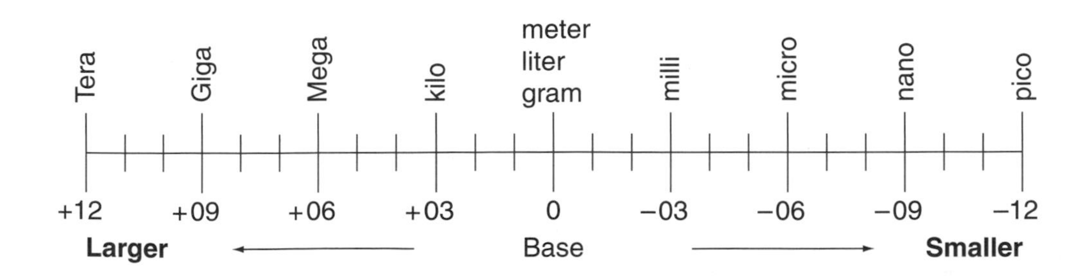

Original Number		Change To	Answer
1.	23.34 m (meters)	mm (millimeters)	23340 mm
2.	0.00267 Mm	m	
3.	51.9876 km	μm	
4.	20 μm	m	
5.	100 m	Mm	
6.	4888.269 mg	kg	
7.	56.9234 Mm	μm	
8.	22 mm	m	
9.	15 mg	g	
10.	100 g	mg	
11.	25789 mm	km	
12.	45.45 Mg	cg	
13.	1.56789 km	μm	
14.	39005.567 mg	Mg	
15.	1 m	μm	
16.	1 km	Mm	
17.	29.456 cm	mm	
18.	999999999.9 μm	Mm	
19.	999999999.9 Mg	μg	
20.	54.545454 m	cm	

FIGURE 4–8 Typical metric conversion calculator.

Converting between metric units and English, or standard, units of measurements normally requires a metric conversion chart included in most automotive reference material (Appendix D). For example, if you were converting 55 miles to kilometers, look in Appendix D to see that miles is multiplied by 1.60935 to get kilometers. The answer is 88.51 kilometers. Another way to perform conversions is to use a metric conversion calculator. Several brands are available for this purpose (Figure 4–8). A metric calculator can be very useful for the automotive technician since you may encounter metric specifications when your tools or test equipment are in standard units of measurement.

The three basic units of electrical measurements we discussed in Chapter 3 are:

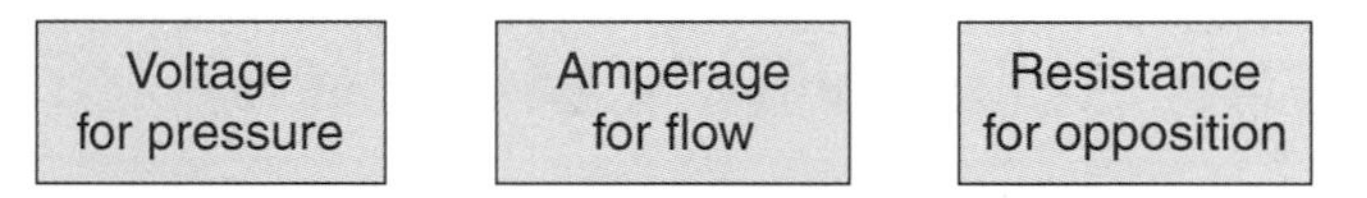

The linear scale we used for metric conversions can be modified for converting basic electrical units (Figure 4–9).

For example, 52.753 A (amps) converted to milliamps (mA) would now be 52,753 mA. For more conversion

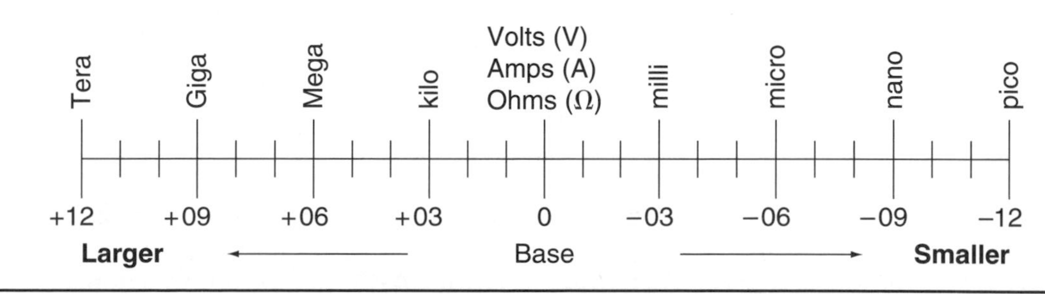

FIGURE 4–9 Linear scale of electrical conversion.

practice, use the linear scale for electrical conversions (Figure 4–9) to solve the following conversions. Write your answers on a *separate* sheet of paper.

Original Number	Change To	Answer
1. 123.25 mV (millivolt)	V (volt)	.12325 V
2. 0.64 kΩ	Ω	
3. 50.244 μV	mV	
4. 12 Ω	μΩ	
5. 0.00078 A	mA	
6. 0.000294 MV	V	
7. 0.0000032473 mA	kA	
8. 3 V	μV	
9. 3 V	mV	
10. 3 V	kV	
11. 3 V	MV	
12. 6,000,000 μA	A	
13. 6 A	kA	
14. 99.9 kΩ	mΩ	
15. 99 MA	mA	
16. 0.034 μV	mV	
17. 0.000055 MA	A	
18. 55,000,000 mΩ	kΩ	
19. 0.000000439 MA	A	
20. 50.54 μV	mV	

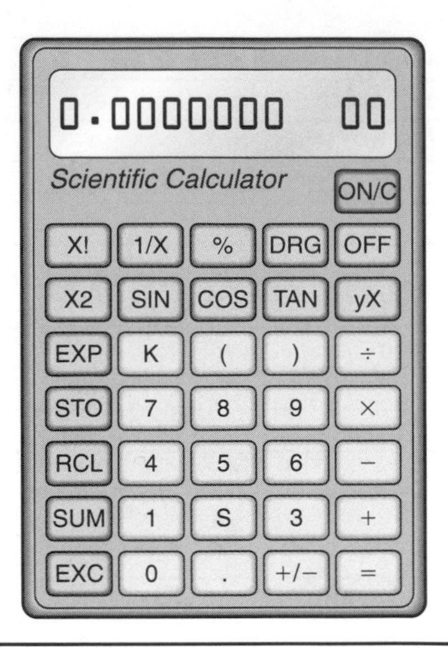

FIGURE 4–10 Typical scientific calculator.

The Scientific Calculator

Automotive technicians, when learning how electrical circuit values and laws interrelate (Ohm's law and Watt's law), may need to practice performing several complex mathematical calculations to determine circuit operations. The use of a scientific calculator can help make this learning fun and easy for two important reasons. First, when you perform these calculations the number unit values *must* be the same. For example, you cannot multiply a kilo-ohm by a milliamp without keeping track of the unit values. Second, sometimes these calculations contain extremely small or large numbers. Since most calculators contain only enough spaces for 8 digits, using the scientific notation features of a calculator permit you to deal with extra large or small numbers. Figure 4–10 is a typical scientific calculator.

Your scientific calculator must have two important features. The first is an EXP feature, which is the power of ten discussed earlier in this chapter and in Table 4–1. Some scientific calculators use an **EE** key for the power of 10 feature instead of the EXP key. The second feature is the **1/X** key. This gives you the reciprocal, or inverse, of a number. The keys for these features are shaded keys on the sample scientific calculator in Figure 4–11. If you are unsure of the actual location of these features on your calculator, refer to the calculator owner's manual. Some of the more expensive scientific calculators have several functions for each key; so you may have to push a **2nd function** key to activate the EXP feature.

FIGURE 4–11 Important function keys for electrical calculations.

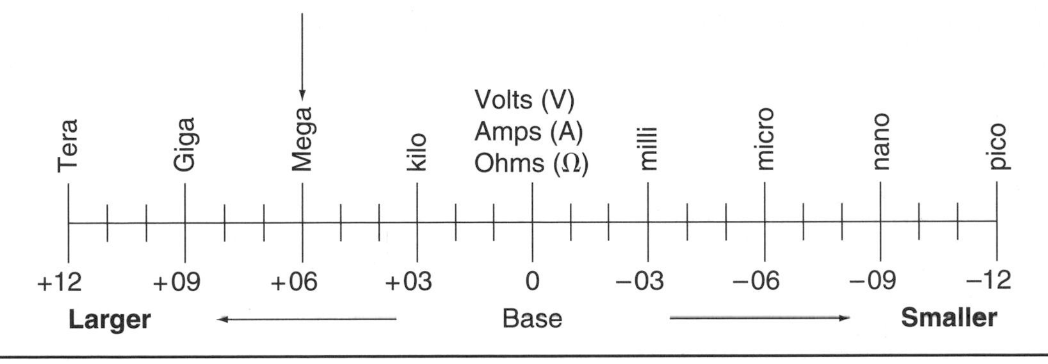

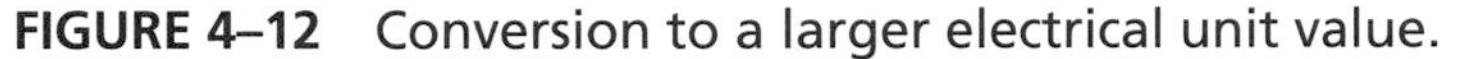
FIGURE 4–12 Conversion to a larger electrical unit value.

The EXP key is normally used for extra large or extra small numbers that do not fit on the standard 8-digit calculator display. When you push the EXP or EE key, you will notice a separate set of zeros that appear offset on the right side of the display. This is the scientific notation feature. As you work with complex problems, this feature can be a big help. For example, the number 125 million would contain more then the 8 digits; so the EXP feature is useful. With a number like 125,000,000 you refer to the linear scale (Figure 4–12) and see that million is +06 power, or mega.

On the calculator you would enter:

125 EXP 6

The calculator should display:

125 ^{06}

When you need to enter a number with a unit value smaller than a whole number, an additional calculator step is necessary. Suppose you had 55 microamps (μA) in your calculation. On the linear scale (Figure 4–13), the unit value is −06 power, or micro. Without the scientific notation feature, a number would have to be converted to a whole unit value by moving the decimal 6 places to the left to get the value 0.000055 A.

On the calculator enter:

55 EXP 6 [+/−]

The calculator should display:

55 $^{-06}$

The additional key used was the **+/− key,** located in the bottom left corner next to the = key in Figure 4–11. The **+/−** key allows you to toggle between the +06 power and −06 power each time you push the key.

Let us try adding a series of numbers to gain experience using the EXP feature. Add the following numbers.

12 kΩ + 23 Ω + 34 mΩ + 45 MΩ + 560 μΩ

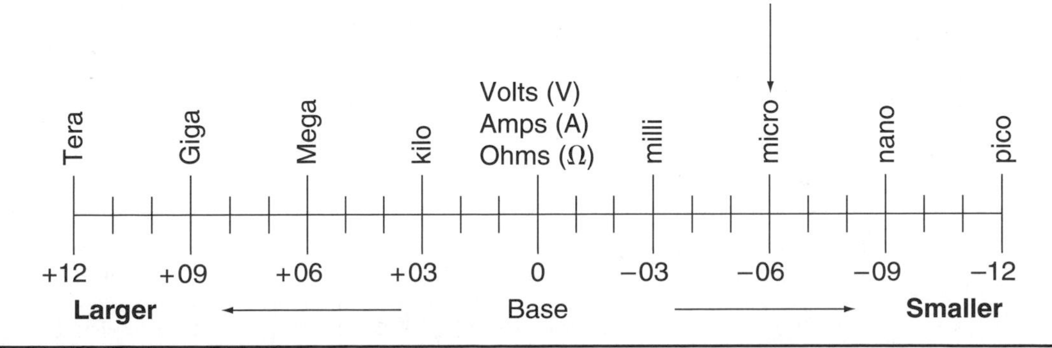

FIGURE 4–13 Conversion to a smaller electrical unit value.

Enter the following steps on the calculator.

12 EXP 3 + 23 + 34 EXP 3 +/−
+ 45 EXP 6 + 560 EXP 6 +/− =

The answer on the calculator may not be displayed in the unit value you are familiar. For example, the answer displayed on the author's calculator is:

4.5012023 07

Moving the decimal one place to the right makes the number 45.012023^{06}, or 45.012023 megaohms (MΩ). Just remember to refer to the linear scale, and change the decimal to a familiar value.

The second important calculator feature that is helpful when performing mathematical calculations is the 1/X key. This key automatically inverts a number, or creates the **reciprocal** of a number. As an example, the whole number 5 is really:

$$\frac{5}{1}$$

The reciprocal or inverse of 5 is the number 0.2.

$$\frac{1}{5} = 0.2$$

In Chapter 8 (Parallel Circuits), we will see that the reciprocal function is a vital step in finding the total resistance of a parallel circuit. Knowing how to use this feature on a scientific calculator makes electrical calculations relatively easy for more complex circuits.

Summary

- ❑ The United States Congress passed the Metric Conversion Act in 1975.
- ❑ The metric system is based on the power of ten principle.
- ❑ Each power of ten has a prefix to identify it.
- ❑ The base units of metric measurement are the meter, liter, and gram.
- ❑ The base units of electrical measurement are the volt, ohm, and amp.
- ❑ The EXP feature on a scientific calculator is for converting extra large or extra small numbers to scientific notation.
- ❑ The +/− key on a scientific calculator allows the user to toggle the unit values of a number from positive to negative numbers.
- ❑ The 1/X feature on a scientific calculator automatically inverts or reciprocates a number.

Key Terms

EE
EXP
giga
gram
kilo
linear scale
mega
meter
micro
milli
1/X
+/− key
power of ten
prefix
reciprocal
scientific notation
2nd function

Review Questions

Short Answer Essays

1. Describe the power of ten principle in the metric measuring system.
2. Explain how to use the 1/X reciprocal feature (1/X key) on a scientific calculator.
3. Explain how to convert between the various metric prefix values without using math formulas.

Fill in the Blanks

1. In 1975, the United States Congress passed the ______________________ in an attempt to get the American public to convert to the ______________________ system.
2. When converting between metric prefix values, 250 grams would be the same as ______________________ kilograms.
3. The ______________________ feature key on a scientific calculator is used for extra large or extra small numbers that do not fit on a standard 8-digit display.

ASE-Style Review Questions

1. Technician A says the base system of measurement for length is the kilometer. Technician B says the millimeter is the base unit of measurement for length. Who is correct?
   A. A only
   B. B only
   C. Both A and B
   D. Neither A nor B

2. Technician A says that 1,000 grams is the same as 1 kilogram. Technician B says that 1,000 grams is the same as 1,000,000 milligrams. Who is correct?
   A. A only
   B. B only
   C. Both A and B
   D. Neither A nor B

3. The linear scale of metric measurement is being discussed. Technician A says that to convert a number to a higher unit of value, the decimal point is moved to the left. Technician B says to move the decimal point to the right to make the unit a value smaller. Who is correct?
   A. A only
   B. B only
   C. Both A and B
   D. Neither A nor B

4. Two technicians are discussing the basic electrical system of measurement. Technician A says the base system of measurement for volts is the kilovolt. Technician B says the milliamp is the base unit of measurement for current. Who is correct?
   A. A only
   B. B only
   C. Both A and B
   D. Neither A nor B

5. The linear scale of electrical measurement is being discussed. Technician A says that to convert a number to a smaller unit of value, the decimal point is moved to the right. Technician B says move the decimal point to the left to make the unit of value larger. Who is correct?
   A. A only
   B. B only
   C. Both A and B
   D. Neither A nor B

6. The EXP feature is being discussed on a scientific calculator. Technician A says this is for numbers that are too large or too small for the calculator. Technician B says this feature is used to reciprocate a number. Who is correct?
   A. A only
   B. B only
   C. Both A and B
   D. Neither A nor B

7. The number 12.5^{03} is shown on the calculator display. The correct whole number is:
   A. 125,000
   B. 0.012500
   C. 1,250
   D. 12,500

8. Technician A says that the number 56.77^{-06} on the calculator display is really 0.00005677. Technician B says that the $^{-06}$ on the display means micro. Who is correct?
   A. A only
   B. B only
   C. Both A and B
   D. Neither A nor B

9. The 1/X feature is being discussed on a scientific calculator. Technician A says this feature is used to invert a whole number to its reciprocal. Technician B says this feature is only for adding or subtracting extra large or extra small numbers. Who is correct?
   A. A only
   B. B only
   C. Both A and B
   D. Neither A nor B

10. Technician A says that 91 MΩ is the same as 91,000,000,000 mΩ. Technician B says that 91 MΩ is the same as 91,000 kΩ. Who is correct?
    A. A only
    B. B only
    C. Both A and B
    D. Neither A nor B

Components of an Electrical Circuit

Introduction

Today's automotive electrical system is a complex maze of systems and subsystems that can be broken down into its various components. A typical modern vehicle has 2,000 feet or more of wiring, hundreds of connectors and terminals, and complex switches and control units. As a vehicle ages, electrical circuits become more prone to failure due to deteriorating conditions like overheating, excessive moisture, road salts, vibration, and improper repairs by unqualified persons. Having the in-depth knowledge of the individual components of electrical circuits and proper repair procedures helps you perform not only a proper diagnosis, but also effective and long-lasting repairs.

Objectives

When you complete this chapter you should be able to:

- ❑ Identify types of primary wire material, insulation properties, and proper wire size for an electrical circuit.
- ❑ Describe the proper procedures for repairing a wire splice.
- ❑ Identify various types of terminals, connectors, and appropriate repair procedures.
- ❑ Identify resistor types and their resistive value from the resistor color bands.
- ❑ Explain the purpose of a circuit protection device and identify the different types commonly used in automotive electrical circuits.
- ❑ Identify the various types of electrical switches and relays used to control electrical circuits.
- ❑ Describe the operation and construction of various types of lamps used in automotive lighting circuits.
- ❑ Explain capacitor applications, ratings, and construction in electrical circuits.

Automotive Wiring

Automotive **primary wire** consists of wiring circuits that carry *low* voltage, normally battery or alternator voltage or less. **Secondary wire** consists of wiring circuits that carry *high* voltages, like spark plug wires. Most primary wire is made of several strands of copper wire that are covered with a durable polyvinyl chloride (PVC) insulation (Figure 5–1).

Stranded wire, instead of solid wire, is normally used in automotive applications for two main reasons. First, the wire needs to be durable enough to withstand the vibrations and flexing that occur constantly. Second, a stranded wire has more surface area then a solid wire. The majority of electrons tend to flow on the outside surface of a wire, so higher current can be used for the same gauge wire when using stranded wire (Figure 5–2).

For a load component to operate properly as designed, the wire size used in primary electrical circuits must handle not only the necessary current, but also any vibration and heat exposure. Two common wire size standards are used to designate a wire size: the **American wire gauge (AWG)** and the metric wire size currently used by most automotive manufacturers.

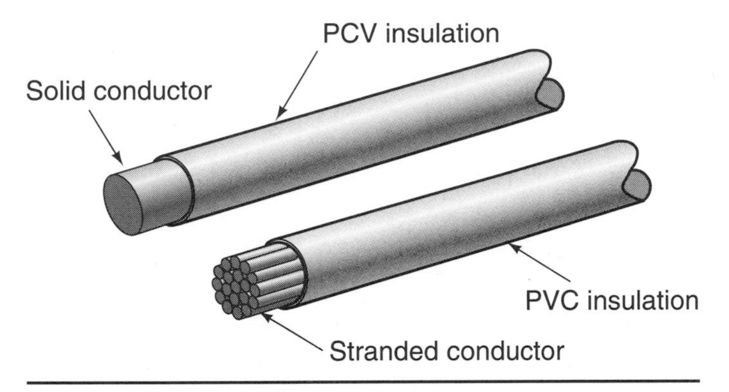

FIGURE 5–1 Typical stranded primary wire compared to solid conductor wire.

The AWG standard assigns a number to a wire based on the diameter of the wire conductor, excluding its insulation. The AWG wire rule states that the *higher* the AWG number is the *smaller* the wire conductor is. For example, an 18 gauge wire is smaller in diameter than a 12 gauge wire. The majority of automotive primary electrical systems use 12 to 18 gauge wires.

The metric system of wire size is determined by the cross-sectional area of the wire, which is expressed in square millimeters (mm^2). In the metric system the *smaller* the number is, the *smaller* the wire conductor is. Figure 5–3 is a cross-reference between AWG and metric wire sizes. Many aftermarket parts suppliers offer wire only in AWG sizes, so it is necessary to convert in most cases.

Another factor in determining the proper wire size for an electrical circuit is the length of the wire. As the wire gets longer, its internal resistance increases. The chart in Figure 5–4 indicates the proper wire gauge and length for a specific amperage load. As an example, on the chart in Figure 5–4, a 16 gauge wire that is 10 feet long is capable of conducting 20 amperes. However, if the 20-ampere current is to be carried for 15 feet, then 14 gauge wire is needed. Changing to 14 gauge wire prevents unnecessary voltage drops in the circuit due to inadequate wire size. Figure 5–4 can be a handy reference chart when installing aftermarket equipment in a vehicle.

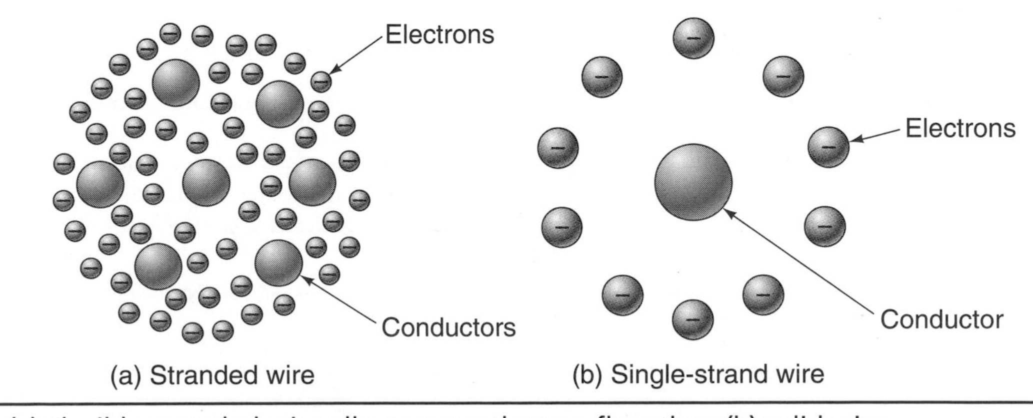

FIGURE 5–2 (a) Flexible stranded wire allows more electron flow than (b) solid wire.

Metric Size (mm²)	AWG (Gauge) Size	Ampere Capacity
0.5	20	4
0.8	18	6
1.0	16	8
2.0	14	15
3.0	12	20
5.0	10	30
8.0	8	40
13.0	6	50
19.0	4	60

FIGURE 5–3 AWG-to-metric cross reference.

Figure 5–5 is a summary of wire sizes and their effect on voltage drops and current-carrying capacity.

Nearly all the wires in an automotive electrical circuit are covered with an insulation that has a **color code** or that is otherwise identified by specific markings. Not all manufacturers use a standard method to mark wires; so it may be necessary to refer to the manufacturer's color code index. Figure 5–6 is an example of the most common abbreviations used today.

Because wiring harnesses normally contain dozens of wires, it is necessary to double up on some of the color abbreviations (Figure 5–7).

In wires with double color codes, the first letters indicate the base color of the insulation, and the second group of letters indicates the color of the tracer. A wire designated as BRN/Y has a brown base color and a yellow tracer stripe.

Total Approximate Circuit Amperes	Wire Gauge (for Length in Feet)								
12 V	3	5	7	10	15	20	25	30	40
1.0	18	18	18	18	18	18	18	18	18
1.5	18	18	18	18	18	18	18	18	18
2	18	18	18	18	18	18	18	18	18
3	18	18	18	18	18	18	18	18	18
4	18	18	18	18	18	18	18	16	16
5	18	18	18	18	18	18	18	16	16
6	18	18	18	18	18	18	16	16	16
7	18	18	18	18	18	18	16	16	14
8	18	18	18	18	18	16	16	16	14
9	18	18	18	18	16	16	16	14	12
10	18	18	18	18	16	16	16	14	12
11	18	18	18	18	16	16	14	14	12
12	18	18	18	18	16	16	14	14	12
15	18	18	18	18	14	14	12	12	12
18	18	18	16	16	14	14	12	12	12
20	18	18	16	16	14	14	12	12	10
22	18	18	16	16	12	12	10	10	10
24	18	18	16	16	12	12	10	10	10
30	18	16	16	14	10	10	10	10	10
40	18	16	14	12	10	10	8	8	6
50	16	14	12	12	10	10	8	8	6
100	12	12	10	10	6	6	4	4	4
150	10	10	8	8	4	4	2	2	2
200	10	8	8	6	4	4	2	2	1

Note: 18 AWG as indicated above this line could be 20 AWG electrically. 18 AWG is recommended for mechanical strength.

FIGURE 5–4 Application load amperage versus wire size and length.

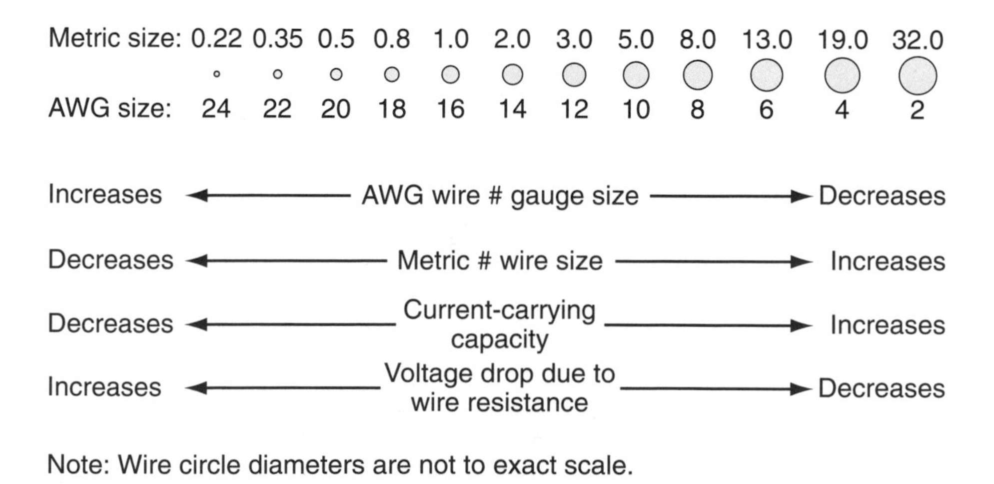

FIGURE 5–5 Summary of wire sizes and their effect on voltage drop and current flow.

Color	Abbreviations		
Aluminum	AL		
Black	BLK	BK	B
Blue (Dark)	BLU DK	DB	DK BLU
Blue	BLU	B	L
Blue (Light)	BLU LT	LB	LT BLU
Brown	BRN	BR	BN
Glazed	GLZ	GL	
Gray	GRA	GR	G
Green (Green)	GRN DK	DG	DK GRN
Green (Light)	GRN LT	LG	LT GRN
Maroon	MAR	M	
Natural	NAT	N	
Orange	ORN	O	ORG
Pink	PNK	PK	P
Purple	PPL	PR	
Red	RED	R	RD
Tan	TAN	T	TN
Violet	VLT	V	
White	WHT	W	WH
Yellow	YEL	Y	YL

FIGURE 5–6 Common color codes used in automotive applications.

Occasionally, a technician needs to replace a wire, and it may be difficult to identify its proper gauge size. The chart and five-step process (Figure 5–8) on page 70 explains the procedure used to determine the gauge size of unknown wires. Use the five steps in Figure 5–8 to determine the wire gauge sizes in the following chart. Assume each wire example has 19 strands, and put your answers on a *separate* sheet of paper.

	Wire Diameter	Gauge Size
1	0.0155	14 gauge
2	0.0385	________
3	0.0101	________
4	0.0244	________
5	0.0900	________
6	0.0429	________

Electrical Connectors

Today's complex wiring circuits use a variety of terminal connectors. Terminals must be able to perform with little or no resistance, or opposition, to current flow. A loose or corroded connection can cause unwanted voltage drops that will result in poor operation of the load

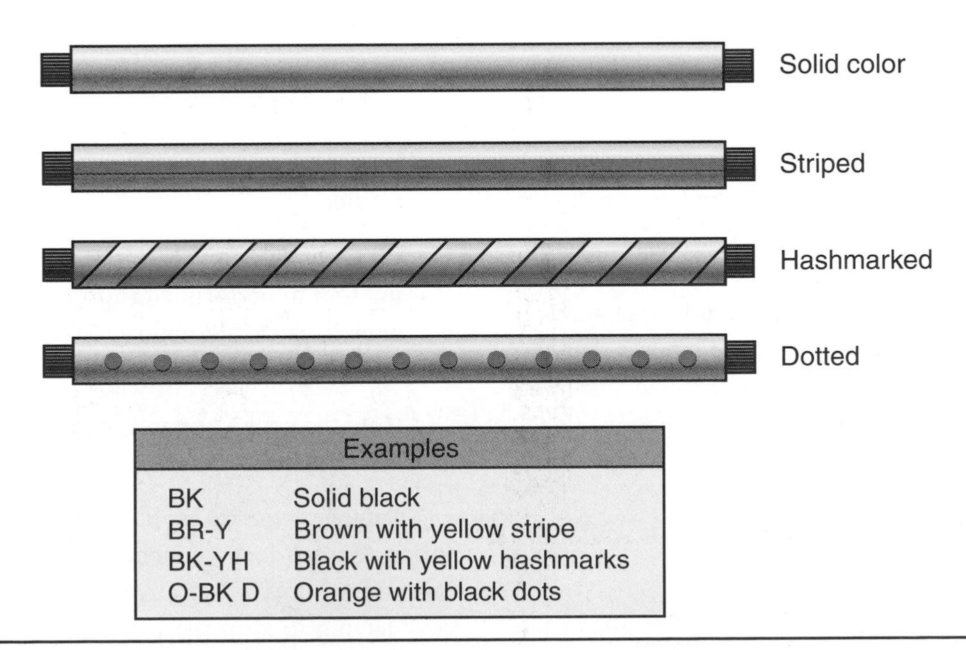

FIGURE 5–7 Four methods that Ford Motor Company uses to color-code wires.

How To Measure the Gauge of Cable

To determine the gauge of a cable, using the table, proceed as follows:

(1) Count the number of strands of wire.

(2) Measure the diameter of a single strand in thousandths of an inch, using a micrometer.

(3) In column A of the table, find the diameter of the wire you have measured, and on the same line, in column C, find its area.

(4) Multiply the area of a single wire by the number of strands, to get the total area.

(5) In column C, find the figure that is closest to the total area obtained by step 4, and on the same line, in column B, note the gauge number of a single wire having that area. This number is the gauge of the cable.

Example: A cable is found to have 19 strands of wire the individual strands (measured by the micrometer) being 0.0112 inch in diameter. The table (column C) shows the circular mil area of each strand to be 127. Multiplying this by the number of strands, 19, results in 2,413 total circular mils. The closest figure in column C is 2,583, and on the same line, in column B, we find that 16 is the nearest cable gauge.

Size and Area of Wire

Wire diameter (inches) (A)	American wire gauge (B)	Circular mil area (C)
.4600	0000	211600
.4096	000	167800
.3648	00	133100
.3249	0	105500
.2893	1	83690
.2576	2	66370
.2294	3	52640
.2043	4	41740
.2893	1	83690
.2576	2	66370
.2294	3	52640
.2043	4	41740
.1620	6	26250
.1285	8	16510
.1019	10	10380
.0808	12	6530
.0640	14	4107
.0508	16	2583
.0403	18	1624
.0319	20	1022
.0284	21	810.1
.0253	22	642.4
.0225	23	509.5
.0201	24	404.0
.0179	25	320.4
.0159	26	254.1
.0142	27	201.5
.0126	28	159.8
.0112	29	126.7
.0100	30	100.5
.0089	31	79.70
.0079	32	63.21
.0070	33	50.13
.0063	34	39.75
.0056	35	31.52
.0050	36	25.00

FIGURE 5–8 Determining the gauge size of unknown wire cable.

component(s). The simplest terminal connector used in automotive circuits, especially older model vehicles, is the **solderless connector** (Figure 5–9).

Crimping a solderless connector is an acceptable method for splicing wires and connectors that are not subjected to dirt or excessive weather elements that can cause corrosion. You must use a specialized crimping tool (Figure 5–10) to obtain a solid connection. Other crimping tools, like slip-joint pliers, do not make a tight enough connection.

The correct procedure for performing a solderless crimp is:

1. Use the correct size hole in the stripping area of the tool to remove enough insulation to allow the wire to completely penetrate the crimping end of the connector.
2. Place the wire into the connector, and use the crimping tool to perform a crimp (Figure 5–11). Place the open area of the connector toward the anvil of the crimping tool, and compress the tool.
3. Hold the connector and give the wire a gentle pull to verify that the crimping is tight enough.

Another type of solderless connector is the **tap splice** (Scotch-lock) connector. This type of connector can add an *additional* circuit to a feed circuit without your having to strip the insulation first. For example, a tap splice is used when a trailer plug is added to a

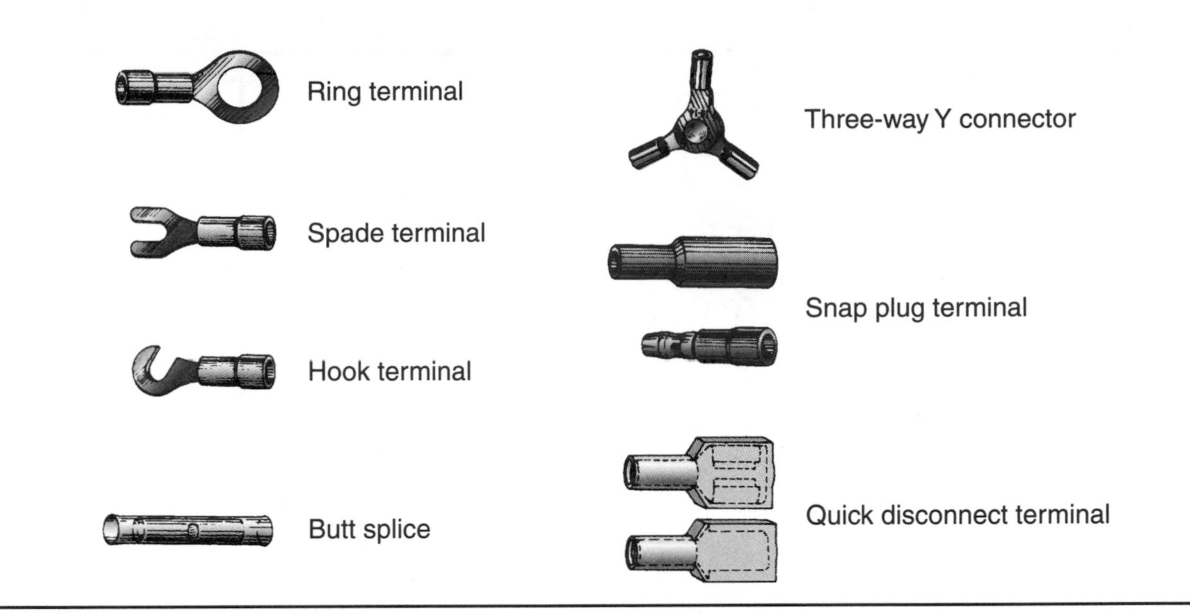

FIGURE 5–9 Common solderless primary wire terminals.

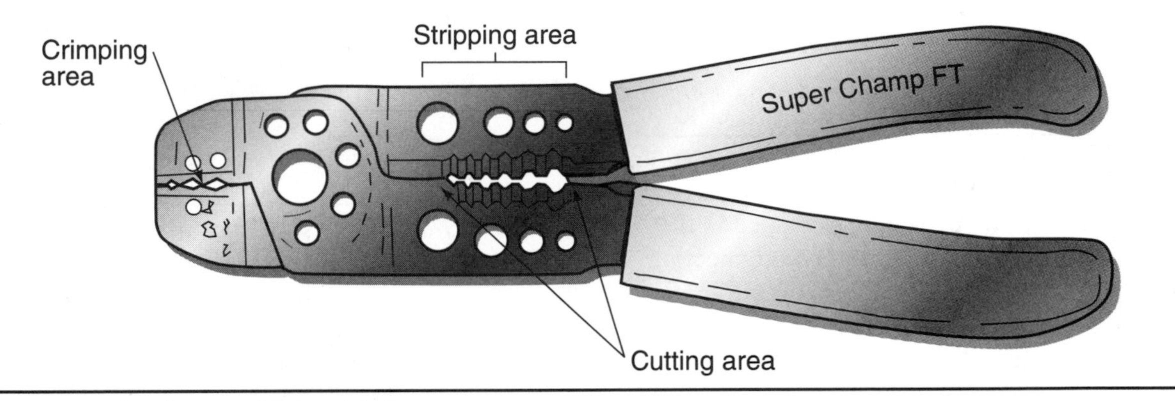

FIGURE 5–10 A crimping tool for solderless connectors.

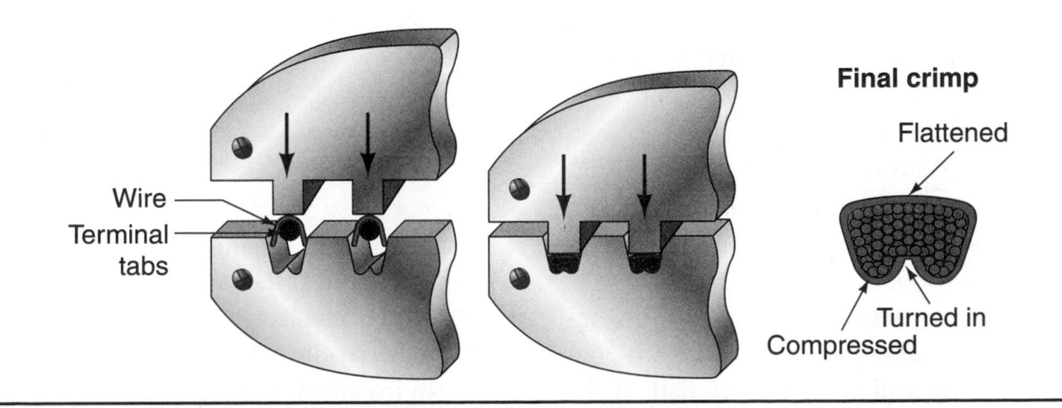

FIGURE 5–11 Using a crimping tool on a solderless connector.

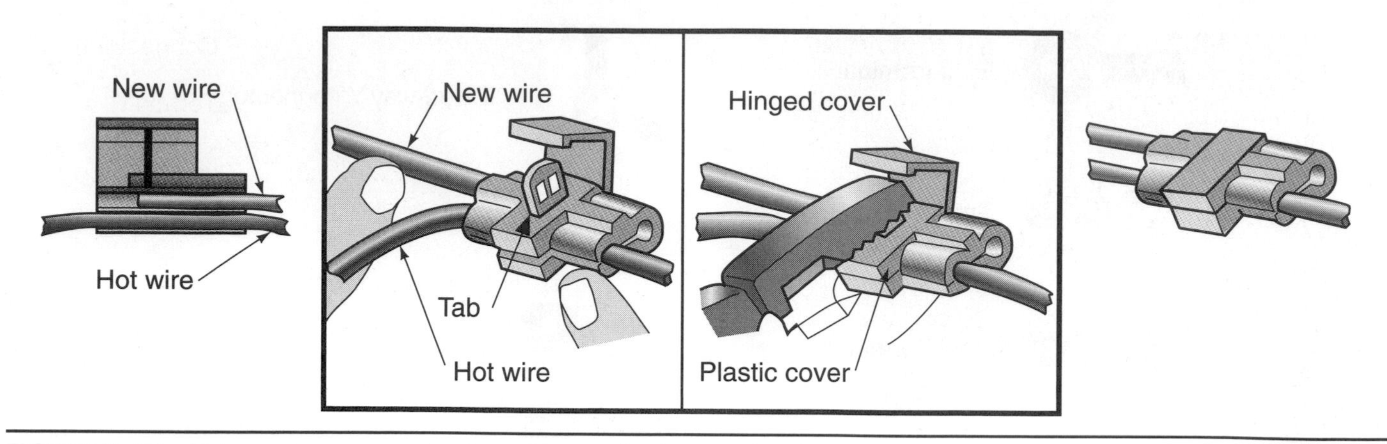

FIGURE 5–12 Using a tap connector to splice a wire into a circuit.

vehicle's brake and tail light circuit. Figure 5–12 shows a tap splice connecting a new wire to an existing hot wire. Extreme care should be exercised in using tap splice because wire strands may be cut during the crimping procedure, and the circuit being tapped may not handle the additional current load, or the connection might turn off poor.

Note: Most automotive manufacturers *do not* recommend this method of wire repair because stranded wires will be cut or damaged in the process. This primitive method of connecting wires is not recommended, but it is still practiced in some low-quality repair shops; so technicians should know about it.

Several other types of connectors have been used over the past several years to ensure weather-tight electrical connections, which are especially important in modern vehicles having computer controls that need a more reliable connector. The first is a **hard shell connector** (Figure 5–13), which can have from one to a dozen wires molded into a connector component.

The next style, a **weather-pack connector** (Figure 5–14), has a rubber seal on the terminal ends and on the covers of the connector halves, which provide for excellent corrosion protection.

The third style is a **metri-pack connector** (Figure 5–15). Similar to a weather-pack connector, this connector does not have the seal on the cover half of the connector, and it is smaller and more compact.

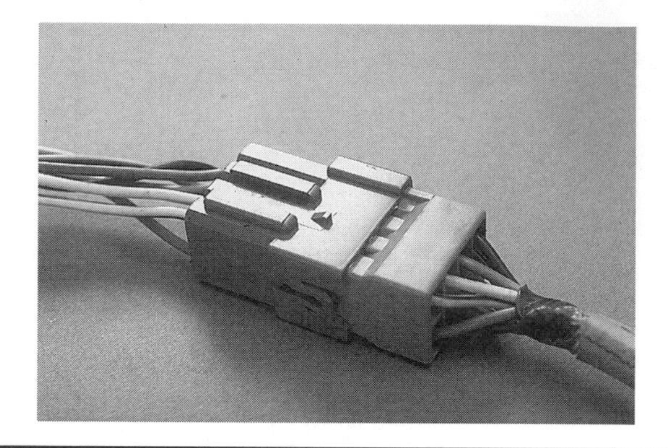

FIGURE 5–13 Hard shell connector with multiple wires.

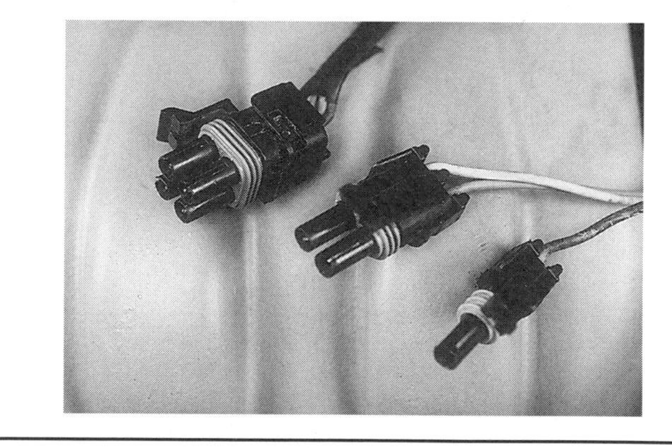

FIGURE 5–14 Weather-pack connector.

Metri-pack connectors use two different methods to fasten the terminal half and connector half. The first is a **push-to-seat** connector and terminal (Figure 5–16). To remove the terminal from the connector requires a special tool to unlock the male locking tang (Figure 5–17) or the female locking tang (Figure 5–18).

FIGURE 5–15 Metri-pack connector.

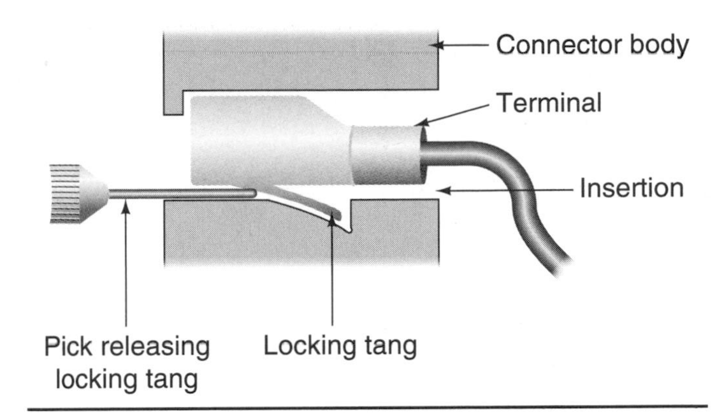

FIGURE 5–16 Typical push-to-seat connector.

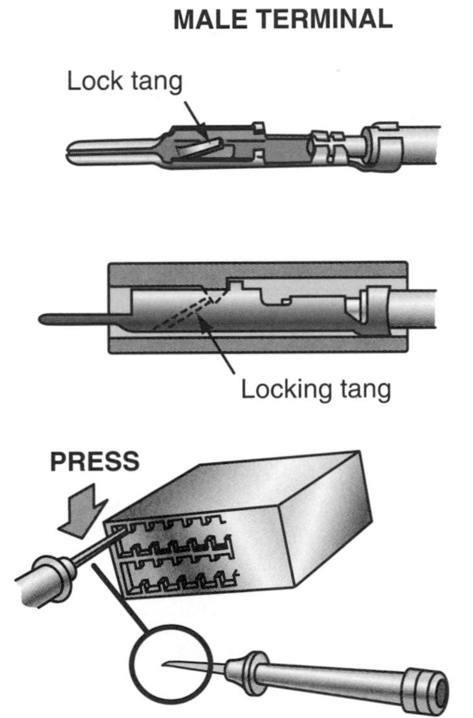

FIGURE 5–17 Using a special pick-tool to unlock the male terminal locking tang.

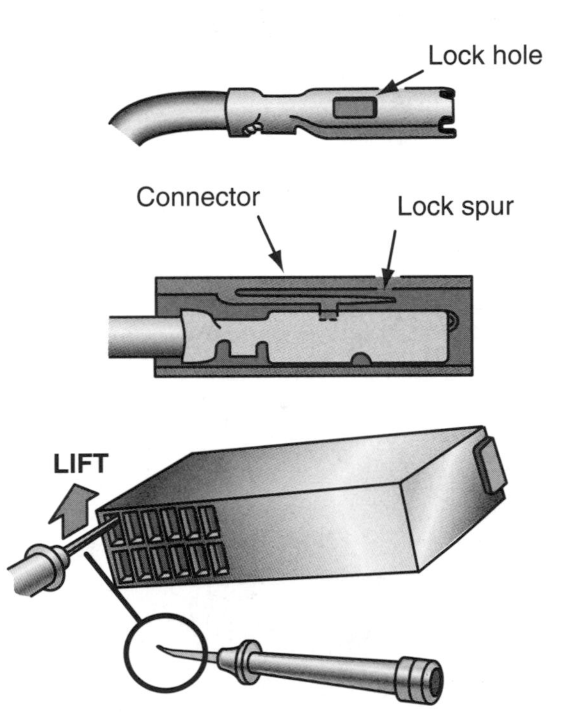

FIGURE 5–18 Using a special pick-tool to unlock the locking tab retainer for a female terminal.

The second method for a metri-pack connector is a **pull-to-seat** connector and terminal (Figure 5–19). The pull-to-seat terminal is removed by inserting a pick-tool into the connector face while prying up on the locking tang. The terminal can then be pushed through the front of the connector (Figure 5–20).

A special tool for unlocking the tang can be purchased at most tool supply stores. Figure 5–21 is an example of a universal wire terminal tool.

A weather-pack connector is similar to a metri-pack in that a tang must be compressed to remove the terminal. Figures 5–22 and 5–23 show the procedure to remove a weather-pack connector.

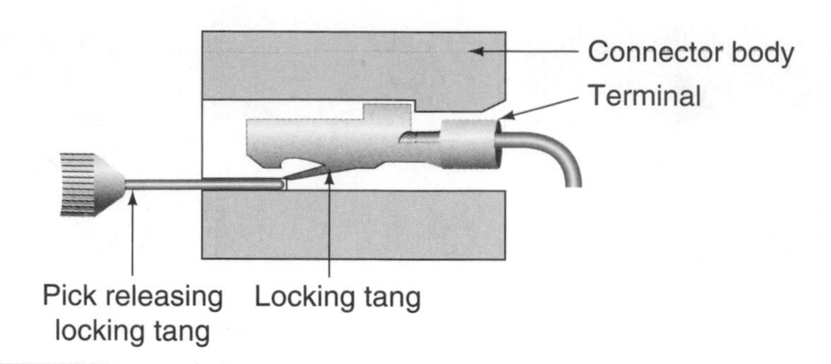

FIGURE 5–19 Typical pull-to-seat connector.

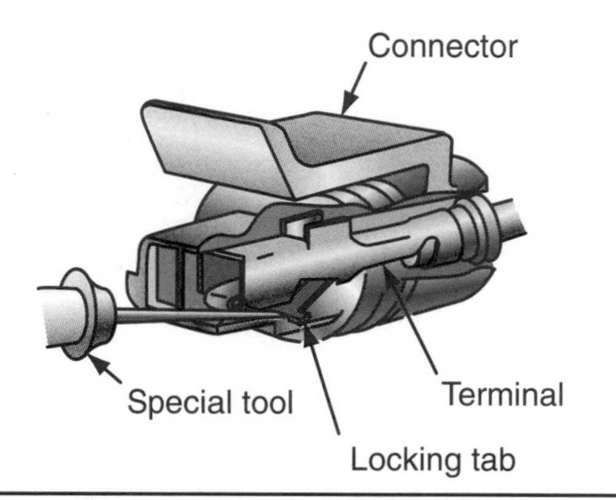

FIGURE 5–20 Pull up on the locking tang to release the terminal from the connector.

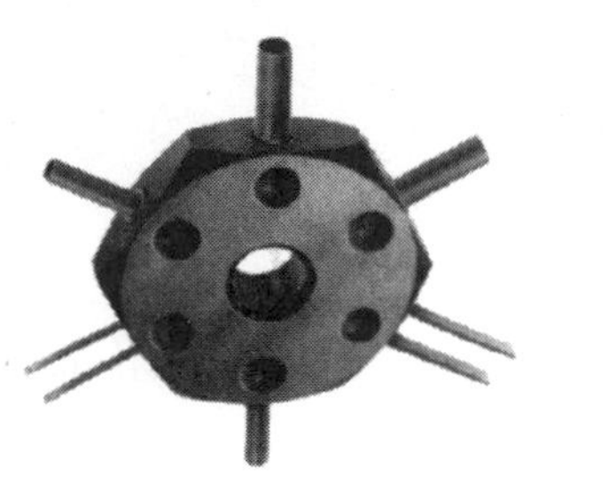

FIGURE 5–21 Universal terminal tang tool.
(Courtesy of the Lisle Corporation)

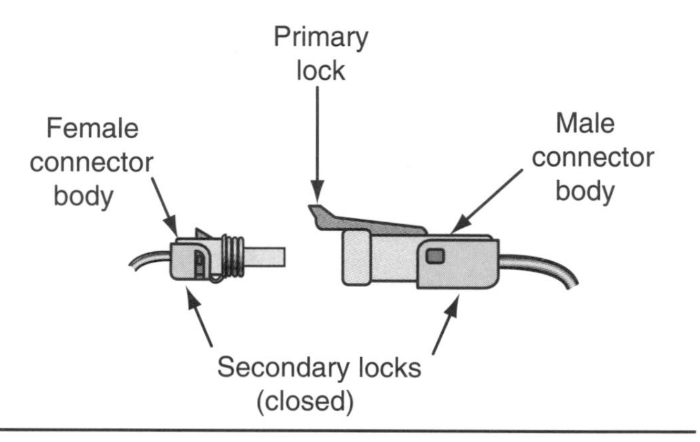

FIGURE 5–22 Primary lock for a weather-pack connector.

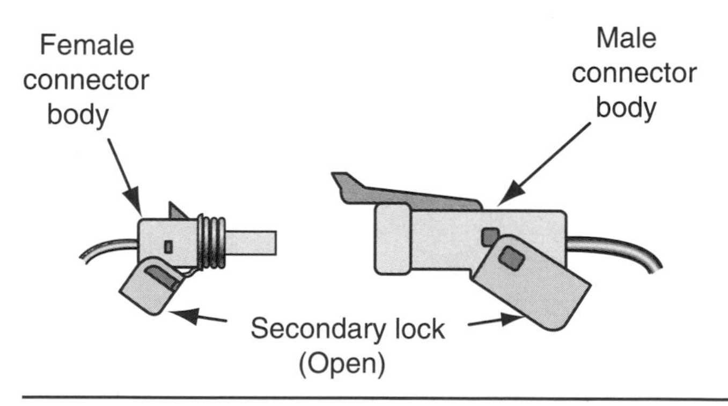

FIGURE 5–23 Secondary lock for a weather-pack connector.

Soldering Procedures for Wire and Connectors

The best way to attach to a terminal or to perform a wire splice is to make a solder connection. Electrical solder is an alloy of lead and tin along with a **rosin core** flux (a cleaning agent). Solder usually comes in a 40/60 mixture (40% tin and 60% lead), but it can also come in a 50/50 or 60/40 mixture. A word of caution about choosing the correct mixture of solder: The higher the tin content is, the higher the heat needs to be to melt the solder. This high heat may damage sensitive electrical circuits. *Do not use* acid core solder for wire or connector repairs. It can cause wire connections to corrode and lead to high resistance. Acid core solder is used for purposes *other* than electrical applications, such as repairing radiators, joining copper pipes, and fusing light metals together. Photo Sequence 1 shows a proper method for soldering.

Photo Sequence 1 demonstrates a proper method for soldering two wires together and for the use of heat shrink tubing to seal the soldered joint afterward.

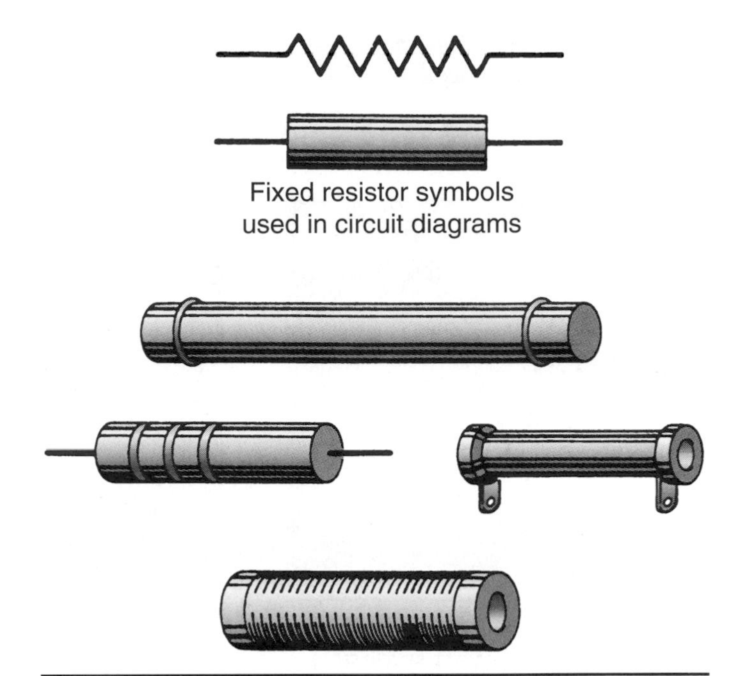

FIGURE 5–24 Several styles of a fixed resistor.

Resistive Devices

Resistors are among the most common components in electrical circuits. Several different types of resistive devices, depending on the functions of the circuit, are in use today. Resistive devices can control a circuit current, produce heat, light a bulb, or provide a designed-in voltage drop. For example, a resistor element made of nickel-chromium wire, or strip, is designed to produce heat in cigar lighters, and heated windows and mirrors. A light bulb, for example, produces light when the fine tungsten resistive element is heated.

The first style of resistor, a **fixed resistor,** is probably the most common resistor used in electrical circuits today (Figure 5–24). Fixed resistors are generally made of a carbon/graphite-based material, mixed with a resin bonding agent. The body is made from a variety of nonconductive enamel insulating materials. Figure 5–25 is a cutaway of a typical carbon film fixed resistor.

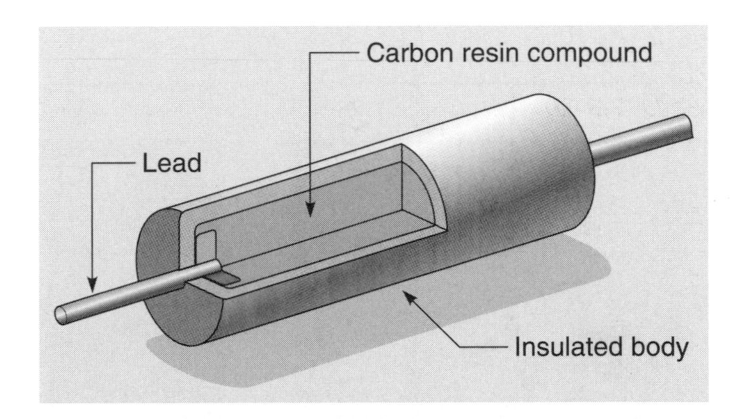

FIGURE 5–25 Typical carbon film fixed resistor.

The size of a resistor does not determine its resistive value, but it does relate to the amount of heat it can dissipate without becoming damaged. This heat production is known as the resistor's power rating, in watts. Typically, the larger the resistor is, the higher the wattage rating will be (Figure 5–26). Fixed resistors with wattage ratings higher that 2 watts are normally wire-wound resistors consisting of a hollow ceramic core to better dissipate excessive heat (Figure 5–27).

Color codes are used to identify the resistive values of carbon film resistors. The most common resistor color code system uses a four-band system (Figure 5–28).

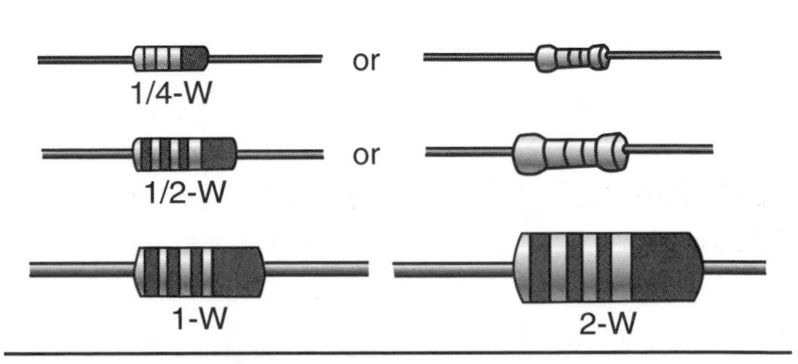

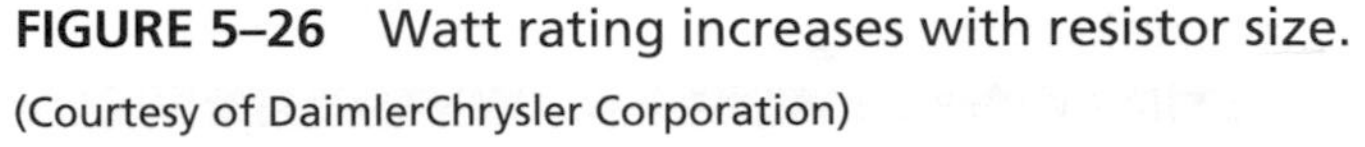

FIGURE 5–26 Watt rating increases with resistor size. (Courtesy of DaimlerChrysler Corporation)

PHOTO SEQUENCE 1

Soldering Two Copper Wires Together

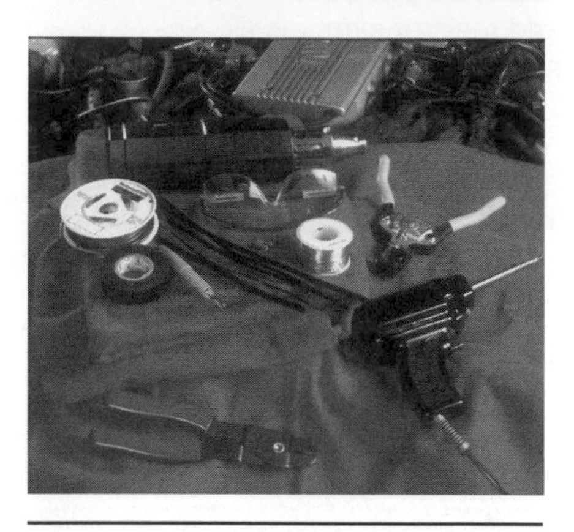

P5–1 Tools required to solder copper wire: 100-watt soldering iron, 60/40 rosin core solder, crimping tool, splice clip, heat shrink tube, heating gun, and safety glasses.

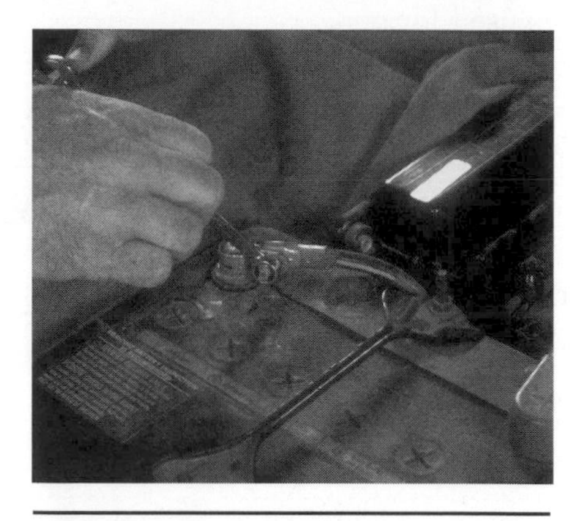

P5–2 Disconnect the fuse that powers the circuit being repaired. *Note:* If the circuit is not protected by a fuse, disconnect the ground lead of the battery.

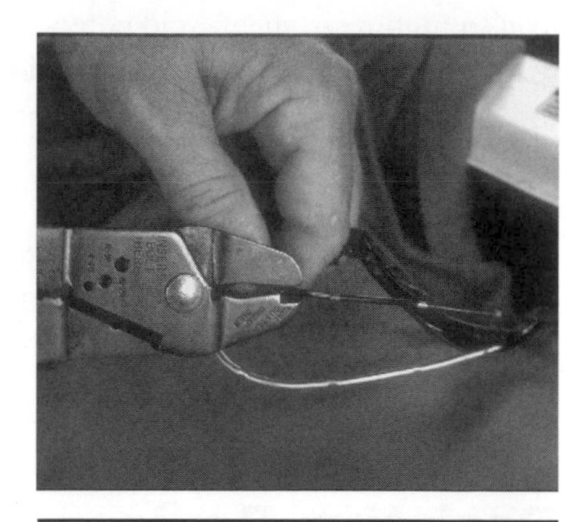

P5–3 Cut out the damaged wire.

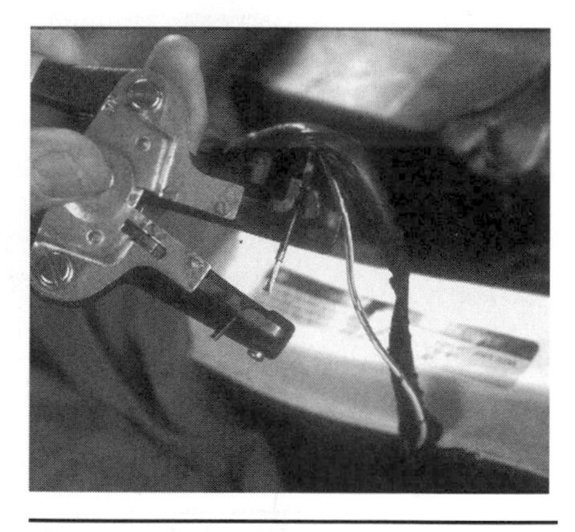

P5–4 Using the correct size stripper, remove about 1/2 inch of the insulation from both wires.

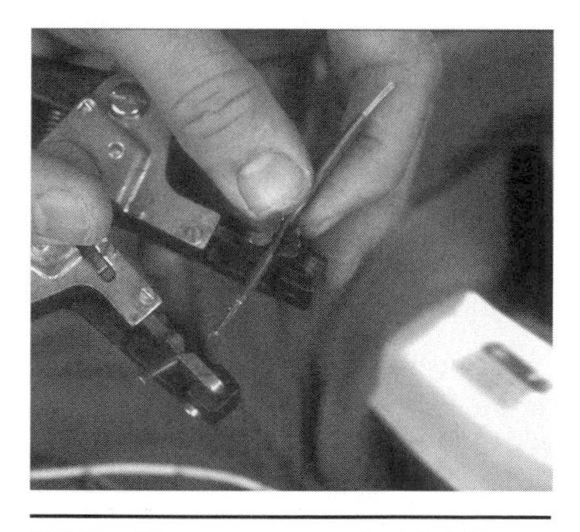

P5–5 Now remove about 1/2 inch of the insulation from both ends of the replacement wire. The length of the replacement wire should be slightly longer than the length of the wire removed.

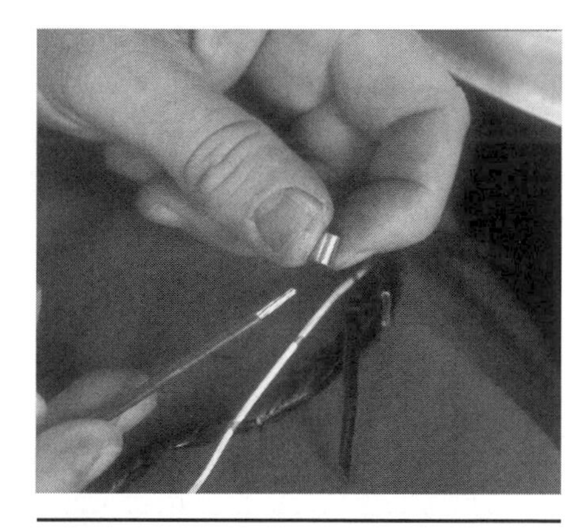

P5–6 Select the proper size splice clip to hold the splice. *Note:* Depending on vehicle manufacturer preferences, wires may be soldered together without a splice clip. If unsure—refer to the vehicle repair manual.

PHOTO SEQUENCE 1

Soldering Two Copper Wires Together, *continued*

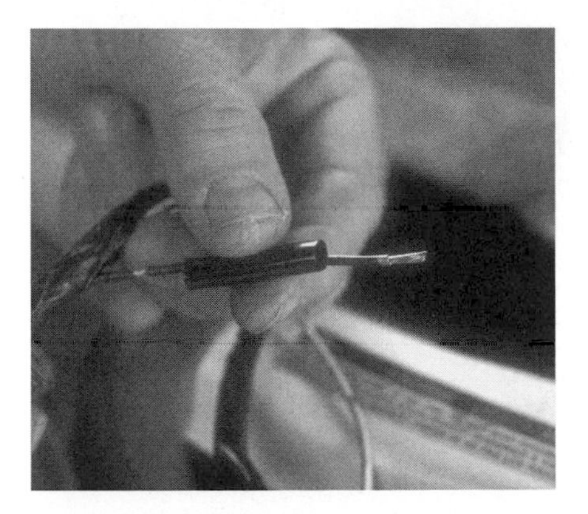

P5–7 Place the correct size and length of heat shrink tube over the two ends of the wire.

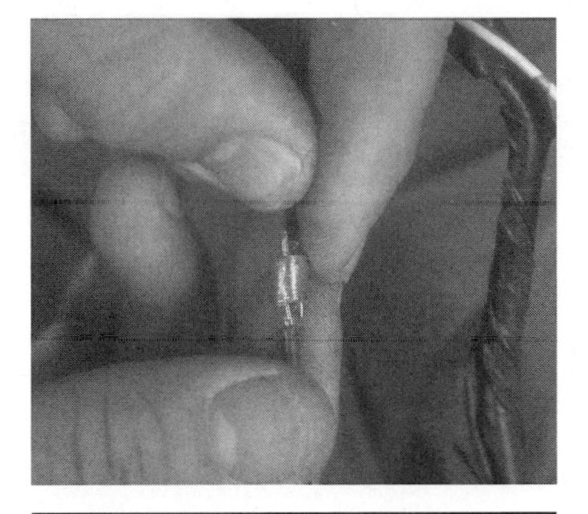

P5–8 Overlap the two splice ends and center the splice clip around the wires, making sure the wires extend beyond the splice clip in both directions.

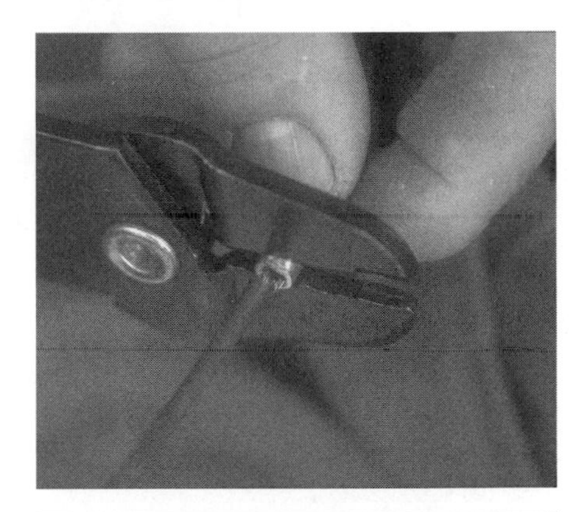

P5–9 Crimp the splice clip firmly in place.

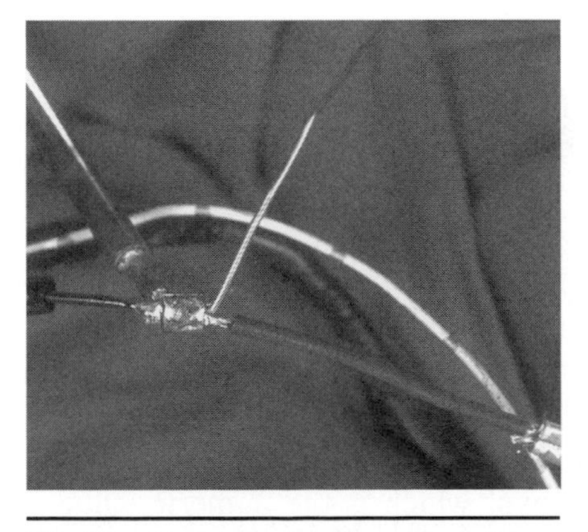

P5–10 Heat the splice clip with the soldering iron while applying solder to the opening of the clip. Do not apply solder to the iron. The iron should be 180 degrees away from the opening of the clip.

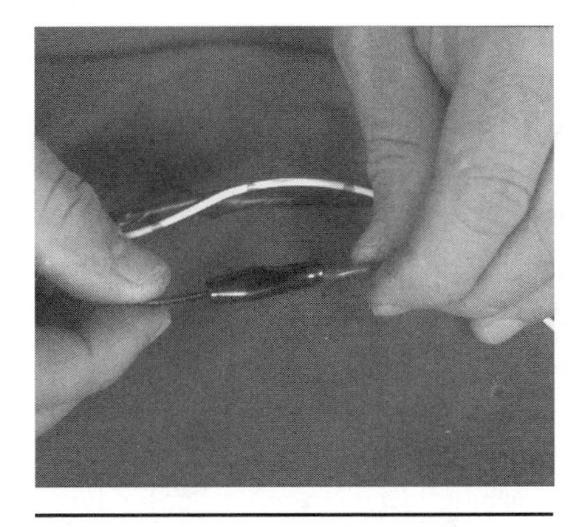

P5–11 After the solder cools, slide the heat shrink tube over the splice.

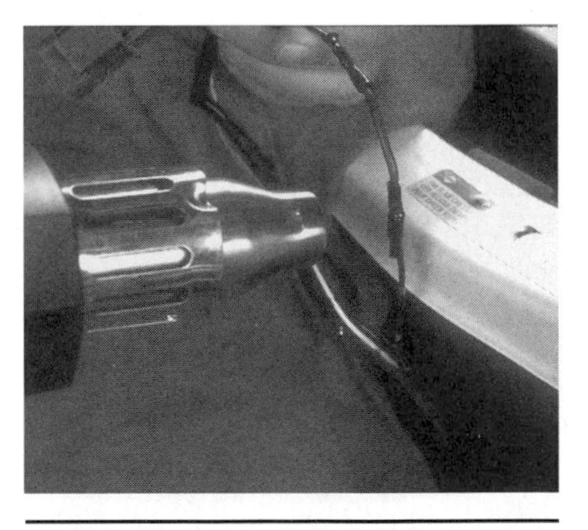

5–12 Heat the tube with the hot air gun until it shrinks around the splice. Do not overheat the heat shrink tube.

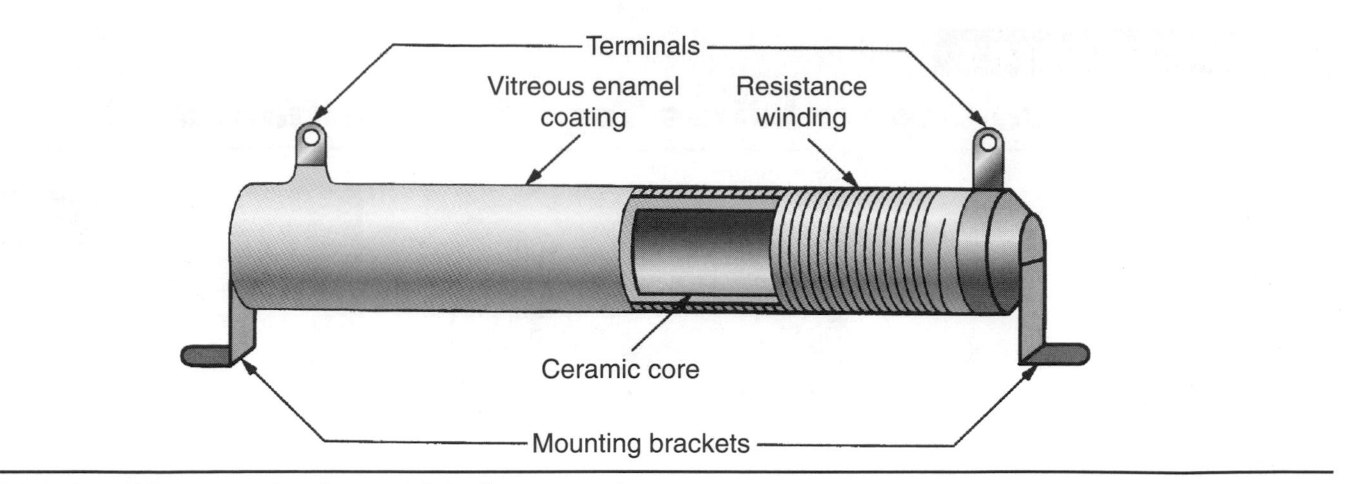

FIGURE 5–27 Wire-wound resistor with hollow ceramic core.

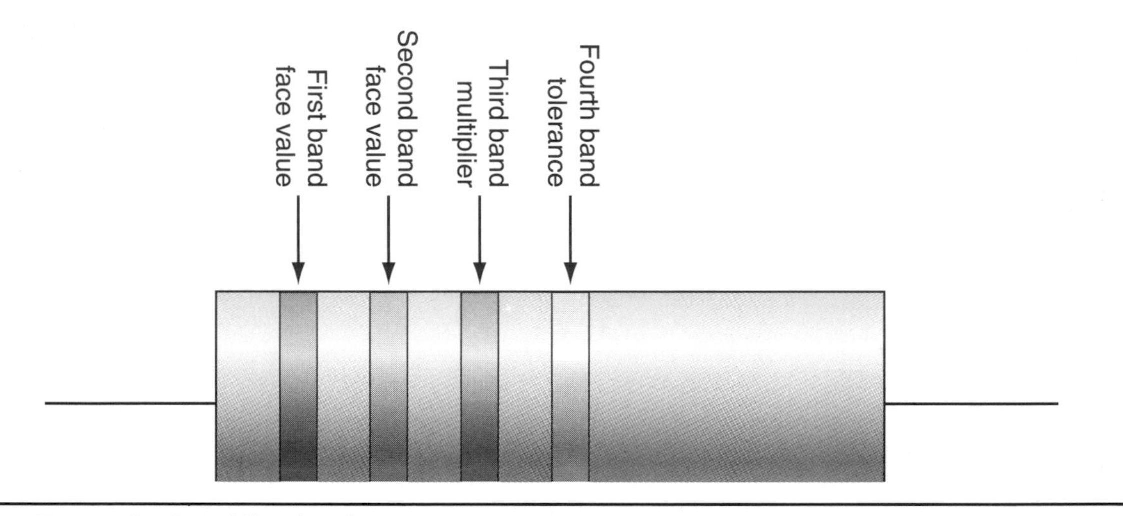

FIGURE 5–28 Four-color band fixed resistor.

The first two color bands represent number values from the color code chart (Figure 5–29). The third color band represents the multiplying value, or the number of zeros to add to the right of the first two color band digits. The fourth color band is the tolerance value of the resistor. A fourth band means the resistor has a 5% tolerance. For example, if a 1,200 Ω resistor has a gold tolerance band, the value limits of this resistor would be $1{,}200 \times 0.05 = 60$. This means the value of the 1,200 Ω resistor can be as high as 1,260 Ω or as low as 1,140 Ω and still be considered a 1,200 Ω resistor. A silver fourth color band means a 10% tolerance value. When there is *no* fourth color band, the tolerance value is 20%.

The two examples given here to show how to determine the resistive value of color-banded fixed resistors.

1. In Figure 5–30, the first color band is red, which corresponds to the numeric value of 2, as shown on the color code chart (Figure 5–29). The second color band is brown, which is the numeric value 1 on the color chart. The third band is the multiplier; it is orange, which corresponds to 3. So adding three zeros to the first two numbers yields 21,000. In other words, the number 21 multiplied by the multiplier 1,000 equals 21,000 Ω. The tolerance color is gold, so it is a 5% resistor. This means the resistor value can be as high as $21{,}000 + 1{,}050 = 22{,}050$ or as low as $21{,}000 - 1{,}050 = 19{,}950$.

2. In Figure 5–31, the first color band is blue, which is 6 on the color code chart (Figure 5–29). The second

4-Band Resistor Color Code Chart			
Color	Number value 1st band	Number value 2nd band	Multiplying value 3rd band
Black	0	0	1
Brown	1	1	10
Red	2	2	100
Orange	3	3	1,000
Yellow	4	4	10,000
Green	5	5	100,000
Blue	6	6	1,000,000
Violet	7	7	10,000,000
Gray	8	8	100,000,000
White	9	9	1,000,000,000

Percentage Tolerance 4th Band	
Brown	± 1%
Red	± 2%
Gold	± 5%
Silver	± 10%
No color	± 20%

FIGURE 5–29 Resistor color code reference chart.

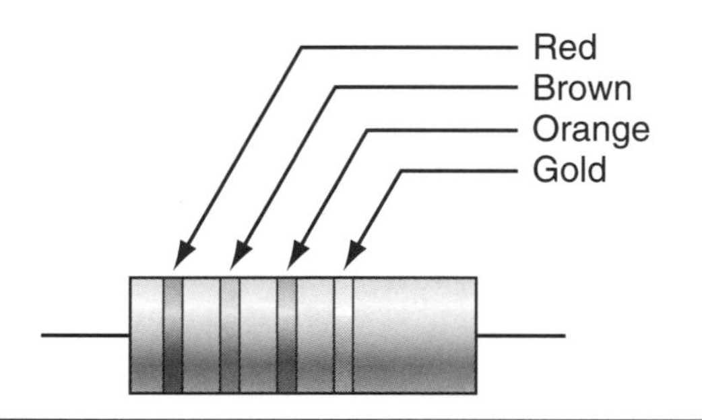

FIGURE 5–30 Resistive value = 21,000 Ω, 5%.

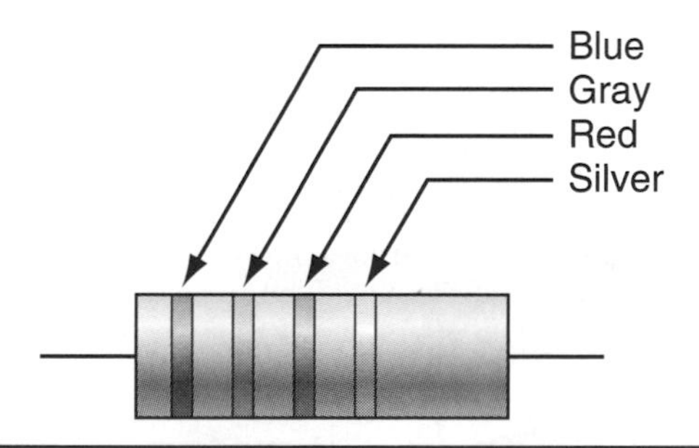

FIGURE 5–31 Resistive value = 6,800 Ω, 10%.

color band is gray, which is 8. The third band is the multiplier and is red, which is 2. Adding two zeros to the first two numbers yields 6,800. The tolerance color is silver; so it is a 10% resistor. The resistor value can be as high as 6,800 + 680 = 7,480 or as low as 6,800 − 680 = 6,120.

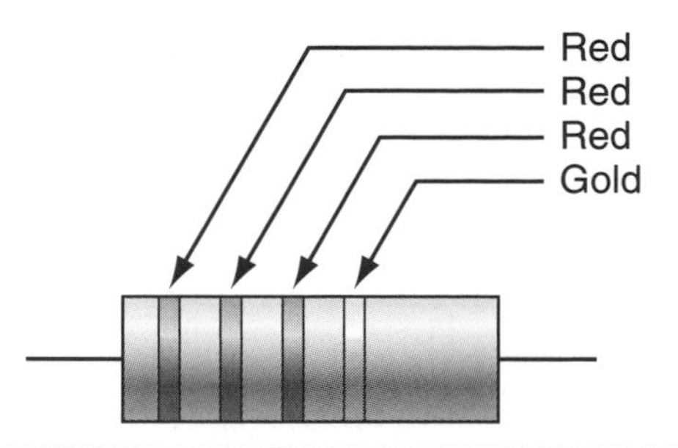

FIGURE 5–32 Resistive value unknown.

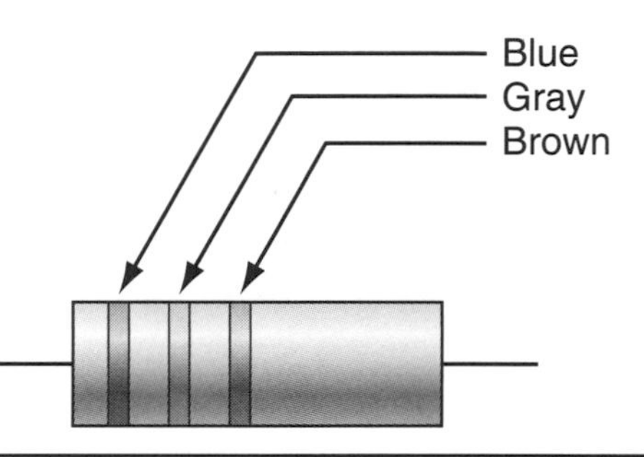

FIGURE 5–33 Resistive value unknown.

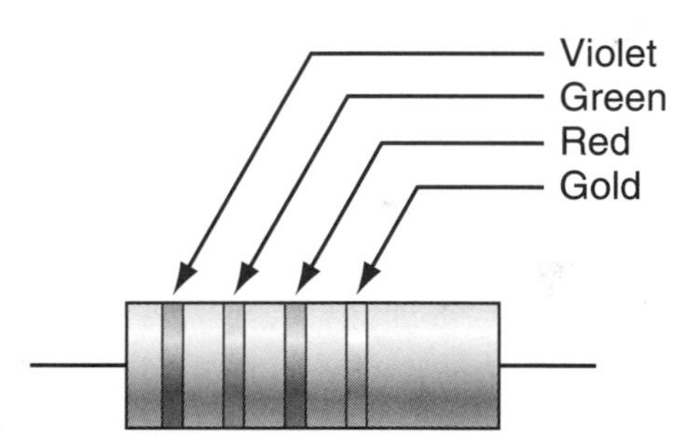

FIGURE 5–34 Resistive value unknown.

Complete the remaining five resistance values by using the color code chart in Figure 5–29. Put your answers on a *separate* sheet of paper.

Resistance value of Figure 5–32 ____________________
Resistance value of Figure 5–33 ____________________
Resistance value of Figure 5–34 ____________________
Resistance value of Figure 5–35 ____________________
Resistance value of Figure 5–36 ____________________

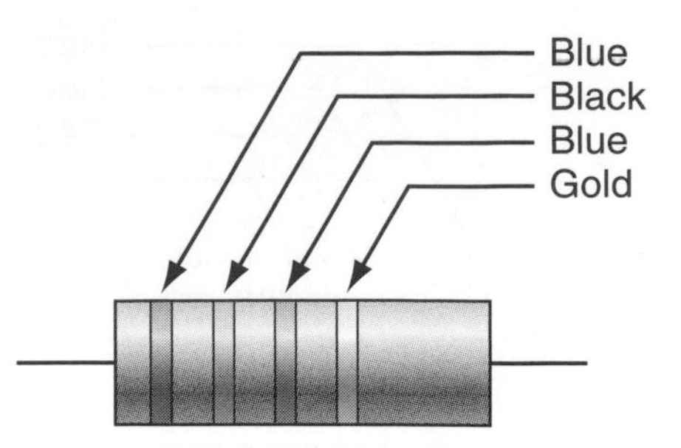

FIGURE 5–35 Resistive value unknown.

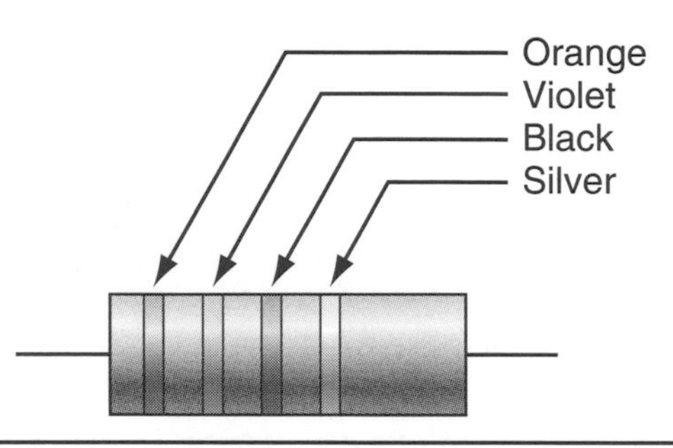

FIGURE 5–36 Resistive value unknown.

The second style of resistor is a **stepped resistor.** For more than 40 years, a stepped resistor has been used in automobiles to control the heater blower fan motor speed (Figure 5–37). If the blower switch is in the low speed position, then the current in the circuit must travel through all three resistors to get to the blower motor. As the switch is moved to the other positions, the current travels through fewer resistors (or less total resistance) (Figure 5–38).

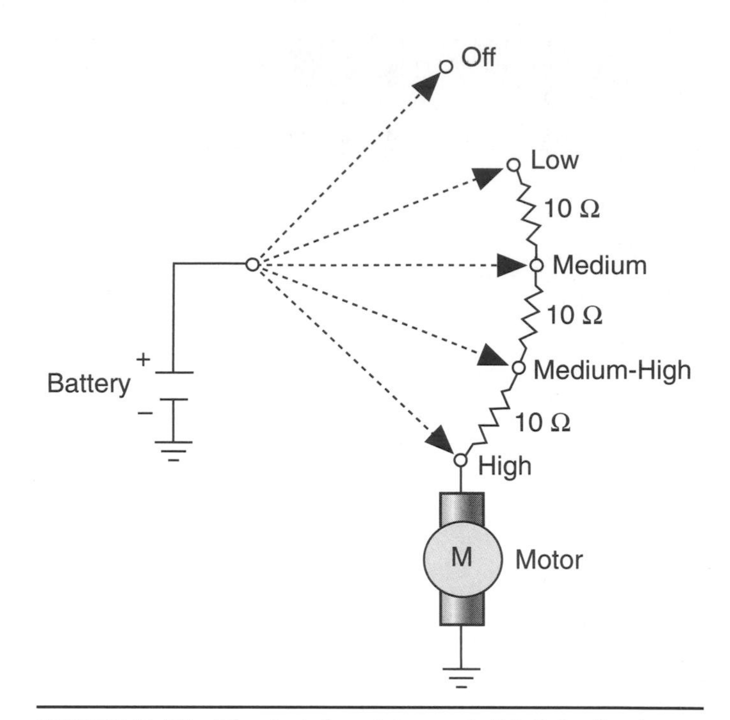

FIGURE 5–37 The total resistance is 30 Ω in the low-speed position, 20 Ω in the medium positions, 10 Ω in the medium-high position, and 0 Ω in the high position.

In newer vehicles with electronic climate control (ECC) systems, the stepped resistor has been replaced with computer modules that control blower speed more precisely (Figure 5–39).

The third style of resistor is a **variable resistor,** which can change the resistance potential of an electrical

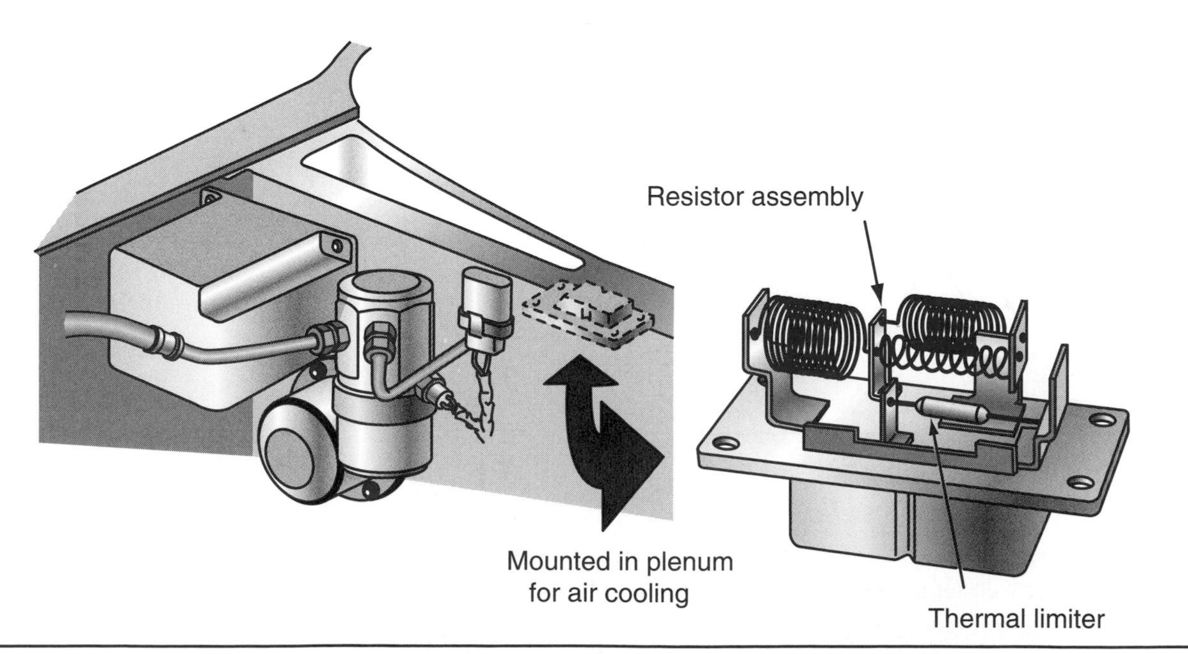

FIGURE 5–38 Typical stepped resistor and location in a manual heating and A/C system.

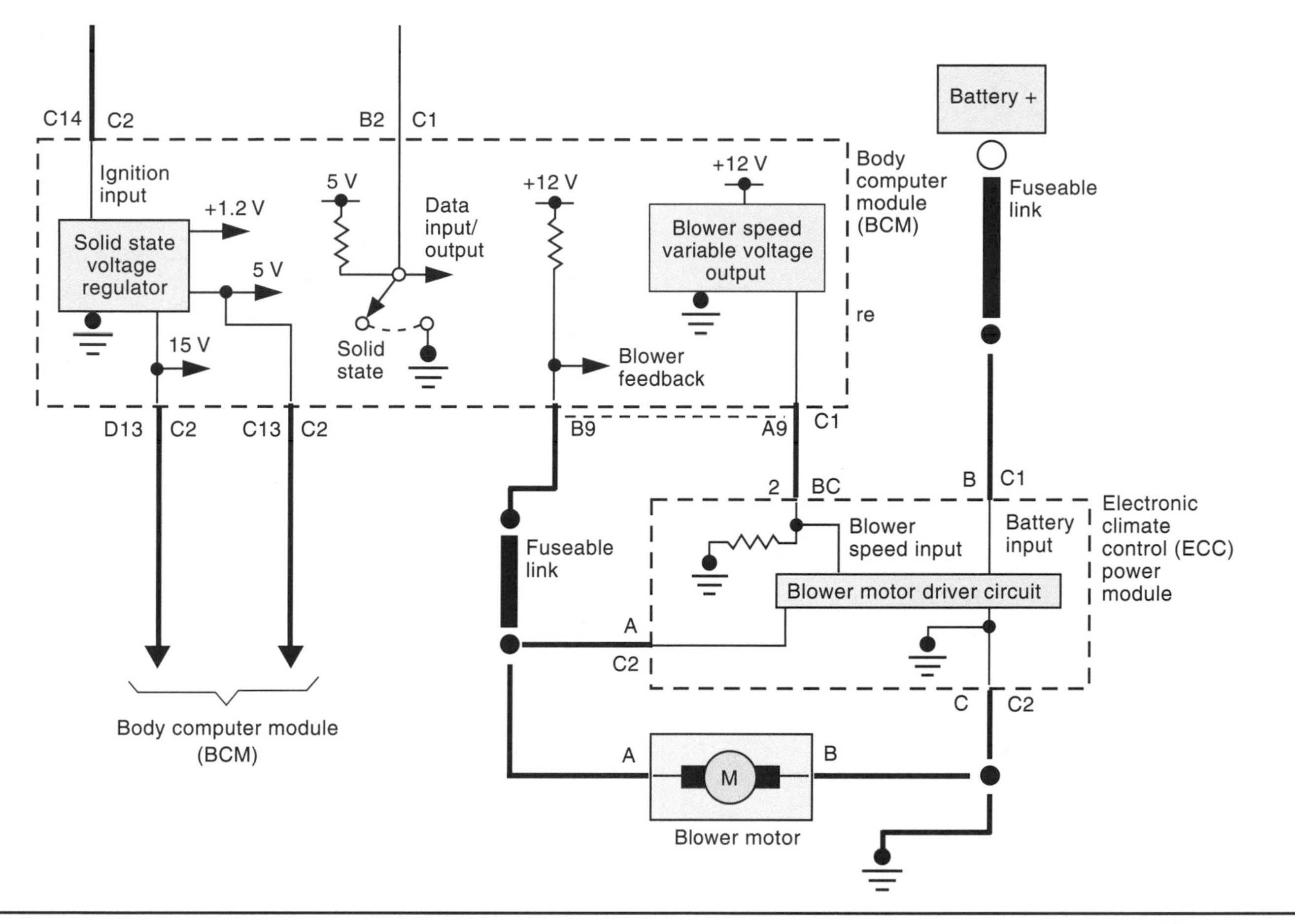

FIGURE 5–39 Blower motor speed controlled by electronic circuits and computer modules.

circuit. The most common types of variable resistors are rheostats, potentiometers, and thermistors. A rheostat has one terminal connected to a fixed end of a resistor and a second terminal connected to a movable contact, called a wiper (Figure 5–40).

The most common **rheostat** in the automobile is the headlight switch with a dimming knob feature for the instrument lights. By changing the position of the wiper on the resistor, the resistance can be increased (bulb dims) or decreased (bulb brightens) to control current (Figure 5–41).

A **potentiometer** (also called a pot) differs from a rheostat in that it has three terminals (or wires) and controls voltage. One terminal at the fixed end of the resistor is connected to a power source. The second terminal of the resistor is connected to ground. The third terminal is connected to a movable wiper contact (Figure 5–42).

The wiper, as it is moved over the resistor, creates a variable voltage drop. Because current is always flowing through the fixed portion of the resistor, the voltage drop generated by the potentiometer is very stable. For this

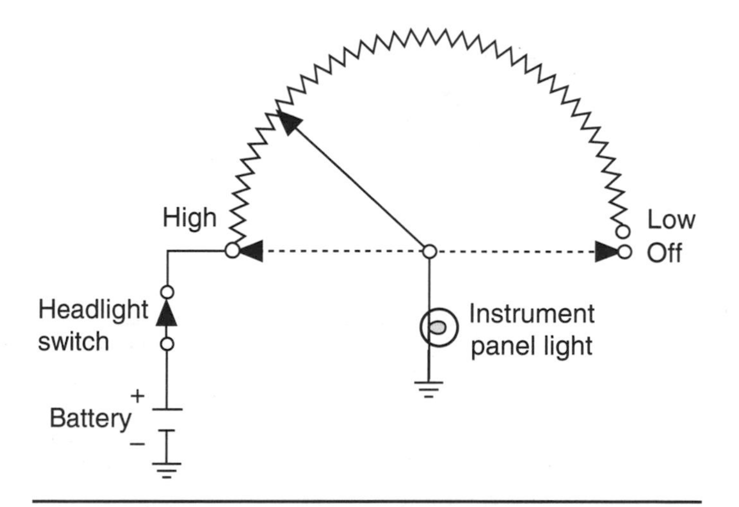

FIGURE 5–40 Using a rheostat to control the brightness of an instrument panel light.

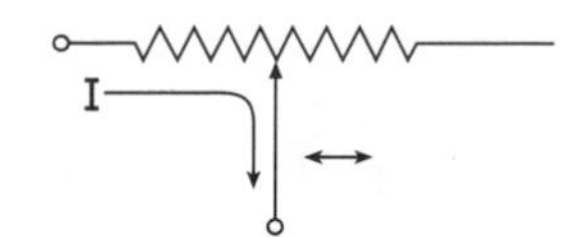

FIGURE 5–41 Wiring diagram for a rheostat.

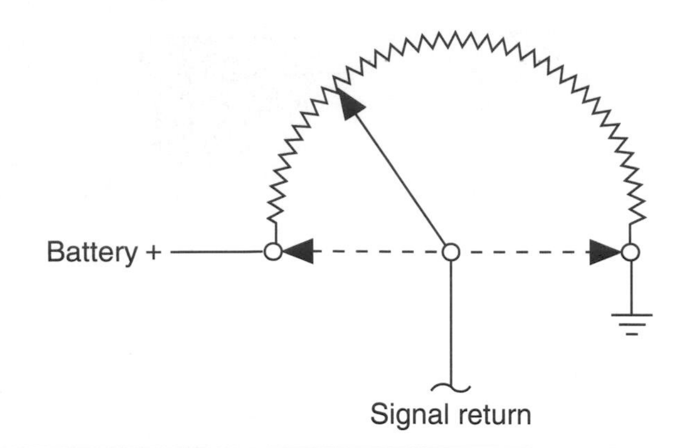

FIGURE 5–42 A typical potentiometer contains three wires.

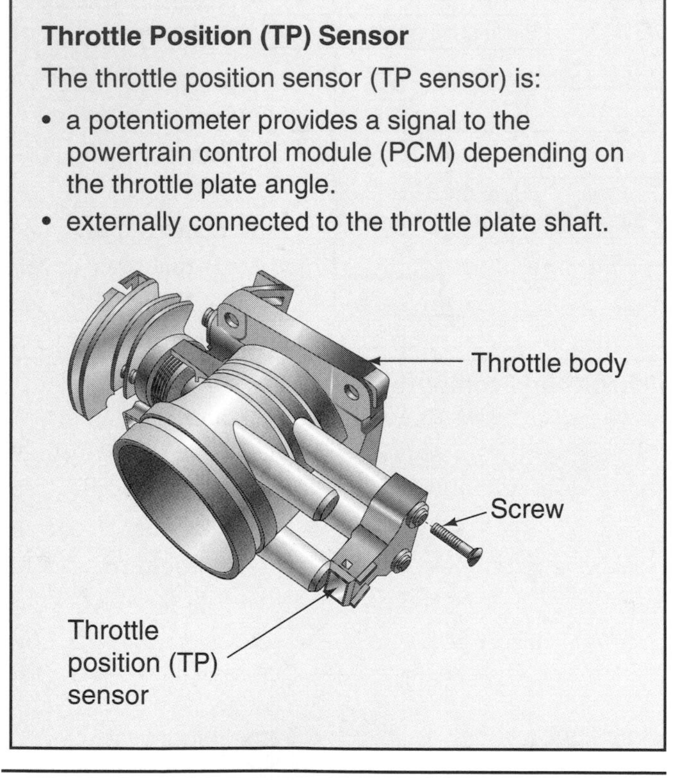

FIGURE 5–43 A throttle position (TP) sensor.

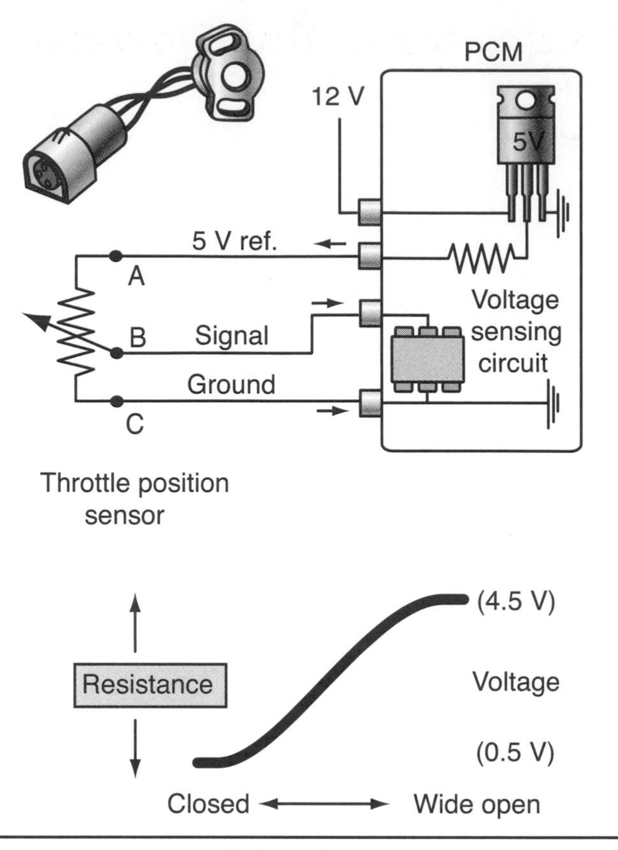

FIGURE 5–44 Typical throttle position sensor circuit.

reason, the potentiometer makes an excellent input sensor for a vehicle's computer system. An example of a potentiometer input sensor is a throttle position sensor (Figure 5–43). The throttle position sensor in Figure 5–44 uses the amount of voltage drop or increase to sense the throttle valve opening angle.

The final style of variable resistor is a **thermistor.** The resistance value of a thermistor changes in proportion to its temperature and is typically used to measure coolant temperature. As coolant temperature rises, the thermistor resistance decreases and the current flow through a temperature gauge increases. When the coolant temperature drops, the thermistor resistance increases, thus decreasing the current flow to the gauge (Figure 5–45).

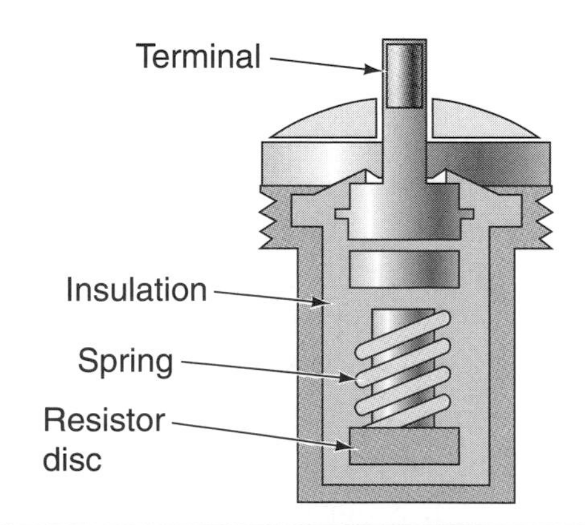

FIGURE 5–45 A thermistor sensor used to control engine temperature gauge.

Circuit Protection Devices

Most automotive electrical circuits are protected from damage from current levels that exceed the designed amperage capacity of the system. The most common **circuit protection device** is the fuse. There are several basic types of quick replacement fuses (Figure 5–46): glass, ceramic, and the newer blade type fuse. Glass and ceramic fuses are found mostly on older vehicles.

Glass fuses are small glass cylinders with metal end caps. The metal strip that connects the two end caps is designed to blow quickly when the current marking indicated on the end cap code is exceeded.

The ceramic fuse is used on many European imports, and it is available in two size codes: GBF (small) and the more common code GBC (large).

There are three different types of blade style fuses (Figure 5–47). These fuses consist of a plastic housing with a metal strip connected between two metal blade-type connectors. The fuse rating is indicated on top of the plastic housing, and the plastic is color-coded for current rating (Figure 5–48). Blade fuses have code ratings of ATC or ATO.

From the late 1970s to mid-1990s, the **auto-fuse** (3–30 amperes) was the most common replacement fuse to protect electrical system components and wiring harnesses from damage. The **mini-fuse** (5–30 amperes) is similar in design to an auto-fuse and has replaced it in most circuits in the last ten years because it is smaller. Using a mini-fuse allows for fitting more individual fuses in the same fuse box. A **maxi-fuse** (20–80 amperes) is larger than an auto-fuse and is a *slow blow* fuse. Common uses for this style of fuse is to place it between the battery and the vehicle's main fuse box.

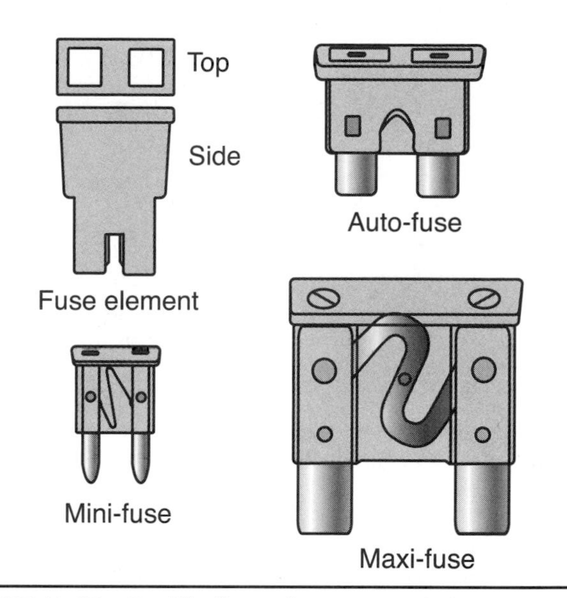

FIGURE 5–47 Pacific fuse element and three types of blade fuses.

The last style of a quick replacement fuse is a **Pacific fuse element** (30–60 amperes) that was developed as a replacement for a **fusible link.** Both are designed to protect the electrical system from a direct short to ground and are normally located in the vicinity of the battery. Figure 5–49 is an example of a fusible link located near the battery. The current capacity of a fusible link is determined by its size. A fusible link is usually four wire sizes smaller than the circuit it protects. For example, a 14 gauge wire would require an 18 gauge fusible link. Remember the rule for AWG wire size: The *smaller* the wire is, the *larger* its number is.

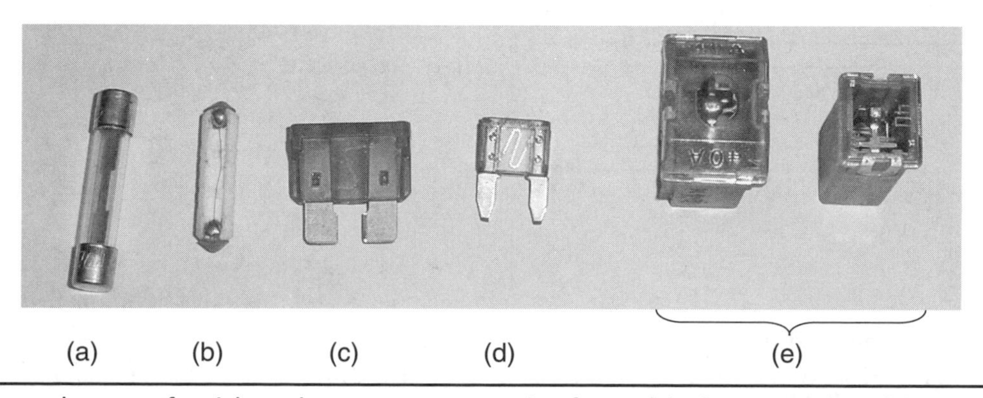

FIGURE 5–46 Several types of quick replacement automotive fuses: (a) glass cartridge, (b) ceramic, (c) auto-fuse, (d) mini-fuse, (e) Pacific fuse.

Auto-fuse

Current rating	Color
3	Violet
5	Tan
7.5	Brown
10	Red
15	Blue
20	Yellow
25	Natural
30	Green

Maxi-fuse

Current rating	Color
20	Yellow
30	Green
40	Amber
50	Red
60	Blue
70	Brown
80	Natural

Mini-fuse

Current rating	Color
5	Tan
7.5	Brown
10	Red
15	Blue
20	Yellow
25	Natural
30	Green

Pacific fuse element

Current rating	Color
30	Pink
40	Green
50	Red
60	Yellow

FIGURE 5–48 Fuse current ratings and corresponding color codes.

A fuse is normally connected in series to an electrical circuit (Figure 5–50) ahead of all the loads in the circuit. However, a fuse may be located at each individual circuit load device (Figure 5–51).

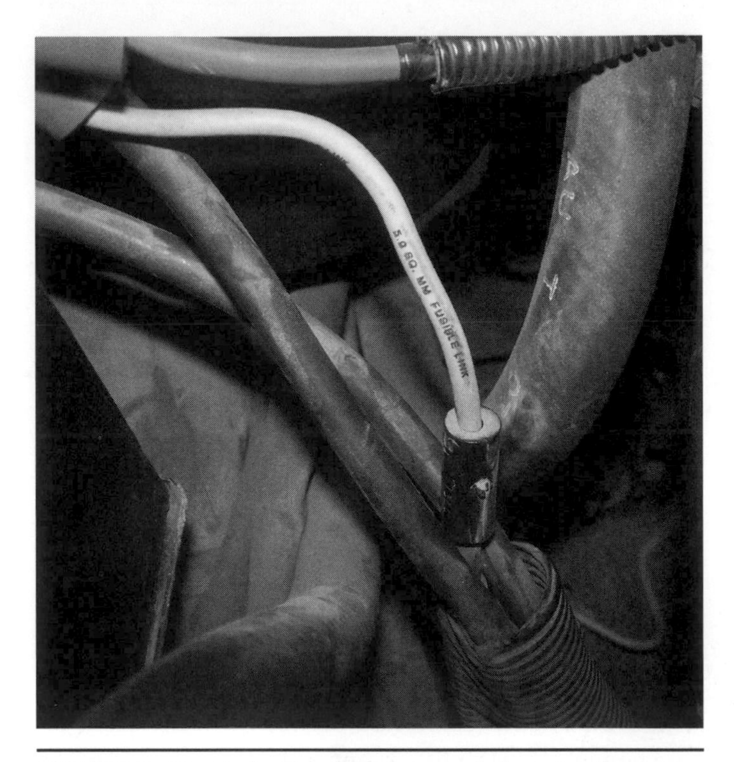

FIGURE 5–49 A fusible link located close to the battery.

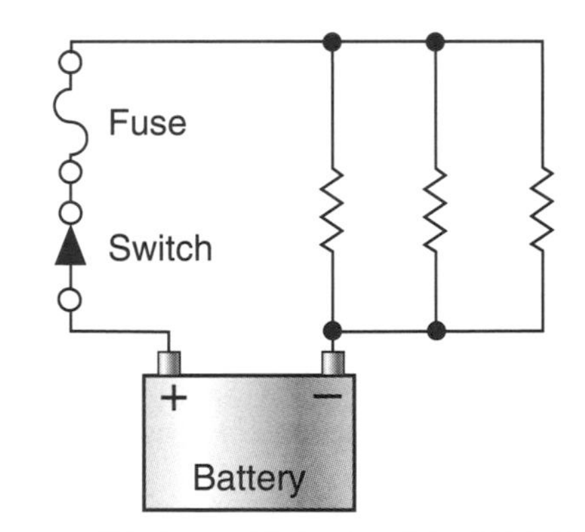

FIGURE 5–50 A fuse connected to protect an entire parallel circuit.

Normally, when fuses reach their current rating, the metal wire inside them melts through, causing the circuit to open (Figure 5–52). Detecting a blown fuse by visual inspection is not always easy. Occasionally, technicians find hidden faults in fuses that look normal (Figure 5–53).

To prevent a visual misdiagnosis, a technician should use an ohmmeter, voltmeter, or test light to properly check a fuse or fusible link. When an ohmmeter

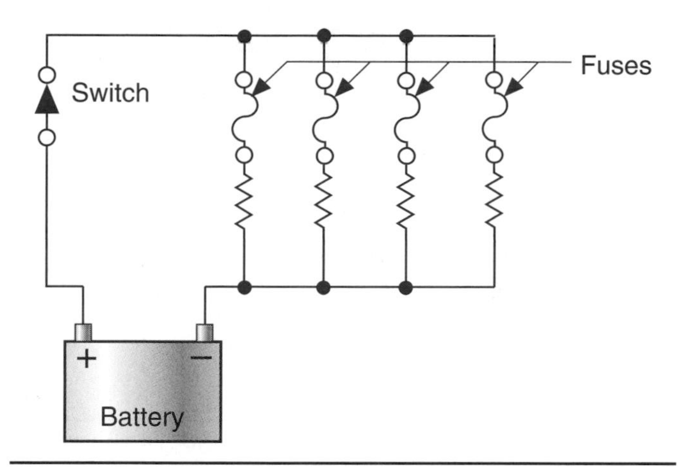

FIGURE 5–51 Individual fuses to protect each branch of a parallel circuit.

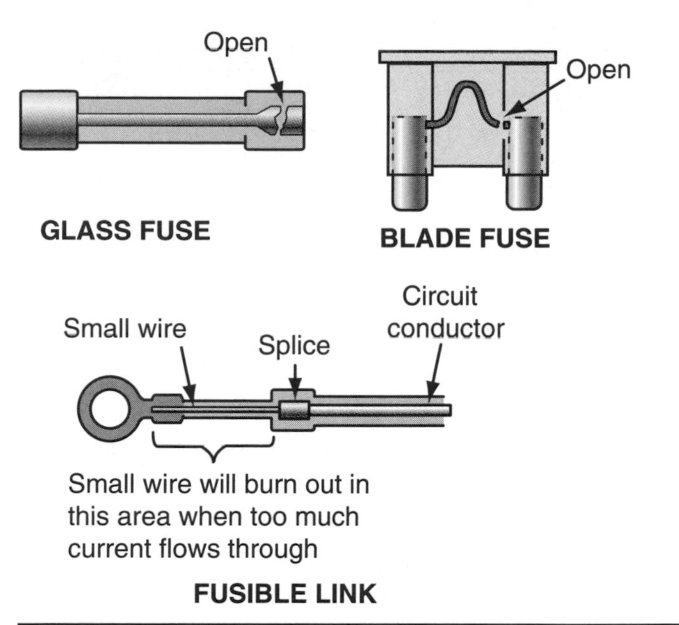

FIGURE 5–53 Hidden faults in a fuse can sometimes go undetected.

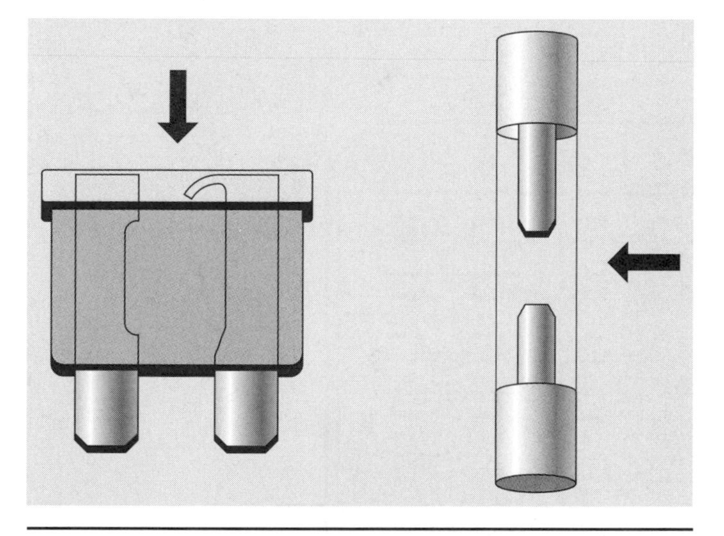

FIGURE 5–52 Typical view of a blown fuse.

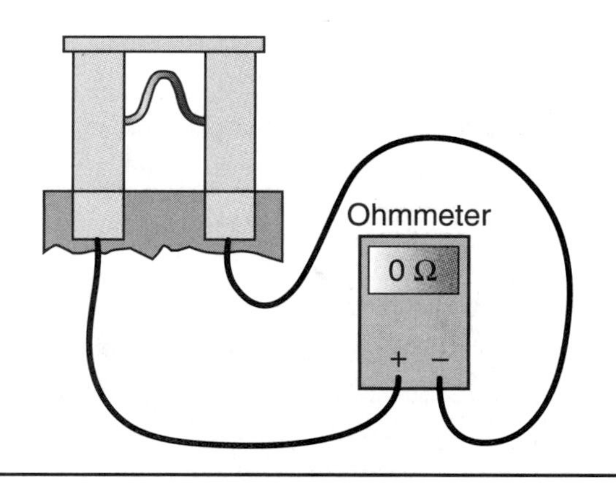

FIGURE 5–54 A good fuse will have zero resistance when tested with an ohmmeter.

is used to check a fuse or fusible link (Figures 5–54 and 5–55), the fuse should first be removed or power removed from the fusible link. The ohmmeter should read 0 to 1 ohms if it is good, and infinite if it is blown. Several manufacturers discourage piercing the insulation on wires. Instead, tug lightly on the fusible link; if the fusible link wire is bad, the insulation easily stretches.

To test a fuse or fusible link with a voltmeter or test light, check for available voltage at both terminals of the unit (Figure 5–56). If the device is good, there will be voltage on both sides. A blown fuse will show no voltage after the fuse.

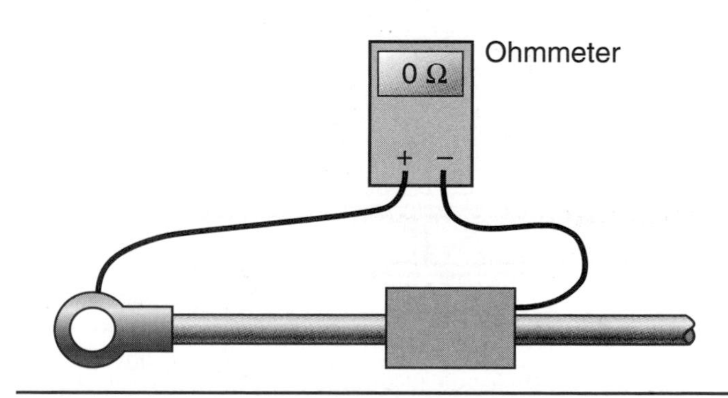

FIGURE 5–55 Testing a fusible link with an ohmmeter once it has been disconnected from the power source.

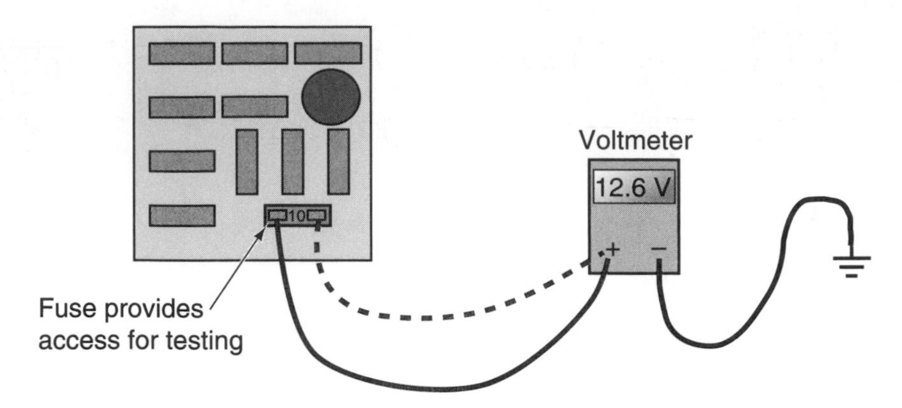

FIGURE 5–56 When testing a fuse with a voltmeter, there should be voltage present on both sides of the fuse.

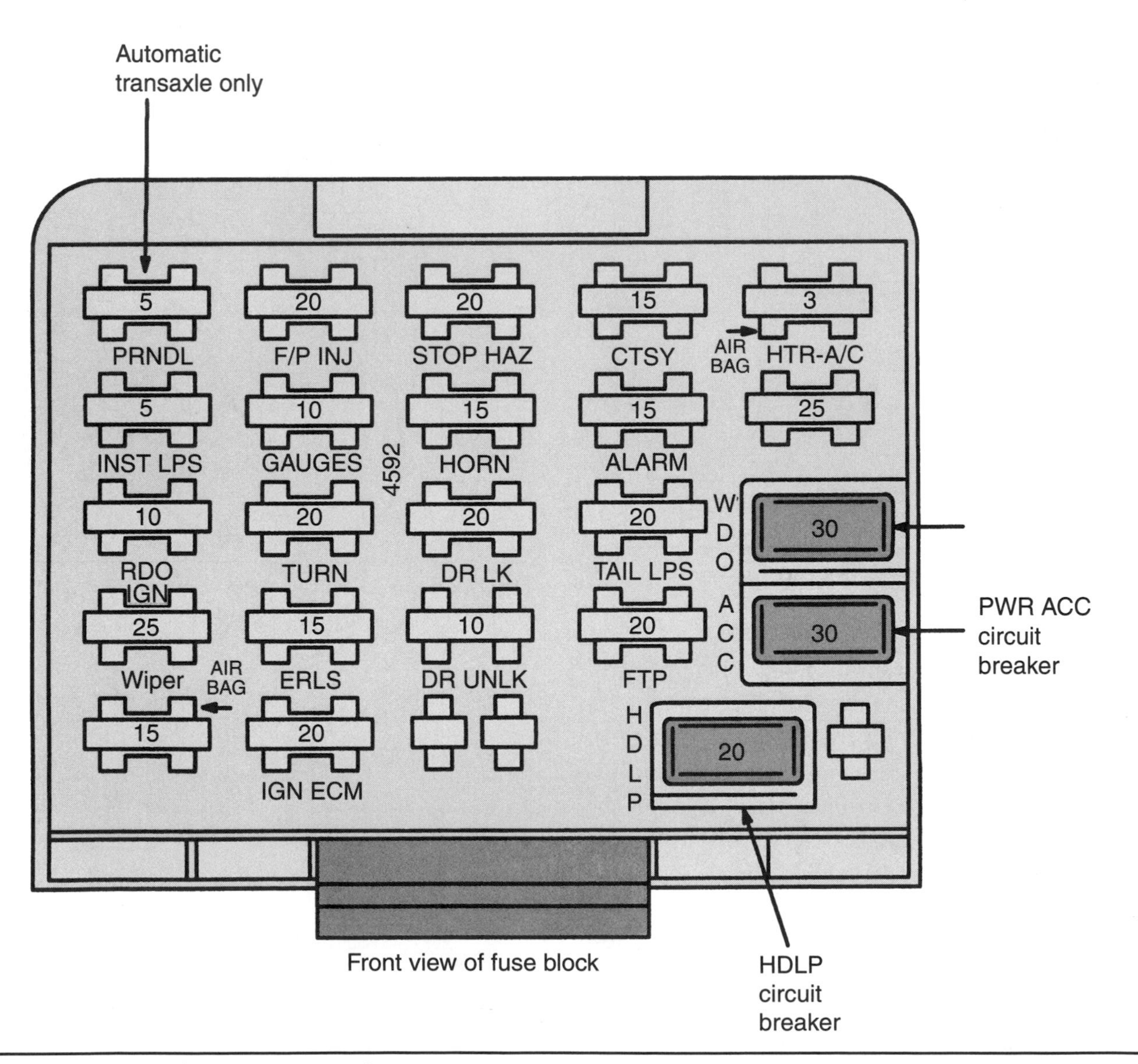

FIGURE 5–57 Typical fuse box or panel.

CAUTION *Fuses are rated by amperage for a specific circuit. Never bypass a fuse, or install a larger-capacity circuit protection device other than what is designed by the manufacturer. Severe circuit damage or user injury could result. A typical fuse box or panel is normally marked with the appropriate fuse size for each circuit (Figure 5–57). On some older vehicles, the printing on the fuse box face may not be readable. In these cases, refer to manufacturer's service manuals for proper fuse sizes.*

Circuit Breakers

Electrical circuits and load devices that are susceptible to overloads on a routine basis are usually protected by a **circuit breaker.** They can be mounted in the fuse panel or installed in-line with the electrical circuit. There are three common types of circuit breakers: the self-resetting circuit breaker (also called the cycling circuit breaker), the manual circuit breaker (also called the resettable circuit breaker), and the positive temperature coefficient (PTC) solid-state circuit breaker.

Figure 5–58 shows an example of a self-resetting circuit breaker. It reacts to a higher-than-designed flow of current by heating a **bimetallic strip** to momentarily open a set of contact points. When the bimetallic strip cools, it automatically closes the contact points and lets current flow in the circuit again.

A typical application for a self-resetting circuit breaker circuit is a power window circuit. Because the window is susceptible to overloading conditions, such as ice or snow buildup, a momentary current overload is possible. When this happens, the circuit breaker opens and closes until the ice or snow buildup is removed.

The second type of circuit breaker, the manual resettable type, is available in two styles. One style is reset by pushing a button on the circuit breaker; the other is reset by removing power from the circuit and letting the circuit breaker cool down (Figure 5–59).

The third type of circuit breaker is a PTC (positive temperature coefficient) solid-state device (Figure 5–60). This type of circuit breaker is usually wired within a load component like a wiper motor or power window motor. As current exceeds its design limits, the resistance

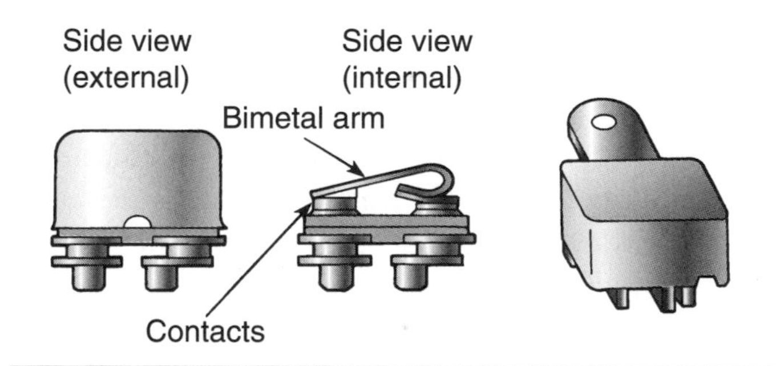

FIGURE 5–58 A self-resetting (cycling) circuit breaker.

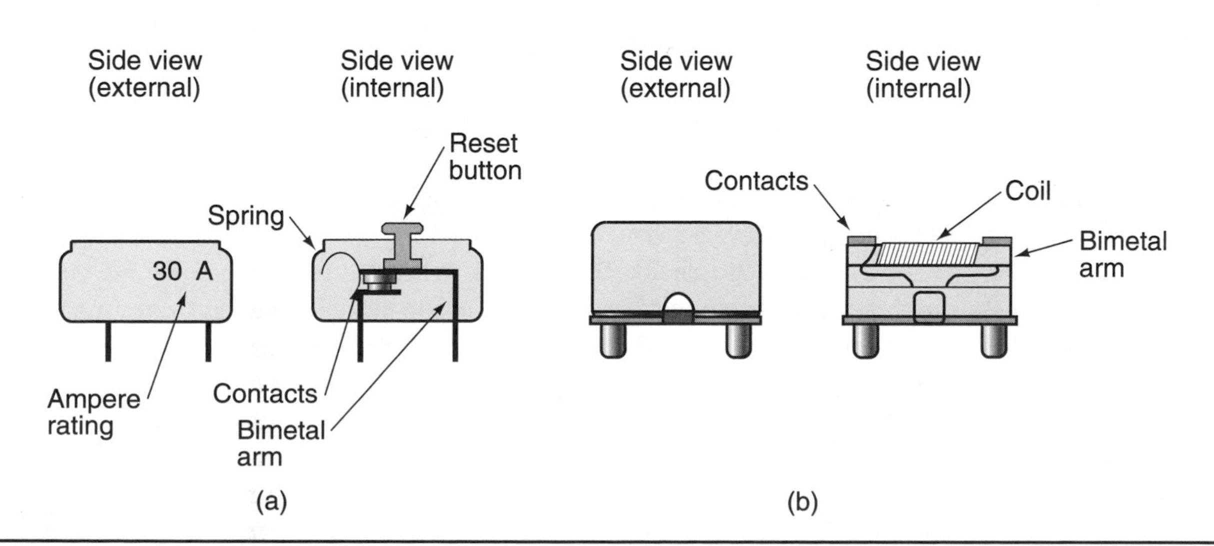

FIGURE 5–59 Typical manual reset circuit breakers: (a) Reset button style, and (b) reset requires removal of power.

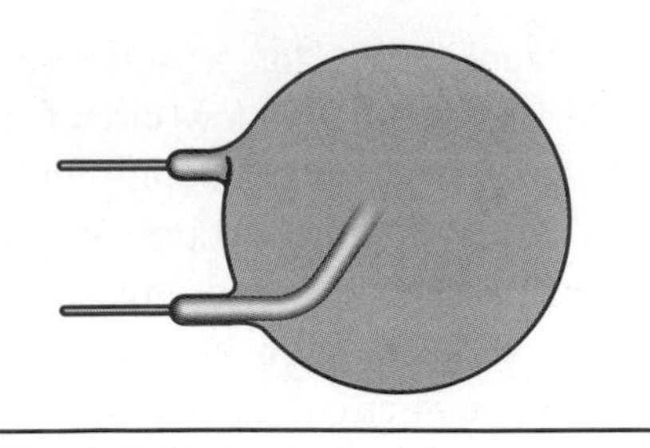

FIGURE 5–60 A PTC circuit breaker.

of the PTC circuit breaker increases until the circuit is opened. The PTC circuit breaker resets only after the high current condition of the circuit no longer exists.

Electrical Switches and Relays

An electrical switch is the most common way of controlling an on/off function or of directing the flow of current in an electrical circuit. The simplest type of switch is the **single-pole/single-throw (SPST) switch** (Figure 5–61). The term *pole* refers to the number of input circuits, and *throw* refers to the number of output circuits.

Some SPST electrical switches are momentary contact switches. For example, a horn switch (button) has a spring-loaded contact that keeps the contacts from closing unless external pressure is applied to the button (Figure 5–62).

Some electrical circuits require a **single-pole/double-throw (SPDT) switch.** A dimmer switch used in a headlight circuit, normally an SPDT switch (Figure 5–63), has one input circuit and two output circuits. Either the high beam or low beam is turned on depending on the position of the switch.

The third type of switch is a **ganged switch,** also referred to as a multiple-pole/multiple-throw (MPMT) switch. An MPMT switch contains multiple wipers that are ganged (hooked) together so that they all move together in unison. An ignition switch (Figure 5–64) is a common example of a ganged switch with five wipers that move together. Several SPST or SPDT switches can be hooked together and called a ganged switch by some manufacturers.

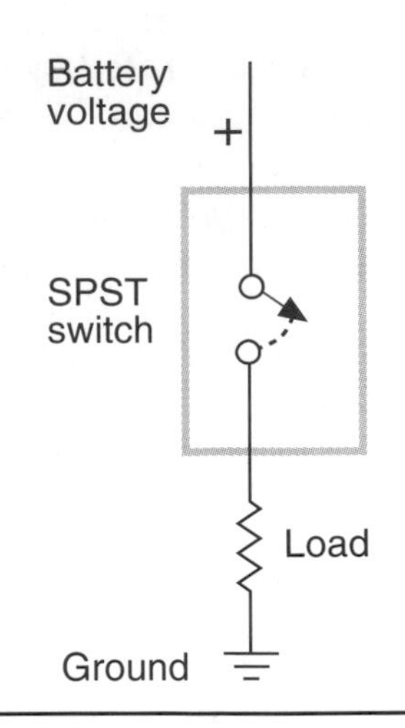

FIGURE 5–61 A simple single-pole/single-throw (SPST) switch.

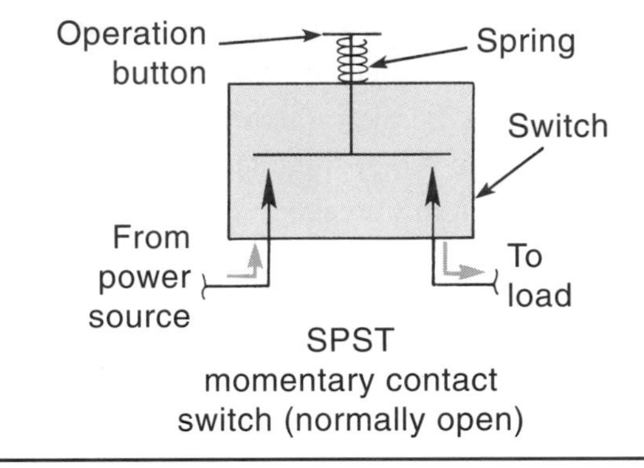

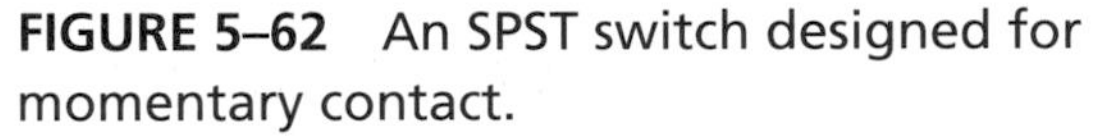

FIGURE 5–62 An SPST switch designed for momentary contact.

In some older vehicle electrical circuits, a **mercury switch** was used to control a light in a luggage or engine compartment. Mercury is used in the switch (Figure 5–65) because it is a good conductor of electricity. A mercury switch is usually mounted to the underside of a hood or trunk lid. When the lid is raised the mercury inside the switch capsule moves to the end, covering the electrical contacts and completing the circuit.

Note: Because mercury has been determined to be a hazardous substance, these switches are no longer used in new production vehicles. However, vehicles older than ten years are still on the road, so be careful when handling this type of switch.

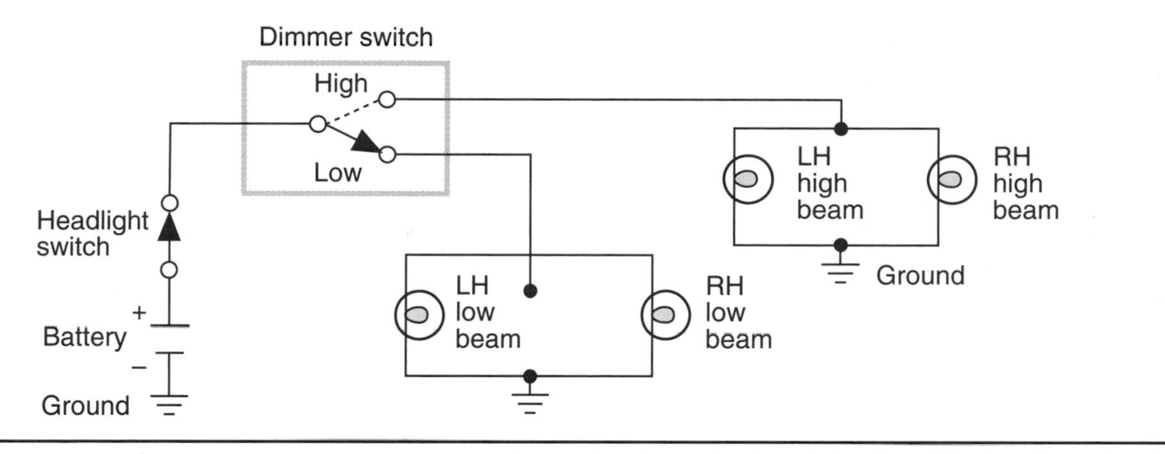

FIGURE 5–63 A simplified schematic of an SPDT dimmer switch in a headlight circuit.

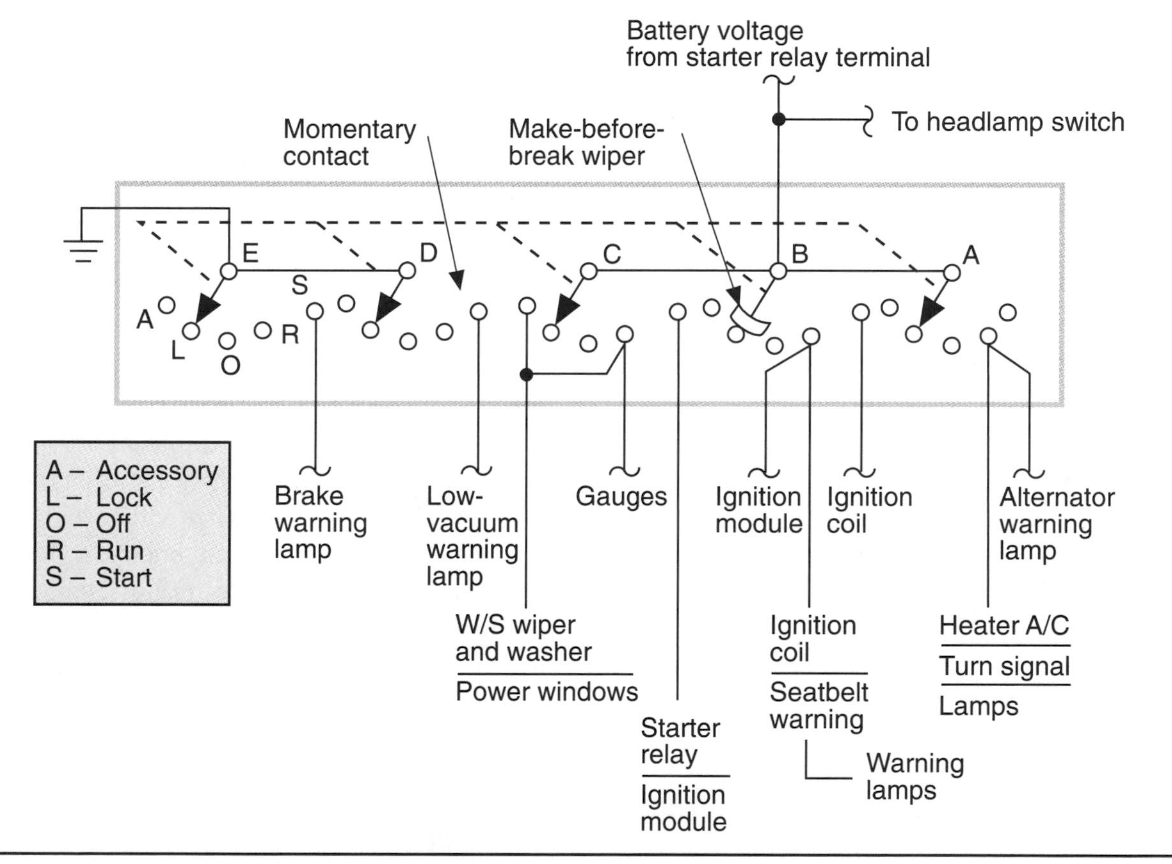

FIGURE 5–64 A typical MPMT ganged ignition switch.

Another type of specialized switch used extensively in vehicles is an **electromagnetic switch,** also called a **relay** (Figure 5–66). This switch is designed to use a small amount of current to control a relay coil that will electromagnetically close a set of contact points to complete a higher amperage circuit. A horn circuit is an example of a relay controlled application (Figure 5–67). The relay coil in the horn circuit in the figure is energized when the horn switch button is closed. The coil develops a magnetic field, which closes the contact points and completes the circuit between the battery and the horn assembly.

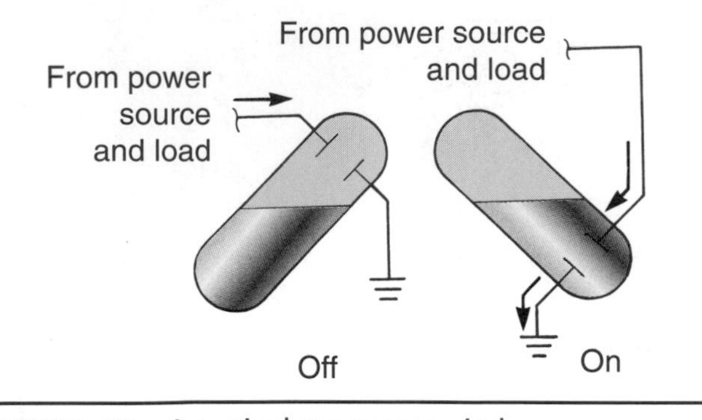

FIGURE 5–65 A typical mercury switch.

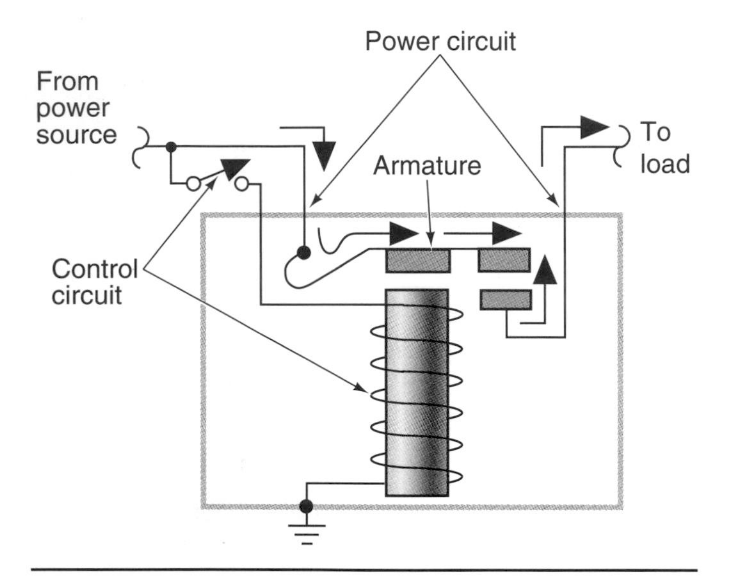

FIGURE 5–66 The relay contacts are closed by electromagnetism created by the coil.

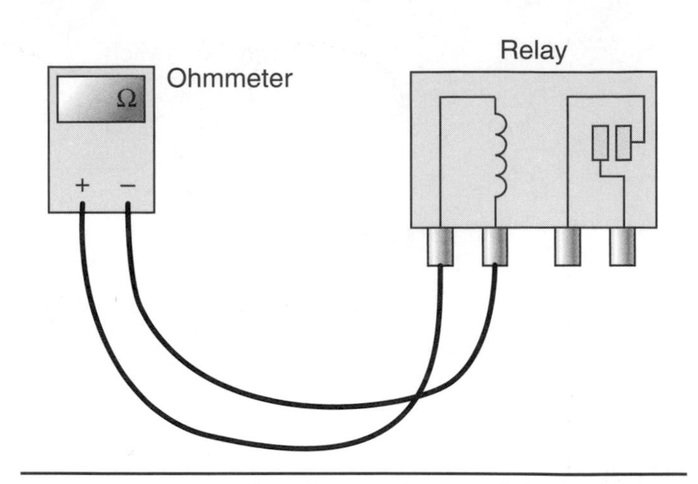

FIGURE 5–68 Testing the continuity of the coil in a relay.

There are normally two tests to perform on a relay. On-vehicle testing of a relay is covered in later chapters. To bench test a relay when it is off the vehicle, first test the relay coil for continuity with an ohmmeter (Figure 5–68). If there is continuity in the coil, the relay winding is good. An open in the coil wiring would be indicated by an infinity reading on the ohmmeter.

Next, test the relay contacts (Figure 5–69) by connecting an ohmmeter to the relay contact terminals. When the relay coil is energized by a suitable battery, the ohmmeter should indicate continuity through the contact points. If the contact points are bad or burned, the resistance reading on the ohmmeter will be high.

There are several different brands, styles, and shapes of relays used by vehicle manufacturers. The most common relay is a normally open (N/O) contact type with a four-pin terminal. The terminal layout and diagram for a four-pin N/O relay is shown in Figure 5–70.

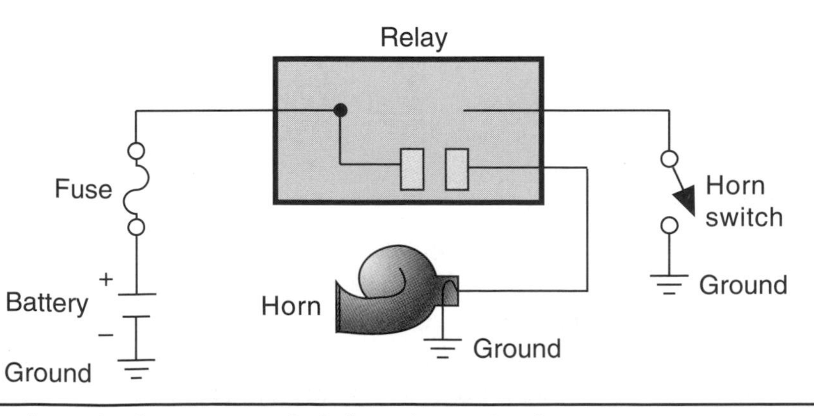

FIGURE 5–67 A relay used as a high current switch for a horn circuit.

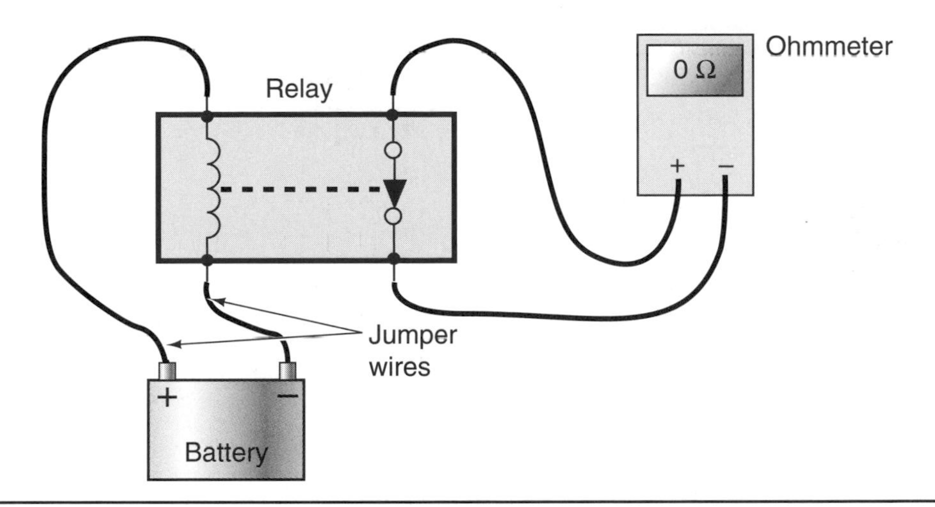

FIGURE 5–69 Checking the continuity of the relay contacts.

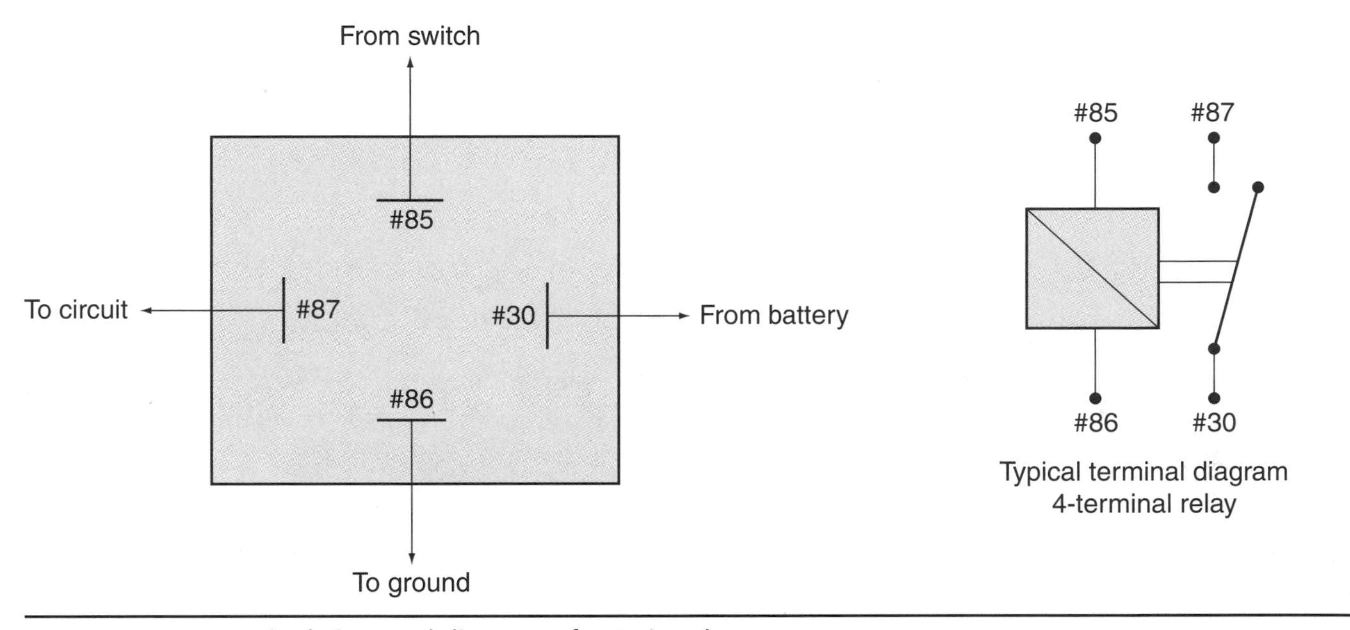

FIGURE 5–70 Terminal view and diagram of a 4-pin relay.

Whenever there is a five-pin relay, the fifth terminal of the relay is most often a normally closed (N/C) contact (Figure 5–71).

Capacitor Construction and Ratings

Some automotive electrical systems use a **capacitor,** or condenser as it is sometimes called, to store electrical charges from a DC circuit. Some of the most commonly used capacitors are shown in Figure 5–72.

A capacitor is a storage device for DC voltage. It is designed to absorb voltage changes in the circuit to control voltage spikes. The majority of capacitors are connected in parallel with the load circuit(s) (Figure 5–73).

When charged, a simple capacitor (Figure 5–74) is constructed of a **positively charged ion** plate which has more protons than electrons, and a **negatively charged ion** plate, which has more electrons than protons. The insulator between the two plates is made of a **dielectric** insulating material like plastic, glass, ceramic, or paper.

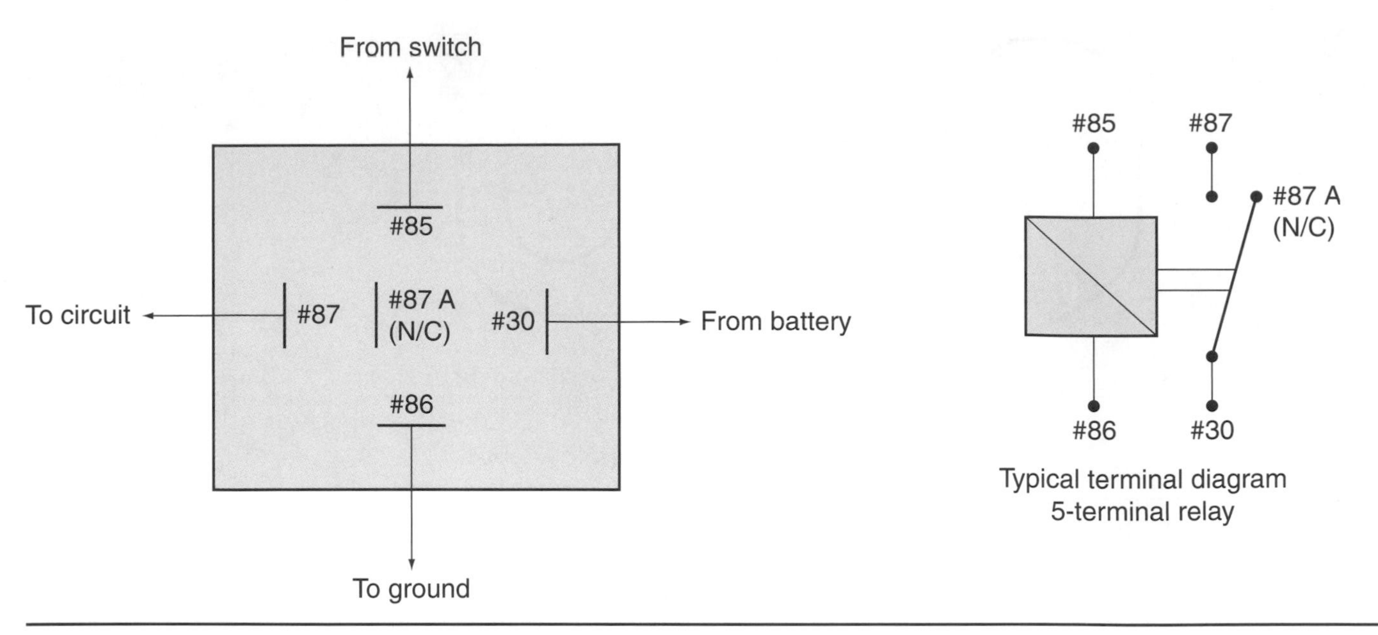

FIGURE 5–71 Terminal view and diagram of a 5-pin relay.

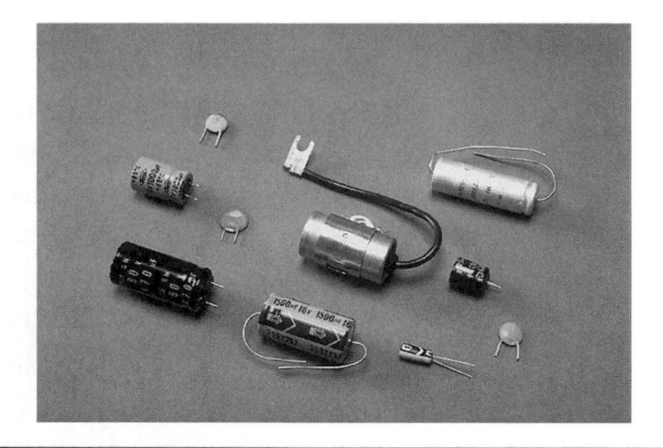

FIGURE 5–72 Common capacitors found in automotive electrical circuits.

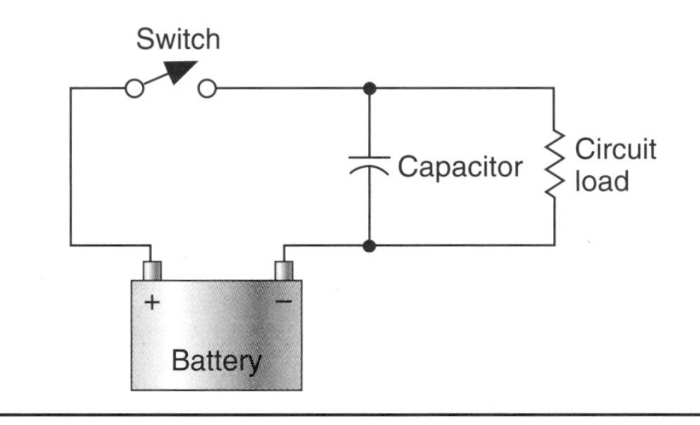

FIGURE 5–73 A capacitor connected in parallel to the circuit load in a simple circuit.

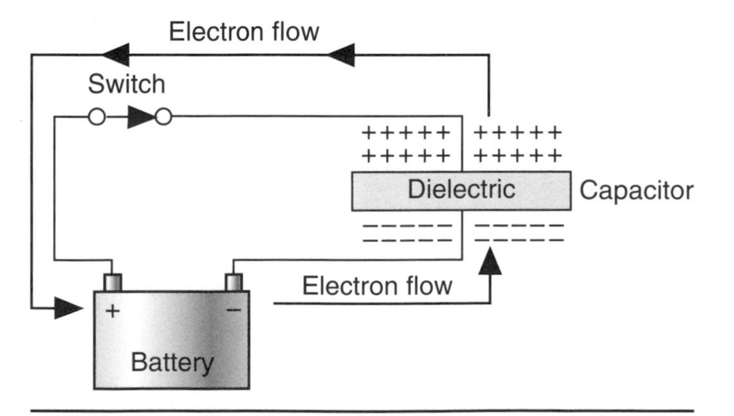

FIGURE 5–74 The flow of electrons in a capacitor.

Capacitors operate on the basic principles that opposite charges attract each other and that there is a potential voltage between any two points of opposite polarity. In Figure 5–75, when the system switch is closed, current flows into the capacitor until the voltage charges are equal to the battery. The effect of electron movement toward the negative plate and away from the positive plate creates the effect of current flow. Electrons do not actually pass through the capacitor's **electrostatic field,** which is between the two opposite charge plates, but are stored on the plates as static electricity.

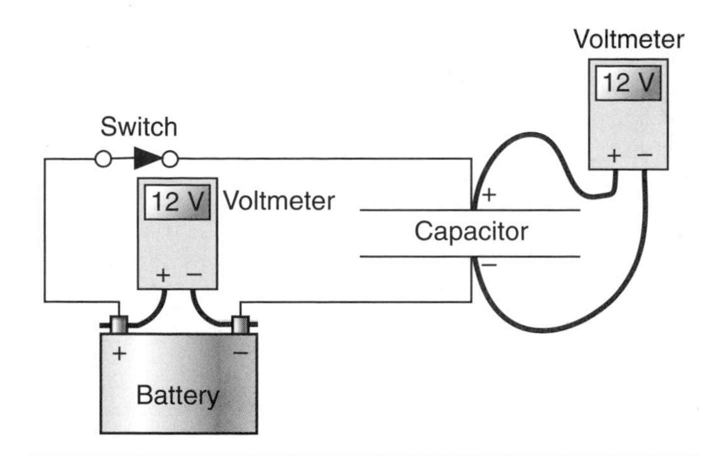

FIGURE 5–75 A capacitor that is fully charged.

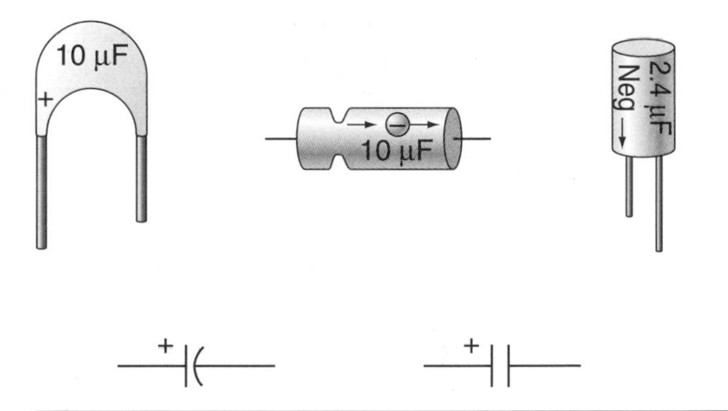

FIGURE 5–77 Common capacitor markings and schematic symbols.

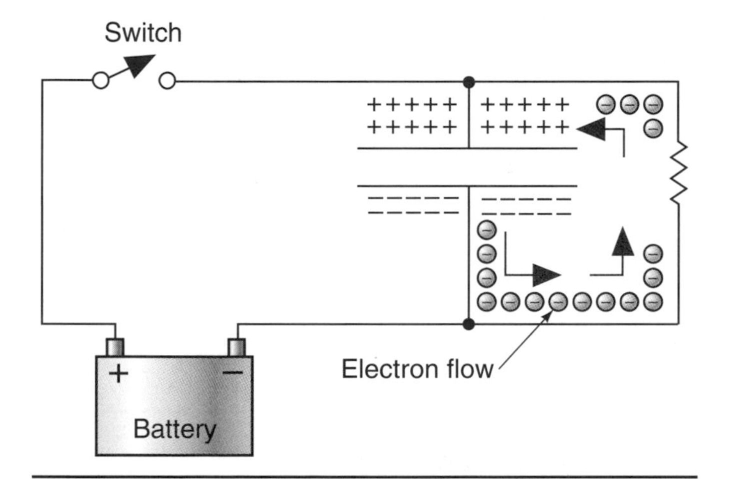

FIGURE 5–76 Current flow with the switch open and the capacitor discharging.

When a DC circuit switch is opened, the capacitor discharges the electrons from the surface of the negative capacitor plate, through the resistive load, and to the positive capacitor plate until a neutral state is reached (Figure 5–76).

Capacitors are normally rated by their maximum voltage rating and capacitance value. The standard capacitance unit value is the **farad.** Common automotive capacitors are usually rated in microfarads (μF), which are one-millionth of a farad. This value may be marked on the capacitor. Common capacitors are made of plastic, ceramic, or electrolytic material. Figure 5–77 is an example of capacitor markings and common schematic symbols.

Automotive Lamps

The lighting circuits of today's vehicles can consist of more than 50 bulbs of different designs and sizes. These bulbs are for an array of applications that include headlights, taillights, parking lights, stop lights, side marker lights, dash lights, courtesy lights, and so on. Knowing the proper size bulb for a specific application is important for proper circuit operation (Figure 5–78).

The simplest lamp design is an **incandescent lamp,** which produces a light as a result of current flowing through a coiled tungsten wire filament inside a glass bulb envelope. Inside the glass bulb is an inert gas that replaces the oxygen and lengthens the life span of the filament. Lamps are normally one of two types: a **single-filament bulb** (Figure 5–79) or a **double-filament bulb** (Figure 5–80). The double-filament bulb is designed to serve more than one electrical function, such as stoplamp and taillamp functions. The stoplamp function uses the lowerresistance filament because it needs to be brighter, and the taillamp function uses the higher-resistance filament.

When replacing a lamp, use the correct lamp for the application. The *Lamp Standard Trade Number* chart in Appendix B is a handy reference to determine the design voltages, amperes, and watts of a specific bulb circuit.

Lamp identification symbols on wiring schematics can vary with the manufacturer of a vehicle (Figure 5–81).

Auto Light Replacement Guide Miniature Bulbs

Description	GE Lamp No.	Replaces
Heavy duty turn signal, stop, tail, parking light	1157	1034
Heavy duty turn signal, parking light	1157NA	1157A, 1034A, 1034NA
Heavy duty turn signal, stop, tail, parking light	2057	-
Heavy duty turn signal, parking light	2057NA	-
Heavy duty back-up turn signal light high mount stop	1156	1073
Back-up light	1141	-
Heavy duty indicator, instrument, side marker, license light	194	158
	168	-
Dome, courtesy light	211-2	211-1
Dome, map, high mount stop	561	-
Turn signal, stop, tail, parking light	*3057	2457, 2358
	*3157	2458, 2359

Description	GE Lamp No.	Replaces
Foreign car interior dome light	DE3022	-
Foreign car interior dome light	DE3175	-
Dome, courtesy light	1004	-
Heavy duty instrument indicator light	1895	57
License, parking light	97	67, 1155
Dome, courtesy light	89	-
Dome, courtesy light	1003	93, 105
Indicator instrument light	1445	53
Dome, map, courtesy light	90	-
Dome, map, high mount stop	912	
	921	
Turn signal, parking light	3057NA	2457NA & 2358NA
	3157NA	3458NA & 2359NA

*Note: The above lamp sockets allow interchangeability of the double filament lamp types (3057, 3157, 3057NA, 3157NA). To comply with the DOT photometric specifications however, it is necessary to replace lamps with the correct type. Double filament lamps (3057, 3157) will fit the single filament socket (3156, 2456, 2356) and can be substituted. They will provide the correct photometric output.

FIGURE 5–78 Common automotive bulb designs.

(Courtesy of General Electric)

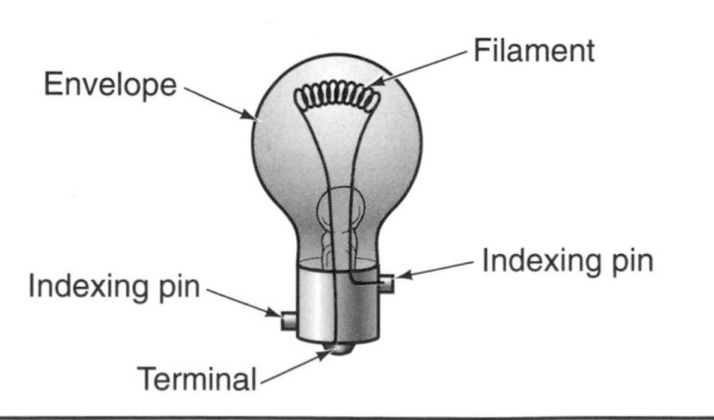

FIGURE 5–79 A single-filament incandescent bulb.

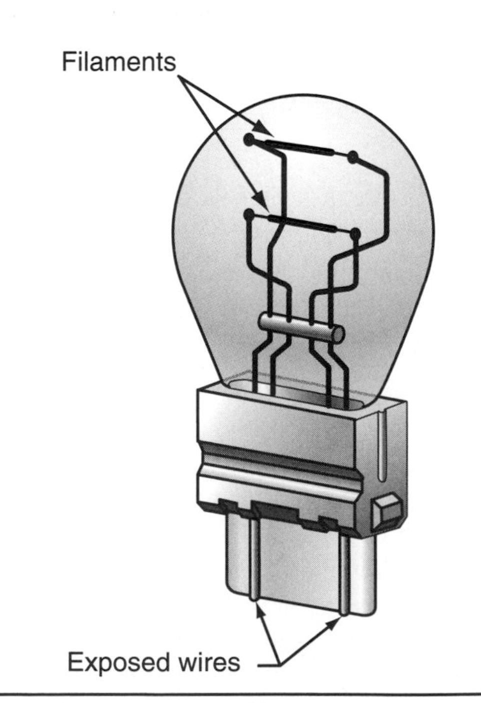

FIGURE 5–80 A double-filament incandescent bulb.

LED Lighting

Over the past ten years, many car and truck manufacturers have been using **light-emitting diodes (LEDs)** in various electrical lighting applications, such as instrument lights and center high-mounted stop lights (CHMSL).

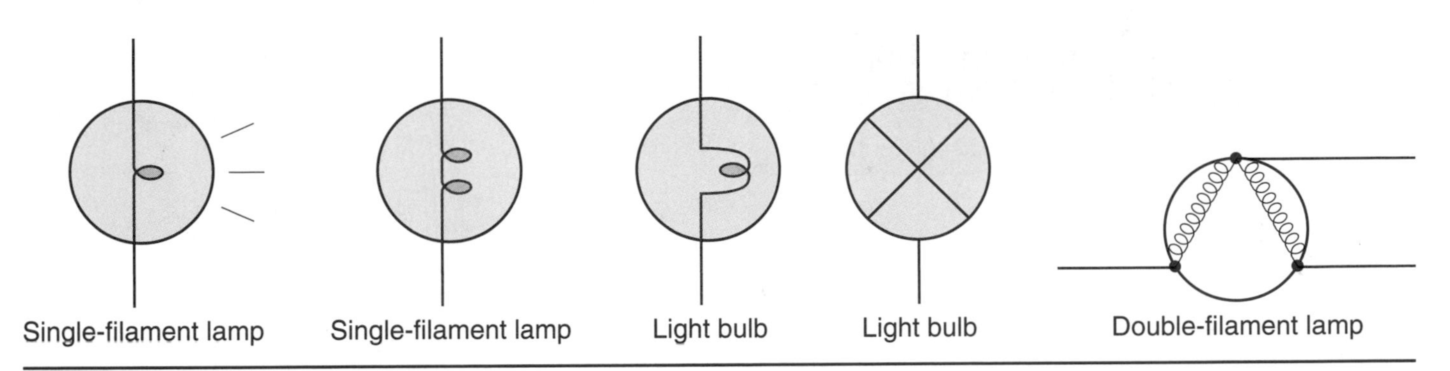

FIGURE 5–81 Common lamp identification symbols.

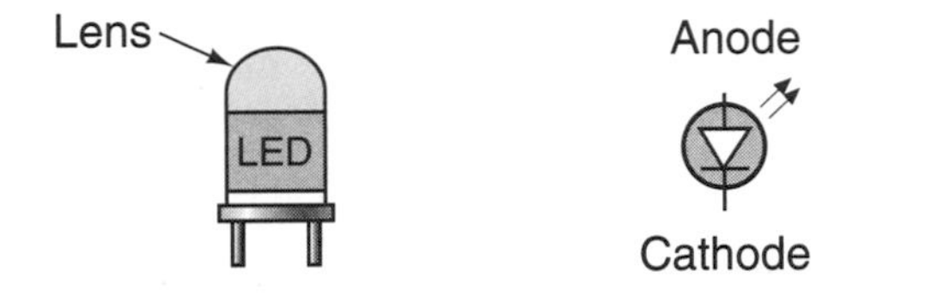

FIGURE 5–82 Typical LED and wiring diagram symbol.

LEDs (Figure 5–82) consist of a crystal that glows like a light bulb when controlled current is passed through them. LED lights have several advantages over the incandescent lamps:

- Quicker on time
- Less amperage draw
- Brighter operation
- Long service life

LED lighting systems have a very fast on time. A typical LED reaches full output brilliance in less than 1 millisecond, whereas incandescent bulbs take about 200 milliseconds or more to reach their full brightness. When LEDs are used in stoplights (Figure 5–83) and CHMSLs, the quicker on time has been proven in Federal studies as helping reduce the number of rear-end collisions.

The current draw in a typical LED lighting system uses significantly less than an incandescent lamp system. Figure 5–84 compares the current draw between LED lighting circuits and incandescent lamps. The actual amperage in an incandescent circuit varies depending on the number of bulbs in the lighting circuit. The start-up surge current is almost nonexistent in LEDs, making this circuit very popular with the trucking industry, which can use dozens of lights on its truck and trailer combinations.

FIGURE 5–83 LED lighting used in a rear light assembly.

Because LEDs are smaller than incandescent lamps, several can be placed behind the lens assembly, giving it overall brighter illumination. In addition, LED lights have an expected life span of 100,000 hours or more, compared to incandescent lamps, which normally last from 1,500 to 10,000 hours depending on heat and vibration factors.

Light circuit	Typical number of incandescent lamps	Operating current incandescent lamps (A)	Operating current LED system (A)
Stop lamps	4–6	8–10	0.75–2
Turn signal	4–6	8–10	0.75–2
Tail lamps	2–4	2–4	0.15–0.75

FIGURE 5–84 Typical current draw comparison between an incandescent lamp system and an LED lighting system.

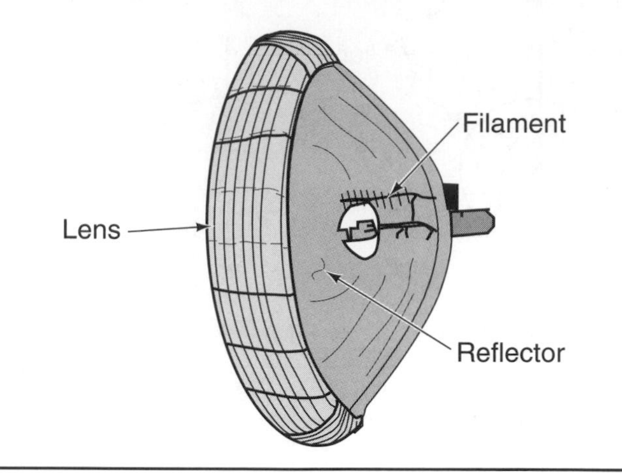

FIGURE 5–85 Sealed beam headlight construction.

Automotive Headlamps

Four types of headlamps are used in vehicles today: (1) sealed beam, (2) halogen sealed beam, (3) composite, and (4) high-intensity discharge (HID).

The first type is the basic round **sealed beam headlamp** that was used extensively between 1939 and 1975. The introduction of the rectangle headlamp in 1975 allowed vehicle manufacturers to streamline the front-end design of their vehicles. Both the round and rectangular sealed beam headlamps use similar construction (Figure 5–85).

In a sealed beam headlamp, the glass lens acts as a prism to direct the light beam from the reflector into a specific pattern (Figure 5–86). Inside the lamp, argon gas prevents the filament from becoming oxidized.

The direction of the light beam is controlled by the location of the filament in relation to the reflector (Figure 5–87). In a double-filament lamp, the lower filament is used for the high beam, and the upper filament is used for the low beam.

The second style of headlamp currently used is the **halogen headlamp** (Figure 5–88). This type of headlamp uses a halogen-filled inner bulb positioned in a glass housing. Its tungsten filament is able to burn approximately 25% brighter than a conventional sealed beam headlamp. If the outer lens is cracked or broken, the filament continues to operate, but the light quality will be reduced.

The third type of headlamp is the **composite headlamp** (Figure 5–89). This style of halogen headlamp allows the vehicle manufacturer to produce any shape of

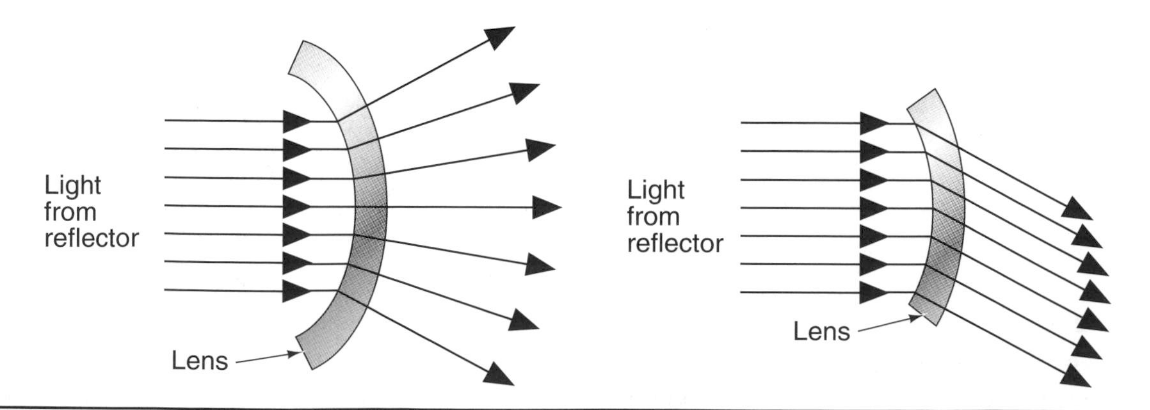

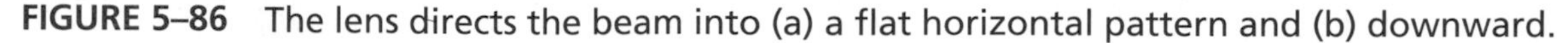

FIGURE 5–86 The lens directs the beam into (a) a flat horizontal pattern and (b) downward.

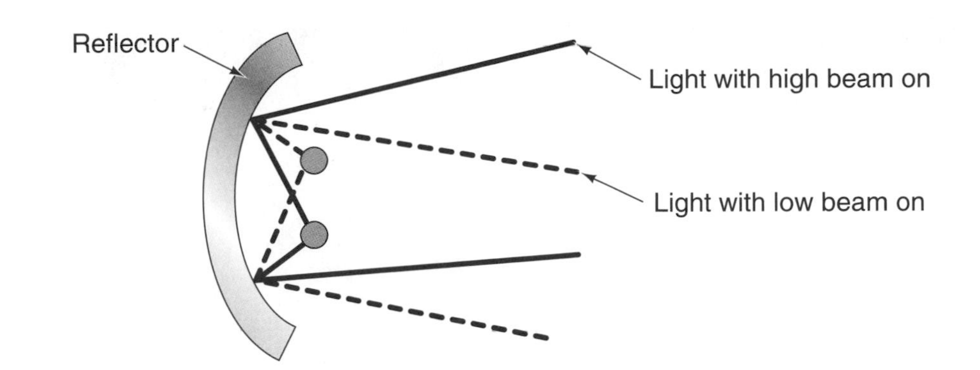

FIGURE 5–87 Filament placement controls projection of the light beam.

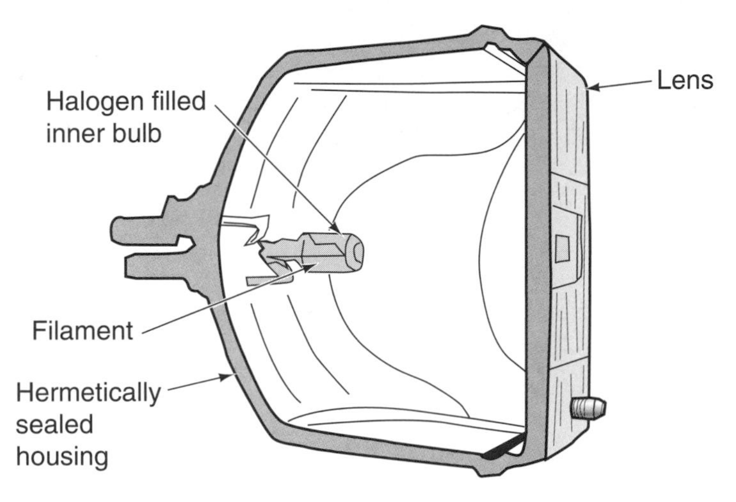

FIGURE 5–88 A sealed beam headlight with a halogen-filled inner bulb.

lens needed for vehicle styling and aerodynamics. Some of these composite headlamps are vented, so that condensation that may naturally develop on the inside of the lens assembly does not affect headlight operation. Refer to a vehicle's service manual if you are unsure whether a particular composite headlight is vented or nonvented. A nonvented composite lamp should be replaced if it collects moisture.

Be careful, and do not touch halogen bulbs when replacing them (Figure 5–90). The oily residue on human skin shortens the life of the bulb.

The remaining type of headlamp used in vehicles today is the **high-intensity discharge (HID),** or arc-discharge, light source (Figure 5–91). Instead of using a tungsten filament found in conventional headlamps, HID headlamps use a high-voltage arc discharge tube that jumps a gap inside a capsule filled with xenon gas. The technology is similar to the electronic flash assembly on a

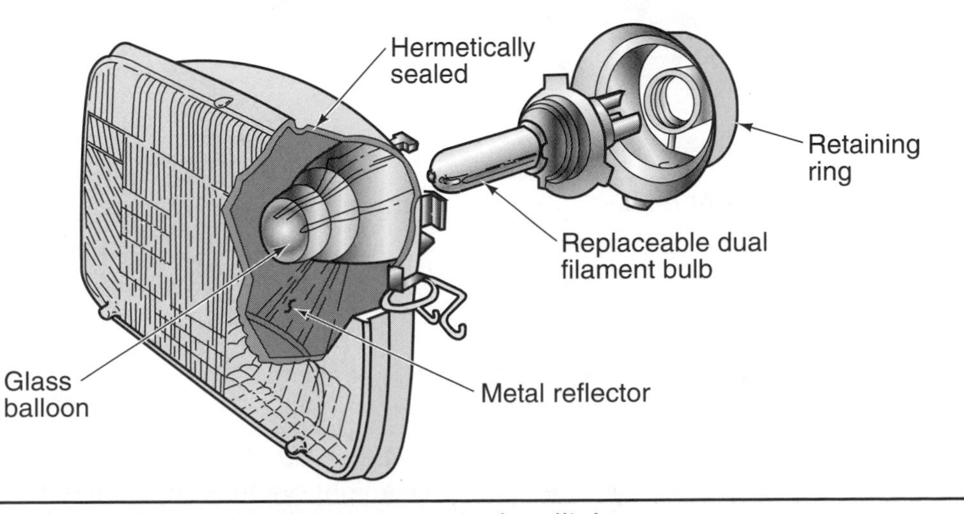

FIGURE 5–89 A replacement halogen bulb in a composite headlight.

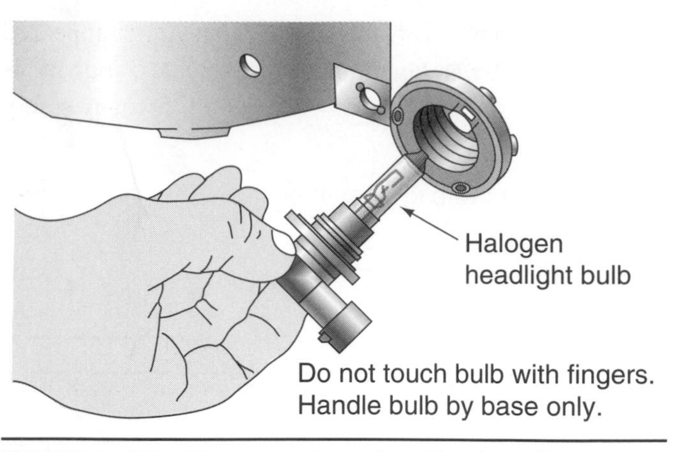

FIGURE 5–90 The correct method for handling a composite bulb during replacement.

FIGURE 5–91 An HID xenon gas light bulb.
(Courtesy of DaimlerChrysler Corporation)

camera. The HID arc produces a bluish-white light that is three to five times brighter than a glowing filament, and it typically lasts 2,000 hours or more (Figure 5–92). HID lighting systems are not available with a double-filament bulb because of the limits of the single-arc discharge tube. An HID system (Figure 5–93) requires a ballast that transforms 12 volts DC battery voltage into 20,000 volts

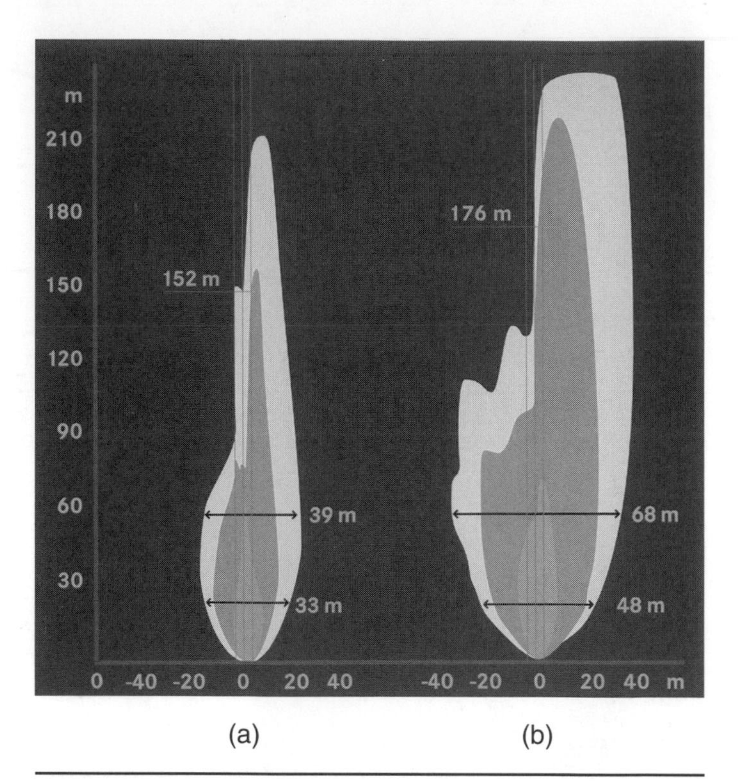

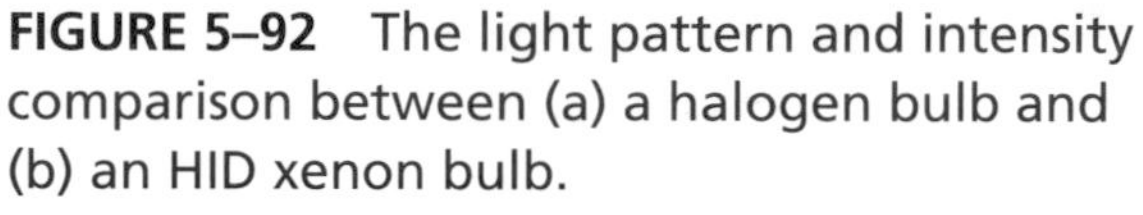

FIGURE 5–92 The light pattern and intensity comparison between (a) a halogen bulb and (b) an HID xenon bulb.
(Courtesy of DaimlerChrysler Corporation)

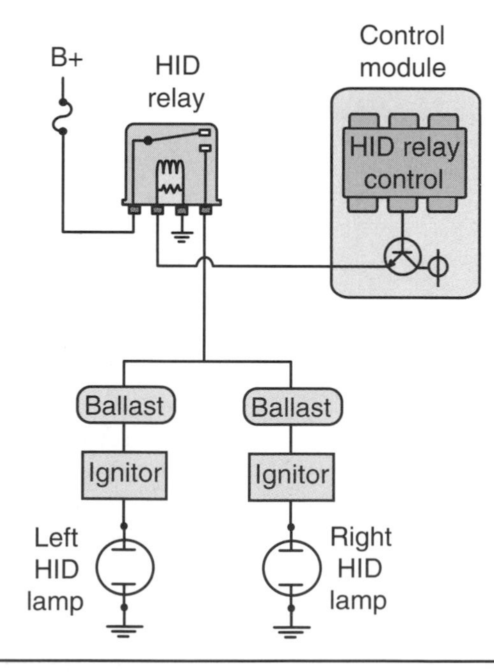

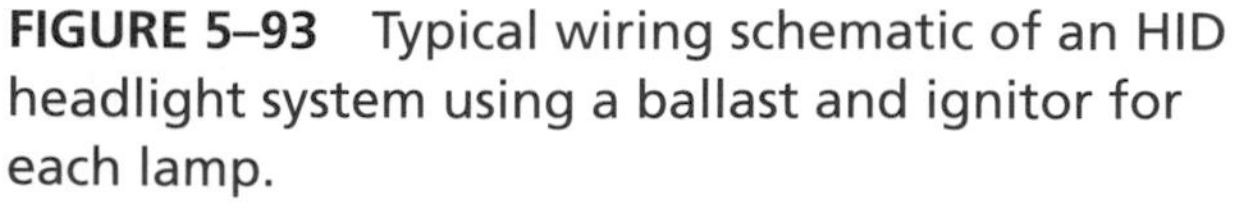

FIGURE 5–93 Typical wiring schematic of an HID headlight system using a ballast and ignitor for each lamp.

AC within 0.3 second for initial start-up. After a short warm-up period, the steady-state light system voltage is around 90 volts AC at about 3 amperes. This makes HID lighting more efficient than typical automotive headlamp systems. This style of headlamp first appeared in the mid-1990s on higher-end vehicles, but it is becoming more popular today because of its many safety advantages.

Hands-On Vehicle Tests

The following two hands-on vehicle checks are included in the NATEF (National Automotive Technician Education Foundation) Task List. Complete your answers to the following questions on a *separate* sheet of paper.

Performance Task 1

Task Description

Inspect and test fusible links, circuit breakers, and fuses; determine necessary action.

Task Objectives

- ❑ Obtain a vehicle that can be used for this task. What model and year of vehicle are you using for this task? ______
- ❑ Identify the following components and list their location on the vehicle.
  1. Fusible link(s) ______
  2. Fuse panel(s) ______
  3. Circuit breaker(s) ______
  4. List the fuses in the fuse panel and their current rating.

a. ______	e. ______	i. ______
b. ______	f. ______	j. ______
c. ______	g. ______	k. ______
d. ______	h. ______	

Circuit Checks

- ❑ Check continuity of a fusible link with an ohmmeter. Disconnect power to the fusible link before checking continuity. What is the result? ______
- ❑ Describe the procedure for replacing a fusible link. ______
- ❑ What is the general rule in determining what size fusible link to use for a circuit? ______
- ❑ Check a fusible link with a voltmeter. What is the result? ______

❑ If you measured a voltage drop across the fusible link, what would this indicate? ________________

____

❑ Check all fuses in the fuse panel with a voltmeter or test light. What are the results if the fuse is good? ____.

What would be the results on a blown fuse? ________________________________

____

❑ Check a fuse that is out of the vehicle with an ohmmeter. What are the results if the fuse is good? ____.

What would be the results if the fuse is blown? ________________________________

❑ Check a circuit breaker for proper operation with an ohmmeter or voltmeter. Determine if the circuit breaker is a self-resetting or a resettable type. Record the results of your findings. ________________

Task Summary

After performing the above NATEF task, what can you determine that will be helpful in future inspection of fusible links, fuses, and circuit breakers? ________________________________

____

Performance Task 2

Task Description

Inspect and test switches, connectors, relays, solenoid solid-state devices, and wires of electrical/ electronic circuits. Perform necessary action.

Task Objectives

❑ Obtain a vehicle that can be used for this task. What model and year of vehicle are you using for this task? ____

❑ Identify at least six of each of the following components and list their location on the vehicle.

1. Switches

   a. ________________ c. ________________ e. ________________

   b. ________________ d. ________________ f. ________________

2. Connectors

   a. ________________ c. ________________ e. ________________

   b. ________________ d. ________________ f. ________________

3. Relays

   a. ________________ c. ________________ e. ________________

   b. ________________ d. ________________ f. ________________

4. Wiring (harnesses)

a. ________________ c. ________________ e. ________________

b. ________________ d. ________________ f. ________________

Circuit Checks

❑ Check the continuity of a switch with an ohmmeter. Disconnect power to the switch before checking continuity. What is the result? ________________

❑ Describe the procedure for replacing a switch. ________________

____

❑ Check a switch with a voltmeter. What is the result? ________________

❑ If you measured a voltage drop across the switch, what would this indicate? ________________

____

❑ Perform a visual inspection of all accessible connectors on the vehicles. Check for loose or corroded connections. What are the results? ________________

❑ Describe how to check an electrical connection for a voltage drop. ________________

____

❑ Perform a bench test of a relay's coil and contact points. Using an ohmmeter, what is the reading on the relay coil? ________________.

❑ Using a battery to energize the relay coil, what is the ohmmeter reading on the relay contact points? ________________

❑ Using a voltmeter, test the proper operation of a relay on the vehicle. What voltage reading do you get on each side of the relay coil?

Input ________________. Output ________________.

❑ What voltage reading do you get on each side of the relay contact points?

Input ________________. Output ________________.

❑ Inspect all accessible wiring harnesses in the vehicle. Check for frayed wires, deteriorated insulation, missing straps, or tie downs. Record results here. ________________

____

Task Summary

After performing the above NATEF task, what can you determine will be helpful in future inspections of fusible links, fuses, and circuit breakers? ________________

____

Summary

- Automotive primary wire is used for low-voltage circuits.
- Automotive secondary wire is used for high-voltage circuits, like spark plug wires.
- Stranded copper wire is used for most automotive wiring applications.
- Wire size is measured by the AWG standard or by metric sizes.
- Metric wire size is measured by determining the cross-sectional area of the wire.
- A solderless connector uses a special crimping tool to connect it to the wire.
- Rosin core flux solder is used when soldering electrical connections.
- A hard-shell connector contains multiple wires in one connector component.
- A weather-pack connector uses a rubber seal on the terminal and connector ends.
- A metri-pack connector is similar to a weather-pack connector except there is no seal on the cover half of the connector.
- Pull-to-seat and push-to-seat are forms of metri-pack connectors.
- Resistors are one of the most common components found in electrical circuits.
- Fixed resistors are generally made of carbon/graphite-based material.
- A stepped resistor is sometimes used to control blower motor speeds in electrical circuits.
- Variable resistors have the ability to change the resistance potential of an electrical circuit.
- A rheostat has two terminals: one to a fixed end and the other to a movable wiper contact.
- A potentiometer has three terminals: one on each fixed end and the third on a movable wiper.
- A thermistor's resistive value changes with temperature.
- Fuses are the most common circuit protection device.
- A glass fuse uses end cap codes of AGA, AGW, or AGC to indicate length and diameter.
- A glass fuse with the end cap code of SFE has the same diameter, but the length varies with the current rating.
- Blade fuses have code ratings of ATC or ATO.
- A maxi-fuse and Pacific fuse element were developed to replace the fusible link.
- Two types of circuit breakers are the self-resetting (cycling) and the resettable.
- A SPST switch has one input pole circuit and one output throw circuit.
- A SPDT switch has one input pole circuit and two output throw circuits.
- A MPMT switch is also known as a ganged switch and contains multiple wiper contacts that are all ganged (hooked) together.

- Mercury switches contain a small vial of mercury to connect two contacts when moved to a specified position.
- A relay is a specialized switch that electromagnetically closes a set of contact points.
- An incandescent lamp uses a tungsten wire filament in a glass bulb envelope.
- LED lighting has several advantages over incandescent lamps: quicker on time, less amperage draw, brighter operation, and longer service life.
- Four types of headlamps are used in vehicles today: sealed beam, halogen sealed beam, composite halogen, and high-intensity discharge.
- A capacitor, or condenser, is used to store electrical charges in a circuit.

Key Terms

American wire gauge (AWG)
auto-fuse
bimetallic strip
capacitor
circuit breaker
circuit-protection device
color code
composite headlamp
dielectric
double-filament bulb
electromagnetic switch
electrostatic field
farad
fixed resistor
fusible link
ganged switch
halogen headlamp
hard-shell connector
high-intensity discharge (HID)
incandescent lamp
light-emitting diode (LED)
maxi-fuse
mercury switch
metri-pack connector
mini-fuse
negatively charged ion
Pacific fuse element
positively charged ion
potentiometer
primary wire
pull-to-seat
push-to-seat
relay
rheostat
rosin core
sealed beam headlamp
secondary wire
single-filament bulb
solderless connector
single-pole/double throw (SPDT) switch
single-pole/single-throw (SPST) switch
stepped resistor
tap splice
thermistor
variable resistor
weather-pack connector

Review Questions

Short Answer Essays

1. Describe the difference between automotive primary wire and secondary wire.
2. Define the difference and use of acid core solder and rosin core solder.
3. Describe how to identify the resistive value of a carbon film resistor.
4. Describe the purpose of a capacitor in an automotive electrical circuit.
5. Describe the difference between a sealed beam headlamp and a halogen headlamp.

Fill in the Blanks

1. The standard rule with AWG wire is the ______________________ the gauge number, the ______________________ the wire conductor.

2. A tap splice connector is sometimes used when a(n) ______________________ circuit is added to a(n) ______________________ circuit.

3. A pacific fuse element was developed as a replacement for a ______________________ ______________________.

4. A ______________________ is used in an electrical circuit to store charges from the DC circuit.

5. A ______________________ ______________________ bulb is designed to serve more than one electrical function.

ASE-Style Review Questions

1. Technician A says that automotive primary wire is used to carry low voltages. Technician B says that spark plug wires are secondary wires than carry high voltages. Who is correct?
   A. A only
   B. B only
   C. Both A and B
   D. Neither A nor B

2. Technician A says that most primary wire is made of several strands of copper wire wound together and covered with an insulation. Technician B says there is more resistance in a stranded wire than in the same gauge solid wire. Who is correct?
   A. A only
   B. B only
   C. Both A and B
   D. Neither A nor B

3. Wire size identification standards are being discussed. Technician A says the rule with AWG standard wire size is the higher the number, the larger the wire conductor. Technician B says in the metric system of determining wire size, the smaller the number, the smaller the wire conductor. Who is correct?
   A. A only
   B. B only
   C. Both A and B
   D. Neither A nor B

4. Wires with double color codes are being discussed. Technician A says the first letters indicate the base color of the insulation. Technician B says the second set of letters indicates the color of the tracer. Who is correct?

A. A only
B. B only
C. Both A and B
D. Neither A nor B

5. Technician A says solderless crimp connectors are used extensively in automotive circuits because they provide excellent protection against corrosion and dirt. Technician B says solderless crimp connectors are normally crimped to the wire with slip-joint pliers. Who is correct?
   A. A only
   B. B only
   C. Both A and B
   D. Neither A nor B

6. Technician A says solder used for electrical use has a rosin core flux. Technician B says solder with a 60/40 mixture means 60% lead and 40% tin. Who is correct?
   A. A only
   B. B only
   C. Both A and B
   D. Neither A nor B

7. Technician A says a hard-shell connector has four wires molded into a connector component. Technician B says a hard-shell connector has six wires molded into a connector component. Who is correct?
   A. A only
   B. B only
   C. Both A and B
   D. Neither A nor B

8. Technician A says a weather-pack connector provides for excellent corrosion protection. Technician B says a weather-pack connector has a rubber seal on the terminal end and on the connector half. Who is correct?
   A. A only
   B. B only
   C. Both A and B
   D. Neither A nor B

9. Technician A says metri-pack connectors have a push-to-seat connector. Technician B says metri-pack connectors have a pull-to-seat connector. Who is correct?
   A. A only
   B. B only
   C. Both A and B
   D. Neither A nor B

10. Technician A says a stepped resistor is used to control temperatures in a circuit. Technician B says a rheostat has one fixed end and one movable wiper contact. Who is correct?
    A. A only
    B. B only
    C. Both A and B
    D. Neither A nor B

11. Technician A says a potentiometer uses three wires: one movable wiper and two fixed ends. Technician B says a potentiometer can sense a variable voltage drop. Who is correct?
    A. A only
    B. B only
    C. Both A and B
    D. Neither A nor B

12. Technician A says glass fuses have end cap codes of AGA, AGW, or AGC. Technician B says glass fuses with the end code SFE have the same length, but their diameter depends on the current rating. Who is correct?
    A. A only
    B. B only
    C. Both A and B
    D. Neither A nor B

13. Technician A says a maxi-fuse is a replacement fuse designed as a fusible link in a modern vehicle. Technician B says a Pacific fuse element is a replacement fuse designed as a fusible link in a modern vehicle. Who is correct?
    A. A only
    B. B only
    C. Both A and B
    D. Neither A nor B

14. Technician A says all circuit breakers in a vehicle's electrical system are self-resetting. Technician B says all circuit breakers in a vehicle's electrical system are cycling. Who is correct?
    A. A only
    B. B only
    C. Both A and B
    D. Neither A nor B

15. Technician A says an SPST switch has a single input and two outputs. Technician B says an SPST switch can be a momentary switch. Who is correct?
    A. A only
    B. B only

C. Both A and B

D. Neither A nor B

16. Technician A says a ganged switch uses multiple wipers hooked together. Technician B says a ganged switch has one input and multiple outputs. Who is correct?

    A. A only

    B. B only

    C. Both A and B

    D. Neither A nor B

17. Technician A says a relay's contact points are energized by a small internal coil. Technician B says a relay's contact points are energized electromagnetically. Who is correct?

    A. A only

    B. B only

    C. Both A and B

    D. Neither A nor B

18. Technician A says a sealed beam headlamp can have a single internal filament. Technician B says a sealed beam headlamp can have two internal filaments. Who is correct?

    A. A only

    B. B only

    C. Both A and B

    D. Neither A nor B

19. Technician A says a halogen headlamp uses a halogen-filled inner bulb sealed in a glass housing. Technician B says in a composite halogen headlamp, the entire headlamp assembly needs to be replaced when burned out. Who is correct?

    A. A only

    B. B only

    C. Both A and B

    D. Neither A nor B

20. Technician A says automotive capacitors generate voltage spikes in a DC electrical circuit. Technician B says a standard capacitor unit is a farad. Who is correct?

    A. A only

    B. B only

    C. Both A and B

    D. Neither A nor B

6 Meters and Measuring Devices

Introduction

An automotive technician working on electrical and electronic circuits must be familiar with the proper use and limitations of instruments that measure electrical properties. When a test instrument is hooked up correctly, it provides the eyes for the technician to see what's inside the circuit. In this chapter, we study several types of electrical measuring devices and their common names, and we look briefly at how each is used.

Objectives

When you complete this chapter you should be able to:

- ❑ Describe the basic design characteristics of an analog multimeter.
- ❑ Describe the basic design characteristics of a digital multimeter.
- ❑ Demonstrate how to properly connect a multimeter to a circuit to measure system voltage.
- ❑ Demonstrate how to properly connect a multimeter to a circuit to measure system amperage.
- ❑ Demonstrate how to properly connect a multimeter to a circuit to measure system resistance.
- ❑ Demonstrate the proper use of a test light to measure for voltage in a circuit.
- ❑ Describe the design characteristics of a digital oscilloscope.
- ❑ Explain the proper use and limitations of a digital oscilloscope.
- ❑ Describe the main characteristics found on most computer scan testers.

Analog Meters

For many years, one of the traditional meters found in automotive shops for basic electrical diagnosis was the **analog meter.** This type of meter uses a pointer needle that moves across the face of a printed scale (Figure 6–1), much the same as a speedometer or standard dash gauges.

Analog meters operate by a **D'Arsonval movement** (Figure 6–2). In this type of movement, also called a **moving coil meter**, a horseshoe style permanent magnet surrounds a small coil of wire attached to a pointer. When current flows through the coil, it interacts with the magnetic field of the magnet and causes the pointer needle to move. The direction of pointer movement is determined by the direction of current flow through the coil.

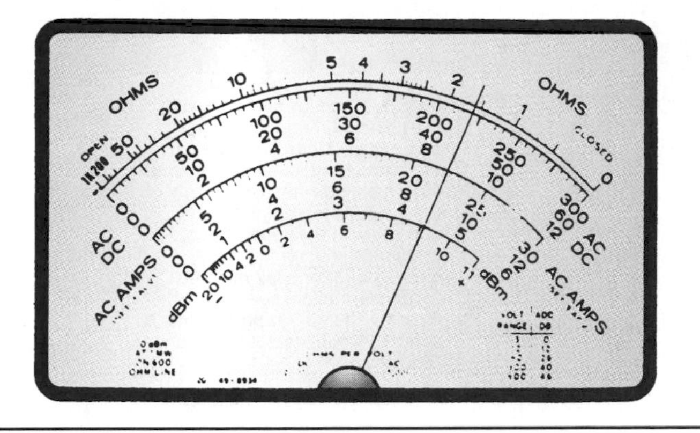

FIGURE 6–1 A typical analog meter face with multiple functions.

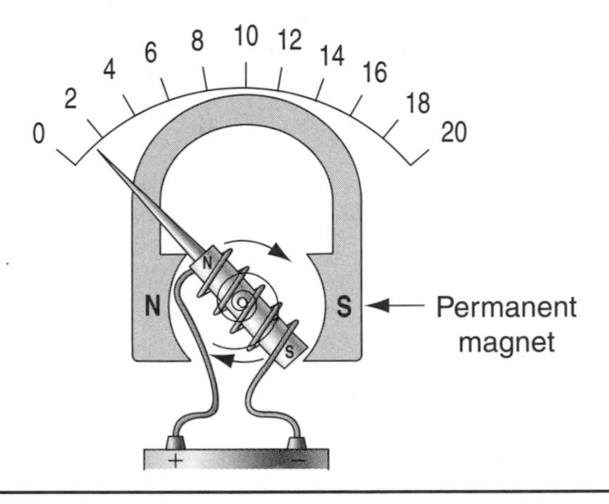

FIGURE 6–2 A D'Arsonval movement used in an analog meter.

Because of the service limitations of an analog style meter, most automotive manufacturers specify using test instruments that have the ability to measure sensitive solid-state circuits like digital dash clusters and engine control sensors. These sensitive circuits require meters with at least 10-megaohm (million) or higher internal resistance. The majority of analog meters *do not* have this high level of internal resistance and should not be used for testing most automotive circuits.

Digital Multimeters

For diagnosing today's automotive electrical and electronic systems, the **digital multimeter (DMM)** has become a necessary tool. Another name for a digital multimeter is a **digital volt-ohm meter (DVOM).** A quality digital meter (Figure 6–3) has a 10-megaohm internal impedance and uses electronic circuitry and a liquid crystal display to provide more precise readings than analog meters. The display face of a quality digital meter (Figure 6–4) is large and easy to read. The various symbols on the display tell the technician which mode the meter has activated and other specific performance features (Figure 6–5).

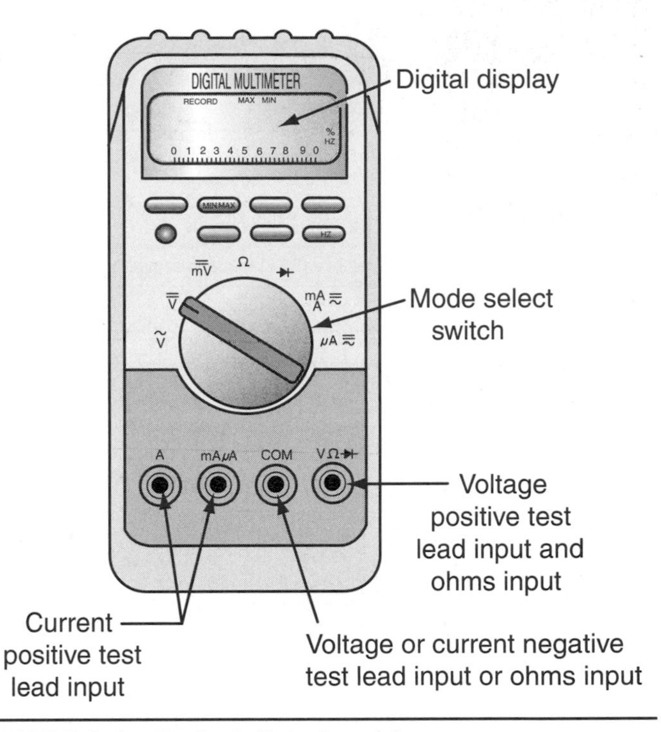

FIGURE 6–3 Typical digital multimeter.

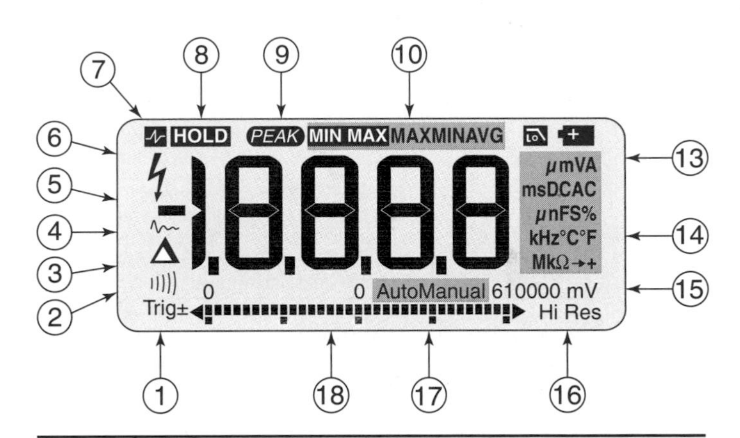

FIGURE 6–4 A digital multimeter liquid crystal display (LCD).

The cost and quality of the digital multimeters available on the market range widely. The most popular DMM types used by technicians have an **autoranging** feature for the mode select switch, which automatically adjusts to the best range for the circuit being measured. A high-quality

Number	Feature	Indication
①	±	Polarity indicator for the analog bar graph.
	Trig ±	Positive or negative slope indicator for Hz/duty cycle triggering.
②	·ı)))	The continuity beeper is on.
③	Δ	Relative (REL) mode is active.
④	∿	Smoothing is active.

Number	Feature	Indication
⑤	–	Indicates negative readings. In relative mode, this sign indicates that the present input is less than the stored reference.
⑥	ϟ	Indicates the presence of a high voltage input. Appears if the input voltage is 30 V or greater (AC or DC). Also appears in low pass filter mode. Also appears in cal, Hz, and duty cycle modes.
⑦	HOLD (Auto)	AutoHOLD is active.
⑧	HOLD	Display Hold is active.
⑨	PEAK	Indicates the Meter is in Peak Min Max mode and the response time is 250 μs (87 only).
⑩	MIN MAX MAX MIN AVG	Indicators for minimum-maximum recording mode.
⑪	LO	Low pass filter mode (87 only). See "Low Pass Filter (87).
⑫	(battery)	The battery is low. ⚠⚠Warning: To avoid false readings, which could lead to possible electric shock or personal injury, replace the battery as soon as the battery indicator appears.

Number	Feature	Indication
⑱	(bar graph)	The number of segments is relative to the full-scale value of the selected range. In normal operation 0 (zero) is on the left. The polarity indicator at the left of the graph indicates the polarity of the input. The graph does not operate with the capacitance, frequency counter functions, temperature, or peak min max. For more information, see "Bar Graph". The bar graph also has a zoom function, as described under "Zoom Mode".
--	OL	Overload condition is detected.
Error Messages		
bAtt	Replace the battery immediately.	
diSC	In the capacitance function, too much electrical charge is present on the capacitor being tested.	
EEPr Err	Invalid EEPROM data. Have Meter serviced.	
CAL Err	Invalid calibration data. Calibrate Meter.	
LEAd	⚠Test lead alert. Displayed when the test leads are in the **A** or **mA/μA** terminal and the selected rotary switch position does not correspond to the terminal being used.	

Number	Feature	Indication
⑬	A, μA mA	Amperes (amps) , Microamp, Milliamp
	V, mV	Volts, Millivolts
	μF, nF	Microfarad, Nanofarad
	nS	Nanosiemens
	%	Percent. Used for duty cycle measurements.
	Ω, MΩ, kΩ	Ohm, Megohm, Kilohm
	Hz, kHz	Hertz, Kilohertz
	AC DC	Alternating current, direct current
⑭	°C, °F	Degrees Celsius, Degrees Fahrenheit
⑮	610000 mV	Displays selected range
⑯	HiRes	The Meter is in high resolution (Hi Res) mode. HiRes=19,999
⑰	Auto	The Meter is in autorange mode and automatically selects the range with the best resolution.
	Manual	The Meter is in manual range mode.

FIGURE 6–5 Display features for Figure 6–4.

DMM also has current jacks that are circuit protected by a fuse if overloaded. Overloading the DMM's current measurement capacity is a common error made by even the most experienced technician when the circuit amperage is unknown.

Measuring Voltage

A voltmeter (analog or digital) is used to measure voltage potential across an electrical circuit's connections, terminals, switches, wiring, and components. Three

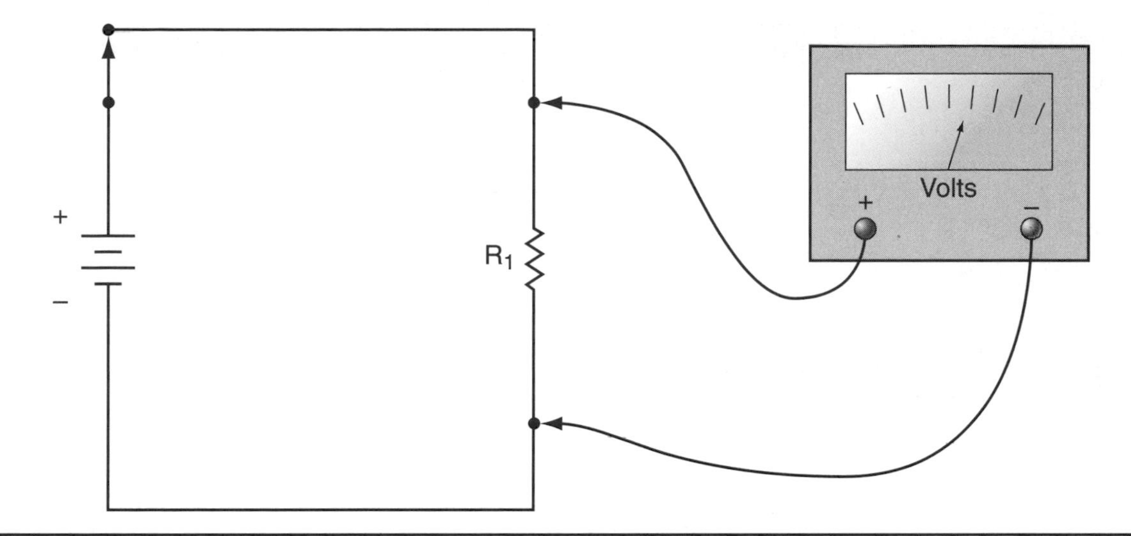

FIGURE 6–6 Meter leads are placed in parallel across the load component of a live circuit.

rules *must* be followed when using a voltmeter to measure voltage potential in a circuit:

Rule 1: Measuring battery voltage or source voltage can be done without energizing a circuit, but an electrical circuit *must* be energized (i.e., be a complete circuit) to measure voltage potential across electrical circuit connections, terminals, switches, wiring, and components with a voltmeter.

Rule 2: A voltmeter is hooked in *parallel* across the component being measured (Figure 6–6).

Rule 3: A voltmeter is *never* intentionally placed in *series* (in-line) with the component being measured (Figure 6–7).

In an electrical circuit, voltage is the pressure that pushes the current through the circuit. Some of the source voltage intended for the circuit components is lost due to resistance through wiring and connections. This loss, called **voltage drop**, can be measured with a voltmeter. In a circuit, the sum of all voltage drop (Figure 6–8) through the connections, wiring and load components *always equals* the source voltage.

The longer and more complex a circuit is, the more chance there is for resistance to develop in connections, thus dropping the voltage to a level that affects the proper operation of the load components. The following general

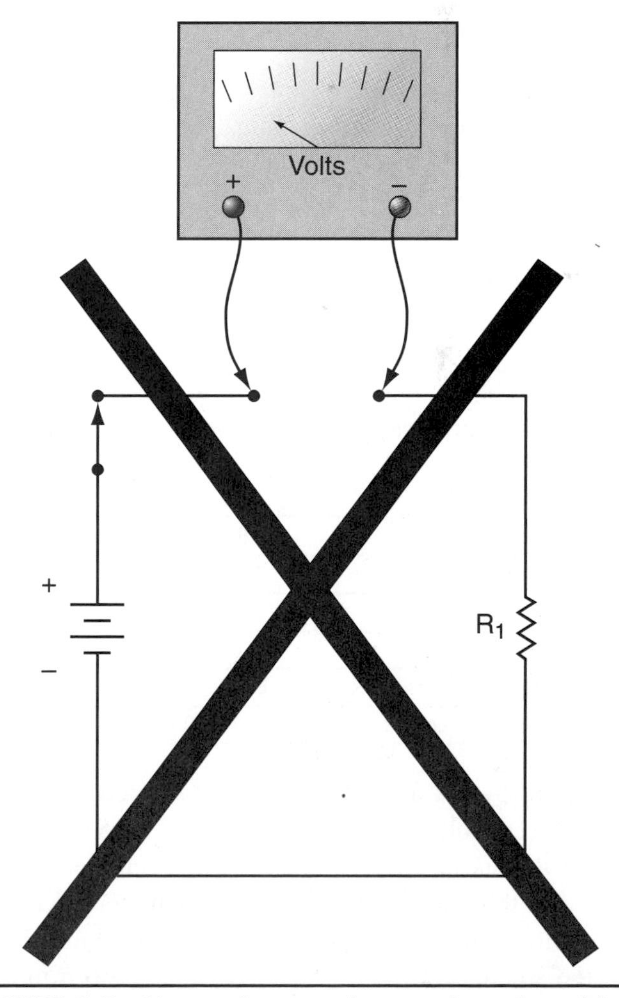

FIGURE 6–7 Never place a voltmeter in series with the load component(s) of a live circuit.

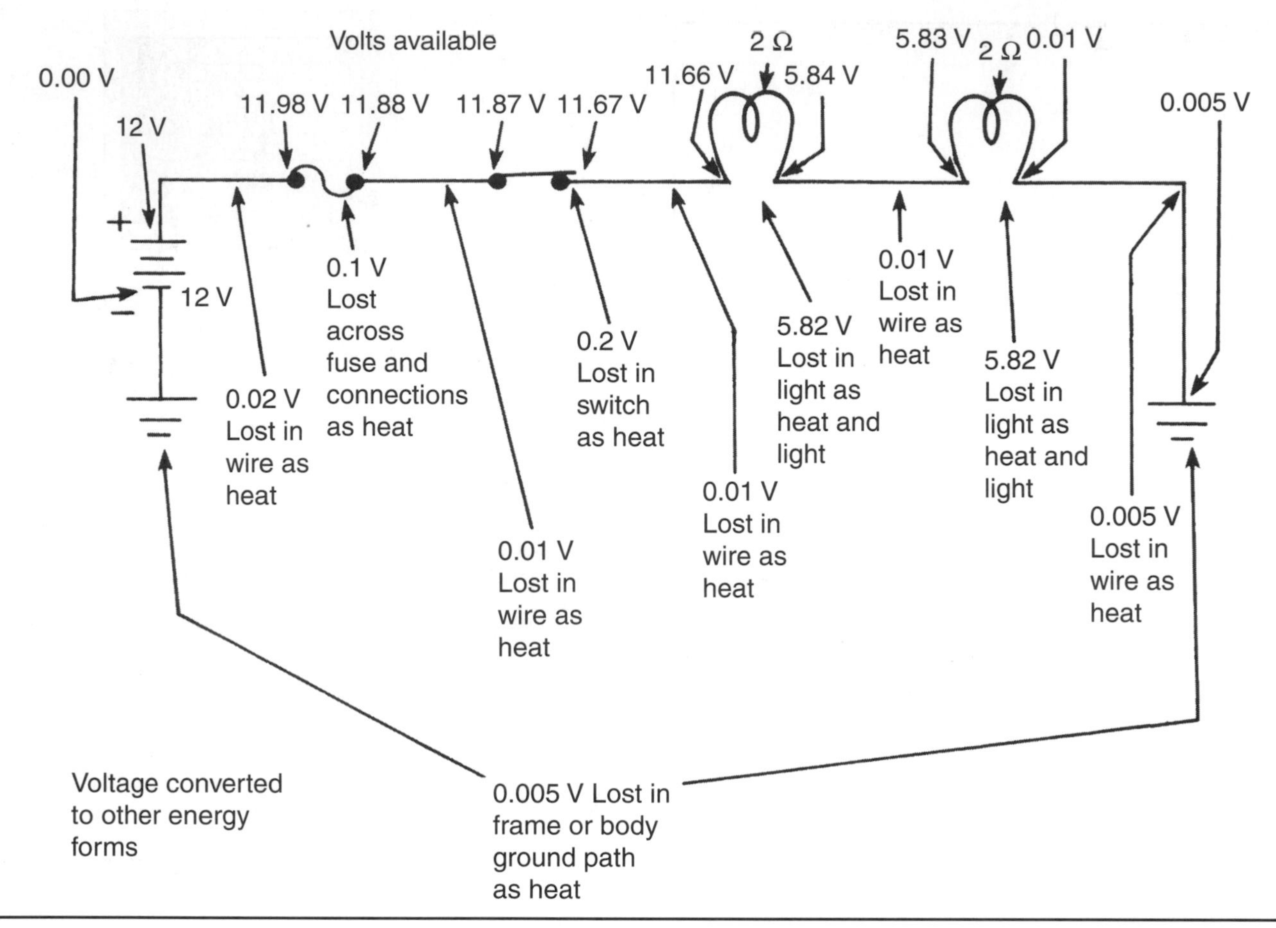

FIGURE 6–8 Measuring voltage drops of connections and components of a circuit.

specifications are used when checking a circuit for voltage drops.

Connection Point	Allowed Voltage Drop (V)
Ground connections	0.1
Switch contacts	0.2
System wiring	0.2
Terminal connections	0.0
Computer sensor connections	0.0 to 0.05

Figure 6–9 is an example of a starter circuit being tested for a voltage drop. This circuit is designed to operate the starter motor on the voltage supplied by the battery. In this starter circuit, there are ten areas to check for a voltage drop: three battery cables, six connections, and the solenoid switch. Excessive voltage drop in any one connection or series of connections could affect the proper operation of the starter motor.

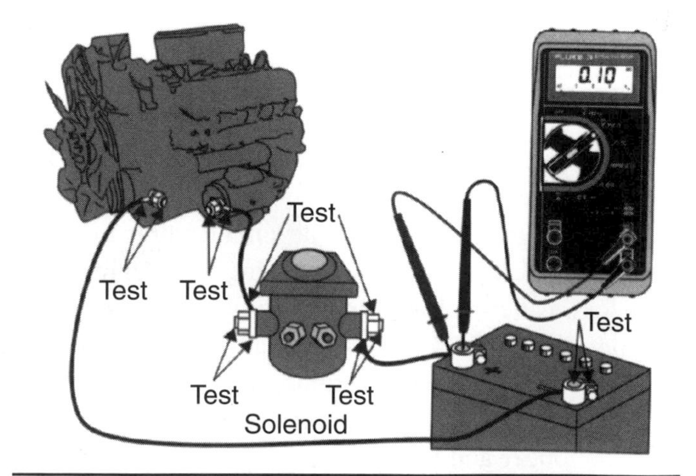

FIGURE 6–9 Testing for excessive voltage drop in a starter circuit.

(Courtesy of The Fluke Corporation)

Measuring Amperage

An ammeter (analog or digital) is used to measure the flow of electrons (current flow) in an electrical circuit. Three rules *must* be followed when the technician is measuring current.

Rule 1: An electrical circuit *must* be energized to measure the current flow.

Rule 2: An ammeter is *always* hooked up in *series* (in-line) with the component or circuit being measured (Figure 6–10).

Rule 3: An ammeter is *never* placed directly in *parallel* with the circuit or component (Figure 6–11). An ammeter connected in parallel will be damaged or blow a fuse.

Because the ammeter must be placed in series with the circuit, normally a wire leading up to or from the component must be disconnected. In some electrical circuits this may not be easy. When the ammeter cannot be placed in series, or when the meter's designed current capacity is exceeded, an **inductive pick-up** adapter (Figure 6–12) senses the amount of current flow by

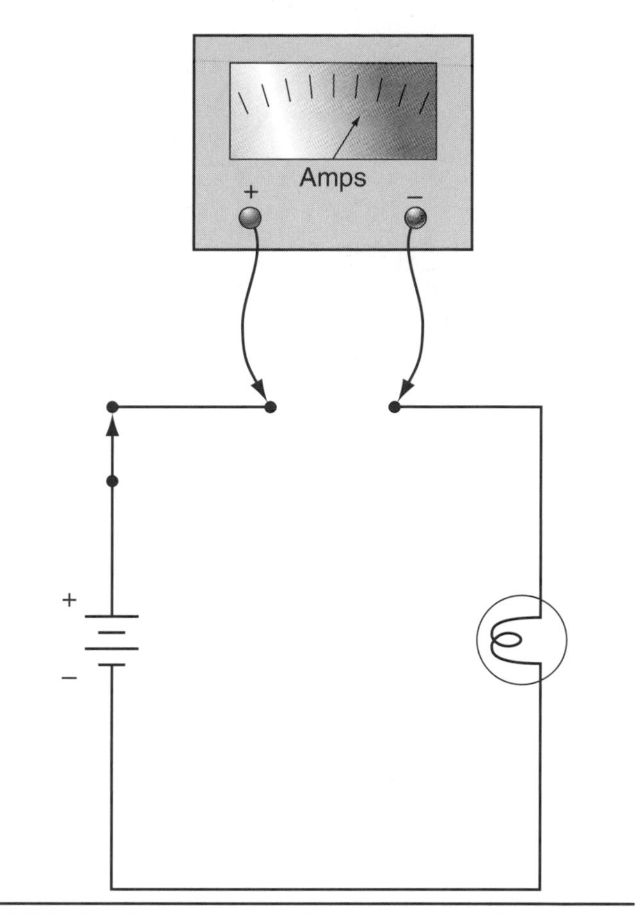

FIGURE 6–10 Ammeter leads are placed in series with the load components of a live circuit.

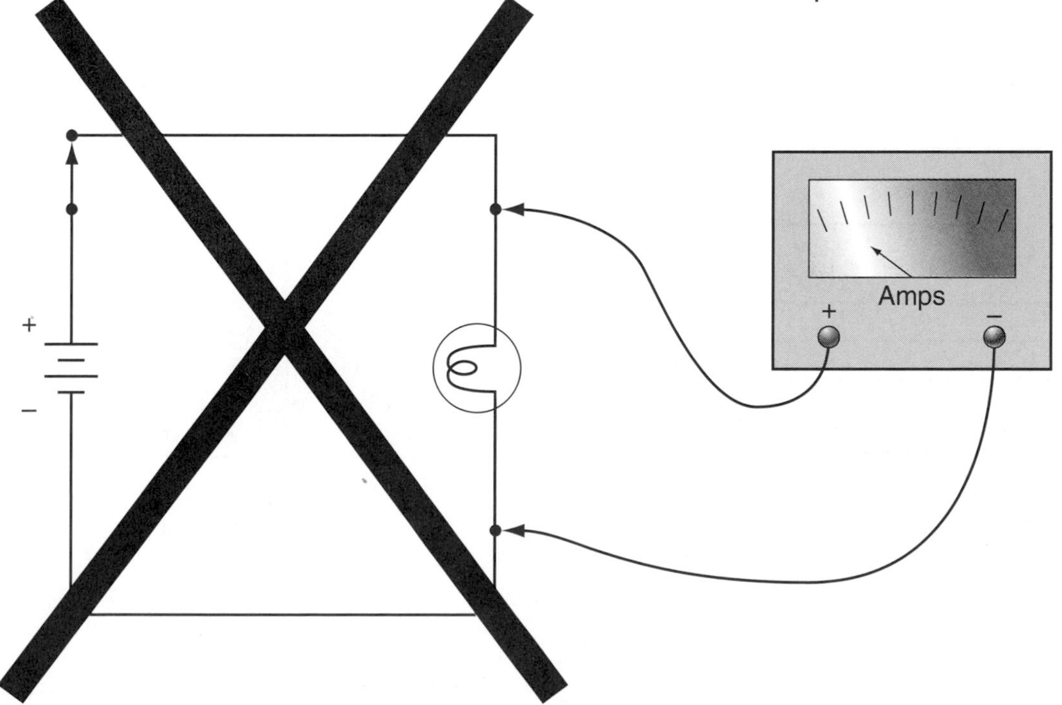

FIGURE 6–11 Never place an ammeter in parallel with the load component(s) of a live circuit.

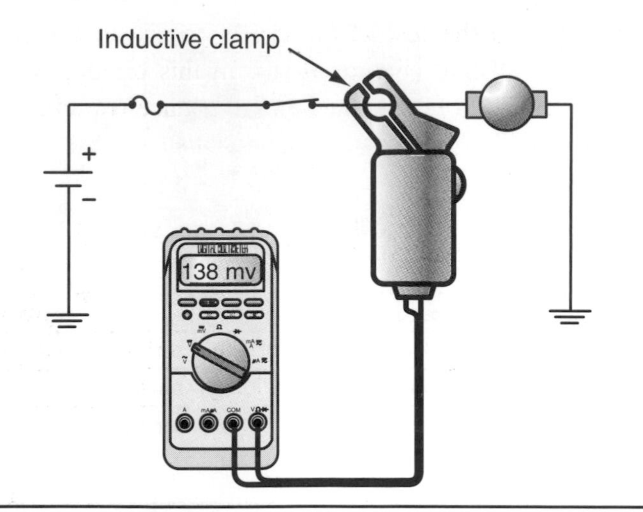

FIGURE 6–12 Using an inductive pickup with a digital meter to measure circuit current.

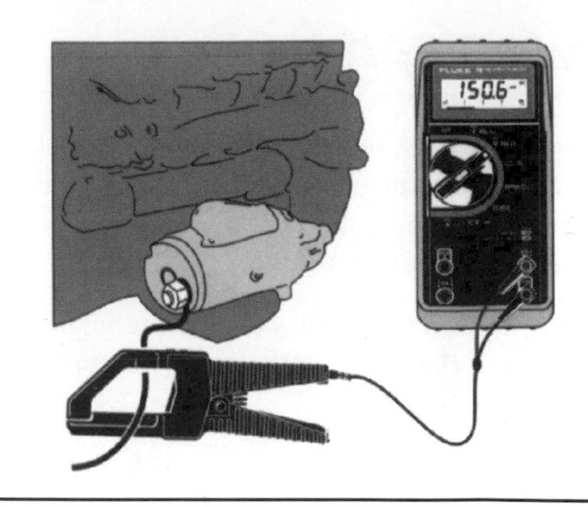

FIGURE 6–13 Measuring starter current draw with an inductive pickup clamp.
(Courtesy of The Fluke Corporation)

measuring the strength of the magnetic field surrounding the conductor.

Most handheld analog or digital multimeters have a maximum current limit of 2 to 20 amperes, so an inductive current clamp is necessary when measuring several hundred amperes, such as the starter circuit (Figure 6–13).

Measuring Resistance

An ohmmeter (analog or digital) is used to measure the resistance, in ohms (Ω), in a electrical circuit. Three rules that *must* be followed when measuring resistance.

Rule 1: An electrical circuit *must* be *disconnected* from its power source to measure the circuit or component resistance. The meter uses its own battery as a power source for measuring resistance.

Rule 2: An ohmmeter is hooked up in *parallel* across the component or circuit being measured (Figure 6–14).

Rule 3: An ohmmeter is *never* placed in *series* (in-line) with the component being measured (Figure 6–15).

When a circuit has multiple load components that share common connections, each component needs to be

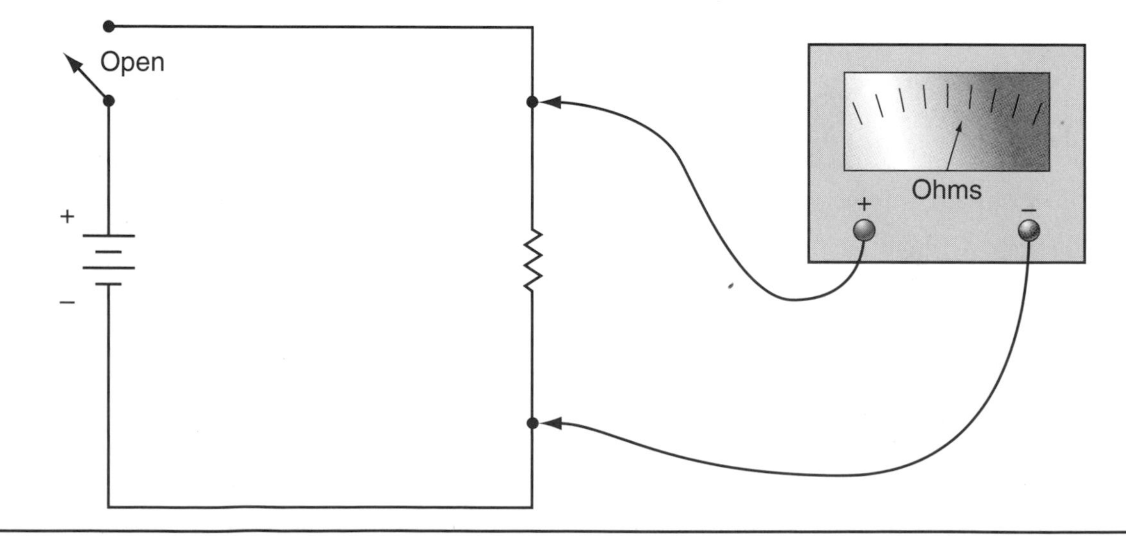

FIGURE 6–14 Ohmmeter leads are placed in parallel across the load component of a dead circuit.

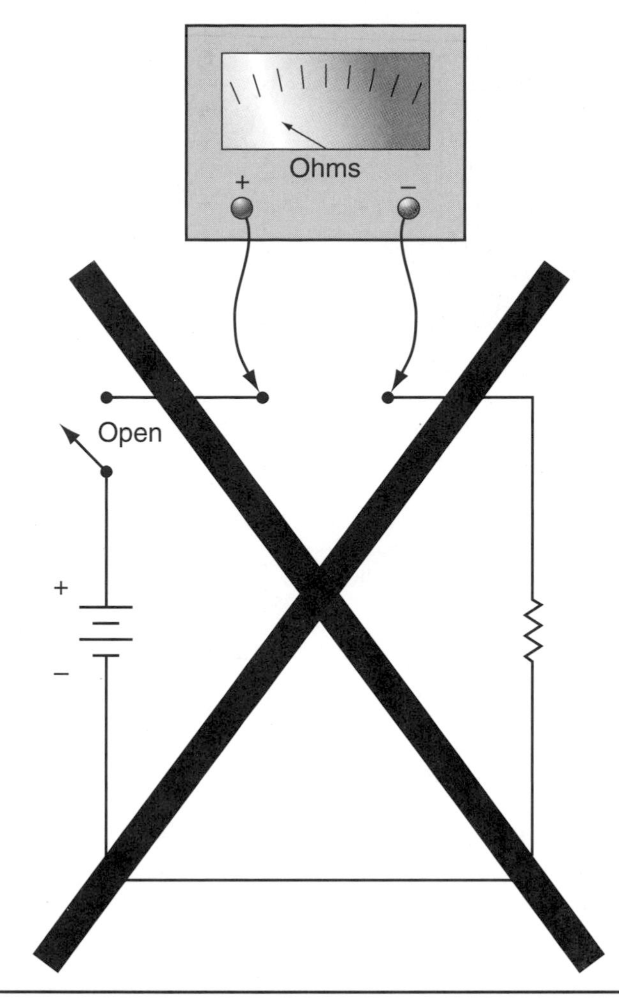

FIGURE 6–15 Never place an ohmmeter in series with the load component(s) of a dead circuit.

isolated from the rest of the circuit to get an accurate resistance reading (Figure 6–16). In this circuit, if the component R_3 is not isolated, the resistance reading will be for all three of the load components.

Summary of Meter Use

Meter	Connection	Do Not	Power
Voltmeter	Place in parallel	Place in series	On
Ammeter	Place in series	Place in parallel	On
Ohmmeter	Place in parallel to isolated component	Place in series	Off

Test Lights

Automotive technicians use three test lights to diagnose simple electrical circuits: a nonpowered test light, a self-powered test light, and a logic probe.

Nonpowered Test Light

A **nonpowered test light** is one of the simplest tools for checking for voltage in an electrical circuit. It consists of a pointed probe, a 12-volt light bulb, and a flexible

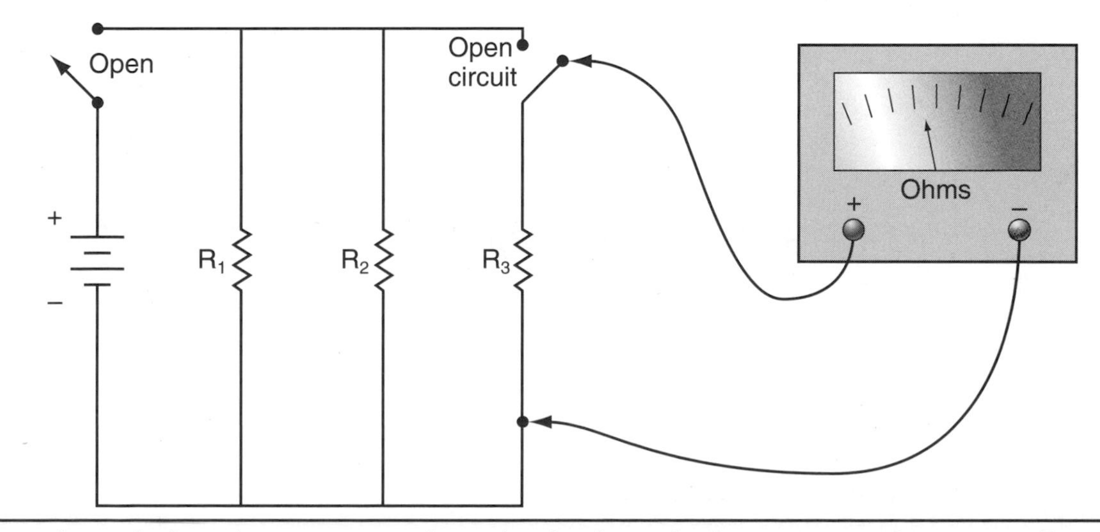

FIGURE 6–16 Always isolate the component from the circuit to obtain a correct resistance reading.

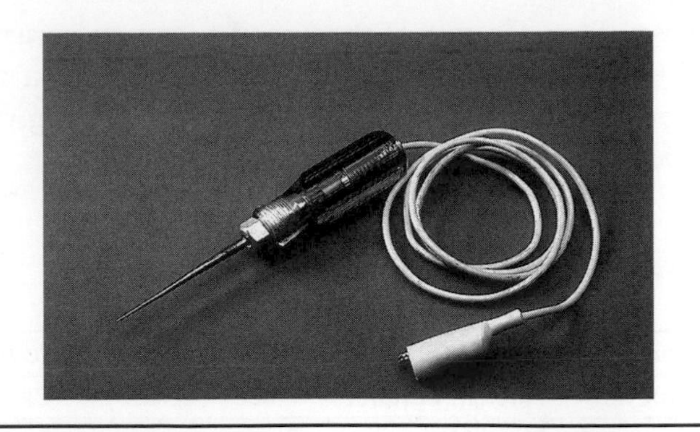

FIGURE 6–17 A typical nonpowered test light.

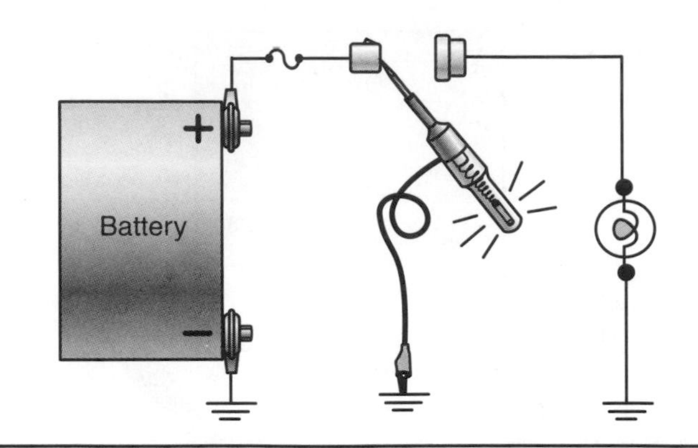

FIGURE 6–18 Nonpowered test light glows when voltage is present.

ground lead (Figure 6–17). It is used merely to check for the presence of voltage in a circuit. When the bulb is lit, a voltage is present (Figure 6–18).

CAUTION *A test light should not be used to test for power in a solid-state electronic circuit. Voltage in a sensitive circuit should be checked only with a voltmeter with a 10-megaohm (10-million-ohm) or higher internal impedance.*

Self-powered Test Light

A **self-powered test light,** also known as a **continuity light,** is used to check the continuity of *open* circuits. It looks similar to a regular test light, except that it contains a small internal battery attached between a light bulb and two leads (Figure 6–19).

With a self-powered test light, if the two leads are touched together or connected through a low resistance circuit (Figure 6–20), the bulb lights because doing so completes the circuit between the bulb and the test light's internal battery. The same caution that applies to a nonpowered test light applies to a self-powered light.

Logic Probes

Most modern vehicles have sensitive computer-controlled circuits and sensors. Standard nonpowered or self-powered test lights should not be used to test these sensitive circuits. The **logic probe** (Figure 6–21) is an electronic test light that is safe to use on computer-controlled circuits. The logic probe consists of *up to three* different light-emitting diodes (LEDs) that determine circuit condition. The *red* LED illuminates when the probe is touched to a power source. The *green* LED illuminates when the probe is touched to a ground or low voltage. The optional *yellow* LED illuminates when the probe is touched to a pulsed voltage, like a fuel injector.

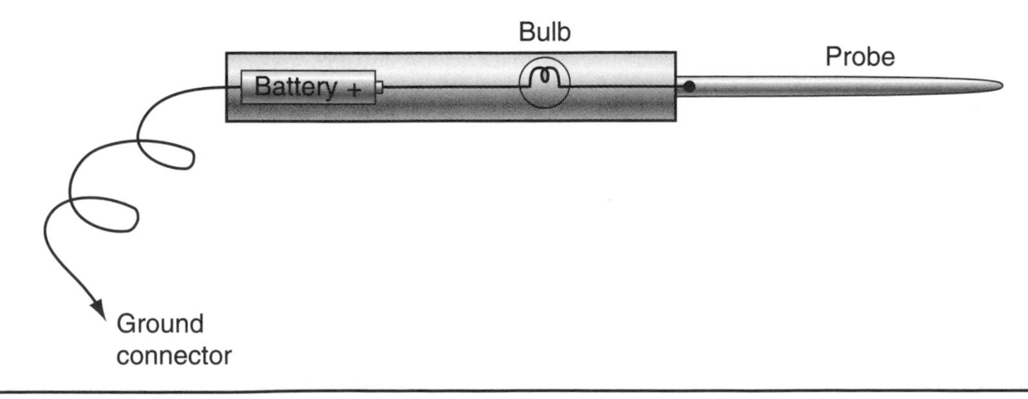

FIGURE 6–19 A typical self-powered test light.

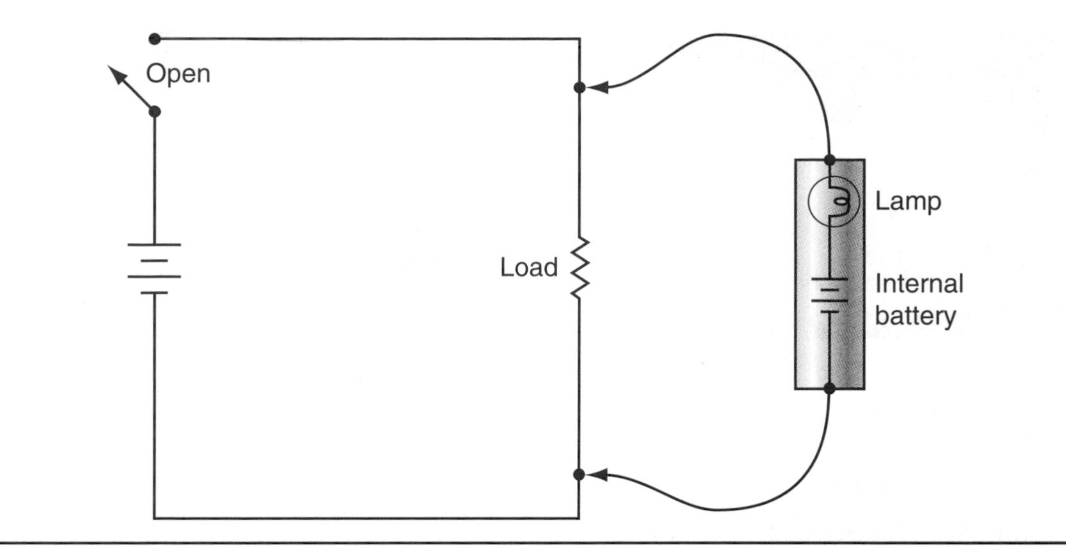

FIGURE 6–20 Using a self-powered test light to check continuity in an open circuit.

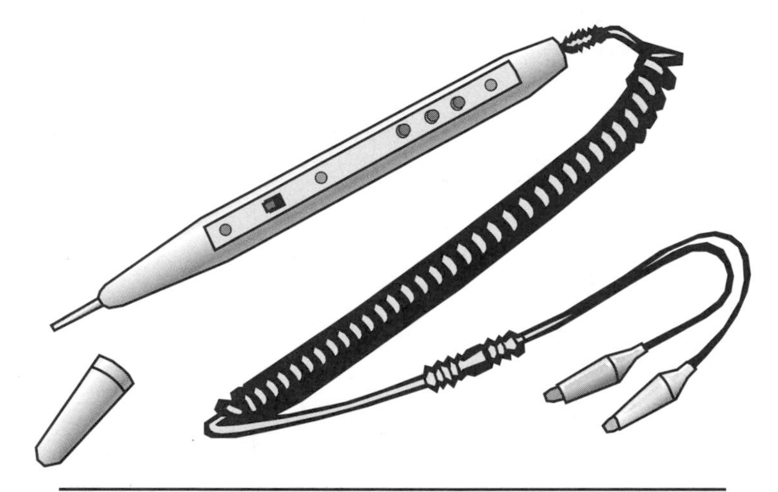

FIGURE 6–21 A typical electronic logic probe.

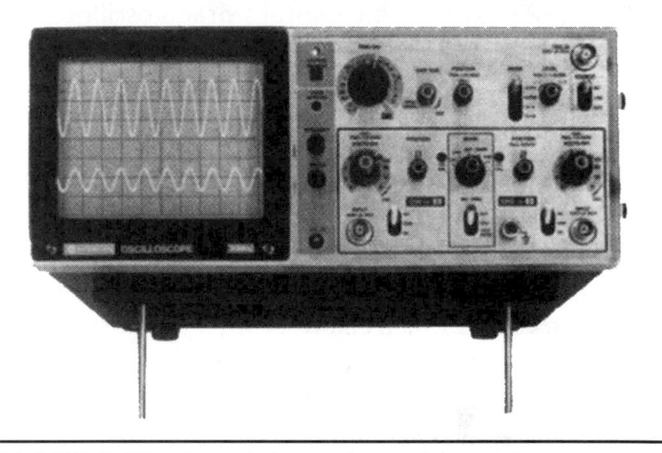

FIGURE 6–22 A typical semiportable analog oscilloscope.

Automotive Oscilloscopes

Over the past several years, oscilloscopes have become a major piece of diagnostic equipment in repair shops for diagnosing ignition or charging system problems. The scope is basically a visual image voltmeter with an extremely fast reaction time to changes in voltage over a period of time. Two types of oscilloscopes are used by technicians to perform electrical diagnosis: analog and digital storage.

Analog Oscilloscope

The **analog oscilloscope**, normally too heavy to be hand carried, reads and displays electrical signals as they occur in the circuit. It is also known as a *real-time scope* because what you see on the screen is presently occurring at the point being measured. This is helpful, and updates are continuous though somewhat slow. Figure 6–22 is an example of a small semiportable analog scope. Some of these scopes are encased in large roll-around cabinets.

Digital Storage Oscilloscope

A **digital storage oscilloscope (DSO)** (Figure 6–23) is highly portable, making it popular with technicians because it can be taken in the vehicle on test drives. A DSO is also known as a scope-meter or lab-scope because it has selections for digital multimeter functions.

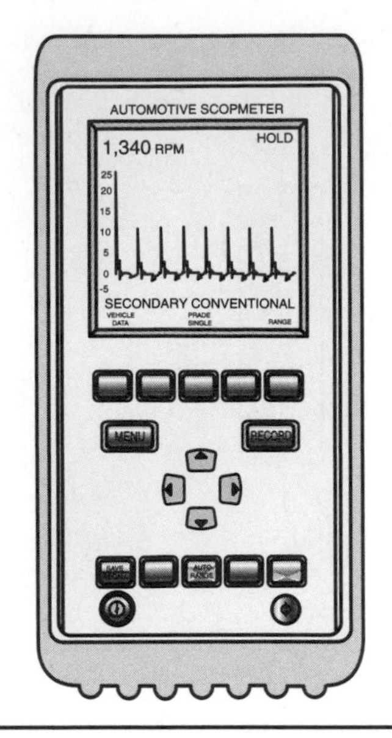

FIGURE 6–23 A portable digital storage oscilloscope (DSO).

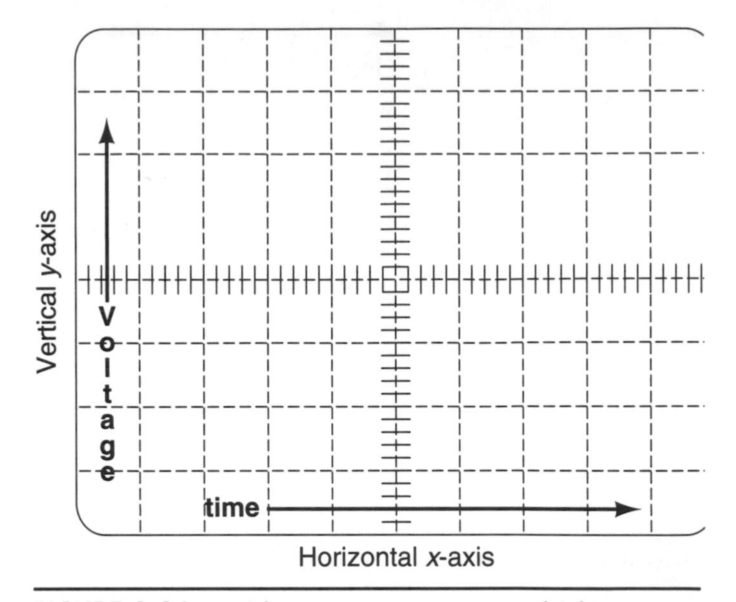

FIGURE 6–24 Grids on a scope screen, which serve as a time and voltage reference.

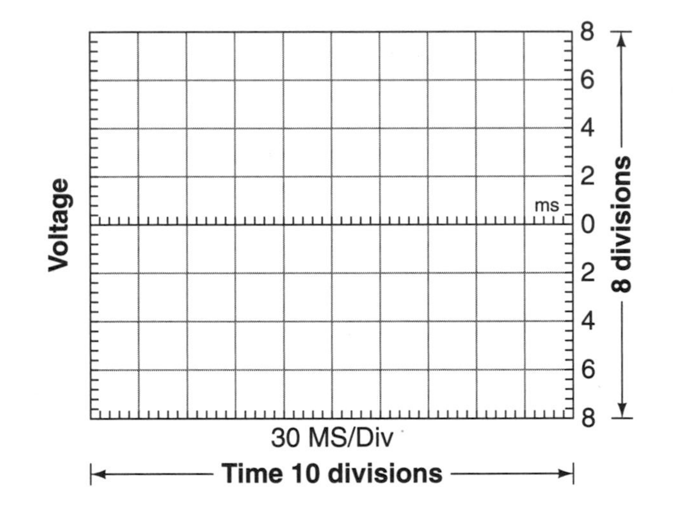

FIGURE 6–25 Setting the appropriate range selection for precise measurement.

A DSO continuously converts the input voltage signal into digital information and then stores it in short-term memory for screen display or for transfer to other diagnostic equipment. Some DSOs sample the input electrical signal at up to 40 billionths of a second, which is 47,000 times faster than analog scopes. The fast capture rate allows the technician to *freeze* the captured signal of input and output sensors or actuators over a period of time for a step-by-step playback and analysis at a later time. The freeze feature, combined with the sampling speed, allows the DSO to display glitches that cannot be seen by analog scopes or multimeters.

The screen on a analog or digital scope is divided into grid squares for measurement purposes. Voltage, or **amplitude,** measurement is represented by the vertical *y*-axis, and time measurement is represented by the horizontal *x*-axis (Figure 6–24).

The range parameter setting on the scope determines the voltage reading for each grid line from the bottom edge to the top, and the time division reading per grid line from the left to the right. For example, in Figure 6–25 the scope is set on 2 volts per grid division, and 30 milliseconds (ms) per time division. The result would be a maximum of 8 volts above the zero line and 8 volts below the zero line, for a total of 16 volts that can be displayed on the screen. Since each of the ten time divisions equal 30 milliseconds, a total of 300 milliseconds would be displayed. The ideal range setting depends on how precise the readings need to be.

The picture produced on a scope screen is called a **waveform** or **trace.** Because a scope displays a voltage measurement over a period of time, the waveform moves from the left to the right on the grid screen. A typical AC (alternating current) **sine wave** is shown in Figure 6–26. One complete sine wave cycle shows the voltage moving from the zero line to its positive peak, then down

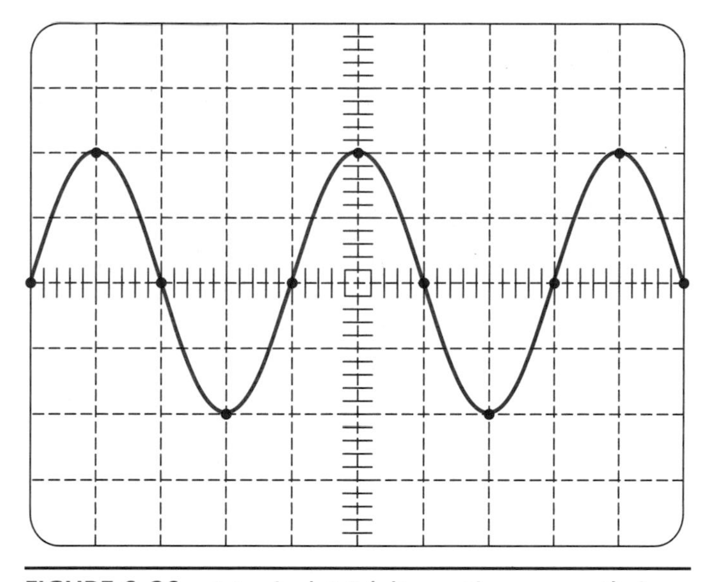

FIGURE 6–26 A typical AC (alternating current) sine wave.

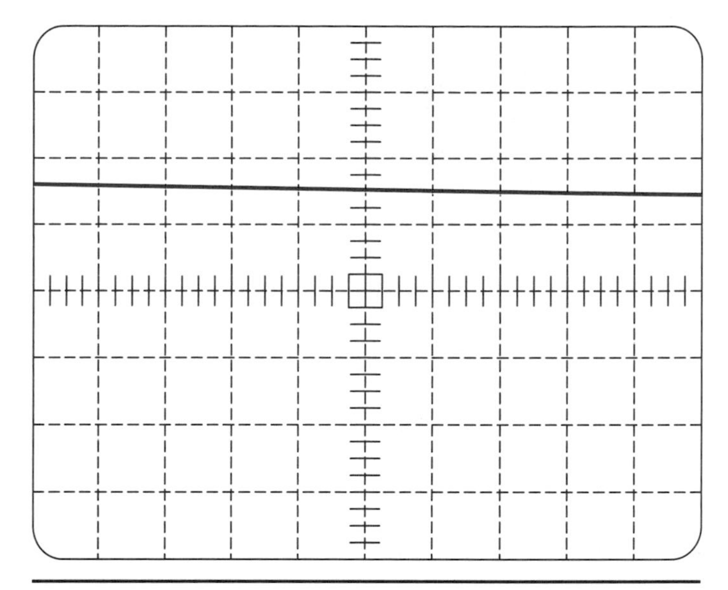

FIGURE 6–27 A typical DC (direct current) constant waveform pattern.

through zero to its negative peak, and back to zero again. If the rise and fall from positive to negative are the same, then the wave cycle is considered to be **sinusoidal.** The number of cycles that occur per second is the frequency of the signal.

A typical DC (direct current) waveform pattern is shown in Figure 6–27. The DC voltage waveform appears as a constant straight line across the screen at a specific voltage. Any changes in the voltage raise or lower the straight line.

Analog or digital oscilloscopes with the ability to display two waveforms at a time are known as **dual-trace** oscilloscopes (Figure 6–28). Observing an abnormal waveform above or below a good waveform can be helpful for efficient diagnosis.

Some of the modern handheld DSOs have technician-friendly controls with preset programs. The DSO in Figure 6–29 can display four voltage waveforms or up to six multimeter functions at a time.

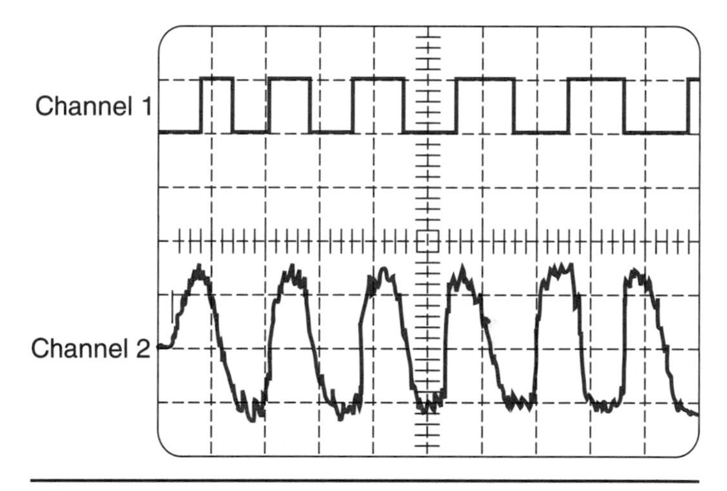

FIGURE 6–28 The benefit of a dual-trace scope is showing two waveforms for comparison.

Analog Voltage Signal

An analog voltage signal produced on a DSO screen shows a variable voltage within a specific range and time period (Figure 6–30). A common vehicle sensor producing an analog signal is an antilock brake (ABS) wheel speed sensor. As the wheel speed increases or decreases, the analog sine wave varies in frequency and amplitude (height) (Figure 6–31).

Digital Voltage Signal

A typical digital voltage signal is either on (high) or off (low). When a signal in an electronic circuit is turned on and off, the wave signal is known as a **square wave** (Figure 6–32). As the voltage is turned on (leading edge), the waveform line increases to a specified on voltage and remains constant until the circuit is shut off

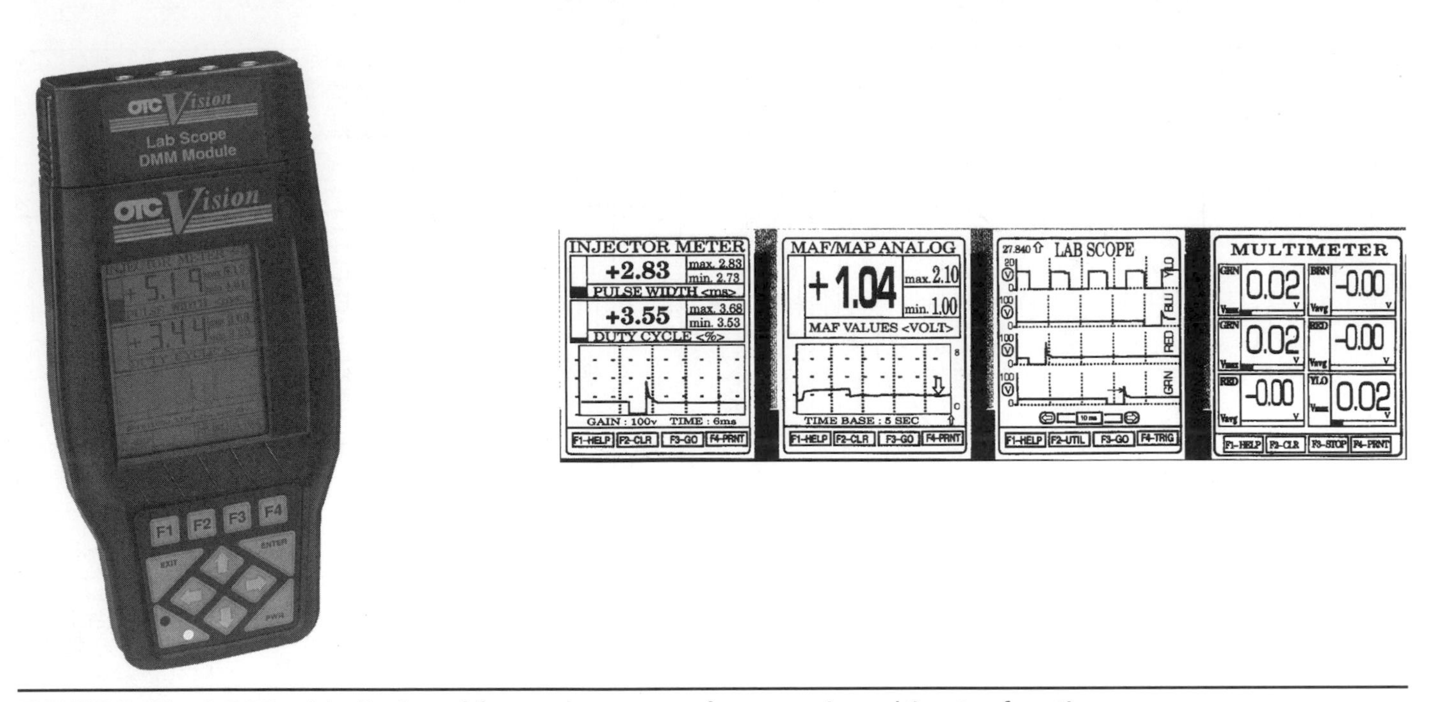

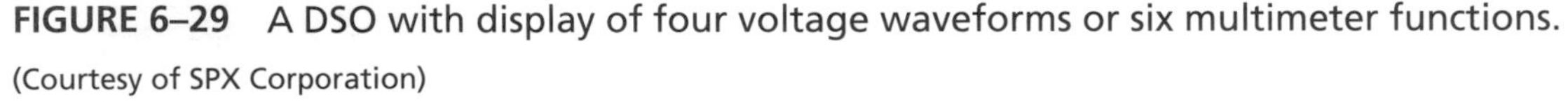

FIGURE 6–29 A DSO with display of four voltage waveforms or six multimeter functions.

(Courtesy of SPX Corporation)

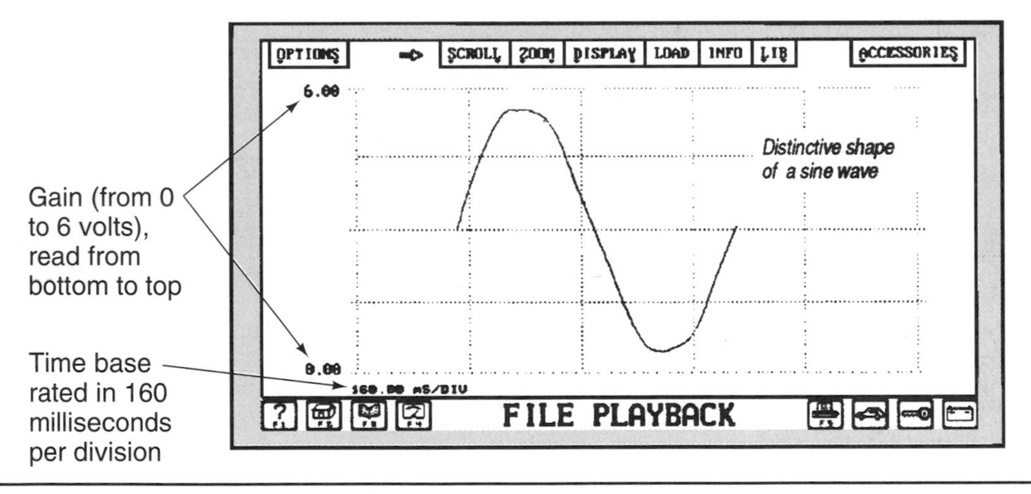

FIGURE 6–30 Analog voltage signal on a DSO screen.

(Courtesy of Snap-On Diagnostics)

(trailing edge), and the voltage decreases (Figure 6–33) to its off voltage.

The amount of time the voltage signal in a digital circuit is on (or off) in a digital circuit is known as the **pulse width.** The distance between the leading edge of one digital signal and the leading edge of the next signal is one cycle. The number of cycles within a defined period of time (usually 1 second) is known as the **frequency** of the circuit. The amount of on time versus the off time in a digital signal is called its **duty cycle.** If the component's on and off time are equal, the duty cycle is 50% (Figure 6–34). If a component has a 75% duty cycle, it is on for 75% of the time and off for 25% of the time. Some computer control outputs vary the pulse width while keeping the frequency constant. This is known as **pulse width modulation.**

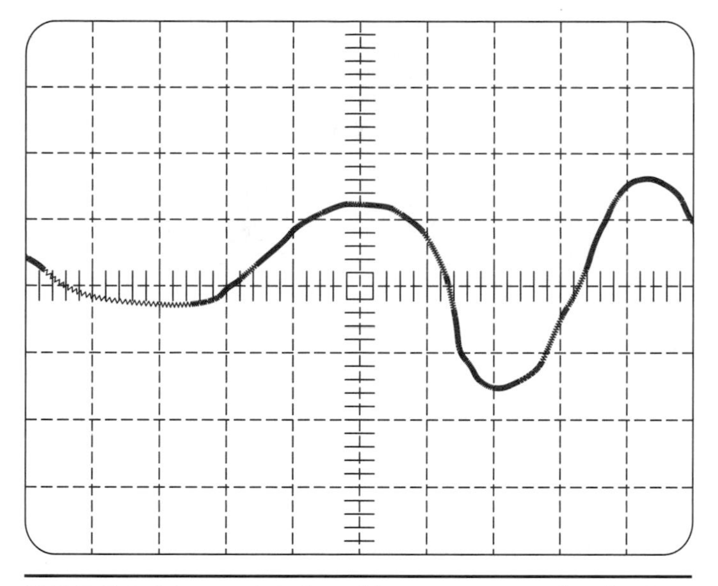

FIGURE 6–31 An AC signal from an ABS wheel spin sensor.

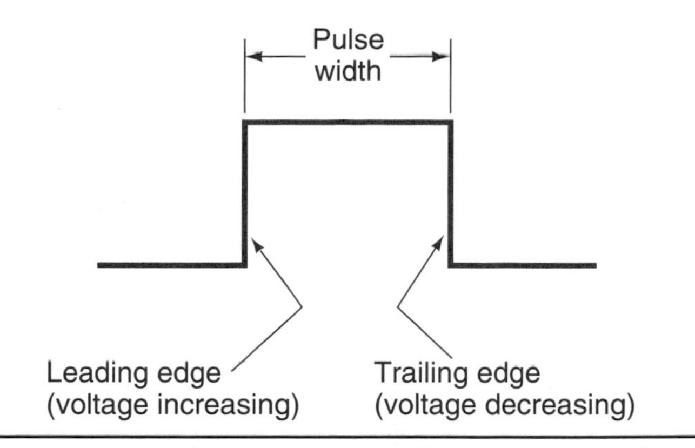

FIGURE 6–33 The component of time in a digital signal is called pulse width.

(Courtesy of Snap-On Diagnostics)

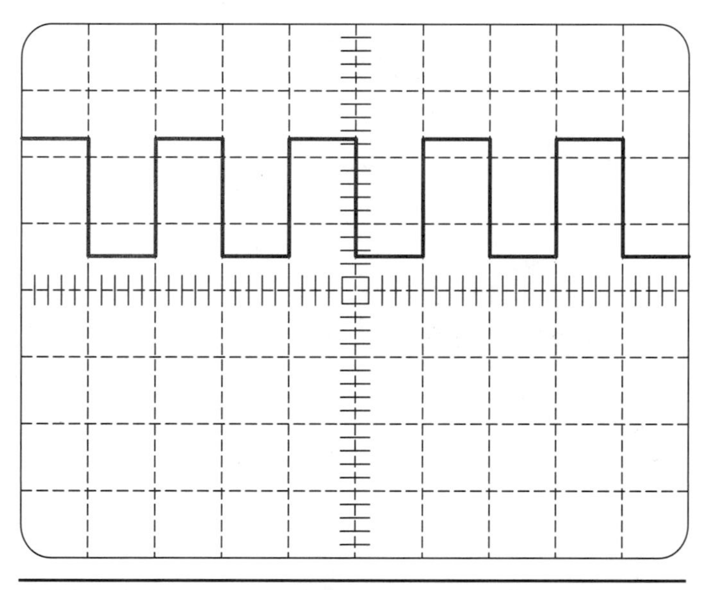

FIGURE 6–32 A typical square wave voltage signal.

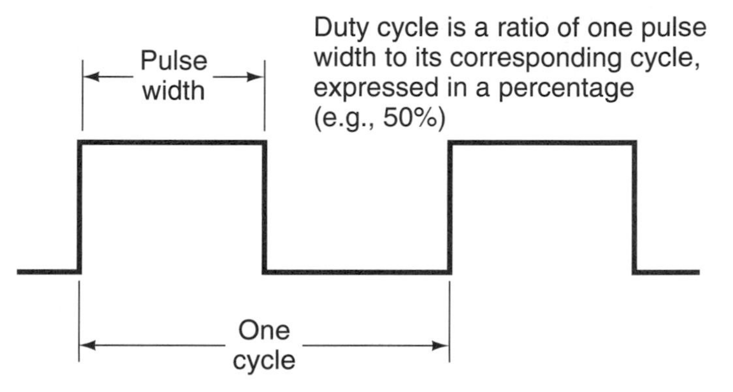

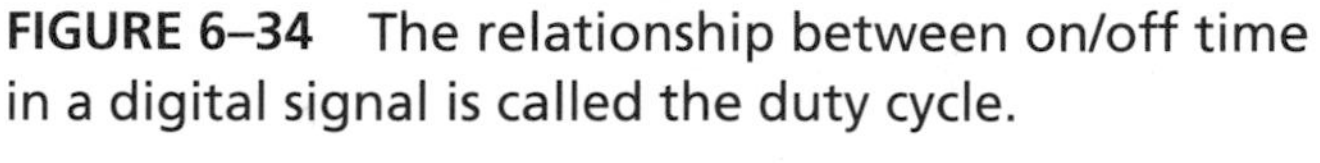

FIGURE 6–34 The relationship between on/off time in a digital signal is called the duty cycle.

(Courtesy of Snap-On Diagnostics)

Computer Scan Testers

The introduction of computer controls for the various vehicle systems listed in Figure 6–35 created the need for diagnostic equipment capable of troubleshooting these complex systems. The best tool to communicate with the vehicle's computer and access stored diagnostic trouble codes is the **scan tool** (Figure 6–36). A scan tool is a portable microprocessor-based instrument designed to connect to the vehicle's computer, through a diagnostic link connector (DLC) (Figure 6–37), to monitor the system's operation. Typical scan tools use a plug-in **programmable read-only memory (PROM)** cartridge that is specific to the vehicle being tested (Figure 6–38).

The benefit of having a portable diagnostic scan tool is being able to take it on a road test with the vehicle (Figure 6–39). The technician can capture a snapshot record of the serial data from the on-board computer at the exact time that fault is occurring. (*Caution:* Depending on traffic and road conditions, a test drive may require a separate driver so that the technician can focus attention on capturing accurate scan tool snapshot data.)

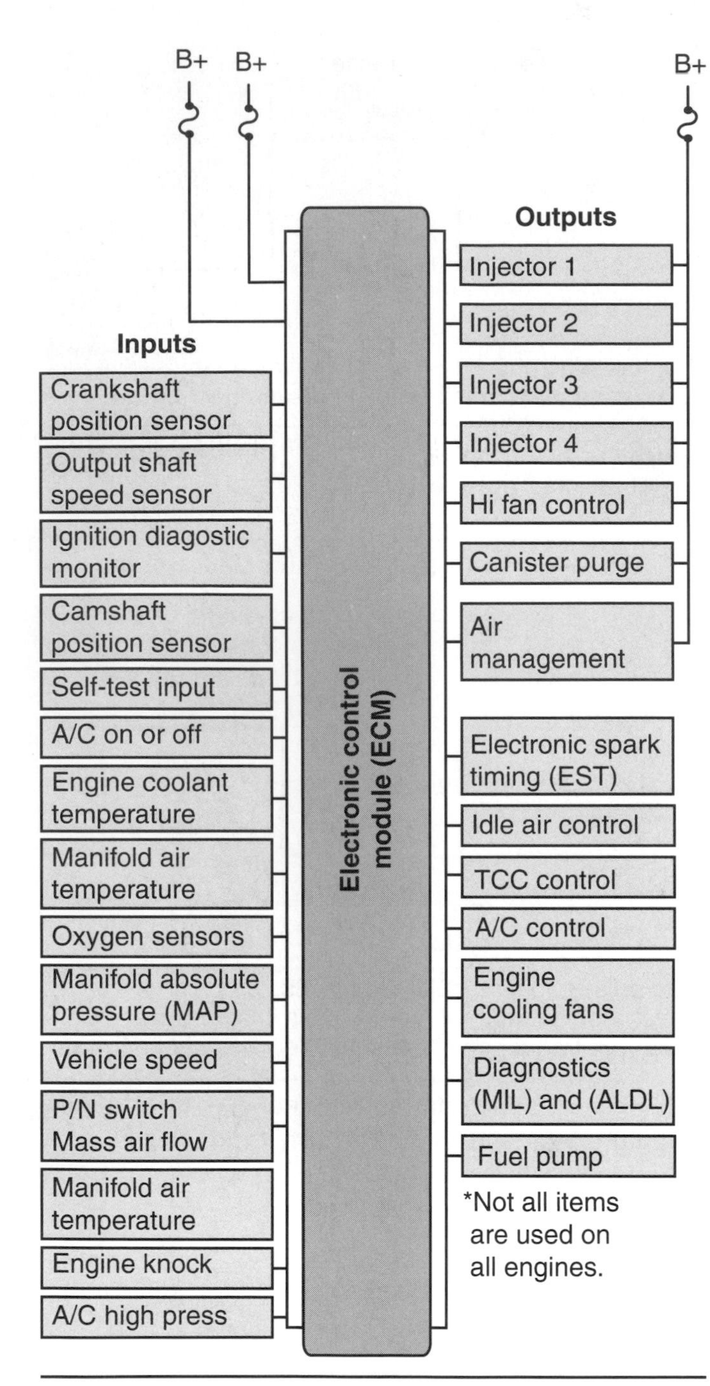

FIGURE 6–35 Typical computer-controlled inputs and outputs.

After the test drive, the information can be viewed on the scan tool's display screen or downloaded to a personal computer (Figure 6–40). With suitable software the technician can view the data and parameters on several sensors at once (Figure 6–41), as well as graph specific data (Figure 6–42).

FIGURE 6–36 Handheld scan tool with an assortment of leads and adapters.

(Courtesy of SPX Corporation)

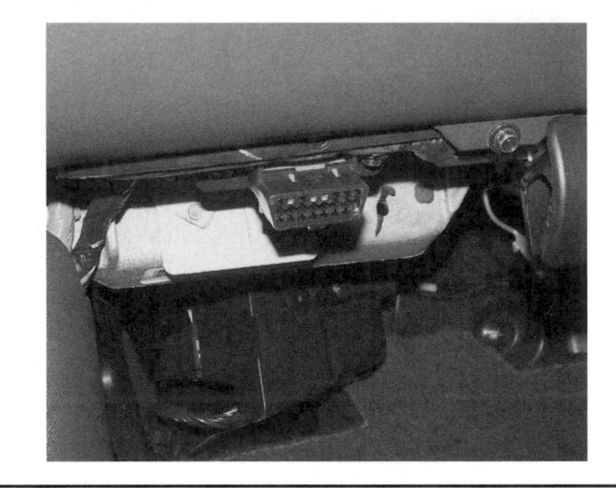

FIGURE 6–37 Typical data link connector (DLC) that is used to connect to a scan tool.

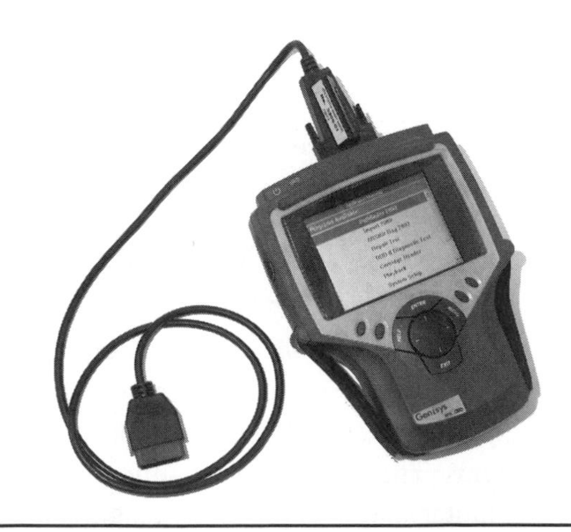

FIGURE 6–38 Scan tool with removable PROM cartridge.

FIGURE 6–39 Using a portable scan tool during a vehicle road test.

(Courtesy of SPX Corporation)

FIGURE 6–40 A scan tool connected to a desktop computer.

TechView® Pro - [Open Coolant Sensor.evn]

File Edit View Tools Data Setup Window Help

New Open Search Save Print Screen Home User Gilles, Tim 5-Gas Engine Scan Camera Import Memo List Notes

Open Coolant Sensor

RO Viewer | Integrated Diagnostics

Record Type	Origin	Date / Time	Technician	Record Note
Snapshot Upload	MTS 3100 Mastertech M	08/29/2002 16:52:58	Gilles, Tim [123]	Mastertech set up to trigger a snapshot if an

Select: All, List, Bar, Digital, Analog, Gauge, Split Line, Multi Line, Custom

Parameter	Value	Units
Engine Speed	722	RPM
TPS (V)	0.80	V
TPS (%)	0	%
ECT (°)	90	°C
ECT (V)	5.08	V
MAP (P)	35	KPa
VAC (P)	64	KPa
Baro (P)	99	KPa
S/C Cancel	Off	
IAT (°)	41	°C
Injector PW	3.00	Msec
MAP (V)	1.36	V
Park/Neutral	P-N	
IAC Motor Posit	17	Steps
Target Idle	800	RPM
P/S Switch	Off	
Battery Volt	12.4	Volts
Chg Sys Goal	13.1	Volts

Parameter	Value	Units
Spark Advance	11	°BTDC
Tot. Knck Ret	0	°
Key On Cycles 1	0	
Knock Sensor	0.68	V
Vehicle Speed	0	MPH
A/C Switch	Off	
S/C Set Speed	0	MPH
S/C On/Off	Off	
S/C Resume	Off	
IAT (V)	2.18	Volts
S/C Set	Off	
Minimum TPS	0.80	Volts
ASD Relay	Enrgzd	
Brake Switch	Off	
Chk Engine Lamp	Enrgzd	
S/C Vent Sol	DeEgzd	
S/C Vacuum Sol	DeEgzd	
Alternator Lamp	DeEgzd	

Parameter	Value	Units
Ignition Sense	High	
Fuel Pump Relay	Enrgzd	
Upstream O2S	0.08	Volts
Upstream O2S ST	Lean	
Dnstream O2S	0.14	Volts
Engine Load	7	%
Fuel Status 1	Closed	
Long Term Adapt	3	%

Frame: +1

Record Note: Mastertech set up to trigger a snapshot if any DTC is read from PCM. DTC P0118 set by an open Coolant Sensor circuit.

2002 Chrysler Pt Cruiser 2.4L Customer: Doe John Repair Order: 123456 Technician: Gilles Tim

Start TechView® Pro - [Open C... 5:40 PM

FIGURE 6–41 Scan tool snapshot data and parameters.

(Reprinted with permission from Bosch Diagnostics)

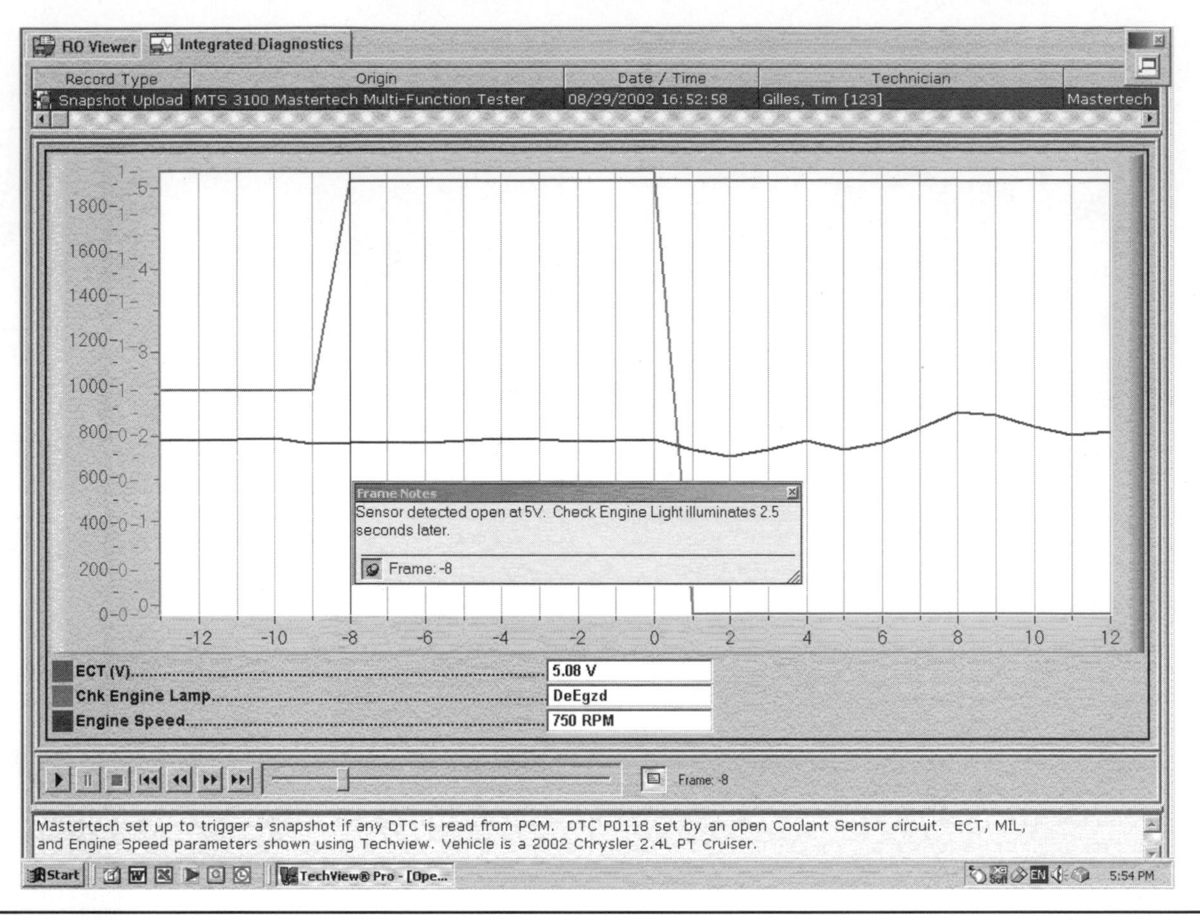

FIGURE 6–42 Snapshot data displayed in graph form.
(Reprinted with permission from Bosch Diagnostics)

Hands-On Vehicle Tests

The following three hands-on vehicle checks are included in the NATEF (National Automotive Technician Education Foundation) Task List. Complete your answers to the below questions on a *separate* sheet of paper for all three tasks.

Performance Task 1

Task Description

Check electrical circuits with a test light; determine the necessary action.

Task Objectives

- ❑ Obtain a vehicle that can be used for this task. What model and year of vehicle are you using for this task? ____________________
- ❑ Using a wiring diagram for the vehicle, determine electrical circuits that are safe to check with a test light. Verify circuits to test with your instructor.

❑ List the circuit(s) to be tested with the test light.

1. ______________________ 4. ______________________

2. ______________________ 5. ______________________

3. ______________________ 6. ______________________

❑ List the procedure(s) used to safely hook up a test light to a vehicle's electrical circuit.

____

____

____

❑ Check the operation of the circuits listed above with the test light. What can you determine about using a test light on each of the circuits?

1. ______________________ 4. ______________________

2. ______________________ 5. ______________________

3. ______________________ 6. ______________________

❑ What would be the benefit(s) of using a test light on electrical circuits compared to other electrical testing equipment?

____

Task Summary

After performing the above NATEF task, what can you determine that will be helpful in using test lights to perform simple electrical circuit troubleshooting? ______________________

____

Performance Task 2

Task Description

Measure source voltage and perform voltage drop tests in electrical/electronic circuits using a voltmeter; determine the necessary action.

Task Objectives

❑ Obtain a vehicle that can be used for this task. What model and year of vehicle are you using for this task? ______________________

❑ Using a wiring diagram for the vehicle, identify an electrical circuit that is suitable for checking voltages and voltage drops. Verify circuits to test with your instructor.

❑ List the circuit(s) to be tested for voltage and voltage drops:

1. ______________________ 3. ______________________

2. ______________________ 4. ______________________

❑ List the procedure(s) used to test for system voltage and voltage drops in an electrical circuit.

____

____

____

____

❑ What is the system voltage of each circuit tested and any voltage drop(s) measured?

System voltage ________. Voltage drops ________, ________, ________, ________.

System voltage ________. Voltage drops ________, ________, ________, ________.

System voltage ________. Voltage drops ________, ________, ________, ________.

System voltage ________. Voltage drops ________, ________, ________, ________.

❑ Do the voltage drops in each circuit tested equal the system total voltage?

____

____

Task Summary

After performing the above NATEF task, what can you determine that will be helpful in knowing how to check for voltage drops in an electrical circuit in reference to system voltage? ________________

____

____

____

____

Performance Task 3

Task Description

Measure current flow in electrical/electronic circuits and components using an ammeter; determine the necessary action.

Task Objectives

❑ Obtain a vehicle that can be used for this task. What model and year of vehicle are you using for this task? ____________________

❑ Using a wiring diagram for the vehicle, determine electrical circuits that are safe to check the current flow with an ammeter. Verify circuits to test with your instructor.

❑ List the circuit(s) to be tested for current flow.

1. ________________ 3. ________________

2. ________________ 4. ________________

❑ List the procedure(s) used to safety hook up an ammeter to a vehicle's electrical circuit.

____

____

____

❑ Check the operation of the circuits listed above with an ammeter. What can you determine about testing for current flow on each of the circuits?

1. ____________________ 3. ____________________

2. ____________________ 4. ____________________

❑ What would be the benefit(s) of using an ammeter on electrical circuits compared to other electrical testing equipment? ____________________

____

____

Task Summary

After performing the above NATEF task, what can you determine that will be helpful in testing a circuit with an ammeter to troubleshoot an electrical circuit? ____________________

____

Summary

❑ An analog meter uses a pointer needle that moves across the face of a printed scale.

❑ A D'Arsonval movement is the moving coil movement in a analog meter.

❑ The internal impedance of a meter is its ability to internally resist any current flow.

❑ A digital multimeter is also known as a DMM or DVOM.

❑ The Greek symbol for resistance is the letter omega, Ω.

❑ An electrical circuit must be energized to measure voltage potential.

❑ A voltmeter is hooked in parallel across the component being measured.

❑ A voltage drop in a circuit is the intended or unintended drop in voltage through a component, wiring, or connection point.

❑ An ammeter is used in an electrical circuit to measure the flow of electrons or current.

❑ An ammeter is hooked in series to the component or circuit.

❑ An inductive pickup may be used to measure current flow of a circuit when the current exceeds the normal capacity of the meter.

❑ An ohmmeter is used to measure resistance in an electrical component.

- An ohmmeter is hooked in parallel across the component or circuit being measured, with the circuit power disconnected.
- A nonpowered test light is used to check for voltage in a simple electrical circuit.
- A self-powered test light is used to check for continuity in an open circuit.
- Nonpowered and self-powered test lights should not be used for voltage sensitive circuits.
- The logic probe is an electronic test light that is safe to use on computer-controlled circuits.
- Analog oscilloscopes read and display electrical signals as they occur in a circuit.
- Digital storage oscilloscopes convert an input voltage signal into digital information.
- Voltage or amplitude measurement is represented by the vertical (y) axis on a scope screen.
- Time measurement is represented by the horizontal (x) axis on a scope screen.
- The picture produced on a scope screen is called a waveform or trace.
- A typical AC sine wave with equal rise and fall is called sinusoidal.
- A typical DC waveform pattern is a straight horizontal line on the scope screen.
- A pulse width is the amount of time the voltage is on per cycle.
- A number of cycles within a period of time is known as the frequency.
- When the pulse width varies in a circuit, but the frequency remains constant is known as pulse width modulation.
- A computer scan tool accesses the serial data from the vehicle computer to determine diagnostic trouble codes and system data.
- A scan tool uses a PROM chip cartridge to read specific vehicle information.

Key Terms

amplitude
analog meter
analog oscilloscope
autoranging
continuity light
D'Arsonval movement
digital multimeter (DMM)
digital storage oscilloscope (DSO)
digital volt-ohm meter (DVOM)
dual-trace
duty cycle
frequency
inductive pick-up
logic probe
moving coil meter
nonpowered test light
programmable read-only memory (PROM)
pulse width
pulse width modulation
scan tool
self-powered test light
sine wave
sinusoidal
square wave
trace
waveform
voltage drop

Review Questions

Short Answer Essays

1. Describe the three service limitations that technicians need to keep in mind when using an analog meter.
2. What are the three rules that must be followed when using a voltmeter to measure voltage potential?
3. What are the three rules that must be followed when using an ammeter to measure current in a circuit?
4. Describe how a typical AC sine wave would look like on a scope screen.
5. Describe the purpose of a computer scan tester in diagnosing today's vehicles.

Fill in the Blanks

1. A voltmeter is always hooked in ________________ across the component being measured.
2. An ammeter is ________________ placed directly in parallel with the circuit or component.
3. The picture produced on a scope screen is called a ________________.
4. A number of cycles within a period of time is known as the ________________ of the circuit.
5. A typical scan tool uses a ________________ cartridge that is specific to the vehicle being tested.

ASE-Style Review Questions

1. Technician A says that analog meters use a pointer that moves across the face of a printed scale. Technician B says that analog meters use complex digital circuitry to measure current flow. Who is correct?
   A. A only
   B. B only
   C. Both A and B
   D. Neither A nor B

2. Technician A says a high-quality DMM has current jacks that are circuit protected by a fuse if overloaded. Technician B says a DMM is more precise than an analog meter. Who is correct?
   A. A only
   B. B only
   C. Both A and B
   D. Neither A nor B

3. Technician A says a quality DMM has an internal resistance of at least 10,000 ohms. Technician B says some DMMs have an autoranging feature to automatically adjust the measured range. Who is correct?
   A. A only
   B. B only
   C. Both A and B
   D. Neither A nor B

4. Technician A says a logic probe is an electronic test light that is safe to use on computer-controlled circuits. Technician B says a logic probe uses three different colored LEDs to determine a circuit's condition. Who is correct?
   A. A only
   B. B only
   C. Both A and B
   D. Neither A nor B

5. Technician A says a voltmeter must be hooked in series with the circuit being measured. Technician B says an electrical circuit must be open to measure the voltage potential of a circuit. Who is correct?
   A. A only
   B. B only
   C. Both A and B
   D. Neither A nor B

6. Technician A says that voltage is the pressure that pushes the current through a circuit. Technician B says resistance at a bad connection in a circuit results in a voltage drop. Who is correct?
   A. A only
   B. B only
   C. Both A and B
   D. Neither A nor B

7. Technician A says that typical terminal connections allow a maximum drop of 0.0 volts per connection. Technician B says that typical ground connections allow a maximum drop of 0.5 volts per connection. Who is correct?
   A. A only
   B. B only
   C. Both A and B
   D. Neither A nor B

8. Technician A says an electrical circuit must be energized to measure the current flow through the circuit. Technician B says the ammeter is hooked in series with the component or circuit being measured. Who is correct?
   A. A only

B. B only

C. Both A and B

D. Neither A nor B

9. Technician A says an inductive pickup is used by a DMM to measure current that exceeds the normal capacity of the meter. Technician B says an inductive pickup is used to measure very small current flow in a circuit. Who is correct?

   A. A only

   B. B only

   C. Both A and B

   D. Neither A nor B

10. Technician A says an electrical circuit must be energized to get an accurate resistance reading with an ohmmeter. Technician B says an ohmmeter is hooked up in parallel across the component or circuit being measured. Who is correct?

    A. A only

    B. B only

    C. Both A and B

    D. Neither A nor B

11. Technician A says a nonpowered test light can be used to check simple solid-state circuits. Technician B says a self-powered test light is used to check an energized circuit. Who is correct?

    A. A only

    B. B only

    C. Both A and B

    D. Neither A nor B

12. Technician A says the analog scope reads and displays electrical signals as they occur in the circuit. Technician B says analog scopes are known as real-time scopes. Who is correct?

    A. A only

    B. B only

    C. Both A and B

    D. Neither A nor B

13. Technician A says a digital storage oscilloscope can sample input electrical signals at up to 40 billionth of a second. Technician B says DSOs have the ability to freeze the captured input or output signals. Who is correct?

    A. A only

    B. B only

    C. Both A and B

    D. Neither A nor B

14. Technician A says the scope waveform for a voltage measurement is represented by the vertical (y) axis. Technician B says the scope waveform for amplitude is represented by the horizontal (x) axis. Who is correct?

    A. A only

    B. B only

    C. Both A and B

    D. Neither A nor B

15. Technician A says the equal rise and fall of an AC sine wave is considered to be sinusoidal. Technician B says one complete sine wave is known as a cycle. Who is correct?

    A. A only

    B. B only

    C. Both A and B

    D. Neither A nor B

16. Technician A says some scopes can display a preprogrammed pattern next to a live pattern as a reference. Technician B says some scopes can show two live patterns of different circuit components at the same time. Who is correct?

    A. A only

    B. B only

    C. Both A and B

    D. Neither A nor B

17. Technician A says a typical square wave digital voltage signal is either on or off. Technician B says one cycle of time in a digital voltage signal is from the leading edge of one signal to the leading edge of the next signal. Who is correct?

    A. A only

    B. B only

    C. Both A and B

    D. Neither A nor B

18. Technician A says the amount of time a voltage is on or off in a digital circuit is the pulse width. Technician B says the number of cycles with a period of time is known as the duty cycle. Who is correct?

    A. A only

    B. B only

    C. Both A and B

    D. Neither A nor B

19. Technician A says a computer scan tool hooks to the vehicle's computer through a data link connector. Technician B says a computer scan tool uses the parallel data from the on-board vehicle computer to determine an intermittent fault. Who is correct?

    A. A only

    B. B only

    C. Both A and B

    D. Neither A nor B

20. Technician A says a computer scan tool can interface with a personal computer to display the findings. Technician B says a computer scan tool uses a plug-in PROM cartridge that is specific to the vehicle being tested. Who is correct?

    A. A only

    B. B only

    C. Both A and B

    D. Neither A nor B

7 Series Circuits

Introduction

In this chapter, you will study the application of voltage, current, and resistance in series circuits that are commonly used in the automotive field. You will be introduced to some electrical circuit laws and formulas that pertain only to series circuits. As you apply these laws and associated formulas to electrical circuits, you will be guided through each new concept step by step and learn the particular behavior of a series circuit. The repetition of the exercises in this chapter are designed to help you remember the series circuit laws and to demonstrate that all series circuits, even complex series circuits, use the same laws and formulas.

Objectives

When you complete this chapter you should be able to:

- ❑ Define a series circuit.
- ❑ Apply series circuit laws for voltage, current, and resistance.
- ❑ Define Kirchoff's voltage law.
- ❑ Apply Ohm's law to series circuit calculations.
- ❑ Apply troubleshooting and testing techniques to series circuits.
- ❑ Perform basic hands-on series circuit exercises on the vehicle.

Series Circuit Explanation

A circuit is a **series circuit** when there is only a *single* path for current to flow through one or more resistive loads in the complete circuit. Because there is only one path for the current to flow, the current is the same at any point in the circuit (Figure 7–1).

Vehicle manufacturers use several series circuits as primary automotive electrical circuits. Some of the most common are heater control circuits, car burglar alarms, and trunk/hatch release solenoids. They are also used as part of more complex automotive electrical circuits that are discussed later. The dependence of individual circuit devices on each other is the main reason for using series circuits for car burglar alarms. Any interruption of the single path of current flow triggers the alarm circuit. For this same reason, a series circuit is not used in a taillight circuit. It would be inconvenient to have all the taillight bulbs

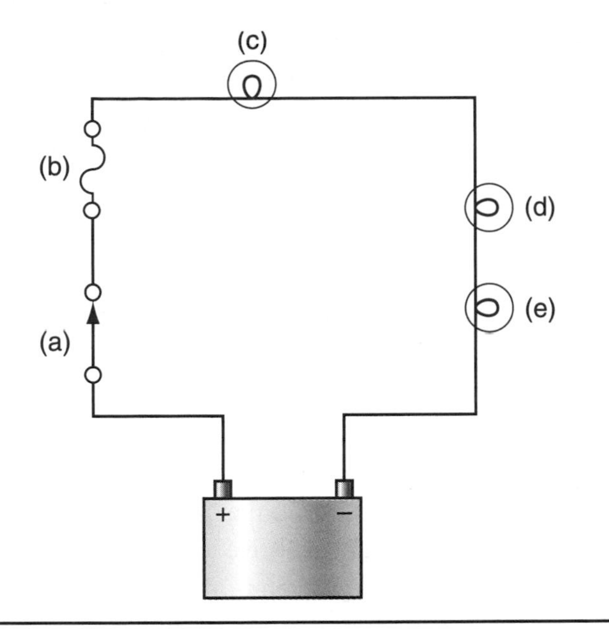

FIGURE 7–1 A basic series circuit including (a) a switch, (b) a fuse, (c) lamp 1, (d) lamp 2, (e) lamp 3.

go out just because one of them burns out. Figure 7–2 is an example of a simple series circuit for a trunk lid release, with a voltage source, fuse protection, control release switch, load device (solenoid), and wiring.

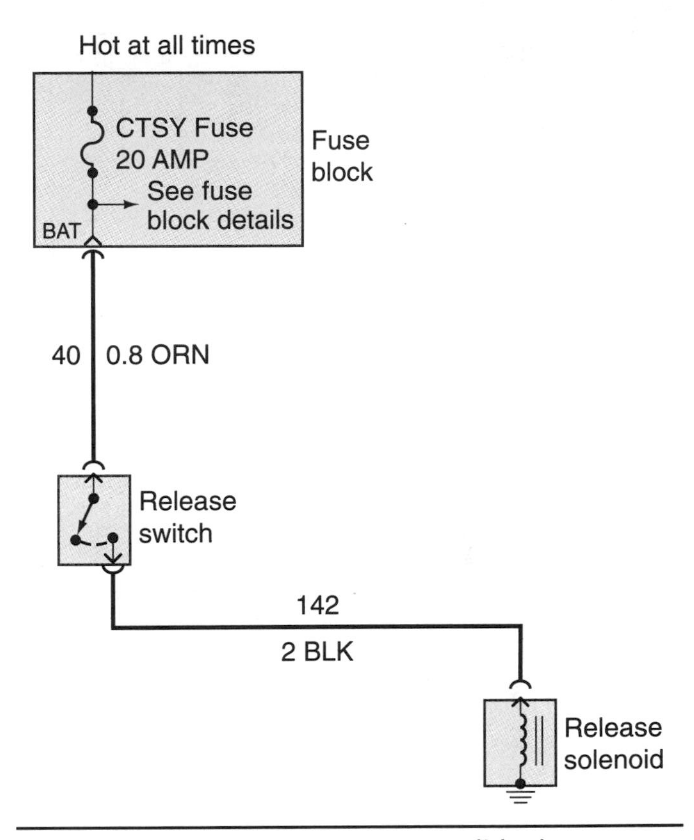

FIGURE 7–2 Luggage compartment lid release.

Circuit Laws

Three **series circuit laws** pertain *only* to series circuits.

Series Law 1

In a series circuit the current is the same at any point in the circuit. Because there is only a single path for the current to flow, the current must pass though each resistive device in the circuit, and each resistive load passes the same current through it. Figure 7–3 is an example of the amperage readings you would get on a simple series circuit with three resistors (R_1, R_2, R_3) wired in series.

In simple math terms, the current total in Figure 7–3 can be expressed as I_T, and I_1, I_2, and I_3 can represent the three resistors as follows:

$$I_T = I_1 = I_2 = I_3 \quad \text{(same or equal)}$$

If you knew the current at any point in the circuit, then you would know the current anywhere else in the circuit. This is important because, if your access to the whole circuit is limited, you might find a convenient point to measure the current.

Series Law 2

The total resistance in a series circuit is equal to the *sum* of the individual resistance loads in the circuit. Remembering Ohm's law—as the resistance goes up, the current flow goes down—then the more resistive loads there are, the lower the current flow will be. In other words, current and resistance are inversely proportional to each other. Figure 7–4 is an example of a series circuit with several resistors. In math terms, R_T represents total resistance, and R_1 through R_6 represent the six resistive loads as follows:

$$R_T = R_1 + R_2 + R_3 + R_4 + R_5 + R_6 \quad \text{(additive)}$$

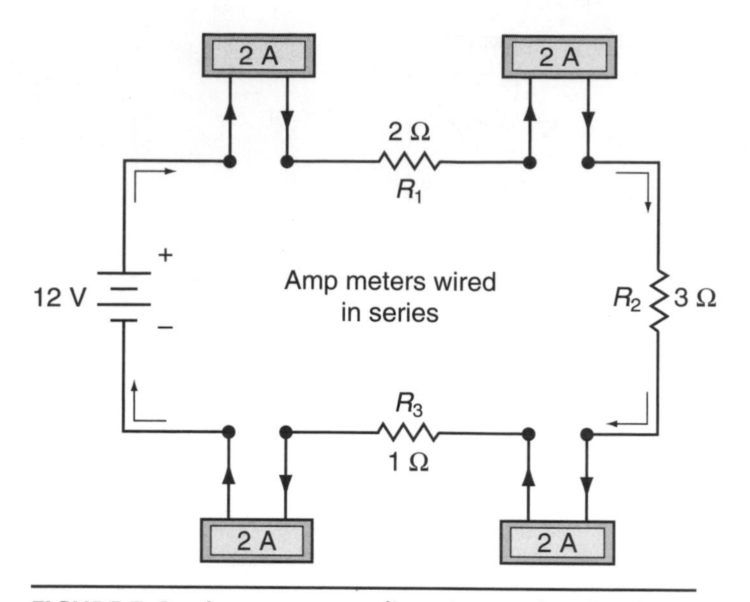

FIGURE 7–3 Amperage readings in a series circuit.

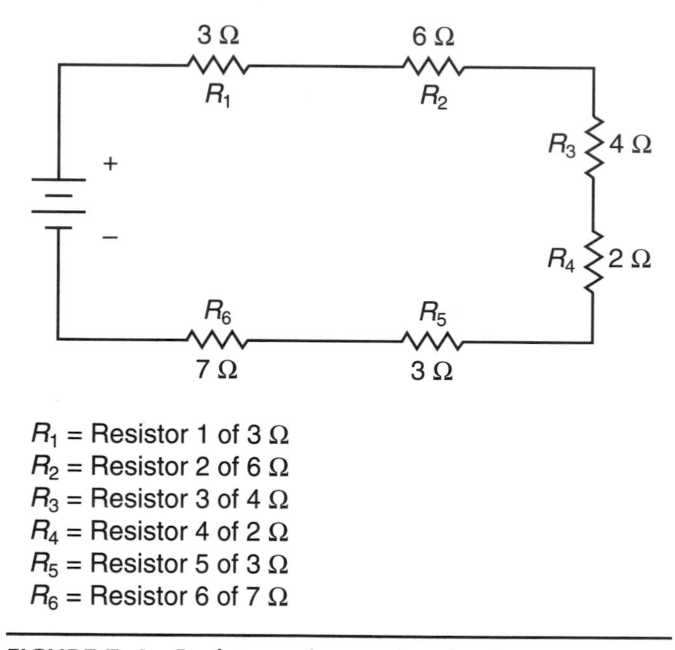

FIGURE 7–4 Resistance in a series circuit.

Using the numbers in Figure 7–4, the total resistance is:

$$R_T = 3\,\Omega + 6\,\Omega + 4\,\Omega + 2\,\Omega + 3\,\Omega + 7\,\Omega$$

$$R_T = 25\,\Omega$$

Series Law 3

The *total* voltage in a series circuit is equal to the *sum* of the individual voltage drops of each resistive load in the circuit. This is according to **Kirchoff's voltage law.** In simple math terms, V_T represents total voltage, and V_1 through V_3 (or more), represent the voltage drop of each resistive device, as follows:

$$V_T = V_1 + V_2 + V_3 + \text{etc.} \qquad \text{(additive)}$$

This principle is also known as the **voltage drop** of a series circuit. Figure 7–5 is an example of a series circuit showing what voltmeters would measure for the voltage drop of each resistive device in the circuit.

When resistive loads differ from one another, the voltage drops for loads also differ. More precisely, voltage drops are proportional to the values of each resistive load in the circuit. If one resistive load is twice as large as another, then the voltage drop is also twice as large. You can see this in Figure 7–5, where the 2-ohm resistive load is twice as large as the 1-ohm value, the voltage drop for the 2-ohm is also twice as large as the 1-ohm resistive load.

There are three primary ways to find the voltage drops in a series circuit. The first is to take the resistance values and use Ohm's law. To do this using the values given in Figure 7–5, first find the resistance total:

$$R_T = R_1 + R_2 + R_3$$
$$R_T = 1\,\Omega + 2\,\Omega + 3\,\Omega$$
$$R_T = 6\,\Omega$$

Having the resistance total, next use Ohm's law to find the amperage total.

12-volt total ÷ 6-ohm resistance total =
2-ampere current total

Because you know the series circuit law for current, you can then use Ohm's law to apply the current total, or 2 amperes, to each resistive device in the circuit to find the voltage drop for each device as follows:

2 amperes × 1 ohm = 2 volts
2 amperes × 2 ohms = 4 volts
2 amperes × 3 ohms = 6 volts

The sum of the individual voltage drops should equal the voltage source or total.

2 volts + 4 volts + 6 volts = 12 volts

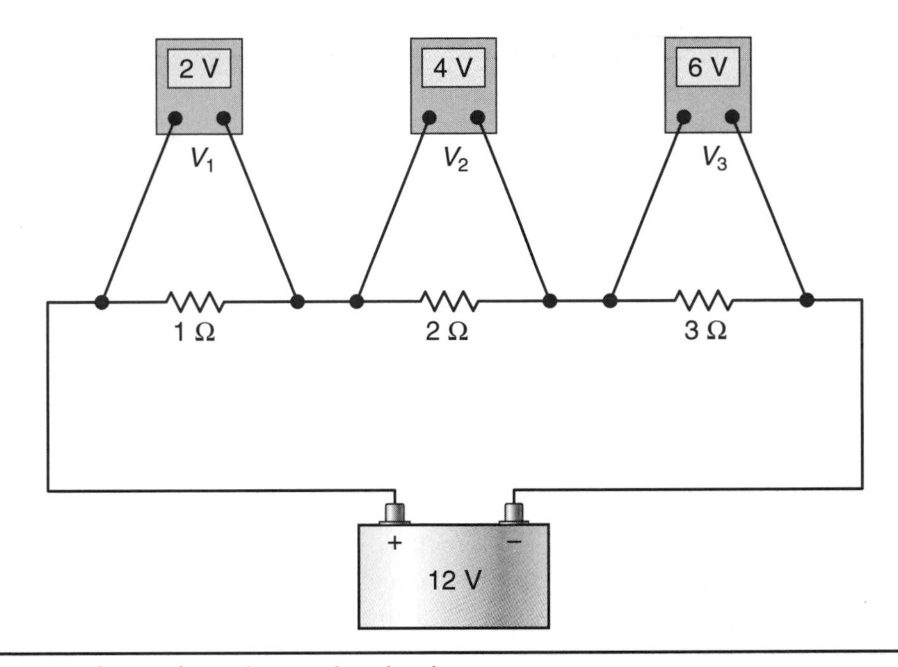

FIGURE 7–5 Measuring voltage drops in a series circuit.

The second way to find voltage drops is to use Kirchoff's voltage law. This method works only if you know the voltage source total and two of the three voltage drops. Then you can take what you know and subtract it from the voltage total to get what you don't know. Using the same values as before, figure the voltage drop of the 3-ohm value.

$$V_T = V_1 + V_2 + V_3$$
$$12\text{ V} = 2\text{ V} + 4\text{ V} + \text{(the unknown volt drop)}$$
$$12\text{ V} - 6\text{ V} = 6\text{ V}$$

The third way to find voltage drops in a series circuit is to use Watt's law. This is not the most efficient way to find voltage drops, but it works if you do not know individual circuit resistances but do know their respective amperages and wattages.

$$W_T = W_1 + W_2 + W_3$$
$$24\text{ W} = 4\text{ W} + 8\text{ W} + 12\text{ W}$$

$$4\text{ W} \div 2\text{ A} = 2\text{ V},$$
$$8\text{ W} \div 2\text{ A} = 4\text{ V},$$
$$12\text{ W} \div 2\text{ A} = 6\text{ V}$$

The individual voltage drops add up to equal the source voltage.

Because the voltage and resistance are both additive in series circuits, the amount of each individual voltage or resistance value is also proportional to its own total. This is known as **voltage and resistance percentage.** Figure 7–6 is an example of a simple series circuit with individual resistive loads that add up to 40-ohm total.

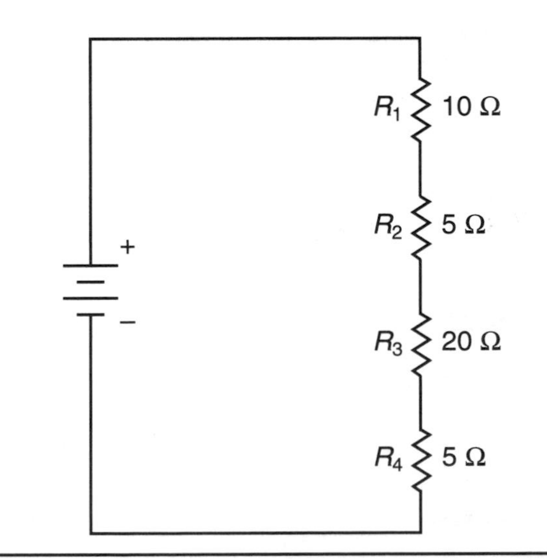

FIGURE 7–6 Simple series circuit with individual resistive loads.

To find the percentage of the resistance total that each individual resistor load takes, use the following calculation where R_n is R_1, R_2, etc.

$$R_n \div R_T \times 100 \text{ (for percent)} =$$

As an example, what percentage of the resistance total is from resistive load 3 (R_3)?

$$20\ \Omega \div 40\ \Omega = 0.50$$
$$0.50 \times 100 = 50\%$$

If you perform this procedure for all the remaining resistive values, the results should add up to 100%. The same procedure can be used to find the voltage percentage of individual resistive loads in a series circuit.

$$V_n \div V_T \times 100 \text{ (for percent)} =$$
Voltage percentage of individual loads

As an example, if the voltage total of Figure 7–6 is 12 volts, then you know that 12 volts ÷ 40 ohms produces a current total of 0.3 amperes. This makes a voltage drop of R_3 equal 6 volts.

$$6\text{ V} \div 12\text{ V} = 0.50$$
$$0.50 \times 100 = 50\%$$

As you can see, the percentage numbers using the resistive values and the voltage drops produce the same values of percentage numbers. This means you can use either set of numbers to find the percentage that an individual resistive load takes in a series circuit.

Here is a brief overview of the series circuit laws to use as a quick reference guide.

Basic Laws of Series Circuit Electricity

	Series
Current	Equal or same $I_T = I_1 = I_2 = I_3 +$ etc.
Voltage	Additive $E_T = E_1 + E_2 + E_3 +$ etc.
Resistance	Additive $R_T = R_1 + R_2 + R_3 +$ etc.

Exercises—Series Circuits

The following cognitive exercises are designed to give you practice using the series laws and methods. The exercises increase in complexity in an effort to test your understanding of the basic series laws and to prepare you for the actual hands-on series circuit lab exercises that follow.

Exercise 1

Instructions Identify the type of circuit in Figure 7–7. On a *separate* sheet of paper make an answer table like Table 7–1. Using the table, fill in the known circuit values from the series diagram. This helpful first step gives you an idea of where to begin and what answers you need. Calculate the answers in the remaining spaces using one of the methods previously learned. Use the following notation system.

R_T = Resistance total I_T = Amperage total
E_T = Voltage total W_T = Watts total

R_1 = Resistor 1 I_1 = Amperage 1
E_1 = Voltage 1 W_1 = Watts 1

R_2 = Resistor 2 I_2 = Amperage 2
E_2 = Voltage 2 W_2 = Watts 2

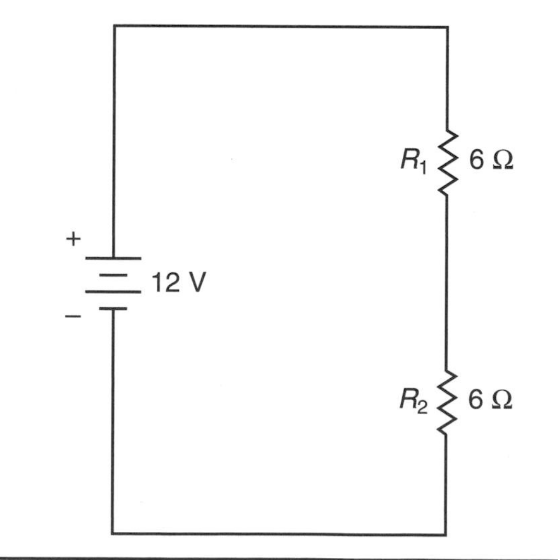

FIGURE 7–7 Basic series circuit.

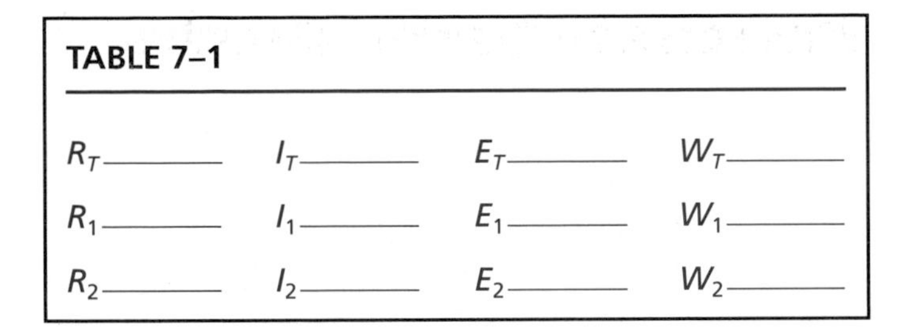

TABLE 7–1

R_T ______	I_T ______	E_T ______	W_T ______
R_1 ______	I_1 ______	E_1 ______	W_1 ______
R_2 ______	I_2 ______	E_2 ______	W_2 ______

TABLE 7–2

R_T 12 Ω	I_T 1 A	E_T 12 V	W_T 12 W
R_1 6 Ω	I_1 1 A	E_1 6 V	W_1 6 W
R_2 6 Ω	I_2 1 A	E_2 6 V	W_2 6 W

Circuit Analysis

To ensure that you are familiar with the series laws, the correct steps and answers for the previous problem follow. If your answers are different, refer to the previous discussion for further study.

Step 1: Known values: $E_T = 12$ V, $R_1 = 6$ Ω, $R_2 = 6$ Ω. Fill them in on a *separate* sheet of paper in a format like that of Table 7–1.

Step 2: Remembering that the resistance law is additive in series circuits, add them up and fill in the resistance total of 12 Ω.

$$R_1 + R_2 = R_T$$
$$6\,\Omega + 6\,\Omega = 12\,\Omega$$

Step 3: $E \div R = I$
Take the voltage total 12 V and divide by the resistance total 12 Ω to get the amps total of 1 A. Remembering that the current law states that the current is the same in series circuits, fill in the other current values with the same 1 A.

Step 4: $I \times R = E$
Next, multiply the individual amperages by the individual resistance values to get the voltage drops of E_1 and E_2.

Step 5: $I \times E = W$
To find the power consumption, use Watt's law. Volts total times amperes total equal watts total, which is 12 W. (If you are unsure, refer to Chapter 3.) Do the same procedure to find W_1 and W_2.

Table 7–2 is a summary of the right answers.

Instructions Follow the instructions used on the previous problem for the remaining series circuit exercises.

Remember to put your answers on a *separate* sheet of paper.

Exercise 2

Note: After you have identified the type of circuit in Figure 7–8, fill in the known values on a table like Table 7–3 on a *separate* sheet of paper.

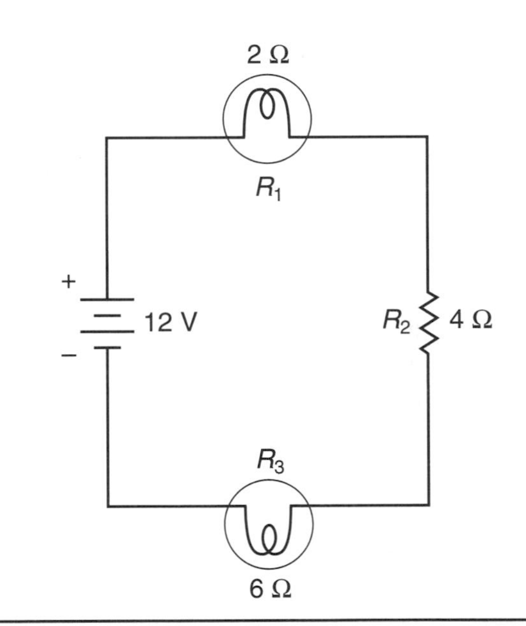

FIGURE 7–8 Identify basic circuit.

TABLE 7–3

R_T ______	I_T ______	E_T ______	W_T ______
R_1 ______	I_1 ______	E_1 ______	W_1 ______
R_2 ______	I_2 ______	E_2 ______	W_2 ______
R_3 ______	I_3 ______	E_3 ______	W_3 ______

Exercise 3

Note: Figure 7–9 is a series circuit that is drawn a little differently. You may find some actual vehicle wiring diagrams drawn like this. Even though the schematic is not drawn from left to right or from top to bottom, the theory of series circuits holds. Identify the circuit by following the wiring from the positive side of the battery, through each resistive load of the circuit, and then to the negative side.

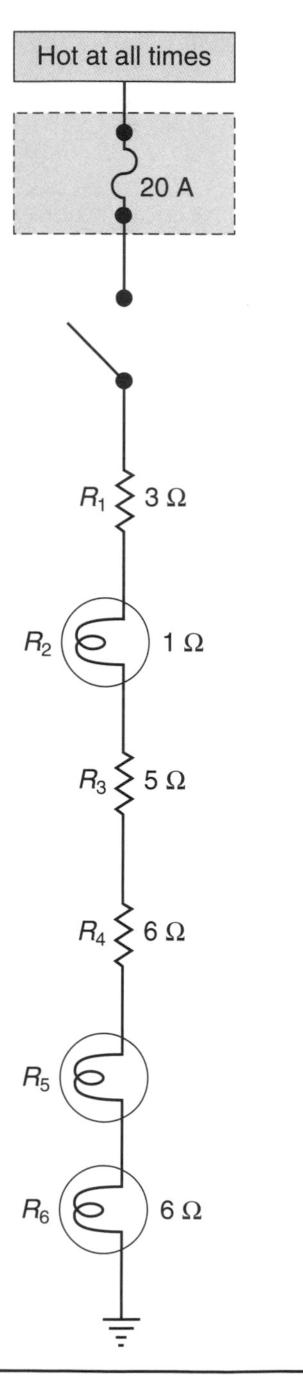

FIGURE 7–9 Basic series circuit.

Complete the answers to the preceding exercise by filling in a table like Table 7–4 on a *separate* sheet of paper.

$$W_T = 0.96 \text{ W}$$

TABLE 7–4

R_T ______	I_T .5 A	E_T ______	W_T ______
R_1 3 Ω	I_1 ______	E_1 ______	W_1 ______
R_2 1 Ω	I_2 ______	E_2 ______	W_2 ______
R_3 5 Ω	I_3 ______	E_3 ______	W_3 ______
R_4 6 Ω	I_4 ______	E_4 ______	W_4 ______
R_5 ______	I_5 ______	E_5 1.5 volts	W_5 ______
R_6 6 Ω	I_6 ______	E_6 ______	W_6 ______

Begin by filling in the known circuit values and then look for the best series law(s) to apply first.

HINT: Remember the series circuit law for current.

Exercise 4

Figure 7–10 is a series circuit with a missing resistive value. Complete your answers to Exercise 4 on a *separate* sheet of paper similar to Table 7–5.

TABLE 7–5

R_T ______	I_T ______	E_T ______	W_T ______
R_1 ______	I_1 ______	E_1 ______	W_1 ______
R_2 ______	I_2 ______	E_2 ______	W_2 ______
R_3 ______	I_3 ______	E_3 ______	W_3 ______
R_4 ______	I_4 ______	E_4 ______	W_4 ______
R_5 ______	I_5 ______	E_5 ______	W_5 ______

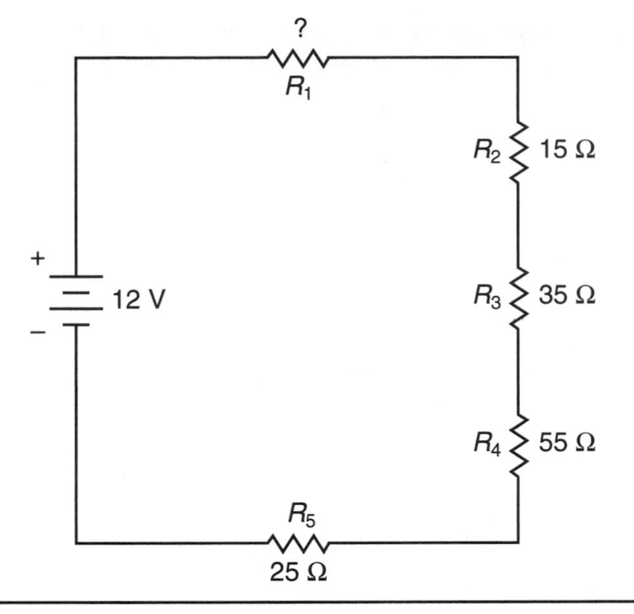

FIGURE 7–10 Basic series circuit.

Now that you have practiced with some basic series circuit examples, let's try a more difficult circuit problem to see how well you know the series circuit laws.

HINT: No matter how complex the circuit looks, the same basic series circuit laws apply. The better you understand the laws applying to series circuits, the easier electrical problem solving becomes.

Exercise 5

Using Figure 7–11 as a reference, use a *separate* sheet of paper and complete Table 7–6.

You've spent some time now learning series circuit theory and laws, and developing the knowledge on how series circuits perform in relation to voltage, amperage, and resistance. So let's apply theory to some real-life case studies and prove the theories you have learned.

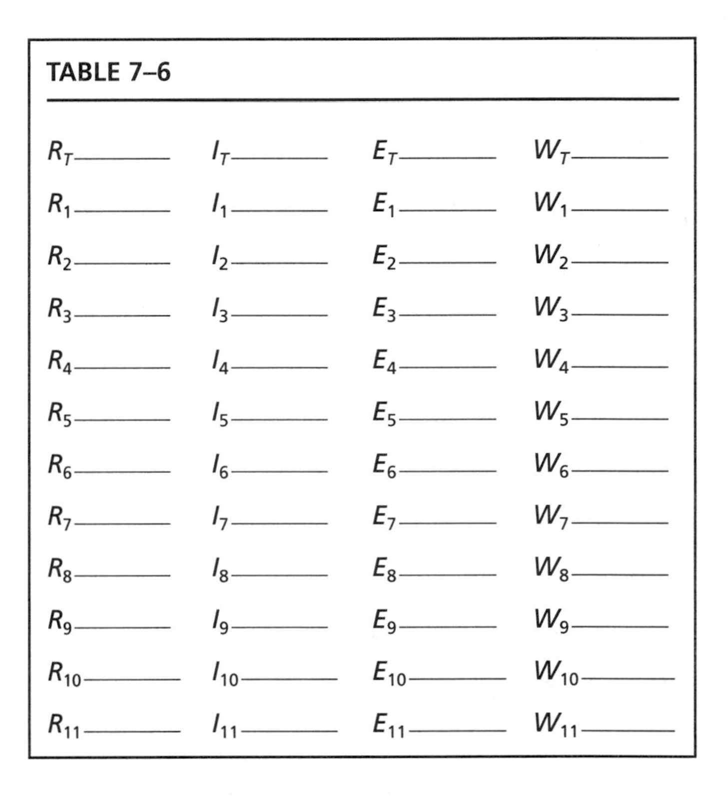

TABLE 7–6

R_T ______	I_T ______	E_T ______	W_T ______
R_1 ______	I_1 ______	E_1 ______	W_1 ______
R_2 ______	I_2 ______	E_2 ______	W_2 ______
R_3 ______	I_3 ______	E_3 ______	W_3 ______
R_4 ______	I_4 ______	E_4 ______	W_4 ______
R_5 ______	I_5 ______	E_5 ______	W_5 ______
R_6 ______	I_6 ______	E_6 ______	W_6 ______
R_7 ______	I_7 ______	E_7 ______	W_7 ______
R_8 ______	I_8 ______	E_8 ______	W_8 ______
R_9 ______	I_9 ______	E_9 ______	W_9 ______
R_{10} ______	I_{10} ______	E_{10} ______	W_{10} ______
R_{11} ______	I_{11} ______	E_{11} ______	W_{11} ______

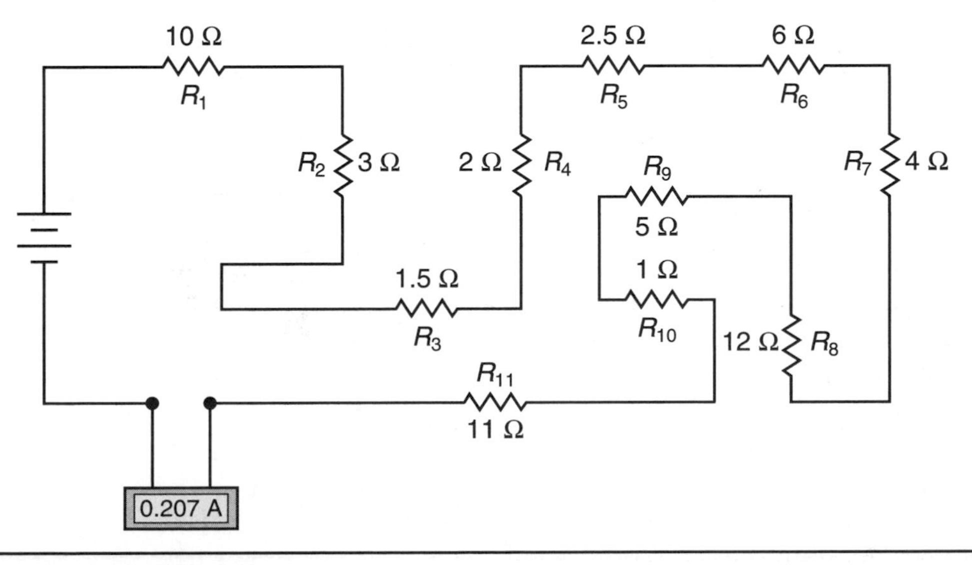

FIGURE 7–11 Complex series circuit.

CASE STUDIES

CASE 1

The following case study is typical of a live series circuit problem that technicians often face. Knowledge of series circuits can help you perform the diagnosis quickly and efficiently, so let's give it a try.

Customer Complaint

A vehicle heater circuit does not work when the blower switch is in the LO and M1 positions, but it does work properly in the M2 and HI positions. Figure 7–12 shows the layout of the blower motor circuit. The customer replaced the blower switch, thinking this was the cause. The results were the same as before.

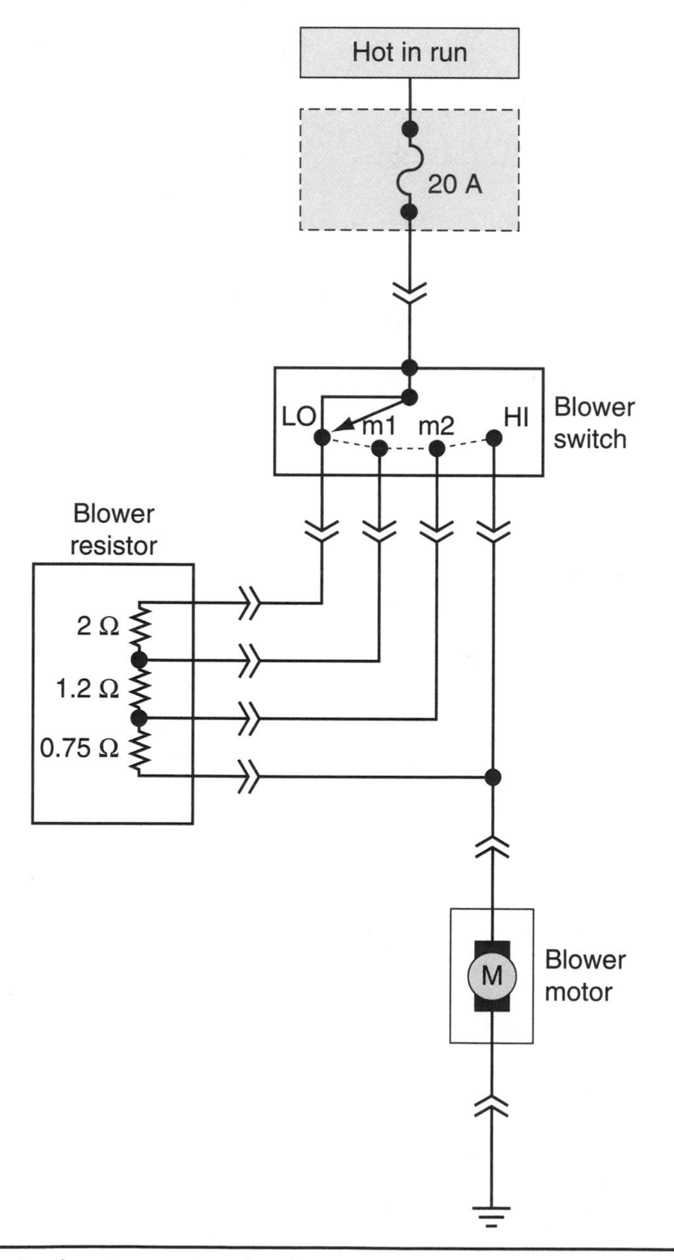

FIGURE 7–12 Typical series heater motor circuit.

Known Information

- ❑ Vehicle operating voltage = 14 volts.
- ❑ Blower switch has been replaced.
- ❑ System wiring and connectors between blower switch and resistor assembly have been checked and are satisfactory.
- ❑ The system fault with LO and M1 happened suddenly.
- ❑ Blower motor has windings with a resistance of 1.5 Ω.
- ❑ All other resistance information is located on wiring diagram Figure 7–12.

Circuit Analysis

Answer the following questions on a *separate* sheet of paper:

1. What is the most likely cause of the inoperative blower circuit in the LO and M1 positions only?
   ______________________________.
2. What is the total circuit resistance of the blower circuit, assuming that the blower switch, connections, and wiring have no measurable resistance? ______________________________.
3. What is the total amperage of the circuit at the following positions if an ampmeter was placed in series with the circuit at the fuse?
   - Amperage reading with blower switch on LO position: ______________________________.
   - Amperage reading with blower switch on M1 position: ______________________________.
   - Amperage reading with blower switch on M2 position: ______________________________.
   - Amperage reading with blower switch on HI position: ______________________________.
4. What is the best way to repair this system? ______________________________.
5. What series circuit law(s) were helpful when troubleshooting this system or explaining it to the customer? ______________________________
   ______________________________.
6. What is your analysis of this case study? ______________________________
   ______________________________.

CASE 2

Using the same circuit diagram, Figure 7–12, solve the following customer complaint.

Customer Complaint

A vehicle heater does not work in any blower switch position. Customer replaced the fuse, but the results were the same.

Known Information

- ❑ Vehicle operating voltage = 14 volts.
- ❑ Twenty-ampere fuse has been replaced.
- ❑ System fault happened suddenly during heater blower operation.

Circuit Analysis

Answer the following questions on a *separate* sheet of paper.

1. If you knew the fuse was good, which system components would you check in the circuit, and in what priority? List the most important component first, as well as the procedure you would use to check the it.

   Component 1: ____________________. Troubleshooting procedure: ____________________.

   Component 2: ____________________. Troubleshooting procedure: ____________________.

   Component 3: ____________________. Troubleshooting procedure: ____________________.

   Component 4: ____________________. Troubleshooting procedure: ____________________.

   Component 5: ____________________. Troubleshooting procedure: ____________________.

2. What series circuit law(s) were helpful when troubleshooting this system or explaining it to the customer? ____

   ____

   ____.

3. What is your analysis of this case study? ____

   ____.

Figure 7–13 is an example of a series heater circuit that includes a blower relay that needs to be energized by the ignition switch in order to provide voltage to the blower circuit.

The blower motor in this circuit is located before the blower resistor instead of after it, as in Figure 7–12. Because in a series circuit there is only one path for the current to flow, the blower motor can be located before or after the blower controls and still operate. The primary difference between the two blower circuits is that the positive side of the blower motor contains the source voltage at all times in Figure 7–12, and it is also on the negative side of the circuit where the resistive speed control block is located. This changes how you would perform your troubleshooting techniques. If you were to do a quick analysis of the circuit in Figure 7–13, you would see the following:

1. The power is *hot* at all times to the front blower motor relay through the 40-ampere fuse in the power distribution center.
2. The front blower motor relay is energized every time the key is in the run position.
3. The ground for the coil side of the relay is supplied on circuit wire Z1.

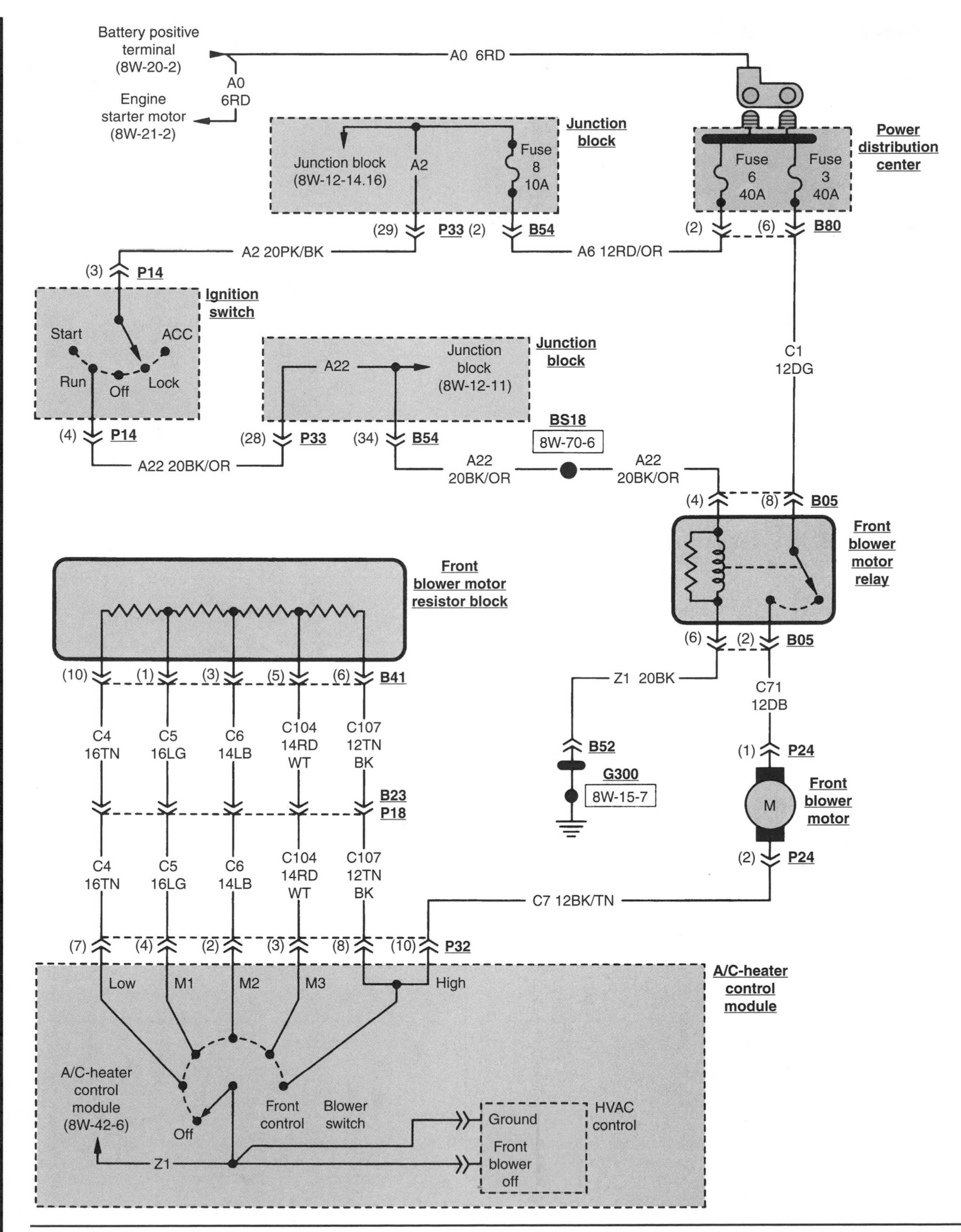

FIGURE 7–13 Heater blower circuit.

(Courtesy of DaimlerChrysler Corporation)

4. The ground for the blower motor is controlled through the blower motor resistor and switch.
5. With the blower motor in the low position, the circuit's current must go through the blower motor, through all four resistors in the resistor block, and then through the low position on the blower switch to ground.

Hands-On Vehicle Tests

The following two hands-on vehicle checks are included in the NATEF (National Automotive Technicians Education Foundation) Task List. Complete your answers to the below questions on a *separate* sheet of paper.

Performance Task 1

Task Description

Check the operation of the parking brake indicator light system.

Task Objectives

- ❑ Obtain a vehicle that can be used for this task. What model and year of vehicle are you using for this task? ________________
- ❑ Find a wiring diagram that shows the parking brake indicator circuit for this vehicle. Does the parking brake circuit follow the characteristics of a series circuit? ________________.
- ❑ Identify the following components and list their location on the vehicle.
    1. Instrument panel cluster: ________________.
    2. Parking brake indicator bulb(s): ________________.
    3. Parking brake switch:________________.
    4. Circuit fuse and fuse number: ________________.
    5. Other electrical circuits that share the same fuse: ________________.
    6. Circuit ground(s): ________________.
    7. Other circuit components: ________________.

    ________________.

Circuit Checks

- ❑ Check operation of parking brake switch. Disconnect wire connections at switch. Set a DVOM to continuity, hook it to the switch terminals, and activate the parking brake. What happens?

  ________________.
- ❑ Reconnect the wires to parking brake switch.
- ❑ Remove the parking brake fuse and hook the DVOM in series in its place. What is the total amperage of the circuit? ________________.

- ❑ With the parking brake off, is there still current flowing through the DVOM? ________________.
- ❑ If the answer is yes to this question, where is the amperage coming from? ________________.
- ❑ What is the total circuit voltage? ________________.
- ❑ Perform a voltage drop test through the circuit components, connectors, and wires. Does the measurable voltage drop(s) add up to the total circuit voltage? ________________.
- ❑ Does this parking brake circuit follow the circuit laws for a series circuit? ________________.

Task Summary

After performing this NATEF task, what can you determine that will be helpful in future parking brake circuit diagnosis? ________________

________________.

Performance Task 2

Task Description

Inspect and test the A/C heater blower, motors, resistors, switches, relays, wiring, and protection devices; perform the necessary action.

Task Objectives

- ❑ Obtain a vehicle that can be used for this task. What model and year of vehicle are you using for this task? ________________.
- ❑ Find a wiring diagram that shows the heater control and motor circuit for this vehicle. Does the heater control circuit follow the characteristics of a series circuit? ________________.
- ❑ Identify the following components and list their location on the vehicle.
  1. Heater motor control switch: ________________.
  2. Blower motor assembly: ________________.
  3. Blower resistor block assembly: ________________.
  4. Blower motor relay (if applicable): ________________.
  5. Blower circuit fuse and fuse number: ________________.
  6. Blower motor relay fuse number (if applicable): ________________.
  7. Other electrical circuits that share the same fuse(s): ________________.
  8. Circuit ground(s): ________________.
  9. Other circuit components: ________________.

     ________________.

Circuit Checks

- ❑ To ensure that all DVOM readings of the circuit are characteristic of the actual circuit operation, the vehicle should be running in a well ventilated area (keeps the battery from discharging).

- ❑ With the blower switch in the low position, measure the voltage available to the circuit. This voltage reading is most accurately taken from the fuse block on the positive side to the ground connection at the end of the circuit. Voltage total of the blower circuit is ________________.
- ❑ Measure the voltage drop across the points listed below. Actual points to record the voltage drops may vary slightly depending on the circuit design.
  1. Across circuit fuse ________________.
  2. From fuse box to blower switch ________________.
  3. From blower switch low to resistor block ________________.
  4. Across each resistor in the blower resistor block assembly. Normally there are at least three resistors in this block assembly. Some vehicles have more or less; so complete as many as necessary.

     Resistor 1 voltage drop: ________________.

     Resistor 2 voltage drop: ________________.

     Resistor 3 voltage drop: ________________.

     Resistor 4 voltage drop: ________________.
  5. Across blower motor ________________.
  6. From blower motor to actual circuit ground connection in vehicle ________________.
- ❑ Do the measured voltage drops add up to equal the measured voltage total? ________________.
- ❑ With the blower switch in the M1 position, measure the voltage available to the circuit in the same way as you did before. Voltage total of the blower circuit: ________________.
- ❑ Measure the voltage drop across the following points. Actual points to record the voltage drops may vary slightly depending on the circuit design.
  1. Across circuit fuse: ________________.
  2. From fuse box to blower switch: ________________.
  3. From blower switch M1 to resistor block: ________________.
  4. Across each resistor in the blower resistor block assembly. Some vehicles have more or less, so complete as many as necessary.

     Resistor 1 voltage drop: ________________.

     Resistor 2 voltage drop: ________________.

     Resistor 3 voltage drop: ________________.

     Resistor 4 voltage drop: ________________.
  5. Across blower motor: ________________.
  6. From blower motor to actual circuit ground connection in vehicle: ________________.

- ❑ Are the voltage drop readings in M1 the same as in the LOW position? ________________________.

  Why? ____.

- ❑ Do the measured voltage drops add up to equal the measured voltage total? ____________________

- ❑ With the blower switch in the M2 position, measure the voltage available to the circuit in the same way as you did before. Voltage total of the blower circuit: ________________________________.

- ❑ Measure the voltage drop across the following points. Actual points to record the voltage drops may vary slightly depending on the circuit design.

  1. Across circuit fuse: ____.
  2. From fuse box to blower switch: ______________________________________.
  3. From blower switch M2 to resistor block: _________________________________.
  4. Across each resistor in the blower resistor block assembly. Some vehicles have more or less, so complete as many as necessary.

     Resistor 1 voltage drop: ____.

     Resistor 2 voltage drop: ____.

     Resistor 3 voltage drop: ____.

     Resistor 4 voltage drop: ____.
  5. Across blower motor: ____.
  6. From blower motor to actual circuit ground connection in vehicle: ____________________.

- ❑ Are the voltage drop readings in M2 the same as in the LOW and M1 positions? ________________.

  Why? ____.

- ❑ Do the measured voltage drops add up to equal the measured voltage total? ____________________.

- ❑ With the blower switch in the HI position, measure the voltage available to the circuit in the way same as you did before. Voltage total of the blower circuit ________________________________.

- ❑ Measure the voltage drop across the points listed below. Actual points to record the voltage drops may vary slightly depending on the circuit design.

  1. Across circuit fuse: ____.
  2. From fuse box to blower switch: ______________________________________.
  3. From blower switch high to resistor block: ________________________________.
  4. Across blower relay (if applicable): ____________________________________.
  5. Across each resistor in the blower resistor block assembly. Some vehicles have more or less, so complete as many as necessary.

     Resistor 1 voltage drop: ____.

     Resistor 2 voltage drop: ____.

Resistor 3 voltage drop: ______.

Resistor 4 voltage drop: ______.

6. Across blower motor: ______.
7. From blower motor to actual circuit ground connection in vehicle: ______.

❑ Are the voltage drop readings in the HI position the same as in the LOW, M1, and M2 positions?

Why? ______.

❑ Does the measured voltage drops add up to equal the measured voltage total? ______.

❑ Turn blower circuit off and install a DVOM in series to measure circuit amperage. Normally, the easiest place to perform this procedure is at the fuse or after the blower motor assembly. What is the circuit amperage with the blower switch in the following positions?

1. LOW: ______
2. M1: ______
3. M2: ______
4. HI: ______

❑ What can you conclude about voltage drops in a series circuit? ______.

______.

❑ What effect do resistors wired in a series line have on the circuit current? ______

______.

Task Summary

After performing this NATEF task, what can you determine will be helpful in future parking brake circuit diagnosis? ______

______.

Summary

❑ In a series circuit there is only one path of flow for the electrons to travel.

❑ In a series circuit the current is the same at any point in the circuit.

❑ In a series circuit the total resistance is the sum of the individual resistance loads in the circuit.

❑ In a series circuit the voltage drops in each system component add up to equal the source voltage, according to Kirchoff's law.

❑ Voltage and resistance are both additive in a series circuit, so the voltage and resistance percentage calculation can be used to find individual percentage loads.

Key Terms

Kirchoff's voltage law
series circuit
series circuit laws
voltage and resistance percentage
voltage drop

Review Questions

Short Answer Essays

1. Describe the circuit law for voltage in a series circuit.
2. Describe the circuit law for amperage in a series circuit.
3. Describe the circuit law for resistance in a series circuit.

Fill in the Blanks

1. The amperage of a load component in a series circuit is the ________________________ as the amperage ________________________.
2. In a series circuit there is only ________________________ path for ________________________ to travel.
3. ________________________ and ________________________ are both additive in a series circuit.

ASE-Style Review Questions

1. Technician A says that in a series circuit the voltage of each circuit load device is the same as the total circuit voltage. Technician B says the amperage of each circuit load is the same throughout the circuit. Who is correct?
   A. A only
   B. B only
   C. Both A and B
   D. Neither A nor B
2. In the following series light circuit the bulbs do not light when the switch is closed. Technician A says the switch can be bad. Technician B says one of the light bulbs could be burned out. Who is correct?
   A. A only
   B. B only
   C. Both A and B
   D. Neither A nor B

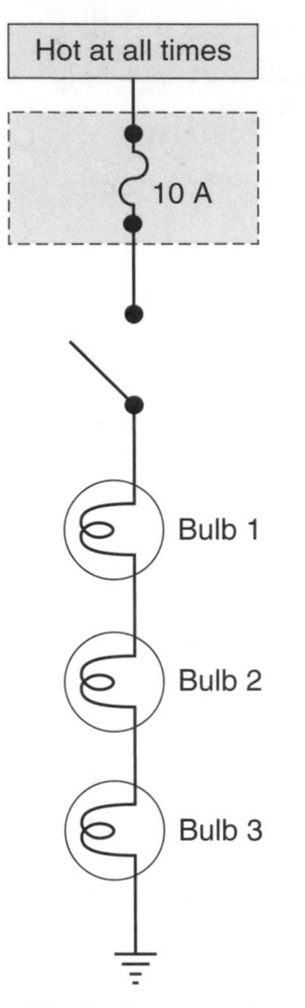

3. A circuit has four resistors wired in series: 10 ohms, 5 ohms, 4 ohms, 2.5 ohms. What is the resistance total? The correct answer is:
   A. 21.5 Ω
   B. 1.05 Ω
   C. 20.5 Ω
   D. None of the above

4. Technician A says the resistance total in a series circuit is always the same as the amperage total. Technician B says the resistance total in a series circuit is always one half of the voltage total. Who is correct?
   A. A only
   B. B only
   C. Both A and B
   D. Neither A nor B

5. Technician A says that adding more series loads to a series circuit increases the total circuit amperage. Technician B says that if a series circuit is opened, the entire circuit will not operate. Who is correct?
   A. A only
   B. B only
   C. Both A and B
   D. Neither A nor B

6. In a typical heater blower circuit, the circuit does not work in the low position but works fine in all other switch positions. Technician A says the switch could be at fault. Technician B says the blower resistor block could be at fault. Who is correct?

   A. A only

   B. B only

   C. Both A and B

   D. Neither A nor B

7. A series circuit has three resistors of 20 Ω each. The voltage drop of each resistor is 4 volts. Technician A says the voltage total is 12 volts. Technician B says the resistance total is 60 Ω. Who is correct?

   A. A only

   B. B only

   C. Both A and B

   D. Neither A nor B

8. Technician A says the individual resistive loads in a series circuit will add up to equal the resistance total. Technician B says voltage drops in a series circuit will add up to equal the voltage total. Who is correct?

   A. A only

   B. B only

   C. Both A and B

   D. Neither A nor B

9. A vehicle blows a fuse in a series blower motor circuit only when the switch is in the HI position. Technician A says the blower motor could be drawing too much amperage in this position for the size of the fuse. Technician B says a wrong amperage fuse could be the problem. Who is correct?

   A. A only

   B. B only

   C. Both A and B

   D. Neither A nor B

10. A total of 150 milliamps is flowing through a circuit that consists of three resistors in series, 30 Ω, 40 Ω, and 10 Ω. Technician A says the source voltage for the circuit is 120 volts. Technician B says the voltage is 1.2 volts. Who is correct?

    A. A only

    B. B only

    C. Both A and B

    D. Neither A nor B

Parallel Circuits

Introduction

In Chapter 7 you were introduced to series circuit laws pertaining to voltage, current, and resistance. In this chapter you will be exposed to electrical circuit laws and formulas pertaining only to parallel circuits. As you apply these electrical laws and formulas to voltage, current, and resistance in parallel circuits, you will be guided through each new concept step by step.

Objectives

When you complete this chapter you should be able to:

- ❑ Define a parallel circuit.
- ❑ Apply parallel circuit laws for voltage, current, and resistance.
- ❑ Define Kirchoff's current law.
- ❑ Perform parallel circuit calculations.
- ❑ Apply troubleshooting and testing techniques to parallel circuits.
- ❑ Perform basic hands-on parallel circuit exercises on a vehicle.

Parallel Circuit Explanation

A **parallel circuit** contains two or more resistive loads with all the *positive terminals* connected to a *common* junction and all *negative terminals* connected to a second *common* junction. The current flows to each branch independent of any other branch.

Note: In Figure 8–1, after the current leaves the positive side of the battery, it splits three ways through the three resistors. The amount of current flowing through each resistive branch load depends on the value of the individual resistor. An open circuit in one resistive load does not affect the operation of the other loads of the circuit.

Remember that Ohm's law states that the higher the resistance is, the lower the current flow will be. Current and resistance are inversely proportional to each other.

An automobile has a lot of simple parallel circuits. Some of these are headlight, backup light, courtesy light, rear window defroster grid, and taillight circuits. Figure 8–2 is an excellent example of a simple exterior lamp parallel circuit. All of the circuit bulbs share a common positive at splice S408, and eventually share a common ground at connection G400.

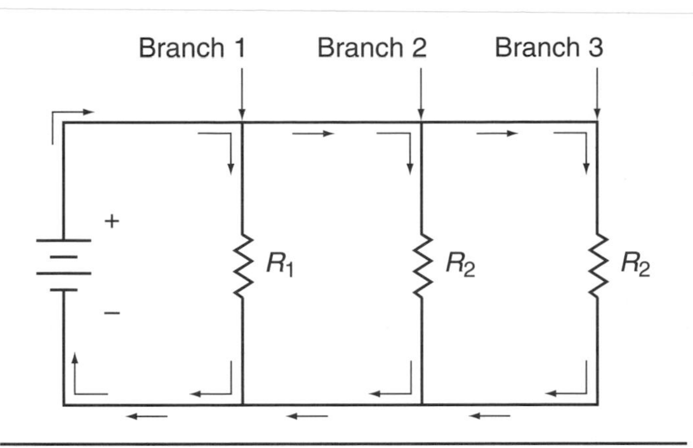

FIGURE 8–1 Simple parallel circuit.

Circuit Laws

Now that we know what a basic parallel circuit is, let's look at the three **parallel circuit laws** that pertain only to parallel circuits.

Parallel Law 1

In a parallel circuit, total current flow is equal to the *sum* of the current flowing through the individual load circuits (**Kirchoff's current law**). In simplified terms, $I_T = I_1 + I_2 + I_3 +$ etc. … (additive). No matter how many resistive loads are connected together at a common junction, Kirchoff's law applies. If one circuit load develops an open circuit, the others continue to operate as designed.

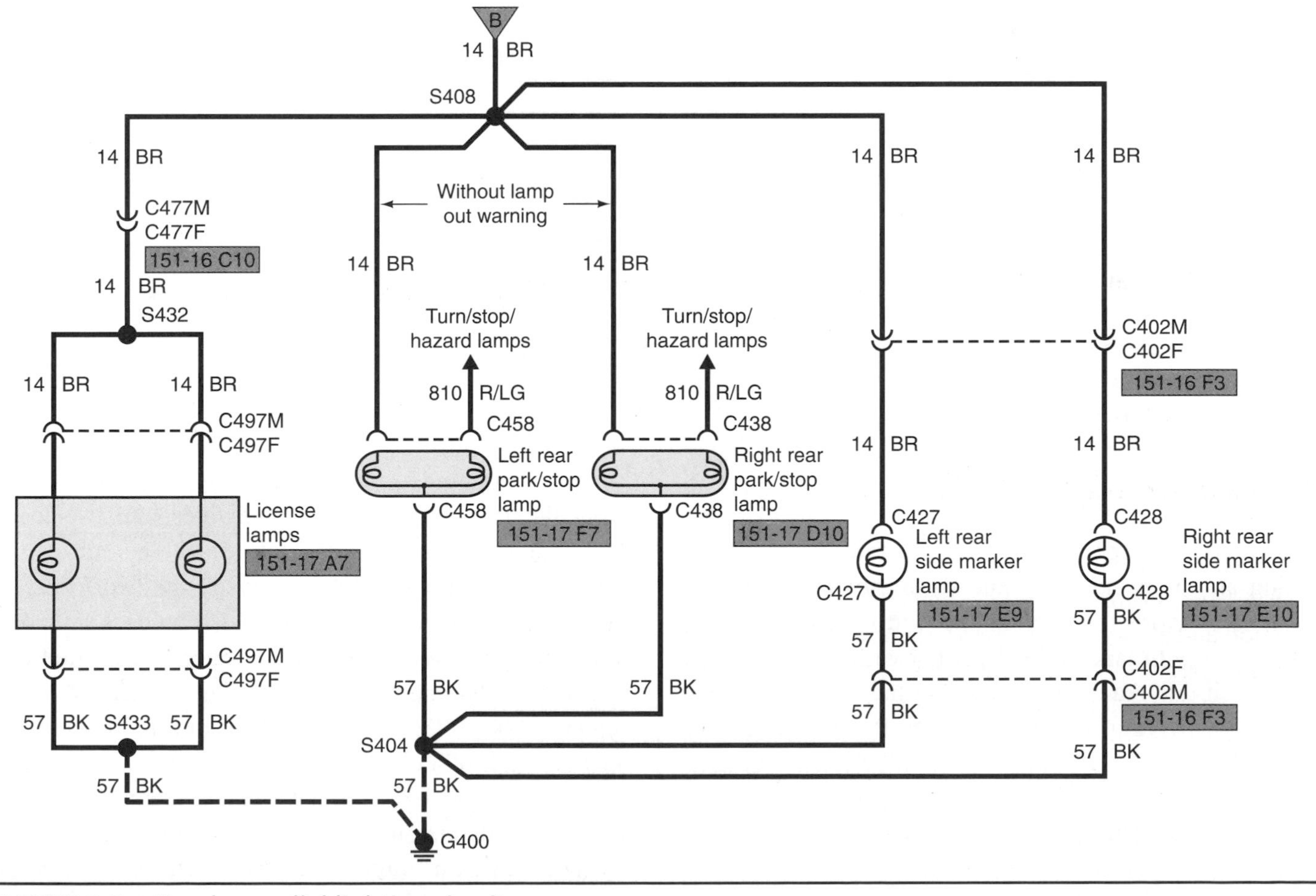

FIGURE 8–2 Simple parallel lighting circuit.

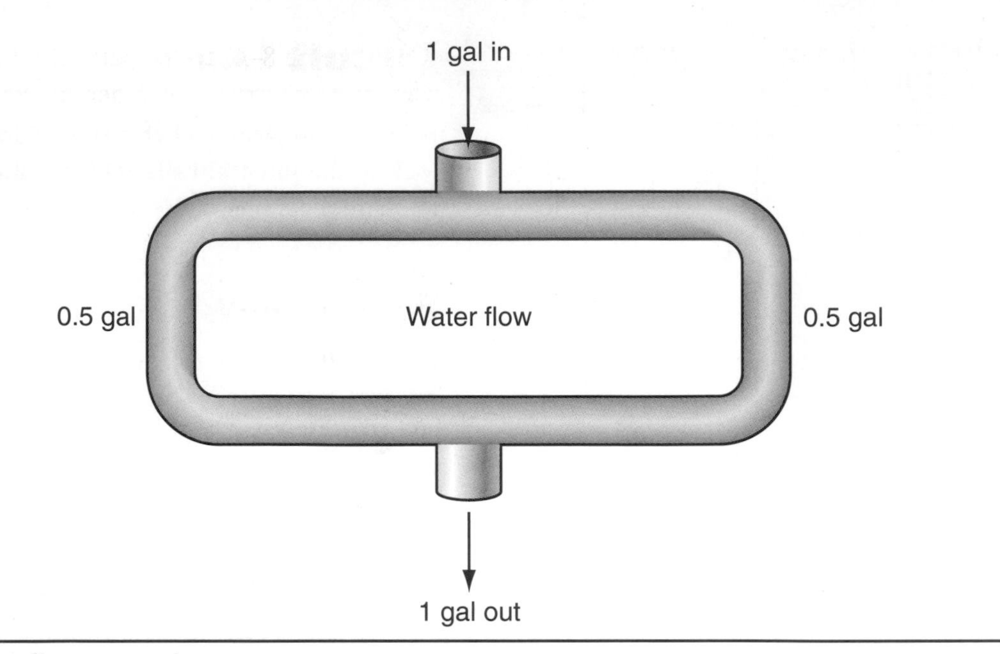

FIGURE 8–3 Water flow example.

If you were to compare this with the series current law in Chapter 7, you will note the difference: $I_T = I_1 = I_2 = I_3 =$ etc. ... Keep in mind that current flow in series circuits is equal throughout the circuit. If one part of the series circuit quits, the whole circuit quits.

Another way of describing the flow of current through a parallel circuit is to use the flow of water through a piping system. If 1 gallon of water is poured into the system (Figure 8–3) and each leg is allowed 0.5 gallon of flow, then the result must be 1 gallon of total output. What goes *into* it must go *out* of it!

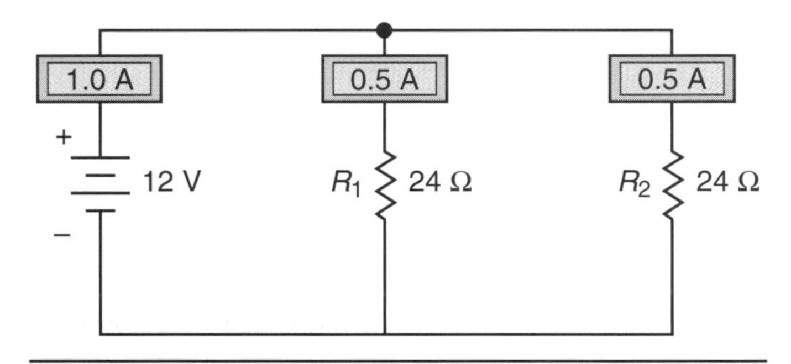

FIGURE 8–4 Parallel circuit flow.

A Simple Experiment

You can try an experiment to demonstrate this theory by taking an empty milk jug and cutting two square half-inch holes close together in the bottom corner. Cover both holes with tape and fill the jug with water. Using a stopwatch that reads in seconds, let the water out of the jug by uncovering *one* of the holes. Record the results in seconds on a slip of paper. Refill the jug with water. Using the stopwatch, let the water out again by uncovering both holes. The results should be approximately one-half of the first reading.

The theory of fluid behind this experiment is directly related to an electrical circuit diagram (Figure 8–4). The circuit in Figure 8–4 has 1 ampere of total current flow from the 12-volt battery that it divides into two loads having equal resistances of 24 Ω each. This means the amperage will divide evenly between the two loads, or 0.5 amps per branch load as indicated by each ammeter wired in series. As you learn the other two parallel circuit laws in the next few pages, you will gain a better understanding of how to calculate current flow in the various loads of a parallel circuit.

For example, Figure 8–5 is a parallel circuit with four resistive loads. If you knew the total current of the circuit and the current of three of the four branches, you could subtract what you know about the three branches from the total to get what you don't know: the fourth branch.

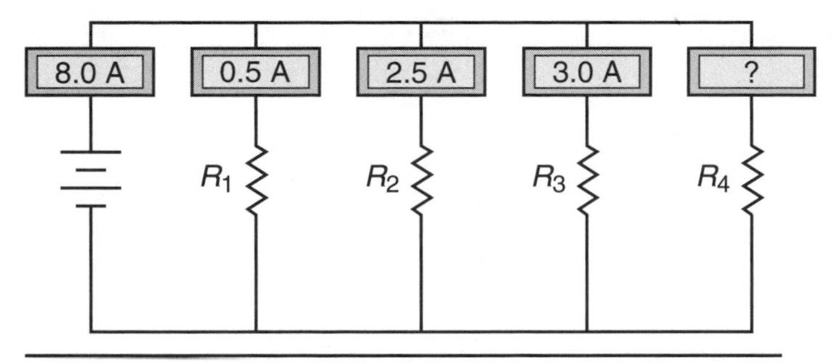

FIGURE 8–5 Parallel circuit with four resistive loads.

$$I_T = I_1 + I_2 + I_3 + I_4$$

$$8\text{ A} = 0.5\text{ A} + 2.5\text{ A} + 3\text{ A} + (\text{the unknown amperage of } I_4)$$

If you know I_1, I_2, and I_3, just subtract them from I_T to get I_4, which is 2 amperes.

Keep this principle in mind because it can assist you in solving future exercises, as well as in solving actual electrical problems.

Parallel Law 2

The source voltage (total) applied to a parallel circuit is the same as the voltage applied to each branch load in the circuit, according to **Kirchoff's voltage law,** where the sum of voltages around any closed circuit is equal to the sum of the applied voltage.

$$E_T = E_1 = E_2 = E_3 = \text{etc. ... (same or equal)}$$

This law is different than the voltage law you learned for the series circuit in Chapter 7: $E_T = E_1 + E_2 + E_3 + \text{etc. ...}$ (additive). Make sure to keep these two laws separate, or you are sure to make some big diagnostic errors!

In Figure 8–6, all the parallel loads are connected to a common positive. The other ends of the loads connect to a common negative. The voltage meters show that the voltage throughout the circuit is the same as the source voltage of 12 volts. Knowing this principle makes it easier to perform calculations and electrical diagnosis on parallel circuits. No matter how many circuit loads there are, as long as there is no circuit resistance between the source voltage and each individual load, and back to the source voltage, then the voltage is *the same* as the source voltage. *Don't forget this principle!*

Parallel Law 3

In a parallel circuit, the total circuit resistance (R_T) is *always* less than the value of the lowest resistive branch in the circuit.

$$R_T = \text{less than } (<) \text{ the lowest resistive branch}$$

The series resistance law $R_T = R_1 + R_2 + R_3 + \text{etc. ...}$ (additive) is quite different from a parallel circuit; so remember to keep them separated.

Computing resistance in a parallel circuit can be simplified depending on the method you use. You can use five different methods to find resistance in a parallel circuit. Since your objective is finding the answer using the quickest and easiest way possible, choose a method comfortable with your math abilities. And remember to use your calculator whenever possible!

Method 1 You can use the **product over sum method** for a parallel circuit with *two* resistive loads only (see Figure 8–7). If there are more than two resistors, then this method can be used in *groups of two* to get the answer.

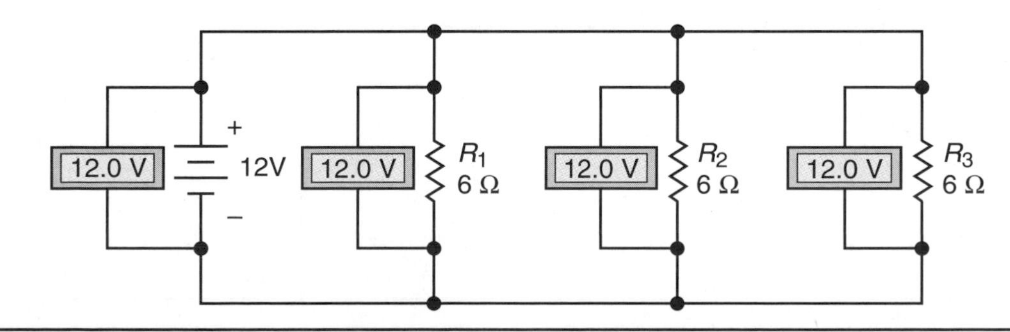

FIGURE 8–6 Voltages of parallel circuit branches.

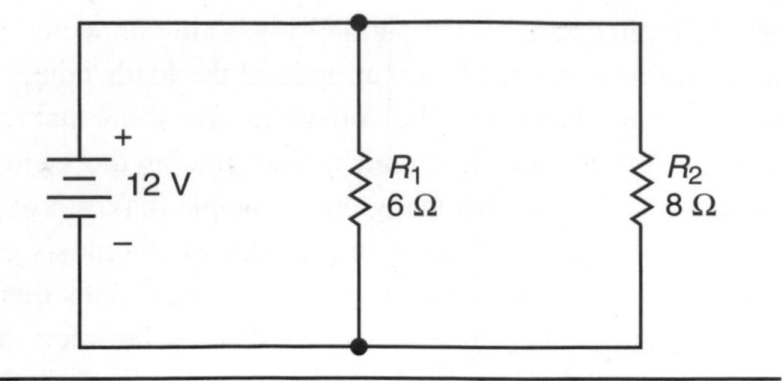

FIGURE 8–7 Parallel circuit with two resistive loads.

HINT: If there are more then two resistive loads, this method is too time-consuming. There is an easier way we will see later in this chapter.

Calculation: $$R_T = \frac{R_1 \times R_2}{R_1 + R_2}$$

To perform the calculation:

Step 1:

$$R_T = \frac{6\ \Omega \times 8\ \Omega}{6\ \Omega + 8\ \Omega}$$

Step 2:

$$R_T = \frac{48\ \Omega}{14\ \Omega}$$

Step 3:

$$R_T = 3.42857\ \Omega\text{, or } 3.43\ \Omega$$

If there are more than two resistive loads in a parallel circuit (Figure 8–8), and you wanted to use Method 1, use the following process to calculate the resistance of the circuit. (Remember you can do only two resistive loads at a time.) Calculation:

$$R_T = \frac{R_1 \times R_2}{R_1 + R_2}$$

Step 1:

$$R_A = \frac{4\ \Omega \times 3\ \Omega}{4\ \Omega + 3\ \Omega}$$

$$R_A = \frac{12\ \Omega}{7\ \Omega}$$

$$R_A = 1.714\ \Omega$$

Next, use R_A (1.714 ohms) as the first of your two numbers and 6 ohms (R_3) as the second.

Step 2:

$$R_B = \frac{1.714\ \Omega \times 6\ \Omega}{1.714\ \Omega + 6\ \Omega}$$

$$R_B = \frac{10.284\ \Omega}{7.714\ \Omega}$$

$$R_B = 1.333\ \Omega$$

Next, use 1.333 ohms as the first number and 8 ohms (R_4) as the second number.

Step 3:

$$R_T = \frac{1.333\ \Omega \times 8\ \Omega}{1.333\ \Omega + 8\ \Omega}$$

$$R_T = \frac{10.664\ \Omega}{9.333\ \Omega}$$

$$R_T = 1.143\ \Omega$$

Even though Method 1 eventually gets the right answer, other methods work better when a parallel circuit has more than two resistive branches.

Method 2 The **same value method** can be used if all the resistive loads in the parallel circuit are of the

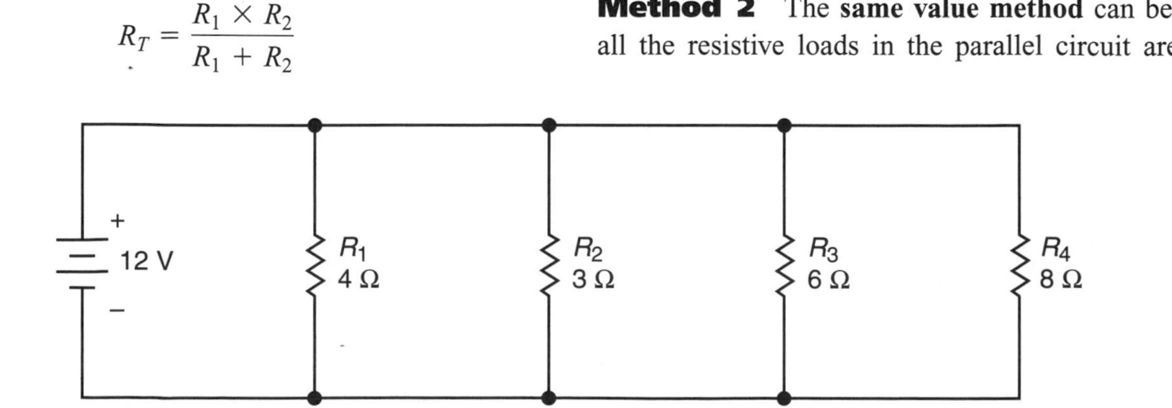

FIGURE 8–8 Parallel circuit with more than two resistive loads.

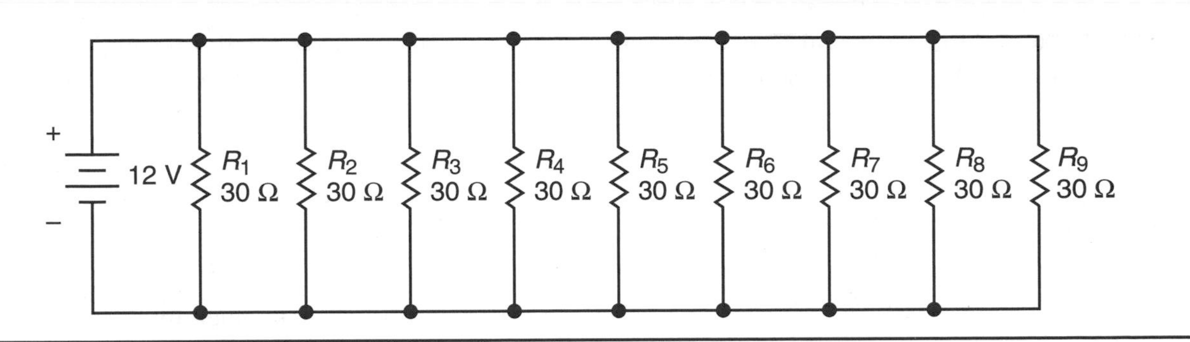

FIGURE 8–9 Parallel circuit with all the same resistance values.

same value. Simply take the value of the resistors and divide by the number of resistors in the circuit. In Figure 8–9, there are nine resistive branches with a 30-ohm load in each branch. Divide 30 by 9 to get R_T, which is 3.33 ohms.

This method applies only when all resistive loads have the same value. If an electrical circuit had some of the loads the same and some different, you could save time by using Method 2 to find the common value of those resistive loads with the same value. Then you could use Method 1 to find the total circuit resistance. This approach takes a considerable amount of time; so we will look at other methods more suited to this type of problem.

Method 3 In the **assumed voltage method,** because the applied voltage is constant in a parallel circuit, we can use Ohm's law to find the resistance. No matter what voltage you use for the calculation, the amperage is proportional to the voltage. The resistance total can be calculated accurately using this method.

For example, use the same four parallel resistances used in Figure 8–8: 4 ohms, 3 ohms, 6 ohms, 8 ohms. And assume a constant voltage of 12 volts. Using Ohms's law, $E \div R = I$. Then:

$$\frac{12\text{ V}}{4\ \Omega} = 3\text{ A}$$

$$\frac{12\text{ V}}{6\ \Omega} = 2\text{ A}$$

$$\frac{12\text{ V}}{3\ \Omega} = 4\text{ A}$$

$$\frac{12\text{ V}}{8\ \Omega} = 1.5\text{ A}$$

Remember that the circuit law for current in a parallel circuit is additive.

$$I_T = I_1 + I_2 + I_3 + \text{etc. } \ldots$$

Add up the answers:

$$3\text{ A} + 4\text{ A} + 2\text{ A} + 1.5\text{ A} = 10.5\text{ A} = I_T$$

Take the assumed voltage total and divide by the amperes total to get resistance total.

$$\frac{12.0\text{ V}}{10.5\text{ A}} = 1.143\ \Omega = R_T$$

R_T equals 1.143 ohms and is the same answer as obtained with Method 1.

Let's try Method 3 again, assuming a source voltage of 38 volts instead to show that this method always works.

$$\frac{38\text{ V}}{4\ \Omega} = 9.5\text{ A} \qquad \frac{38\text{ V}}{3\ \Omega} = 12.67\text{ A}$$

$$\frac{38\text{ V}}{6\ \Omega} = 6.33\text{ A} \qquad \frac{38\text{ V}}{8\ \Omega} = 4.75\text{ A}$$

Add the amperes as before.

$$9.5\text{ A} + 12.67\text{ A} + 6.33\text{ A} + 4.75\text{ A} = 33.25\text{ A} = I_T$$

$$\frac{38.0\text{ V}}{33.25\text{ A}} = 1.143\ \Omega = R_T$$

Note: As you can see, an R_T of 1.143 ohms is the same for both problems. If you used any other value for the voltage, the answer would still come out the same. Remember that voltage and amperage are proportional to each other. Try a few voltage examples of your own to show that the answer will be the same (1.143 ohms).

Method 3 works, but it can be very time-consuming in finding a resistance total. Other methods are better suited for this type of circuit.

Method 4 The **reciprocal formula method** is most often used in textbooks and taught by electrical educators. For consistency, let's use the same values as in the previous parallel circuit example: 4 ohms, 3 ohms, 6 ohms, 8 ohms. The formula for Method 4 is:

$$R_T = \frac{1}{\frac{1}{R_1} + \frac{1}{R_2} + \frac{1}{R_3} + \frac{1}{R_4}}$$

Step 1:

$$R_T = \frac{1}{\frac{1}{4\ \Omega} + \frac{1}{3\ \Omega} + \frac{1}{6\ \Omega} + \frac{1}{8\ \Omega}}$$

At this point, you can calculate the fractions either by finding their lowest common denominator or by converting the reciprocal numbers to their decimal equivalents. Let's do it both ways. First, we will find the lowest common denominator.

Step 2: Fraction style. The lowest common denominator in this problem is 24.

$$1/4 = 6/24 \qquad 1/3 = 8/24$$
$$1/6 = 4/24 \qquad 1/8 = 3/24$$

Step 3:

$$R_T = \frac{1}{\frac{6}{24} + \frac{8}{24} + \frac{4}{24} + \frac{3}{24}}$$

Step 4:

$$R_T = \frac{1}{21/24}$$

Step 5:

$$R_T = 1 \times \frac{24}{21}$$

Step 6:

$$24 \div 21 = R_T = 1.143\ \Omega$$

This is the same answer as in the previous two methods.

Now let's try using Method 4 again, but this time converting the same resistance values to decimals.

Step 1:

$$R_T = \frac{1}{\frac{1}{4\ \Omega} + \frac{1}{3\ \Omega} + \frac{1}{6\ \Omega} + \frac{1}{8\ \Omega}}$$

Step 2: Divide each number into 100 to get decimal equivalents.

$$\frac{100}{4} = 25 \qquad \frac{100}{3} = 33.3$$
$$\frac{100}{6} = 16.67 \qquad \frac{100}{8} = 12.5$$

Step 3: Add up all the decimal answers to find the total decimal value.

$$25 + 33.3 + 16.67 + 12.5 = 87.47$$

Step 4: Because of the numerator 1 over the original calculation (in step 1), we need to divide 87.47 into 100.

$$\frac{100.00}{87.47} = 1.143\ \Omega$$

This is the same answer as before.

Note: As you recall, the *primary* goal when finding the resistance total of a parallel circuit is to get a correct answer the fastest and easiest way possible. Let's do that now.

Method 5 The **scientific calculator method** really performs all the calculations of Method 4, but you are letting the calculator do all the work. If a calculator is available, use it!

Step 1: Identify the 1/X key on your scientific calculator. On some calculators you need to push the 2nd function key first because the 1/X key is on the lower printed portion of the calculator face. Consult your calculator reference guide if you are unsure of location of the 1/X key.

Step 2: Use the same four parallel circuit load values as before: 4 ohms, 3 ohms, 6 ohms, 8 ohms.

Step 3: Press the keys exactly as shown:

4 1/X + 3 1/X + 6 1/X + 8 1/X = 1/X

The answer on the calculator should be 1.1428571. Round off to 1.143 ohms.

Note: This is the *same* answer as we got using the previous methods.

By entering this series of numbers and functions, the calculator performed the reciprocal calculations for you. A common mistake is to forget to push the 1/X key one final time after you push the equals (=) key. No matter how complex the parallel resistance numbers are, the process in Method 5 is still the same. If a circuit has a mixture of ohms (Ω), kiloohms (kΩ), and megaohms (MΩ), then you would need to apply the EXP function key to the process.

For example, suppose the parallel circuit has resistive loads of 12 kiloohms, 2.5 megaohms, 640 ohms, and 275 kiloohms. Enter the keys as follows:

12 EXP 3 1/X + 2.5 EXP 6 1/X +
640 1/X + 275 EXP 3 1/X = 1/X

Your answer should be 606.11 ohms.

That last example made the process look simple. Using the calculator properly makes Method 5 the simplest method when finding the resistance total in a parallel circuit. From now on, all discussions and practice examples in this text dealing with parallel circuit resistance calculations will use primarily Method 5. Having an understanding of parallel circuit laws and methods will enable you to focus more on circuit and electrical diagnosis than on performing complex math calculations.

Let us start with a few easy exercises in parallel circuits to verify that you know the circuit laws and calculations, then move on to more complex circuits with real electrical diagrams. Through repetition of these parallel exercises, you will build a solid foundation of parallel circuit laws that will aid you in diagnosing complex circuits.

Before we continue, though, here is a brief overview of series and parallel circuit laws. Mixing up the different circuit laws could be frustrating; so use this overview as a quick reference guide.

Review of Basic Laws of Electricity

	Series	Parallel
Current	Equal or Same $I_T = I_1 = I_2 = I_3$ = etc.	Additive $I_T = I_1 + I_2 + I_3$ + etc.
Voltage	Additive $E_T = E_1 + E_2 + E_3$ + etc.	Equal or same $E_T = E_1 = E_2 = E_3$ = etc.
Resistance	Additive $R_T = R_1 + R_2 + R_3$ + etc.	Less than $R_T = <$ lowest number of resistor

Exercises—Parallel Circuits

The following exercises are designed to help give you practice using the parallel laws and methods previously explained. The exercises increase in complexity in an effort to test your understanding of the basic parallel laws and to prepare you for the actual hands-on parallel circuit lab exercises that follow.

Exercise 1

Instructions Identify the type of circuit in Figure 8–10. On a *separate* sheet of paper make an answer table like Table 8–1. Using the table, fill in the known circuit values from the parallel circuit diagram. This

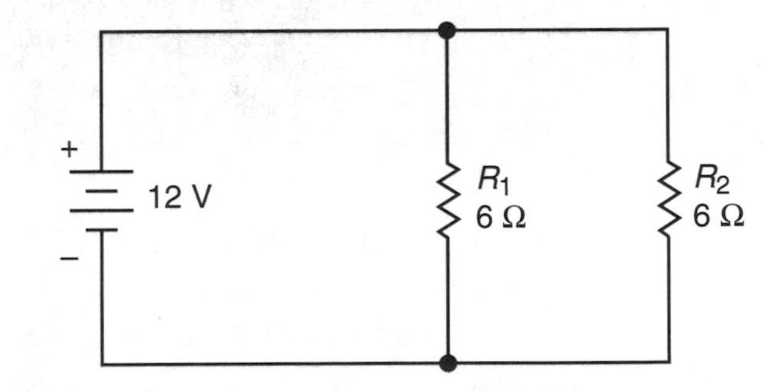

FIGURE 8–10 Parallel circuit.

TABLE 8–1

R_T ______	I_T ______	E_T ______	W_T ______
R_1 ______	I_1 ______	E_1 ______	W_1 ______
R_2 ______	I_2 ______	E_2 ______	W_2 ______

helpful first step gives you an idea of where to begin and what answers you need. Then calculate the answers and write them in the remaining spaces using one of the methods previously learned.

R_T = Resistance total, I_T = Amperage total
E_T = Voltage total, W_T = Watts total

R_1 = Resistor 1, I_1 = Amperage 1
E_1 = Voltage 1, W_1 = Watts 1

R_2 = Resistor 2, I_2 = Amperage 2
E_2 = Voltage 2, W_2 = Watts 2

Circuit Analysis Here are the correct steps and answers for Exercise 1. If your answers are different, refer to the discussion section at the front of this chapter for further study.

Step 1: Known values: E_T = 12 V, R_1 = 6 Ω, and R_2 = 6 Ω. Remember that the voltage law states that the voltage is constant in parallel circuits, so fill in the voltage for E_1 and E_2 with 12 V.

Step 2: To find total resistance, use Method 1, 2, or 5. We use Method 5 because it gives you practice using the calculator's 1/X key. Begin by entering the following into the calculator: 6 1/X + 6 1/X = 1/X. The resistance total should be 3 Ω. Add this to your table.

Step 3: Take the voltage total of 12 V and divide it by the resistance total of 3 Ω to get the amperes total of 4 A. Divide E_1 (12 V) by R_1 (6 Ω) to get I_1, which equals 2 A. Do the same with E_2 and R_2 to get I_2. To verify the correct answer make sure the amps add up to equal the total amperes, or $I_T = I_1 + I_2$.

Step 4: To find the watts total, use Watt's law: Voltage total times amperage total equals the watts total, or 48 W (if unsure, refer back to Chapter 3). Use the same procedure to find W_1 and W_2.

Table 8–2 is a summary of the correct answers.

TABLE 8–2

R_T 3 Ω	I_T 4 A	E_T 12 V	W_T 48 W
R_1 6 Ω	I_1 2 A	E_1 12 V	W_1 24 W
R_2 6 Ω	I_2 2 A	E_2 12 V	W_2 24 W

Note: Follow the steps used in Exercise 1 for the remaining parallel circuit exercises. Remember to put your answers on a *separate* sheet of paper.

Exercise 2

Instructions Make sure you always look at the known values! In Figure 8–11 the voltage was changed from 12 volts to 24 volts. Complete your answers on a table like Table 8–3.

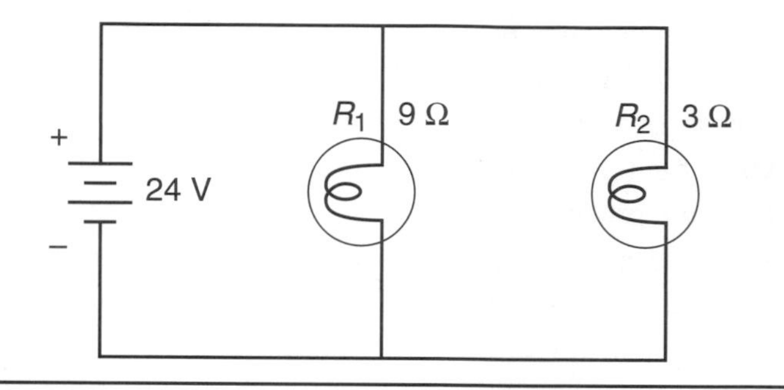

FIGURE 8–11 Simple parallel light circuit.

TABLE 8–3

R_T ____	I_T ____	E_T 24 V	W_T ____
R_1 9 Ω	I_1 ____	E_1 ____	W_1 ____
R_2 3 Ω	I_2 ____	E_2 ____	W_2 ____

Exercise 3

Instructions After you have identified the type of circuit in Figure 8–12, fill in the known values on a table like Table 8–4.

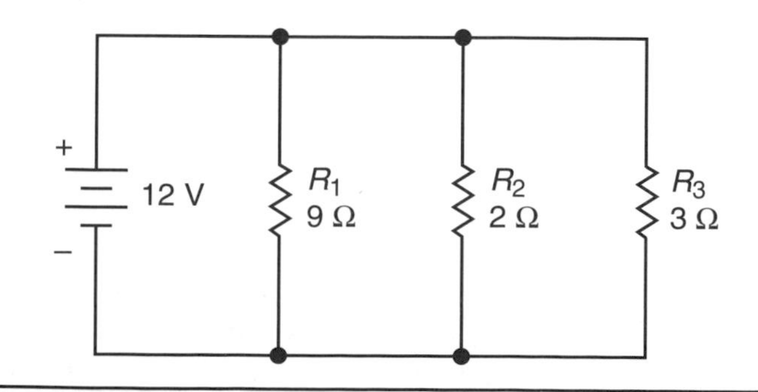

FIGURE 8–12 Simple parallel circuit.

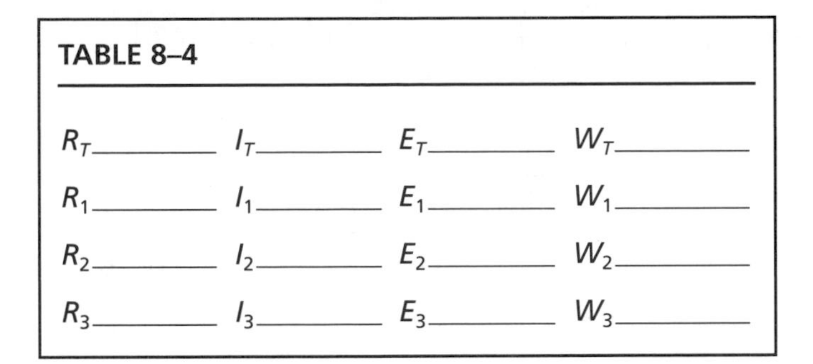

TABLE 8–4

R_T ______	I_T ______	E_T ______	W_T ______
R_1 ______	I_1 ______	E_1 ______	W_1 ______
R_2 ______	I_2 ______	E_2 ______	W_2 ______
R_3 ______	I_3 ______	E_3 ______	W_3 ______

Exercise 4

Instructions The circuit in Figure 8–13 has been drawn differently than in the previous exercises. If you look closely, you can see that the circuit starts at the top, then splits three ways depending on the resistive value of each light bulb. The circuit then recombines to form a common ground at the bottom of the page. Complete the answers to this parallel circuit on a table like Table 8–5.

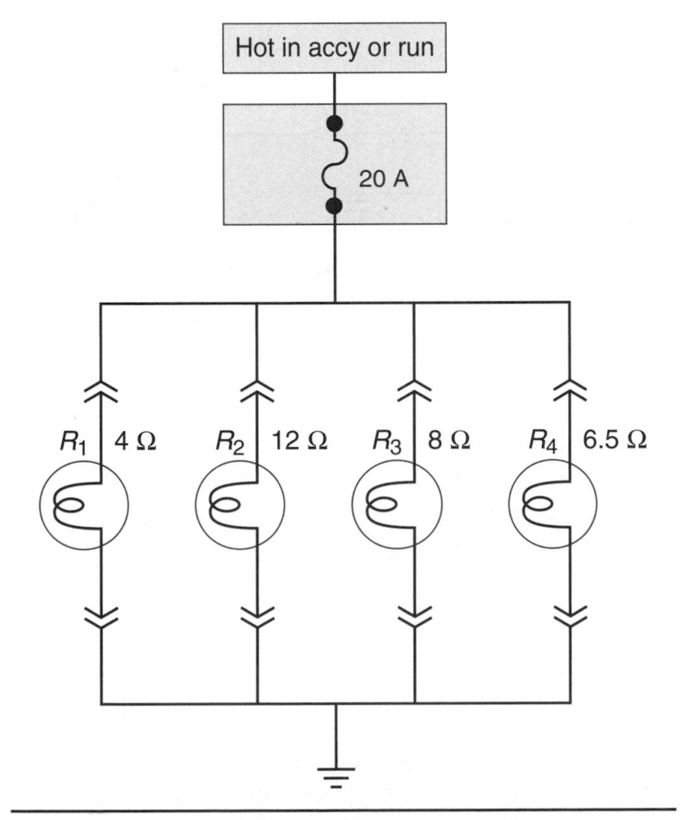

FIGURE 8–13 Typical parallel light circuit.

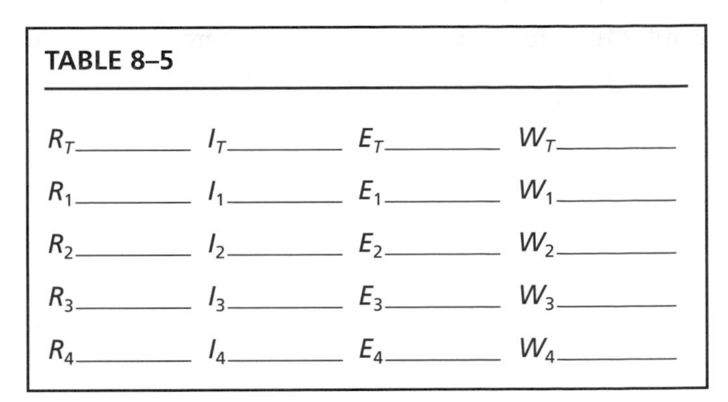

TABLE 8–5

R_T ______	I_T ______	E_T ______	W_T ______
R_1 ______	I_1 ______	E_1 ______	W_1 ______
R_2 ______	I_2 ______	E_2 ______	W_2 ______
R_3 ______	I_3 ______	E_3 ______	W_3 ______
R_4 ______	I_4 ______	E_4 ______	W_4 ______

Exercise 5

Instructions For this exercise the voltage is the unknown property (see Figure 8–14). After you find R_T you can then use I_T and Ohm's law to find the voltages. Once you have all the voltages, you can get the remaining individual amperages. Complete your answers on a table like Table 8–6.

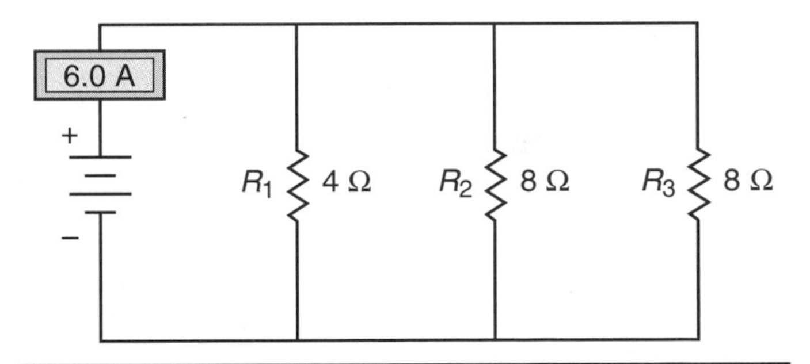

FIGURE 8–14 Parallel circuit with an unknown voltage.

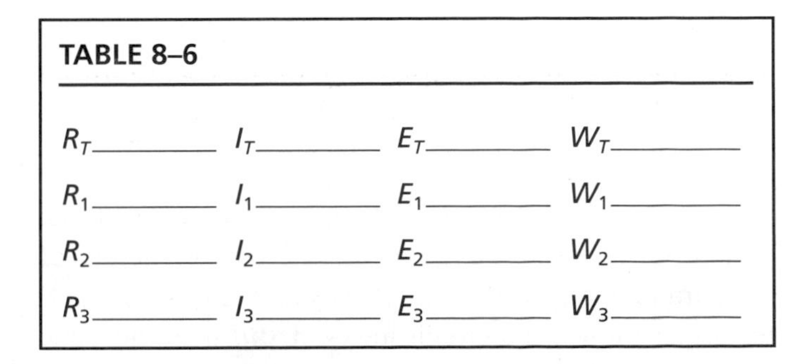

TABLE 8–6

R_T ______	I_T ______	E_T ______	W_T ______
R_1 ______	I_1 ______	E_1 ______	W_1 ______
R_2 ______	I_2 ______	E_2 ______	W_2 ______
R_3 ______	I_3 ______	E_3 ______	W_3 ______

Exercise 6

Instructions Figure 8–15 shows a parallel circuit drawn a little differently. Even if the schematic is not drawn from left to right or from top to bottom, the theory of parallel circuits is the same. Identify the circuit by following the wiring from the positive side of the battery to each resistive branch independently of each other. This can also be done from the negative side of each branch to the battery negative. This circuit can be redrawn in more simplified forms like the three examples in Figure 8–16.

Complete the answers to Exercise 6 by filling in a table like Table 8–7.

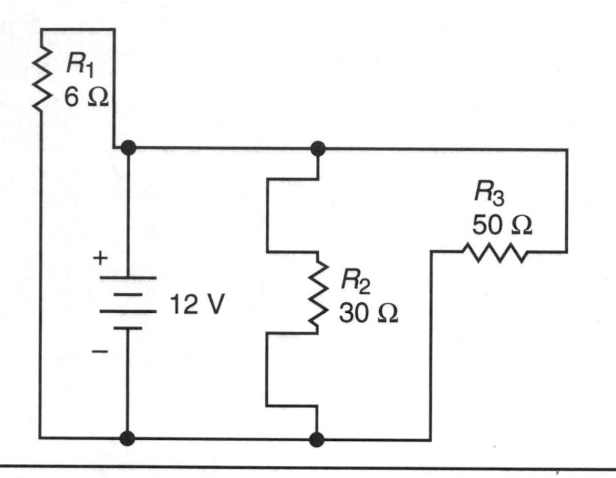

FIGURE 8–15 Simple parallel circuit.

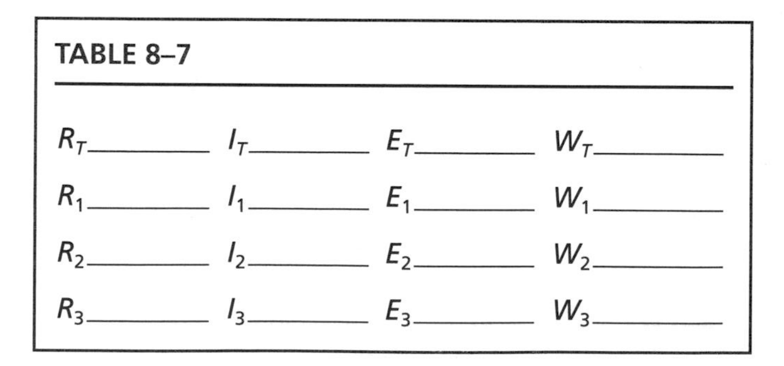

TABLE 8–7

R_T ______	I_T ______	E_T ______	W_T ______
R_1 ______	I_1 ______	E_1 ______	W_1 ______
R_2 ______	I_2 ______	E_2 ______	W_2 ______
R_3 ______	I_3 ______	E_3 ______	W_3 ______

Now that you have practiced with examples of basic parallel circuits, it is time to increase the difficulty of the parallel circuit exercises. Always remember, no matter how complex the circuit looks, it *still* uses the same basic parallel circuit laws. The better you understand the laws applying to parallel circuits, the easier electrical problem solving will become.

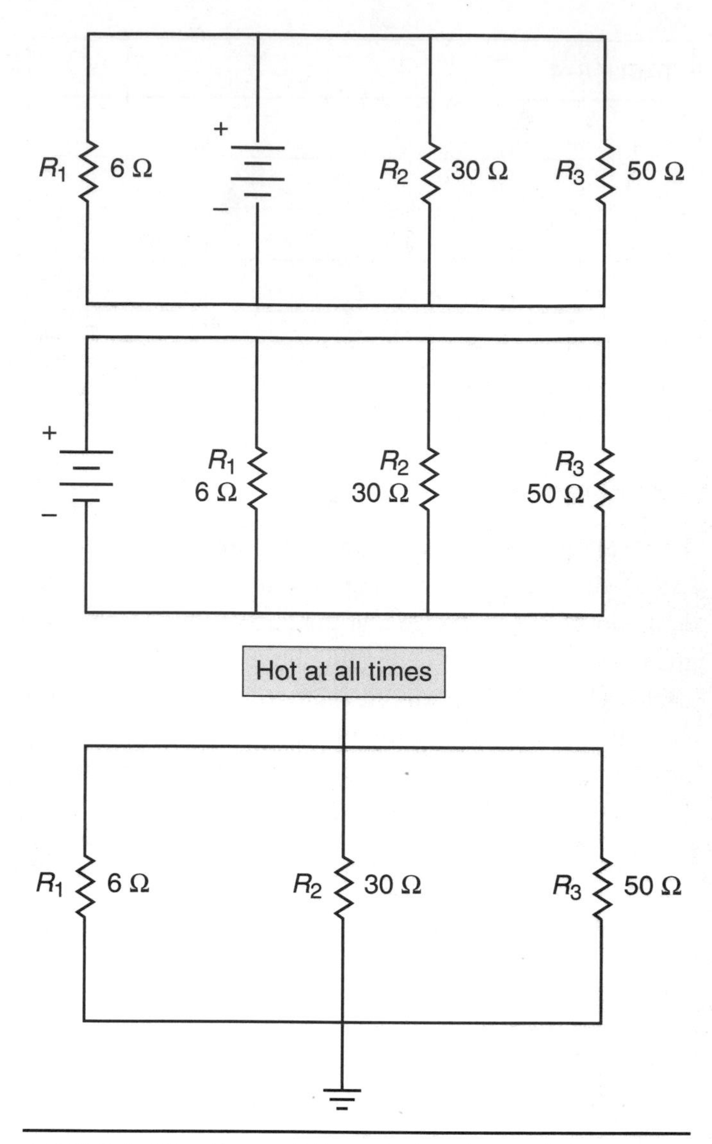

FIGURE 8–16 Three different drawings of the same parallel circuit.

Exercise 7

Identify the type of circuit in Figure 8–17. Using the known values for this circuit from the Table 8–8, complete the answers for this exercise on a *separate* sheet of paper.

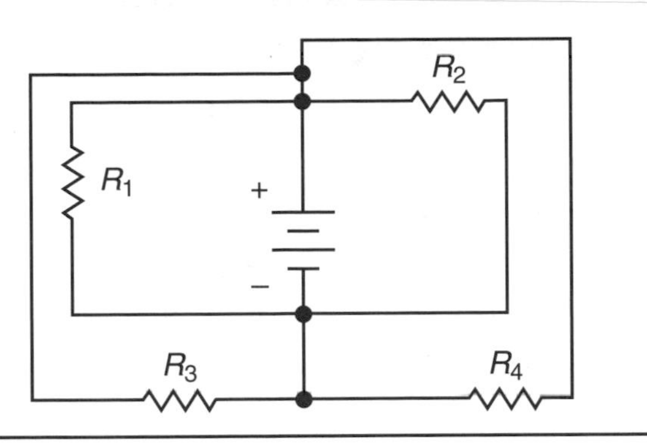

FIGURE 8–17 It looks different, but it is still a parallel circuit.

TABLE 8–8

R_T	I_T 3.75 A	E_T	W_T
R_1	I_1 1.50 A	E_1	W_1
R_2	I_2 1.00 A	E_2	W_2
R_3	I_3	E_3 18 V	W_3
R_4	I_4 .75 A	E_4	W_4

Exercise 8

Using the circuit in Figure 8–18 as a reference, use a *separate* sheet of paper and complete Table 8–9.

Stop If you completed Exercise 8 using the parallel circuit laws, you made an *error!* If you follow the circuit from the battery positive, then back to the battery negative you will see that it is a *series* circuit. Make sure you always know what type of circuit you are working on. If you noticed this right away, congratulations, you are very observant!

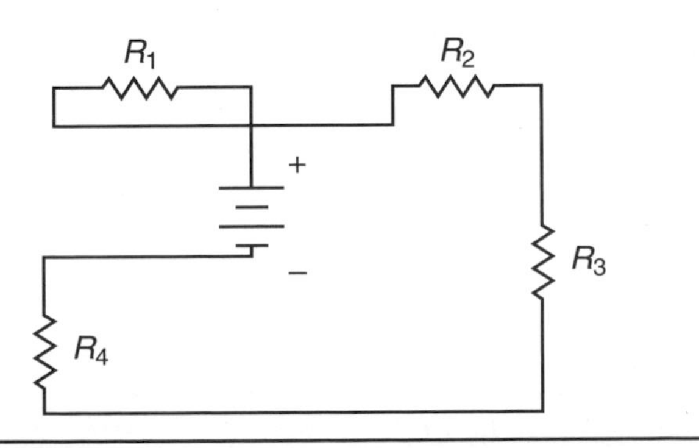

FIGURE 8–18 Identifying unknown electrical circuit.

TABLE 8–9

R_T	I_T	E_T	W_T
R_1 2 Ω	I_1	E_1	W_1
R_2 4 Ω	I_2	E_2 12 V	W_2
R_3 3 Ω	I_3	E_3	W_3
R_4 6 Ω	I_4	E_4	W_4

Exercise 9

The parallel circuit shown in Figure 8–19 has an operating voltage of 14 volts. Because resistance in a bulb cannot be measured directly (due to hot bulb resistance), refer to the operating amperages listed in the Lamp Standard Trade Number Chart in Appendix B. The two high-level stoplight bulbs are GE-194 bulbs, which use 0.27 amperes each at 14 volts. The six stoplights are GE-2057 bulbs, which use 2.297 amperes each at 14 volts (2.10 amperes each at 12.8 volts). On a *separate* sheet of paper, complete the calculations to complete Table 8–10.

Exercise 10

Complete your answers to Exercise 10 on a *separate* sheet of paper using a table similar to Table 8–11. Use 12 volts (as shown in Figure 8–20) as the circuit operating voltage.

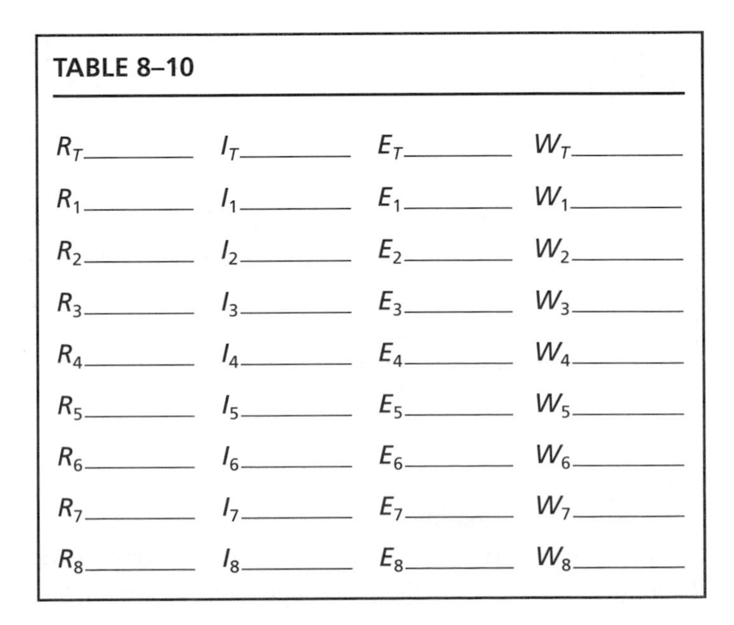

TABLE 8–10

R_T	I_T	E_T	W_T
R_1	I_1	E_1	W_1
R_2	I_2	E_2	W_2
R_3	I_3	E_3	W_3
R_4	I_4	E_4	W_4
R_5	I_5	E_5	W_5
R_6	I_6	E_6	W_6
R_7	I_7	E_7	W_7
R_8	I_8	E_8	W_8

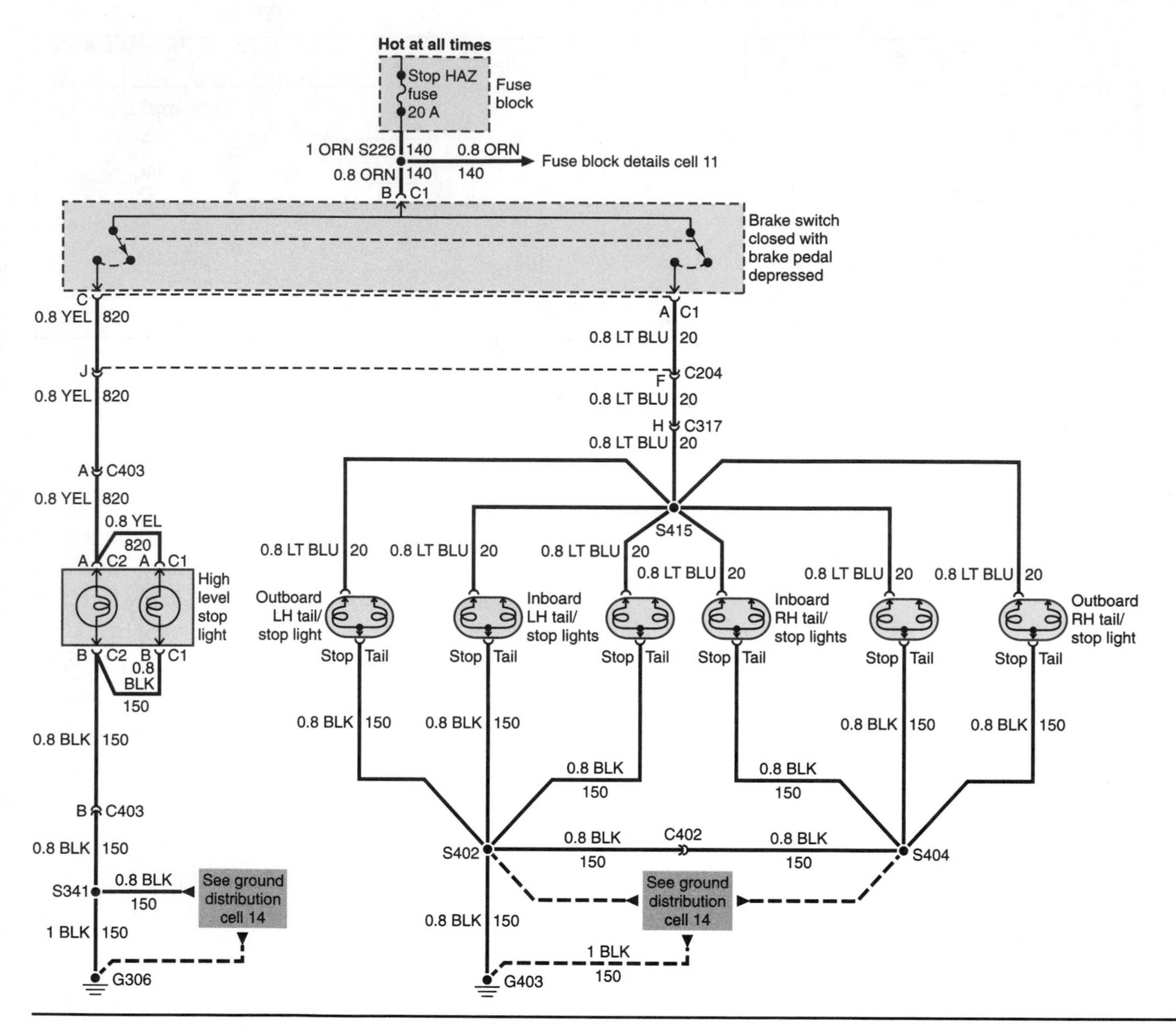

FIGURE 8–19 Exterior stoplight circuit.

Method 5 comes in handy for finding the resistance total in this exercise. Because of the mixture of kiloohms and ohms, enter the following on your calculator to find the resistance total.

6 EXP 3 1/X + 3.5 EXP
3 1/X + 10 EXP 3 1/X + 500

1/X + 6 EXP 3 1/X +
3 EXP 3 1/X 2.5 EXP 3 1/X = 1/X

The answer is 289.66 ohms.

In Exercise 10, what percentage of the current flows through I_2? To answer this, you need to start with the

TABLE 8–11

R_T ____	I_T ____	E_T ____	W_T ____
R_1 ____	I_1 ____	E_1 ____	W_1 ____
R_2 ____	I_2 ____	E_2 ____	W_2 ____
R_3 ____	I_3 ____	E_3 ____	W_3 ____
R_4 ____	I_4 ____	E_4 ____	W_4 ____
R_5 ____	I_5 ____	E_5 ____	W_5 ____
R_6 ____	I_6 ____	E_6 ____	W_6 ____
R_7 ____	I_7 ____	E_7 ____	W_7 ____

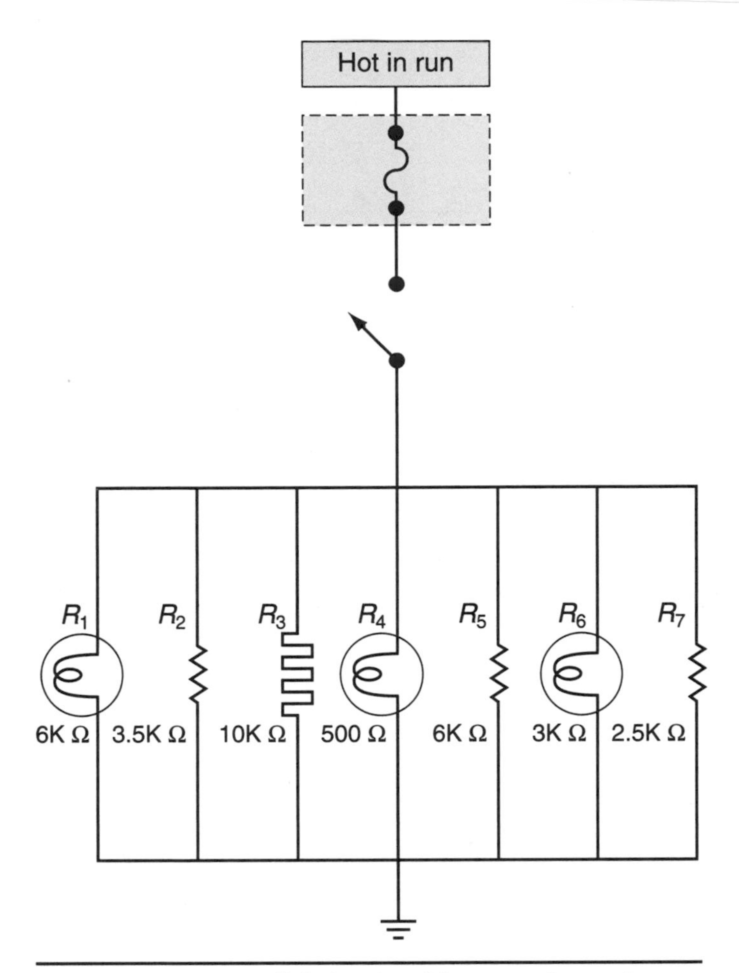

FIGURE 8–20 Parallel circuit with operating voltage of 12 volts.

following calculation: $I_A \div I_T$. You then multiply this result by 100 to get a percentage.

$$\frac{I_2}{I_T} = \frac{0.0034 \text{ A}}{0.0414 \text{ A}} = 0.082 \qquad 0.082 \times 100 = 8.2\%$$

You can determine the remaining six circuit loads by using the preceding calculations.

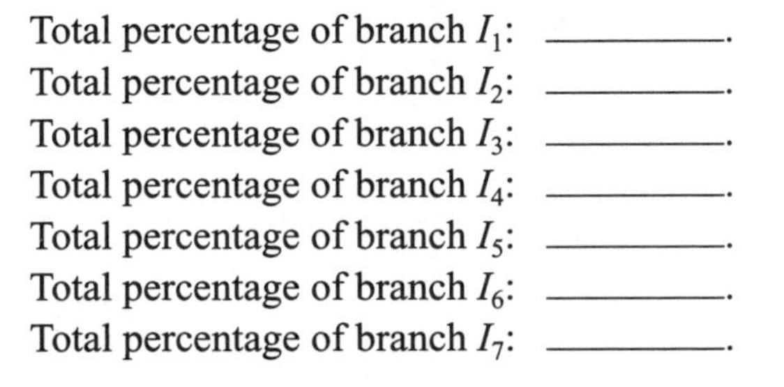

Total percentage of branch I_1: ________.
Total percentage of branch I_2: ________.
Total percentage of branch I_3: ________.
Total percentage of branch I_4: ________.
Total percentage of branch I_5: ________.
Total percentage of branch I_6: ________.
Total percentage of branch I_7: ________.

The total of all the branch percentages should equal 100%. Does it?

In Exercise 10, what percentage of the resistance total is R_2? To answer this, you need to first use the following calculation: $R_T \div R_n$. You then multiply this result by 100 to get a percentage.

$$\frac{R_T}{R_2} = \frac{289.66\ \Omega}{3500\ \Omega} = 0.082 \qquad 0.082 \times 100 = 8.2\%$$

You can determine the remaining six resistive loads by using the above calculations.

Total percentage of branch R_1: ________.
Total percentage of branch R_2: ________.
Total percentage of branch R_3: ________.
Total percentage of branch R_4: ________.
Total percentage of branch R_5: ________.
Total percentage of branch R_6: ________.
Total percentage of branch R_7: ________.

The total of all the branch percentages should equal 100%. Does it? ________.

If you look at the answers to the current and resistance percentage of Exercise 10, you will notice that they are the *same*. The reason for this is that resistance and current in parallel circuits are not constant, but are divided proportionally into their own totals. As a result, you could use either current or resistance to determine the same percentage answers. Remember this for future problem solving.

Exercise 11

HINT: The parallel circuit in Figure 8–21 appears more complicated than it really is. Redraw it, if necessary, to see that each load shares a common positive and that each negative shares its own common. Just remember Ohm's law, Watt's law, and the parallel circuit laws.

Complete your answers to Exercise 11 on a separate sheet of paper similar to Table 8–12.

Exercise 12

HINT: When you look at the number values given in Figure 8–22 and at which ones are similar, keep in mind the circuit laws you have learned and that amperage and resistance are inversely

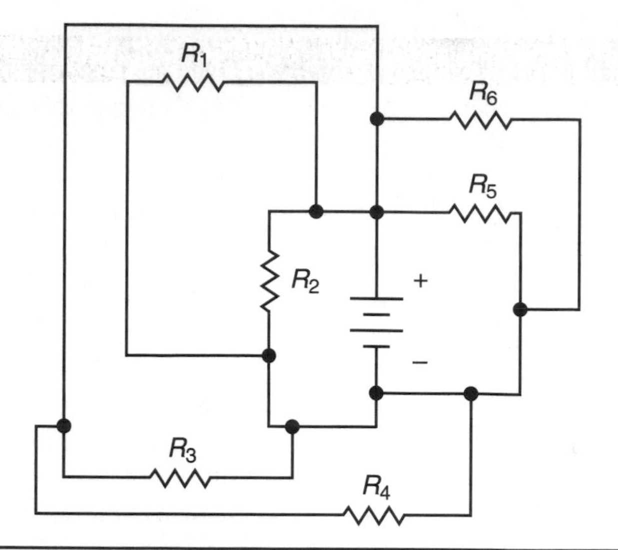

FIGURE 8–21 Complex parallel circuit.

proportional to each other. Complete your answers to this exercise on a *separate* sheet of paper using a table similar to Table 8–13.

You have spent some time learning the parallel circuit theory laws. Even though you most likely will not be standing next to a vehicle electrical problem with a calculator in hand, by performing these parallel exercises, you should have the basic knowledge to know how a parallel circuit is supposed to perform. Just remember to keep the series and parallel circuit laws separate.

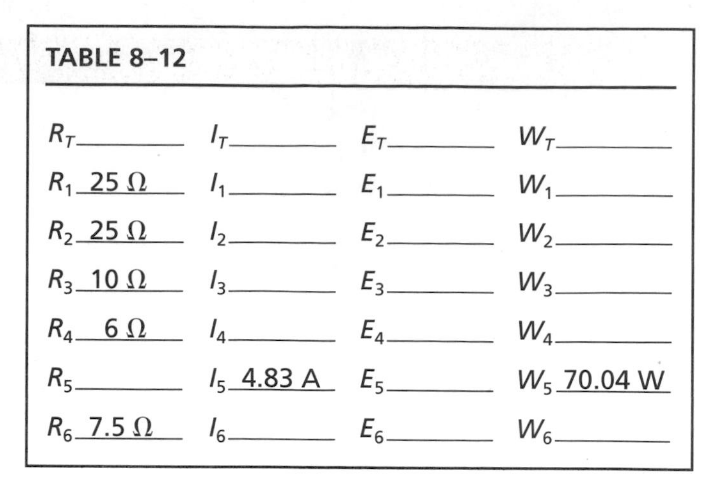

TABLE 8–12

R_T	I_T	E_T	W_T
R_1 25 Ω	I_1	E_1	W_1
R_2 25 Ω	I_2	E_2	W_2
R_3 10 Ω	I_3	E_3	W_3
R_4 6 Ω	I_4	E_4	W_4
R_5	I_5 4.83 A	E_5	W_5 70.04 W
R_6 7.5 Ω	I_6	E_6	W_6

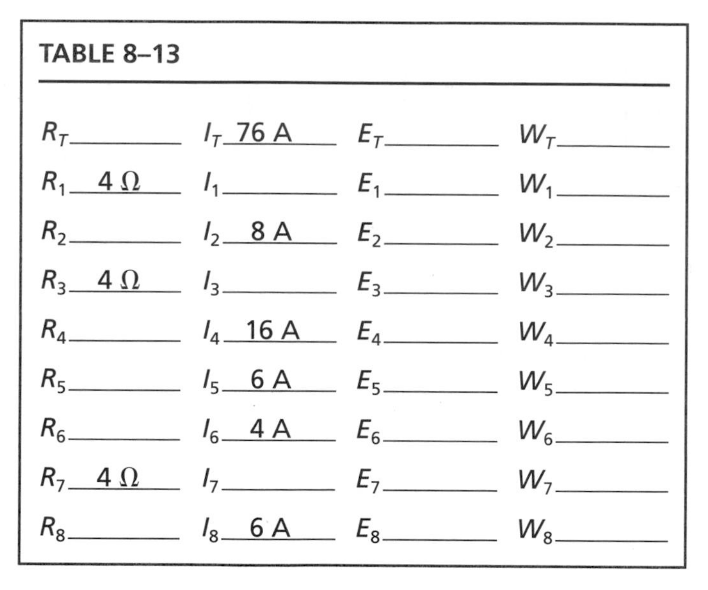

TABLE 8–13

R_T	I_T 76 A	E_T	W_T
R_1 4 Ω	I_1	E_1	W_1
R_2	I_2 8 A	E_2	W_2
R_3 4 Ω	I_3	E_3	W_3
R_4	I_4 16 A	E_4	W_4
R_5	I_5 6 A	E_5	W_5
R_6	I_6 4 A	E_6	W_6
R_7 4 Ω	I_7	E_7	W_7
R_8	I_8 6 A	E_8	W_8

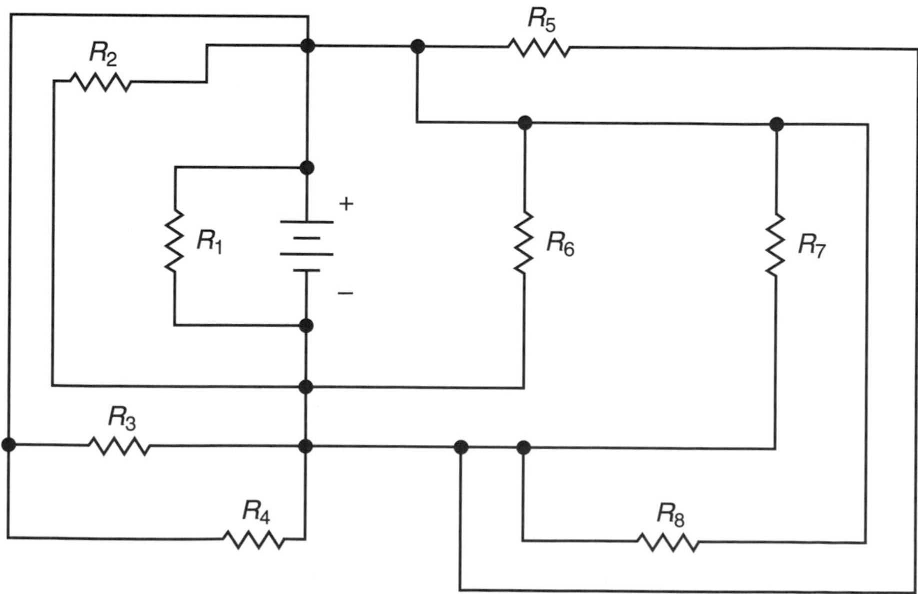

FIGURE 8–22 Complex parallel circuit.

CASE STUDIES

CASE 1

The following case study is typical of a live parallel circuit problem that technicians often face. Knowledge of parallel circuits can help you perform the diagnosis quickly and efficiently.

Customer Complaint

The vehicle's parking lamps and taillamps work fine when not hooked up to a camping trailer. As soon as the owner hooks up a trailer, the 10-amp taillamp fuse blows. The customer thinks there must be a short circuit in the trailer lights and wants it fixed. Figure 8–23 indicates the location of the lights on the vehicle and trailer.

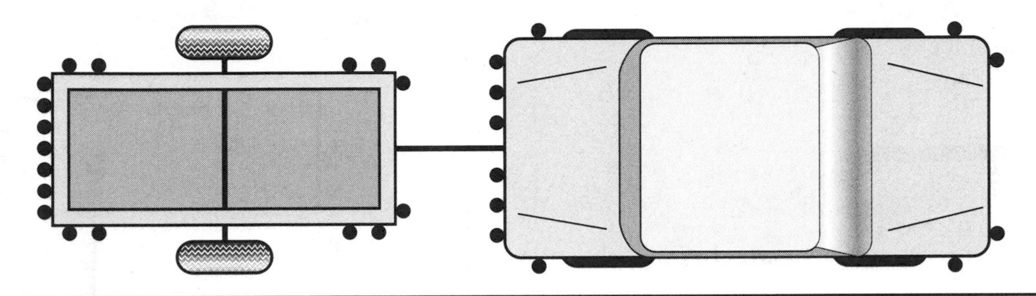

FIGURE 8–23 Vehicle and trailer lighting layout.

Unless you have had considerable experience with exactly the same circuits, first assumptions about the wiring layout could lead to a misdiagnosis. Figure 8–24 provides a clearer picture for the technician to determine which type of circuit it is and which circuit laws apply.

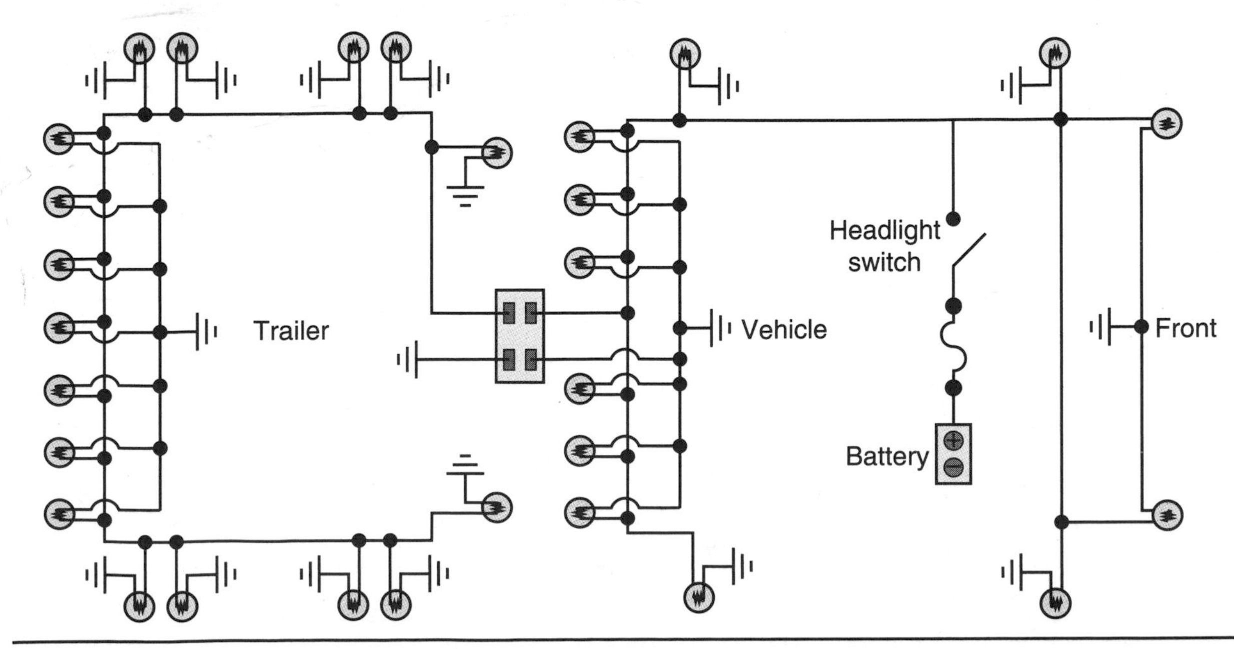

FIGURE 8–24 Vehicle and trailer schematic diagram.

Known Information

- ❑ The vehicle operating voltage equals 14 volts.
- ❑ The vehicle's front parking lamps (#1157 bulb) have a design amperage of 0.59 amperes each and a hot bulb resistance of 23.73 Ω each.
- ❑ The vehicle's side marker lamps (#194 bulb) have a design amperage of 0.27 amperes each and a hot bulb resistance of 51.85 Ω each.
- ❑ The vehicle's rear taillamps (#1157 bulb) have a design amperage of 0.59 amperes each and a hot bulb resistance of 23.73 Ω each.
- ❑ The trailer's side and front marker lamps (#57 bulb) have a design amperage of 0.24 amperes each and a hot bulb resistance of 58.33 Ω each.
- ❑ The trailer's rear taillamps (#1157 bulb) have a design amperage of 0.59 amperes each and a hot bulb resistance of 23.73 Ω each.

Circuit Analysis

Answer the following questions on a *separate* sheet of paper:

1. What is the total amperage of the vehicle's parking lamp/taillamp circuit? ______________________.
2. What is the total hot-bulb resistance of the vehicle's parking lamp/taillamp circuit? ______________.
3. Is the 10-ampere fuse designed for the circuit suitable for the vehicle's amperage load only?

   ____.
4. What is the hot-bulb resistance of the trailer parking lamp/taillamp circuit? ____________________.
5. With the trailer hooked to the vehicle, what is the total resistance of the parking lamp/taillamp circuit? ____.
6. What is the amperage total of the circuit with the trailer hooked to the vehicle? ________________.
7. Is the rated 10-ampere circuit fuse satisfactory for the vehicle and trailer combined? ____________.
8. What are at least three ways to repair this problem?
   1. ____.
   2. ____.
   3. ____.
9. What circuit laws were helpful when troubleshooting this system or when explaining the problem to the customer?

   ____.
10. What is your analysis of Case Study 1? ____

    ____

    ____.

CASE 2

The backup lights in a modern sedan do not work (Figure 8–25).

Because this is a sedan, the station wagon part of the diagram should be disregarded. If you were to do a quick analysis of the diagram, you would see the following:

1. The turn signal flasher shares the fuse as the backup lights.
2. The backup lights should work only in the reverse position.
3. All the backup lights share a common ground at G403.
4. The voltage to the backup bulbs should be the source voltage of 12 volts.
5. There are other circuits that share the G403 common ground.
6. A further study of the ground distribution chart in Figure 8–26 shows all common links at G403 ground.

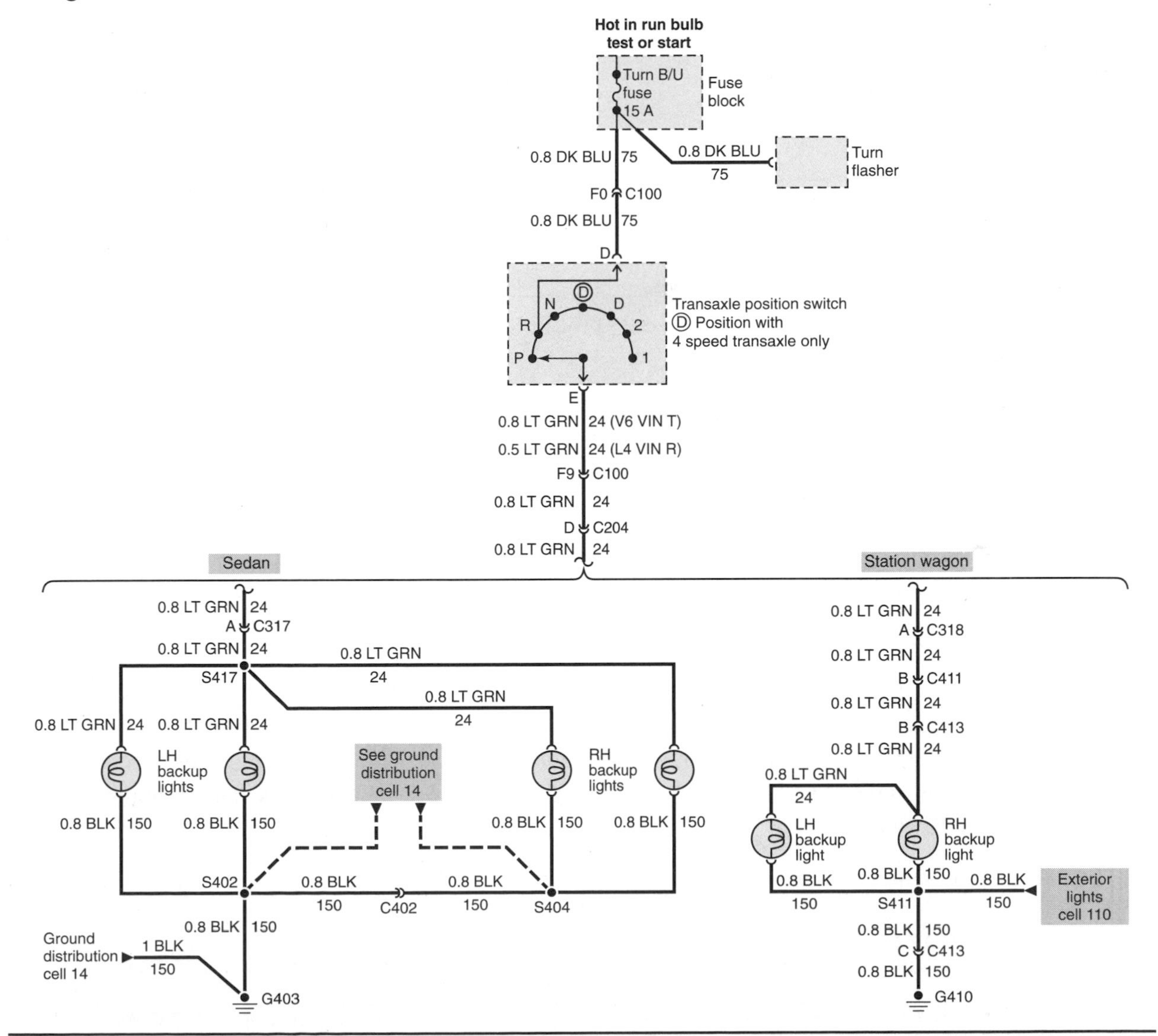

FIGURE 8–25 Backup light circuit.

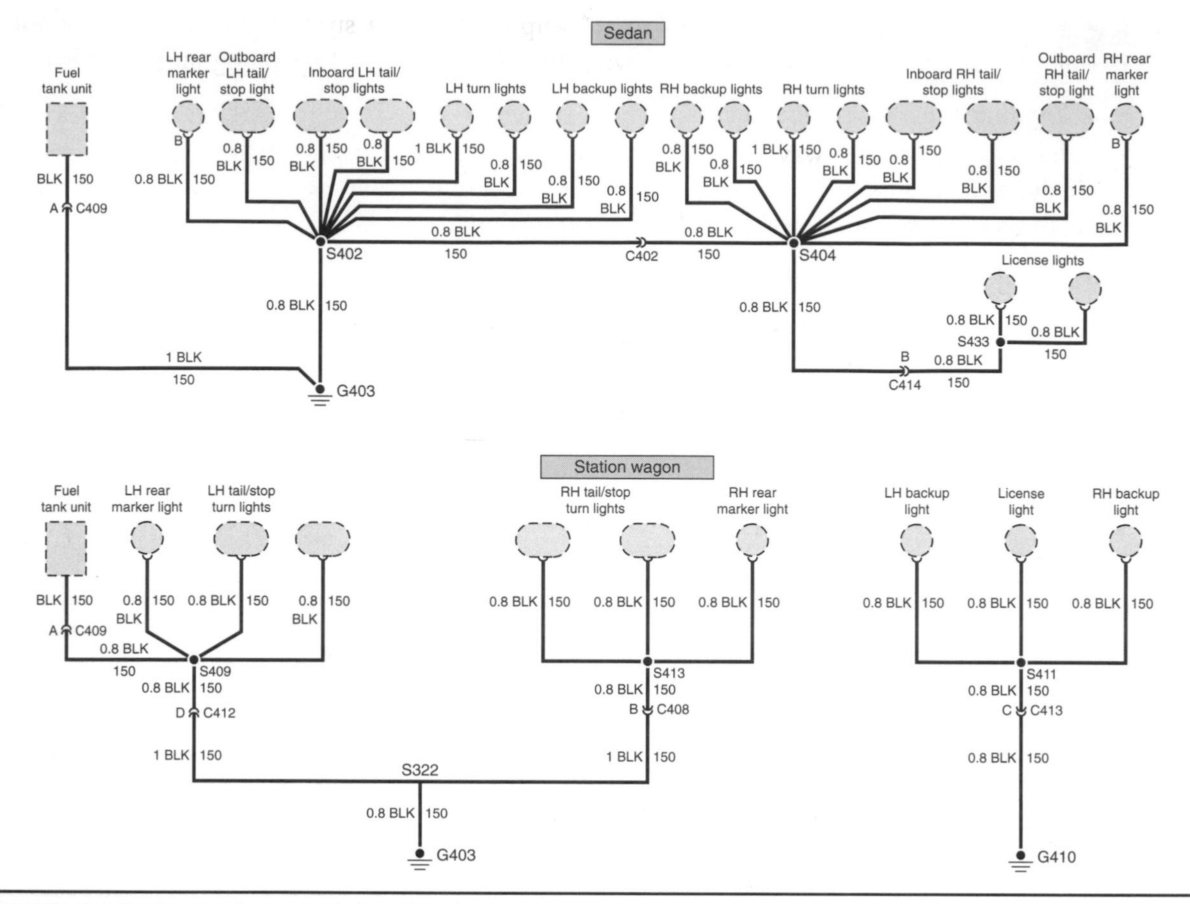

FIGURE 8–26 Typical ground distribution system.

To quickly isolate the problem, you could refer to the wiring diagrams in Figure 8–25 and Figure 8–26, and do the following.

1. Operate the turn signals. If they work, the fuse is OK.
2. If the taillights and fuel gauge work fine, the G403 ground is OK.
3. Put transmission selector in reverse, there should be 12 volts at each backup light.

More extensive checks would consist of the following.

1. Check for voltage at the transaxle position switch (terminals D and E). There should be source voltage at both connection points. If there is, the switch is OK. If there is voltage only at the inlet side (D), then suspect a bad switch. The switch can be double checked with an ohmmeter by disconnecting both leads to the switch, and hooking the ohmmeter across the switch contacts. There should be continuity through the switch.
2. If there is source voltage at the backup light bulbs, check the bulbs to see if all are burned out. It could happen!
3. If the bulbs are OK, check the circuit ground wiring leading up to the G 403 connection.
4. If the bulbs were a #1156 bulb with a designed amperage of 2.10 amperes, what would the circuit amperage be if all four bulbs were operating as designed?

5. Knowing the theoretical amperage total is helpful in case a suspected resistance at a connection point may cause a voltage drop. Removing the circuit fuse and installing an ammeter in its place could tell you the total circuit amperage.

6. What are at least three ways to quickly isolate a backup light circuit problem? ____.

7. What is your analysis of this study? ____.

CASE 3

The right rear and left rear side marker lamps do not work on a sedan (Figure 8–27). The circuit begins at the top of Figure 8–27 at point C which is a continuation of another circuit on "page 92-1" (Figure 8–28).

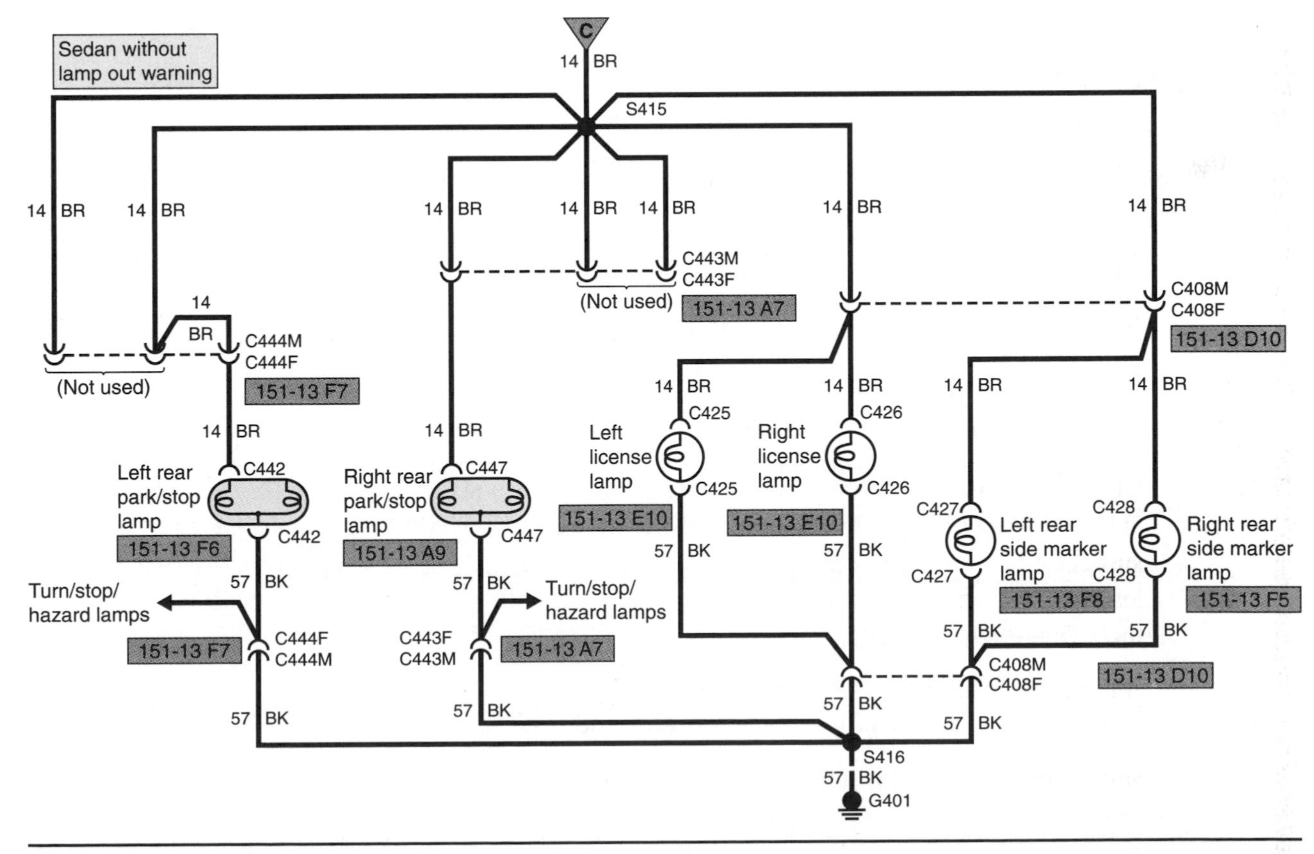

FIGURE 8–27 Exterior lamp circuit.

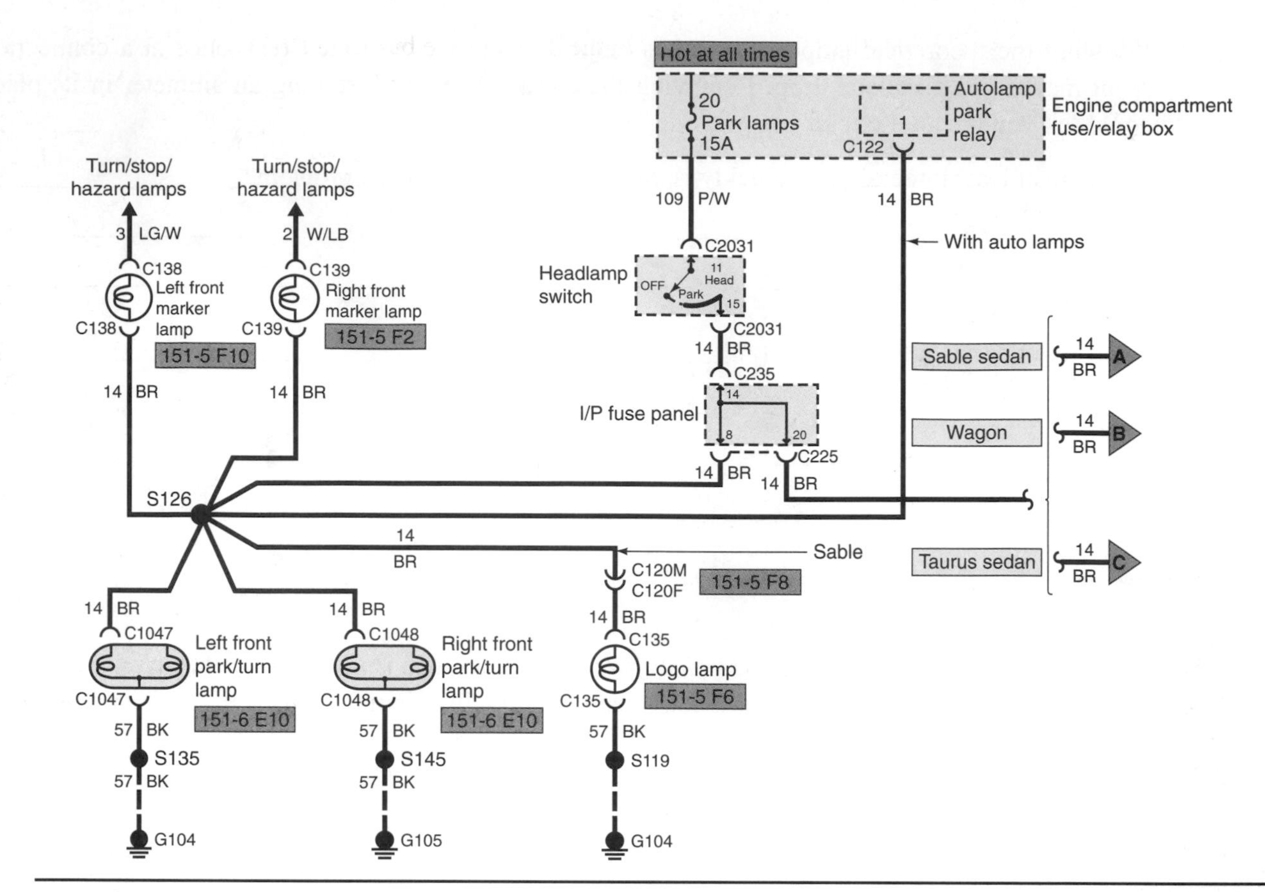

FIGURE 8–28 Exterior lamp circuit.

If you were to do a quick analysis of the diagram, you would see the following.

1. The source power starts at the 15-ampere park lamp fuse and goes through the headlamp switch. The circuit then splits in the I/P fuse panel.
2. One part of the circuit goes to the front lights, the other part goes to point C, where the circuit we need begins.
3. The circuit divides into seven parallel subsections at the S415 connection. Three of these connections are not used in this model vehicle. The remaining subsections go to lamps.
4. Two of the subsections further divide into parallel branch circuits.
5. All the lamps recombine at S416 and then share a common ground at G401.

To quickly isolate the problem, you could do the following.

1. Making sure all the remaining lamps work in the circuit would tell you that there is power coming in at point C and that the grounds at G401 are good.
2. Since the left rear and right rear side marker lamps both work off the same parallel branch, check the connection at point C408 coming into the bulbs.
3. If there is power at C408, check the C408 ground side connection.

4. If both of these are OK, then the problem is in the bulbs or the bulb sockets.

What is your analysis of this study? ______________________________

____

______________________________.

Hands-On Vehicle Tests

The following two hands-on vehicle checks are included in the NATEF (National Automotive Technician Education Foundation) Task List. Complete your answers to the following questions on a *separate* sheet of paper.

Performance Task 1

Task Description

Check the operation of the brake stoplight system; determine the necessary action.

Task Objectives

❑ Obtain a vehicle that can be used for this task. What model and year of vehicle are you using for this task? ______________________________

❑ Find a wiring diagram that shows the stoplight circuit for this vehicle.

❑ Does the stoplight circuit follow the characteristics of a parallel circuit? ______________________________

❑ Identify the following components and list their location on the vehicle.

1. Left stoplight bulb(s): ______________________________.
2. Right stoplight bulb(s): ______________________________.
3. Stoplight switch: ______________________________.
4. Circuit fuse and fuse number: ______________________________.
5. Circuit ground(s): ______________________________.
6. Other circuit components: ______________________________

______________________________.

Circuit Checks

❑ Check the operation of the stoplight switch. Disconnect wire connections at the switch. Connect a DVOM, set on continuity, to the switch terminals and then depress the brake pedal. What happens?

____

❑ Remove the stoplight fuse and connect the DVOM, in series, in place of the fuse. What is the total amperage of the circuit? ______________________________

- ❑ Although not a normal diagnostic procedure, leave the DVOM connected up to read amperage so that you can verify that Kirchoff's law, which you used in the previous parallel exercises, is true. Remove each stoplight bulb and record the amperage difference between the old amperage total and each new total amperage. Replace each bulb before removing the next.

  Bulb 1: ________ Bulb 2: ________ Bulb 3: ________ Bulb 4: ________

  Bulb 5: ________ Bulb 6: ________ Bulb 7: ________ Bulb 8: ________

- ❑ Do the totals of all branch amperages just measured equal the circuit amperage total?

  ____

- ❑ If amperages do not add up to the same as the total, where are the remainder amps going?

  ____

Task Summary

After performing the above NATEF task, what can you determine will be helpful in future stoplight circuit diagnosis? ____

____

____.

Performance Task 2

Task Description

Diagnose incorrect heated glass, mirror, or seat operation; determine the necessary action.

Task Objectives

- ❑ Obtain a vehicle that can be used for this task. What model and year of vehicle are you using for this task? ____.
- ❑ Find a wiring diagram that shows the rear window heated glass circuit for this vehicle.
- ❑ Does this heated glass circuit follow the characteristics of a parallel circuit? ________________.
- ❑ Identify the following components and list their location on the vehicle.

  1. Rear window grid: ________________________________.
  2. Control switch: ________________________________.
  3. Relay (if applicable): ________________________________.
  4. Circuit fuse and fuse number: ________________________________.
  5. Circuit ground(s): ________________________________.
  6. Circuit indicator light: ________________________________.
  7. Circuit timer assembly (if applicable): ________________________________.
  8. Other circuit components: ________________________________.

Circuit Checks

- ❑ Check the operation of the indicator light. If operating, proceed to rear window heating grid. If the indicator is not operating, check to see if the rear window grid feels warm. If not, is the circuit fuse OK? ______________________________.
- ❑ Check for circuit source voltage at the rear window heater grid. DVOM voltage reading: __________.
- ❑ What is the voltage from the positive end of the heating grid to the negative end? ______________.
- ❑ If you leave the DVOM connected to the negative side and slowly move the positive lead along a grid line, what happens to the voltage readings? ____________________. Why? ______________.
- ❑ Why are rear window grid lines wired in parallel? ______________________________.
- ❑ Remove the circuit fuse and install the DVOM, in series, in its place. What is the amperage of the circuit? ______________________________.
- ❑ What happens to the total circuit amperage as the rear heating grid starts to heat up? ____________.

Task Summary

After performing the above NATEF task, what can you determine will be helpful in future rear window heating grid circuit diagnoses? ______________________________

____

______________________________.

Summary

- ❑ In a parallel circuit there is more than one path for the flow of electrons to travel.
- ❑ In a parallel circuit, the voltage is the same at any point in the circuit.
- ❑ In a parallel circuit, the branch amperages are added to equal the total amperage.
- ❑ In a parallel circuit, the resistance total is always less than the lowest-value resistive device in the circuit.
- ❑ The product over sum method can be used to compute the parallel resistance total for two resistances at a time.
- ❑ The same value method can be used to compute the parallel resistance total for multiple resistors with the same value.
- ❑ The assumed voltage method can be used to compute the parallel resistance total for circuits when the voltage is unknown.
- ❑ The reciprocal formula method is the mathematical formula for computing the parallel resistance total when there are multiple resistors with different values.
- ❑ The scientific calculator method can be used to compute parallel resistance total using the 1/X key on a scientific calculator.

Key Terms

assumed voltage method
Kirchoff's current law
Kirchoff's voltage law
parallel circuit
parallel circuit laws
product over sum method
reciprocal formula method
same value method
scientific calculator method

Review Questions

Short Answer Essays

1. Describe the circuit law for voltage in a parallel circuit.
2. Describe the parallel circuit law for amperage in a parallel circuit.
3. Describe the parallel circuit law for resistance in a parallel circuit.
4. Explain how to use the 1/X feature on a scientific calculator when computing resistance total in a parallel circuit.

Fill in the Blanks

1. The amperage in a load component in a parallel circuit is ________________ to ________________ the total amperage.
2. In a parallel circuit, the voltage ________________ is the ________________ as the voltage ________________ each branch load in the circuit.
3. When current ________________ in a parallel circuit increases, the resistance ________________.

ASE-Style Review Questions

1. Technician A says that in a parallel circuit the voltage across each circuit load adds up to the total circuit voltage. Technician B says the amperages of each parallel circuit load are equal because they share a common positive and a common negative. Who is correct?
   A. A only
   B. B only
   C. Both A and B
   D. Neither A nor B

2. In the following parallel light circuit the bulbs do not light when the switch is closed. Technician A says the switch can be bad. Technician B says both light bulbs could be burned out. Who is correct?
   A. A only
   B. B only
   C. Both A and B
   D. Neither A nor B

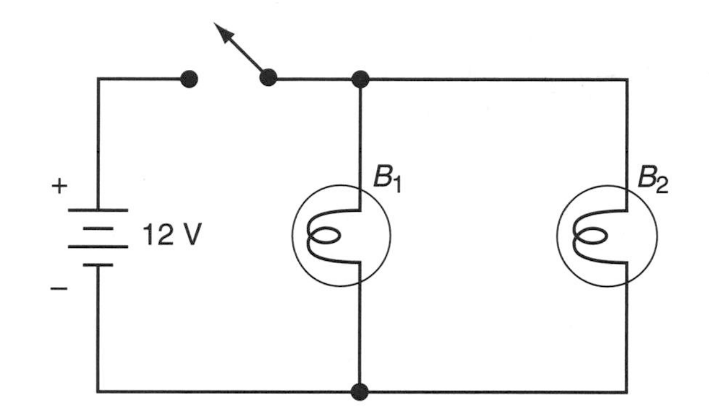

3. A circuit has the following four fixed resistors wired in parallel: 10 Ω, 5 Ω, 10 Ω, and 2.5 Ω. What is the resistance total of the circuit?
   A. 27.5 Ω
   B. 1.25 Ω
   C. 125 Ω
   D. None of the above

4. Technician A says the resistance total in a parallel circuit is always less then the smallest resistive branch in the circuit. Technician B says the parallel circuit laws are the same as series circuit laws. Who is correct?
   A. A only
   B. B only
   C. Both A and B
   D. Neither A nor B

5. Technician A says that adding resistive branches to a parallel circuit increases the total circuit amperage. Technician B says that, if a parallel branch is opened, the circuit source voltage is lowered. Who is correct?
   A. A only
   B. B only
   C. Both A and B
   D. Neither A nor B

6. In a rear window heating grid, some of the grid lines are not working. Technician A says the grid lines have a break in them creating an open. Technician B says some of the grid line will continue to work because they are wired in parallel. Who is correct?
   A. A only
   B. B only
   C. Both A and B
   D. Neither A nor B

7. In the circuit below, the lights do not operate, but the relay can be heard making a clicking sound when the switch is energized. Technician A says the relay could be bad. Technician B says the bulb ground could be bad. Who is correct?
   A. A only
   B. B only
   C. Both A and B
   D. Neither A nor B

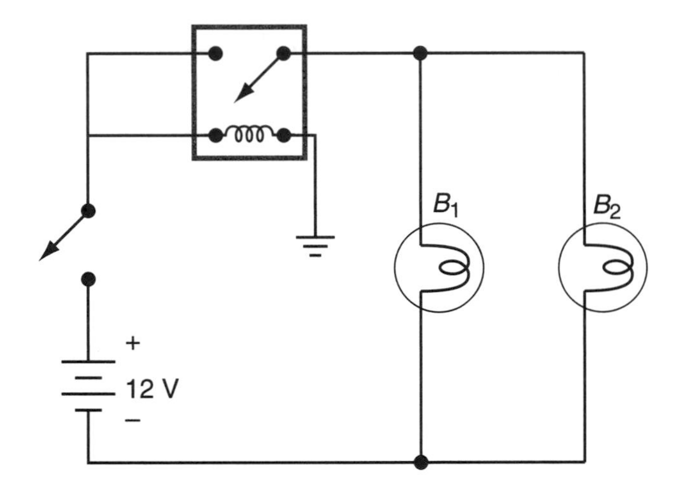

8. Technician A says the proper way to find the resistance of a hot bulb parallel circuit is to measure the amperage and voltage of the circuit, then calculate the circuit resistance total. Technician B says that the hot bulb resistance total of a parallel circuit can be determined by taking the hot bulb resistance of each circuit bulb and finding the reciprocal of the sum of each parallel branch. Who is correct?
   A. A only
   B. B only
   C. Both A and B
   D. Neither A nor B

9. A vehicle has six taillights wired in parallel with 0.75 ampere per bulb. When a trailer is hooked up to the vehicle, an additional four lights with 0.50 ampere per bulb is added to the circuit. What is the amperage total of this parallel circuit?

   A. 4.5 amperes total

   B. 6.0 amperes total

   C. 1.25 amperes total

   D. None of the above

10. A total of 2.4 mA is flowing through a circuit that consists of three resistors in parallel: 20 kΩ, 20 kΩ, and 10 kΩ. Technician A says the source voltage of the circuit is 1.2 V DC. Technician B says the voltage is 120 V DC. Who is correct?

    A. A only

    B. B only

    C. Both A and B

    D. Neither A nor B

Series/Parallel Circuits

Introduction

In this chapter you will apply the individual series and parallel circuit laws learned in previous chapters to a combination circuit consisting of some components connected in series and some in parallel. Series/parallel electrical circuits seem complicated, but they are in fact fairly simple to understand if you remember which circuit laws apply to each circuit load component. Most vehicle electrical circuits used today contain several series/parallel circuits or portions of a series/parallel circuit.

Objectives

When you complete this chapter you should be able to:

- ❑ Define a series/parallel electrical circuit.
- ❑ Identify the individual series and parallel circuit loads of a compound series/parallel circuit.
- ❑ Perform series/parallel circuit calculations for voltage, current, and resistance.
- ❑ Apply troubleshooting and testing techniques to series/parallel circuits.
- ❑ Perform basic hands-on series/parallel exercises on the vehicle.

Series/Parallel Circuit Explanation

A **series/parallel circuit** is defined as a circuit that contains both series circuits and parallel circuits. This type of circuit is also known as a **combination circuit** (Figure 9–1). The simple circuit in Figure 9–1 has resistor 1 (a) in series from the battery, then splits into two parallel branches [resistor 2 (b) and resistor 3 (c)] before recombining and returning to the battery.

No specific law or formula pertains to the whole series/parallel circuit for voltage, amperage, and resistance. Instead, you must determine which branch loads of the circuit are in series and which are in parallel, simplify the circuit where possible, and use the circuit laws that apply to each of these branches to find the

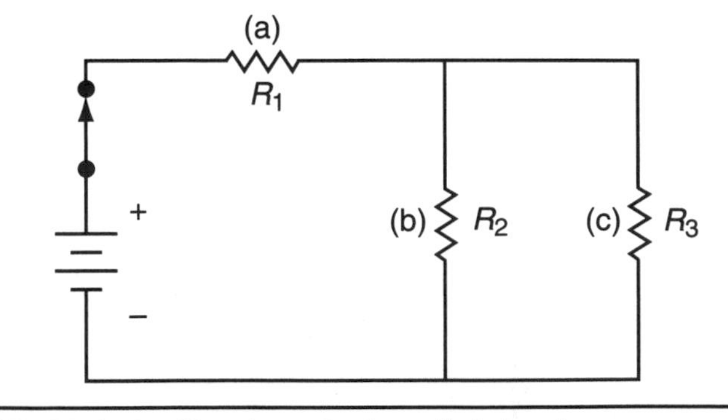

FIGURE 9–1 Simple series/parallel circuit.

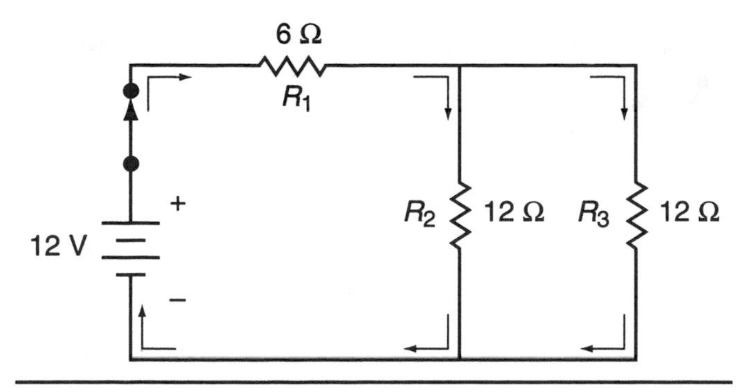

FIGURE 9–2 Simple series/parallel circuit.

value totals. Figure 9–2 is an example of a series/parallel circuit with resistor 1 (6 ohms) in series, and resistors 2 and 3 (12 ohms each) in parallel.

To find the **series/parallel resistance** total for the circuit in Figure 9–2, perform the following steps.

1. Resistance total equals resistor 1 + the reciprocal sum of resistor 2 and resistor 3.
2. $R_T = R_1 +$ (reciprocal sum of R_2 and R_3)
3. $R_1 = 6\ \Omega$
4. R_2 and $R_3 = 12$ 1/X + 12 1/X = 1/X
5. $R_T = 6\ \Omega$ (series) + $6\ \Omega$ (sum parallel)
6. $R_T = 12\ \Omega$

As you can see by these steps, it was necessary to simplify the series/parallel circuit to find the resistance total. Figure 9–3 shows how the circuit is eventually converted to a series-only circuit to calculate the resistance total. Remember that the series circuit law for resistance, $R_T = R_1 + R_2 + R_3 +$ etc. is additive; so once the sum of the parallel branch loads is calculated, it is reduced to a single value.

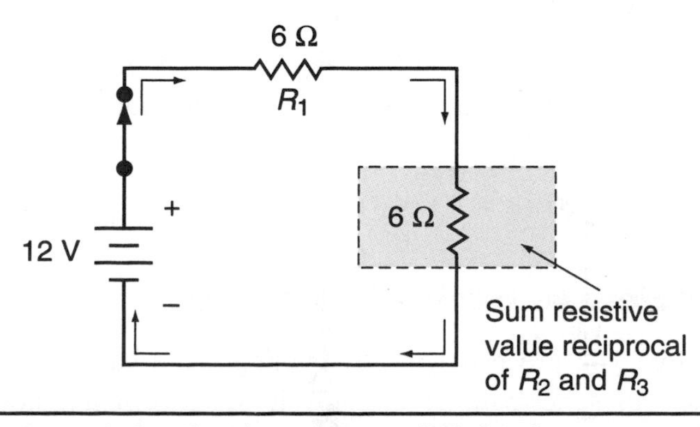

FIGURE 9–3 Simple series/parallel circuit.

To find the **series/parallel current** total of Figure 9–2, perform just one step. Use Ohm's law to find current total.

$$\begin{aligned} I &= E \div R \\ &= 12\ V_T \div 12\ \Omega_T \\ &= 1\ \text{A total} \end{aligned}$$

With these circuit totals, you can now find the **series/parallel voltage** drops by performing the following steps:

1. Because resistor 1 (6 Ω) is in series, the current is the same as the current total of 1 ampere.

   $$\begin{aligned} V &= I \times R \\ &= 1\ \text{A} \times 6\ \Omega \\ &= 6\ \text{V} \end{aligned}$$

   The voltage drop across resistor 1 is 6 volts. If you subtract the voltage drop from the voltage total of 12 volts, the remaining voltage available for the parallel circuit is 6 volts.

2. Because the voltage is constant in a parallel circuit, resistors 2 and 3 share the same 6 volts.
3. Find the current of R_2 using Ohm's law.

   $$\begin{aligned} I &= E \div R \\ &= 6\ \text{V} \div 12\ \Omega \\ &= 0.5\ \text{A} \end{aligned}$$

4. Find the current of R_3 using Ohm's law.

   $$\begin{aligned} I &= E \div R \\ &= 6\ \text{V} \div 12\ \Omega \\ &= 0.5\ \text{A} \end{aligned}$$

Note: Even though the parallel resistance values in these calculations were identical, and the current total could have been divided by two, performing this step was important because the values are not always going to be the same.

The procedure for finding circuit values is the same even if the circuit's series and parallel branches are located in different places, such as in Figure 9–4.

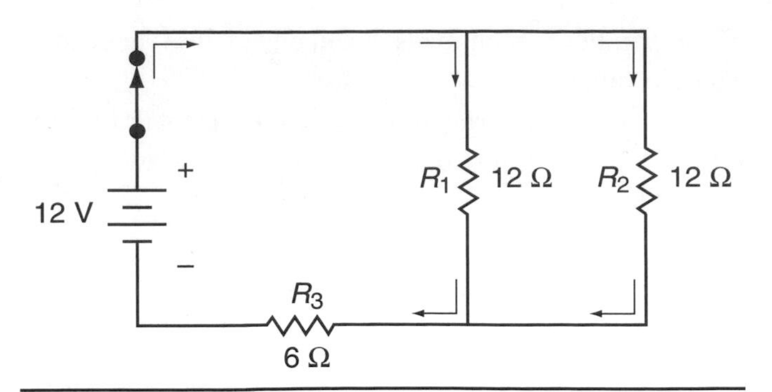

FIGURE 9–4 Simple series/parallel circuit.

In Figure 9–4, resistors 1 and 2 are in parallel, and resistor 3 is in series. The values are the same as in the previous problem because the voltage drop at the series load device determines the remaining voltage for the parallel circuit.

Figure 9–5 is a series/parallel circuit that consists of **multiple circuits** that in turn consist of more than one circuit using the same circuit laws.

In Figure 9–5, resistors 1 and 4 are in series with the 12-volt source, and resistors 2 and 3 are in parallel. To find the resistive values of this circuit, perform the following steps:

1. Resistance total equals resistor 1 + resistor 4 + the sum reciprocal of resistors 2 and 3.
2. $R_T = R_1 + R_4 +$ (sum reciprocal of R_2 and R_3)
3. $R_1 + R_4 = 3\ \Omega + 1\ \Omega$
4. Sum parallel of R_2 and R_3 = 6 1/X + 3 1/X = 2 Ω

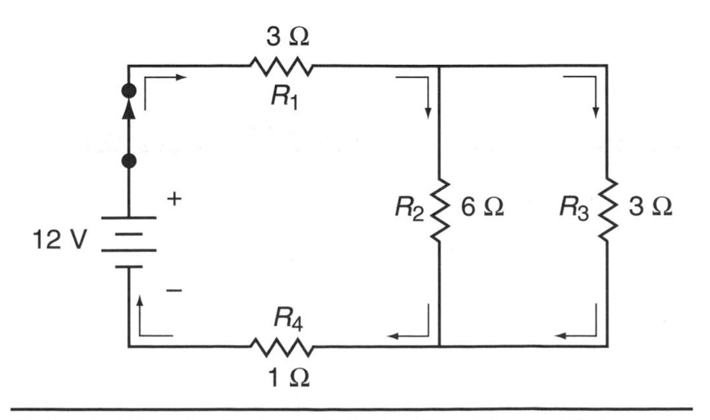

FIGURE 9–5 Series/parallel circuit with multiple series loads.

5. $R_T = 3\ \Omega + 1\ \Omega$ (series) + 2 Ω (sum parallel)
6. $R_T = 6\ \Omega$

To find the current total of Figure 9–5, perform the following step.

12 V ÷ 6 Ω = 2 A total

With the circuit totals of 12 volts, 6 ohms, and 2 amperes, it is now possible to find the voltage drops of the series values (R_1 and R_4) and the remaining parallel voltage by performing the following steps.

1. $V = I \times R$, resistor 1 voltage
   $= 3\ \Omega \times 2$ A
   $= 6$ V
2. $V = I \times R$, resistor 4 voltage
   $= 1\ \Omega \times 2$ A
   $= 2$ V
3. Add the voltage drops of R_1 and R_4 = 6 V + 2 V = 8 V.
4. Subtract the series voltage drops from the voltage total to get the parallel voltage.

   12 V − 8 V = 4 V
5. Because the voltage is constant in a parallel circuit, both resistor 2 and resistor 3 share the same 4 volts.
6. Find the current of R_2 using Ohm's law.

   $I = E \div R$
   $= 4$ V ÷ 6 Ω
   $= 0.667$ A
7. Find the current of R_3 using Ohm's law.

   4 V ÷ 3 Ω = 1.333 A
8. Add the current of R_2 and R_3, which should then be the same as each series load and the total circuit current.
9. R_2 (0.667) + R_3 (1.333) = 2 A

Review of Basic Circuit Laws

Before you proceed with performing some basic exercises in series/parallel circuits, let's perform a quick snapshot calculation review of the differences between

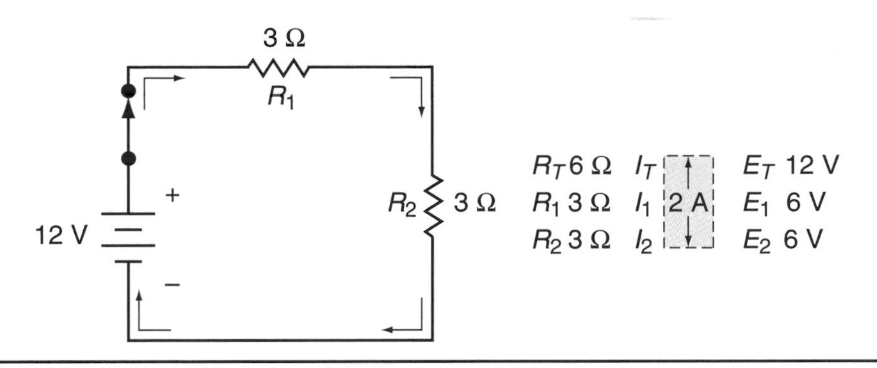

FIGURE 9–6 Basic series circuit.

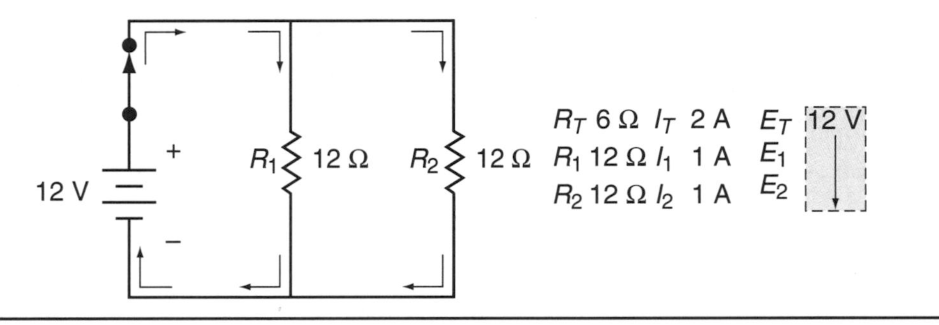

FIGURE 9–7 Basic parallel circuit.

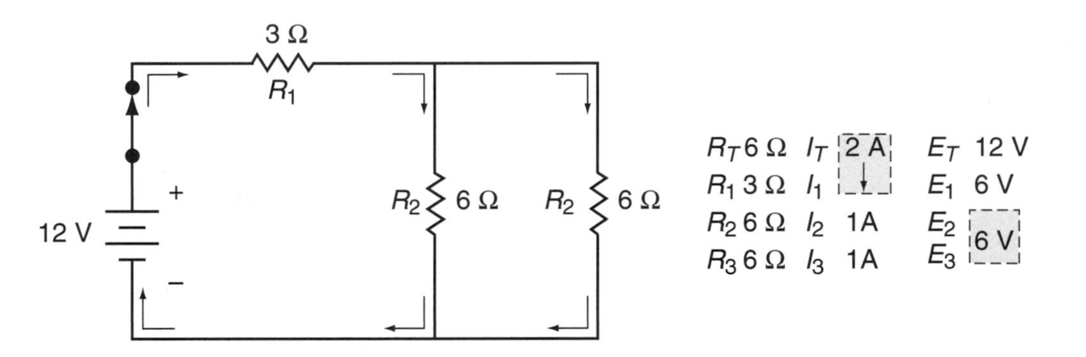

FIGURE 9–8 Basic series/parallel circuit.

series circuits, parallel circuits, and series/parallel circuits. Figure 9–6 indicates how 12 volts total, 2 amperes total, and 6 ohms resistance total are distributed throughout a basic series circuit. In comparison, in Figure 9–7 the basic parallel circuit distributes the same total volts, amperes, and resistance in accordance with the parallel circuit laws. Figure 9–8 is a basic series/parallel circuit that identifies how the series circuit laws and parallel circuit laws remain specific to how each resistive load is wired into the circuit.

Exercises—Series/Parallel Circuits

The following exercises are designed to give you practice using the examples just explained. The exercises increase in complexity in an effort to test your understanding of the mixture of series and parallel circuit laws and to prepare you for the actual hands-on series/parallel electrical exercises later in this chapter.

Exercise 1

Instructions Identify the type of circuit in Figure 9–9. On a *separate* sheet of paper, make an answer table like Table 9–1. Using the answer table, fill in the known circuit values from Figure 9–9. This is a helpful first step to give you an idea of where to begin and what answers you need. Calculate the answers in the remaining spaces using one of the methods previously learned.

R_T = Resistance total, I_T = Amperage total
E_T = Voltage total, W_T = Watts total

R_1 = Resistor 1, I_1 = Amperage 1
E_1 = Voltage 1, W_1 = Watts 1

R_2 = Resistor 2, I_2 = Amperage 2
E_2 = Voltage 2, W_2 = Watts 2

R_3 = Resistor 3, I_3 = Amperage 3
E_3 = Voltage 3, W_3 = Watts 3

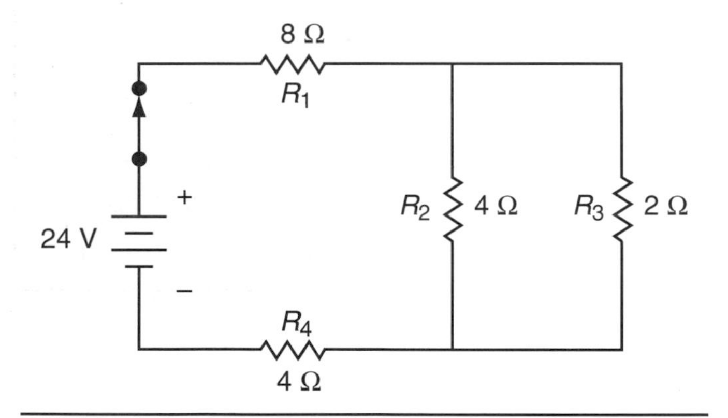

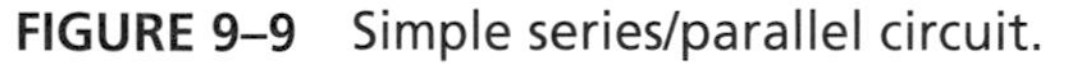

FIGURE 9–9 Simple series/parallel circuit.

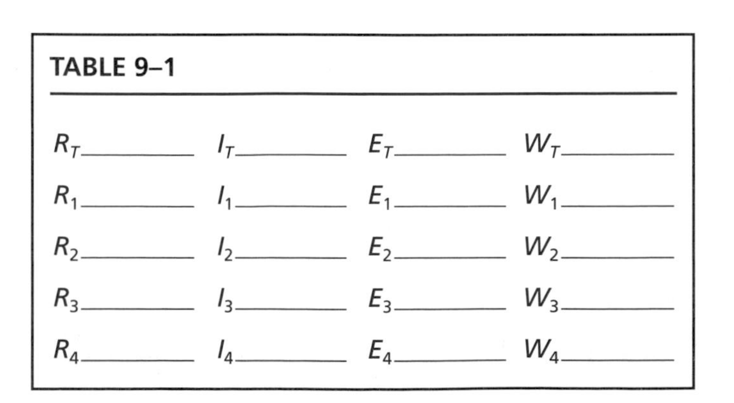

TABLE 9–1

R_T	I_T	E_T	W_T
R_1	I_1	E_1	W_1
R_2	I_2	E_2	W_2
R_3	I_3	E_3	W_3
R_4	I_4	E_4	W_4

R_4 = Resistor 4, I_4 = Amperage 4
E_4 = Voltage 4, W_4 = Watts 4

To verify that you are on the right track for finding the circuit values of series/parallel circuits, look at Table 9–2; which is a summary of the correct answers.

TABLE 9–2

R_T 1.33 Ω	I_T 1.80 A	E_T 24 V	W_T 43.2 W
R_1 8 Ω	I_1 1.80 A	E_1 14.4 V	W_1 25.92 W
R_2 4 Ω	I_2 0.6 A	E_2 2.4 V	W_2 1.44 W
R_3 2 Ω	I_3 1.2 A	E_3 2.4 V	W_3 2.88 W
R_4 4 Ω	I_4 1.80 A	E_4 7.2 V	W_4 12.96 W

Instructions Follow the instructions used in the previous problem for the remaining series/parallel exercises. Remember to put your answers on a *separate* sheet of paper.

Exercise 2

Note: Use Figure 9–10 to fill in Table 9–3. The voltage is not listed on the circuit diagram, but the amperage total is shown on the ammeter mounted in series with the battery.

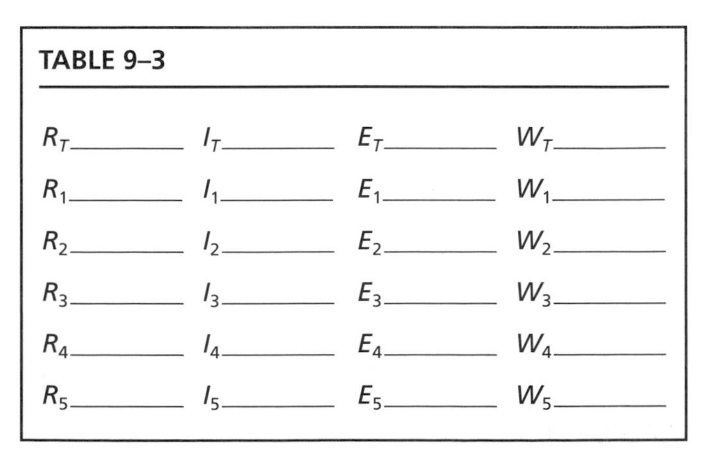

TABLE 9–3

R_T	I_T	E_T	W_T
R_1	I_1	E_1	W_1
R_2	I_2	E_2	W_2
R_3	I_3	E_3	W_3
R_4	I_4	E_4	W_4
R_5	I_5	E_5	W_5

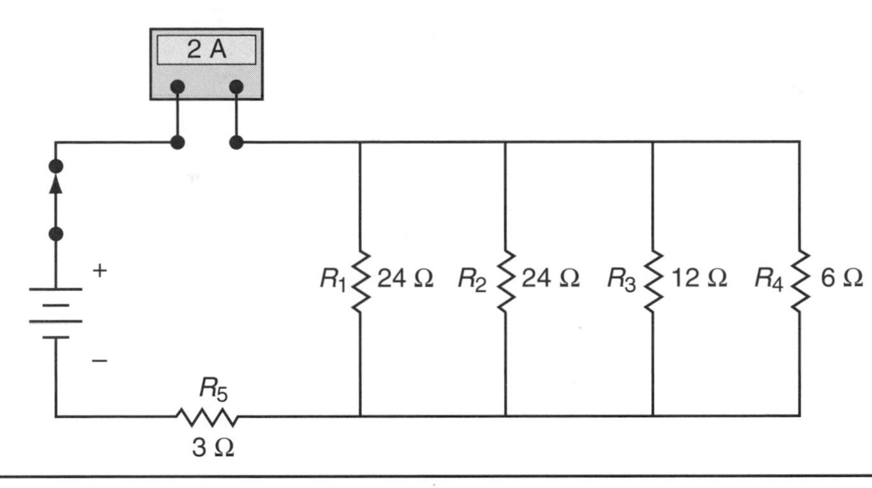

FIGURE 9–10 Measuring current total on a series/parallel circuit.

Exercise 3

The series/parallel circuit in Figure 9–11 consists of several series resistors before and after a parallel circuit. Identify the circuit by following the wire from the positive side of the battery through resistors 1 and 2 (in series), and then the circuit splits three ways through resistors 3, 4, and 5 (in parallel) before going through resistors 6 and 7 (in series). Write your answers in a chart like Table 9–4.

TABLE 9–4

R_T ______	I_T ______	E_T ______	W_T ______
R_1 ______	I_1 ______	E_1 ______	W_1 ______
R_2 ______	I_2 ______	E_2 ______	W_2 ______
R_3 ______	I_3 ______	E_3 ______	W_3 ______
R_4 ______	I_4 ______	E_4 ______	W_4 ______
R_5 ______	I_5 ______	E_5 ______	W_5 ______
R_6 ______	I_6 ______	E_6 ______	W_6 ______
R_7 ______	I_7 ______	E_7 ______	W_7 ______

Exercise 4

Note: The series/parallel circuit in Figure 9–12 is drawn a little differently than previous circuits. Identify the circuit by following the wire from the positive side of the battery through resistor 1 and resistor 2 (in series), and then as it splits four ways through resistors 3, 4, 5, and 7 (in parallel) before going through resistor 6 (in series). Write your answers in a chart like Table 9–5.

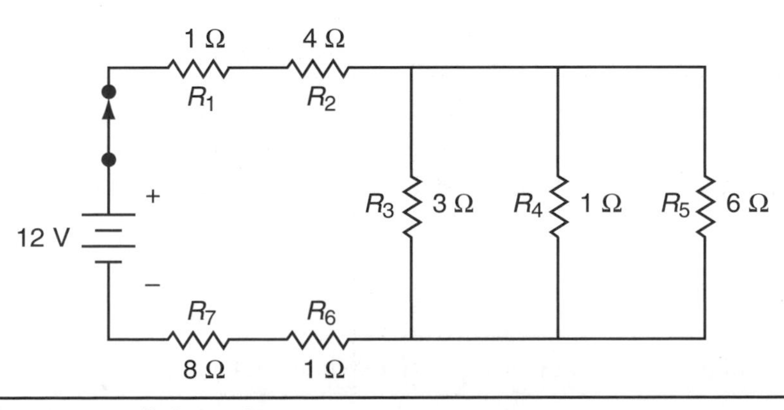

FIGURE 9–11 Expanded series/parallel circuit.

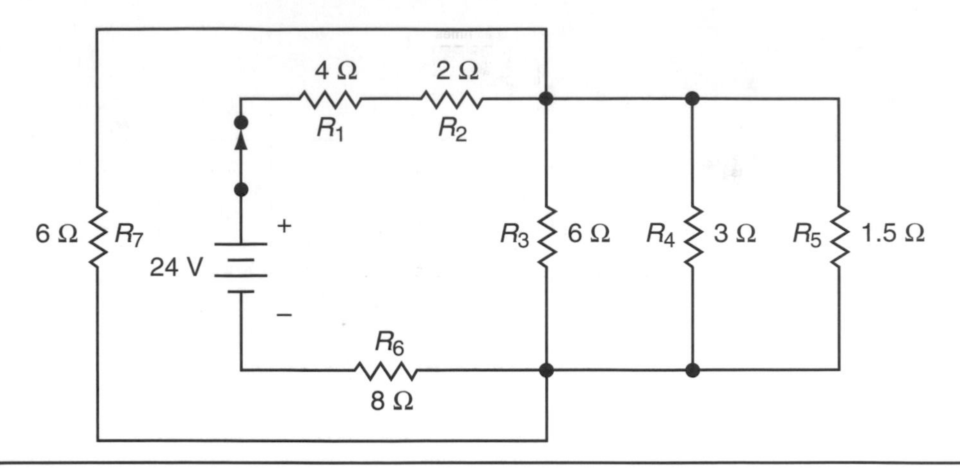

FIGURE 9–12 Complex series/parallel circuit.

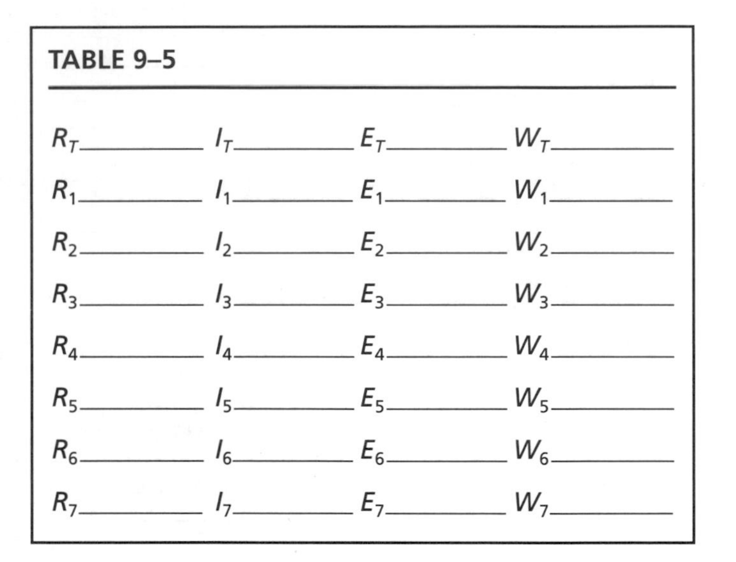

TABLE 9–5

R_T ______	I_T ______	E_T ______	W_T ______
R_1 ______	I_1 ______	E_1 ______	W_1 ______
R_2 ______	I_2 ______	E_2 ______	W_2 ______
R_3 ______	I_3 ______	E_3 ______	W_3 ______
R_4 ______	I_4 ______	E_4 ______	W_4 ______
R_5 ______	I_5 ______	E_5 ______	W_5 ______
R_6 ______	I_6 ______	E_6 ______	W_6 ______
R_7 ______	I_7 ______	E_7 ______	W_7 ______

Now that you have learned how to find the values for the combined series and parallel circuits, it is time to test your abilities on real-life case studies. Just remember to trace out the circuit to determine what part of the circuit is in series and what part is in parallel.

CASE STUDIES

CASE 1

The case study in Figure 9–13 and Figure 9–14 are typical of a live series/parallel interior dash light circuit that technicians often encounter. Knowledge of series/parallel circuits can help you verify the values of circuit components quickly and efficiently and perform an accurate diagnosis.

Known Information

- ❑ The system operating volts = 14.0 volts.
- ❑ The light switch dimmer rheostat resistance on maximum (HI) or DIM position = 15 ohms.
- ❑ The light switch dimmer rheostat resistance on maximum (LOW) or BRIGHT position = 5 ohms.

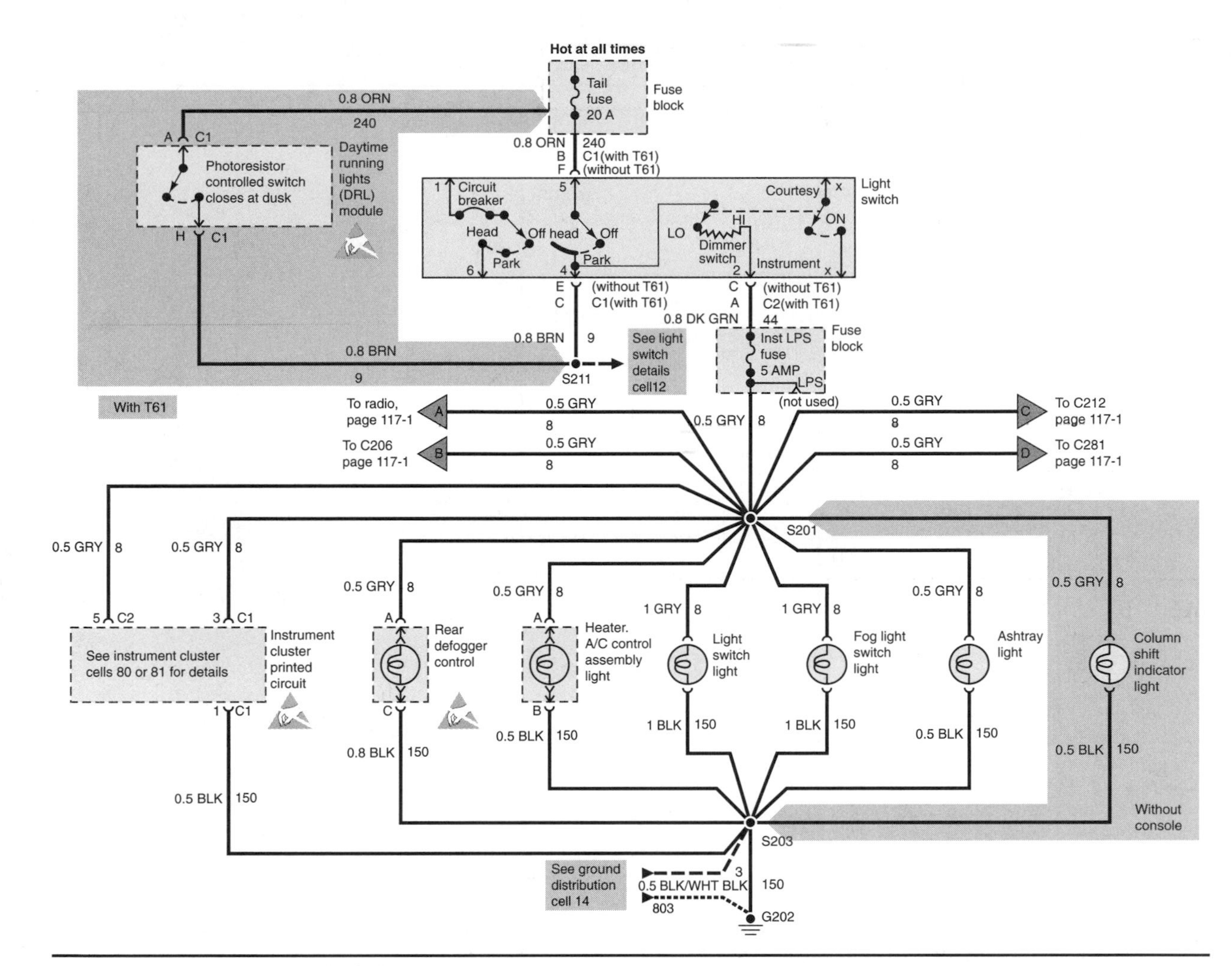

FIGURE 9–13 Interior dash light circuit.

- ❑ The interior dash light bulb specifications = #194 bulb, design amperes 0.27 amperes, hot bulb resistance = 51.85 ohms (see Appendix A).
- ❑ Five bulbs are not specifically shown on Figure 9–13, located in the Instrument Cluster Printed Circuit.

Circuit Analysis

Answer the following questions on a *separate* sheet of paper:

1. What component(s) of the circuit are in series? ______.
2. What component(s) of the circuit are in parallel? ______.
3. What is the total number of #194 bulbs in this circuit? ______.
4. What is the resistance total of bulbs *only?* ______.
5. What is the resistance total of the bulbs and dimmer switch in the LOW position? ______.

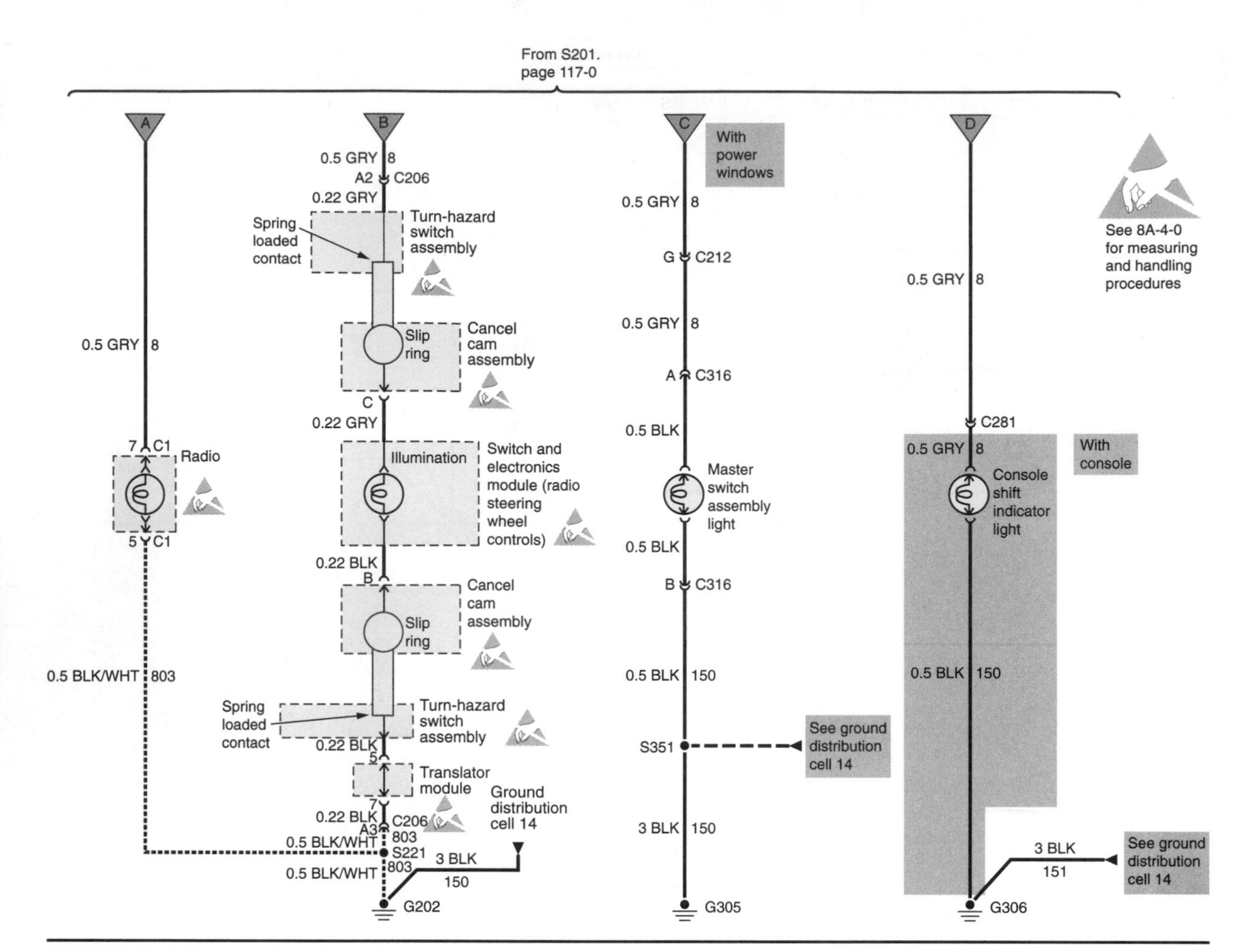

FIGURE 9–14 Interior dash light circuit.

6. What is the resistance total of the bulbs and dimmer switch in the HI position? ________________.
7. What is the amperage total of the circuit with the dimmer switch in the LOW position? __________.
8. What is the amperage total of the circuit with the dimmer switch in the HI position? ____________.
9. What is your analysis of this case study? ____

____.

CASE 2

Using the same circuit diagrams as in Case 1 (Figure 9–13 and Figure 9–14), diagnose the following customer complaint.

Customer Complaint

Interior dash lights do not work in any position. Customer replaced the 20-ampere tail fuse and the 5-ampere instrument lamp fuse, but the results were the same.

Known Information

- ❑ The vehicle operating voltage = 14 volts.
- ❑ Both circuit fuses have been replaced.
- ❑ The bulbs are not burned out.

Circuit Analysis

Answer the following on a *separate* sheet of paper.

1. If you know both fuses are good, which system components would you check in the circuit and in what order? List the most important component first, with the procedure you would use to perform the check.

   Component 1: ____________________. Troubleshooting procedure: ____________________.

   Component 2: ____________________. Troubleshooting procedure: ____________________.

   Component 3: ____________________. Troubleshooting procedure: ____________________.

   Component 4: ____________________. Troubleshooting procedure: ____________________.

2. Which circuit laws were helpful when troubleshooting this system or explaining it to the customer?

   ____.

3. What is your analysis of this case study? ____

   ____.

CASE 3

The case study in Figure 9–15 is typical of a backup light circuit that operates parallel lights from a series fuse and switch.

Customer Complaint

Backup lights do not operate when position switch is put in the reverse (R) position.

Known Information

- ❑ Vehicle operating voltage = 14 volts.
- ❑ Turn signals operate normally.
- ❑ Twenty-ampere turn-backup fuse is OK.
- ❑ This is an automatic transmission vehicle.
- ❑ Backup lamps, #1156 bulb, design amperes 2.10, hot bulb resistance = 6.10 ohms each.

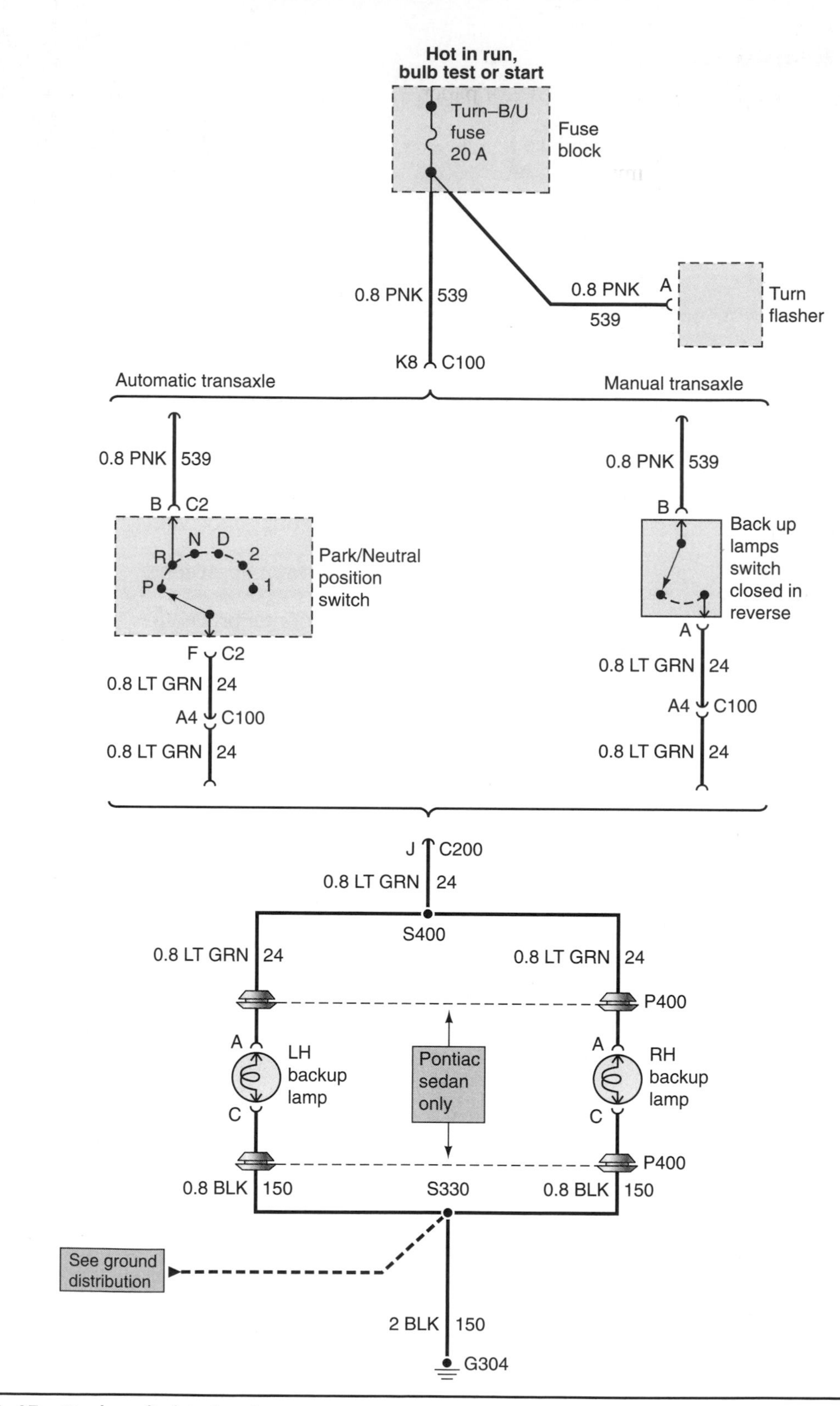

FIGURE 9–15 Backup light circuit.

Circuit Analysis

Answer the following on a *separate* sheet of paper.

1. If you know the fuse is good, which system components would you check in the circuit and in what order? List the most important component first, with the procedure you would use to perform the check.

   Component 1: ____________________. Troubleshooting procedure: ____________________.

   Component 2: ____________________. Troubleshooting procedure: ____________________.

   Component 3: ____________________. Troubleshooting procedure: ____________________.

   Component 4: ____________________. Troubleshooting procedure: ____________________.

2. What is the resistance total of this circuit? ____________________.
3. What is the amperage total of the circuit? ____________________.
4. What is the amperage of each bulb in the circuit? ____________________.
5. What is your analysis of this case study? ____________________

   ____________________.

CASE 4

The circuit in Figure 9–16 operates correctly, except for the righthand seat roof vanity mirror.

Known Information

- ❑ The vehicle operating voltage is available at points A and B at the top of circuit diagram (12 volts).
- ❑ Righthand (RH) seat roof vanity mirror lamps are OK.
- ❑ The power available at the righthand (RH) seat roof vanity mirror = 12 volts.
- ❑ Righthand (RH) seat roof vanity mirror dimmer rheostat has a resistance of 5 ohms in the maximum bright position and 20 ohms in the maximum dim position.
- ❑ All other vanity mirrors work satisfactorily.

Circuit Analysis

Answer the following on a *separate* sheet of paper.

1. If you know there is 12 volts at connection A going to the righthand (RH) vanity mirror, what would you check in the circuit and in what order? List the most important component check first, with the procedure you would use to check the item.

   Component 1: ____________________. Troubleshooting procedure: ____________________.

   Component 2: ____________________. Troubleshooting procedure: ____________________.

   Component 3: ____________________. Troubleshooting procedure: ____________________.

   Component 4: ____________________. Troubleshooting procedure: ____________________.

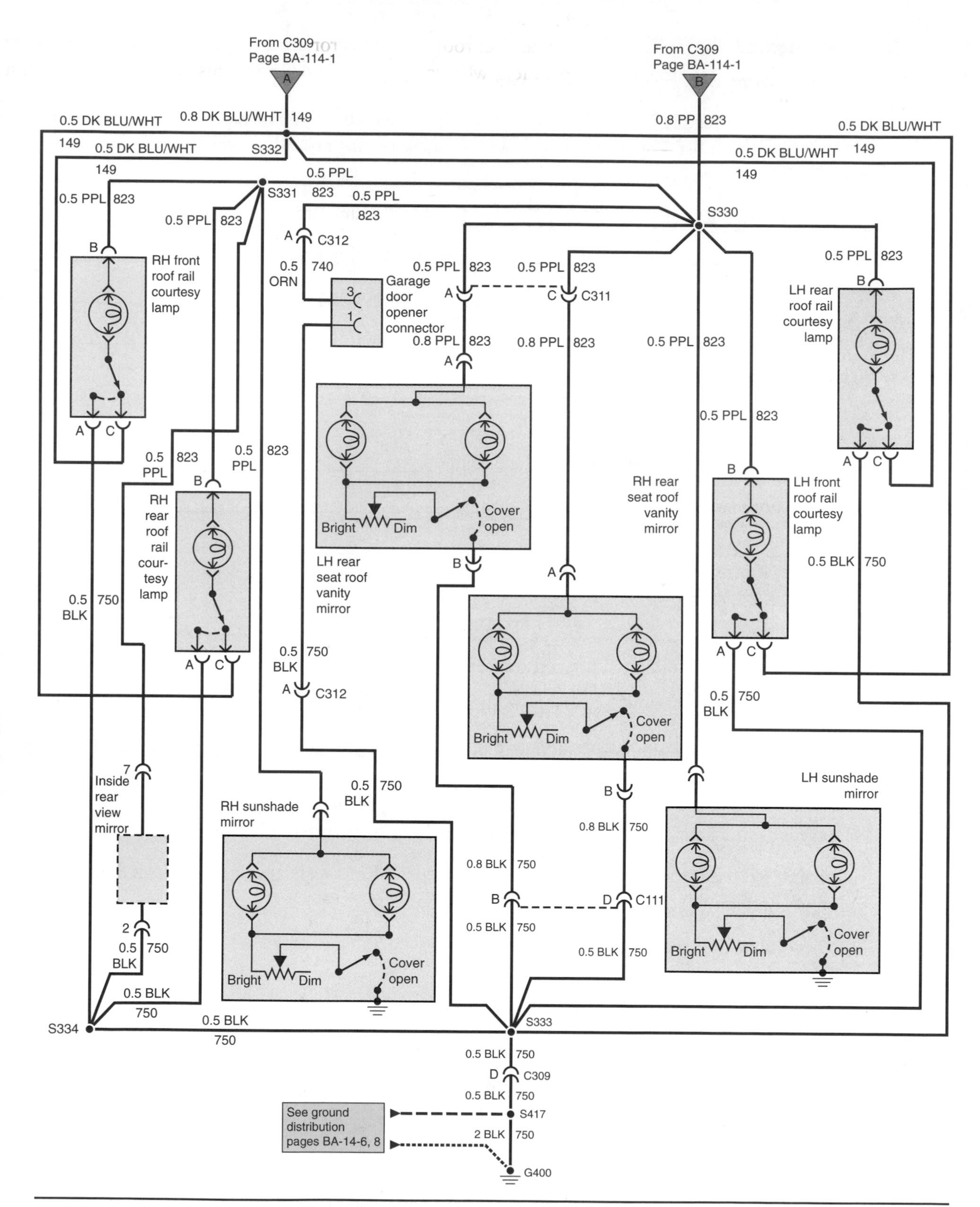

FIGURE 9–16 Interior vanity light circuit.

2. If the lamps for the righthand (RH) rear seat roof vanity mirror are 51 ohms each, and the dimmer rheostat is at the maximum bright position, what is the resistance total of this portion of the interior lamp circuit?

____.

3. What is the amperage going to the righthand (RH) rear seat roof vanity mirror at connector point A?

____.

4. What is the amperage coming from the righthand (RH) rear seat roof vanity mirror at connector point B?

____.

5. What is your analysis of this case study? ____

____.

Hands-On Vehicle Tests

The following hands-on vehicle check is included in the NATEF (National Automotive Technician Education Foundation) Task List. Complete your answers to the following questions on a *separate* sheet of paper.

Performance Task 1

Task Description

Diagnose the cause of brighter-than-normal, intermittent, dim, or no light operation; determine necessary action.

Task Objectives

- ❑ Obtain a vehicle that can be used for this task. What model and year of vehicle are you using for this task? ____.
- ❑ Find a wiring diagram that shows the interior dash light circuit for this vehicle.
- ❑ Does the dash light circuit follow the characteristics of a series/parallel circuit? ____________________.
- ❑ Identify the following components and list their location on the vehicle:
  1. Light switch assembly: ____.
  2. Circuit fuse(s) and fuse numbers: ____.
  3. Other electrical circuits that share the same fuse: ____.
  4. Number of instrument panel dash lights: ____.

5. Circuit ground(s): ______________________________.
6. Other circuit components: ______________________________.

Circuit Checks

- ❑ Check the operation of the headlight switch. Rotate the dimmer switch throughout its full range from maximum DIM to maximum BRIGHT position. Does it operate as designed? ______________.
- ❑ Identify the inlet and outlet wires to the dimmer switch. Connect a DVOM, set on voltage, in parallel with the inlet and outlet wires to the dimmer switch. Measure the voltage drop at maximum DIM ______________, and maximum BRIGHT ______________.
- ❑ What is the available voltage remaining for the parallel dash lights? ______________.
- ❑ Remove the circuit fuse and install an ammeter (in series) in place of the fuse. Measure the amperage total with the dimmer switch in the fully DIM position ______________.
- ❑ Measure the amperage total with the dimmer switch in the maximum BRIGHT position ______________.
- ❑ What would happen to the dash light circuit if the dimmer switch became inoperative?

  ______________________________.
- ❑ What would happen to the dash light if one of the lights burned out? ______________.

Task Summary

After performing this NATEF task, what can you determine will be helpful in future light circuits that operate other than as intended? ______________________________

______________________________.

Summary

- ❑ In a series/parallel circuit, there are individual series load components and parallel load components.
- ❑ In a series/parallel circuit, the resistance total is computed by adding the series resistance to the sum reciprocal of the parallel branches.
- ❑ In a series/parallel circuit, voltage drops exist at the series loads, with the remaining voltage available for the parallel load branch(s).
- ❑ A series/parallel circuit is also called a combination circuit.
- ❑ Series/parallel circuits can have several series load branches and several parallel load branches.
- ❑ An interior dash light circuit uses a series dimmer rheostat to adjust the parallel lamp circuit brightness.
- ❑ In a series/parallel circuit, the failure of a main series load device will disable the entire circuit.

Key Terms

combination circuit
multiple circuits
series/parallel circuit
series/parallel current
series/parallel resistance
series/parallel voltage

Review Questions

Short Answer Essays

1. Describe how to find the current total in a compound circuit.
2. Describe how to find the resistance total in a compound circuit.
3. Describe how to find the voltage total in a compound circuit.

Fill in the Blanks

1. The ________________________ of a series load component in a compound circuit is the same as ________________________ total.
2. In a series/parallel circuit, the resistance total is computed by ________________________ the series load components with the sum of the ________________________ load components.
3. A compound electrical circuit with a series load of 10 ohms and two parallel loads of 10 ohms each, has a resistance total of ________________________.

ASE-Style Review Questions

1. Series/parallel theory is being discussed. Technician A says that a compound circuit contains both series and parallel load components. Technician B says a series parallel circuit is also known as a combination circuit. Who is correct?
   A. A only
   B. B only
   C. Both A and B
   D. Neither A nor B

2. In the following simple series/parallel circuit, the bulbs do not light when the switch is closed. Technician A says the switch could be bad. Technician B says both bulbs could be burned out. Who is correct?
   A. A only
   B. B only
   C. Both A and B
   D. Neither A nor B

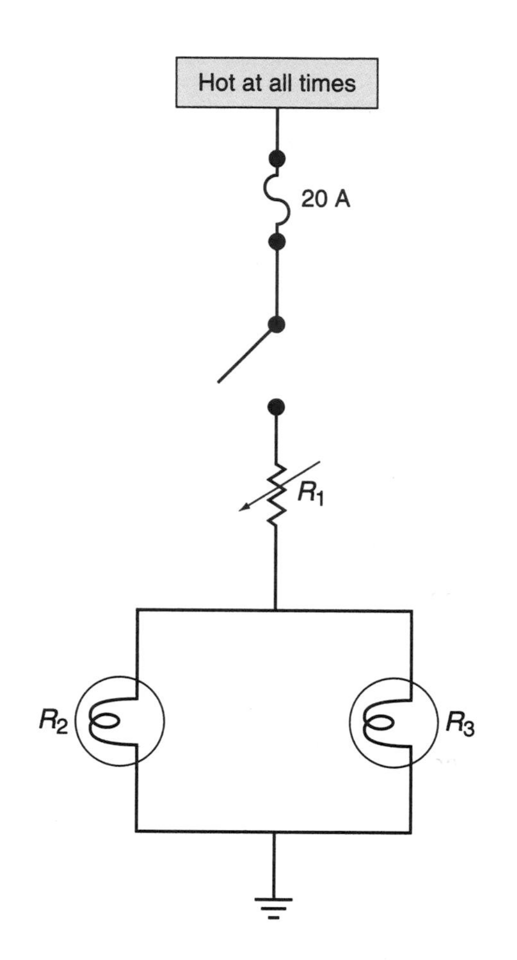

3. Two technicians are discussing circuit laws that pertain to a series/parallel circuit. Technician A says that in a compound circuit the resistance total is computed by adding the series load devices with each individual parallel load device. Technician B says that in a compound circuit the circuit voltage is the same throughout the whole circuit. Who is correct?
   A. A only
   B. B only
   C. Both A and B
   D. Neither A nor B

4. Technician A says the amperage of each series load in a compound circuit is the same as the amperage total. Technician B says the amperage of each parallel load in a compound circuit is the same as the amperage total. Who is correct?

   A. A only

   B. B only

   C. Both A and B

   D. Neither A nor B

5. A series/parallel circuit has two series resistors, 3 ohms and 4 ohms, and four parallel resistors, 6 ohms, 6 ohms, 3 ohms, and 1.5 ohms. What is the resistance total?

   A. 23.5 Ω

   B. 7.75 Ω

   C. 0.5217 Ω

   D. None of the above

6. Two technicians are discussing a simple series/parallel circuit. Technician A says if a series load component fails in a compound circuit, the entire circuit will quit working. Technician B says if a parallel load component fails in a compound circuit, the rest of the circuit will continue to operate as designed. Who is correct?

   A. A only

   B. B only

   C. Both A and B

   D. Neither A nor B

7. The amperage total is being measured in a series/parallel circuit. Technician A says to remove a series load device and install an ammeter in its place to read the amperage total. Technician B says the ammeter is installed in series in place of the circuit fuse to read amperage total. Who is correct?

   A. A only

   B. B only

   C. Both A and B

   D. Neither A nor B

8. A series/parallel circuit has a series load device with an amperage reading of 8 amperes, and three parallel load devices with readings of 4.5 amperes, 1.5 amperes, and 2.0 amperes. What is the amperage total of this circuit?

   A. 8 amps

   B. 16 amps

   C. 12.5 amps

   D. None of the above

9. Voltage total of a simple series/parallel circuit is being discussed. There are two series load devices with voltage drops of 6 volts each and four load devices in parallel with a voltage reading across each device of 12 volts. What is the voltage total of this circuit?
   A. 12 volts
   B. 24 volts
   C. 60 volts
   D. None of the above

10. The following simple series/parallel lamp circuit does not work when the switch is closed. Technician A says the circuit will not work if only lamp C is burned out. Technician B says the circuit will not work if only lamp B is burned out. Who is correct?
   A. A only
   B. B only
   C. Both A and B
   D. Neither A nor B

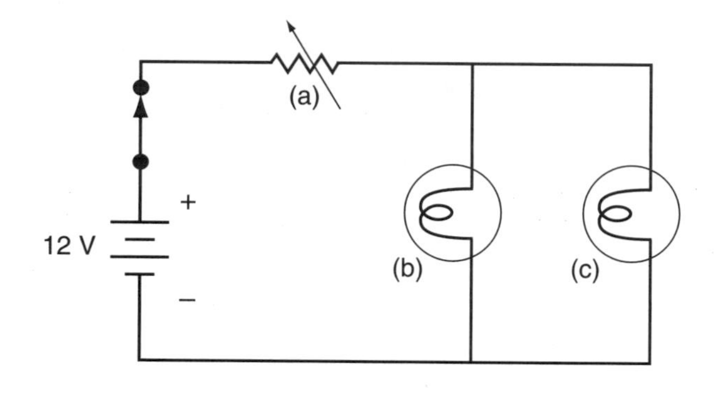

10 Basic Troubleshooting Techniques and Tips

Introduction

Automotive technicians working on electrical circuits must be familiar with commonly used troubleshooting techniques for locating circuit faults. Successful troubleshooting of a circuit begins with the knowledge of electrical theory and the circuit laws you learned in previous chapters. In this chapter you apply this knowledge to find faults in an electrical circuit.

Objectives

When you complete this chapter you should be able to:

- ❑ Define an open circuit.
- ❑ Troubleshoot an electrical circuit to find an unintentional open.
- ❑ Define a short-to-voltage.
- ❑ Troubleshoot an electrical circuit to find a short-to-voltage.
- ❑ Define a short-to-ground.
- ❑ Troubleshoot an electrical circuit to find a short-to-ground.
- ❑ Define excessive resistance in an electrical circuit.
- ❑ Troubleshoot an electrical circuit to find excessive resistance.
- ❑ Measure unwanted voltage drops in electrical circuits.
- ❑ Diagnose the cause of a key-off parasitic battery drain.
- ❑ Define a sneak circuit and its characteristics.

Circuit Faults

Open Circuits

An **open circuit** is a break in the current flow of a complete circuit. When an electrical circuit is open, the current does not flow and the load component does not work. Two types of open circuits affect current flow in a circuit. The first type is an **intentional open circuit** (Figure 10–1). The so-called open is designed as a circuit control device, like a switch or relay that turns on the circuit load when needed by the driver. Almost all electrical circuits have a switching device for on/off control.

The second type of open circuit is an **unintentional open circuit.** This type of open circuit can be created

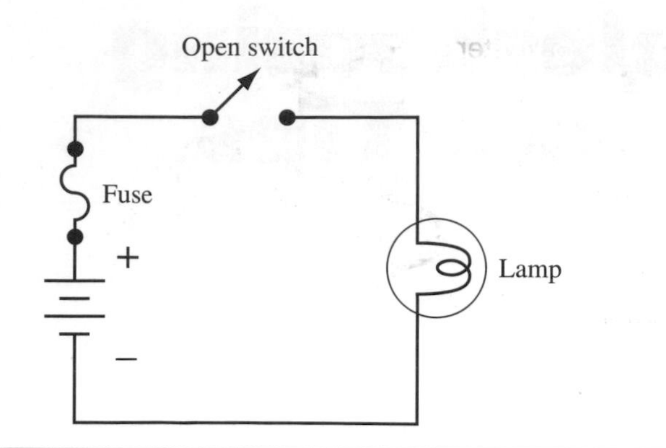

FIGURE 10–1 A switch is an intentional open circuit.

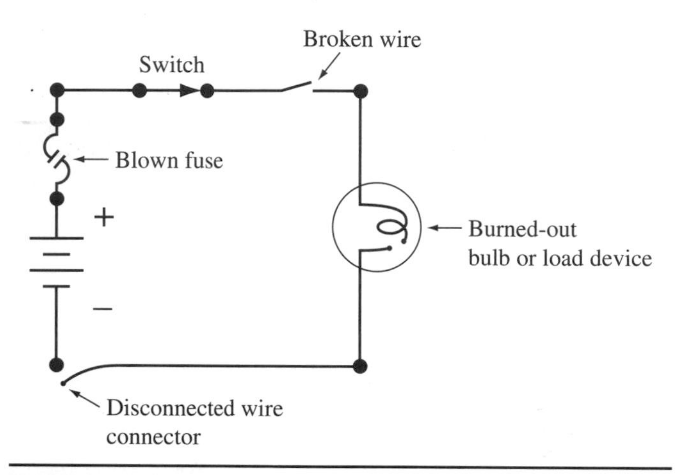

FIGURE 10–2 Several faults could cause an unintentional open circuit: a blown fuse, a broken disconnected wire, or a burned out bulb or load device.

when there is an electrical problem in the circuit (Figure 10–2). The current flow will stop when a bulb (or other load device) burns out, a wire is broken or disconnected, or a circuit protection device (fuse or circuit breaker) blows. Poor electrical connection are normally the most common causes of open circuits because they are often hidden from view within a connector. Open circuits caused by a blown fuse are normally associated with other circuit problems that create enough excessive current to blow the fuse.

Depending on the electrical circuit being tested and the accessibility of the components that need testing, several acceptable devices can be used to troubleshoot: an ohmmeter, voltmeter, test light, self-powered test light, and a fused jumper wire. To test a circuit, the technician must know its correct operation. This is where your knowledge of series and parallel circuit laws, as well as Ohm's law, comes in handy.

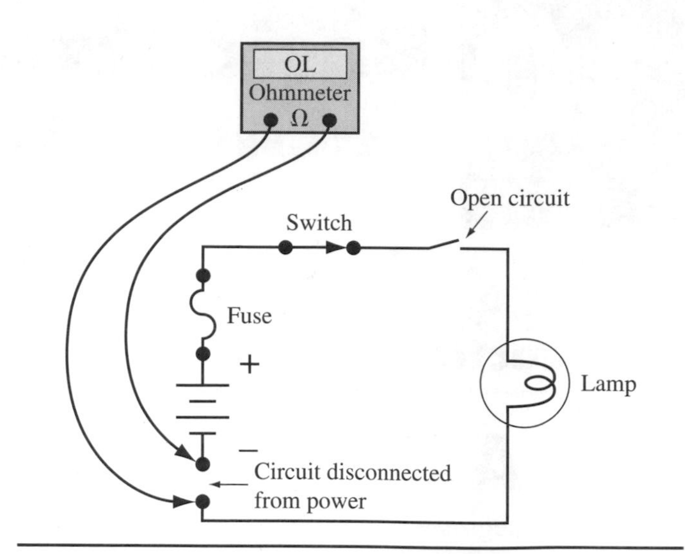

FIGURE 10–3 Testing for an open circuit using an ohmmeter.

To test for an open circuit using an ohmmeter (Figure 10–3), the circuit *must* be disconnected from the power source before installing the ohmmeter in parallel. A circuit with an open will read over limit (OL) or infinite. A complete (not open) circuit would read the designed circuit resistance (usually less than 100 ohms).

To test for an open circuit using a voltmeter (Figure 10–4), first hook up the meter in parallel as close to the circuit load as possible, then work backward through the circuit until the open circuit is found. The voltage reading in the circuit will be zero *after* the open circuit, and it will be the source voltage *before* the open circuit, as long as the rest of the circuit has no other problems.

Note: If the open circuit is after the circuit load, then the voltage reading is the source voltage all the way up to the open circuit and zero after the open circuit. This reading could mean that the open circuit could be at the end (or ground connection) of the circuit.

The more complex a circuit is, the harder it may be to diagnose an open circuit. This is why it is important to remember the circuit laws for series and parallel circuits, so that you know how the circuit is supposed to operate.

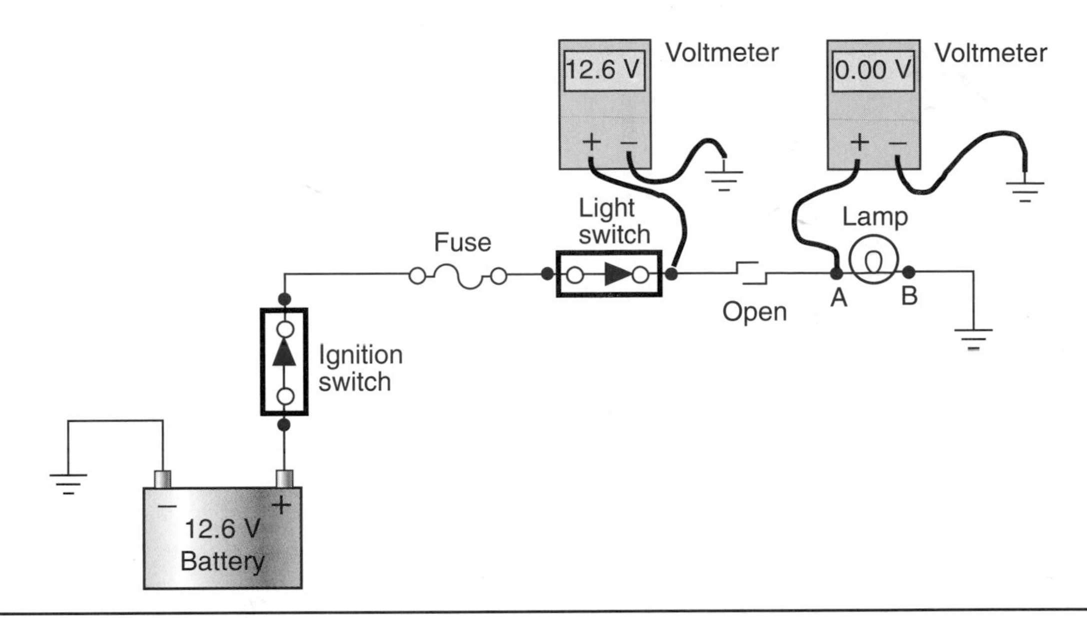

FIGURE 10–4 An open circuit is located between the points where a voltage is measured and where no voltage is found.

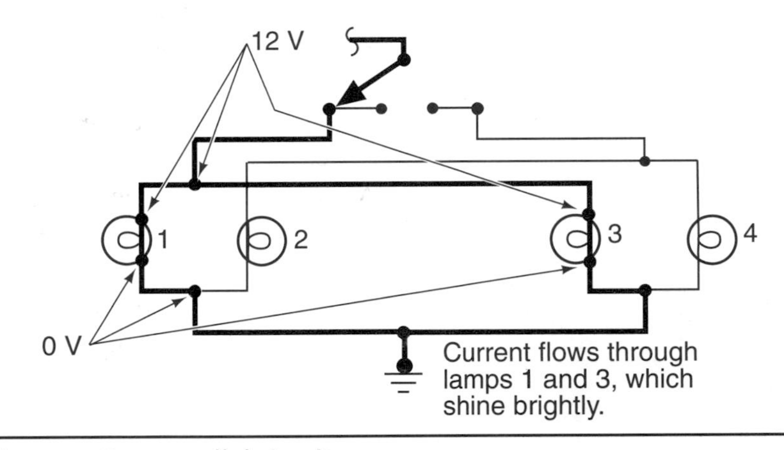

FIGURE 10–5 A properly operating parallel circuit.
(Courtesy of DaimlerChrysler Corporation)

In Figure 10–5, lamps 1 and 3 are parallel to each other with 12 volts going to each bulb and with a common ground. When bulb 1 develops an open circuit (Figure 10–6) on the ground side, bulbs 2 and 4 become wired in series with bulb 1, causing them to glow dimly. Bulb 3 continues to glow brightly, as designed, because the ground is unaffected by the circuit's open.

To test for an open circuit using a test light (Figure 10–7), hook up the ground lead of the test light to a known good ground to prevent any misdiagnosis of the circuit. Ensure that the lamp is working properly by first touching a known good power connection, and then probe along the circuit until the test lamp goes out. This is the point at which an open circuit exists.

CAUTION *Test lights and self-powered test lights should not be used around electronic or computer-controlled circuits. Damage may result to the sensitive circuit due to excessive current flow from the test light.*

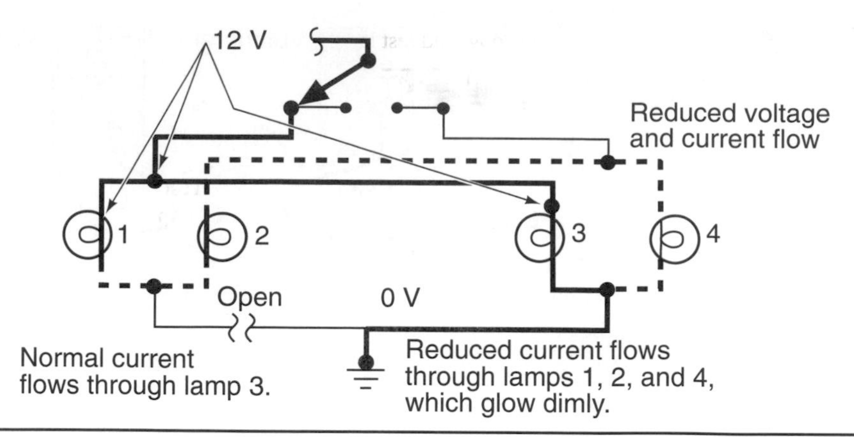

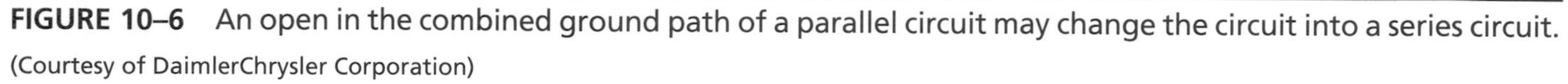
FIGURE 10–6 An open in the combined ground path of a parallel circuit may change the circuit into a series circuit. (Courtesy of DaimlerChrysler Corporation)

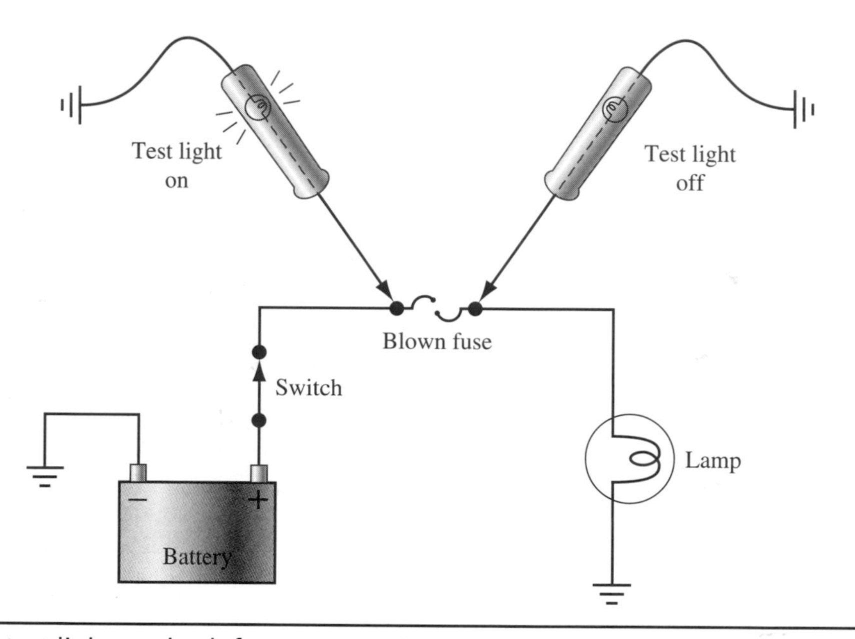

FIGURE 10–7 Using a test light to check for an open circuit.

A self-powered test light to test for an open circuit can be a useful tool for checking for circuit continuity. Because the test light uses its own internal battery, all it needs is a completed path of current flow through a circuit to energize the test light (Figure 10–8). The test lamp lights as long as there is continuity through the circuit being tested. When an open is found with the test lamp probe, the test lamp does not light.

On electrical circuits that contain sensitive electronic or computer controls, the use of a logic probe (Figure 10–9) is considered safe because of its high internal resistance. Similar to a test light in operation, the logic probe provides the technician with an easy-to-use tool. The logic probe contains three LED lights and is powered by the vehicle battery positive and ground connections, which provide the probe circuitry with a reference voltage. Touching the probe tip to a ground will light the green LED, and touching the probe tip to a power source will light the red LED. The yellow LED lights only when a pulsed voltage is sensed, such as a fuel injector circuit.

Individual circuit components that are removed from the circuit can be checked for continuity or for an open

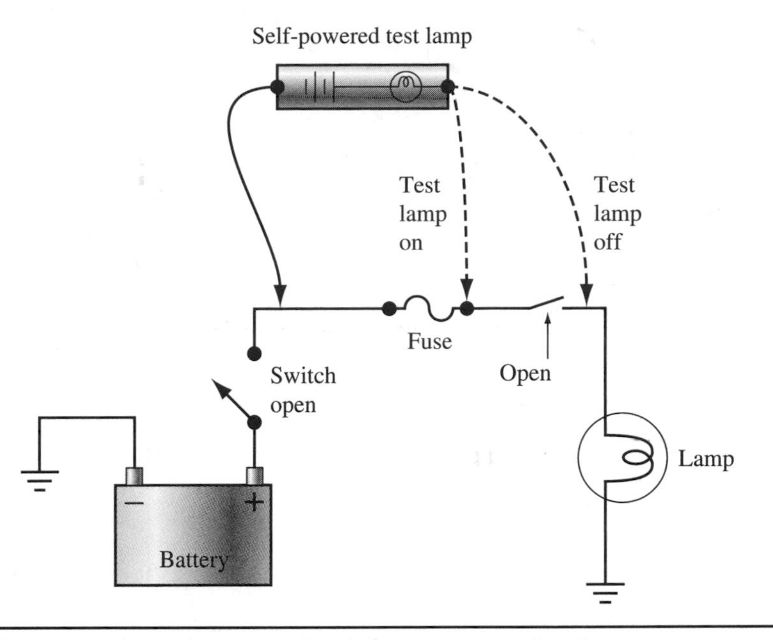

FIGURE 10–8 Using a self-powered test lamp to check for an open circuit.

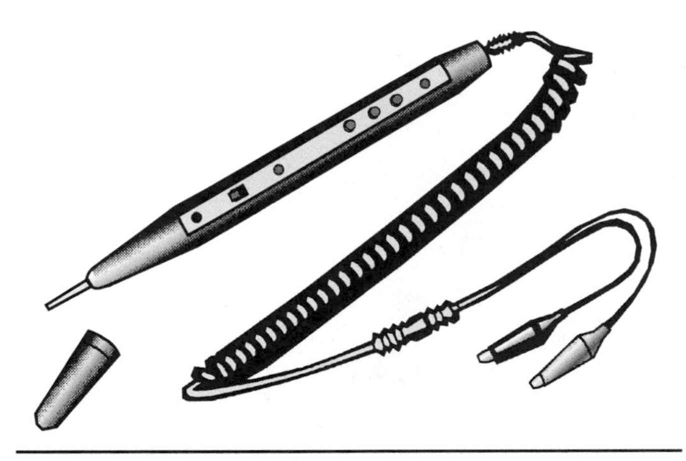

FIGURE 10–9 Typical logic probe.

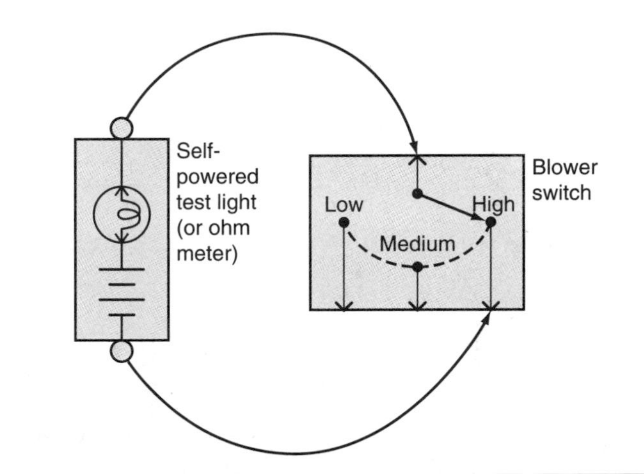

FIGURE 10–10 Checking a blower switch for continuity using a self-powered test light.

with a self-powered test light. Figure 10–10 shows a blower switch HI contact being checked for continuity. If an open exists, the lamp will not light.

Testing for an open circuit using a fused jumper wire can be a simple way of diagnosing circuits when the load control switch is suspected to be faulty. Using an in-line fuse along with the jumper wire ensures circuit safety if the jumper wire is connected incorrectly. In Figure 10–11, a fused jumper wire is being used to bypass a suspected bad light switch. Jumper wires can also be used on the ground side of the circuit load to diagnose opens in the ground.

CAUTION *Never use a fuse in the jumper wire with a rating higher than the fuse that protects the circuit being tested.*

Short Circuits

A **short circuit** is an unwanted or accidental bypass of the current in a circuit. Two types of short circuits can affect the current flow in a circuit. The first is a **short-to-voltage** (Figure 10–12). In this type of short, there is an unwanted copper-to-copper connection

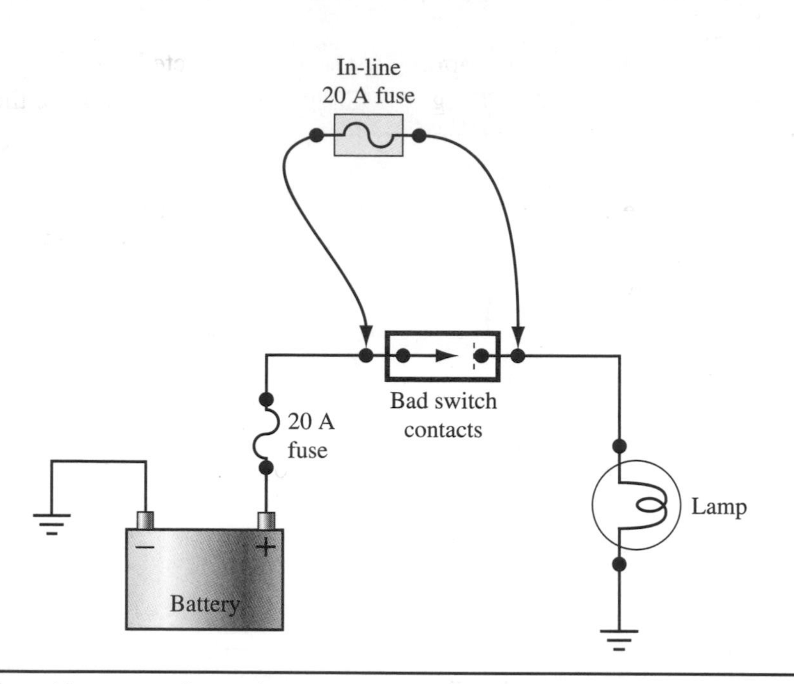

FIGURE 10–11 Using a fused jumper wire to bypass a suspected bad switch.

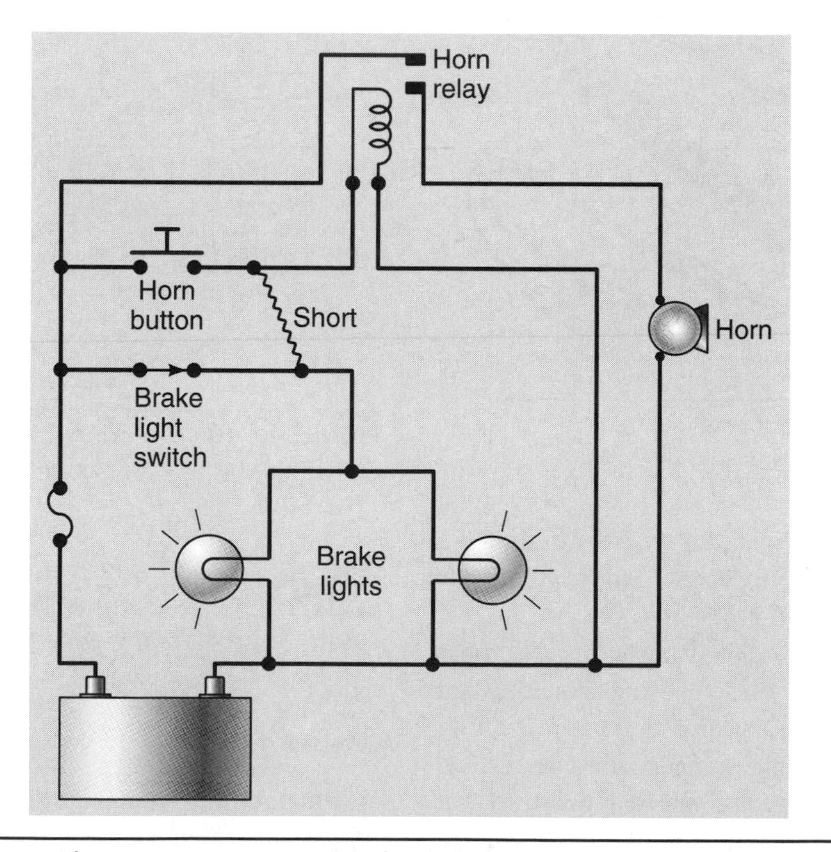

FIGURE 10–12 An unwanted copper-to-copper wire short circuit.

between two separate circuits, causing one or more circuits to operate when not needed. In the case of Figure 10–12, every time the brake light switch is pushed, the horn honks, or if the horn button is pushed, the brake lights go on. Sometimes, with this type of short, the circuit fuse may blow if the designed amperage flow exceeds the fuse limits.

Locating a short-to-voltage can sometimes be very difficult. If a visual check of the wiring, insulation, connectors, and circuit components does not show the cause of the short, then isolating the circuit is necessary. An excellent, home-built diagnostic tool that can be used to help find a short-to-voltage is a 12-volt buzzer with an in-line fuse (Figure 10–13).

If the affected circuits share a common fuse (see Figure 10–12), then remove the fuse and install a buzzer that has been fitted with terminals to fit across the fuse holder terminals (Figure 10–14). Using a buzzer is helpful when you isolate circuit components through a process of elimination by disconnecting them. When the affected spot is reached, the buzzer stops.

Similarly, activate the circuit that the buzzer is connected to and disconnect all of the loads that are supposed to be activated by each switch. Then disconnect the wire connectors at the circuit that go from the load(s) back to the switch(es). If the buzzer stops when a connector is disconnected, the short-to-voltage is in that portion of the circuit.

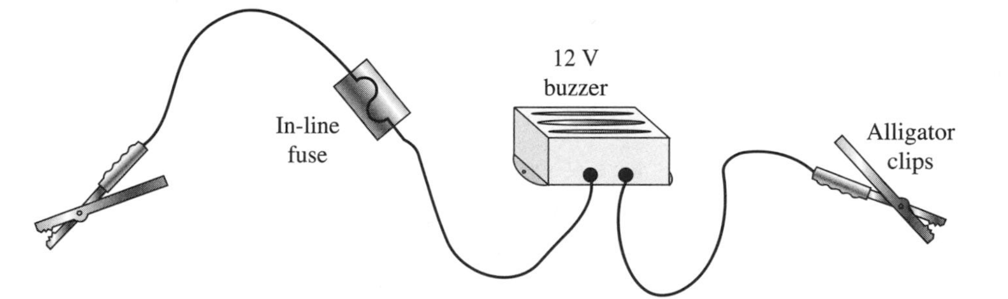

FIGURE 10–13 Buzzer project: old seat belt buzzer, or Radio Shack part #273-055, or equivalent; in-line glass fuse holder, Radio Shack part #270-1217 or equivalent, or blade fuse holder, Radio Shack part #270-1213 or equivalent; alligator clips, Radio Shack part #270-375 or equivalent.

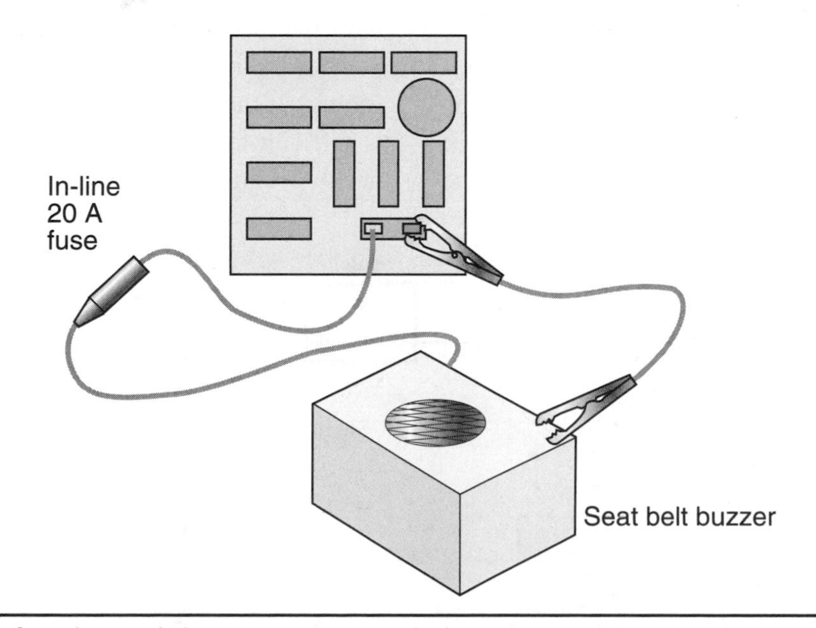

FIGURE 10–14 Using a fused 12-volt buzzer in series to help isolate a short-to-voltage.

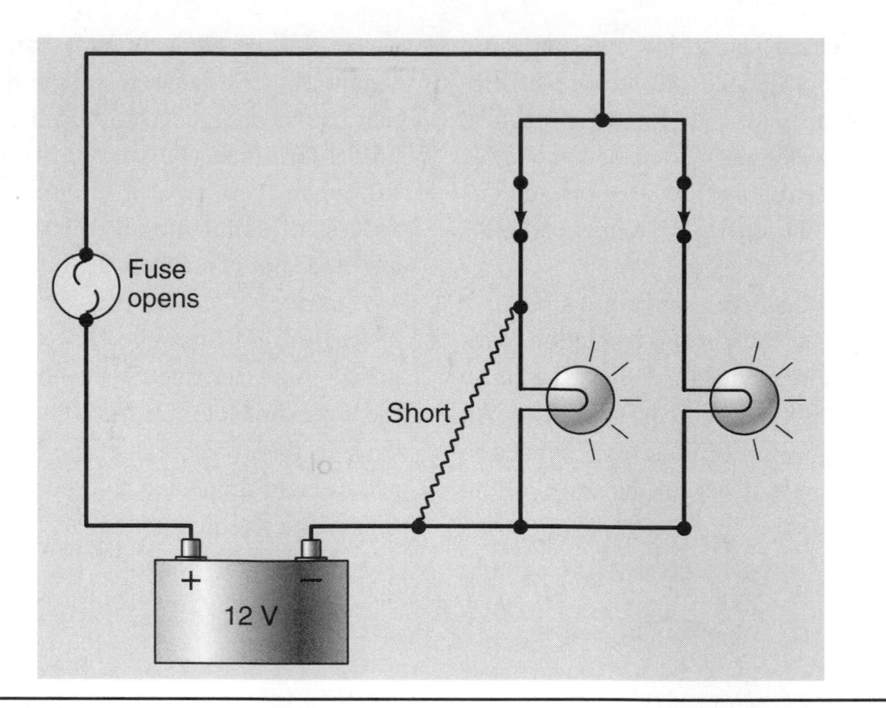

FIGURE 10–15 A typical short-to-ground in a parallel circuit.

The second type of short circuit, is a **short-to-ground** (Figure 10–15), also known as a **grounded circuit.** In this type of short, there is an unwanted path for current to flow directly to ground. This normally occurs when wire insulation breaks down from vibration or unwanted rubbing, and the wire touches a ground. This problem provides a low-resistance path for the current to flow, which most likely blows the circuit protection fuse or overheats wiring insulation and connectors.

Where a short-to-ground is located within a circuit can affect the circuit differently depending on where the control switch and circuit load device(s) are located. In Figure 10–16, if the short-to-ground occurs before the circuit switch and load, the fuse protection device blows even if the switch is open.

In Figure 10–17, the short-to-ground is located after the circuit switch but before the load. The fuse blows only after the switch is closed.

If the short-to-ground is located after the load but before the control switch, then the load remains on all the time (Figure 10–18). This type of short does not blow the fuse but may run the circuit battery down.

Several methods can be used to diagnose a short-to-ground. If the circuit fuse blows, a test light or voltmeter is difficult to use. To prevent this from occurring, connect a cycling (self-resetting) circuit breaker in place of

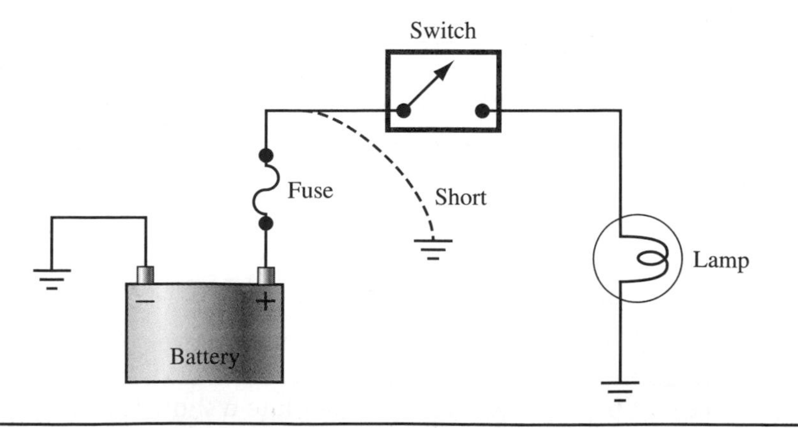

FIGURE 10–16 A short-to-ground located before the circuit control switch.

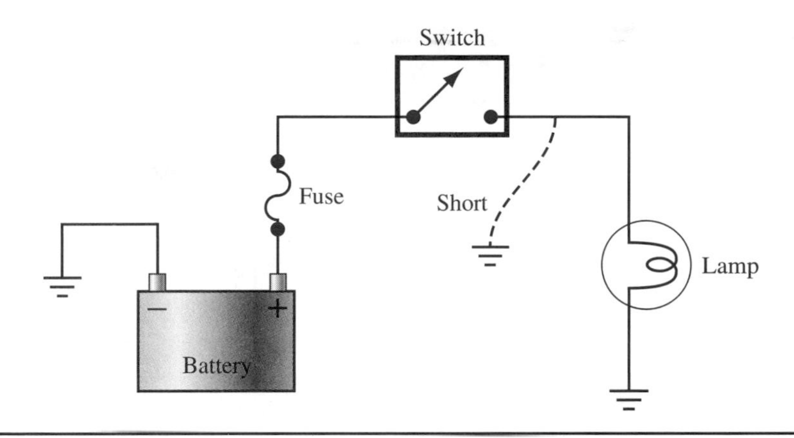

FIGURE 10–17 A short-to-ground located after the circuit control switch.

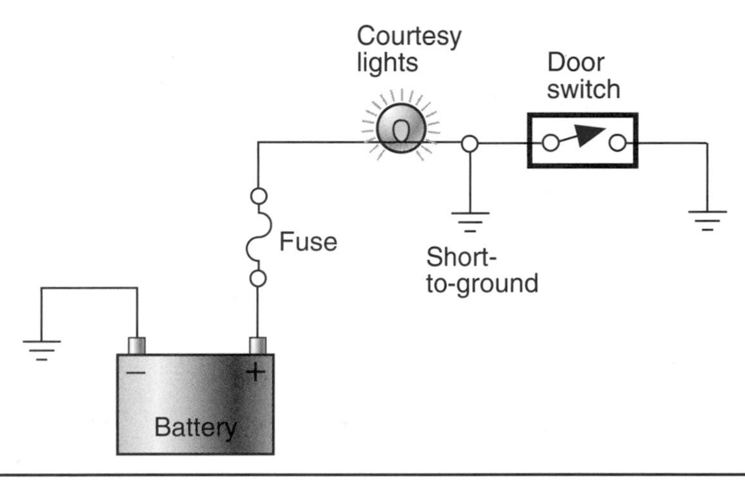

FIGURE 10–18 A short-to-ground located after the load, but before the circuit control switch.

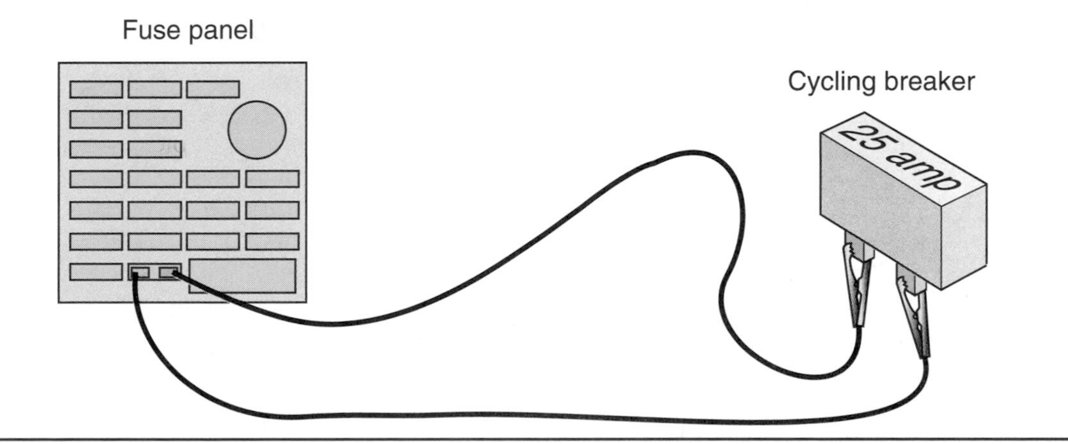

FIGURE 10–19 To protect the circuit while checking for a short-to-ground, install a cycling circuit breaker in place of the circuit fuse.

the fuse (Figure 10–19). The circuit breaker cycles open and closes to let the technician test for voltage.

With a cycling circuit breaker installed (Figure 10–20), current initially flows through the circuit, creating a magnetic field surrounding the wire while the breaker is closed. When the short-to-ground causes the circuit breaker to open, the magnetic field collapses. When the breaker resets itself, the cycling repeats continuously. When you hold a compass or gauss gauge over the circuit wiring, the needle fluctuates rapidly from the

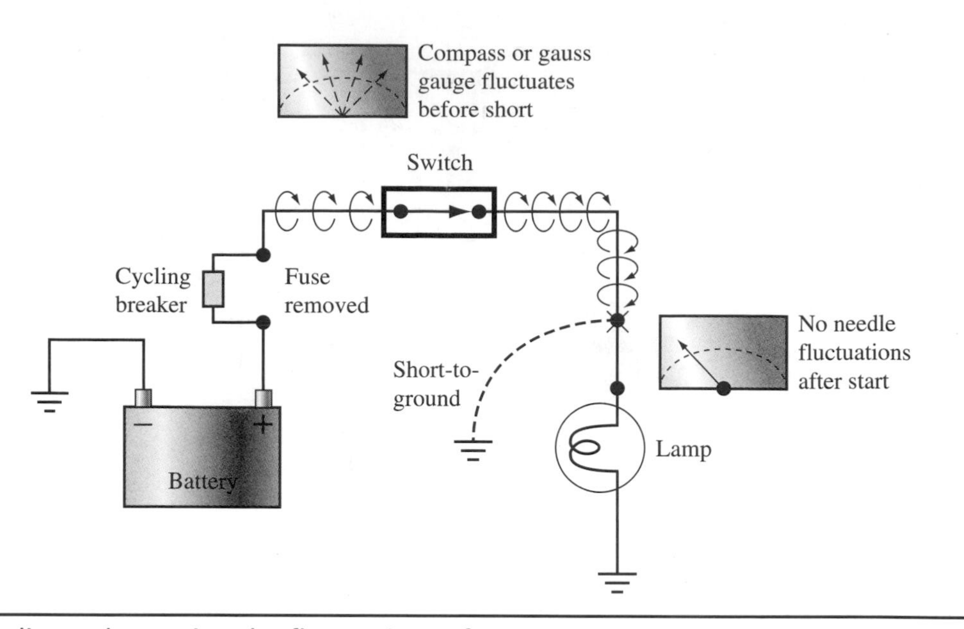

FIGURE 10–20 Finding a short using the fluctuations of a compass or gauss meter.

pulsing magnetic field *up to* the location of the short. *After* the short, there is no current flow, and the compass or gauge needle does not fluctuate. This method works well even when the wire is covered by the vehicle's trim.

You can test for a short-to-ground with a test light or buzzer (Figure 10–21), which should be used in series with a cycling circuit breaker for protection. The test light or buzzer provides resistance in the circuit and allows you to isolate the short by disconnecting the circuit components and connectors until the problem is found. If the short-to-ground is intermittent, this method is helpful because you can wiggle the harness every few inches until the test light glows or the buzzer sounds, indicating the location of the short.

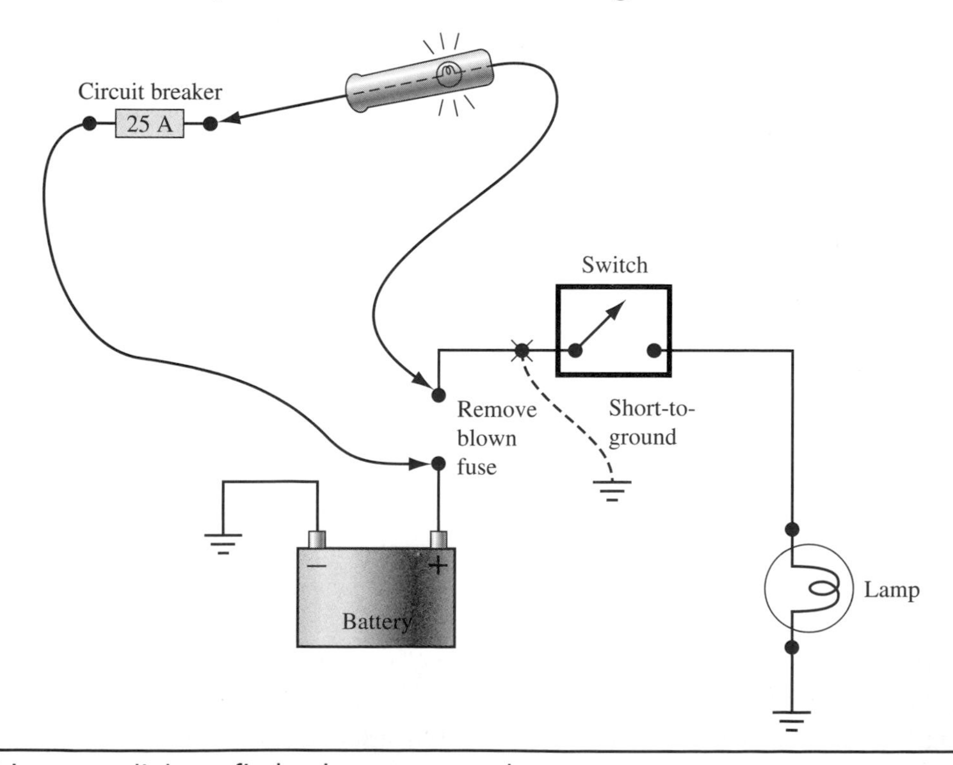

FIGURE 10–21 Using a test light to find a short-to-ground.

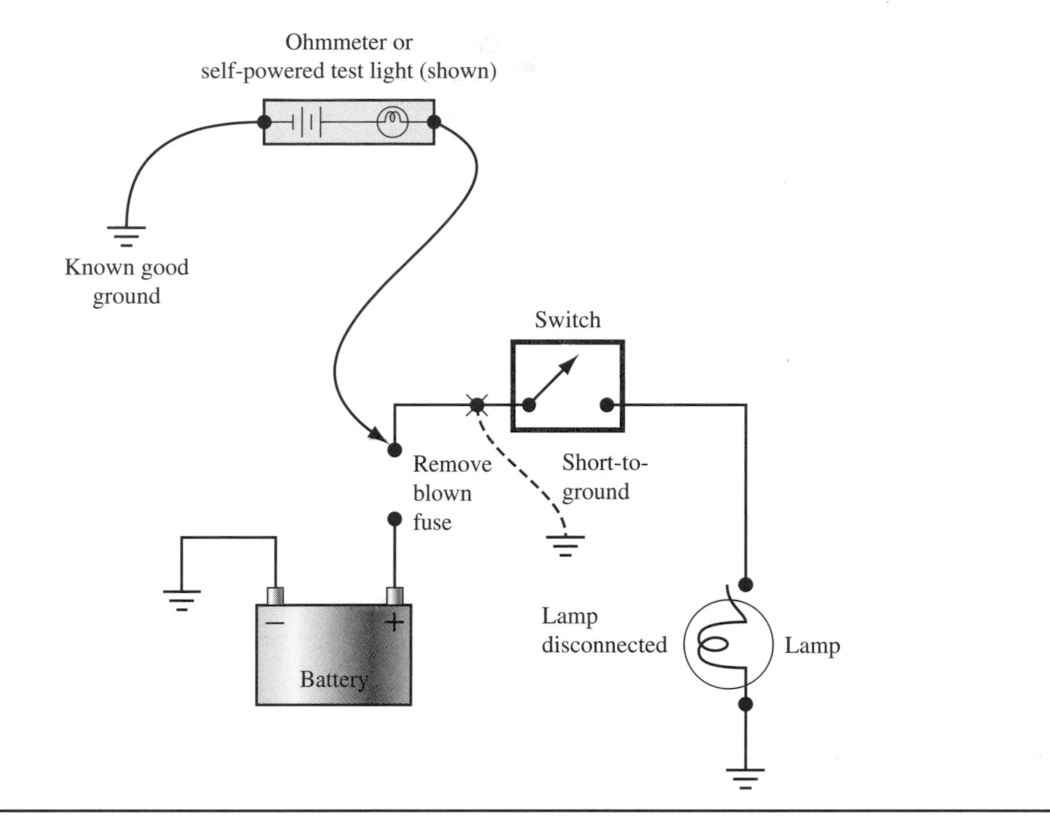

FIGURE 10–22 Using a self-powered test light to find a short-to-ground.

A self-powered test light or ohmmeter can also be used in finding a short-to-ground (Figure 10–22). Remove the blown fuse from the shorted circuit, and disconnect the battery and the circuit load component(s) before hooking up the self-powered test light. The test lamp glows or the ohmmeter records a reading if there is a short-to-ground. To locate the short, wiggle the harness every few inches until the test lamp goes out.

CAUTION *Be careful when using test lights and self-powered test lights around electronic circuits like air bags or computer-controlled circuits. Damage may occur to sensitive circuits from the current flow of the test light.*

Voltage Drops

When an electrical circuit develops problems that cause, for example, dim lights or a slowly operating starting motor, the cause is normally unwanted excessive resistance in the circuit. **Excessive resistance** is a higher-than-normal resistance that causes the current flow to be lower. Remember Ohm's law concerning resistance and current being inversely proportional to each other.

The best way to test for excessive resistance in a complete circuit is an **unwanted voltage drop test,** which is the portion of the applied circuit voltage that is used up at points of the circuit other than by the designed load components. The resistance in the circuit reduces the available electrical pressure beyond the excessive resistance point(s). Voltage drops may occur on the positive side or the ground (negative) side of the circuit. Whenever a voltage drop is suspected, both sides of a circuit should be checked for excessive resistance. In Figure 10–23, a voltmeter is being used to check for the cause of the dim light in the load of a simple series circuit.

If you recall that the series circuit law for voltage states that $V_T = V_1 + V_2 + V_3 +$ etc., then you know that the voltage across the lamp in Figure 10–23 should be 12 volts. In this case it reads 6 volts; so unwanted resistance exists somewhere in the circuit. Figure 10–24 identifies where corrosion has caused the voltage loss in the circuit, causing the lamp to operate dimly.

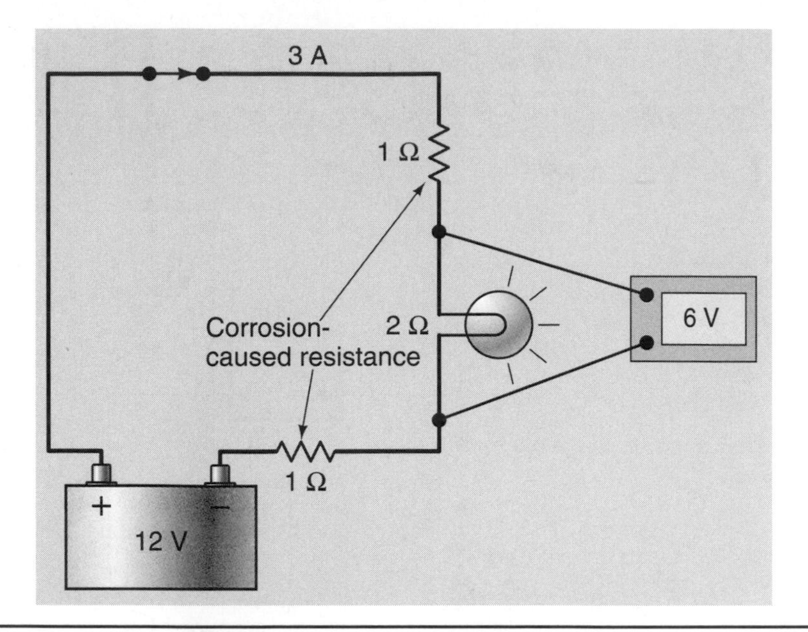

FIGURE 10–23 Checking a simple series circuit for an unwanted voltage.

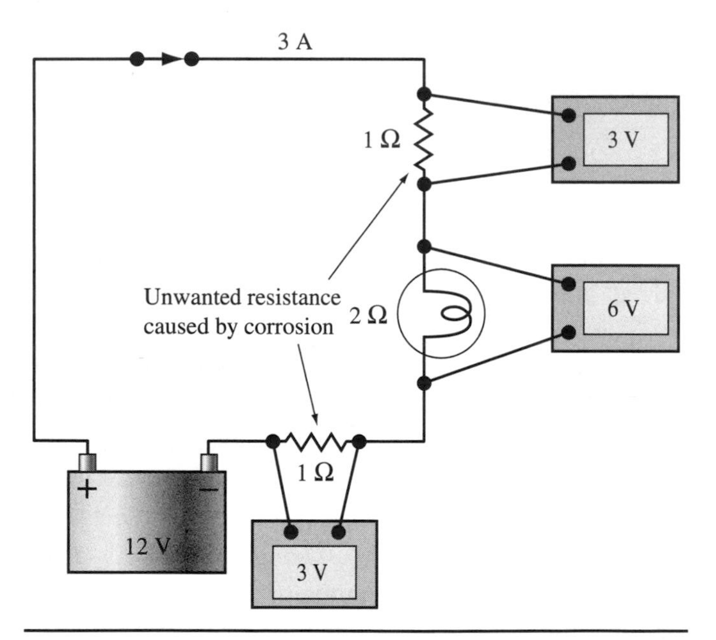

FIGURE 10–24 Using a voltmeter to find unwanted voltage drops in a circuit.

The following specifications are general rules for permitted voltage drops in circuits when manufacturers' specifications are not available.

Connection Point	Allowed Voltage Drop (V)
Ground connections	0.1
Switch contacts	0.2
System wiring	0.2
Terminal connections	0.0
Computer sensor connections	0.0 to 0.05

The more connections and wiring a complex circuit contains, the greater the chance is for unwanted resistance to eventually develop that will cause a voltage drop. In Figure 10–25, several small voltage drops add up to over 0.3 volt. Even though this is a minimum amount overall, the voltage drops could eventually increase and cause the lamps to operate dimly.

Figure 10–26 is an example of an older starter circuit being tested for a voltage drop. This circuit is designed to operate the starter motor on the entire 12 volts supplied by the battery. In this starter circuit, you should check ten areas for unwanted resistance that would cause a voltage drop: three battery cables, six connections, and the solenoid switch. A voltage drop in any one of these connections or a series of connections could effect the proper operation of the starter motor. Large current circuits like a starter motor must have very low resistance (voltage drop) in order for the 12-volt source to push as much as 200 amperes through the circuit.

Parasitic Drains

When customers complain about the vehicle battery going dead if the vehicle has been sitting for a few days or longer, and if the battery has been tested and ruled out as the cause, an unwanted key-off current drain can be suspected. A **parasitic drain** is defined as a current drain present in the electrical system after the ignition switch is in the key-off position. Of the several possible

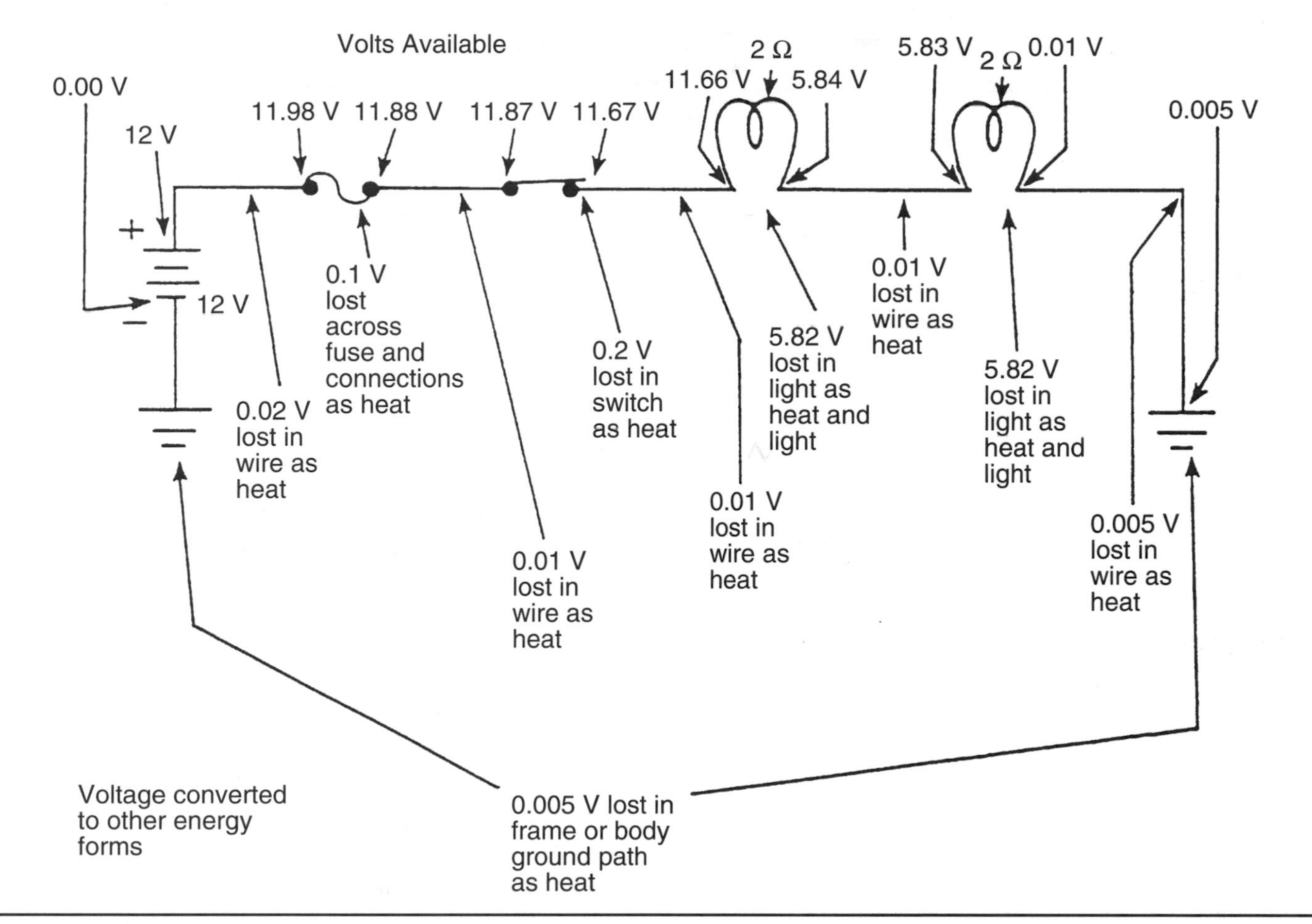

FIGURE 10–25 Adding up all the unwanted voltage drops in a circuit.

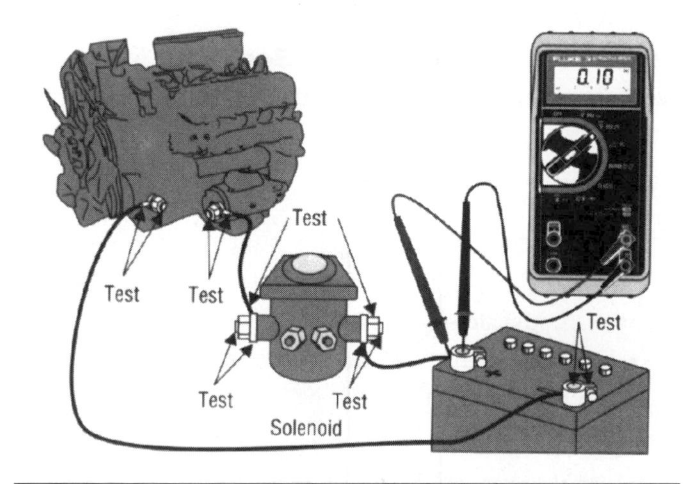

FIGURE 10–26 Checking a starter circuit for a voltage drop.

(Courtesy of The Fluke Corporation)

causes of an excessive parasitic drain, the most common is a light not being turned off, such as a trunk light, engine compartment light, or glove box light.

Acceptable limits of parasitic loads depend on the vehicle manufacturer and on the amount of electrical accessories on the vehicle, but most consider the maximum limit at 35 to 50 milliamperes (0.035 to 0.050 ampere). Modern vehicles with several on-board computers and electrical accessories may have as much as 200-milliampere (0.2-ampere) parasitic drain designed into the system that can take several minutes to shut off or to reach the sleep mode. Some of the more common acceptable parasitic drains are listed in Table 10–1.

Normal Electrical Parasitic Drains

The preferred procedure for performing a parasitic drain test may vary with according to the manufacturer due to specific vehicle electrical applications or complexity. Six methods have been identified to find an unwanted parasitic drain. The first three methods are *not* considered reliable on newer, computerized vehicles and are included here only to give you a complete understanding

TABLE 10–1

Load Device	Parasitic Drain (mA)
ECM	3–10
PCM	5–3
ABS computer	1–3
Digital clock	2–5
Electronic radio	3–8
CD changer	3–5
Memory seats	2–4
Alternator	0.5–2
Door lamp/chime module	0.5–1
Horn module	0.25–0.5
Twilight sentinel module	3–8

of common troubleshooting practices from the past. Methods four and five have been the most popular methods used in the last 10 years because of their accuracy in reading the actual parasitic amperage drain. The sixth method is becoming more popular with computerized vehicles because it does not require the disconnection of the battery cable as part of the measurement process.

Method 1: Voltmeter in Series This method is also known as the voltage draw test, which is *not* a reliable test on newer computerized vehicles or when a high-impedance digital voltmeter is used. An *analog* voltmeter is connected in series with the negative battery cable (Figure 10–27), and a reading higher than 6 volts indicates a parasitic drain. A digital voltmeter reads the source voltage of 12 volts even if there is not a parasitic drain, which makes this method unreliable with a digital voltmeter.

Method 2: Test Light in Series This method uses the test light connected in series in the same way as the voltmeter is used in method 1. This method is also *not* a reliable test. The brightness of the test light is used to determine whether a parasitic drain exists. However, it is hard to determine how much parasitic drain exists by looking at how bright the test lamp is.

Method 3: Ohmmeter in Parallel This method uses an ohmmeter connected in parallel to the vehicle's electrical circuit and in place of the battery (Figure 10–28). This method is *not* a reliable test because some computer circuits have to be energized before going into a standby or sleep mode, and the ohmmeter battery will not operate computer switching circuits.

Method 4: Ammeter in Series This method uses an ammeter in series with the vehicle electrical

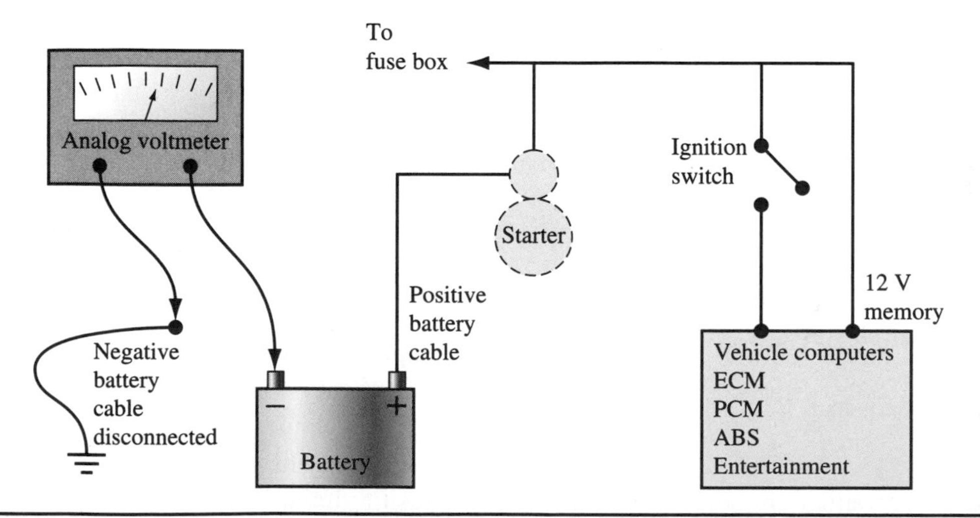

FIGURE 10–27 An analog voltmeter connected in series to the negative cable to find a parasitic drain.

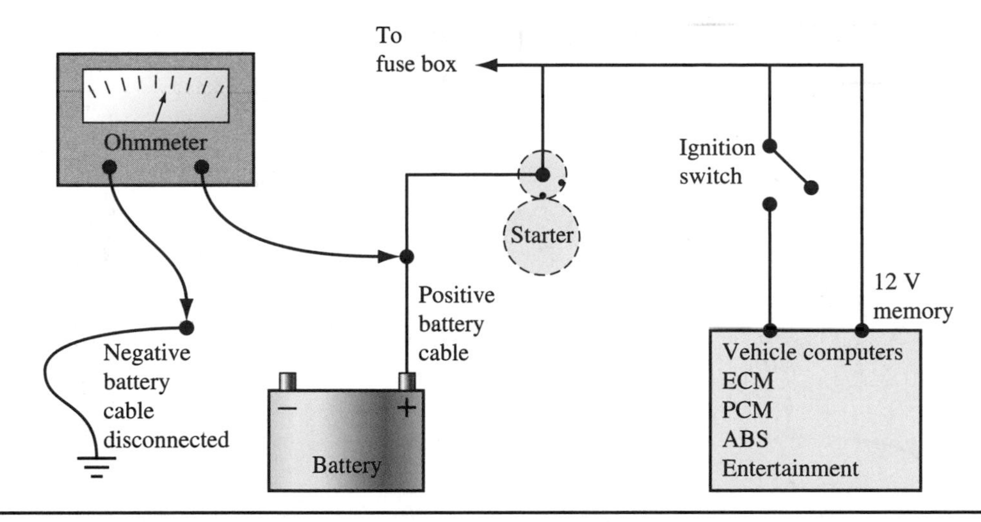

FIGURE 10–28 With the battery disconnected, an ohmmeter is connected in parallel to the circuit to check for a parasitic drain.

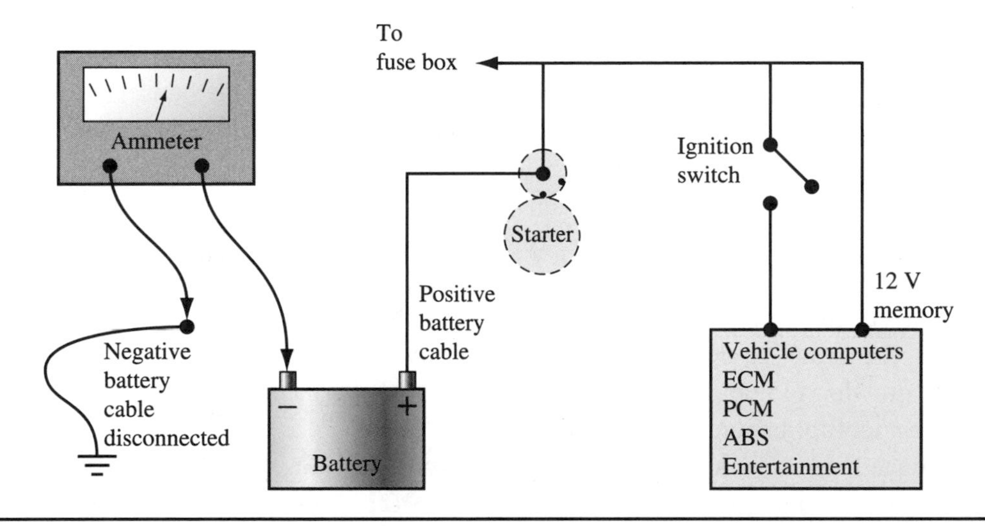

FIGURE 10–29 An ammeter is connected in series to the negative cable to find a parasitic drain.

circuit, placed between the negative battery post and the negative battery cable (Figure 10–29). This method is a *reliable* test that is used widely by technicians because it shows the exact current draw on the electrical system. However, take a precaution when using this method: The ammeter has a current capacity limit that should never be exceeded when checking for parasitic drains. Most quality digital ammeters are fuse protected to prevent overloading the meter and can handle up to 10 amperes satisfactorily. Typical parasitic drains are less then 5 amperes, unless major circuit problems exist.

Method 5: Voltage Drop This method is very effective in testing for parasitic drains in a circuit under actual voltage conditions. Some computer circuit faults may be voltage-sensitive-related and show up only when stressed under normal working conditions. In this method, a fused 1-ohm, 10-watt resistor is put in series between the negative battery post and the negative battery cable (Figure 10–30). A digital voltmeter is connected in parallel across the 1-ohm resistor. If there is a parasitic drain present in the electrical system, the current has to pass through the 1-ohm resistor and develop a

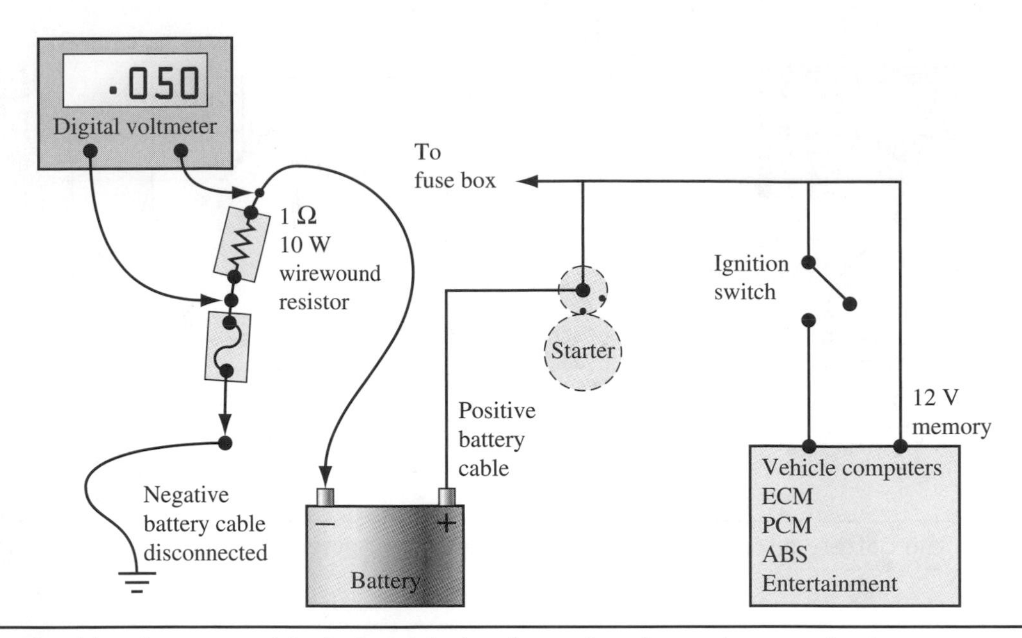

FIGURE 10–30 Checking for a parasitic drain with the dynamic voltage drop method.

voltage drop across it. Using Ohm's law ($E = I \times R$), the amount of voltage drop across the resistor is directly related to the amount of current flowing in the circuit.

For example, if the voltage reading across the 1-ohm resistor is 0.050 volts, this would be the same as 0.050 amperes (0.050 volts ÷ 1 ohm = 0.050 amperes). If you remember the conversion method from Chapter 3, then 0.050 amperes is the same as 50 milliamperes. Because this method is growing in popularity, making a parasitic drain, 1-ohm test tool for under $10.00 can be a handy diagnostic addition to have a tool kit. Figure 10–31 is a sample project that can be made with simple parts from your local electronic parts supply store.

For diagnostic methods 1–5, the procedure used to determine the circuit or electrical device that is causing the parasitic drain, the *fuse pull technique,* is the most common. In Figure 10–32, the fuses are being removed one at a time while the tester views the ammeter for any change in its reading. If all the fuses have been removed and the parasitic drain is still present, then the most likely suspect is an excessively dirty battery case or an internal alternator problem.

Method 6: Ignition-Off Parasitic Drain (IOPD) This method is gaining popularity because in modern vehicles a parasitic drain may be due to an

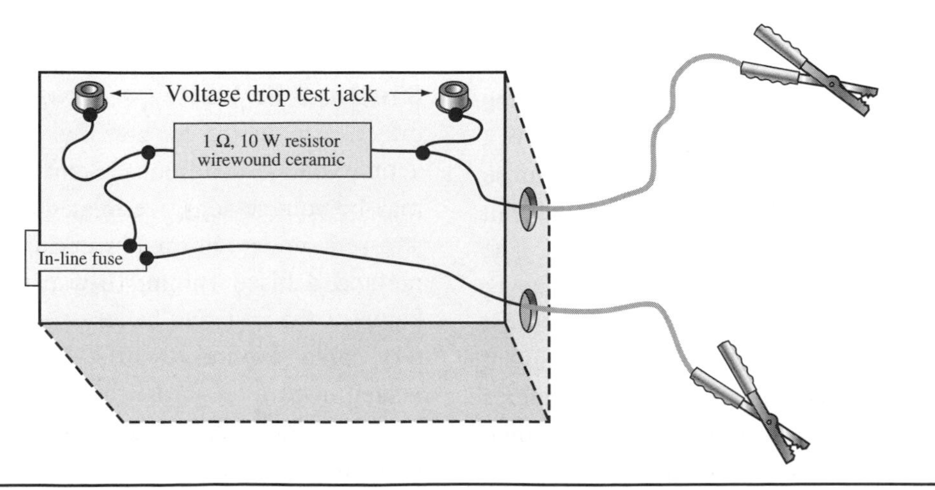

FIGURE 10–31 A parasitic drain test tool project.

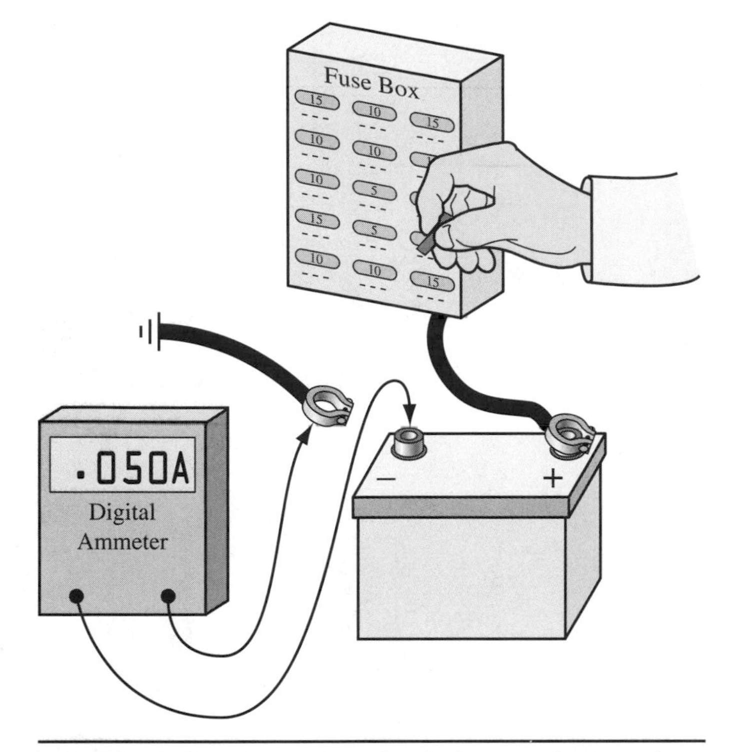

FIGURE 10–32 Removing circuit fuses to isolate the parasitic drain.

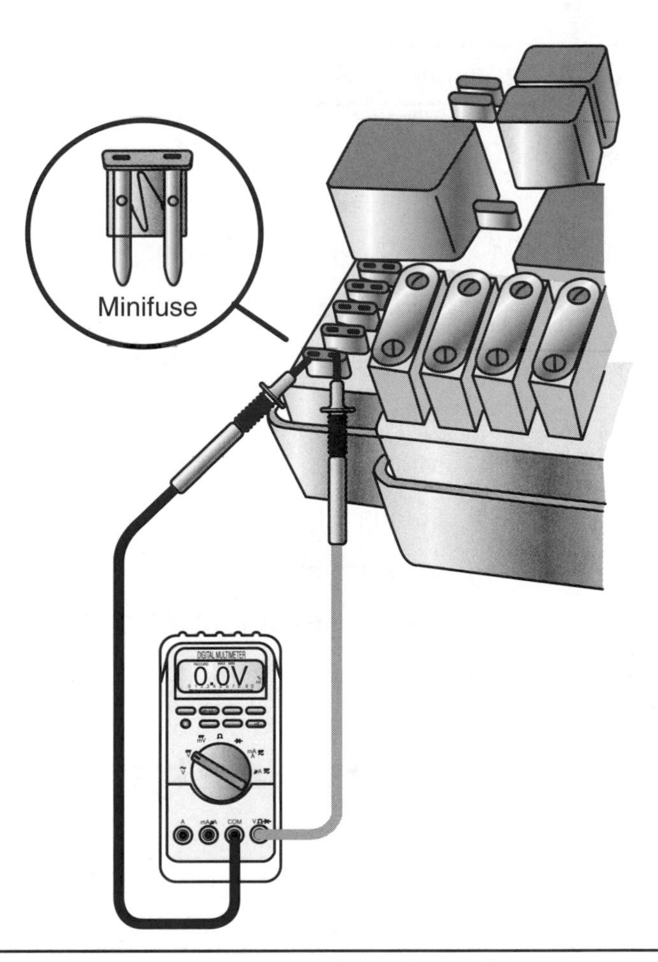

FIGURE 10–33 Measuring for voltage drop across the fuse with a digital voltmeter set on the millivolts setting.

electrical circuit control module that is not powering down (i.e., going to sleep). The control module parasitic drain may be caused by a hardware or software problem. If a technician uses one of the previous five diagnostic parasitic drain methods, main battery power may be lost in the process, possibly temporarily powering down the module but not identifying the problem. Some modules take several minutes or more to go to sleep; so isolating a module by pulling a fuse does not work.

To determine current flow in a circuit without removing a fuse, the technician can simply use the IOPD method by measuring the voltage drop across a circuit fuse (Figure 10–33) with a digital voltmeter set on the millivolts scale. During normal operation, the fuse in a circuit gets hot and has a small resistance that can be measured in millivolts. Remember Ohm's law, the current is directly proportional to the voltage and inversely proportional to the resistance. This rule applies in this method. If the current flow in the circuit is minimal or none, then the millivolts drop measured across the fuse is close to zero also. If there is a higher-than-normal parasitic current drain in the circuit, then the millivolts drop measured across the fuse indicates the proportionate voltage reading. The amount of current flowing through the circuit can be calculated using the reference chart in Table 10–2 after measuring the millivolts drop across the fuse.

For example, the technician reads an 11.87-millivolt drop across a 20-ampere standard blade fuse. Using Table 10–2, the Division Number for a 20-ampere standard blade fuse is 5.52. The technician then performs the following calculation: 11.87 ÷ 5.52 = 2.15 amperes current draw.

Parasitic Draw Test Switch Some technicians have also been successful locating parasitic drains by using a parasitic draw test switch (Figure 10–34) in combination with a high-capacity (at least 20-ampere) ammeter. When vehicle on-board computer modules take several minutes to go to sleep and the technician wants to perform a variety of tests to verify proper module operation, then a parasitic draw test switch can be used. This test switch is connected in series between the

TABLE 10–2 Ignition-Off Parasitic Drain

Fuse Value (A)	Fuse Type	Division Number
5	Miniblade	18.107
10	Miniblade	8.89
15	Miniblade	6.67
20	Miniblade	5.71
25	Miniblade	4.33
30	Miniblade	3.51
3	Standard blade	35.81
5	Standard blade	20.48
7.5	Standard blade	12.59
10	Standard blade	10.17
15	Standard blade	6.69
20	Standard blade	5.52
25	Standard blade	4.13
30	Standard blade	3.98
20	Maxiblade	6.73
30	Maxiblade	3.89
40	Maxiblade	3.65
50	Maxiblade	2.78
60	Maxiblade	2.65

Instructions:

Step 1: Determine the fuse type and amperage rating.

Step 2: Measure the millivoltage (mV) drop across the fuse using a digital voltmeter.

Step 3: Determine the division number for fuse type and amperage rating.

Example: The circuit has a 20-ampere standard blade fuse, and the millivolt drop measured in the circuit is 11.87 mV.

$$11.87 \div 5.52 = 2.15 \text{ A current draw}$$

negative battery post and the negative battery cable, and it allows the vehicle to operate as normal when the switch is in the ON position. With the ammeter connected across the test switch terminals, the technician waits for all delay circuits to shut off (this can take several minutes or longer depending on manufacturer). When the test switch is turned to the OFF position, the ammeter is now in series with the electrical circuit and reading the total amperage of the vehicle electrical system (this is why a high-capacity ammeter is needed). The amount of current flowing can now be recorded, and the isolation process can proceed by removing fuses one at a time or by disconnecting individual modules or circuits to locate the parasitic drain. Remember to close the test switch before attempting to operate any of the vehicle's electrical systems. This switch is intended for testing parasitic drains only, and it is not to be used as a battery on/off isolation switch.

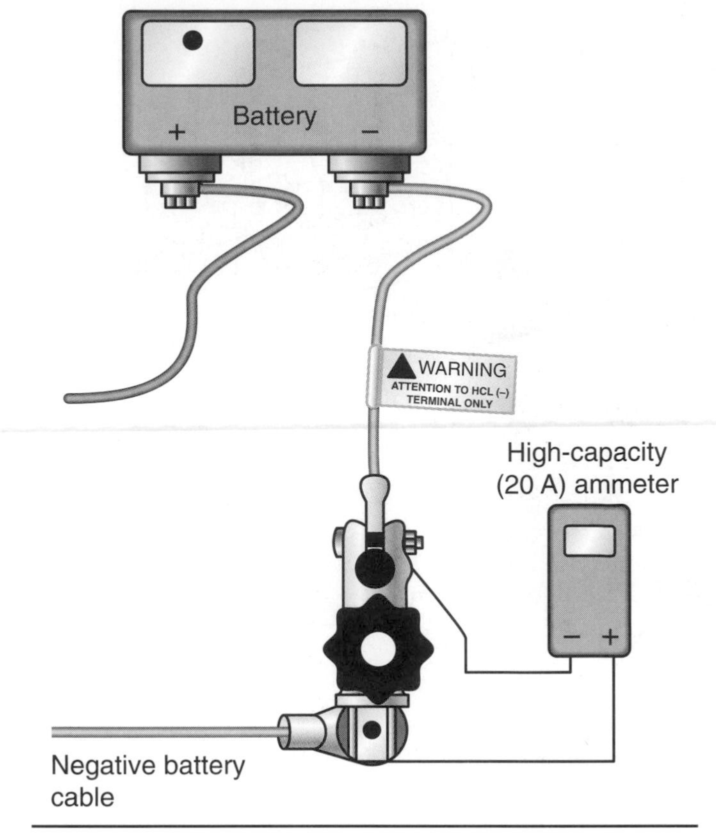

FIGURE 10–34 Checking for a parasitic battery drain using a parasitic draw test switch.

Sneak Circuits

A **sneak circuit** is an unwanted backfeed of current from one circuit through another that shares the same power feed or ground path. This can happen when a shared fuse is blown or a shared ground is disconnected. Figure 10–35 is an example of a sneak circuit where

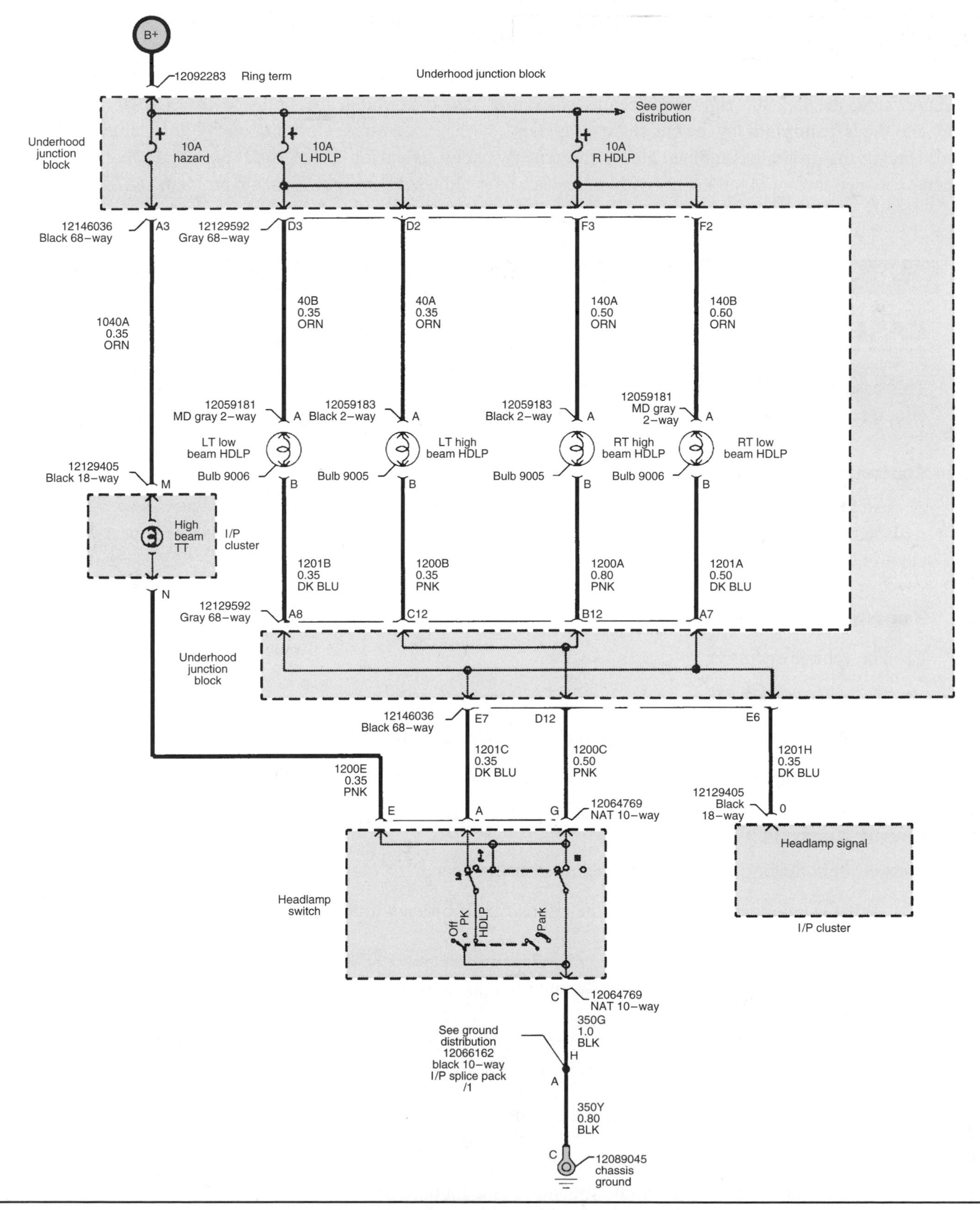

FIGURE 10–35 Locating a sneak circuit in a headlamp schematic.

(Courtesy of The Saturn Corporation)

there is a 10-ampere fuse for the left low and high sealed beam headlamps and a separate 10-ampere fuse for the right low and high sealed beam headlamp circuit. On the ground side, the left and right high beams share a ground, and the left and right low beams share a separate ground. Having the grounds combined allows the driver to operate between low and high beams with one switch. When either circuit fuse blows, a series circuit is created and the headlamps on the side with the blown fuse dimly glow because voltage from the good headlamp fuse *sneaks* through the bulbs to the headlamp switch and then to ground. Customers may not realize a sneak circuit is taking place, and they replace the headlamp bulbs, only to find the same problem.

CASE STUDIES

CASE 1

This case study is typical of an open circuit problem that technicians often encounter. Knowledge of the circuit and the simple test equipment needed can help you perform the diagnosis quickly and efficiently.

Customer Complaint

A customer brings his car into the shop because the left rear taillamp does not work. The customer replaced the left rear taillamp, but the lamp still does not light. The rest of the rear lamps appear to operate properly (Figure 10–36).

Known Information

- ❑ The vehicle operating voltage is 14 volts.
- ❑ The circuit fuse is OK.
- ❑ All taillamps, license lamp, and marker lamp bulbs are OK.
- ❑ The headlamp switch is OK.

Circuit Analysis

Answer the following questions on a *separate* sheet of paper:

1. What is the most likely cause of the left rear taillamp's not illuminating? ____________________

   ____________________.

2. What should the voltage be at connector A of the 0.8 BRN wire at the LH taillamp? ____________________

   ____________________.

3. Can the 0.8 BLK wire from the LH taillamp to the S317 connector be open? ____________________

   Why? ____________________

   ____________________.

4. How could you verify the circuit problem using a jumper wire? ____________________

   ____________________.

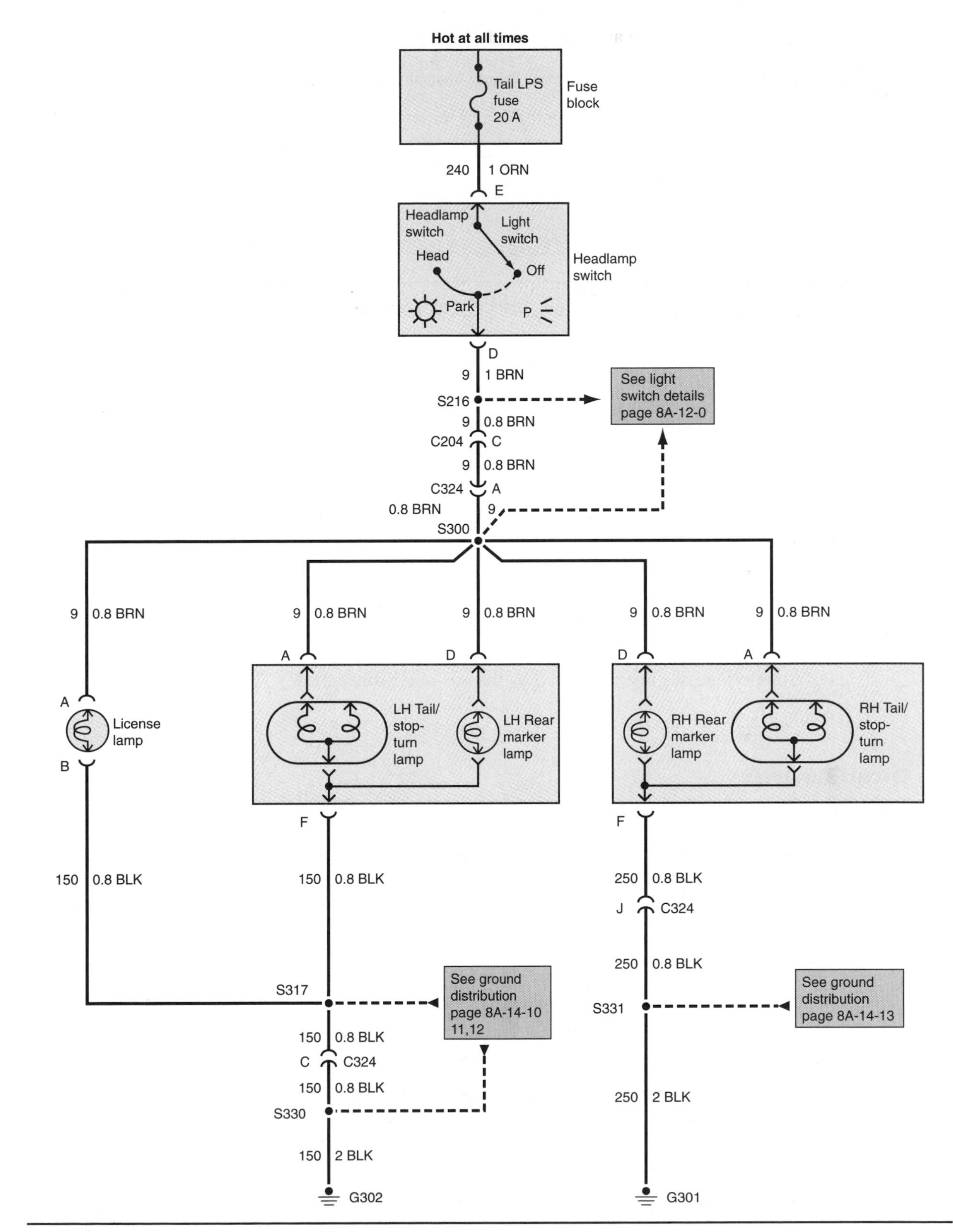

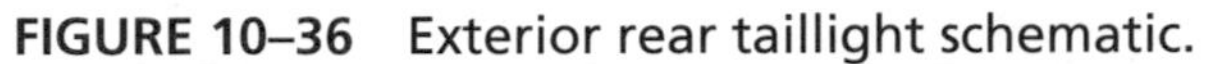

FIGURE 10–36 Exterior rear taillight schematic.

5. What is the best way to repair this system? ______.
6. What diagnostic steps were helpful when troubleshooting this system or explaining it to the customer? ______
______.
7. What is your analysis of this case study? ______
______
______.

CASE 2

Using the same circuit diagram (Figure 10–36), complete the following customer complaint.

Customer Complaint

The 20-ampere taillamp fuse blows when the headlamp switch is turned on to the park or headlamp position. The customer replaced the fuse four times, but it blows instantly.

Known Information

- ❑ The vehicle operating voltage = 14 volts.
- ❑ Twenty-ampere fuse has been replaced by the customer four times.
- ❑ The headlamp switch is OK.

Circuit Analysis

Answer the following questions on a *separate* sheet of paper:

1. What is the most likely cause of the taillamp fuse to blow when the headlight switch is turned on?
______.
2. What diagnostic steps and tools would you use to find the suspected problem in this taillamp circuit?
______
______.
3. If you disconnected the 0.8 BRN wire at C324 and the fuse still blows, but it does not blow when the 0.8 BRN wire is disconnected at C204, what is the suspected problem in the circuit? ______
______.
4. How could you verify the circuit problem using a jumper wire? ______.
5. What is the best way to repair this system? ______
______.

6. What diagnostic steps were helpful when troubleshooting this system or explaining it to the customer? ______________________________

______________________________.

7. What is your analysis of this case study? ______________________________

____

______________________________.

CASE 3

The following case study, shown in Figure 10–37, is typical of a short circuit problem that technicians often encounter. Knowledge of the circuit and use of simple diagnostic test tools can help you perform the diagnosis quickly and efficiently.

Customer Complaint

A customer complains that the coolant temperature gauge reads hot any time the ignition switch is on or the vehicle is running. The customer replaced the temperature gauge and the temperature gauge sensor, but the problem still exists.

Known Information

- ❑ The vehicle operating voltage = 14 volts.
- ❑ All other instrument cluster gauges and indicator lights operate properly.
- ❑ The ignition switch is OK.

Circuit Analysis

Answer the following questions on a *separate* sheet of paper:

1. What is the most likely cause for the temperature gauge's reading hot if the gauge and sending unit have both been replaced? ______________________________

   ______________________________.

2. What should happen to the temperature gauge reading if the coolant temperature gauge sensor is disconnected while the engine was running? ______________________________.

3. If you disconnect the temperature gauge sensor, and the gauge still shows hot, but moves back to cold when the 0.35 DK GRN is disconnected at the C205 connector, what is the most likely cause of this circuit problem? ______________________________

   ______________________________.

4. What is the best way to repair this system? ______________________________.

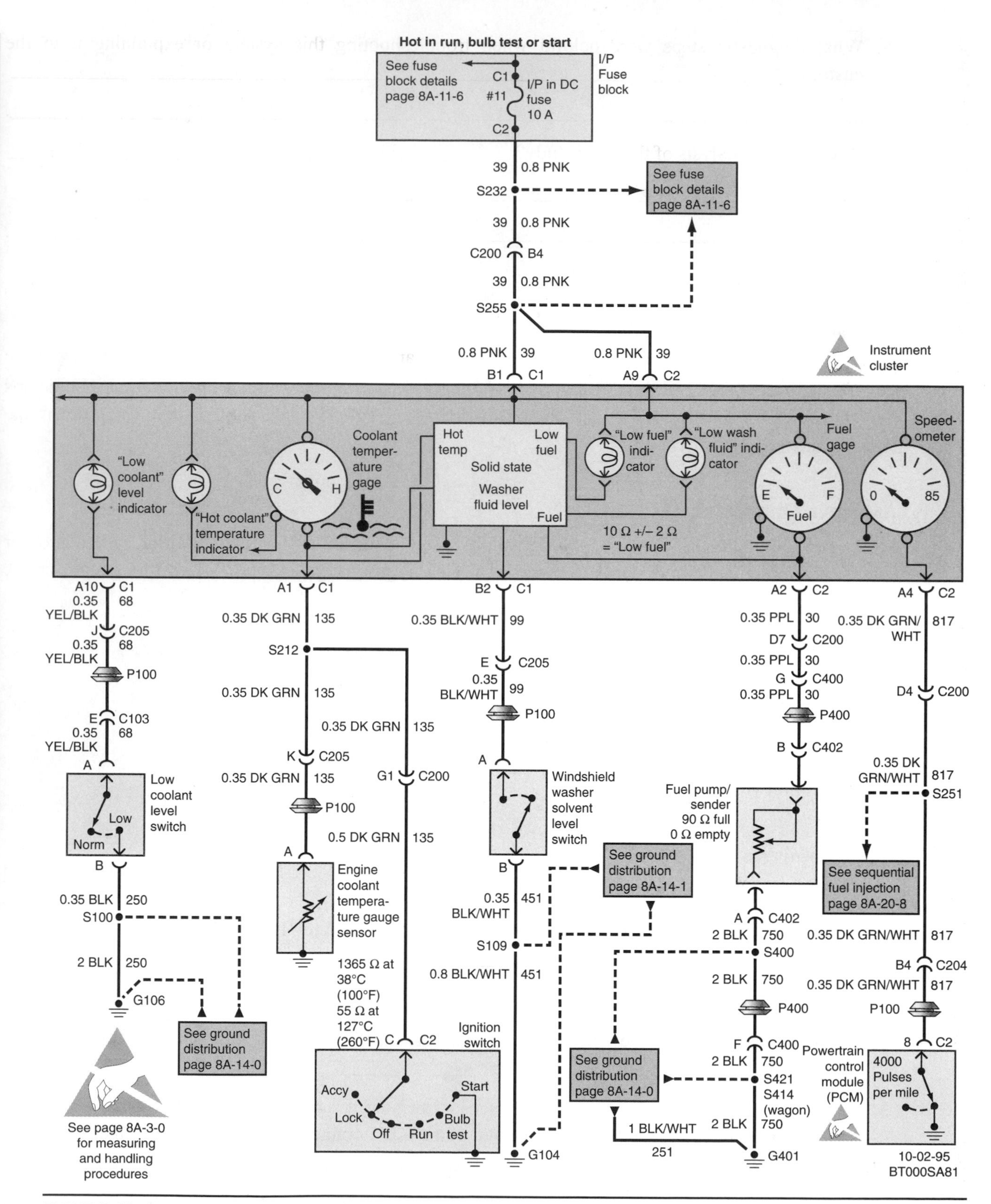

FIGURE 10–37 Instrument cluster and control circuit schematic.

5. What diagnostic steps were helpful when troubleshooting this system or explaining it to the customer?____.

6. What is your analysis of this case study? ____.

CASE 4

The following case study (Figure 10–38) is typical of a parasitic drain problem that technicians often are required to diagnose. Knowledge of the circuit, test procedures, and use of simple diagnostic test tools can help you diagnose the circuit safely and efficiently.

Customer Complaint

A customer complains that the vehicle battery runs down after the vehicle sits for a few days. The battery has been replaced twice, and the alternator tested satisfactorily.

Known Information

- ❑ The vehicle operating voltage = 14 volts.
- ❑ The battery and alternator both test satisfactory.
- ❑ A parasitic drain is suspected.

Circuit Analysis

Answer the following questions on a *separate* sheet of paper:

1. What diagnostic steps and test tools would you use to check this vehicle for a parasitic drain? ____.

2. The vehicle was drawing 2 amperes of parasitic current until the 10-ampere RH DR (righthand door) fuse was pulled out. The parasitic drain then went down to 35 milliamperes. What would be the *most likely* cause for the excessive parasitic drain in the system? _________________________.

3. How can you verify that the circuit problem exists in the identified problem component? ____.

4. What is the best way to repair this system? ____.

5. What diagnostic steps were helpful when troubleshooting this system or explaining it to the customer? ____.

6. What is your analysis of this case study? ____.

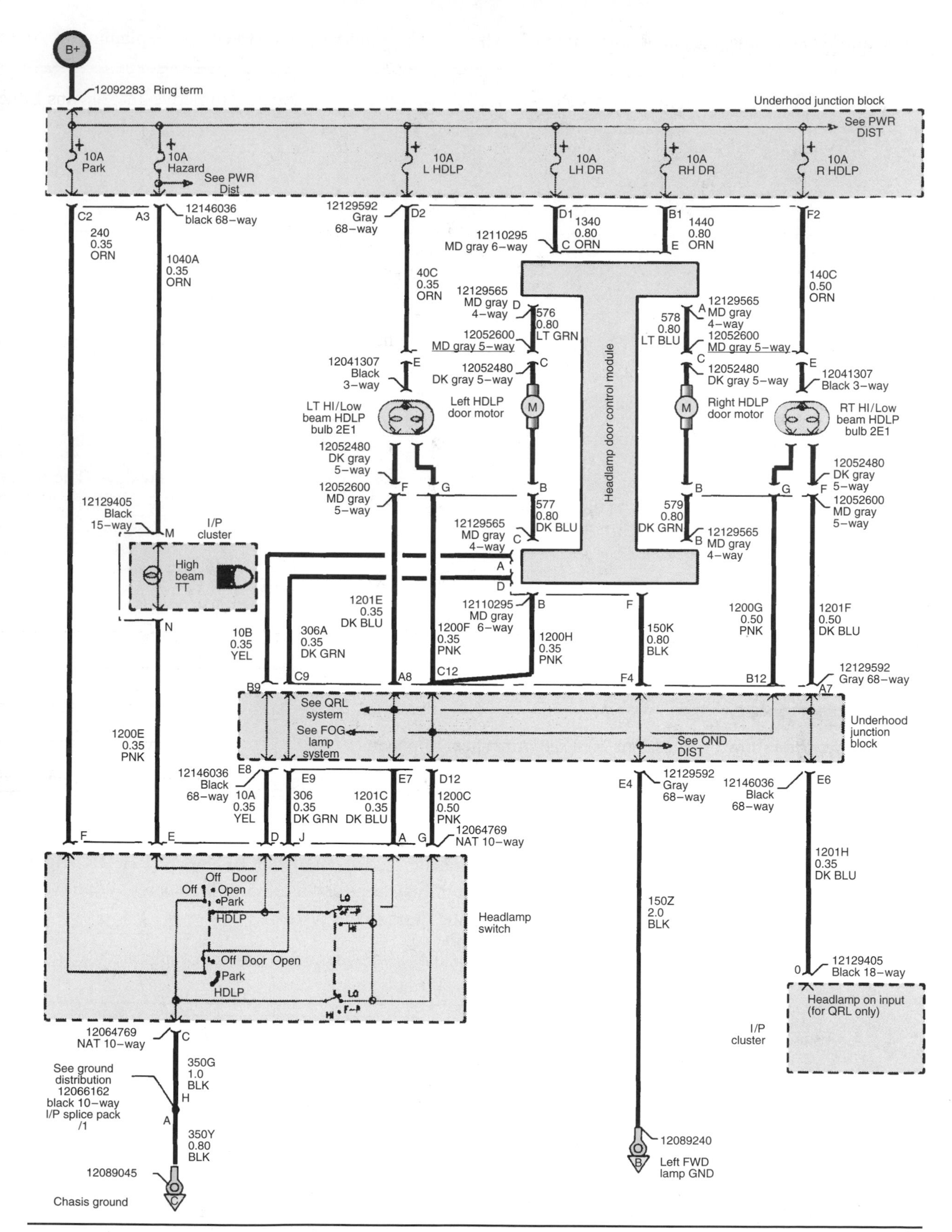

FIGURE 10–38 High/low beam headlamp schematic.

(Courtesy of The Saturn Corporation)

Hands-On Vehicle Tests

The following seven hands-on vehicle checks are included in the NATEF (National Automotive Technicians Education Foundation) Task List. Complete your answers to the following questions on a *separate* sheet of paper.

Performance Task 1

Task Description
Use wiring diagrams during the diagnosis of electrical circuit problems.

Task Objectives

- ❑ Obtain a vehicle that can be used for this task. What model and year of vehicle are you using for this task? ______________________________.
- ❑ What electrical circuit problem(s) exist with this vehicle? ______________________________

  ____

  ______________________________.
- ❑ Locate a wiring diagram(s) that will show the electrical circuit identified.
- ❑ Using the wiring diagram(s), identify areas in the circuit such as connectors, wires, fuses, and components, that can cause the problem.

Identified Area	Symptom	Cause	Test Procedure

- ❑ Perform the diagnostic steps and procedures identified in the previous question. Does this help identify the circuit problem(s)?______________________________.
- ❑ What diagnostic steps were helpful when troubleshooting this system? ______________________________.
- ❑ What is your analysis of using wiring diagrams to help diagnose electrical circuit problems? ______

  ______________________________.

Performance Task 2

Task Description
Check electrical circuits with a test light. Determine the necessary action.

Task Objectives

- ❑ Obtain a vehicle that can be used for this task. What model and year of vehicle are you using for this task? ______________________________.

❑ Using a wiring diagram for the vehicle, determine electrical circuits that are safe to check with a test light. Verify circuits to test with your instructor.

❑ List the circuit(s) to be tested with the test light.

1. ______________________ 4. ______________________

2. ______________________ 5. ______________________

3. ______________________ 6. ______________________

❑ List the procedure(s) used to safely connect a test light to a vehicle's electrical circuit.

____

____

____

❑ Check the operation of the circuits you listed with the test light. What can you determine about using a test light on each of the circuits?

1. ______________________ 4. ______________________

2. ______________________ 5. ______________________

3. ______________________ 6. ______________________

❑ What are the benefit(s) of using a test light on electrical circuits compared to other electrical testing equipment?____

____.

Task Summary

After performing the preceding NATEF task, what can you determine will be helpful in using test lights to perform simple electrical circuit troubleshooting?______________________________

____.

Performance Task 3

Task Description

Measure source voltage and perform voltage drop tests in electrical/electronic circuits using a voltmeter; determine the necessary action.

Task Objectives

❑ Obtain a vehicle that can be used for this task. What model and year of vehicle are you using for this task? ____.

❑ Using a wiring diagram for the vehicle, identify an electrical circuit that is suitable for checking voltages and voltage drops. Verify circuits to test with your instructor.

❑ List the circuit(s) to be tested for voltage and voltage drops.

1. ____________________ 4. ____________________

2. ____________________ 5. ____________________

3. ____________________ 6. ____________________

❑ List the procedure(s) used to test for system voltage and voltage drops in an electrical circuit.

____

____

____

❑ What is the system voltage of each circuit tested, as well as any voltage drop(s) measured?

System voltage: ________ Voltage drops: ________, ________, ________, ________.

System voltage: ________ Voltage drops: ________, ________, ________, ________.

System voltage: ________ Voltage drops: ________, ________, ________, ________.

System voltage: ________ Voltage drops: ________, ________, ________, ________.

❑ Do the voltage drops in each circuit tested equal the system total voltage? ____________________

____.

Task Summary

After performing the preceding NATEF task, what can you determine will be helpful in knowing how to check for voltage drops in an electrical circuit in reference to system voltage? ____________________

____.

Performance Task 4

Task Description

Check current flow in electrical/electronic circuits and components using an ammeter. Determine the necessary action.

Task Objectives

❑ Obtain a vehicle that can be used for this task. What model and year of vehicle are you using for this task? ____.

❑ Using a wiring diagram for the vehicle, determine electrical circuits that are safe to check the current flow with an ammeter. Verify circuits to test with your instructor.

❑ List the circuit(s) to be tested for current flow.

1. ____________________ 3. ____________________

2. ____________________ 4. ____________________

- ❑ List the procedure(s) used to safely connect an ammeter to a vehicle's electrical circuit.

  ____

  ____

  ____

- ❑ Check the operation of the preceding circuits with an ammeter. What can you determine about testing for current flow on each of the circuits?

  1. ____________ 3. ____________

  2. ____________ 4. ____________

- ❑ What would be the benefit(s) of using an ammeter on electrical circuits compared to other electrical testing equipment? ______________________________

  ____

  ______________________________.

Task Summary

After performing the preceding NATEF task, what can you determine will be helpful in testing a circuit with an ammeter to troubleshoot an electrical circuit? ______________________________

______________________________.

Performance Task 5

Task Description

Check electrical circuits using fused jumper wires; determine the necessary action.

Task Objectives

- ❑ Obtain a vehicle that can be used for this task. What model and year of vehicle are you using for this task? ______________________________.
- ❑ Obtain a wiring diagram of the horn circuit for the vehicle. Determine that the circuit is safe to use with a jumper wire to bypass circuit components. Verify circuit components to bypass with your instructor.
- ❑ List the circuit components to be bypassed with the jumper wire.

  1. ____________ 3. ____________

  2. ____________ 4. ____________

- ❑ List the procedure(s) used to safely connect the jumper wire to bypass circuit components.

  ____

Task Summary

After performing the preceding NATEF task, what can you determine will be helpful in knowing how to use a jumper wire as an effective tool in diagnosing a horn circuit problem or other electrical circuit problems? ______________________________

______________________________.

Performance Task 6

Task Description

Locate shorts, grounds, opens, and resistance problems in electrical/electronic circuits. Determine the necessary action.

Task Objectives

Short-to-Voltage

- ❑ Obtain a vehicle that can be used for finding a short-to-voltage in an electrical system. What model and year of vehicle are you using for this task? ______________________________.
- ❑ Determine which electrical circuit has a short-to-voltage. ______________________________.
- ❑ What symptoms are present in this short-to-voltage? ______________________________.
- ❑ Obtain a wiring diagram of the circuit that has the short-to-voltage. List the components, connectors, and wiring that can be isolated to determine the cause of the short. ______________________________

  ______________________________.
- ❑ List the diagnostic procedure(s) and test equipment used to safely locate the short-to-voltage.

  ______________________________.
- ❑ Where is the short-to-voltage in the circuit, and what caused it to happen? ______________________________

  ______________________________.
- ❑ What diagnostic steps were helpful when troubleshooting for a short-to-voltage? ______________________________

  ______________________________.
- ❑ What is your analysis of finding a short-to-voltage in an electrical circuit? ______________________________

  ______________________________.

Short-to-Ground

- ❑ Obtain a vehicle that can be used for finding a short-to-ground in an electrical system. What model and year of vehicle are you using for this task? ______________________________.
- ❑ Determine which electrical circuit has a short-to-ground. ______________________________.
- ❑ What symptoms are present in this short-to-ground? ______________________________.

- ❑ Obtain a wiring diagram of the circuit that has the short-to-ground. List the components, connectors, and wiring that can be isolated to determine the cause of the short. ______________________________
- ❑ List the diagnostic procedure(s) and test equipment used to safely locate the short-to-ground. ______________________________
- ❑ Where is the short-to-ground in the circuit, and what caused it to happen? ______________________________
- ❑ What diagnostic steps were helpful when troubleshooting for a short-to-ground? ______________________________
- ❑ What is your analysis of finding a short-to-ground in an electrical circuit? ______________________________

Open Circuit

- ❑ Obtain a vehicle that can be used for finding an open in an electrical system. What model and year of vehicle are you using for this task? ______________________________.
- ❑ Determine which electrical circuit has an open. ______________________________.
- ❑ What symptoms are present in this open? ______________________________.
- ❑ Obtain a wiring diagram of the circuit with the open. List the components, connectors, and wiring that can be isolated to determine the cause of the open. ______________________________
- ❑ List the diagnostic procedure(s) and test equipment used to safely locate the open. ______________________________
- ❑ Where is the open in the circuit, and what caused it to happen? ______________________________
- ❑ What diagnostic steps were helpful when troubleshooting for an open? ______________________________
- ❑ What is your analysis of finding an open in an electrical circuit? ______________________________

High-Resistance Circuit

- ❑ Obtain a vehicle that can be used for finding high resistance in an electrical system. What model and year of vehicle are you using for this task? ______________________________.
- ❑ Determine which electrical circuit has a high resistance. ______________________________.
- ❑ What symptoms are present in this high resistance? ______________________________.

❑ Obtain a wiring diagram of the circuit with the high resistance. List the components, connectors, and wiring that can be isolated to determine the cause of the high resistance. ____________________.

❑ List the diagnostic procedure(s) and test equipment used to safely locate the high resistance. ______

____.

❑ Where is the high resistance in the circuit, and what caused it to happen? ____________________

____.

❑ What diagnostic steps were helpful when troubleshooting for high resistance? ____________________

____.

❑ What is your analysis of finding a high resistance in a electrical circuit? ____________________

____.

Task Summary

After performing the preceding NATEF task, what can you determine will be helpful in testing a circuit with an ammeter to troubleshoot an electrical circuit? ____________________

____.

Performance Task 7

Task Description

Measure and diagnose the cause(s) of excessive key-off battery drain (parasitic draw); determine the necessary action.

Task Objectives

❑ Obtain a vehicle that can be used for finding a parasitic drain (key-off battery drain). What model and year of vehicle are you using for this task? ____________________.

❑ Describe the procedure you would use to find an unwanted current drain on the electrical system.

____.

❑ Which electrical circuit did you find with the unwanted parasitic drain? ____________________.

❑ How much current is flowing through this parasitic drain? ____________________.

❑ What is the maximum amount of parasitic drain allowed by this vehicle manufacturer? __________.

❑ Using a wiring diagram for the vehicle, determine which electrical circuits are part of the circuit that you isolated as causing the parasitic drain? ____________________

____.

❑ Check the operation of each circuit that you identified by the wiring diagram for proper operation. Which circuit can be isolated as the cause of the parasitic drain? ____________________.

- ❑ List the procedures that you would use to repair this electrical circuit to manufacturer standards.

____

____.

Task Summary

After performing the preceding NATEF task, what can you determine will be helpful in knowing how to check for parasitic key-off drains in an electrical circuit? _______________________

____.

Summary

- ❑ An open circuit is defined as a break in the current flow of a complete circuit.
- ❑ An intentional open circuit is designed into the circuit, such as a switch or relay for on/off control.
- ❑ An unintentional open circuit is created when there is a problem in the circuit, such as a broken wire, tripped circuit breaker or blown fuse, or loose connection.
- ❑ A poor electrical connection is the most common cause of open circuits.
- ❑ An ohmmeter can be used to test for an open circuit. A circuit with an open reads over limit or infinite.
- ❑ A voltmeter can be used to test for an open circuit. Voltage readings should be zero after the open and at the circuit voltage before the open.
- ❑ A test light can be used to test for an open circuit. The test light lights before the open and goes out after the open.
- ❑ A self-powered test light can be used to test for an open circuit. The test light remains lit until the test lamp probe goes beyond the open.
- ❑ Test lights and self-powered test lights should not be used around electronic or sensitive computer-controlled circuit.
- ❑ Testing for an open circuit with a jumper wire is an excellent way to bypass bad circuit switches.
- ❑ A short circuit is an accidental bypass of the current flow in a circuit.
- ❑ A short-to-voltage is an unwanted copper-to-copper connection between two separate circuits.
- ❑ A short-to-ground is an unwanted path for current to flow directly to ground.
- ❑ A short-to-ground blows the circuit protection fuse.
- ❑ A voltage drop is defined as the portion of the applied circuit voltage that is lost at points of the circuit other than at the designed load component(s).
- ❑ A parasitic drain is defined as an unwanted current drain still present in the electrical system after the ignition switch is turned to the key-off position.

- ❑ Acceptable parasitic drains vary with manufacturer, but most allow up to 35 to 50 milliamperes of current drain.
- ❑ The most reliable method for testing for parasitic drains is to use an ammeter or to perform a dynamic voltage drop test.
- ❑ Some electrical circuits take several minutes to shut off or go into the sleep mode, possibly giving the technician erroneous information when diagnosing a parasitic drain.
- ❑ An ignition-off parasitic drain (IOPD) test can accurately measure current draw in an electrical circuit due to a control module that is not powering down (or going to sleep).
- ❑ A sneak circuit is an unwanted backfeed of current from one circuit through another that shares the same power feed or ground path.

Key Terms

excessive resistance
grounded circuit
intentional open circuit
open circuit
parasitic drain
short circuit
short-to-ground
short-to-voltage
sneak circuit
unintentional open circuit
unwanted voltage drop

Review Questions

Short Answer Essays

1. Describe how you would test for an unintentional open circuit using a voltmeter.
2. Describe a short circuit in an electrical circuit.
3. What are the general specifications allowed for voltage drops in an electrical circuit.
4. Describe the six methods used to test an electrical circuit for a parasitic drain.
5. Describe how an electrical circuit can develop a sneak circuit.

Fill in the Blanks

1. An ____________________ circuit is defined as a ____________________ in the current flow of a complete circuit.
2. An ohmmeter will read ____________________ or ____________________ when connected in parallel to an open circuit.
3. Current flow through an unwanted path to ground is called a ____________________.
4. Dim lights or a slow turning motor could be caused by ____________________ resistance.
5. Maximum parasitic loads normally allowed for a vehicle are ____________________.

ASE-Style Review Questions

1. Technician A says an open circuit could be caused by a switch. Technician B says a bad relay could be the cause of an open circuit. Who is correct?
   A. A only
   B. B only
   C. Both A and B
   D. Neither A nor B

2. A simple series 12-volt light circuit has a short-to-ground. Technician A says the current flow will be higher than normal through the lamp. Technician B says the switch will not turn off the lamp. Who is correct?
   A. A only
   B. B only
   C. Both A and B
   D. Neither A nor B

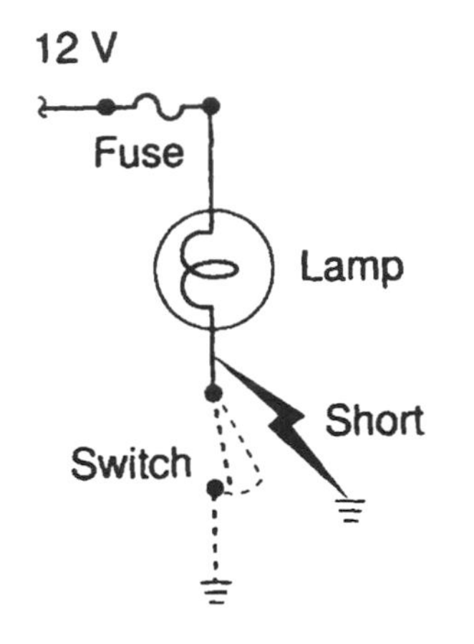

(Courtesy of DaimlerChrysler Corporation)

3. Technician A says an intentional open circuit is caused by a switch or relay. Technician B says an unintentional open circuit could be caused by a blown fuse. Who is correct?
   A. A only
   B. B only
   C. Both A and B
   D. Neither A nor B

4. Two technicians are discussing the use of a test light to check for an open circuit. Technician A says the test lamp will light when probing the circuit from the battery positive connector up to the open,

not after the open. Technician B says the test light will only light after the open connection. Who is correct?

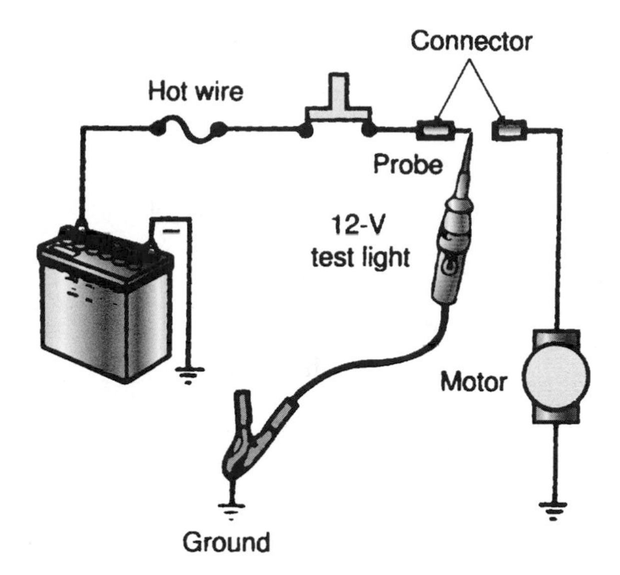

A. A only
B. B only
C. Both A and B
D. Neither A nor B

5. Technician A says a fused jumper wire can be used to bypass a switch to determine whether a circuit is working. Technician B says a self-powered test light can be a useful tool for checking for circuit continuity. Who is correct?
   A. A only
   B. B only
   C. Both A and B
   D. Neither A nor B

6. Technician A says a short-to-voltage is an unwanted copper-to-copper connection between two circuits. Technician B says a short-to-voltage may not blow the circuit fuse. Who is correct?
   A. A only
   B. B only
   C. Both A and B
   D. Neither A nor B

7. A test light is being used to test for a short. Technician A says the test light will not light with the circuit fuse removed as shown. Technician B says the test light should not be used to check this circuit. Who is correct?
   A. A only
   B. B only

C. Both A and B

D. Neither A nor B

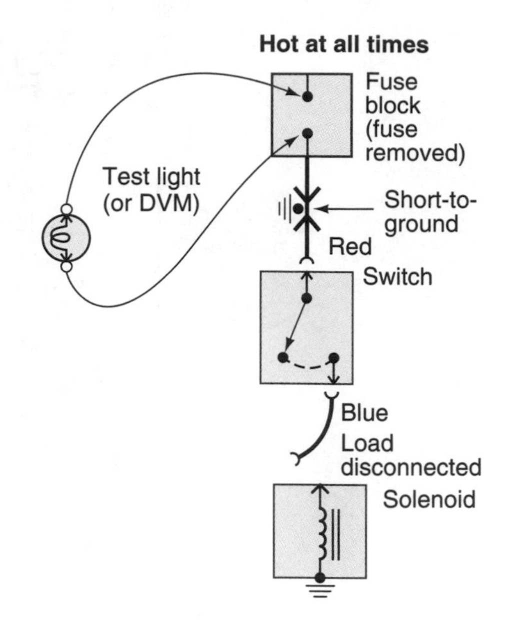

8. Technician A says a short-to-ground is always located after the circuit load device(s). Technician B says a short-to-ground can occur when wire insulation breaks down from vibration. Who is correct?
   A. A only
   B. B only
   C. Both A and B
   D. Neither A nor B

9. Technician A says a cycling circuit breaker can be used in place of the circuit fuse when diagnosing the cause of a short-to-ground. Technician B says a short-to-ground always causes the circuit fuse to blow. Who is correct?
   A. A only
   B. B only
   C. Both A and B
   D. Neither A nor B

10. Technician A says a test light can be connected in series with a cycling circuit breaker to test for a short-to-ground. Technician B says a buzzer can be connected in series with a cycling circuit breaker to test for a short-to-ground. Who is correct?
    A. A only
    B. B only
    C. Both A and B
    D. Neither A nor B

11. Technician A says that test lights and self-powered test lights should never be used around sensitive electronic circuits. Technician B says computer control circuits have fail-safe controls that make the use of test lights safe to use. Who is correct?
    A. A only
    B. B only
    C. Both A and B
    D. Neither A nor B

12. Technician A says excessive circuit resistance in connectors could be the cause of an unwanted voltage drop. Technician B says bad wiring could be the cause of an unwanted voltage drop. Who is correct?
    A. A only
    B. B only
    C. Both A and B
    D. Neither A nor B

13. Technician A says that voltage drops of 0.5 volt at ground connections are satisfactory. Technician B says that voltage drops at terminal connections of 0.2 volt are satisfactory. Who is correct?
    A. A only
    B. B only
    C. Both A and B
    D. Neither A nor B

14. Technician A says a parasitic drain of 1 ampere is the maximum allowed by most manufacturers. Technician B says some parasitic drains may take several minutes to shut off or reach the sleep mode after the key is turned off. Who is correct?
    A. A only
    B. B only
    C. Both A and B
    D. Neither A nor B

15. Technician A says a sneak circuit is when current can backfeed through a circuit when two circuits share a common power feed. Technician B says a sneak circuit is when current can backfeed through a circuit when two circuits share a common ground path. Who is correct?
    A. A only
    B. B only
    C. Both A and B
    D. Neither A nor B

11 Lighting Circuits

Introduction

In previous chapters you learned basic electrical theory, concepts, circuit laws, and troubleshooting techniques. In this chapter, you can apply what you have learned to a variety of exterior and interior lighting circuits. Electrical lighting circuits of modern vehicles can consist of complex circuit boards, hidden wiring looms and connectors, several lamps, circuit protectors, relays, and control switches. In addition, after-market accessories may have been installed over the car's life by an unqualified technician. All of this can make the job of diagnosing circuits that do not work challenging.

Objectives

When you complete this chapter you should be able to:

- Explain the operation and diagnosis of the following exterior light systems: parking and side marker lamps, headlights and dimmer switches, flash-to-pass, taillights, brake and collision avoidance lights, turn signal/hazard lights and flashers, fog lights, and backup lights.
- Explain the operation and diagnosis of the various interior light systems, including courtesy lights, dome lights, and reading/map/vanity lights.

Exterior Lighting Circuits

Exterior lighting circuits of today's vehicles are largely regulated by federal laws. Even though there are differences among manufacturers in lighting circuit designs and style, the result is that a vehicle must meet or exceed all vehicle safety requirements and laws.

In addition, the neon, fiber optic, or LED lighting technology that used to be reserved for high-end vehicles only has become standard in most new vehicles produced today. As a result, today's technician needs to be familiar with the various lighting circuit spanning over the last 20 or 30 years, depending on location.

Headlight Systems

Although we discussed headlamp types and design features in Chapter 5, we now apply them to various headlight circuits common in vehicles over the past 30 years. A basic headlight circuit normally consists of a headlight switch, dimmer switch, low and high beam headlights, high beam indicator, and circuit wiring and connectors. Figure 11–1 is an example of a headlight circuit with the dimmer switch directing the battery voltage *only* to the low beam circuit.

When the dimmer switch is changed to the high beam position (Figure 11–2), the battery voltage is directed *only* to the high beams and the high beam indicator lamp.

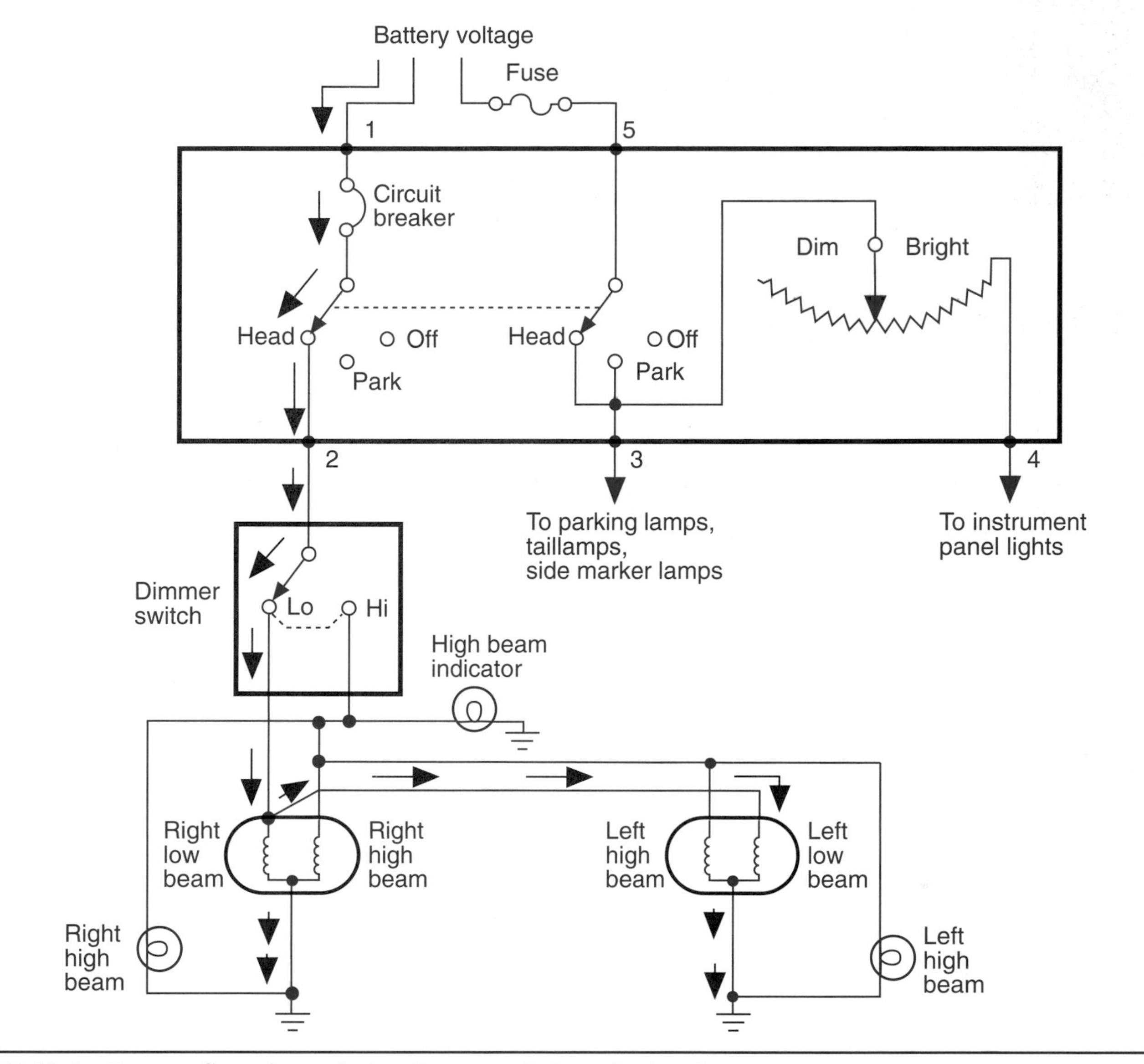

FIGURE 11–1 Current flow through the dimmer switch to the low beams.

In the headlight circuit just discussed, the headlight switch and dimmer switch are both on the positive side of the circuit, and the headlamps complete the path to ground. In many import vehicles the headlamps are on the positive side of the circuit, and the headlight switch and dimmer switch complete the path to ground (Figure 11–3).

Headlight Switch Regardless of the type of headlight system, a headlight switch controls the operation of the circuit. The typical headlight switch could be mounted either in the dash or instrument panel (Figure 11–4) or in the steering column (Figure 11–5).

On older vehicles, the most common headlight switch is mounted in the dash panel. The switch combined the ON/PARK/OFF functions with a ceramic rheostat used to control dimming of the instrument panel lights (Figure 11–6). When this switch has to be removed for troubleshooting or replacement, the control knob can be removed from the switch by pushing on the release button located on the bottom of the switch (Figure 11–7). Knowing about the location of the release button is important. Some switches are hard to see and are not easily accessible, so the technician has to rely on feel to locate the release button and remove the control knob.

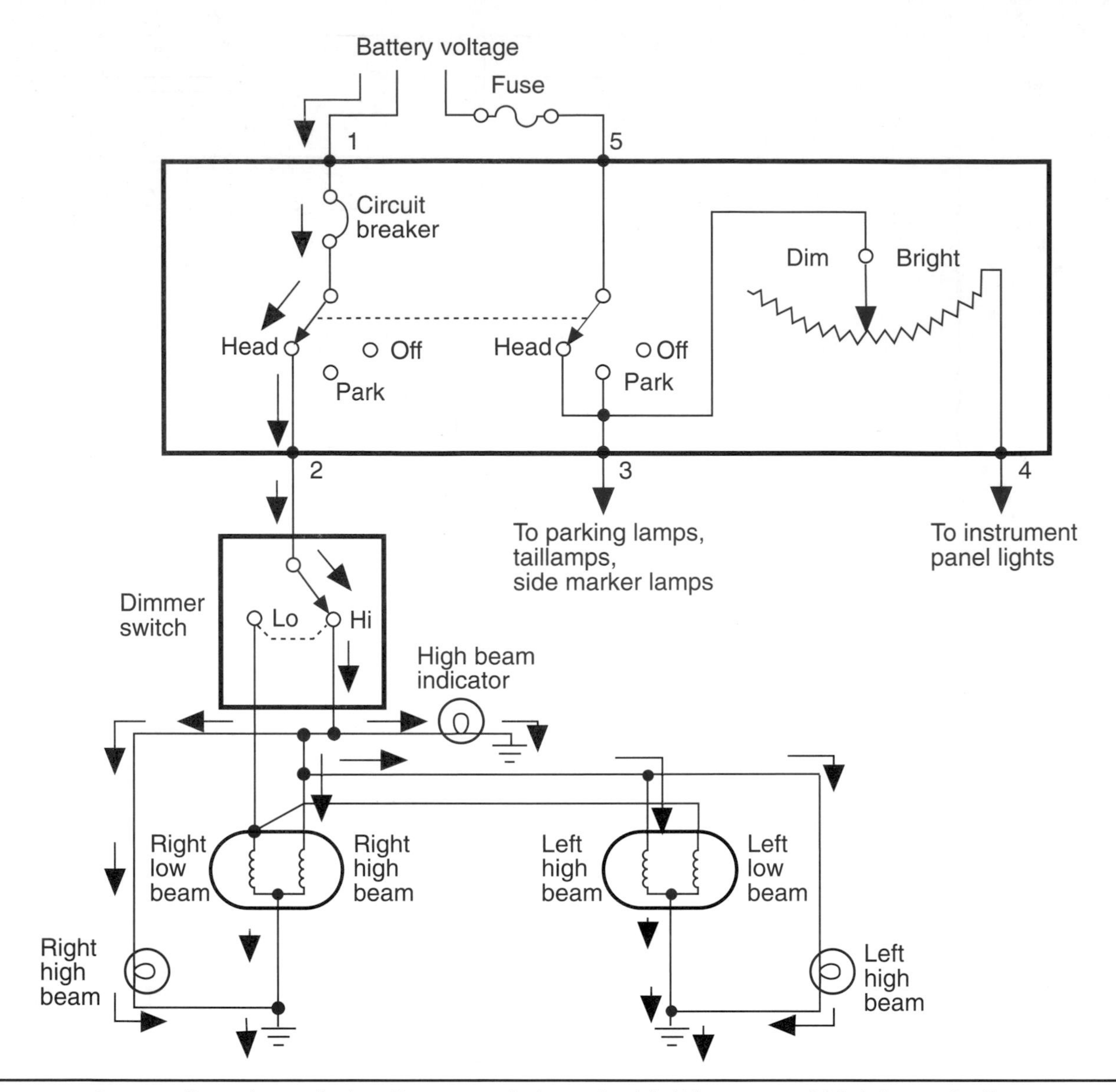

FIGURE 11–2 Current flow through the dimmer switch to the high beams and high beam indicator.

In a typical three-position headlight switch, two terminals normally have battery voltage. This allows the headlight switch to operate the lights independently of the circuits controlled by the ignition switch (Figure 11–8).

When the headlight switch is pulled to the park or headlight position (Figure 11–9), the battery voltage present at terminal 5 is directed to the parking lamps, taillamps, side marker lamps, license plate lamp, and (through a ceramic dimming rheostat) the instrument cluster lamps, but *not* the headlights.

When the switch is pulled to the headlight position (Figure 11–10), the battery voltage present at terminal 1 is directed through the circuit breaker to the headlamps. The circuit breaker is designed so that it does not totally disable the headlights when temporary overloads exist in the system. The voltage at terminal 5 continues to feed battery voltage to the park circuit.

Dimmer Switches In older vehicles, the dimmer switch is located next to the left kick panel on the floor board (Figure 11–11). This location subjects the switch to water, dirt, and other contaminants that shorten the service life of the switch. Newer vehicles locate the dimmer switch on the steering column as a separate switch (Figure 11–11) or as part of a **multifunction switch** (Figure 11–12), which combines the dimmer switch with

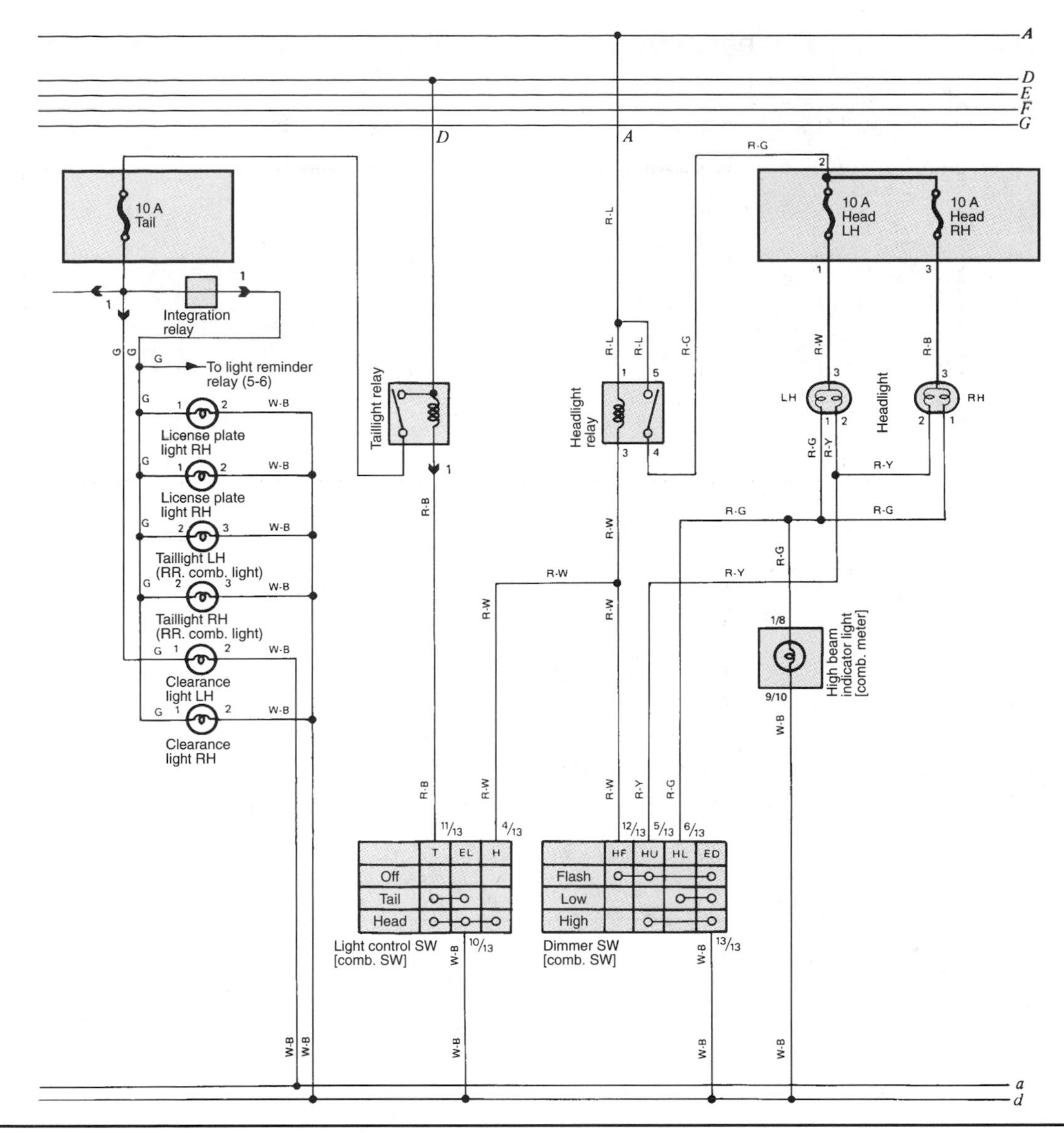

FIGURE 11–3 Headlight circuit with the control switches on the ground side.

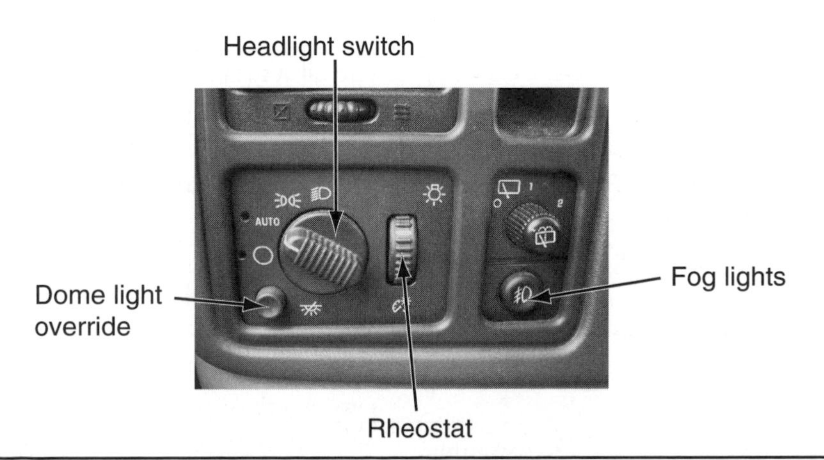

FIGURE 11–4 Typical dash-mounted light switch.

(Courtesy of Tim Gilles)

FIGURE 11–5 Typical headlight switch mounted on the steering column, with a separate control on the dash for dimming the instrument panel lights.

FIGURE 11–6 Typical dash-mounted headlight switch used on older vehicles.

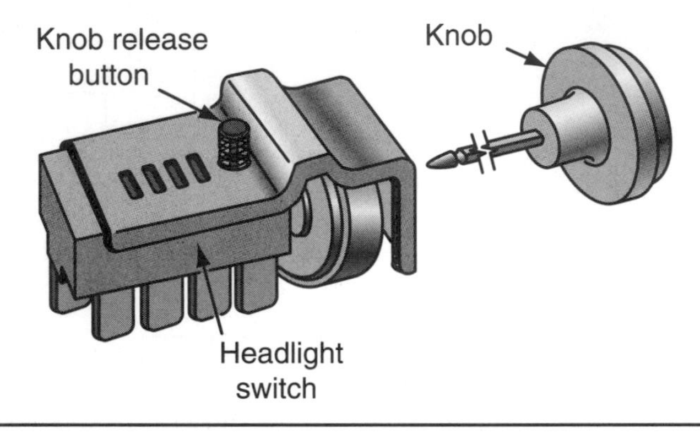

FIGURE 11–7 Typical location of the headlight switch knob release button.

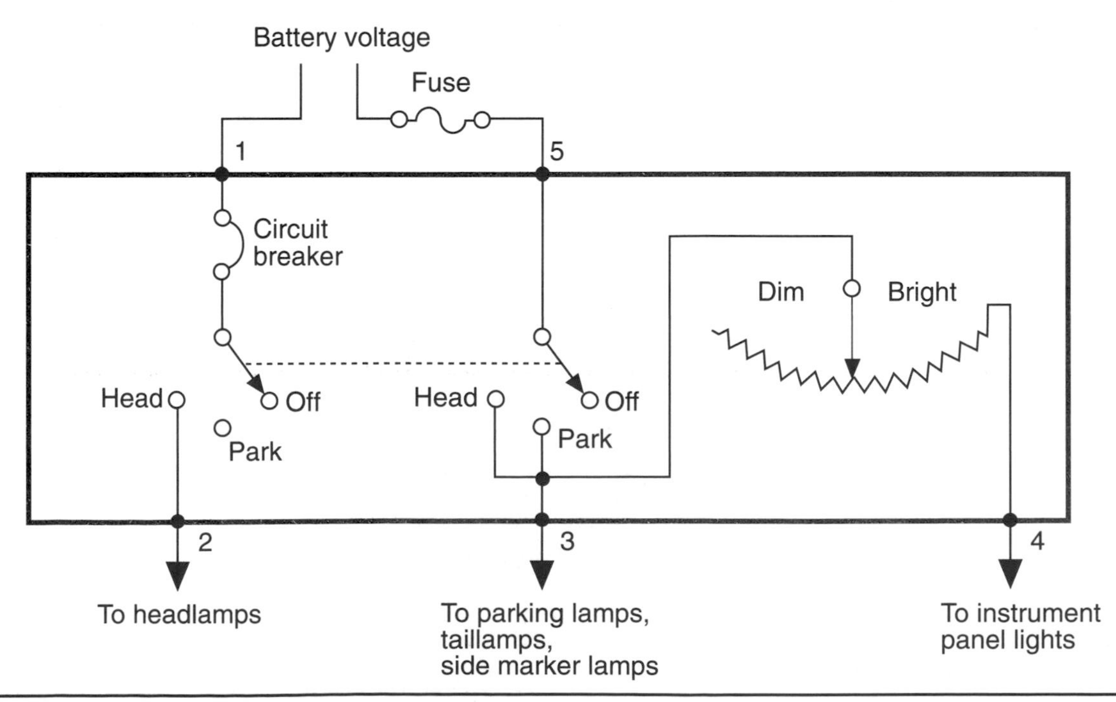

FIGURE 11–8 A headlight switch in the off position.

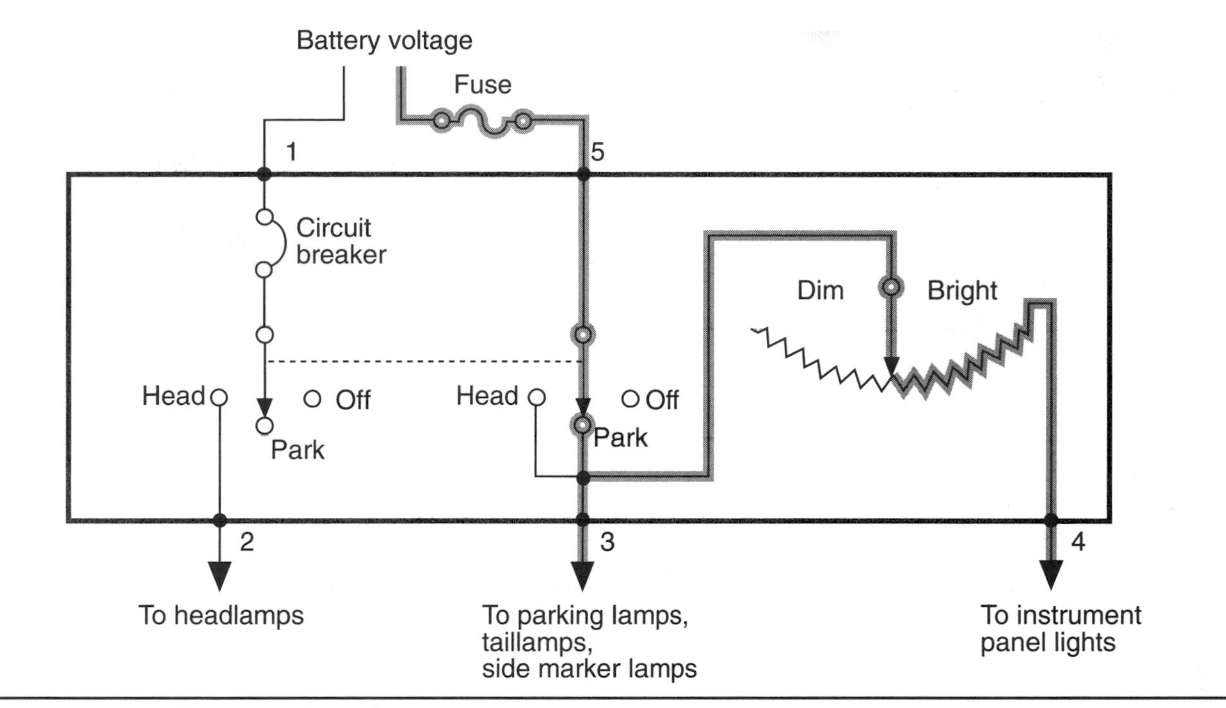

FIGURE 11–9 A headlight switch in the park position.

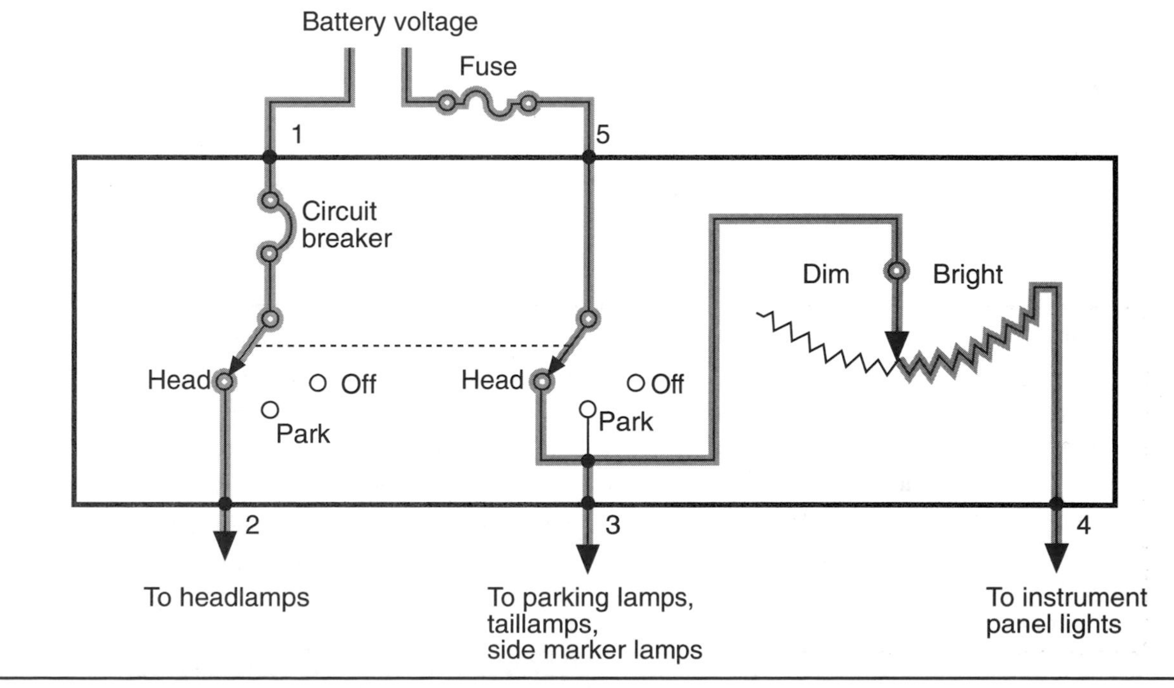

FIGURE 11–10 A headlight switch in the headlight position.

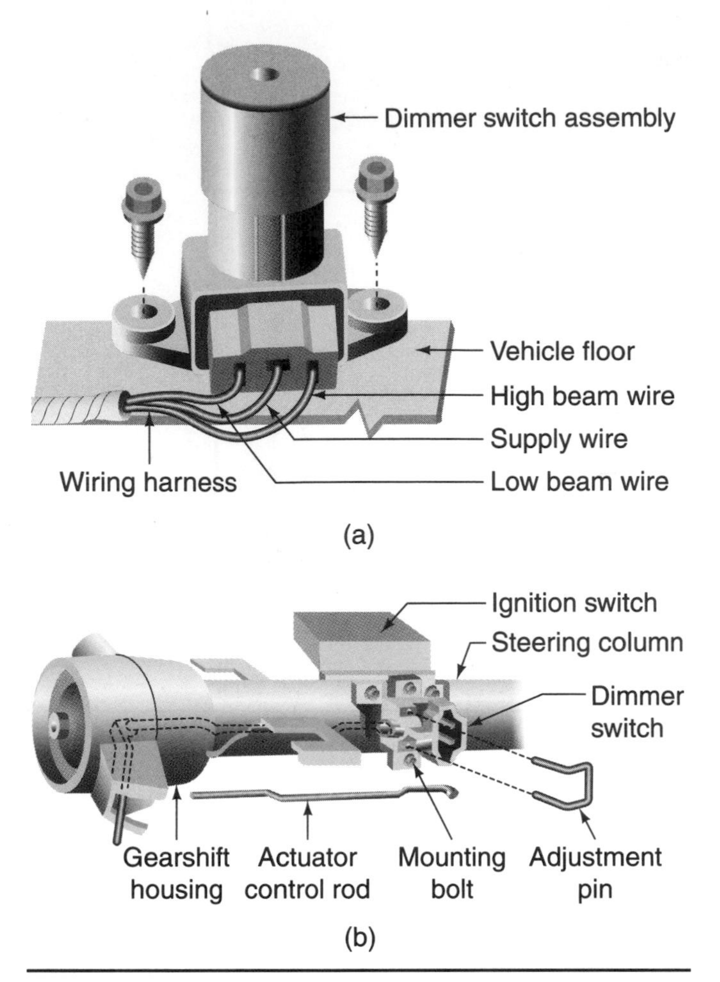

FIGURE 11–11 (a) Floor-mounted dimmer switch, and (b) steering column–mounted dimmer switch.

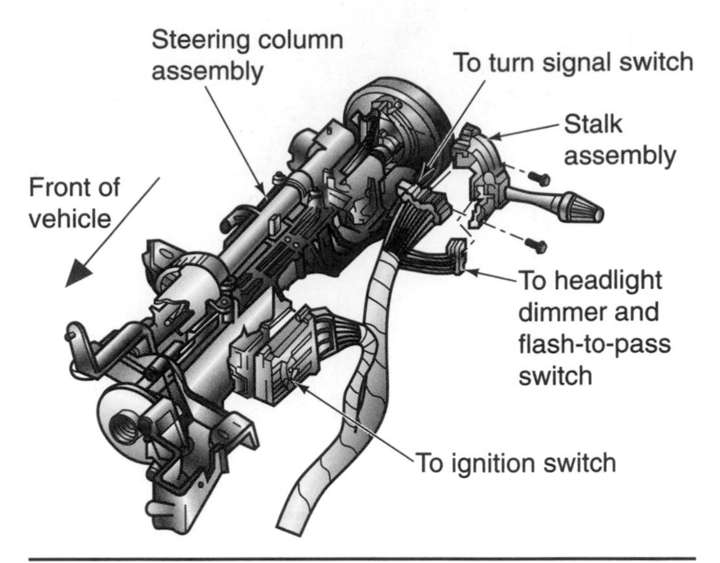

FIGURE 11–12 A multifunction switch with dimmer switch incorporated into it.

one or more of the following switches: turn signal, headlight, hazard, horn, and flash-to-pass.

To diagnose a suspected malfunctioning dimmer switch, a fused jumper wire can be used to bypass the switch to either low or high beams (Figure 11–13). The dimmer switch is typically connected in series with the headlight circuit; so, when the switch is bypassed, the low or high beams go on unless something else is wrong in the circuit.

On dimmer switches that are part of the multifunction switch (Figure 11–14), you might be able to test the switch without removing it from the steering column. Normally, a multiconnector at the base of the column can be used for testing the switch. If not, you may have to remove the column covers to gain access. Refer to specific vehicle manufacturer guidelines for this procedure to prevent damaging access covers or connectors.

Some dimmer switches have an additional feature: **flash-to-pass,** which switches on the high beams and, in some vehicles, the low beams, even if the headlight switch is off. The circuit in Figure 11–15 supplies battery voltage from the B1 terminal at the headlight switch to the dimmer switch. This allows battery voltage at the dimmer switch even if the headlight switch is in the off or park position. When the driver activates the flash-to-pass feature, the dimmer switch completes the circuit to the high beams and, in some vehicles, to the low beams as well.

On many luxury vehicles the dimmer switch circuit contains a photoelectric cell and dimmer module (Figure 11–16) that automatically dim the headlights when the shine of oncoming headlights approach and then return to high beams after the vehicle passes. The photoelectric cell is normally located in the front grill area, the front part of the dashboard, or an area that will pick up the shine of oncoming lights.

Concealed Headlight Systems On several older vehicles (and a few newer ones), in an effort to combine vehicle styling with aerodynamics, headlights are concealed behind movable doors (Figure 11–17). When the headlight switch is off, the headlight doors are closed,

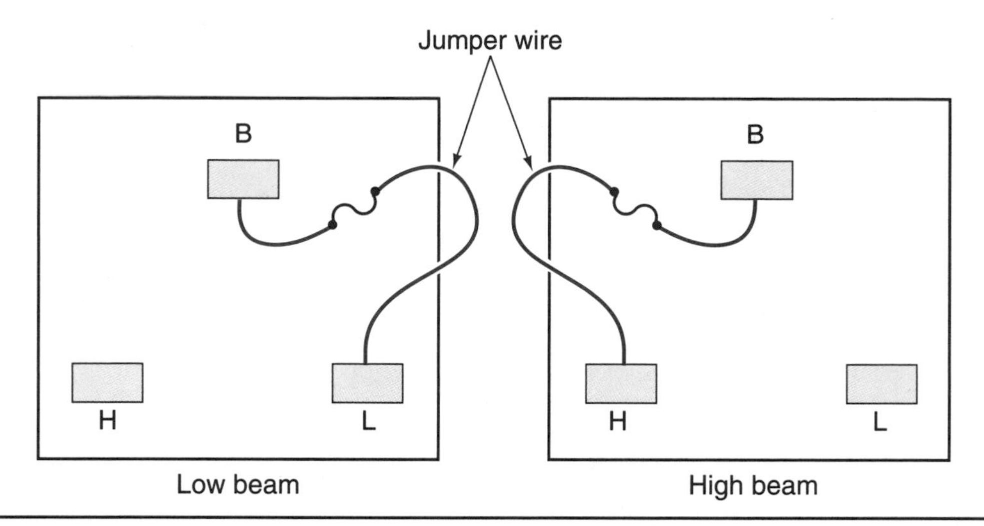

FIGURE 11–13 Using a jumper wire to bypass a suspected malfunctioning dimmer switch.

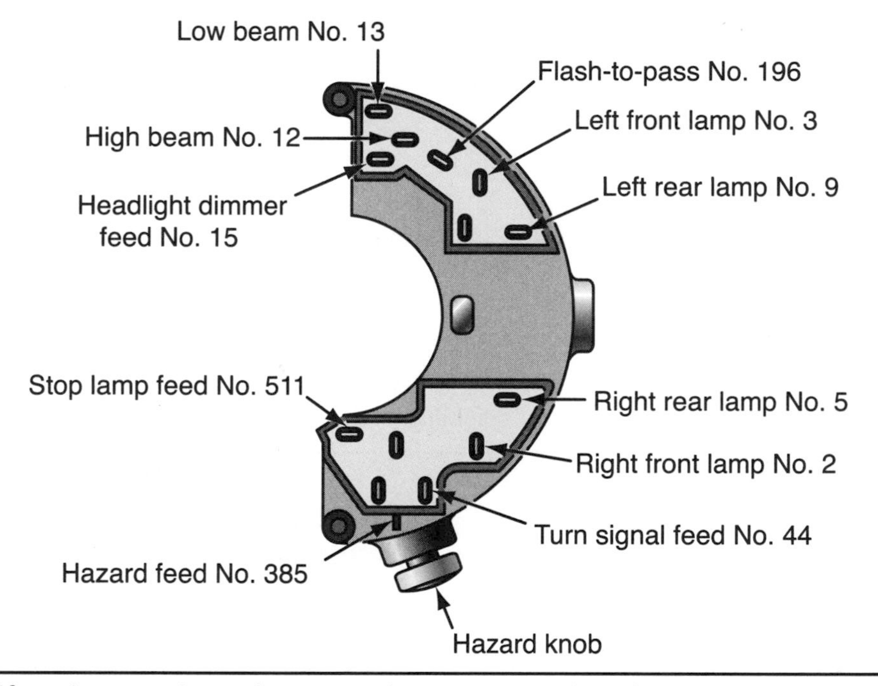

FIGURE 11–14 Multifunction steering column switch connection identification.

but they automatically open when the headlight switch is turned to the headlight position (but not the park position).

The headlight doors are controlled by either a vacuum circuit with vacuum motors or an electric circuit with electric motors. In Figure 11–18, a vacuum distribution valve is attached to the back of an older headlight switch to vent the vacuum from the motors that hold the doors closed (Figure 11–19). If vacuum is used to close the headlight doors and springs are used to open them as the vacuum is vented, this assures the doors will be open if vacuum to the system is lost.

In electrically operated headlight doors, a door motor is normally located at each headlight door (Figure 11–20); or, when only one motor is used, a torsion bar connects both doors (Figure 11–21).

In both vacuum-controlled motors and electric motor systems, a means of manually opening the doors is provided in the event the system fails. In Figure 11–22, a

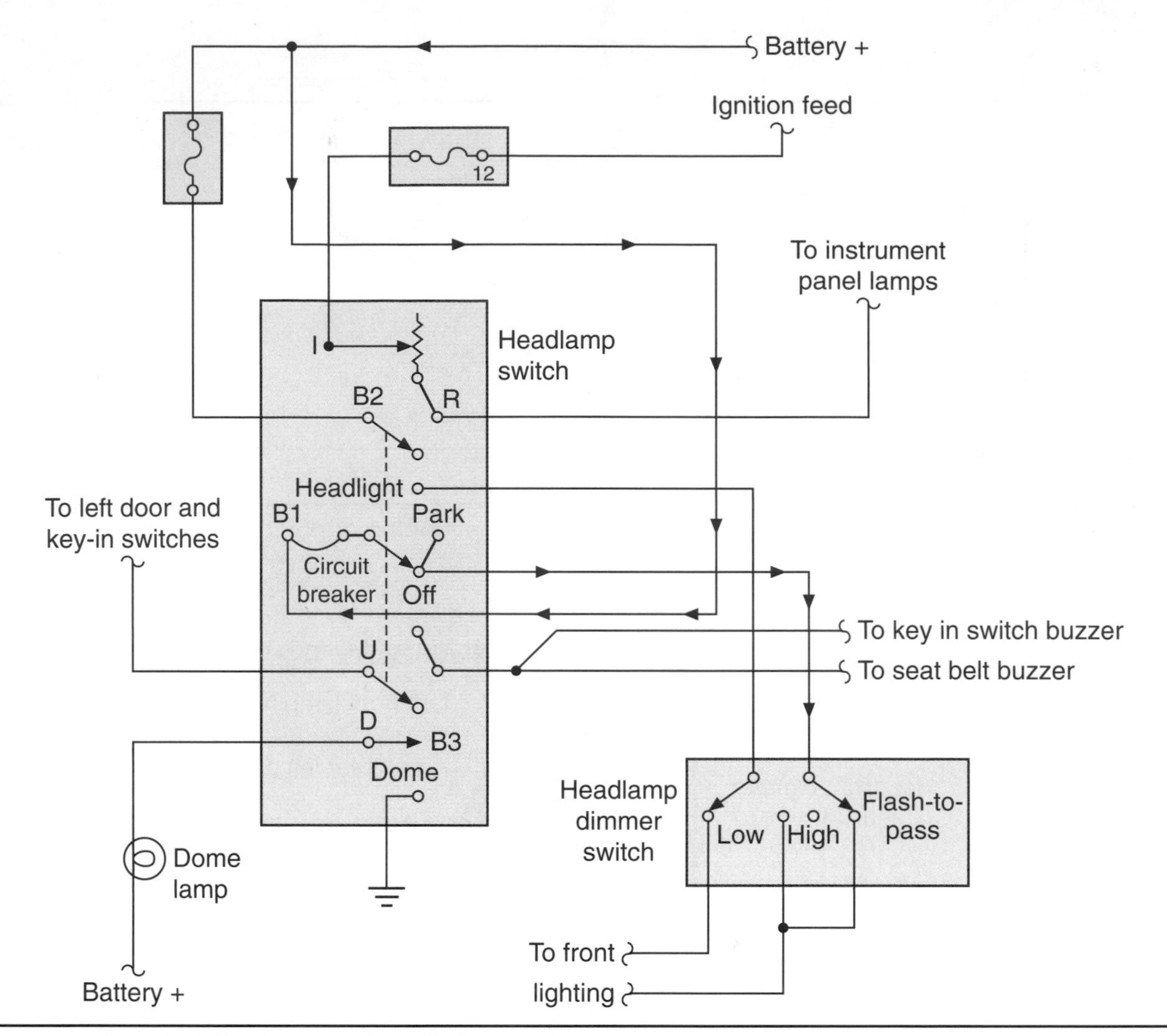

FIGURE 11–15 A typical flash-to-pass headlight circuit.

bypass valve supplies vacuum directly to the doors from the engine. In Figure 11–20, the manual knob allows the doors to be turned up manually. Remember to check appropriate service information before attempting manual procedures.

Over the last several years, vehicle headlight design has improved dramatically. Instead of using concealed headlight systems with complex door mechanisms to obtain aerodynamics, headlights have become an integral part of the vehicle body contour (Figure 11–23).

CAUTION *Remember to keep hands clear of door mechanisms when performing testing or repair. Doors may suddenly open with a snap due to a bind or malfunction. Play it safe! Pay attention!*

Daylight Running Lights (DRL) In an effort to increase vehicle safety, Canadian and U.S. federal laws require that vehicles be equipped with **daylight running lights (DRL)** for daylight safety. Depending on the system, the low beam or high beam lights operate any time the vehicle is running, but usually at a reduced voltage and intensity to extend headlamp life. In Figure 11–24, the DRL relay is connected directly to battery voltage. When the vehicle is running during daylight hours and the parking brake is off, the DRL control module in the instrument cluster microprocessor grounds the DRL relay coil, which then energizes the low beam lamps in *series* with each other. This drops the voltage to approximately 6 volts per low beam bulb. When the headlight switch is activated for night or poor light conditions, the DRL system is deactivated and the lamps operate at regular brightness.

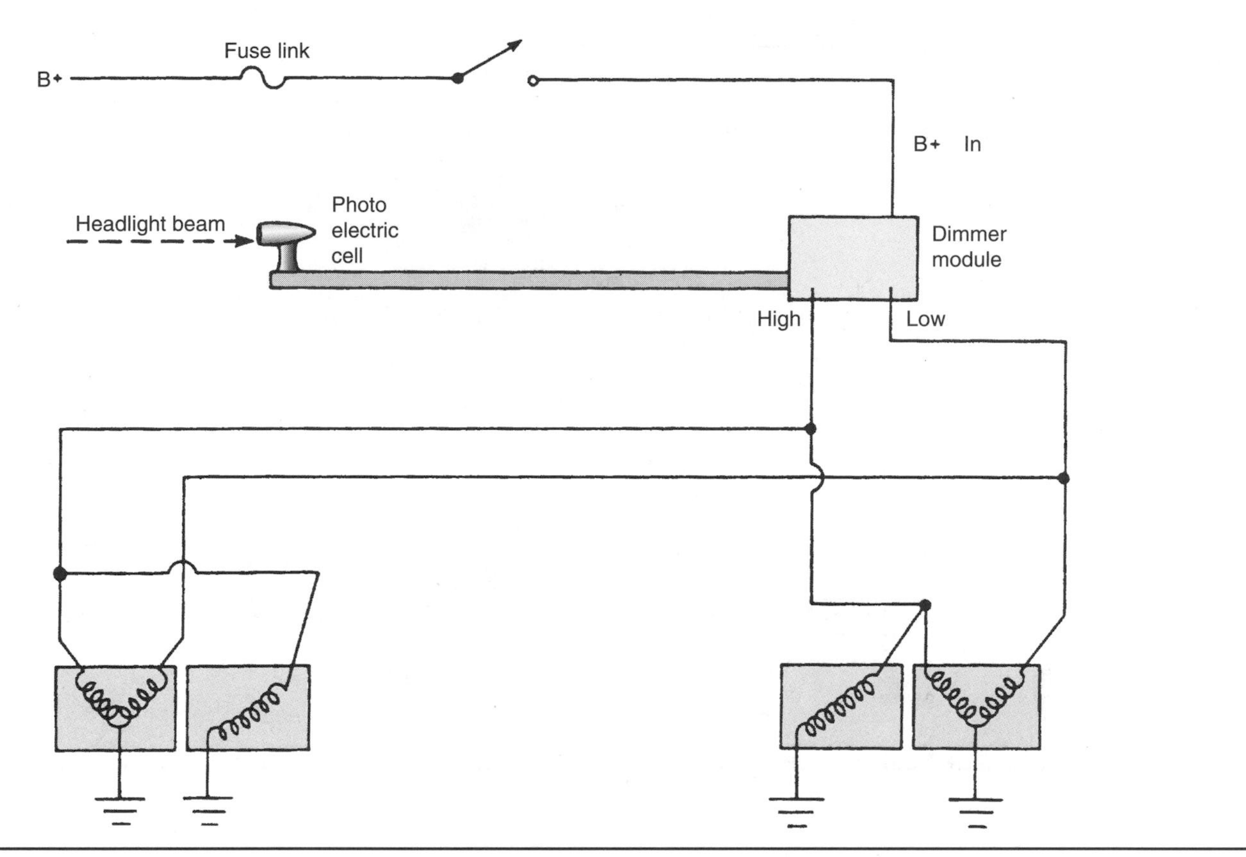

FIGURE 11–16 Typical photoelectric headlight dimming circuit.

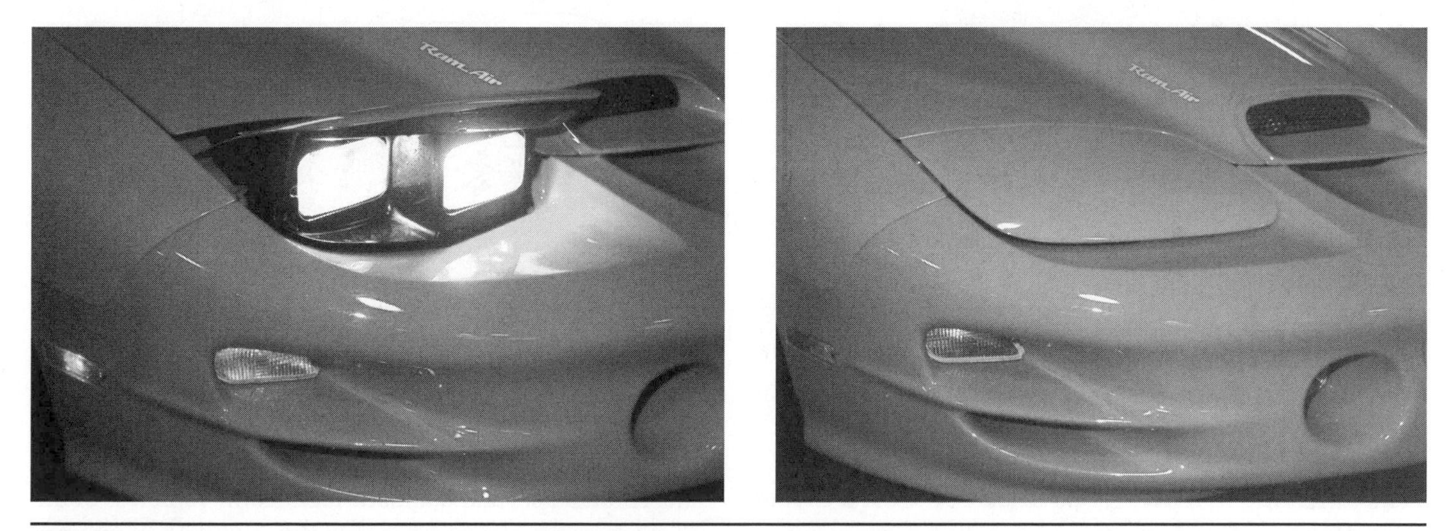

FIGURE 11–17 Concealed headlights designed for styling and aerodynamics.

FIGURE 11–18 An older style headlight switch combined with a vacuum distribution valve.

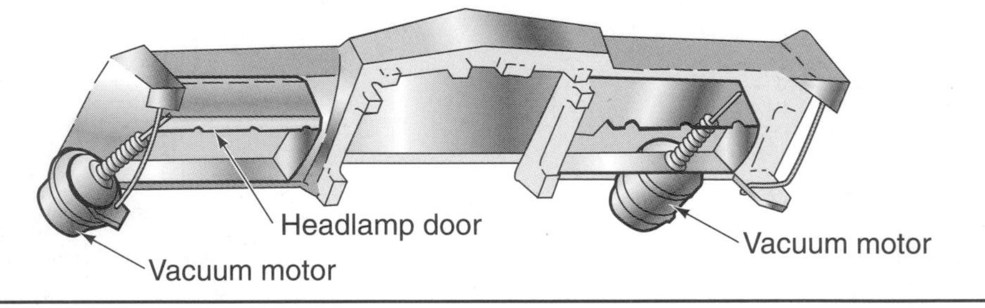

FIGURE 11–19 Typical headlight doors connected to vacuum control motors.

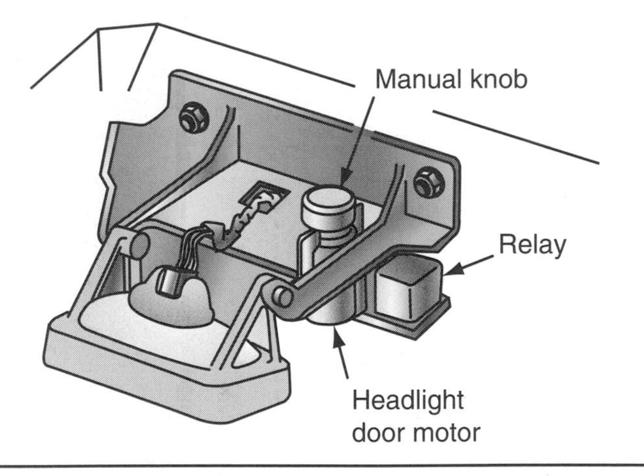

FIGURE 11–20 Typical electrically operated headlight doors with manual over-ride knob.

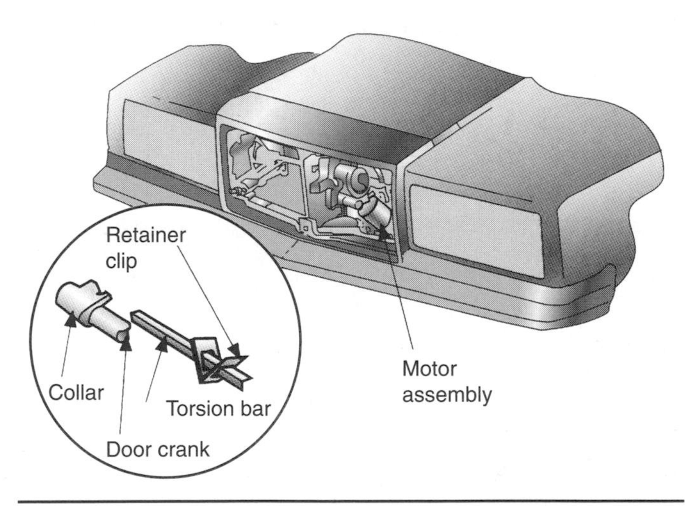

FIGURE 11–21 Torsion bar–operated headlight door system with single motor.

Automatic and Computer-Controlled Headlight Systems Manufacturers have made several refinements to headlight circuits in the last several years. In an effort to increase driving safety by providing the best lighting possible, manufacturers have incorporated several standard features in vehicles. Some of these are automatic headlight high/low dimming, automatic headlight washers, automatically timed delay ON/OFF or twilight sentinel, and adaptive headlight systems.

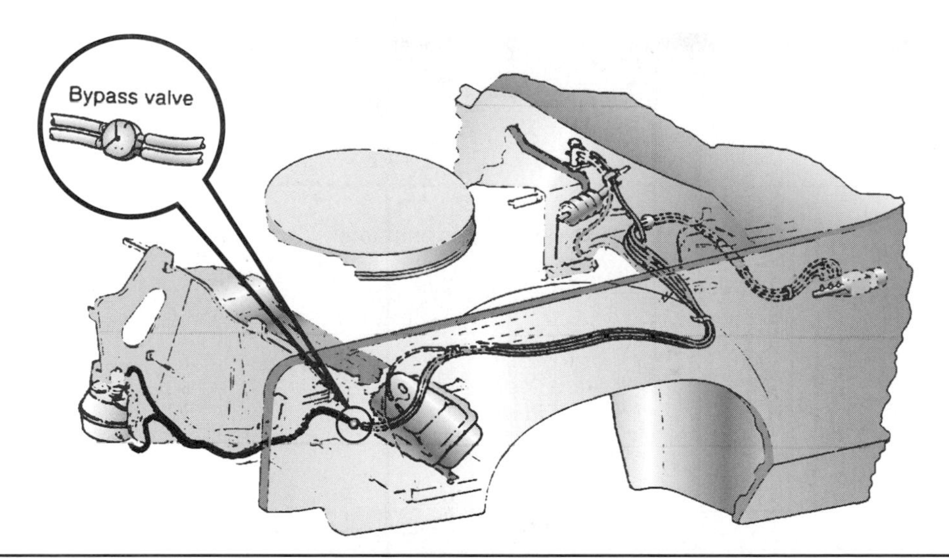

FIGURE 11–22 A bypass valve used to open headlight doors in the event of a system failure.
(Courtesy of DaimlerChrysler Corporation)

FIGURE 11–23 Aerodynamic design of newer style headlights.
(Courtesy of American Honda Motor Co.)

A typical twilight sentinel headlight system (Figure 11–25) activates the headlights automatically when the photocell senses a decrease in the ambient light to a predetermined level. In addition, the twilight sentinel also controls the amount of time the headlights remain on after the vehicle is turned off. This delay time can normally be adjusted from zero to a maximum of 3 minutes, depending on the location of the control switch.

In a typical computer-controlled headlight system, the headlight switch provides an input to a body control module (BCM) (Figure 11–26), which then controls activation of the headlight circuit through a series of relays and switches.

An adaptive headlight system is probably the highest-tech headlight system available on high-end vehicles today (Figure 11–27). This system uses headlights that swivel up to 15 degrees through bidirectional motors at the headlight base (Figure 11–28), depending on several input signals. Depending on the manufacturer, this system uses a combination of inputs, such as vehicle speed, steering wheel angle sensor, yaw rate, and in some cases global positioning system (GPS) navigation data to inform the BCM of the best headlight angle.

Headlight Aiming Suspension changes can alter vehicle ride height, or a load condition can change a vehicle's headlight aim. For various reasons, when headlights are replaced headlights should be adjusted to meet all laws and regulations. Headlights that are misaimed even 1 degree downward can reduce the vision distance by almost 160 feet. Some federal, state, and local laws specify requirements for headlamp aim that *must* be followed. Before headlights are aimed, the following preparation should be done to the vehicle.

- Remove any excess dirt, mud, snow, etc., from the vehicle that will affect the ride height.

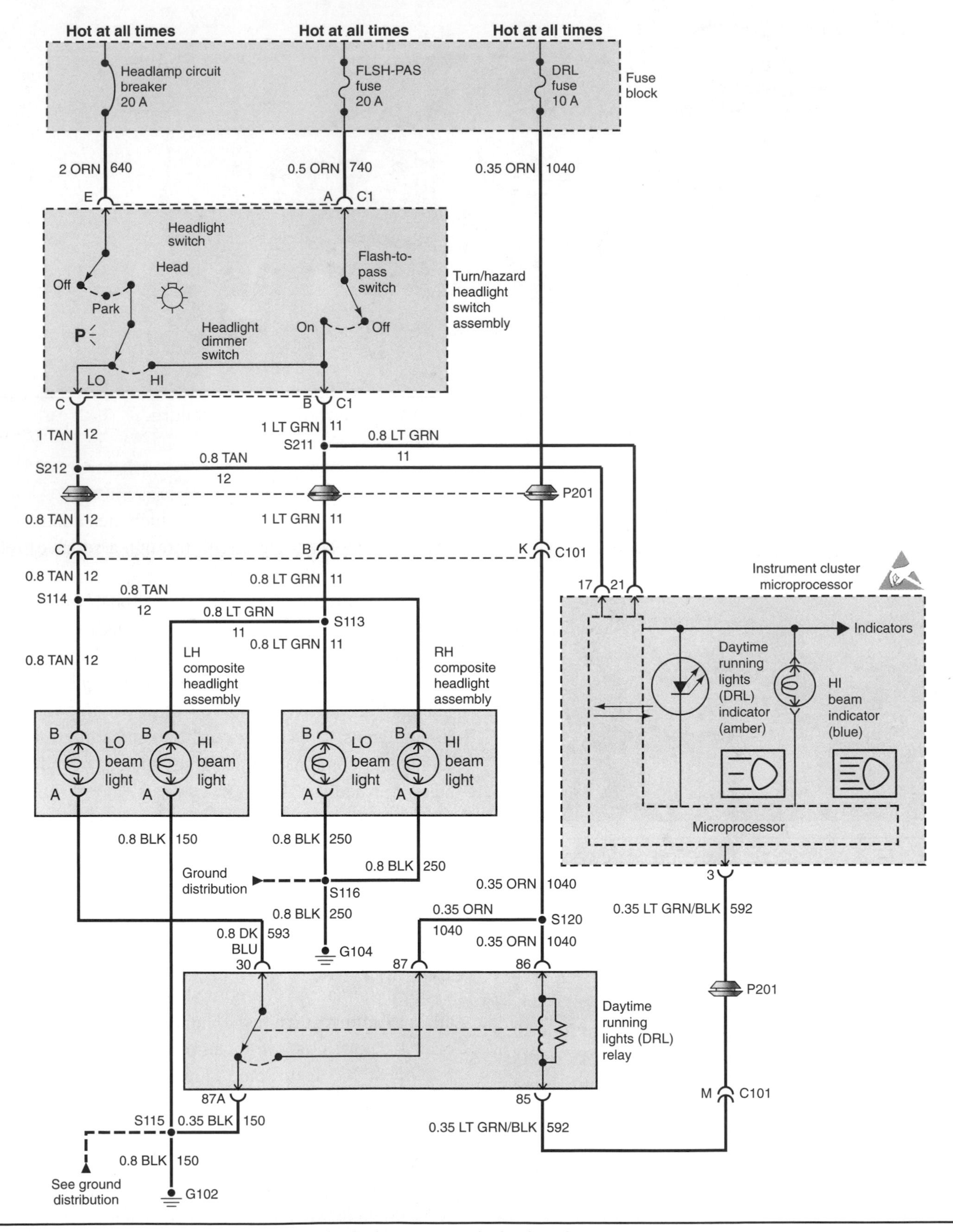

FIGURE 11–24 Headlight system with daylight running light (DRL) circuit.

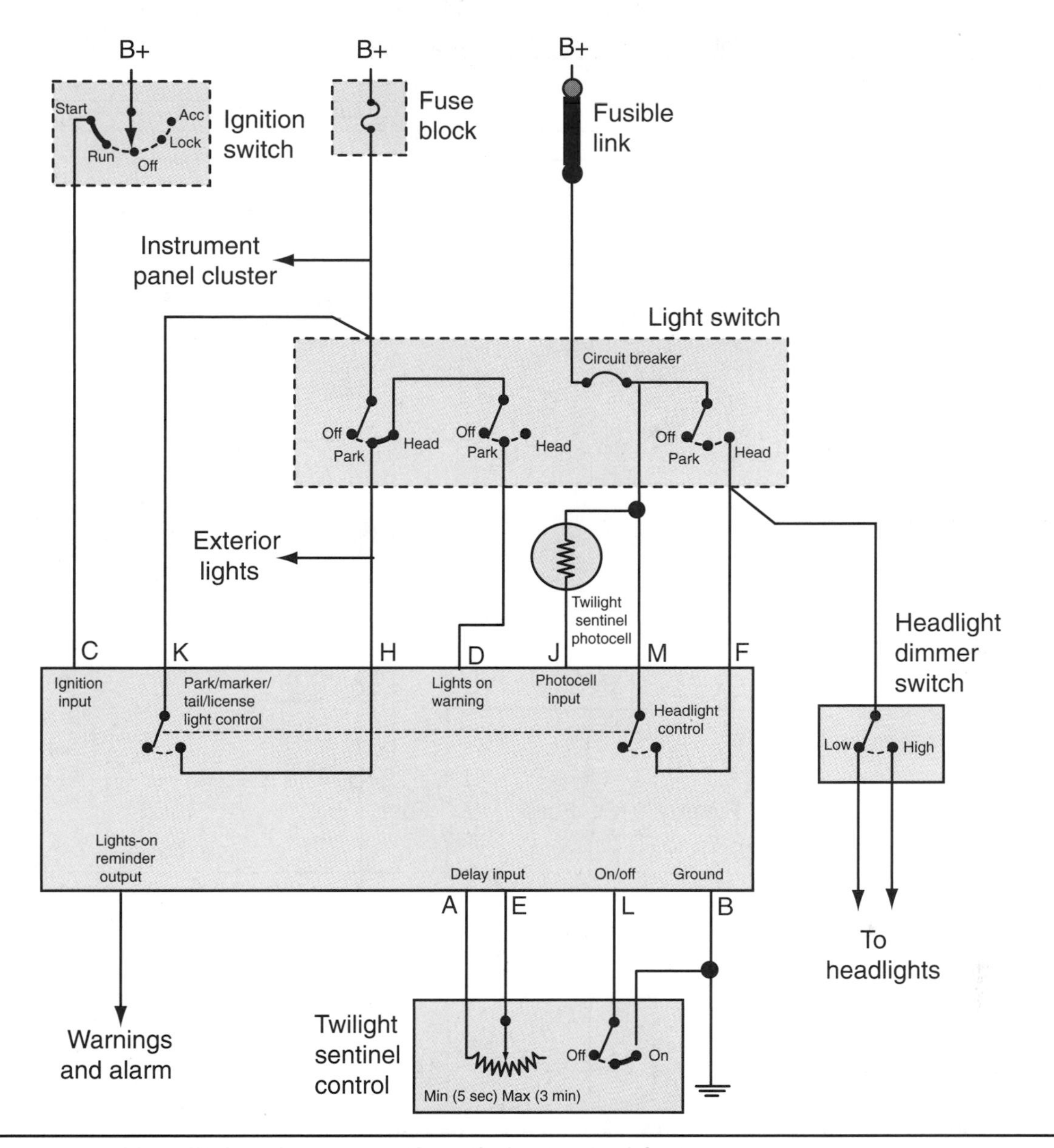

FIGURE 11–25 Typical headlight schematic with Twilight Sentinel feature.

- Consult with the customer about load conditions on the vehicle. A vehicle continually loaded with heavy material in the luggage compartment could cause the light to shine into oncoming traffic.
- Place the vehicle on a flat, level surface. If such a surface is not available, the floor slope must be compensated for with the aiming tools.
- Jounce the vehicle suspension so that it rests in its normal driving plane.
- Make sure the fuel tank is at least half full. Some manufacturers recommend the fuel tank be filled completely for the most accurate headlight adjustment.
- Adjust tire pressures to the recommended settings.
- Determine the weight distribution of vehicle occupants and cargo. This step could be important on the ride height of smaller vehicles with weaker suspension systems.

Although not preferred or recommended, another method is possible. In the absence of professional headlight aiming equipment, vehicle headlights can be checked for basic accuracy by projecting the intensity or

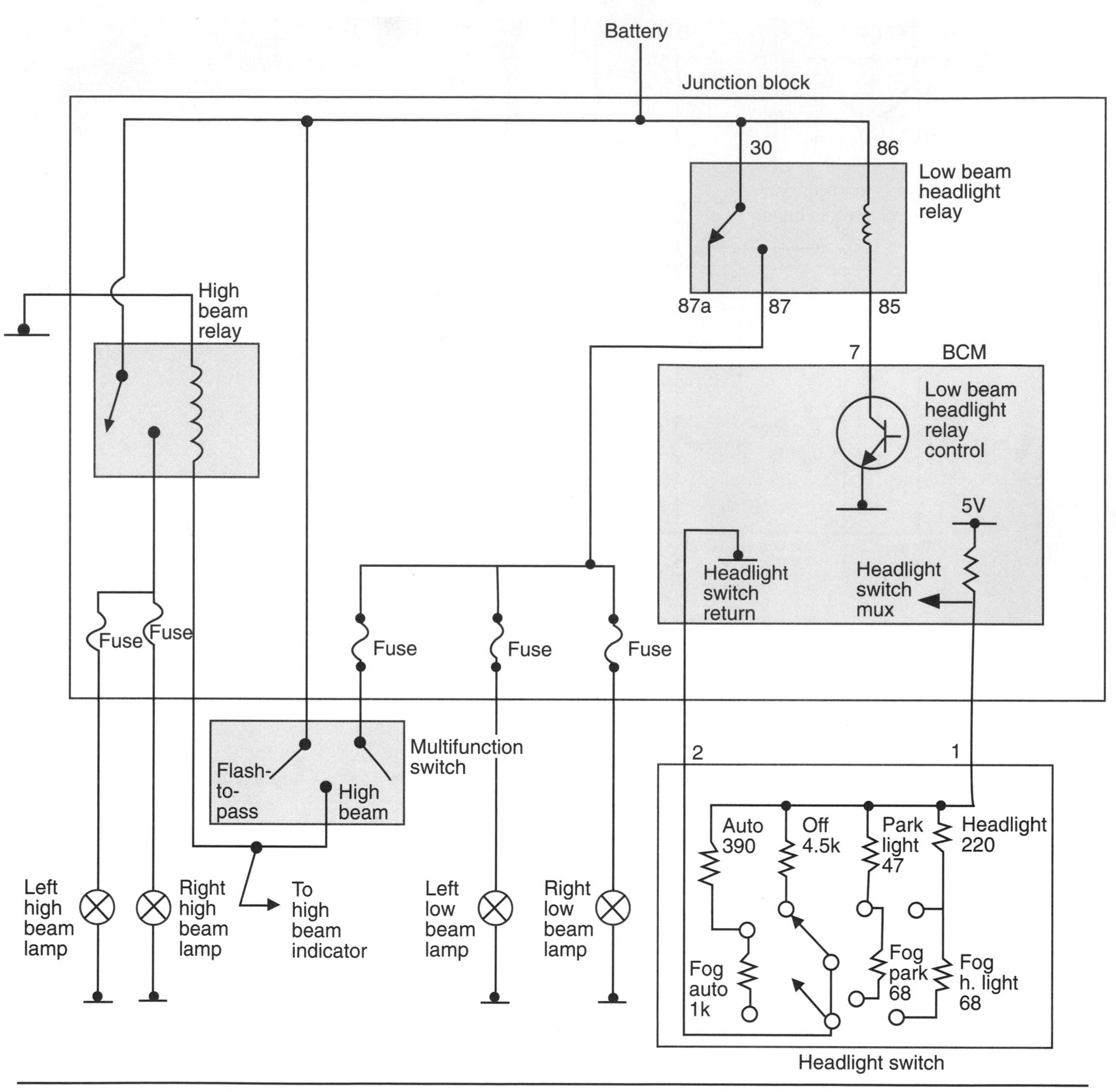

FIGURE 11–26 Typical computer-controlled headlight system schematic.

brightness of the headlights on a vertical surface that is 25 feet in front of the vehicle (Figure 11–29). The vehicle must be positioned exactly perpendicular to the vertical surface to be reasonably accurate. Again, this method is used only to check the basic aim of the headlights when professional equipment is not available.

For accuracy, most shops use portable headlight aiming equipment and adapters (Figure 11–30). The aiming units are fit to the headlight aiming pads (Figure 11–31) by adapters that accommodate each style of headlight. Composite headlights have additional adapters that adjust to the aerodynamic contour of the headlight lens (Figure 11–32).

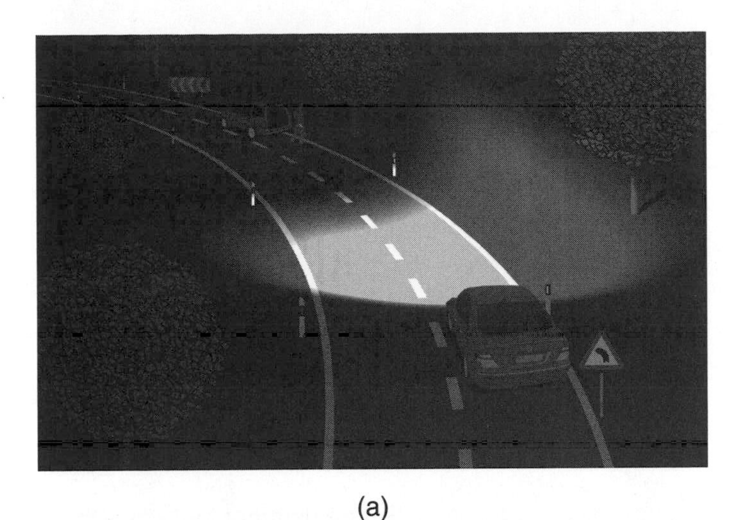

(a)

(b)

FIGURE 11–27 Illumination comparison between (a) a conventional headlight system and (b) an adaptive headlight system.
(Courtesy of DaimlerChrysler Corporation)

FIGURE 11–28 An adaptive headlight assembly.
(Courtesy of DaimlerChrysler Corporation)

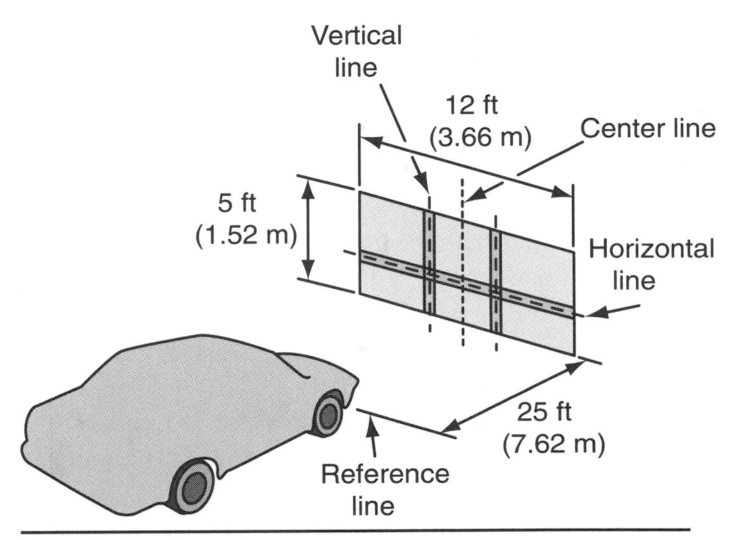

FIGURE 11–29 Projecting headlight beam patterns on a wall to determine basic alignment.

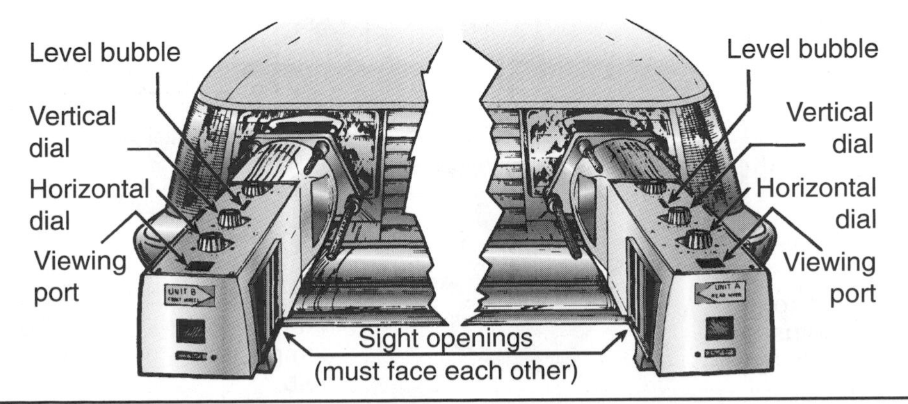

FIGURE 11–30 Mechanical headlight aiming equipment attached to a vehicle.
(Courtesy of Hopkins Mfg. Co.)

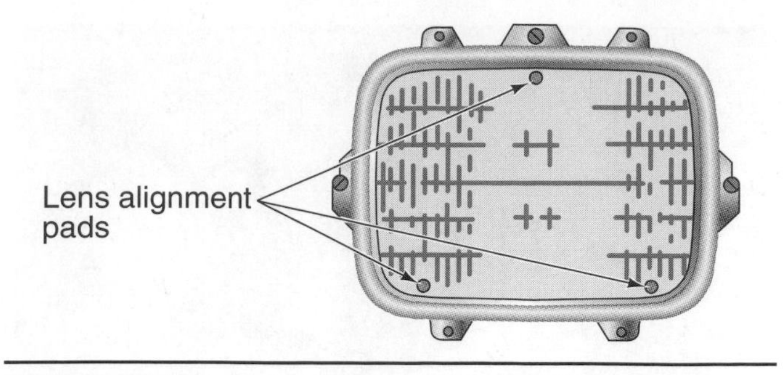

FIGURE 11–31 Lens alignment pad location on a headlight.

The vertical adjustment on the headlight is accomplished by adjusting the vertical adjustment screw (Figure 11–33). The vertical indicator on the headlight aiming unit is a *spirit bubble* that is centered when adjustment is complete (Figure 11–34).

The horizontal adjustment on the headlight is accomplished by adjusting the horizontal adjustment screw (Figure 11–33). The horizontal indicator on the headlight aiming unit is a *split image* that is aligned when the adjustment is complete (Figure 11–35).

Many newer vehicles incorporate composite headlight assemblies with the bubble level (also called a spirit level) built into them (Figure 11–36). This feature allows the vehicle driver to make necessary adjustments based on the vehicle load.

Headlight Voltage Drop Test Dimmer-than-normal headlights are most often the result of a voltage drop caused by excessive resistance in the system due to a bad connection. Other common causes of dim headlights could be low alternator output, the wrong type of lamp, or excessive electrical loads on the system. Headlights do not wear out and get dimmer over time unless the light has been damaged or chipped. To test the headlight system for voltage drop from excessive resistance, determine the number and location of connections, switches, and components from a wiring diagram and component locator diagram. Using a voltmeter, measure each point in the system that has been identified as a possible cause of the voltage drop (Figure 11–37). Follow the manufacturer specification for voltage drop, or use the general table in Chapter 10.

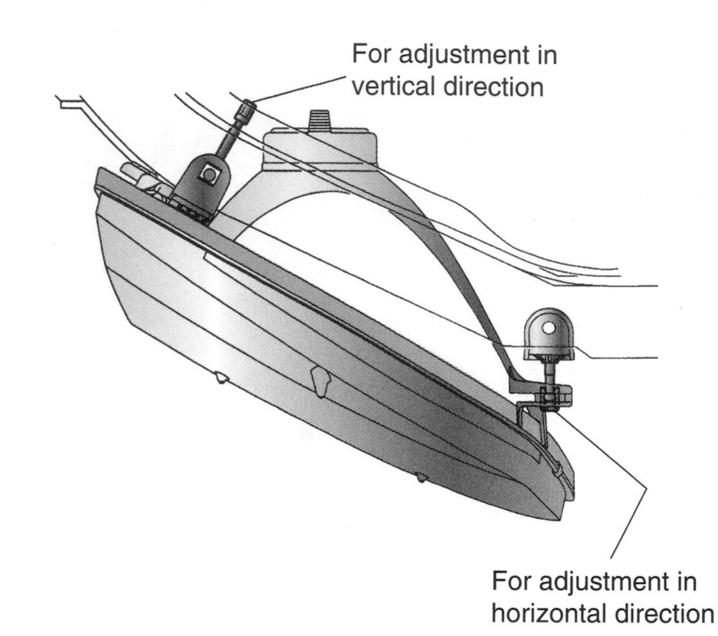

FIGURE 11–33 Typical location of headlight vertical and horizontal adjusting screws.
(Courtesy of Toyota Motor Sales, U.S.A., Inc.)

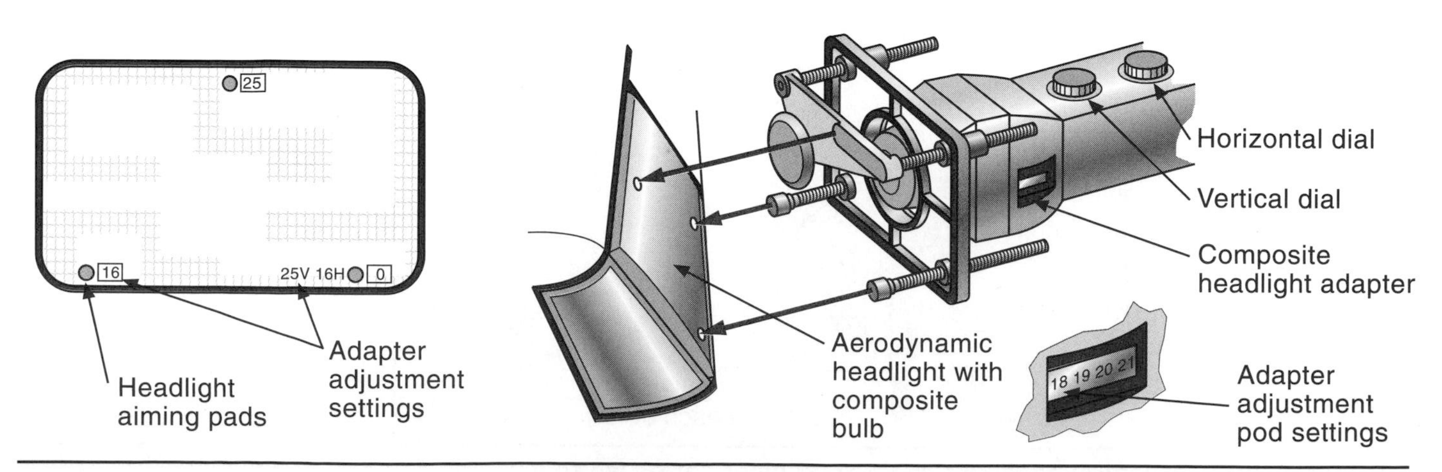

FIGURE 11–32 Special aiming adapters and adjustment settings for composite headlights.

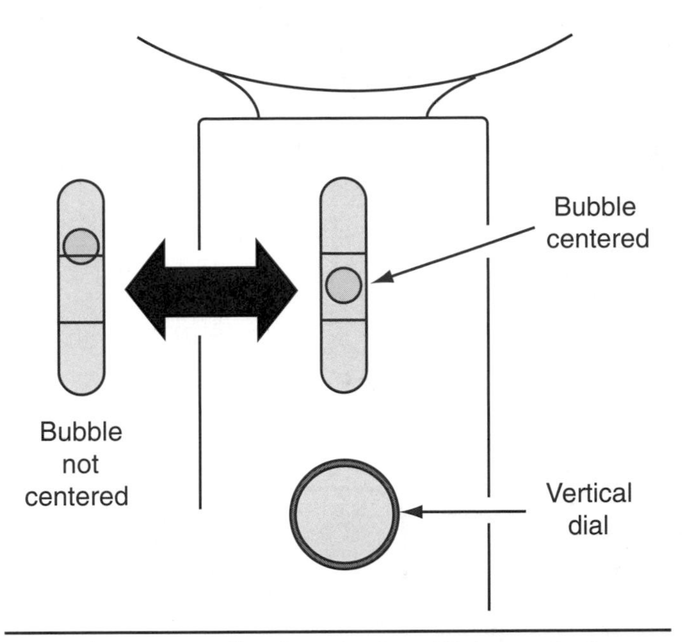

FIGURE 11–34 Adjusting the headlight vertical setting while viewing the spirit bubble.

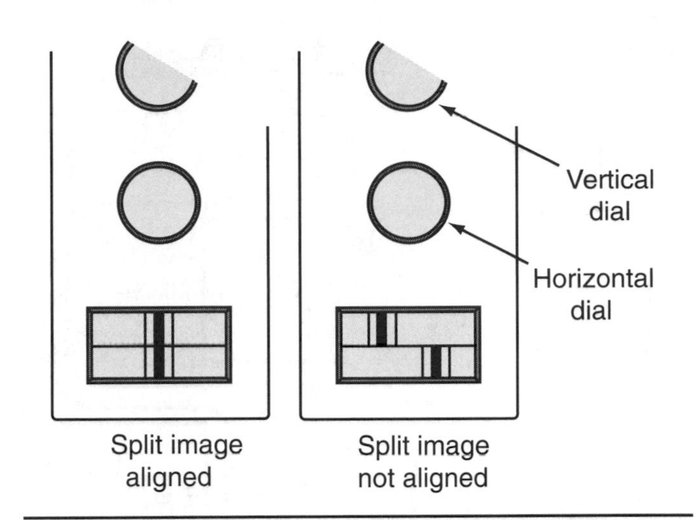

FIGURE 11–35 Adjusting the headlight horizontal settings while viewing the split image target.

Parking and Taillight Circuits

Parking and taillight circuits are normally controlled by the headlight switch. When the headlight switch is pulled to the first detent position, the parking lamps, license lamp, marker lamps, and taillamps turn on, but the headlights remain off. Figure 11–38 is an example of a two-bulb taillight circuit. In this circuit a dual-filament bulb is used on each side of the circuit for taillights and brake lights. If a single-filament bayonet-based bulb is mistakenly forced into a socket designed for a dual-filament bayonet-based bulb, the single contact may short across the dual filament contacts in the socket, causing problems in the operation of the attached circuits (Figure 11–39).

Figure 11–40 is an example of a three-bulb taillight circuit with individual control for each bulb circuit. Most imported vehicles and several newer domestic vehicles use this type of system to prevent lighting problems that might occur when joining two separate systems.

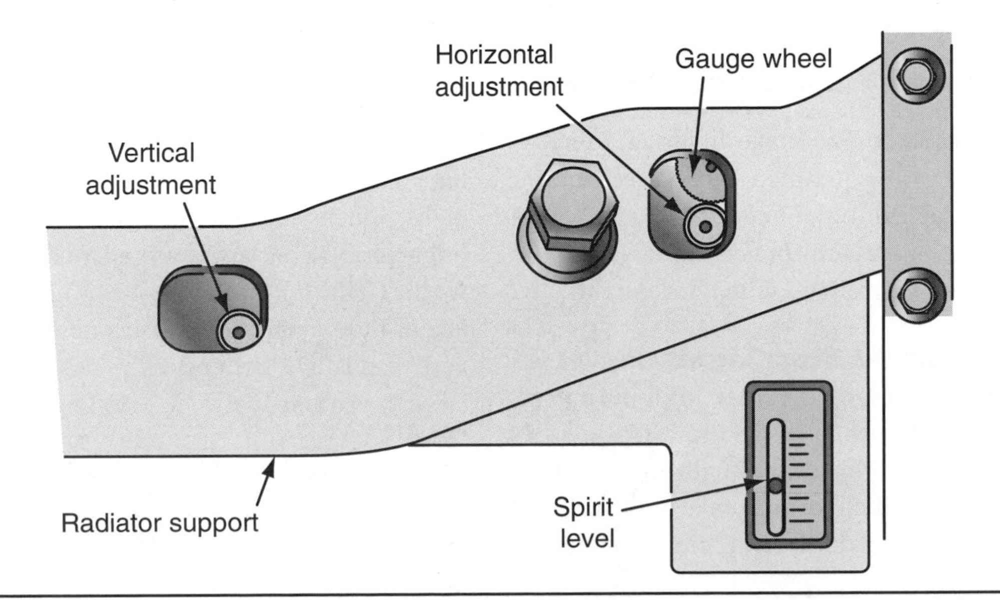

FIGURE 11–36 A typical composite headlight assembly equipped with a spirit level.

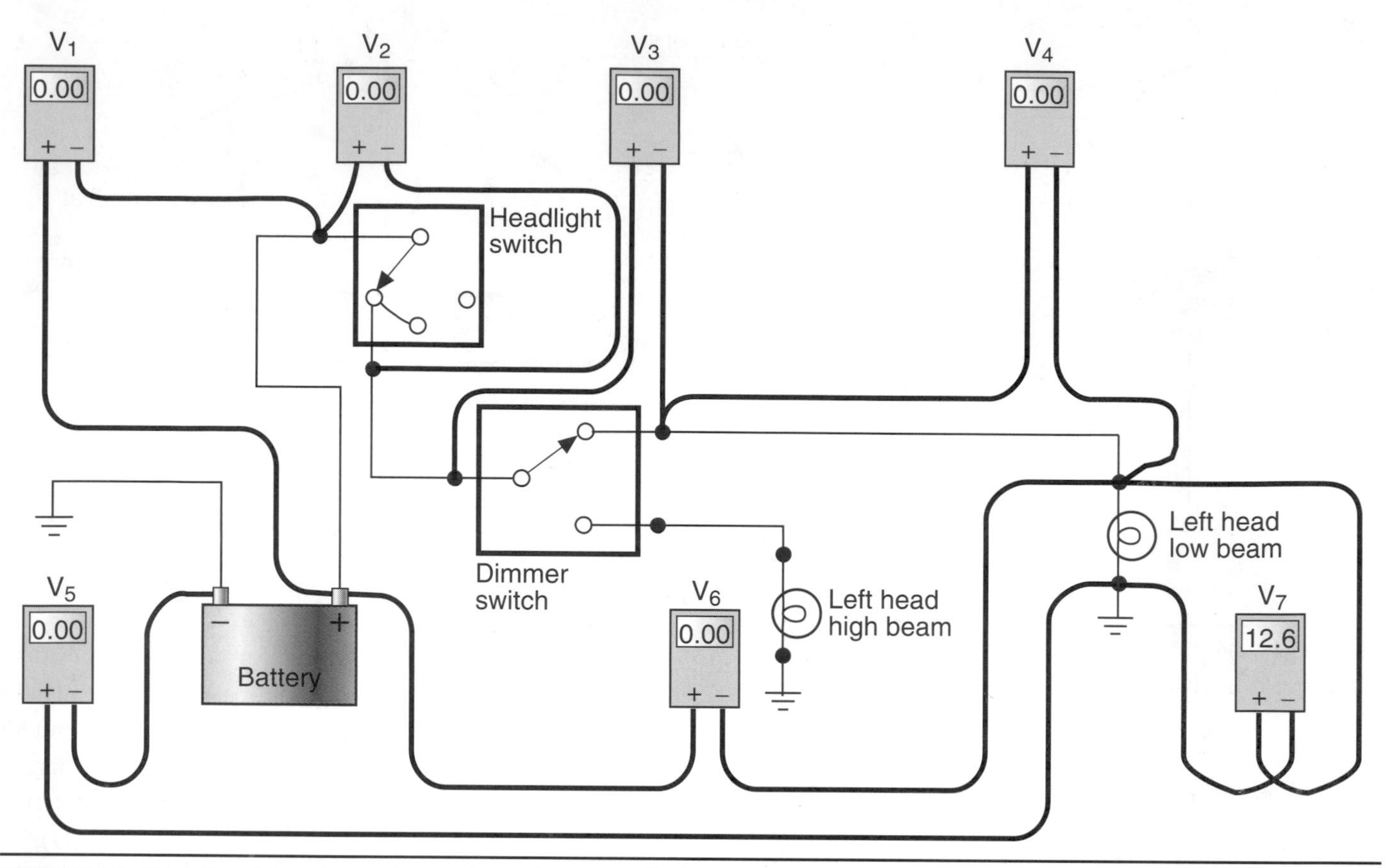

FIGURE 11–37 Checking for voltage drops in a headlight circuit.

Some late model vehicles are also using more LED lighting in tail light circuits instead of incandescent filament bulbs. This technology (discussed in Chapter 5) offers many advantages for all exterior lighting circuits.

Brake Lights and Turn Signal Circuits

Brake Light Switch The brake lights in a three-bulb taillight circuit are controlled by a mechanical brake light switch attached to the brake pedal arm (Figure 11–41). When the brakes are applied, the switch contacts close and complete the circuit to the brake lights.

Center High-Mounted Stop Light Beginning in 1986, all vehicles must have a **center high-mounted stop light (CHMSL).** This light is also referred to as a **collision avoidance light.** In a three-bulb system, the CHMSL is wired in parallel to the brake light circuit because this circuit is separate from the turn signal circuit (Figure 11–42).

In a two-bulb brake light system, the CHMSL can be wired in two common configurations. The first is to wire the CHMSL to the brake light circuit after the brake light switch but before the turn signal switch (Figure 11–43), increasing the amount of wiring needed in the circuit. The second is to install diodes in the system (Figure 11–44) that isolate the CHMSL from the turn signal functions.

Several late model vehicles utilize LED technology specifically for the CHMSL because of the quicker ON time of less than 1 millisecond, compared to incandescent bulbs, which take 200 milliseconds or more to reach full brightness. Neon lamps are also used on some vehicles for the CHMSL (Figure 11–45). Although the neon lamps take a little longer (3 milliseconds) to reach full brightness than LEDs, both offer the same advantage of giving an early warning to a vehicle following close behind.

Turn Signal Switch The turn signal switch is an important safety feature in any vehicle. The turn signal switch on a two-bulb system determines which brake light is isolated from the brake light circuit. In Figure 11–46, the rectangular bars are stationary contacts that are connected to wires. The triangles are conductive

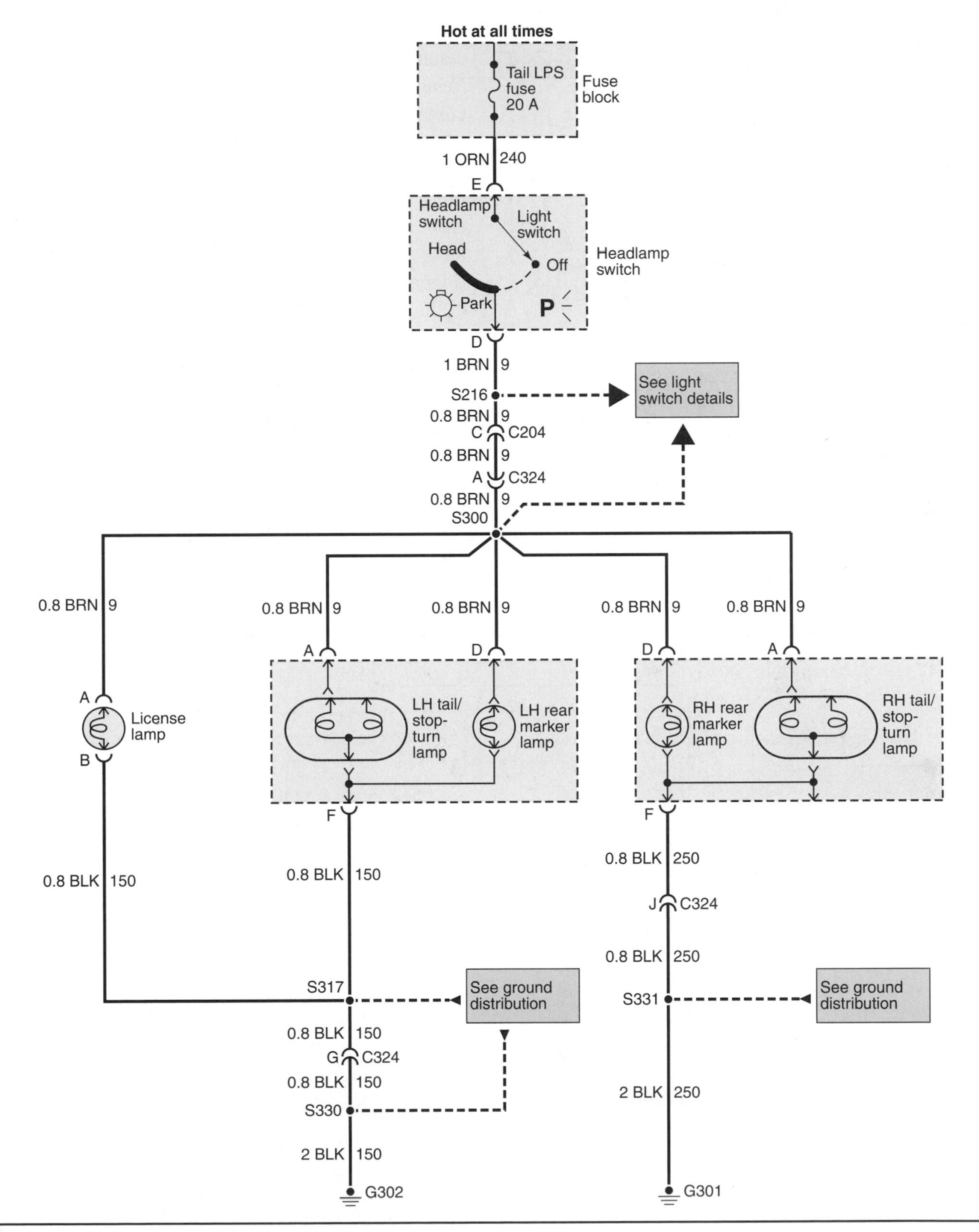

FIGURE 11–38 Typical two-bulb taillight circuit.

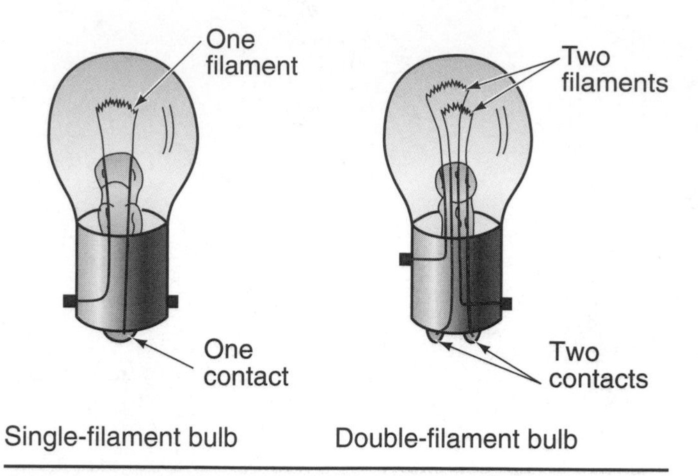

FIGURE 11–39 Single-filament and double-filament bulbs.

pads that slide over the bars to direct the current flow where indicated.

Figure 11–47 is an example of a turn signal switch that is integrated into a multifunction switch, which also includes the headlight switch, wiper/washer switch, and the air bag clock spring. Refer to the specific vehicle manufacturer for instructions on removal and repair. Vehicles with air bags or that have electronic function controls in the steering wheel need special precautions when servicing the turn signal switch.

Figure 11–48 is an illustration of a turn signal circuit in the neutral or rest position. No current flows past the turn signal flasher to the turn signal switch. When the turn signal switch is moved to the left turn position (Figure 11–49), current flows from the turn signal flasher to the designated left turn circuit. When the turn signal switch is moved to the right turn position (Figure 11–50), the current flows to the designated right turn circuit.

Turn Signal Flasher An integral part of the turn signal and hazard signal circuits is the flasher unit, which is located on or around the fuse panel (Figure 11–51). On most vehicles, the flasher unit for the turn signal circuit is separate from the hazard flasher. Some vehicles, though, may incorporate it into one unit.

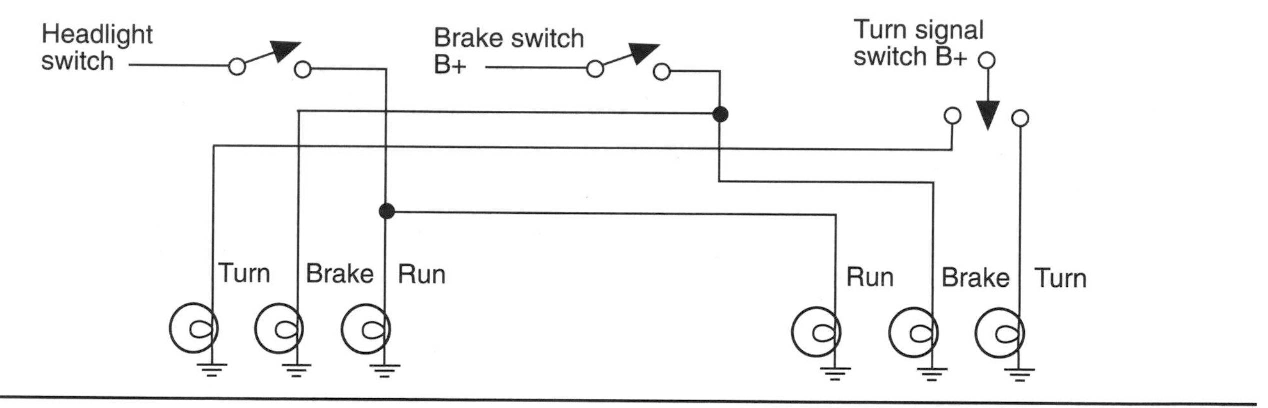

FIGURE 11–40 Typical three-bulb taillight circuit.

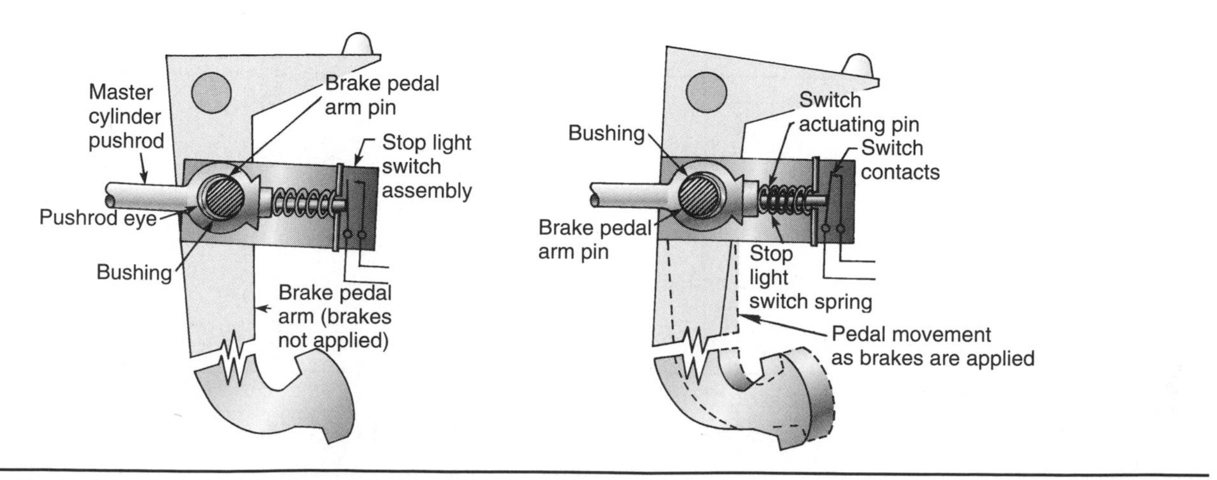

FIGURE 11–41 Typical operation of a brake light switch.

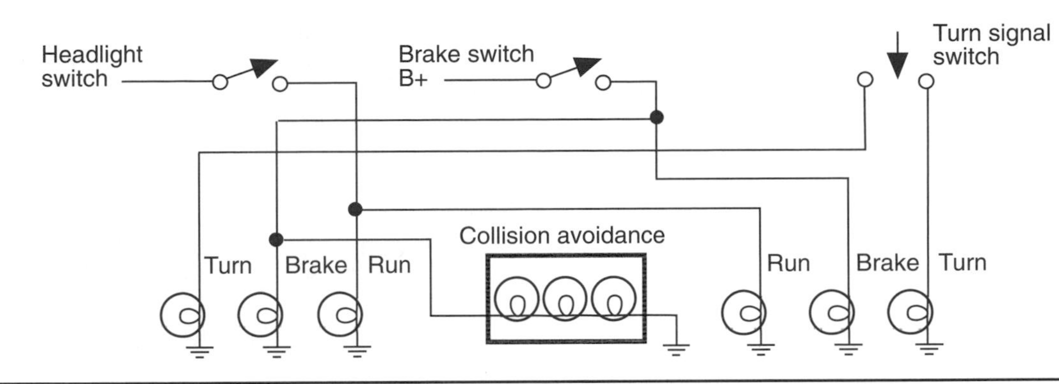

FIGURE 11–42 Brake light circuit with a center high-mounted stop light (CHMSL).

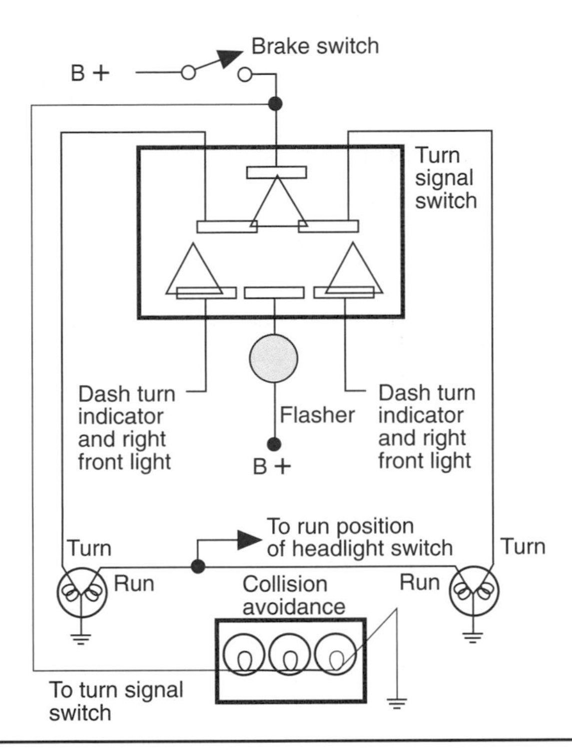

FIGURE 11–43 The CHMSL wired after the brake light switch, but before the turn signal switch.

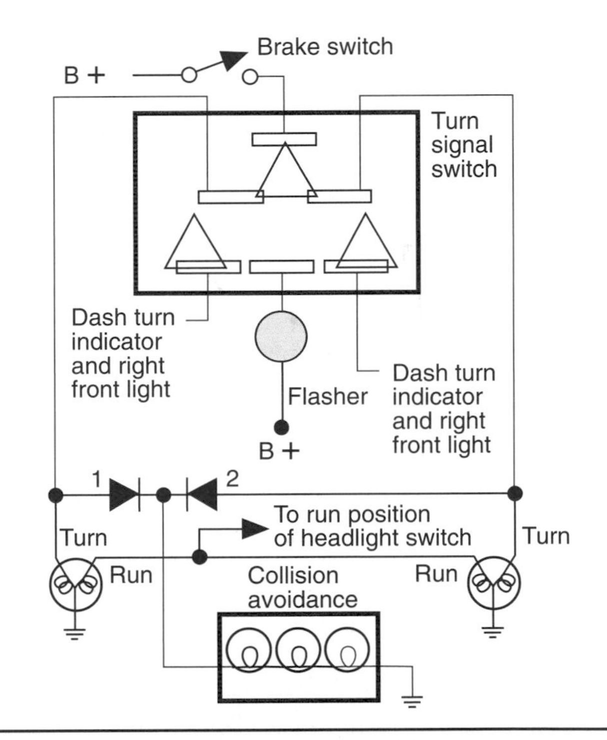

FIGURE 11–44 A CHMSL installed into the two-bulb taillight circuit with isolating diodes.

There are two types of flasher devices: a bimetallic strip flasher and an electronic controlled flasher that is used in most modern vehicles. The first flasher type (Figure 11–51) consists of a bimetallic strip and heating element wired in series with the turn signal circuit to control the current flow to the turn signal lamps. When the bimetallic strip bends from the heat, the contact points open the turn signal circuit. The absence of current allows the strip to cool and close the circuit. Electronic flasher units (Figure 11–52) use solid-state circuitry and transistors to cycle the circuit by timing the current flow through the turn signal system.

FIGURE 11–45 A neon lamp used for a CHMSL.

(Courtesy of BMW of North America Inc.)

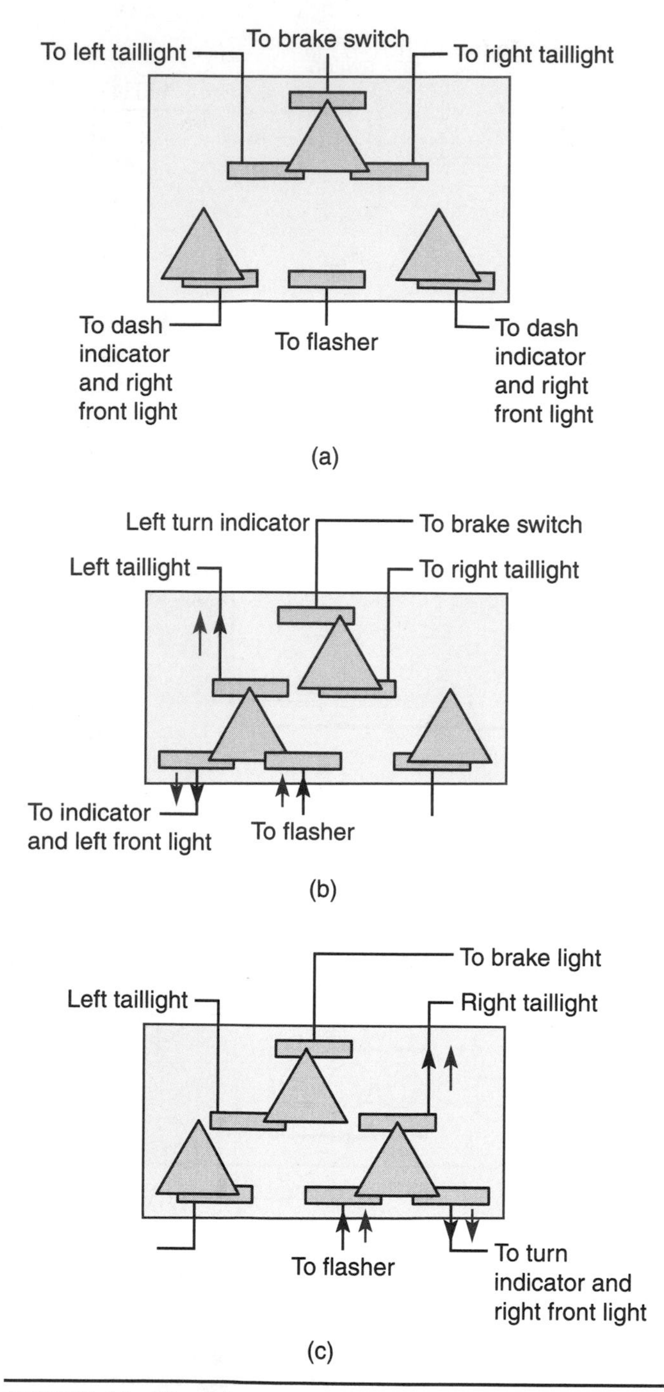

FIGURE 11–46 (a) Turn signal switch in the off position, (b) left turn position, (c) right turn position.

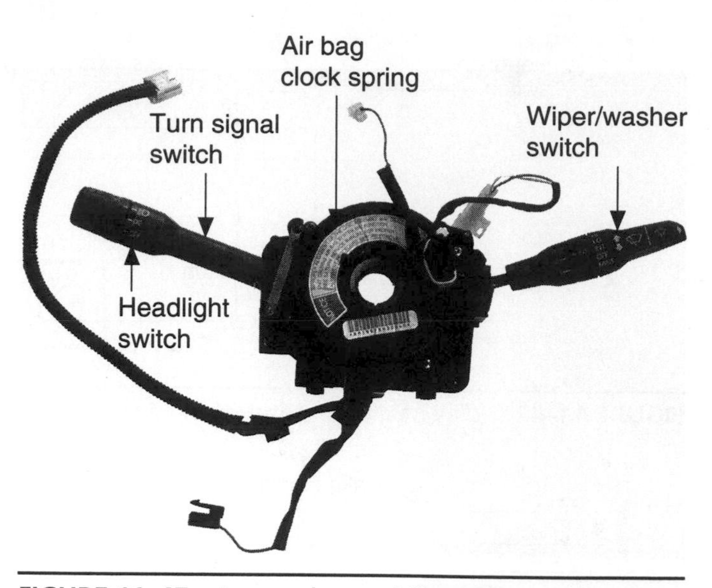

FIGURE 11–47 A complex multifunction switch that includes the turn signal switch.

(Courtesy of Tim Gilles)

When flashers fail to flash as fast as originally designed or flash too fast, a bulb may be burned out. Check all the bulbs for proper operation before replacing the flasher unit. When a bulb burns out on a system that uses a bimetallic flasher, the lights normally flash more slowly than normal. When a bulb burns out on a system with an electronic flasher, the lights flash more rapidly than normal. When trailers are hooked up to a vehicle, the trailer lights are wired in parallel with the vehicle light systems. This connection can raise the current flow through the flasher unit and cause it to blink faster than normal. Flashers are designed to operate a specific number of bulbs. To correct this problem, heavy-duty flasher units are available to correct the blink rate (Figure 11–53).

Side Marker Lights The side marker lights have been a safety requirement on all vehicles sold in North America since 1968. When a vehicle enters traffic from the side, the marker lamps enable others to see the vehicle. The front side marker lens is amber, and the rear is always red. Many manufacturers wire the side marker lamp circuit in parallel to the parking lamp circuit so that they will work anytime the park lights or headlights are on (Figure 11–54).

Some vehicle manufacturers wire the side marker lamps across the parking light and turn signal light circuits (Figure 11–55). In this circuit, the side marker lamp flashes when the turn signals are used. If the parking lights are on, the side marker lamp uses the turn signal lamp as a ground to complete the circuit (Figure 11–56). The turn signal lamp does not light because the voltage drop across the side marker lamp is too high.

If the parking lamps are off when the turn signals are used, the side marker lamp flashes with the turn signal lamp (Figure 11–57). The side marker lamp uses the parking lamp as a ground to complete the circuit.

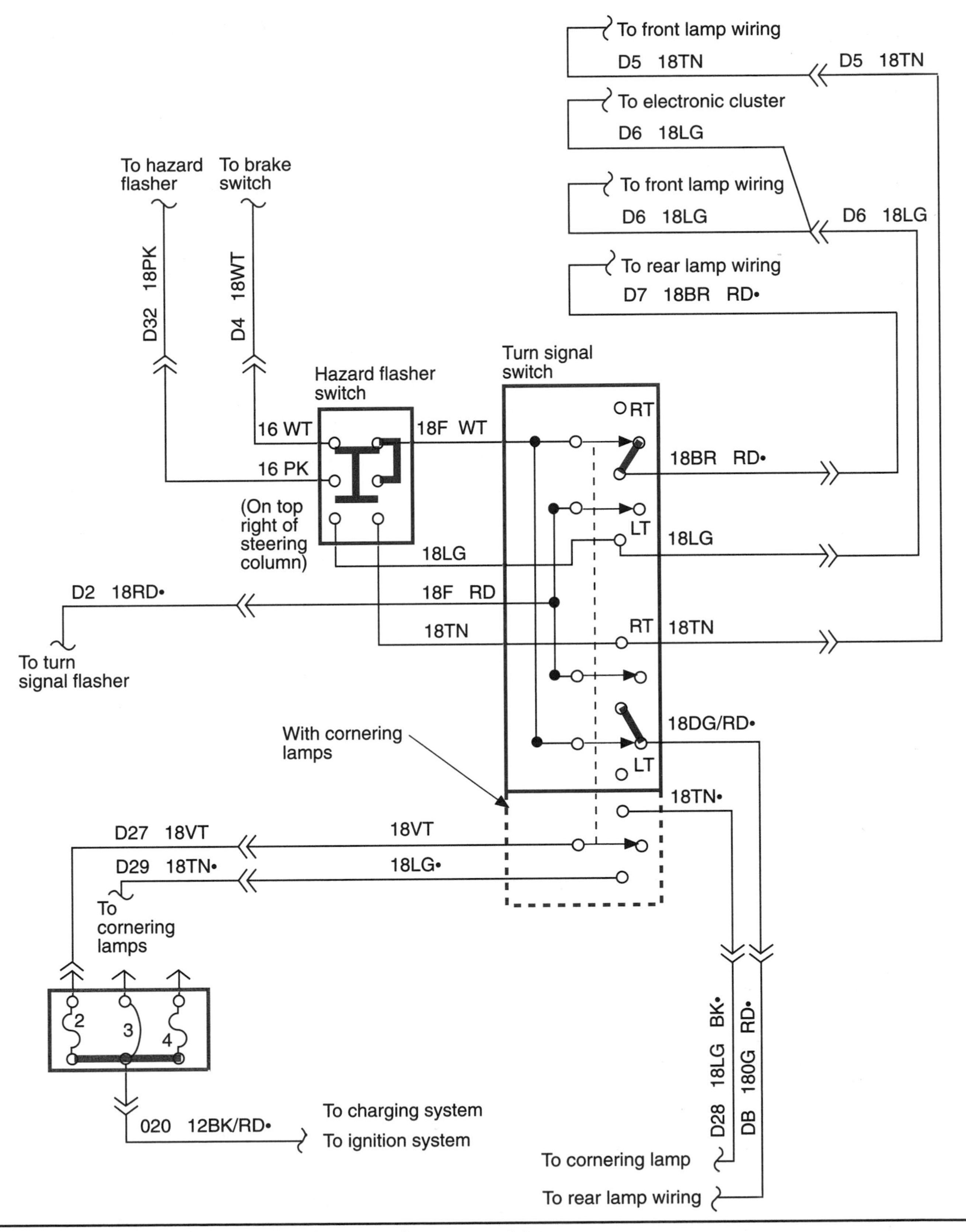

FIGURE 11–48 Turn signal circuit in the neutral or rest position.

(Courtesy of DaimlerChrysler Corporation)

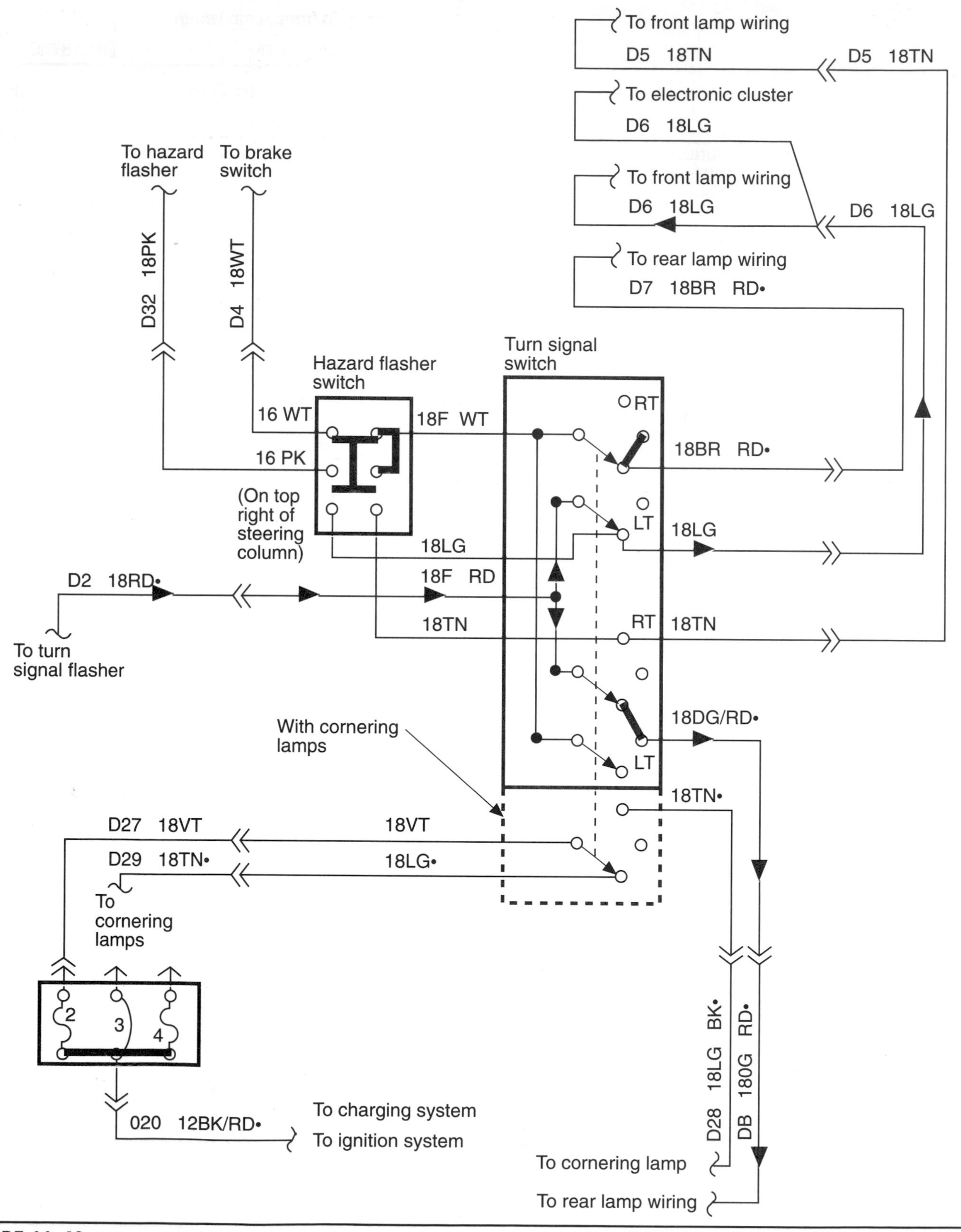

FIGURE 11–49 Turn signal circuit in the left turn position.

(Courtesy of DaimlerChrysler Corporation)

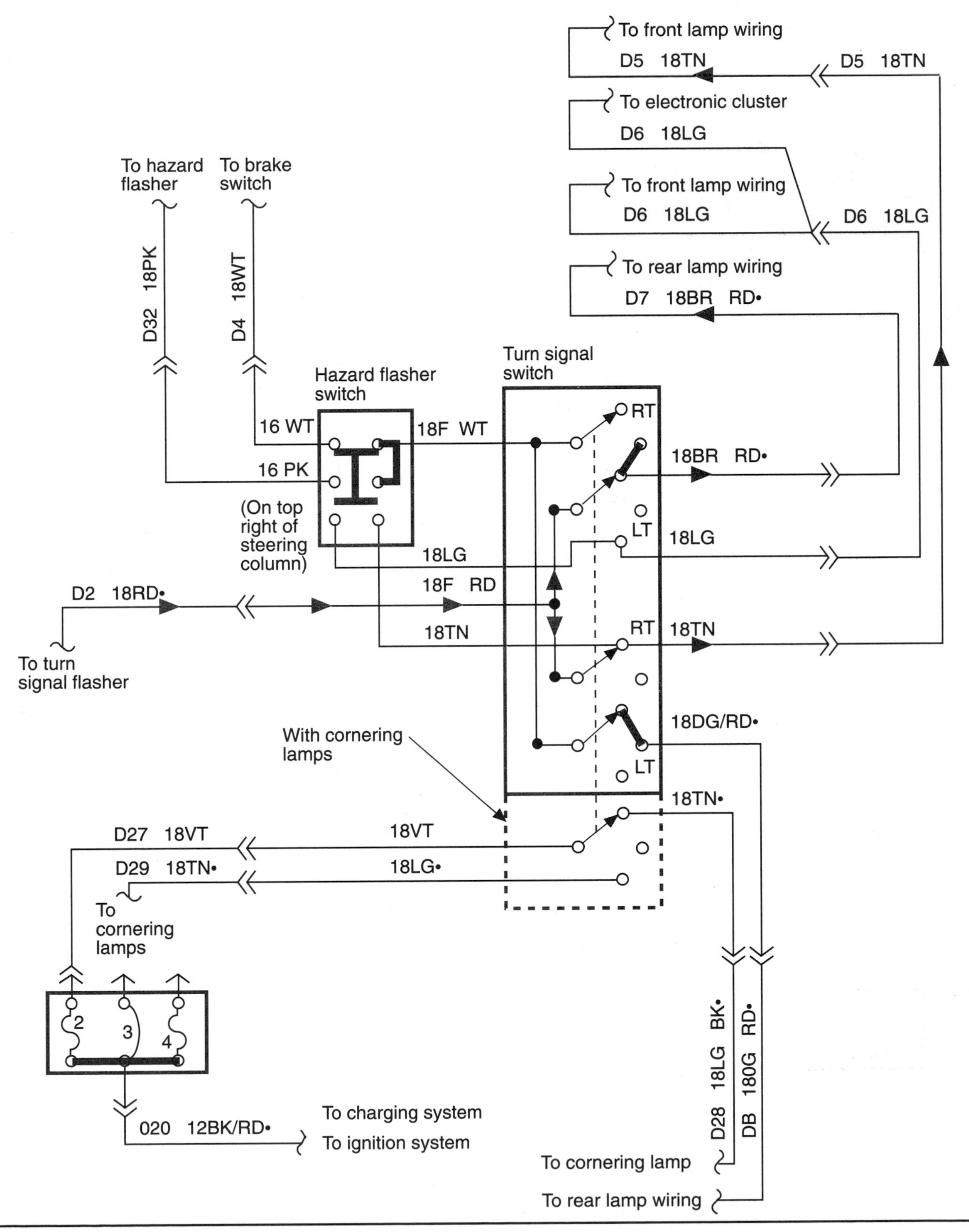

FIGURE 11–50 Turn signal circuit in the right turn position.

(Courtesy of DaimlerChrysler Corporation)

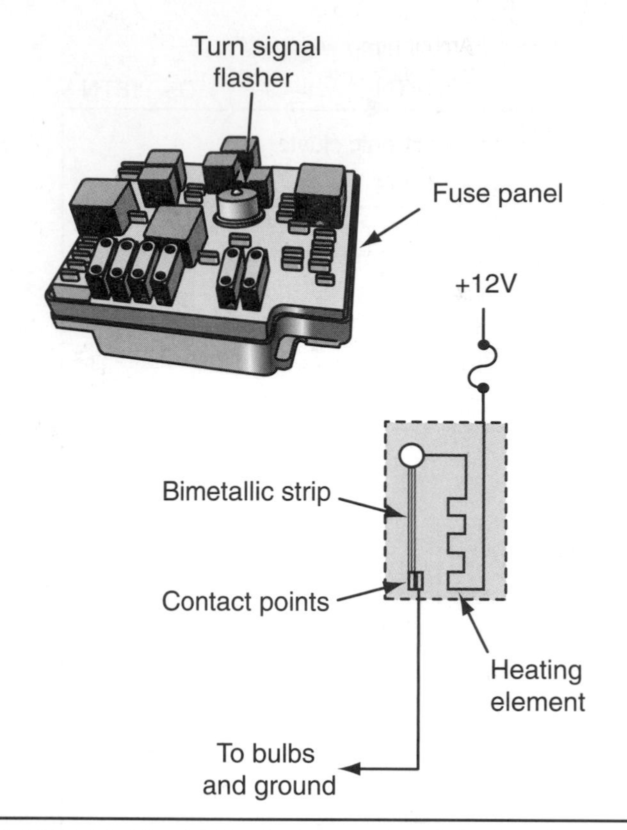

FIGURE 11–51 (a) Typical turn signal flasher location and (b) flasher bimetallic strip and contacts.

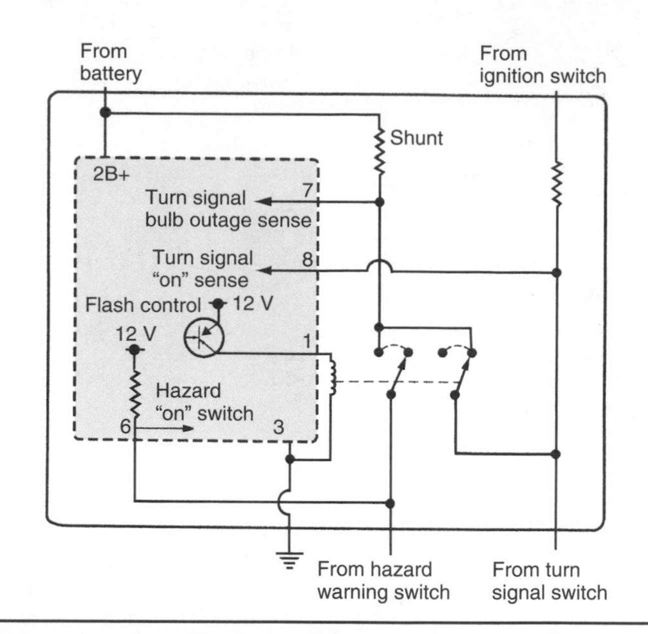

FIGURE 11–52 Typical solid-state electronic flasher unit.

(Courtesy of DaimlerChrysler Corporation)

FIGURE 11–53 Comparison between a standard duty and a heavy duty flasher.

(Courtesy of Tim Gilles)

Because of the high voltage drop across the side marker lamp, the parking lamp does not light.

When the parking lamps are on and the turn signals are activated, the side marker lamp flashes alternately with the turn signal (Figure 11–58). This is due to having equal voltage on both sides of the side marker lamp until the turn signal flasher opens. When this happens, the side marker operates (Figure 11–56) until the flasher closes and provides voltage to the turn signal lamp again.

Because the parking lamp circuit and turn signal circuit are connected through the side marker lamp, feedback of the current flow can occur if a ground connection goes bad in one of the circuits. This is important to remember when diagnosing a lighting circuit problem.

Fog Light Circuit For additional safety under certain driving conditions, some vehicles have fog lights as optional equipment from the factory or as an aftermarket add-on. The fog lamp circuit normally has a relay because of the high amperage required by the fog lamps. Figure 11–59 is an example of a fog lamp circuit that can operate only if the headlight switch is in the park or headlight position. Some off-road or sport utility vehicles have the fog light circuit wired directly to battery voltage to allow the circuit to operate independently of the headlight circuit. Refer to specific vehicle wiring diagram before attempting any repairs.

Backup Light Circuit The backup light circuit is designed to illuminate the road behind the vehicle. Figure 11–60 illustrates a backup light circuit for an

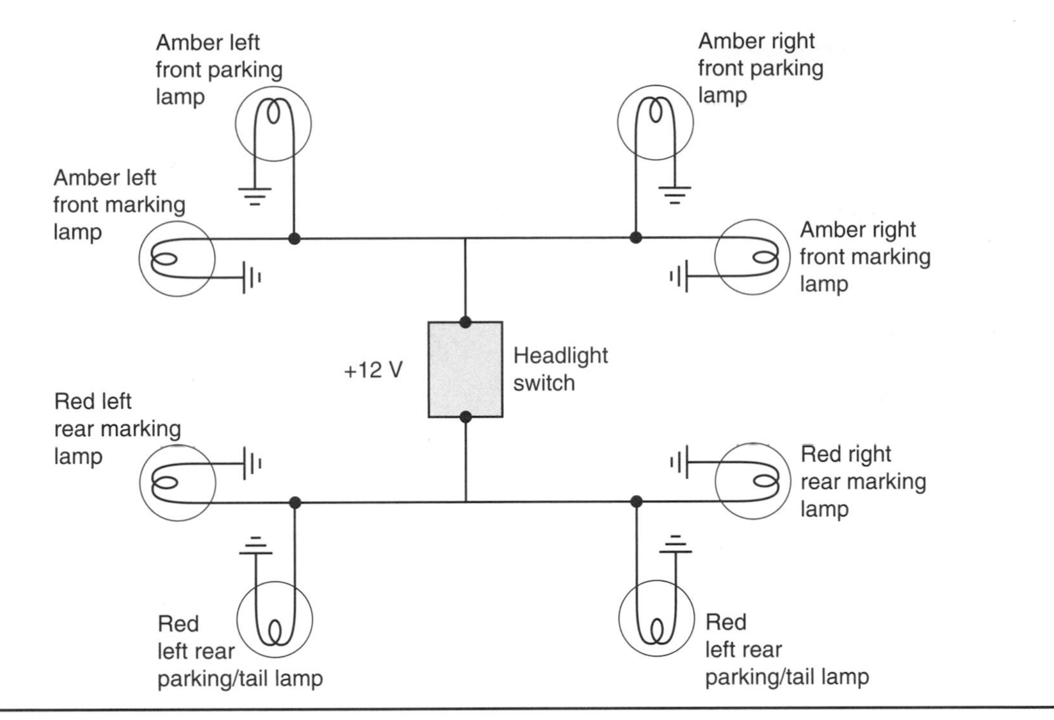

FIGURE 11–54 Typical side marker light circuit.

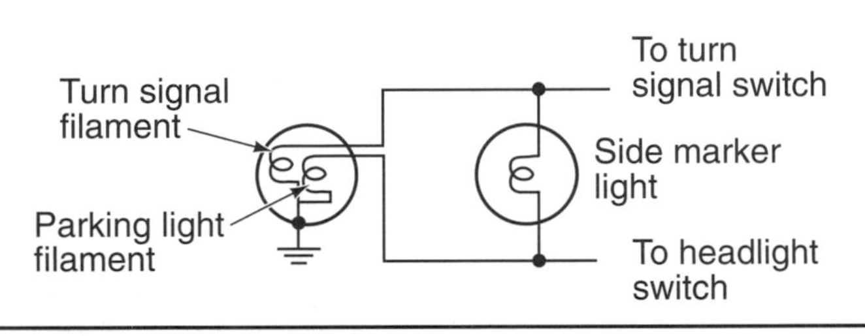

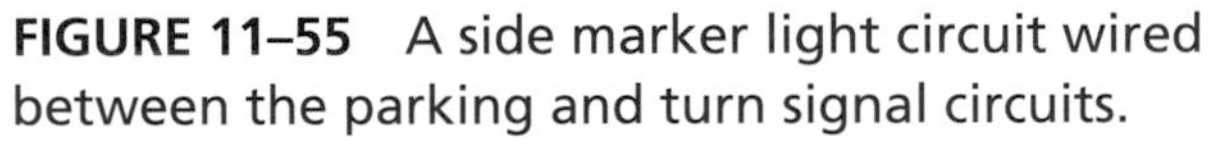
FIGURE 11–55 A side marker light circuit wired between the parking and turn signal circuits.

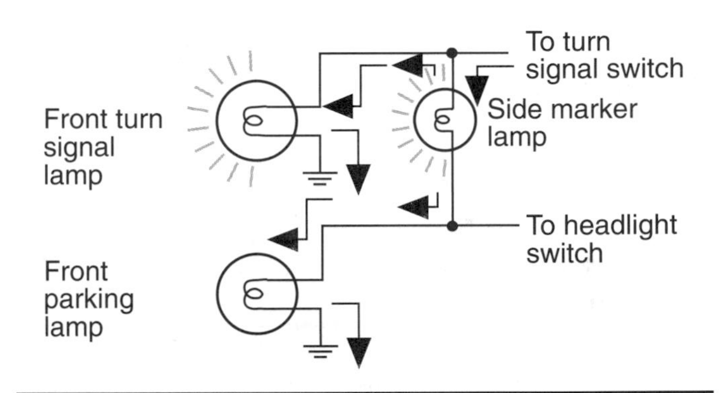

FIGURE 11–57 Current flow through the side marker lamp with the turn signal circuit on and the parking light switch off.

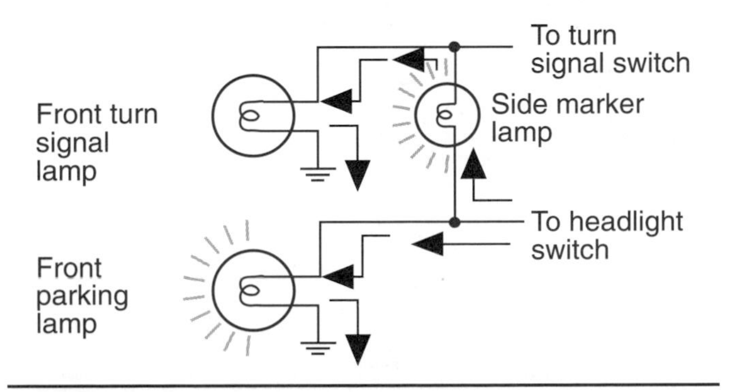

FIGURE 11–56 Current flow through the side marker lamp with the parking light switch on.

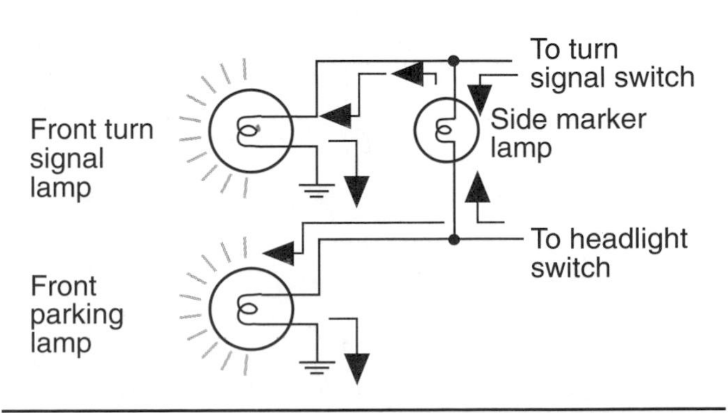

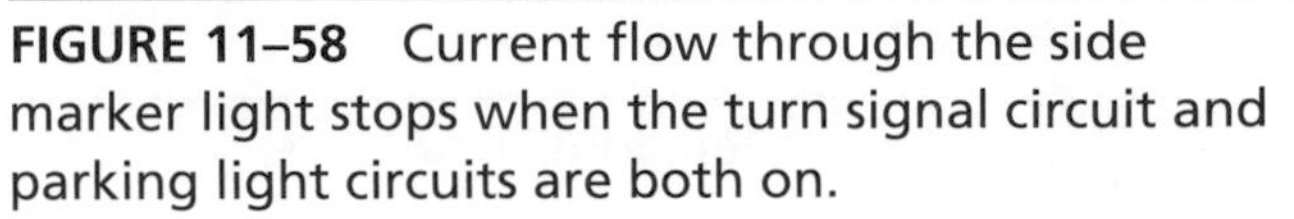
FIGURE 11–58 Current flow through the side marker light stops when the turn signal circuit and parking light circuits are both on.

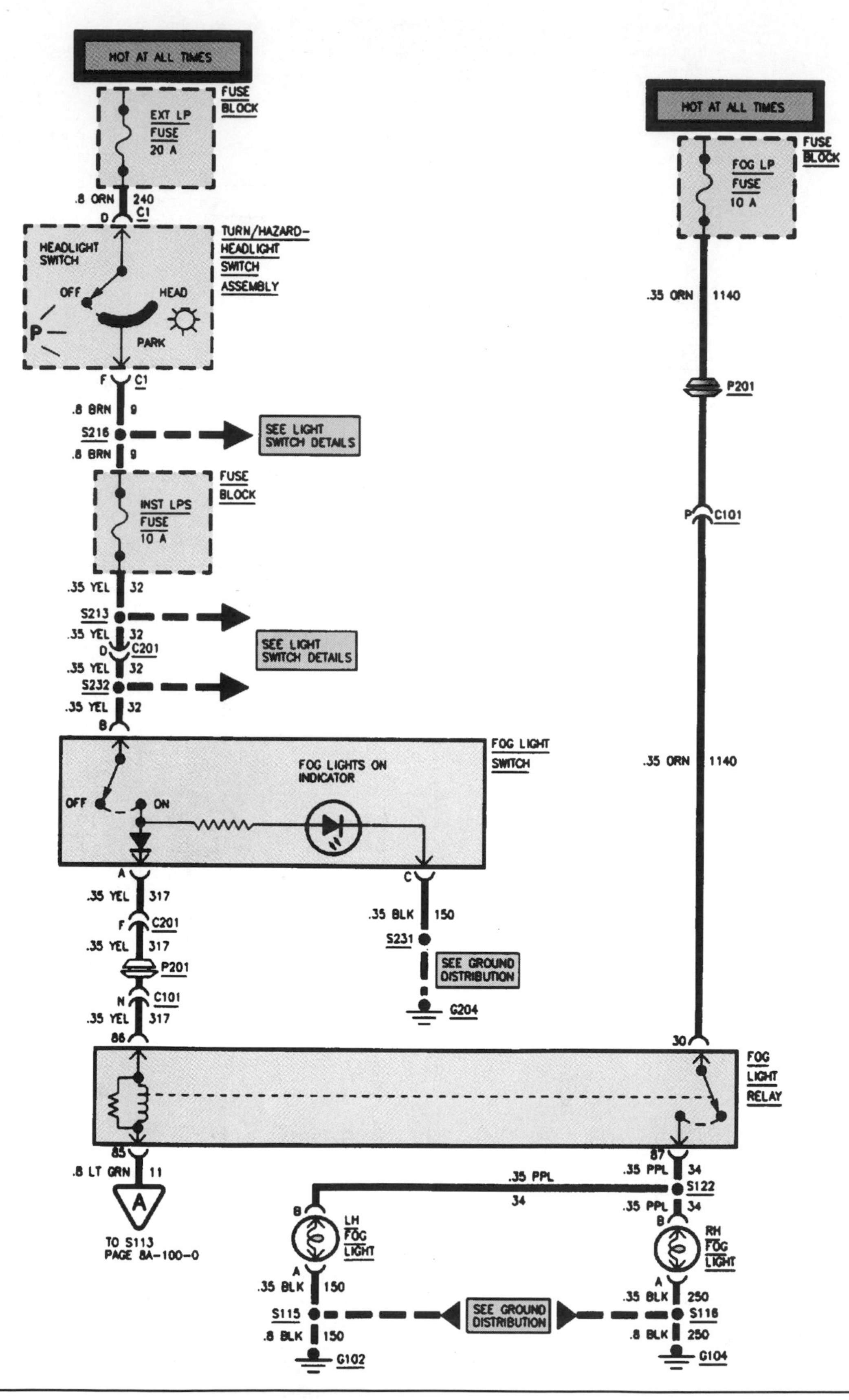

FIGURE 11–59 Typical fog light circuit.

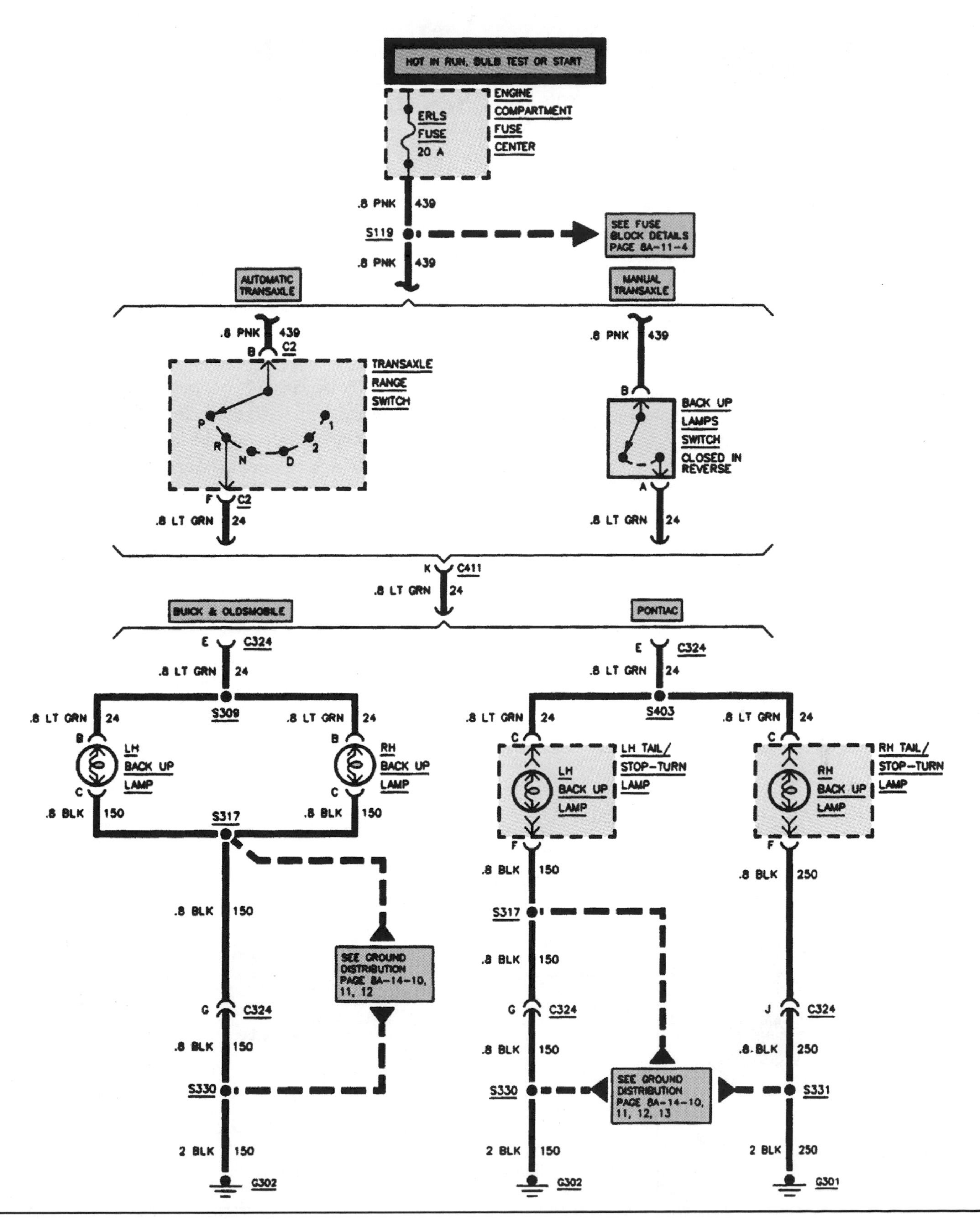

FIGURE 11–60 Backup light circuit.

automatic transaxle vehicle and a manual transaxle vehicle. Power is supplied to a fuse only when the ignition switch is in the run position. This prevents the backup lamps from operating by mistake if the ignition is off. The circuit bulbs are energized from battery voltage when the transmission is put into reverse with the key on.

The backup light system is relatively easy to diagnose by performing the following checks:

- Check the system fuse. If other accessories operate off the same fuse, check them for proper operation.
- If power to the switch and the switch check out satisfactorily, check the backup lamps for burned-out elements, and replace them if necessary.
- If the lamps check out OK, then check the bulb sockets, connectors, and associated wiring.
- Check for voltage input at the backup light switch. If voltage is present, bypass the switch with a jumper wire and check the backup lights for operation. If the system works with the jumper wire, then the switch is faulty and should be replaced.

Interior Lighting Circuits

Interior Light Systems

The interior lighting circuits in today's vehicles vary with the style of vehicle and the amount of optional equipment. Basic interior lighting includes courtesy lights, reading lights, map lights, and instrument panel lights. Luxury vehicles tend to have more complex interior lighting circuits, which can operate through another system, such as a body control module (BCM). In addition, the use of fiber optics for lighting control knobs and panels is gaining in popularity in more than just high-end vehicles because of their ability to illuminate several objects with a single bulb (Figure 11–61). In a fiber optic system, light rays are transmitted by internal reflection through strands of polymethacrylate plastic encased in a sheath that keeps the light rays from escaping until reaching the end object. The strands of fiber optics are very flexible, making them perfect for lighting objects several feet away.

Courtesy Light Circuit Several types of courtesy lights are on vehicles today. Some courtesy lights are located in the front and rear door trim panels, in floor

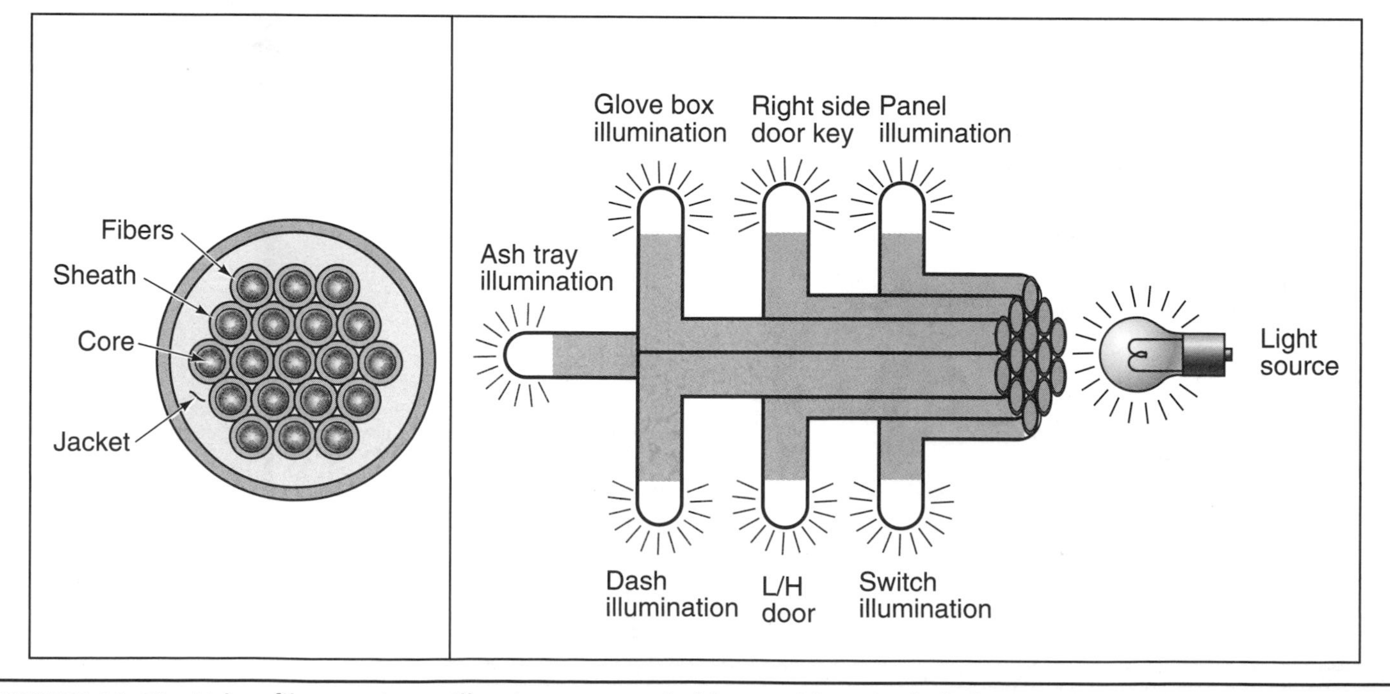

FIGURE 11–61 Using fiber optics to illuminate several objects with a single light source.

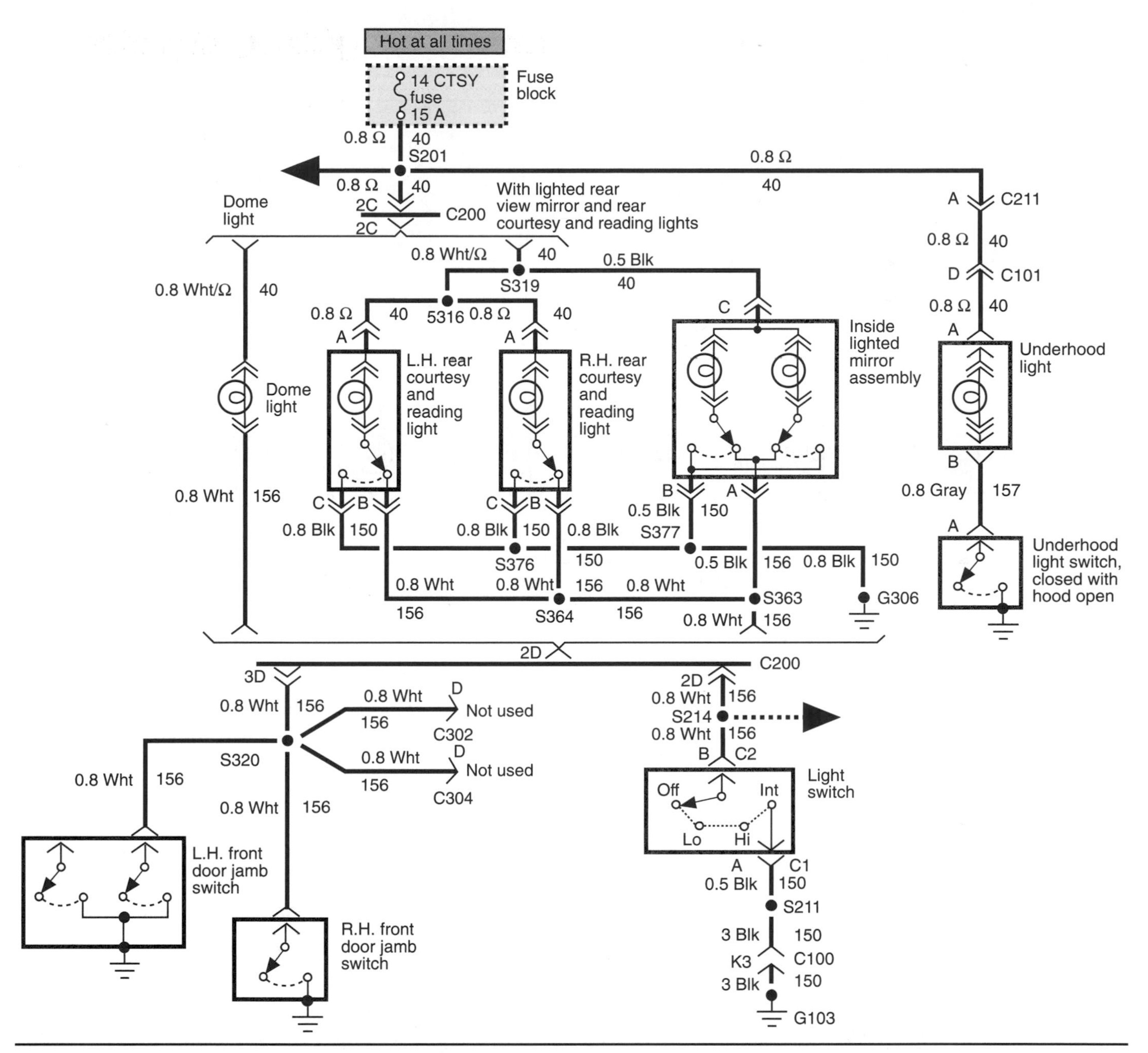

FIGURE 11–62 Typical courtesy light circuit with control switches on the ground side.

kick panels, under the instrument panel, and in the center of the headliner. Figure 11–62 is a courtesy light circuit with the door jamb switches on the ground side of the circuit. The courtesy light circuit completes the path to ground when either of the doors is opened or when the headlight switch is turned to the interior light position.

The courtesy light circuit can also be designed with the door jamb switches on the battery *positive side* of the circuit (Figure 11–63). These switches are insulated from the ground side of the circuit (chassis), and they control the battery voltage to the courtesy lights.

The courtesy light circuit may share its system fuse with other circuits in the vehicle. This may cause a sneak circuit feedback, which is when the current in one circuit seeks an alternate path of flow when the original path is interrupted. In turn, this effect can cause lighting circuits or other accessories to operate when not intended. Figure 11–64 is an example of a circuit that can experience

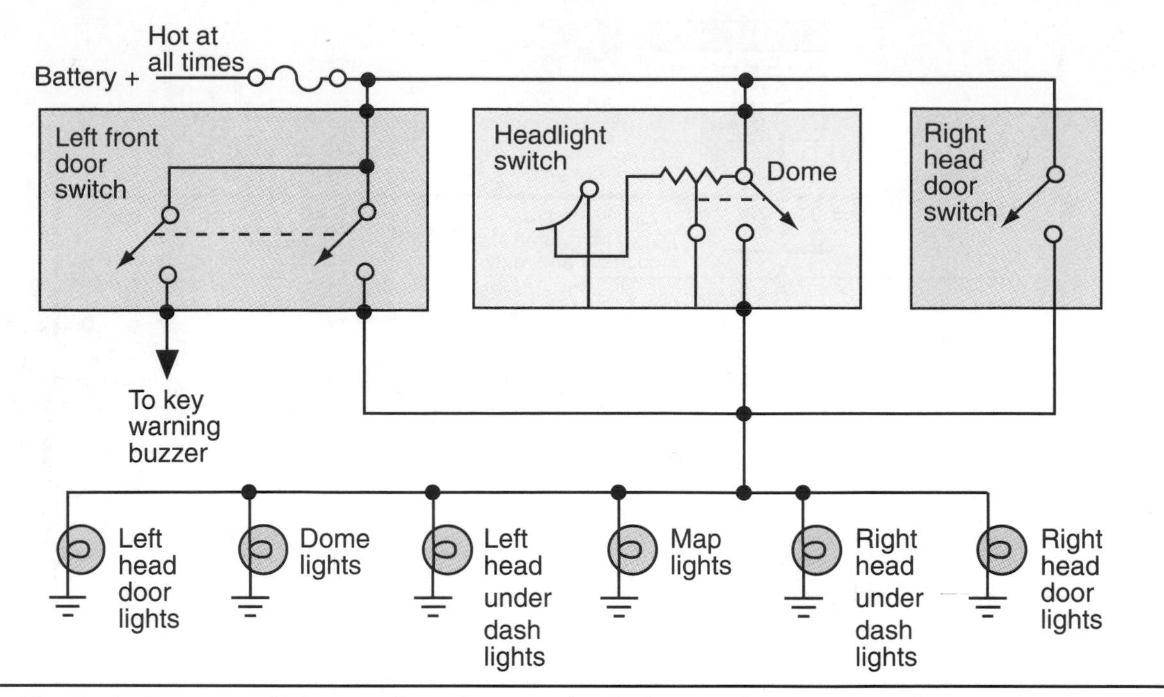

FIGURE 11–63 Typical courtesy light switch with control switches on the positive side.

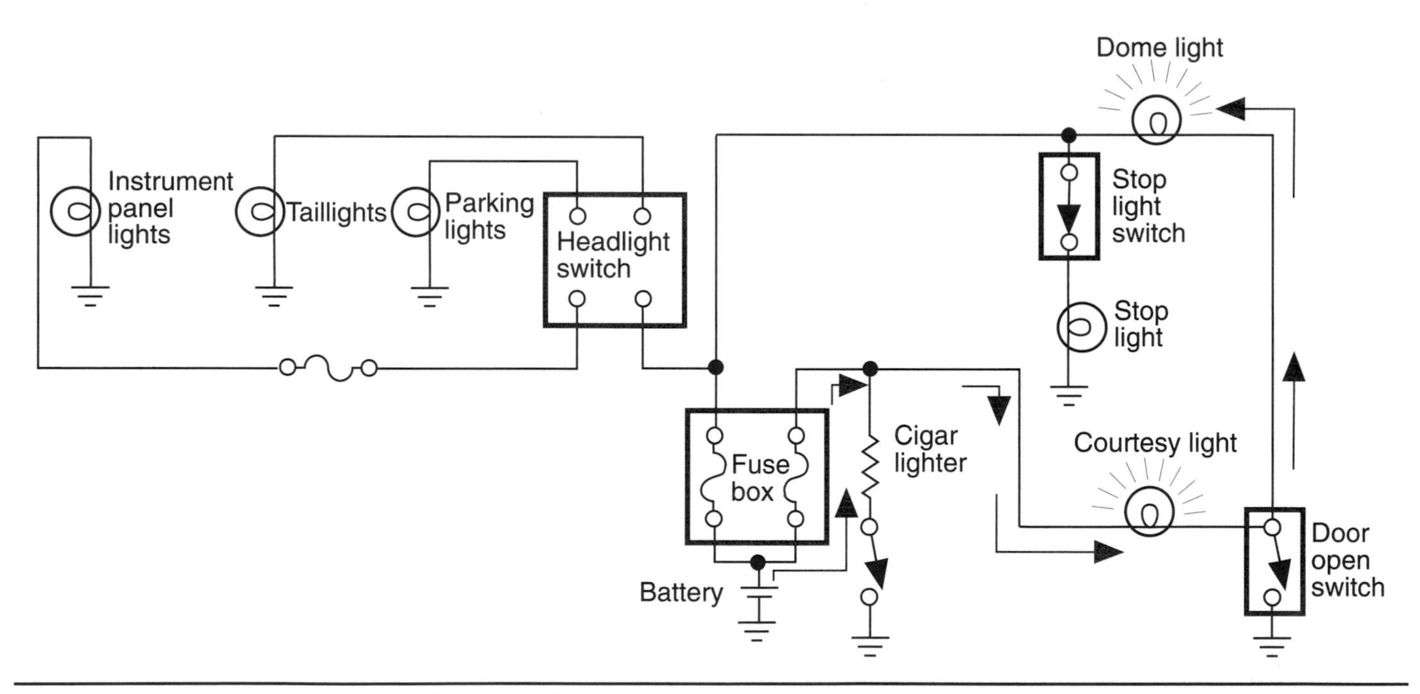

FIGURE 11–64 Circuit feedback if the tail/dome/brake light fuse is blown and the stop light switch is closed.

a feedback condition. The dome light, taillight, and brake light share a fuse, and the cigar lighter and courtesy light share a fuse. If the dome/tail/brake fuse is blown and the headlight switch is off, whenever the stoplight switch is closed, the courtesy light, dome light, and stop light all glow dimly because these lights are in series.

Reading, Map, and Courtesy Light Circuits

The interior lighting circuit in most vehicles have some branch lighting circuits that are controlled by individual switches. Figure 11–65 is an example of a reading/vanity light circuit. Battery voltage is supplied at connection B from the interior light circuit fuse. Each reading lamp

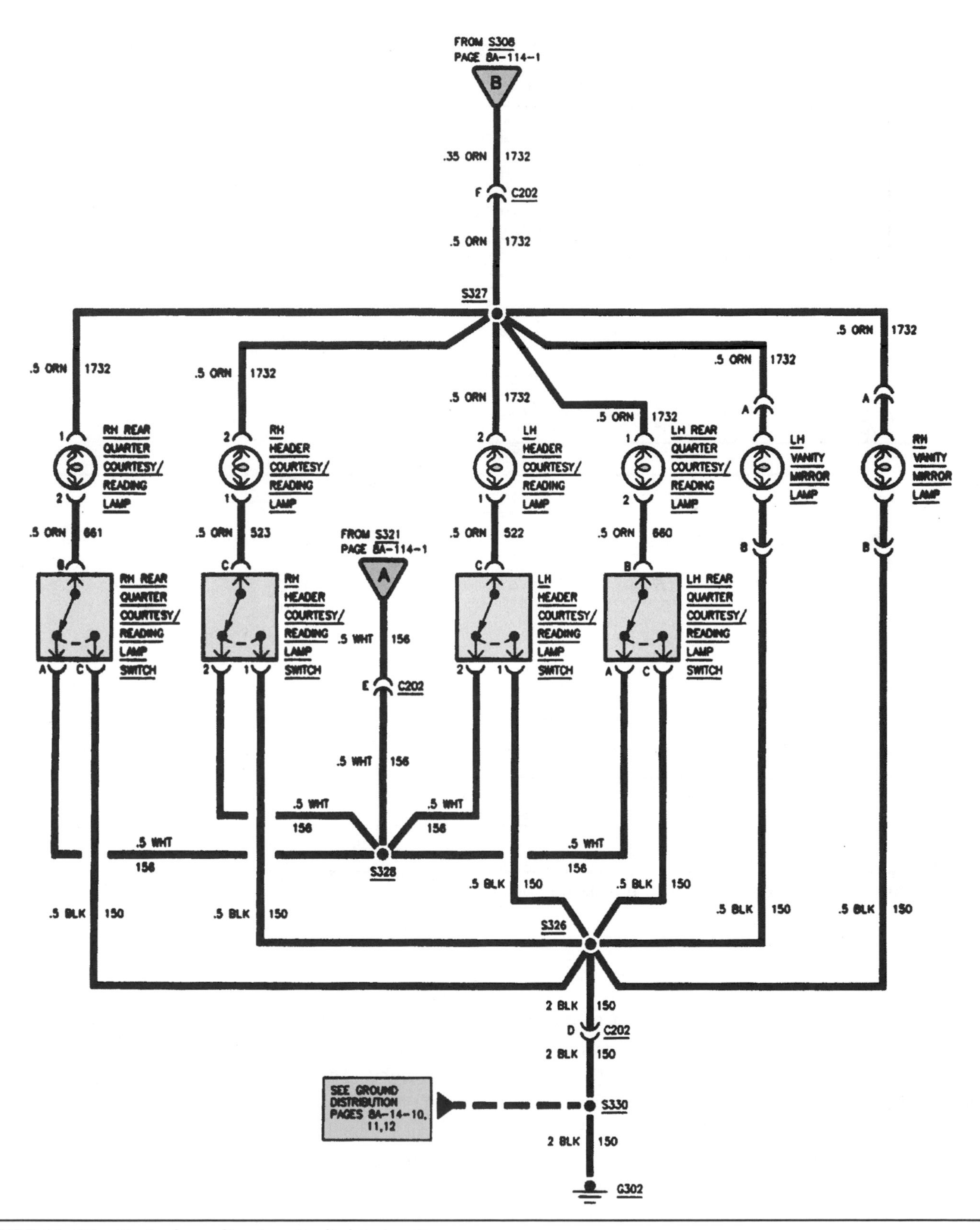

FIGURE 11–65 Typical reading/vanity light circuit.

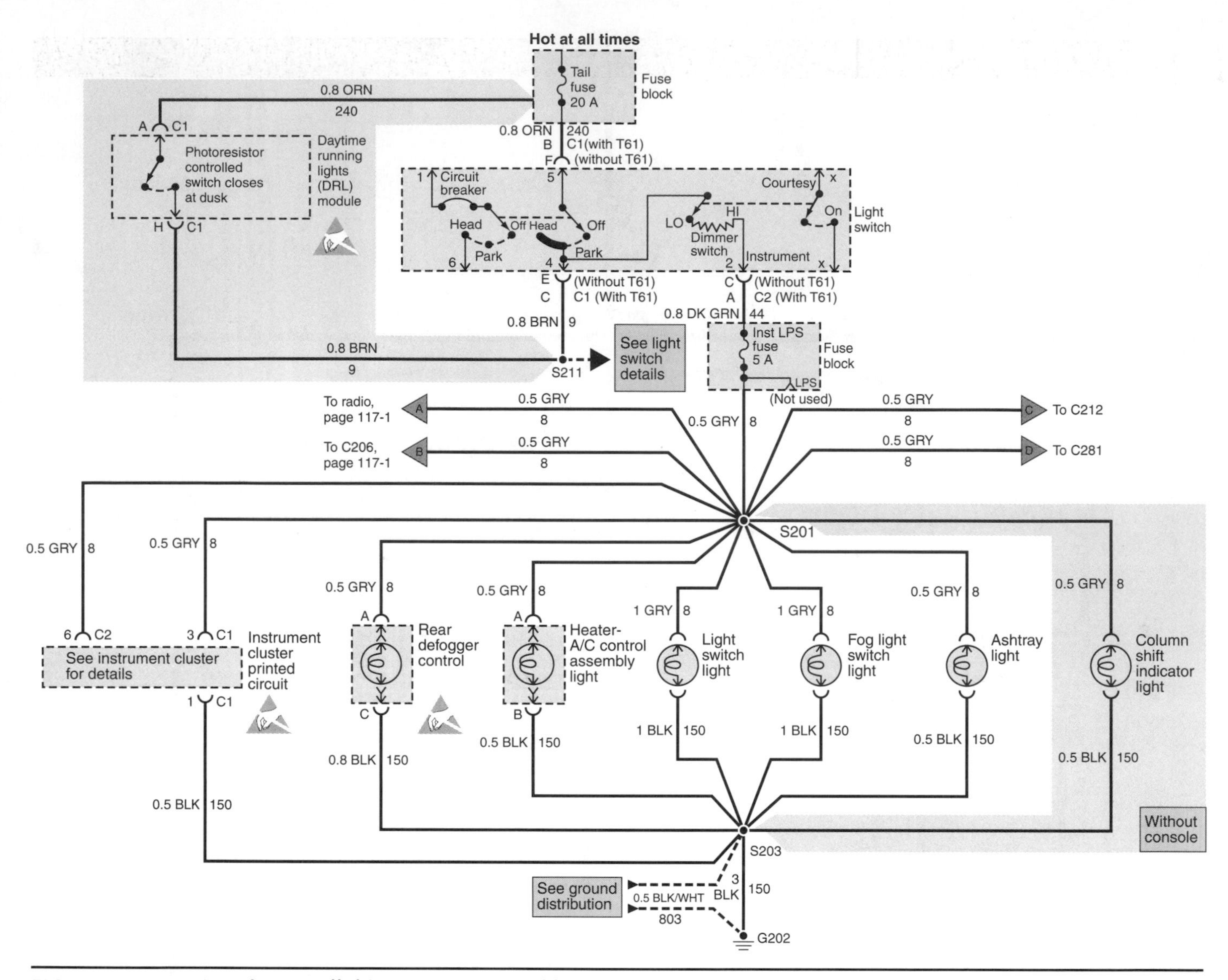

FIGURE 11–66 A series/parallel instrument panel light circuit.

has a control switch to complete the circuit to ground. The vanity mirror lamps have an internal switch that completes the circuit to ground whenever the mirror cover is opened.

Instrument Panel Light Circuit A simple instrument panel light circuit is a perfect example of a series/parallel or compound circuit. In Figure 11–66, the rheostat is connected in a series line to the instrument panel lights. The voltage available at the parallel instrument panel lights depends on the setting of the rheostat control knob. Because the lamps are in parallel, if one lamp burns out, the rest remain on.

CASE STUDIES

CASE 1

The following case study is typical of a headlight circuit that technicians often encounter. Knowledge of the circuit and simple test procedures can help you perform the diagnosis quickly and efficiently.

Customer Complaint

A customer brings a vehicle into the shop because the headlights go out when they are switched from low to high beams. The customer replaced both headlamps, thinking the high beam bulb elements were burned out, but the problem remained (Figure 11–67).

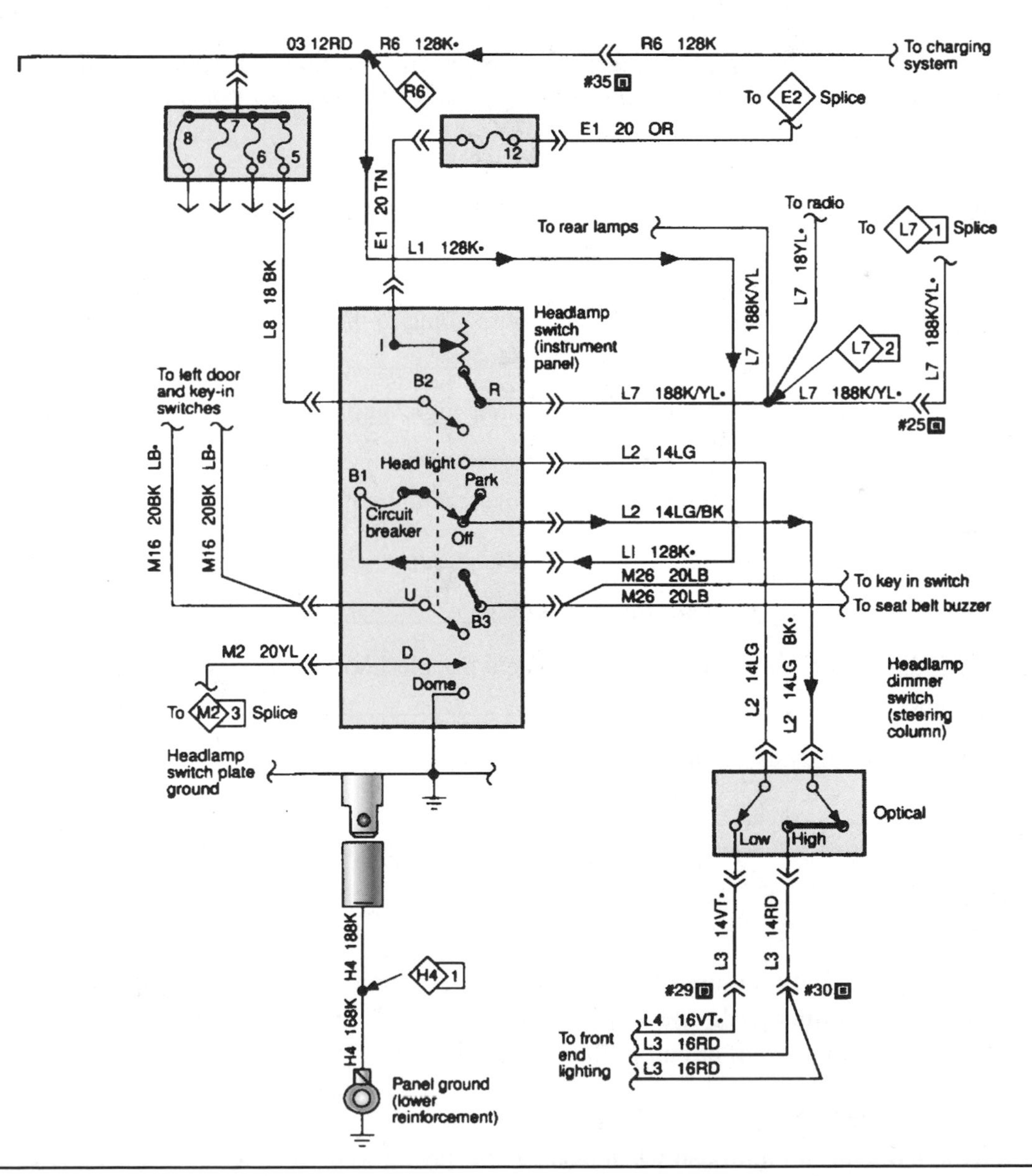

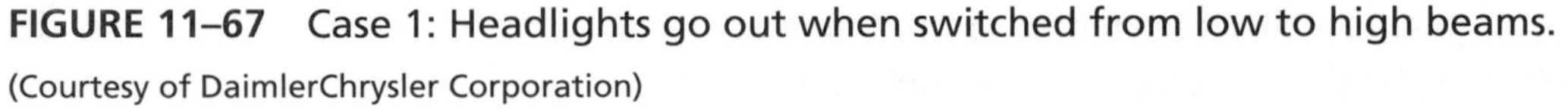

FIGURE 11–67 Case 1: Headlights go out when switched from low to high beams.

(Courtesy of DaimlerChrysler Corporation)

Known Information

- ❑ The vehicle operating voltage = 14 volts.
- ❑ All circuit fuses are OK.
- ❑ Both headlamps have been replaced.
- ❑ The headlamp switch is OK.

Circuit Analysis

Answer the following questions on a *separate* sheet of paper:

1. If the bulbs go out when the beam select switch is switched from low to high beams, what problems could be preventing the high beams from illuminating? ______________________.
2. If the high beam bulbs operate when a jumper wire is connected from battery power to the high beam circuit connector #30, what would be the *most likely* cause of the problem? ______________________ ______________________.
3. What should the voltage be of the L2 14LG wire low connection at the headlamp dimmer switch? ______________________.
4. What is the purpose of the L2 14LG BK wire going to the high connection at the headlamp dimmer switch? ______________________.
5. If there is no voltage at the dimmer switch high output, what is your diagnosis of this circuit problem? ______________________.
6. How could you verify the circuit problem using a jumper wire? ______________________ ______________________.
7. What is the best way to repair this system? ______________________.
8. What diagnostic steps were helpful when troubleshooting this system or explaining it to the customer? ______________________.
9. What is your analysis of this case study? ______________________ ______________________.

CASE 2

The following case study (Figure 11–68) is a turn signal circuit problem. Depending on the wiring design, knowledge of how turn signals operate and where the circuit components are located can help the technician perform the diagnosis of circuit problems quickly and efficiently.

Customer Complaint

The turn signal bulbs do not flash when needed but stay on when the switch is moved to either the left or the right turn positions. The customer replaced a flasher relay located on the fuse panel located under the dash, assuming it was the flasher relay, but the problem remained.

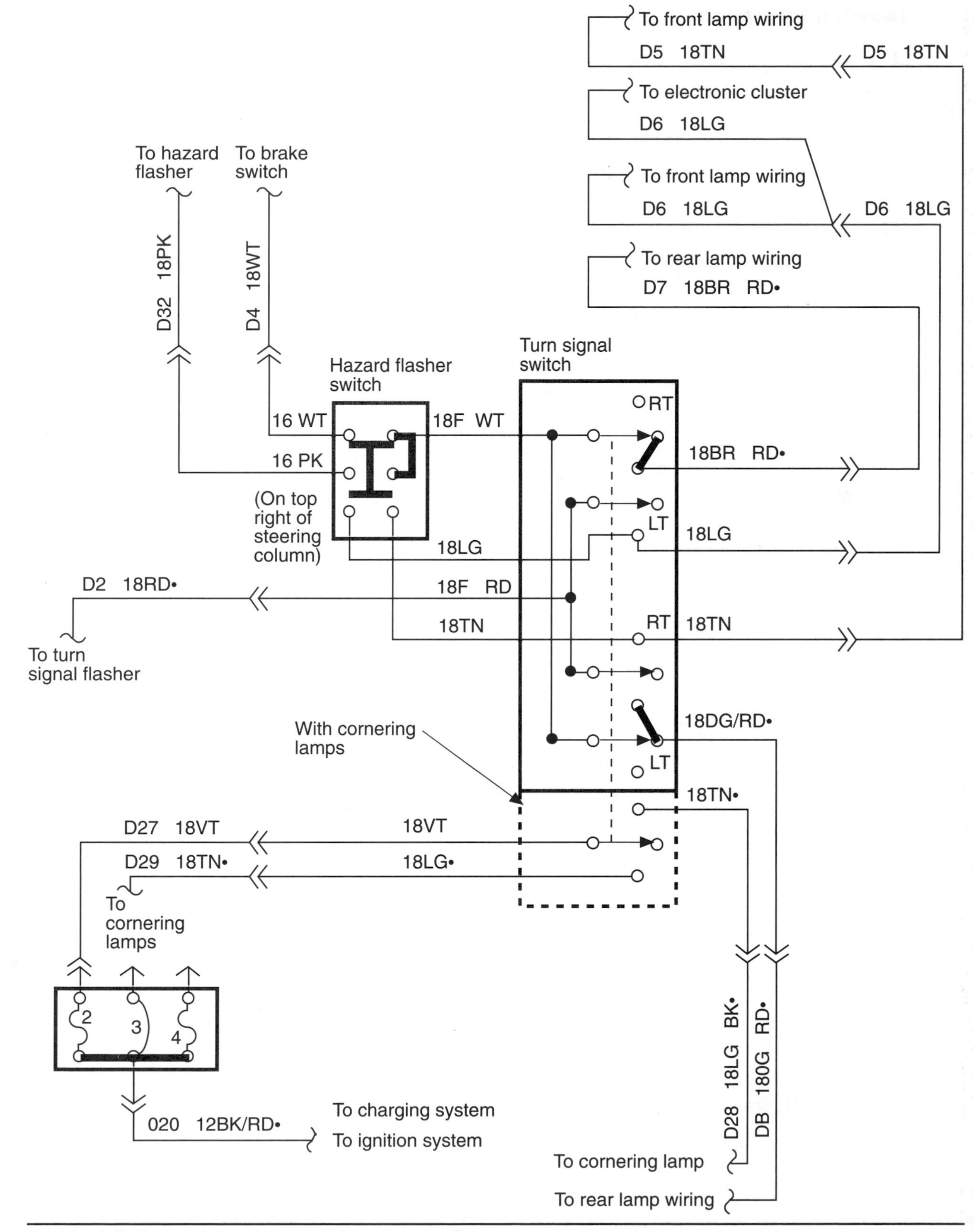

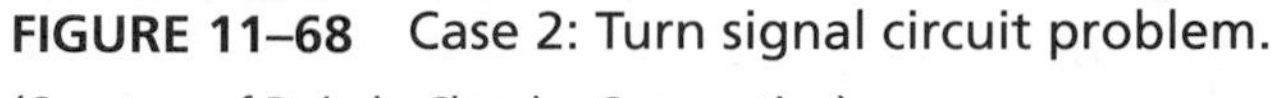
FIGURE 11–68 Case 2: Turn signal circuit problem.

(Courtesy of DaimlerChrysler Corporation)

Known Information

- ❑ The vehicle operating voltage = 14 volts.
- ❑ The hazard flasher circuit operates OK.
- ❑ All system bulbs are OK.

Circuit Analysis

Answer the following questions on a *separate* sheet of paper:

1. What is the *most likely* cause of the turn signals to stay on? ______________________________
______________________________.
2. How could the technician quickly determine the difference between the turn signal flasher relay and the hazard flasher relay? ______________________________.
3. What should the voltage be from the turn signal flasher to the turn signal switch? ______________.
4. What diagnostic steps are helpful when troubleshooting a turn signal system? ______________
______________________________.
5. What is your analysis of this case study? ______________________________
______________________________.

CASE 3

The following case study (Figure 11–69) is typical of a backup light circuit problem that technicians often encounter. Knowledge of the circuit and simple test procedures can help you diagnose the circuit safely and efficiently.

Customer Complaint

The backup lights do not work on a customer's car. The two backup lamps were both replaced, but the problem remained.

Known Information

- ❑ The vehicle operating voltage = 14 volts.
- ❑ Both backup bulbs have been replaced.
- ❑ The circuit fuse is OK.

Circuit Analysis

Answer the following questions on a *separate* sheet of paper:

1. What is the *most likely* cause of the backup lights not working? ______________________________
______________________________.
2. What should the voltage be at each of the backup light bulbs? ______________________________.

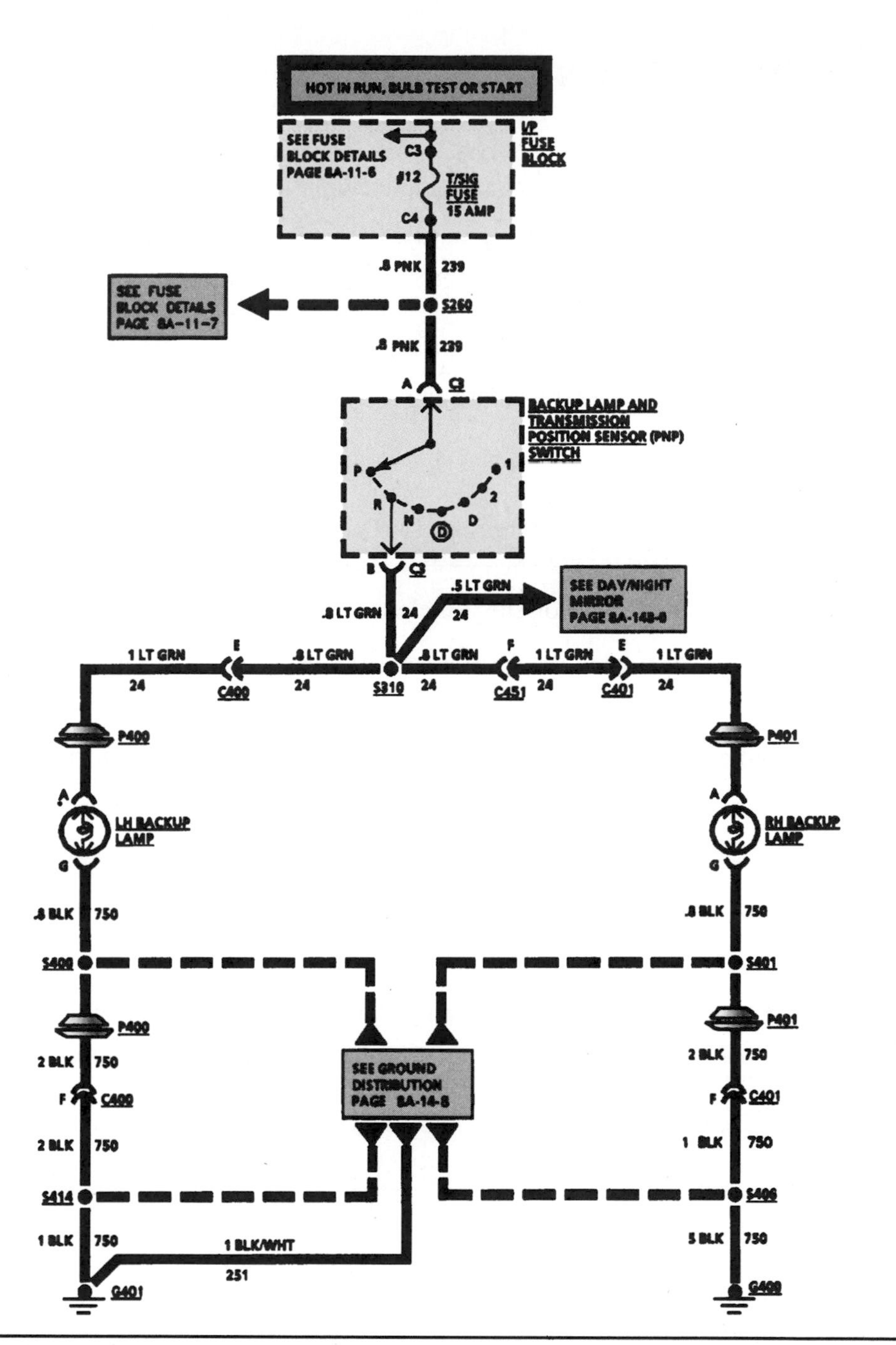

FIGURE 11–69 Case 3: Backup light circuit problem.

3. How could you verify that the backup lamp switch is functioning properly using a jumper wire?

____.

4. What diagnostic steps are helpful when troubleshooting this system or explaining the diagnosis to the customer? ____.

5. What is your analysis of this case study? ____

____.

Hands-On Vehicle Tests

The following two hands-on vehicle checks are included in the NATEF (National Automotive Technicians Education Foundation) Task List. Complete your answers to the following questions on a *separate* sheet of paper.

Performance Task 1

Task Description

Inspect, replace, and aim headlights and bulbs.

Task Objectives

- ❑ Obtain a vehicle that can be used for this task. What model and year of vehicle are you using for this task? ____________________.
- ❑ What type of headlights are on the vehicle you are using? ____________________.
- ❑ What are the prealignment checks to be made to the vehicle before adjusting the headlights?
  ________, ________, ________, ________, ________, ________, ________, ________.
- ❑ Describe the procedures used when aligning headlights with mechanical headlight aiming equipment.
  ____
  ____
  ____________________.
- ❑ How is the horizontal adjustment accomplished on the vehicle headlights? ____________________
  ____________________.
- ❑ How is the vertical adjustment accomplished on the vehicle headlights? ____________________
  ____

Task Summary

After performing the preceding NATEF task, what can you determine will be helpful when aligning headlights with mechanical aiming equipment? ____________________

____________________.

Performance Task 2

Task Description

Inspect and diagnose incorrect turn signal or hazard light operation; perform the necessary action.

Task Objectives

- ❑ Obtain a vehicle that can be used for this task. What model and year of vehicle are you using for this task? ____________________.

❑ Locate a wiring diagram for the vehicle that shows the turn/hazard light circuit.

❑ Using the wiring diagram(s), identify areas in the circuit such as connectors, wires, fuses, and components that can cause a possible problem.

Identified Area	Symptom	Cause	Test Procedure

❑ Perform the diagnostic steps and procedures identified in the previous question. Does this help identify the circuit problem(s)? ______________________________.

❑ What diagnostic steps were helpful when troubleshooting this system? ______________________.

❑ What is your analysis of using wiring diagrams to help diagnose turn signal circuit problems?

____

____.

Task Summary

After performing the preceding NATEF task, what can you determine that will be helpful in knowing how to check turn signal and hazard flasher circuits for proper operation?

____

____.

Summary

❑ A dimmer switch is used to change headlights from low to high beam.

❑ Typical headlight switches can be mounted in the dash board or on the steering column.

❑ A typical headlight switch is a three-position style, with off, park, and headlight positions.

❑ A multifunction switch combines the dimmer switch with other function switches in the steering columns.

❑ The flash-to-pass feature of a headlight circuit allows the high beams and/or the low beams to light when the headlight switch is off.

❑ For styling and aerodynamics, some concealed headlights are hidden behind vacuum or electrically controlled doors.

- ❑ Daytime running lights (DRL) operate the headlights at a reduced voltage during daylight hours for vehicle safety.
- ❑ A twilight sentinel headlight system activates the headlights automatically.
- ❑ In a computer-controlled headlight system, the headlight switch provides input to a BCM, which then controls the activation of the headlight circuit.
- ❑ Adaptive headlights swivel up to 15 degrees depending on input from several sensors.
- ❑ A common way to align headlights is with mechanical headlight aiming equipment.
- ❑ In a two-bulb tail/park lamp circuit, a dual-filament bulb is used on each side of the circuit for taillights and brake lights.
- ❑ In a three-bulb taillight circuit, an individual bulb is used for each taillight, brake light, and turn signal function.
- ❑ Some late model vehicles use LED lighting in taillights instead of incandescent filament bulbs.
- ❑ All brake light circuits since 1986 are required to have a CHMSL (center high-mounted stop light).
- ❑ Several CHMSLs use LED or neon lamps because of the quick ON time to full brightness.
- ❑ In a two-bulb system, the turn signal switch determines which brake light is separated out of the circuit.
- ❑ A typical turn signal or hazard flasher consists of a bimetallic strip and a heating element wired in series to the circuit.
- ❑ Electronic flasher units use solid-state circuitry and transistors to cycle the turn signal circuit.
- ❑ Side marker lights have been a safety requirement on all vehicles sold in North America since 1969. The front marker lens is always amber, and the rear marker lens is always red.
- ❑ In a courtesy light circuit with the door jamb switch on the ground side of the circuit, the circuit is complete when the door is opened.
- ❑ In a courtesy light circuit with the door jamb switch is on the positive side of the circuit, the switches are insulated from the ground side of the circuit.
- ❑ Circuit feedback occurs when the current in one circuit seeks an alternate path of flow when the original path is interrupted.
- ❑ A typical instrument panel light circuit uses a rheostat to control the intensity of the lamps.
- ❑ In a fiber optic system, a single-filament bulb can illuminate several objects through a fiber optic stranded cable.

Key Terms

center high-mounted stoplight (CHMSL)
collision avoidance light
daylight running lights (DRL)
flash-to-pass
multifunction switch

Review Questions

Short Answer Essays

1. Describe some of the combined features of a multifunction steering column switch.
2. In a concealed headlight system, when the headlights are turned on, how are the headlight doors opened?
3. Describe the purpose of a CHMSL in a brake light system.
4. What causes a bimetallic turn signal flasher to operate as designed?
5. Describe the most common way instrument panel lights are dimmed.

Fill in the Blanks

1. A front side marker lamp lens is always ______________________ in color.
2. When a bulb burns out in a bimetallic turn signal flasher system, the remaining lights flash ______________________ than normal.
3. A brake light switch is usually attached to the ______________________.
4. Concealed headlight doors are opened by either ______________________ or ______________________.
5. Dimmer headlights are most often the result of ______________________.

ASE-Style Review Questions

1. Technician A says the dimmer switch controls the current to the low beams only in a headlight circuit. Technician B says the dimmer switch controls the current to the high beams only in a headlight circuit. Who is correct?
   A. A only
   B. B only
   C. Both A and B
   D. Neither A nor B

2. Technician A says a headlight switch could be mounted in the dash panel or the steering column. Technician B says all headlight switches have a ceramic rheostat that is used to control the dimming of the instrument lights. Who is correct?
   A. A only
   B. B only
   C. Both A and B
   D. Neither A nor B

3. The courtesy and dome light circuit has the dome light on even when the doors are closed. Technician A says the dome light circuit is grounded between the dome light and the door jamb switches. Technician B says a door switch is grounded. Who is correct?
   A. A only
   B. B only
   C. Both A and B
   D. Neither A nor B

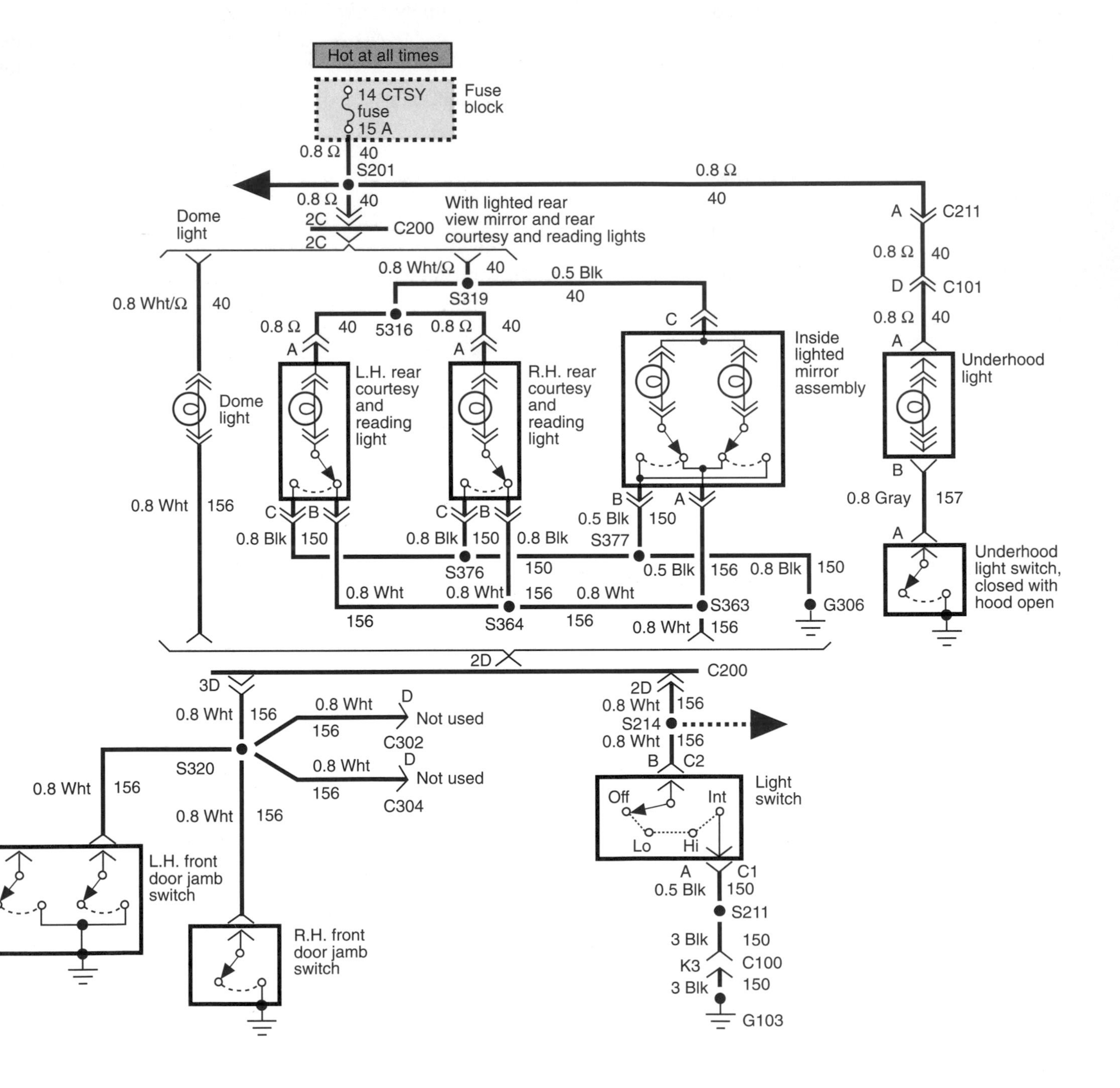

4. Two technicians are discussing a multifunction steering column switch. Technician A says the dimmer switch can be part of a multifunction steering column switch. Technician B says the horn switch could be part of a multifunction steering column switch. Who is correct?
   A. A only
   B. B only
   C. Both A and B
   D. Neither A nor B

5. A flash-to-pass feature is being discussed. Technician A says this system will flash the high beam bulbs at driver command. Technician B says this feature will not work when the headlight switch is in the off position. Who is correct?
   A. A only
   B. B only
   C. Both A and B
   D. Neither A nor B

6. Technician A says concealed headlight doors can be controlled by vacuum motors. Technician B says headlight doors can be opened manually. Who is correct?
   A. A only
   B. B only
   C. Both A and B
   D. Neither A nor B

7. Daylight running lights (DRL) are being discussed. Technician A says the headlights in a DRL system run at maximum designed voltage for brightness. Technician B says a DRL system cannot be overridden when the headlight switch is turned on. Who is correct?
   A. A only
   B. B only
   C. Both A and B
   D. Neither A nor B

8. Technician A says that mechanical headlight aiming equipment uses a spirit bubble for adjusting the headlights horizontally. Technician B says the vehicle should be on a flat level surface to properly adjust the headlights. Who is correct?
   A. A only
   B. B only
   C. Both A and B
   D. Neither A nor B

9. Technician A says a typical three-bulb taillight circuit has individual control for each bulb circuit. Technician B says the mechanical brake light switch attached to the brake pedal arm controls the brake light circuit in a three-bulb tail light circuit. Who is correct?
   A. A only
   B. B only
   C. Both A and B
   D. Neither A nor B

10. Technician A says a turn signal switch isolates a brake light from the brake light circuit. Technician B says a turn signal flasher opens to let the current flow to the circuit bulbs. Who is correct?
    A. A only
    B. B only
    C. Both A and B
    D. Neither A nor B

11. The backup light circuit in the figure does not work when the back-up switch is closed. Technician A says both bulbs could be burned out. Technician B says the fuse could be blown. Who is correct?
    A. A only
    B. B only
    C. Both A and B
    D. Neither A nor B

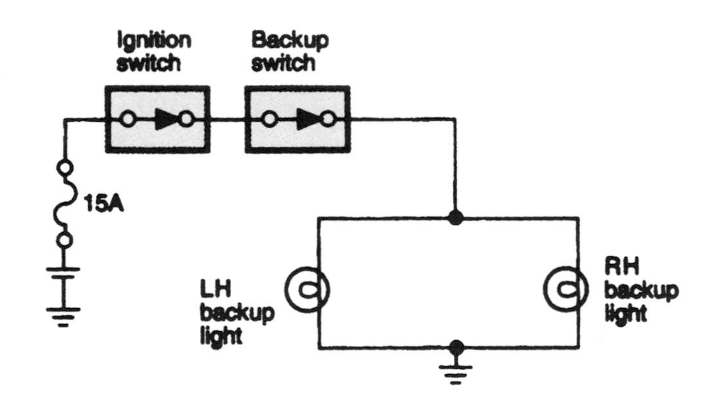

12. Technician A says that a sneak circuit feedback occurs when the current in one circuit seeks an alternate path of flow because the original path is interrupted. Technician B says a sneak circuit feedback can occur *only* when the insulation breaks down between the wires of two separate circuits. Who is correct?
    A. A only
    B. B only
    C. Both A and B
    D. Neither A nor B

13. Technician A says that when one turn signal bulb is burned out, the turn signal flasher may be faulty. Technician B says that the hazard signal circuit and the turn signal circuit normally operate from the same flasher unit. Who is correct?

    A. A only

    B. B only

    C. Both A and B

    D. Neither A nor B

14. A taillight bulb is glowing dimly. A voltmeter across the bulb reads 9.7 volts. Technician A says this is normal in some taillight circuits. Technician B says there is an unwanted voltage drop elsewhere in the circuit. Who is correct?

    A. A only

    B. B only

    C. Both A and B

    D. Neither A nor B

15. Technician A states that most instrument panel lamps are wired in series to allow them to dim evenly. Technician B says that the voltage available to the instrument panel lamps depends on the setting of the rheostat control knob. Who is correct?

    A. A only

    B. B only

    C. Both A and B

    D. Neither A nor B

12 Basics of Electronics and Computers

Introduction

Over the past several years, electronic controls and computer circuits have taken over the control and monitoring of most vehicle functions. Some of these functions are engine operation, climate control systems, antilock brakes, electronic suspension systems, electronically shifted transmissions, and charging system regulation. Some textbooks dedicate their entire content to their study of computer sensors and troubleshooting. This chapter covers the basic concepts of electronics and sensors in the operation of the automotive computer, giving you the basic background necessary to understand vehicle service manual diagnostic procedures.

Objectives

When you complete this chapter you should be able to:

- ❑ Explain the purpose of electronics and computers in automobiles.
- ❑ Explain the purpose of semiconductors, diodes, and transistors in simple electronic circuits.
- ❑ Describe the difference between analog and digital voltage signals.
- ❑ Describe how the binary code system works in computers.
- ❑ Describe the purpose of amplifiers and converters in computers.
- ❑ Describe the different types of memory in a microprocessor.
- ❑ Explain the purpose and operation of various logic gates in computer circuits.
- ❑ Describe the operation of input and output sensors in an automotive computer control circuit.
- ❑ Identify proper safeguard procedures when working with electronic circuits.

Basics of Electronics

Understanding the basics of electronics begins with knowing the materials used to build the components commonly used in electronic circuits. The material most often used is called a **semiconductor.** As mentioned in Chapter 3, a semiconductor is a stable element with exactly four valence electrons in its structure. The two typical semiconductor materials are silicon (Si) and germanium (Ge), of which silicon is the more commonly used.

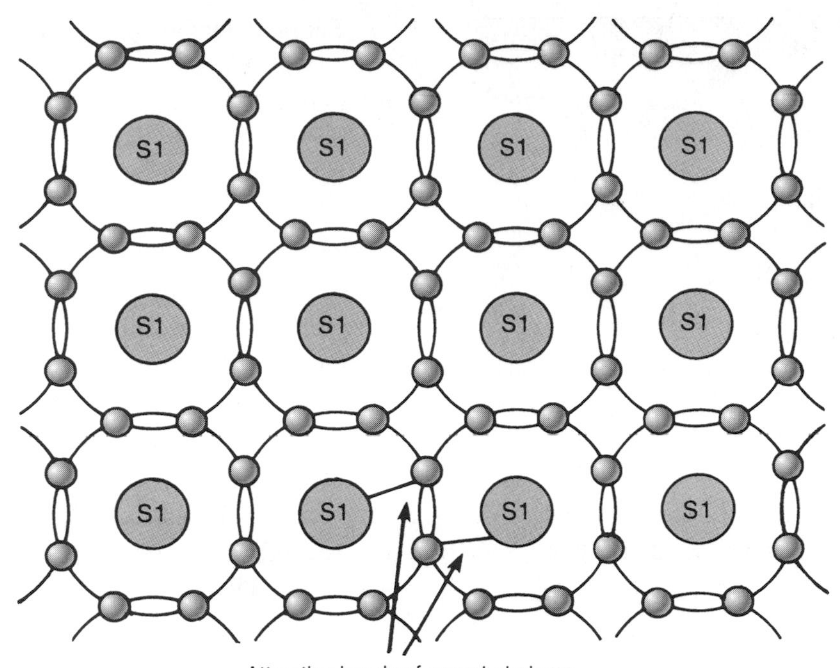

FIGURE 12–1 Silicon crystal covalent bond.

Semiconductor materials have a crystal structure called a **covalent bond,** where atoms share the valance electrons of neighboring atoms (Figure 12–1).

To function as a semiconductor, a material has to be doped, that is, a small amount of trace element impurities are added to the crystal structure. Two types of semiconductor materials are produced by doping. The first is an **N-type** semiconductor, which has loose or excess electrons, and a negative charge (Figure 12–2). N-type semiconductors are produced by adding an impurity with five electrons in the outer ring (called pentavalent atoms such as arsenic, antimony, or phosphorus). Four of these electrons fit into the crystal structure, but the fifth is free. The excess electrons produce a negative charge and enables the semiconductor to carry a current.

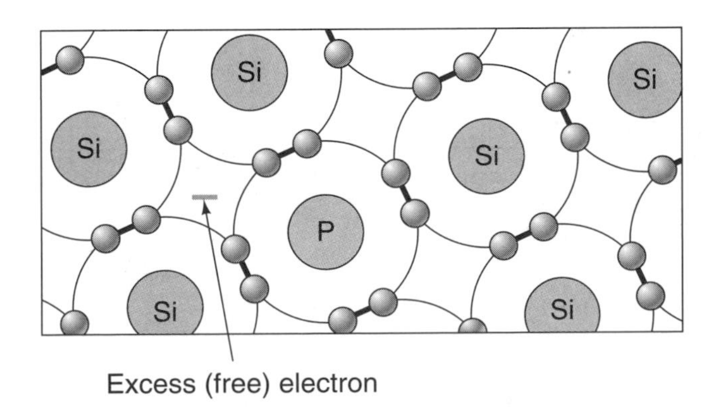

FIGURE 12–2 N-type silicon semiconductor atomic structure.

The second type of semiconductor material is a **P-type** semiconductor, which has positively charged material that carries a current (Figure 12–3). P-type semiconductors are produced by adding an impurity with three electrons in the outer ring (called trivalent atoms, such as aluminum, indium, gallium, or boron). When this element is added to silicon or germanium, the three outer electrons fit into the pattern of the crystal, leaving a hole where the fourth would fit. This hole is actually a positive charged empty space, that carries the current in the P-type semiconductor because it attracts electrons. Although the electrons cannot be freed from their atom, they can rearrange their pattern and fill a hole in a nearby atom. Whenever this occurs, the electron leaves a hole, which is filled by

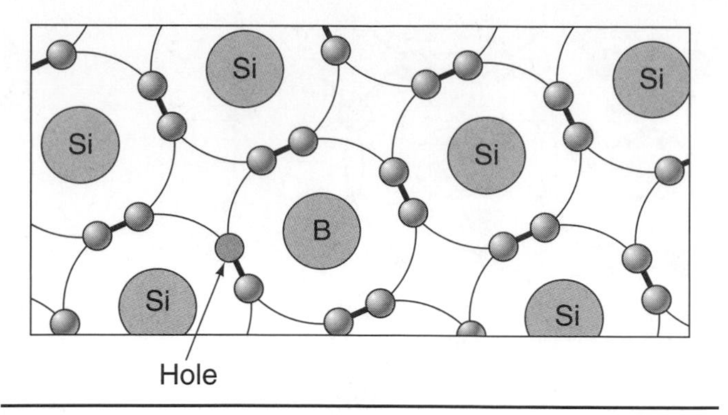

FIGURE 12–3 P-type silicon semiconductor atomic structure.

another electron, and so on. The electrons move toward the positive side of the structure, and the holes move toward the negative side.

Semiconductors can function as both a conductor and an insulator. Because of this, semiconductors serve as excellent switching and circuit control devices. Two common semiconductor devices used in solid-state circuits are diodes and transistors.

Diodes A **diode** is the simplest semiconductor device made. It is an electrical one-way check valve that allows current to flow in one direction only. It is formed by joining P-type semiconductor material with an N-type material (Figure 12–4). The positive side (P) is called the anode, and the negative side (N) is called the cathode. The area where the anode and cathode material join is called the **PN junction** or **depletion zone** (Figure 12–5).

A diode allows the current to flow in one direction only. In a DC circuit, when a diode has the positive

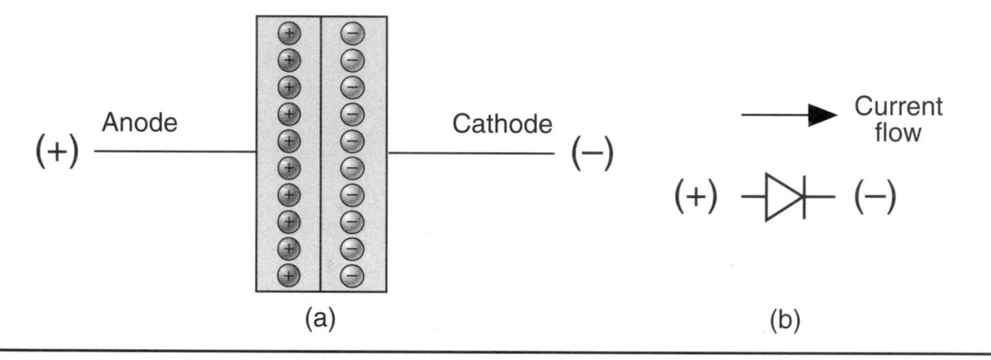

FIGURE 12–4 (a) A diode and (b) its electrical symbol.

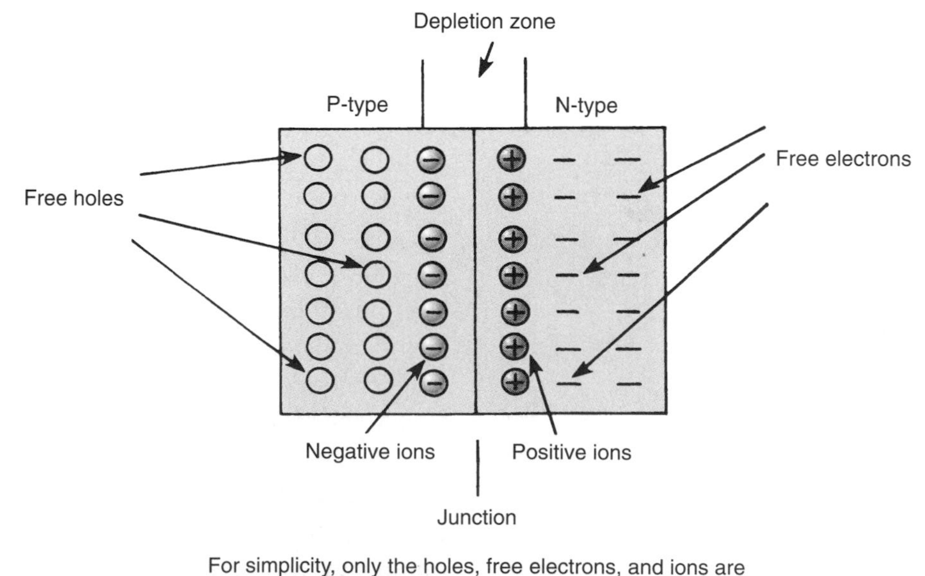

FIGURE 12–5 PN junction of a diode.

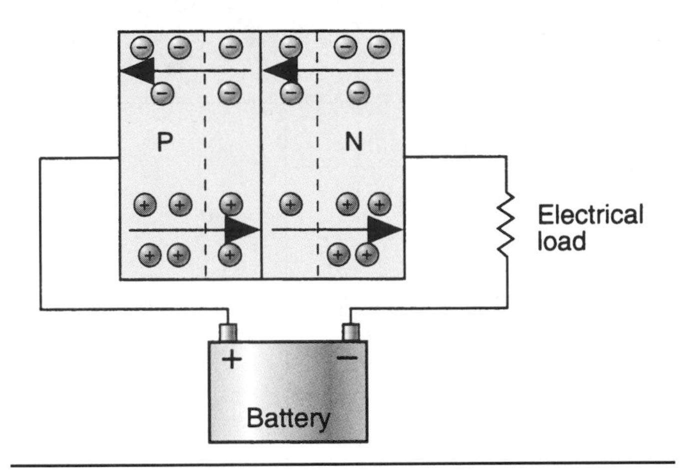

FIGURE 12–6 Diode is forward biased, causing current flow.

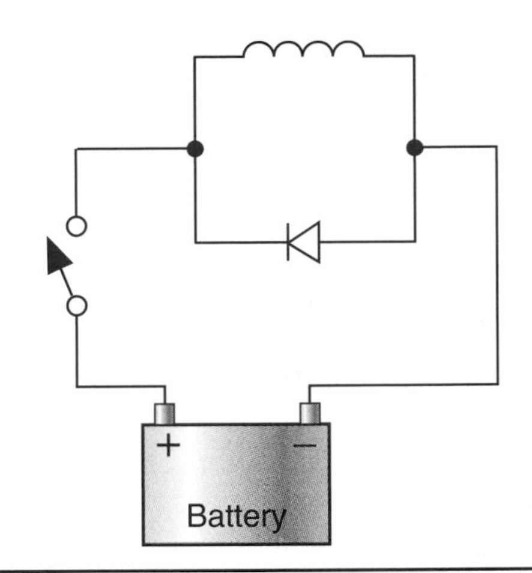

FIGURE 12–8 A clamping diode wired in parallel to a coil prevents voltage spikes when the switch is opened.

voltage on the P-side or anode, the diode is considered to be in a **forward bias,** and current flows through it (Figure 12–6). When a positive voltage is present on the N-side or cathode, the diode is in a **reverse bias,** and current flow is prevented (Figure 12–7).

In an AC circuit, diodes are used to separate the positive and negative voltages. A good example is an AC generator or alternator in a vehicle. The function of diodes in converting AC current to DC current in alternators is covered in depth in Chapter 16.

An application of a basic diode used as a control device in a circuit is a **clamping diode** (Figure 12–8). When the current flow through a relay coil or solenoid is discontinued, a voltage surge or spike is produced as the collapse of the magnetic field moves across the coil windings. If a reversed-biased diode is wired in parallel to the circuit coil, a bypass through the clamping diode is provided for the electrons in the circuit when the control switch is opened. Figure 12–9 is an example of a clamping diode installed in parallel with an air conditioning compressor clutch coil to prevent voltage spikes. Without the clamping diode, when the compressor clutch control switch is opened, a voltage spike can reach several hundred volts positive or negative for a split second before

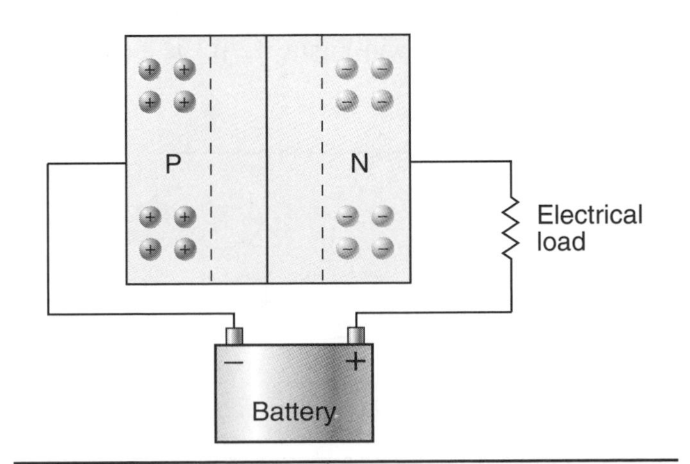

FIGURE 12–7 Diode is reverse biased, preventing current flow.

FIGURE 12–9 A clamping diode connected in parallel across the terminals of an air conditioning compressor clutch.

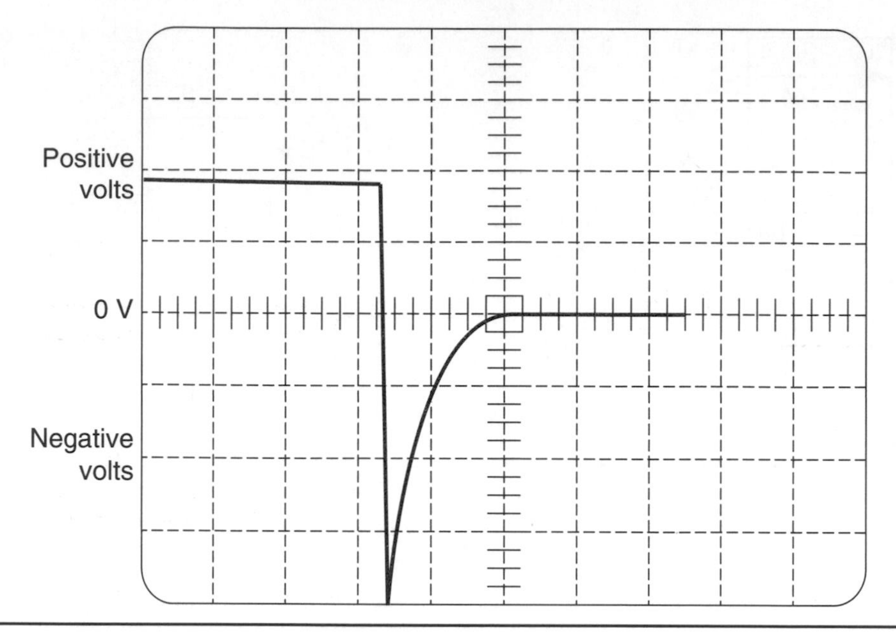

FIGURE 12–10 A typical negative voltage spike created for a split second when a compressor clutch without a clamping diode is turned off.

returning to 0 volts (Figure 12–10). This voltage spike is similar to the static electricity discussed in Chapter 1 and has been known to damage sensitive electronic control circuits.

Another type of diode that is similar in operation to a regular diode is a **light-emitting diode (LED).** An LED has a small lens built into it and emits light when forward biased with a voltage of 1.5 to 2.5 volts (Figure 12–11). The light from an LED is not heat energy, as is the case with other lights; it is electrical energy. LEDs are used in instrument cluster digital displays, triggering devices in some ignition systems, as the lighting source in center high-mounted brake lights, and taillights.

A **zener diode** is a complex type of diode used to regulate voltage in it a circuit by installing it reversed biased (backward) in a circuit. The depletion zone or PN junction is much narrower in a zener diode, and its barrier voltage becomes very intense when a reverse bias voltage is applied. When a zener diode's specified voltage level is reached, a breakdown occurs within the PN junction zone and current flows across the diode in reverse direction. Figure 12–12 shows the common symbol for a zener diode, and Figure 12–13 is a simplified

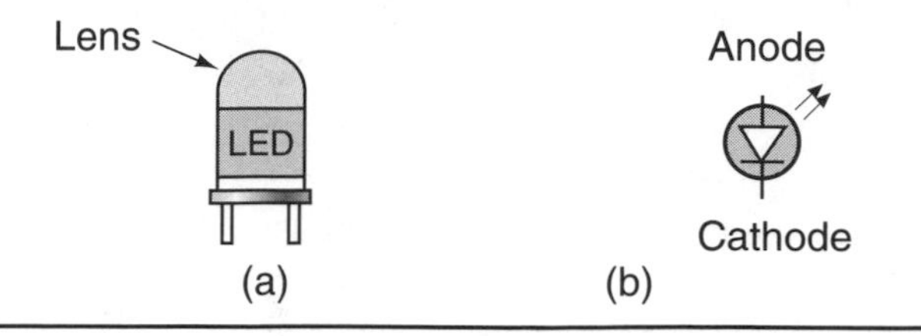

FIGURE 12–11 (a) Light-emitting diode (LED) and (b) its symbol.

FIGURE 12–12 Typical zener diode symbol.

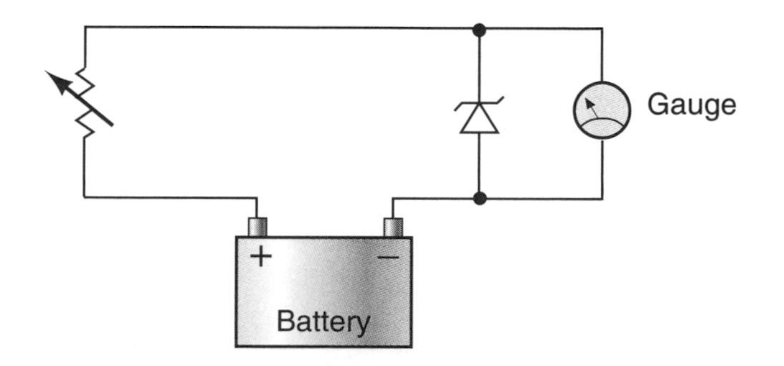

FIGURE 12–13 Zener diode used in a simple gauge circuit to maintain a constant gauge voltage.

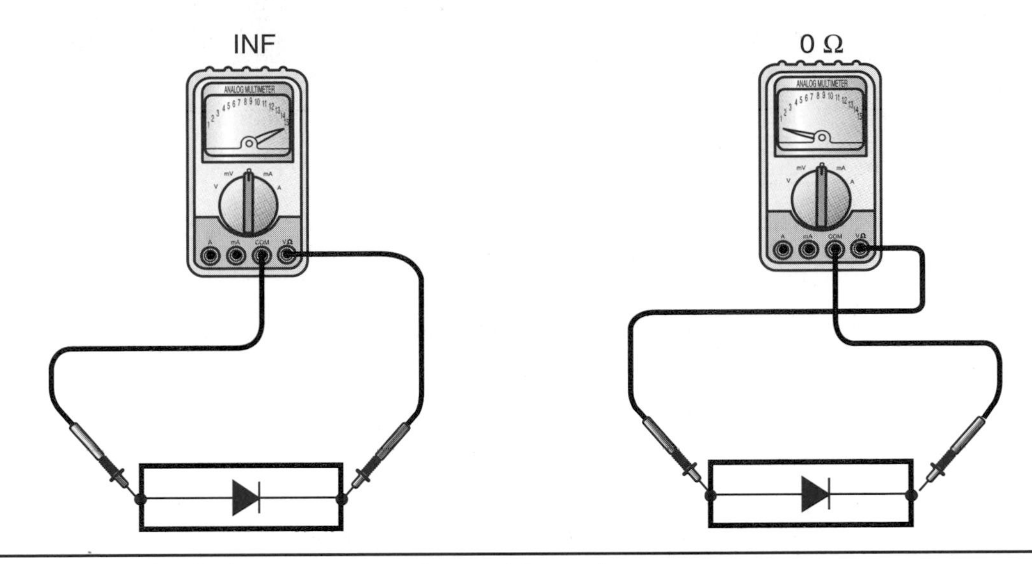

FIGURE 12–14 Testing a diode for an open or short with an ohmmeter.

gauge circuit with a zener installed to maintain a constant voltage to the gauge. If the gauge voltage must be limited to 9 volts, the 9-volt-specified zener would break down and conduct reverse current when the system voltage reached 9 volts. The amount of voltage to the gauge remains 9 volts because the zener diode causes the variable resistor to drop the excess voltage to maintain the voltage limit.

Diode Tests A regular diode can be tested with an ohmmeter or digital multimeter (DMM) equipped with a diode test feature. To test a regular diode for shorted and open conditions with an ohmmeter, connect the meter leads across the diode (Figure 12–14). Observe the ohmmeter reading, then reverse the ohmmeter leads and observe the reading again. There should be a reading close to zero in one direction and infinite (OL) in the other direction. A *shorted* defective diode reads close to zero in both directions. An *open* defective diode reads infinite in both directions.

Some of the less expensive high-impedance digital multimeters (DMM) can give a false diode test reading when a diode is checked for a forward bias using the ohmmeter function. Many diodes allow current flow through them only when the voltage is more than 0.6 volt, resulting in a diode that shows *open,* when in fact it may not be. Because of this problem, several DMMs are equipped with a diode test function that increases the voltage at the test leads (Figure 12–15).

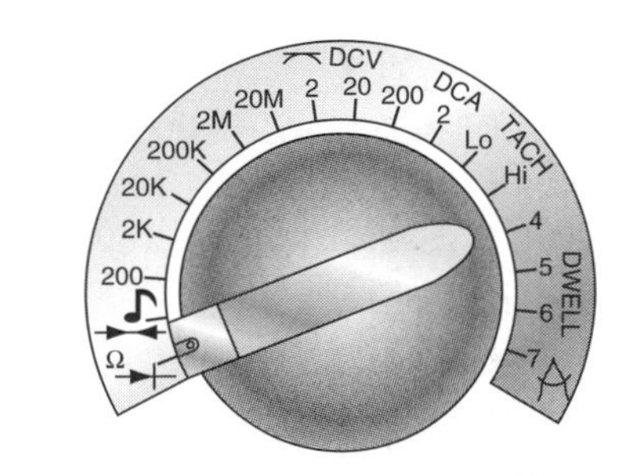

FIGURE 12–15 Diode test function on a DMM.

Transistors The **transistor** is another semiconductor commonly used in solid-state circuits throughout the automotive industry. The transistor devices do not have any moving parts and, when used in place of mechanical switches and relays to turn circuits ON and OFF, give trouble-free operation. A transistor consists of three layers of semiconductor materials joined together that are designated as **emitter, collector,** and **base.** In effect, a transistor can be thought of as two diodes that share a common center layer (Figure 12–16).

Two types of transistors are commonly used in electronic circuits. The first, an **NPN transistor,** consists of two N-type materials and one P-type material (Figure 12–17). In an NPN transistor, a positive control

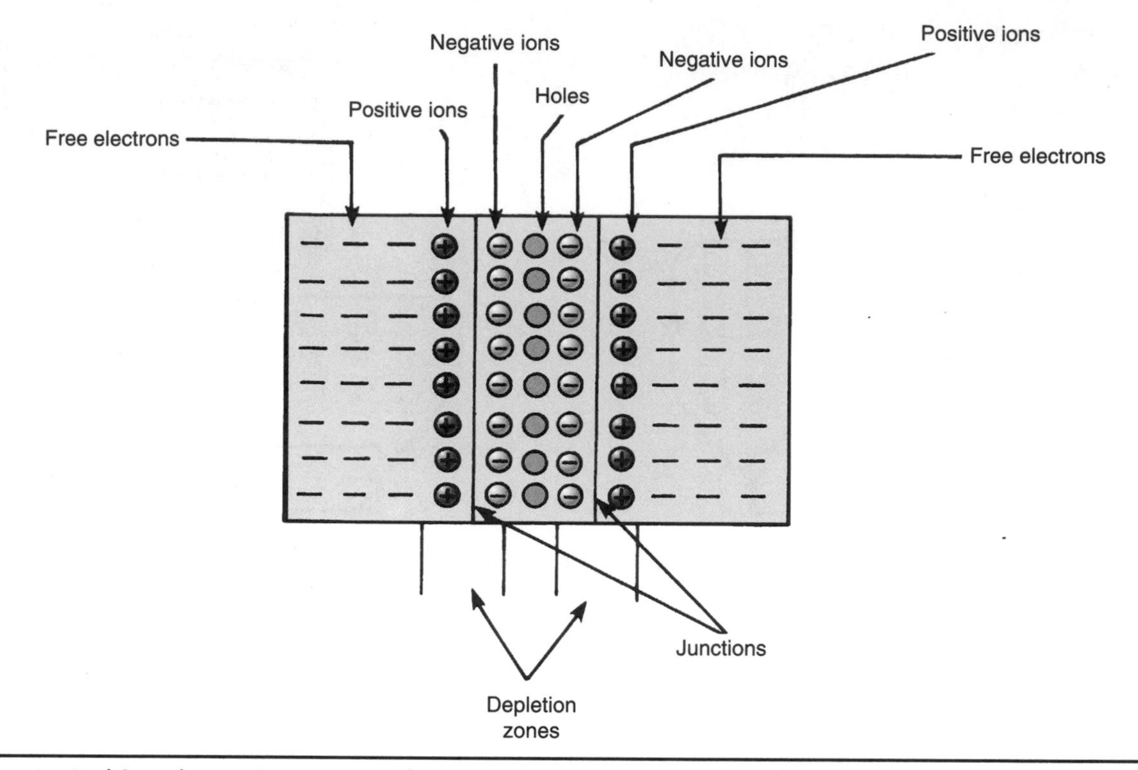

FIGURE 12–16 Unbiased transistor consisting of two diodes that share a common center material.

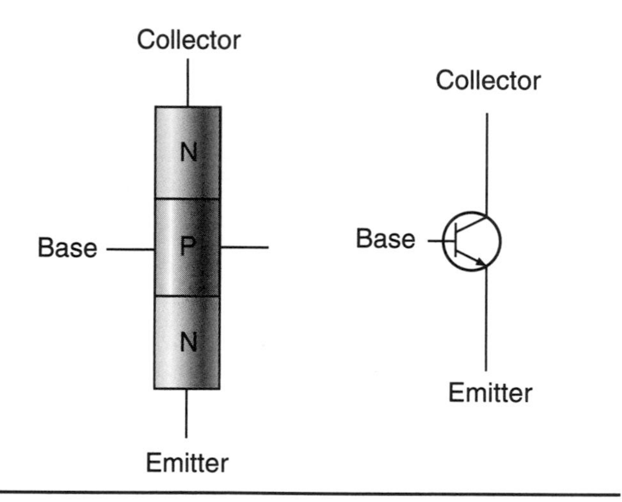

FIGURE 12–17 NPN transistor and its schematic symbol.

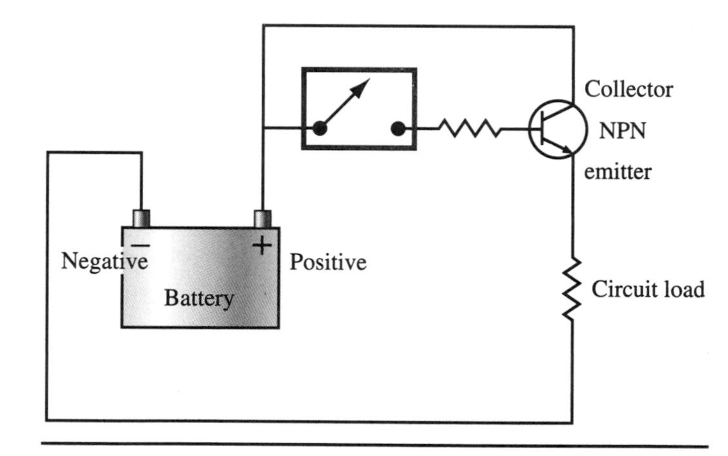

FIGURE 12–18 Simple circuit with NPN transistor.

voltage is applied through a current limiting resistor to the base of the transistor to turn the transistor on. In a *typical* transistor-controlled circuit, current flows through the collector and emitter and then through the load device before returning to the negative side of the battery (Figure 12–18).

The second type of transistor, a **PNP transistor,** consists of two P-type materials and one N-type material (Figure 12–19). For a PNP transistor, a negative voltage is applied through a current-limiting resistor to the base of the transistor to turn the transistor on. Current then flows from the emitter through the transistor to the collector, then through the load device before returning to the battery (Figure 12–20).

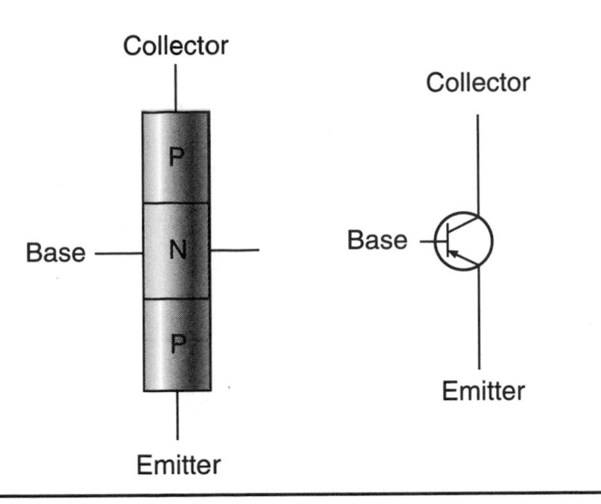

FIGURE 12–19 PNP transistor and its schematic symbol.

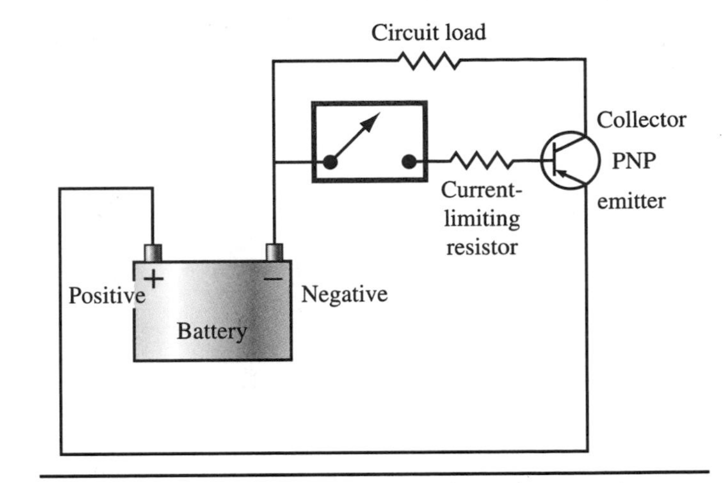

FIGURE 12–20 Simple circuit with a PNP transistor as a load current switch.

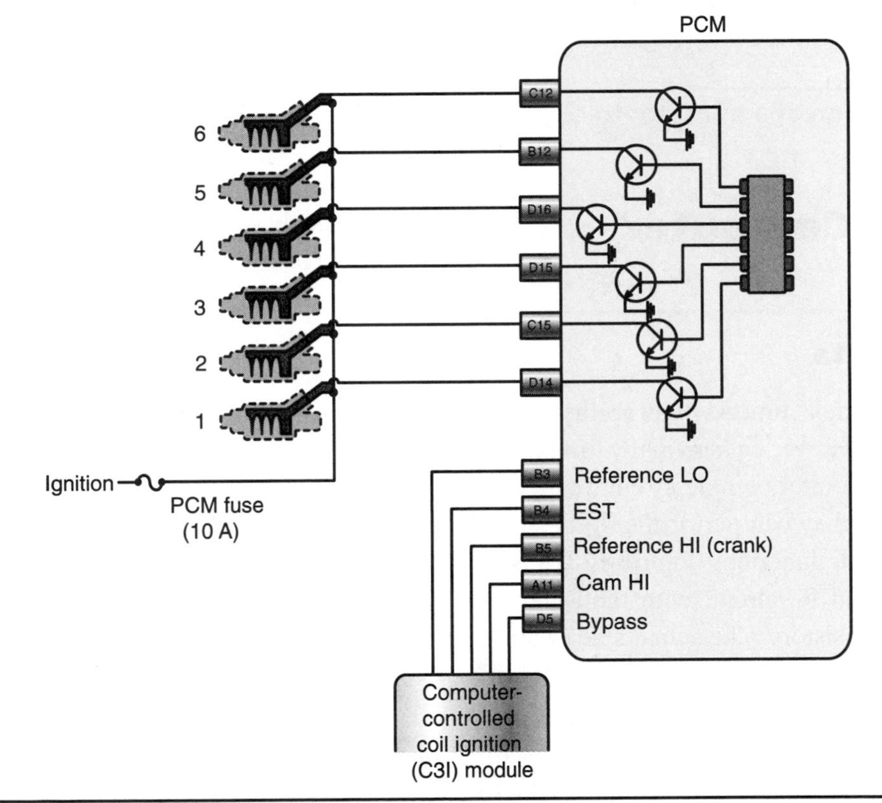

FIGURE 12–21 Typical use of transistors in an injector control circuit.

Figure 12–21 is an example of the use of transistors to control the ground circuit of a sequential fuel injection system. In this system the **powertrain control module (PCM)** receives input from the ignition module and other sensors. Then, through several transistors, the PCM controls the amount of ON/OFF time of each injector. The transistor is the actual switching device that the current flows through to complete the injector circuit.

The use of transistors in combination with mechanical relays and switches is also very common. In Figure 12–22 the cooling fan control circuit is using a transistor located in the PCM to control the ground circuit of the small coil in a standard mechanical relay. The mechanical relay is the actual device that current flows through to operate the fan motor.

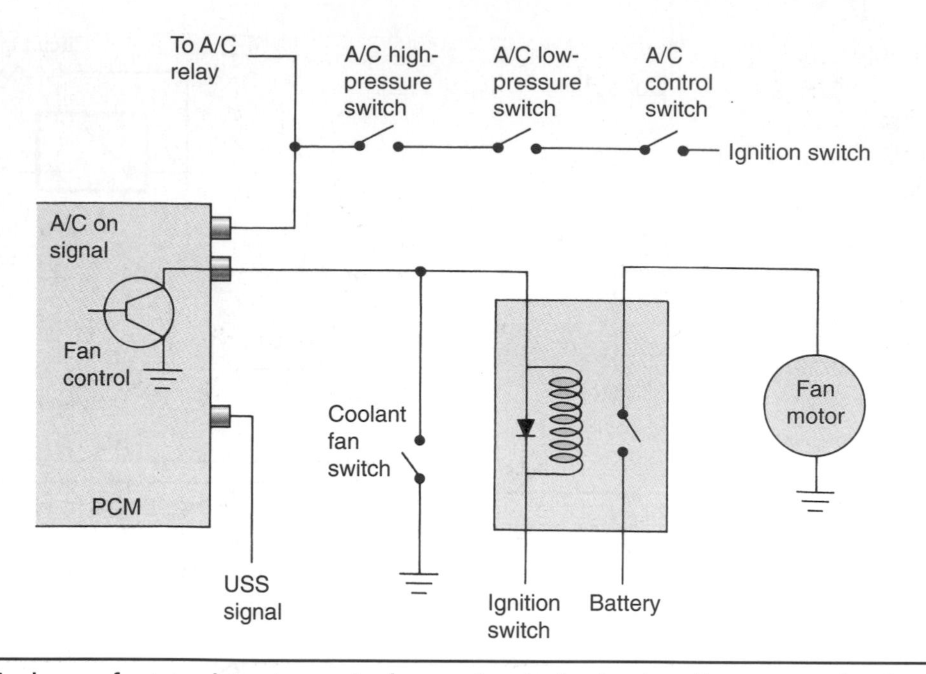

FIGURE 12–22 Typical use of a transistor to control a mechanical relay in a fan motor circuit.

Automotive Computer Components

Integrated Circuits

A single transistor or diode is limited in its ability to perform complex tasks. However, when many transistors and diodes are combined into a single circuit, called an **integrated circuit (IC),** they can perform several simultaneous complex electronic functions. Figure 12–23 is an enlarged illustration of an IC circuit with thousands of transistors, diodes, and resistors. The actual size of an IC can be less than 4 millimeters square (Figure 12–24). IC chips are constantly shrinking in size and being designed into vehicles to make logical decisions and issue commands to various sensors and control systems. Some of these systems are the ignition module, fuel injection control, transmission shifting, antilock brakes, active suspension system, climate controls, and entertainment. With a basic knowledge of the common electronics and IC chips has used in today's vehicles, the technician the diagnostic ability to find faulty electronics.

Basic Voltage Signals

Analog Voltage Signal An **analog voltage signal** is usually produced by an input sensor that has a continuously variable output within a certain voltage range (Figure 12–25). For example, if a rheostat is used to manually pulse a 5-volt bulb on and off, then the corresponding sine wave the rheostat produces would be an analog voltage signal with a 5-volt range. Some of the sensors in automotive computer systems produce analog voltages.

Digital Voltage Signal If an on/off switch is connected *before* a 5-volt bulb and the switch is off, then 0 volts are available at the bulb. When the switch is turned on, a 5-volt signal is sent to the bulb to light it to full brilliance. Switching this circuit on and off rapidly produces a **digital voltage signal,** also known as a **square wave signal** (Figure 12–26).

The digital voltage signal is reversed if the on/off switch and a scope are connected *after* a 5-volt bulb (Figure 12–27). When the switch is off (open), 5 volts are available through the bulb to the scope and to the switch. When the switch is turned on (closed), the bulb glows and 0 volts are available to the scope after the bulb.

Digital voltage signals are used to control various relays and components in a vehicle. The computer can vary the length of time the digital signal is high or low for precise component control (Figure 12–28). Depending on the circuit, some digital systems monitor a circuit before a load device and some after a load device.

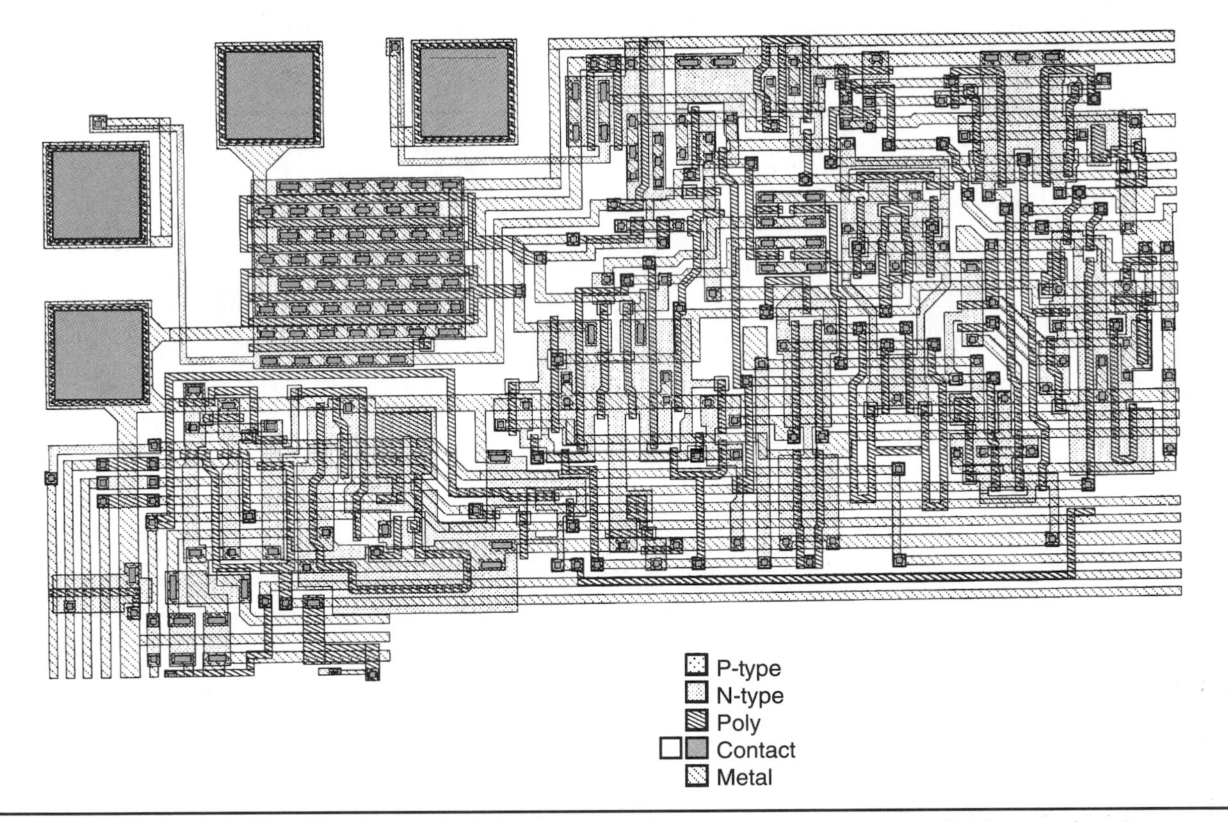

FIGURE 12–23 Enlarged illustration of an IC circuit with thousands of transistors, diodes, and resistors.

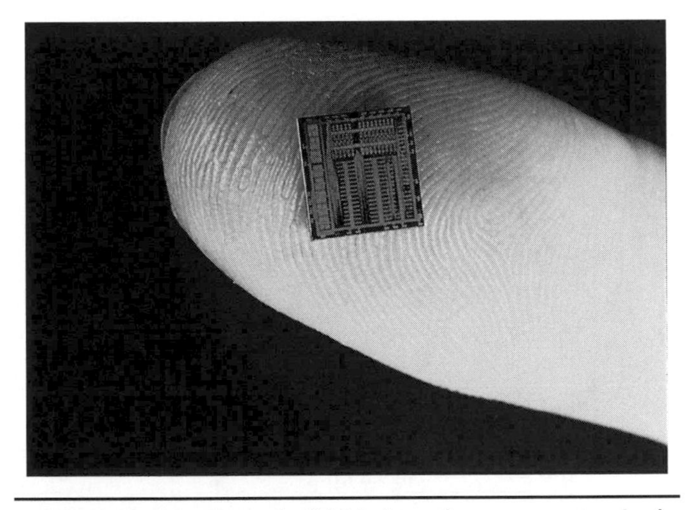

FIGURE 12–24 A typical IC is less than a quarter inch square.

(Courtesy of Texas Instruments)

Binary Code When a digital signal is ON (or OFF), it is considered HIGH (or LOW). A digital low signal is assigned the numeric value of 0, and a digital high signal is given the numeric value of 1. The assignment of numeric values to digital signals is called **binary code** (Figure 12–29). In an automotive computer, all communications are transmitted in binary code form. Each 1 or 0 represents a **bit** of information. The 16 possible combinations of a group of 4 bits is represented by the numbers 0 (0000) to 9, and the letters A to F (1111). Eight bits of information are equal to a **byte.**

Amplification Most of the input signals sent to the computer are analog signals, which have to be converted to a digital signal to be understood by the computer's software program (Figure 12–30). Some of these input sensors, like the exhaust gas oxygen sensor, produce a very low output voltage of less than 1 volt, along with a low current flow. This type of input signal must be **amplified** through an amplification circuit to increase the strength of the signal going to the **analog-to-digital (A/D) converter** before the computer can interpret the information received. Other sensors, like a throttle position sensor, have an analog signal high enough that amplification of the signal is not necessary.

The A/D converter continually scans the input signals and assigns numeric values to these voltages. Voltages that are above a given value are converted to a 1, and voltages below a given value are assigned a

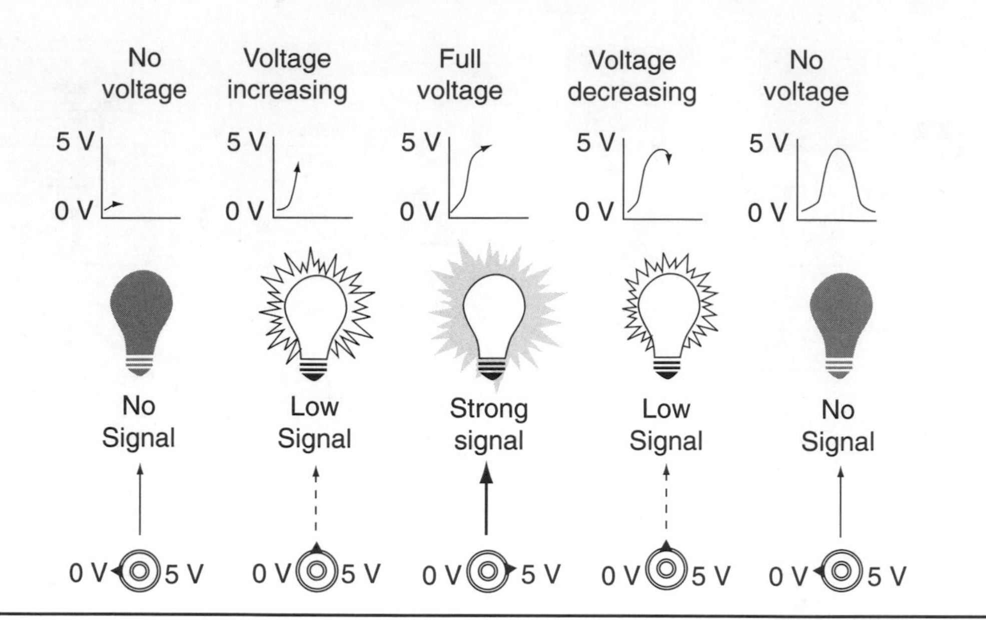

FIGURE 12–25 Typical analog voltage signal.

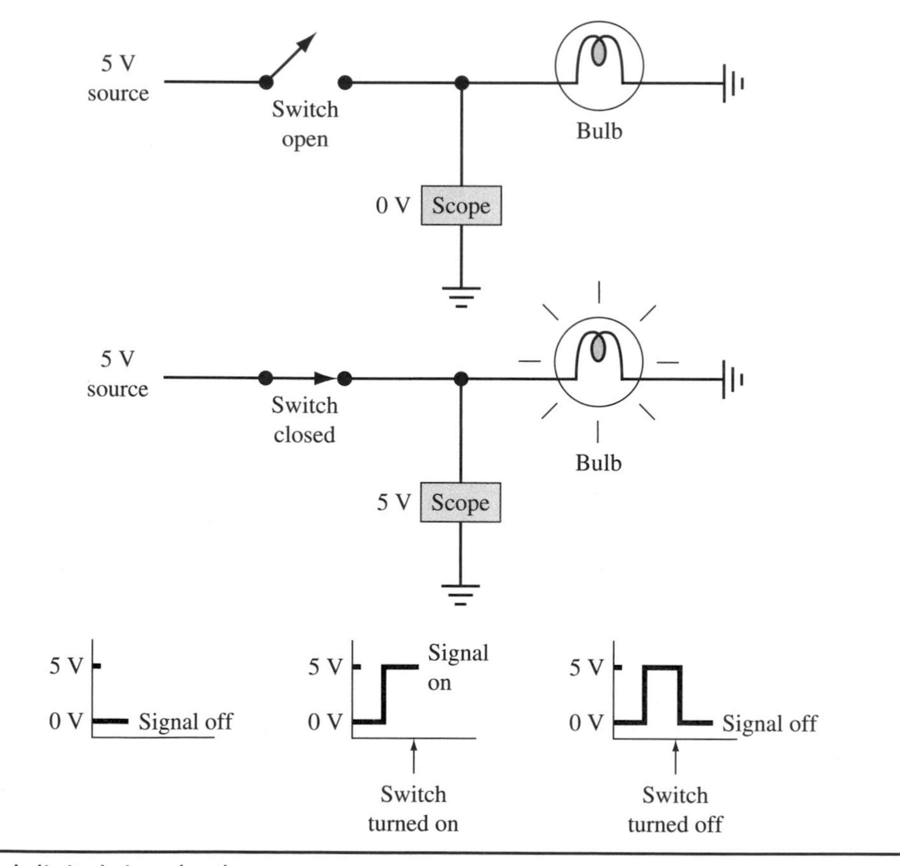

FIGURE 12–26 Typical digital signal voltage.

value of 0. In Figure 12–31, the A/D converter changes a series of signals to a binary number made up of 1s and 0s. When 8 bits are combined in a specific sequence, they form a byte, or word, that makes up the basis of the computer's software.

Microprocessors

In an automotive computer, the **microprocessor** is the decision-making chip in the computer, and it contains a large number of transistor and diode switches in an IC

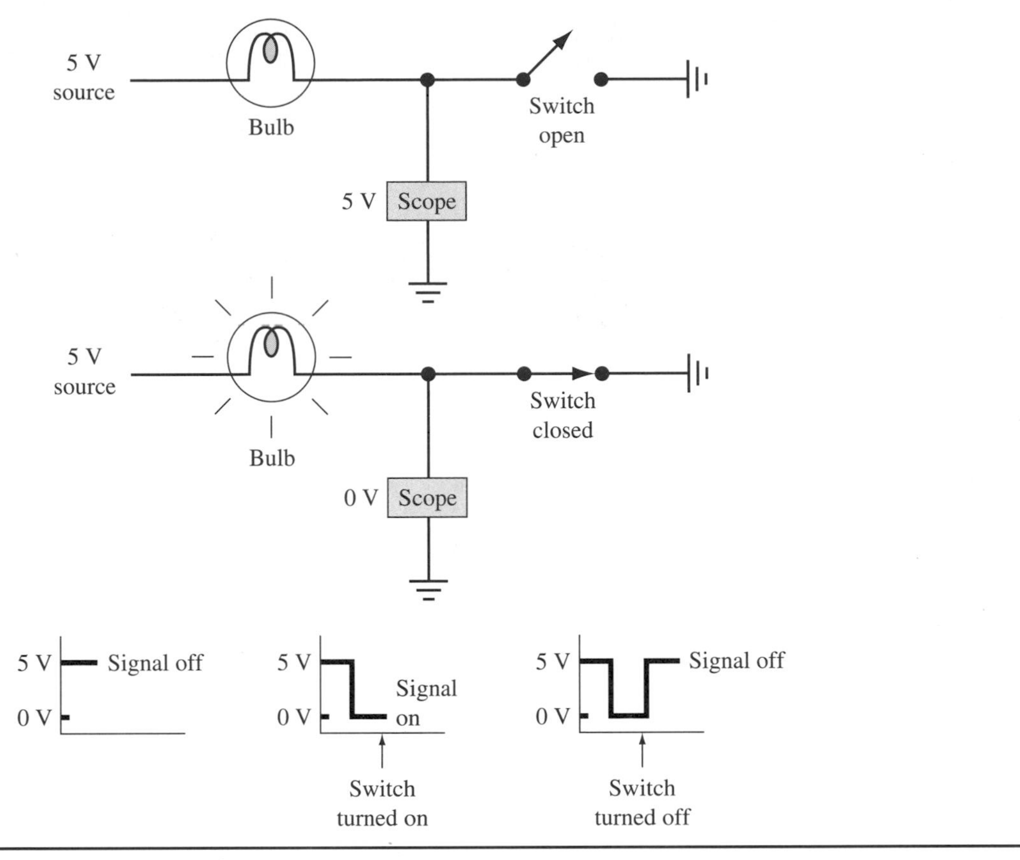

FIGURE 12–27 Reversed digital signal voltage.

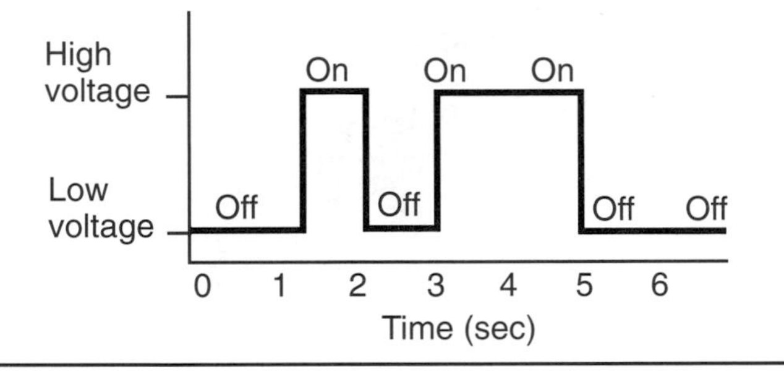

FIGURE 12–28 Variable digital signal voltage.

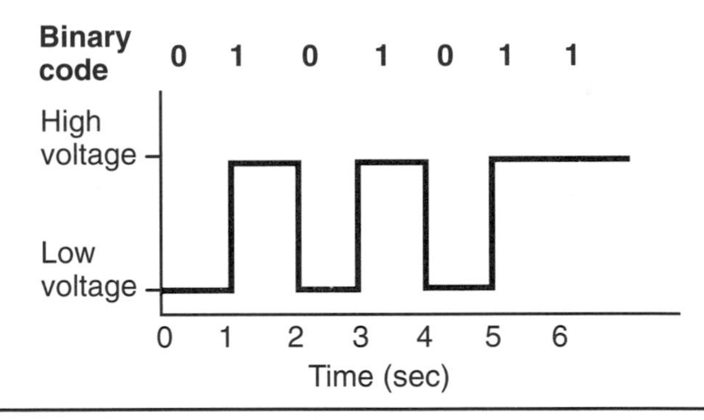

FIGURE 12–29 Assigning a binary code to a digital signal voltage.

that is capable of producing many digital voltage signals per second. The chip containing the silicon IC is mounted in a flat rectangular protective package with metal connecting pins (Figure 12–32).

Computer Memory

When the microprocessor interprets input information, performs calculations, and makes decisions, it requires the use of three specific types of computer memories (Figure 12–33).

The first, **random access memory (RAM),** is the note pad for a computer. Information that requires temporary storage is transferred to the RAM for use by the microprocessor to read, write, and erase information in relationship to the various operating conditions of the vehicle (Figure 12–34). There are two types of RAM: volatile and nonvolatile. A volatile RAM maintains information as long as it is connected to a battery power

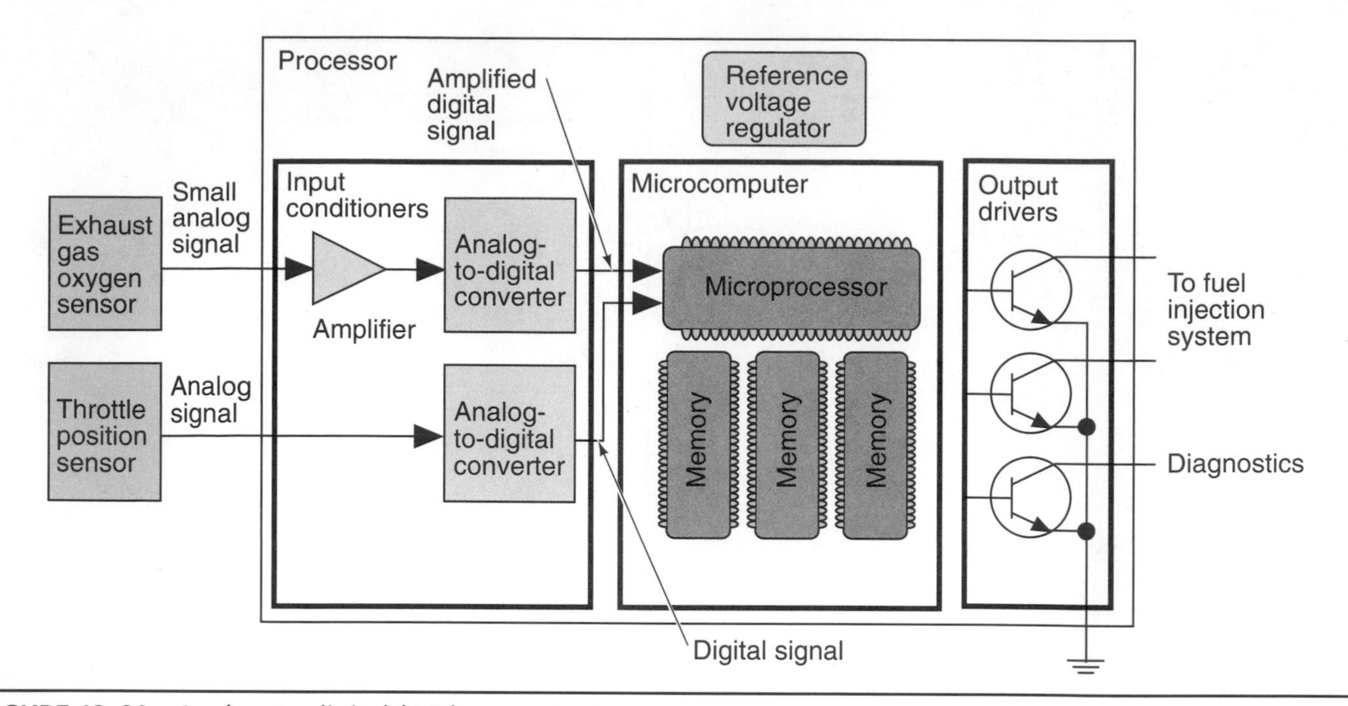

FIGURE 12–30 Analog-to-digital (A/D) converter in a computer.

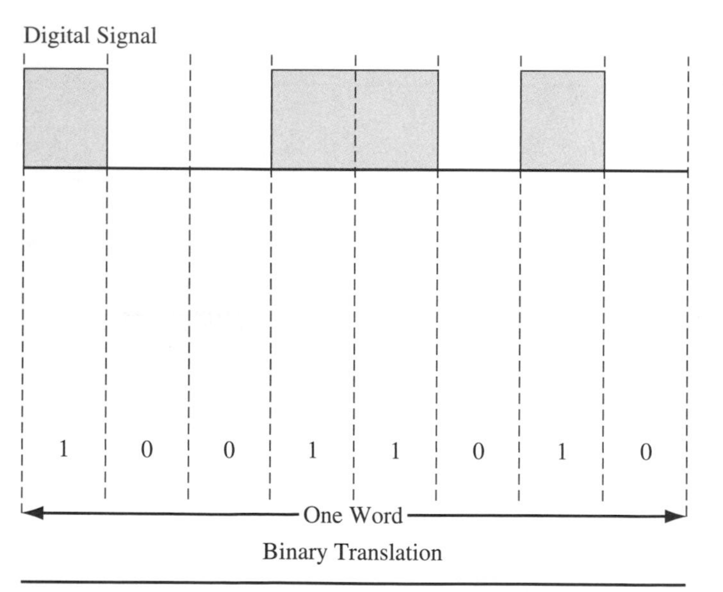

FIGURE 12–31 A/D conversion of a signal and assigning binary code numbers.

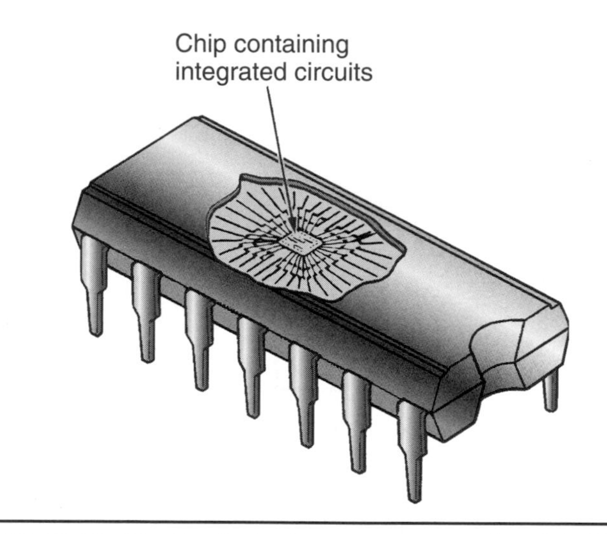

FIGURE 12–32 Typical microprocessor chip with IC mounted in a protective package.

source. The only time the RAM loses its stored information is when the power source is disconnected. It is also referred to as a **keep alive memory (KAM)** (Figure 12–35). KAM allows variations in the options a vehicle uses, such as electronic transmission shifting patterns, adaptive fuel trim for air/fuel ratios, and idle air control. A nonvolatile RAM does *not* lose its stored memory if the source power is lost. Nonvolatile RAM is typically used to store information on vehicles with digital odometers.

The second type of memory, **read only memory (ROM),** contains information like a textbook that is permanently stored for use by the microprocessor to determine what to do with the input data (Figure 12–36). The ROM contains specifications and calibration information of various components and systems under a variety of

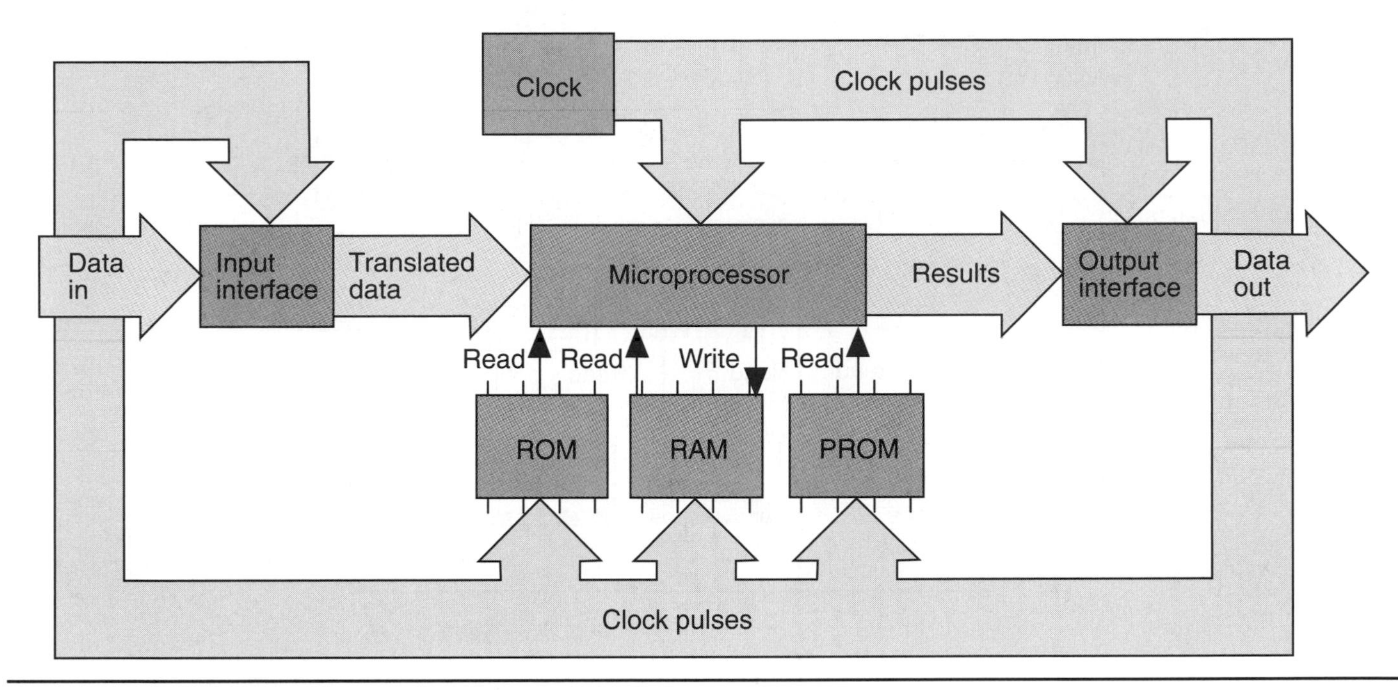

FIGURE 12–33 Various computer memories and their interaction with a microprocessor.

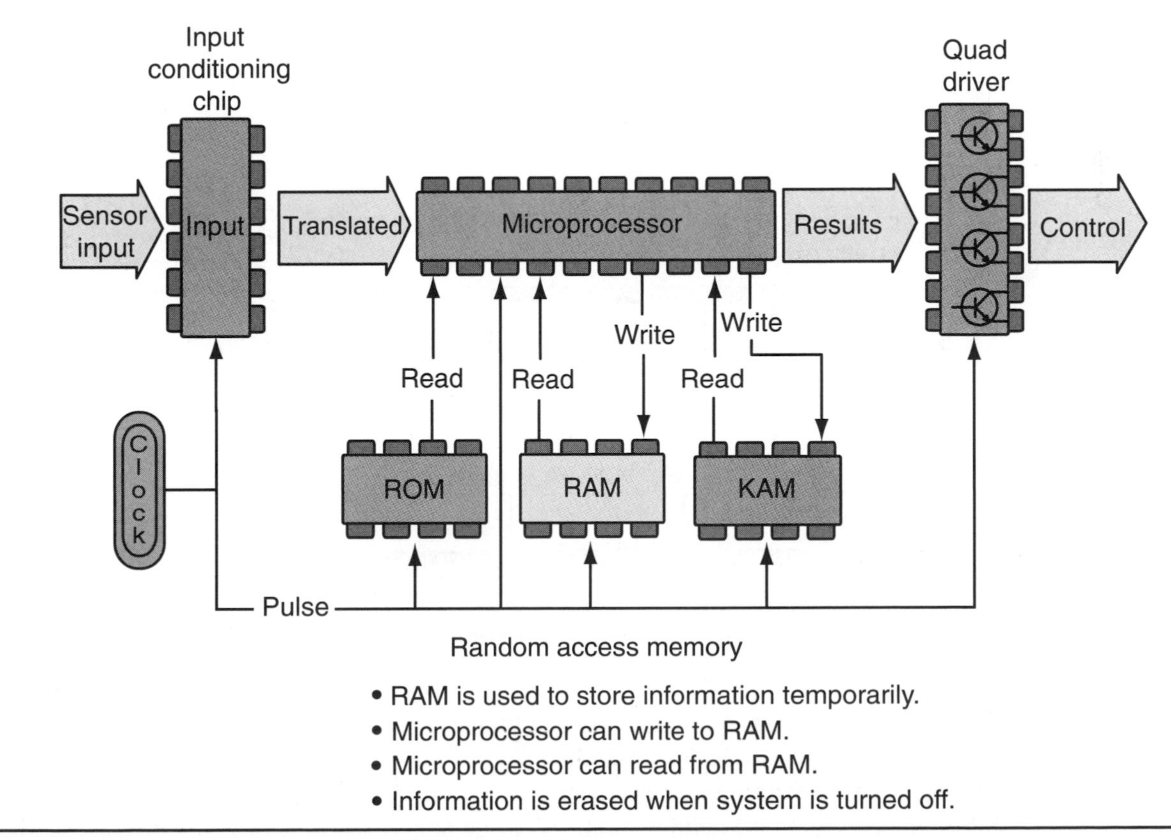

FIGURE 12–34 Random access memory (RAM).

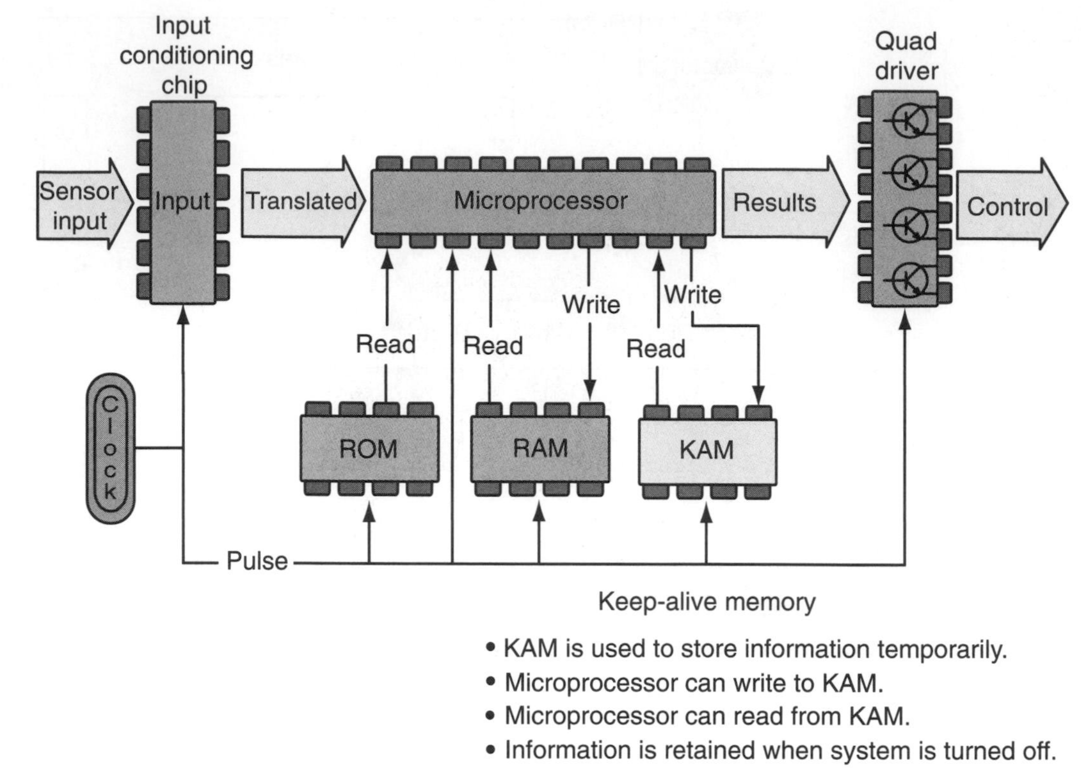

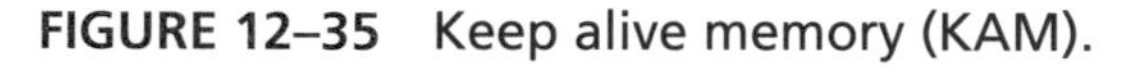
FIGURE 12–35 Keep alive memory (KAM).

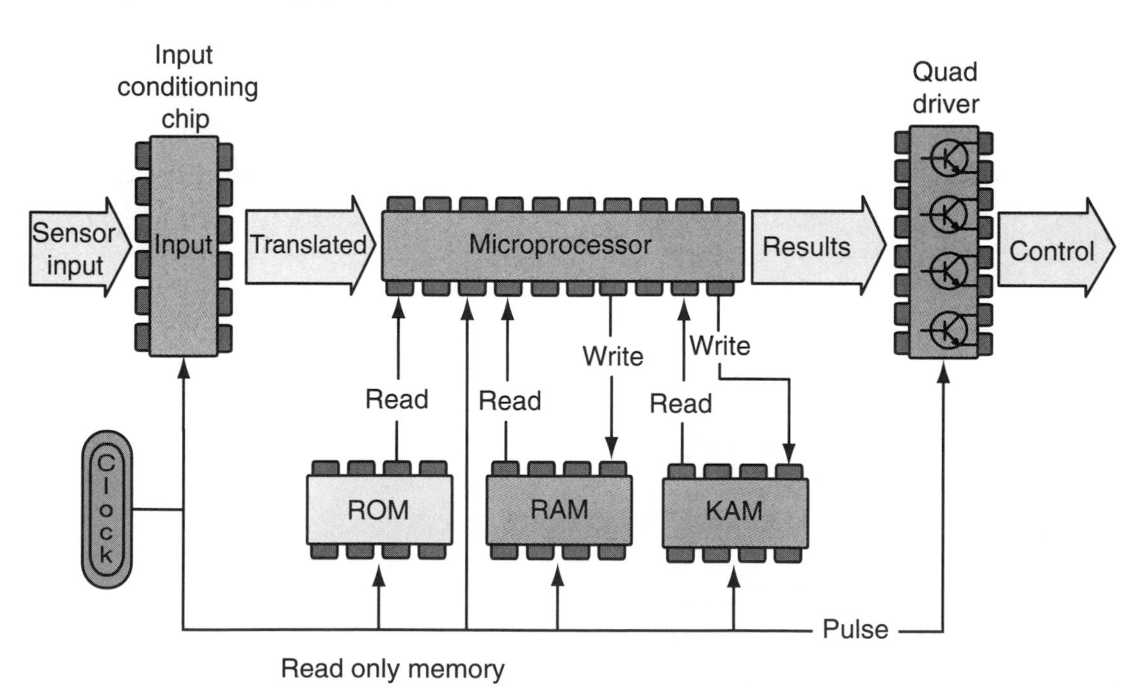

- ROM is used to store information permanently.
- Microprocessor can read from ROM.
- Microprocessor cannot write to ROM.
- Information is retained even when battery power is disconnected.

FIGURE 12–36 Read only memory (ROM).

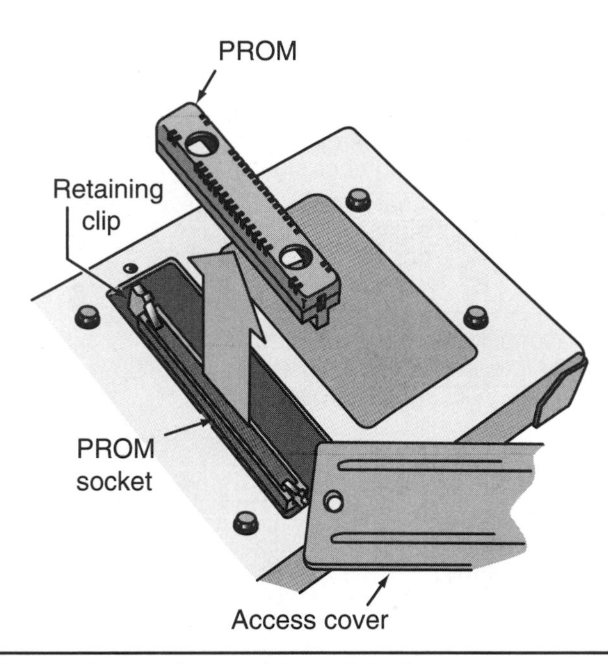

FIGURE 12–37 PROM location in an electronic control module (ECM).

engine operating conditions. The microprocessor compares sensor inputs to the preprogrammed look-up tables and specifications stored in the ROM and then makes decisions for controlling various engine functions.

Some computers have a removable **programmable read only memory (PROM),** which contains specific calibration programs that pertain to a vehicle model (Figure 12–37). The PROM can be changed easily if revised information is available from the manufacturer. On some of the newer computers the PROM has been changed to an **electronic erasable programmable read only memory (EEPROM).** This type of memory chip can be reprogrammed only with special test equipment available from the vehicle manufacturer.

Figure 12–38 shows a basic overview of the systematic processing and conversion of input information into digital signals that control various circuits.

Logic Gates

For the computer's microprocessor to make the most informed decisions regarding the operation of various systems, sensor inputs may be sent through various logic gates, which incorporate thousands of **field effect transistors (FET)** into the computer's circuitry. A **logic gate** controls the output voltage signal depending on the different combinations of input signals.

Five common logic gates are used to perform the processing functions in the computer.

1. **NOT Gate** A NOT gate simply reverses binary 1s and 0s and vice versa (Figure 12–39). A high input to a NOT gate results in a low output. A NOT gate is sometime referred to as an inverter.

2. **AND Gate** The AND gate has at least two inputs and one output. The operation of the AND gate is similar to two switches in a series line to a load (Figure 12–40). The only way the light turns on is when both switches are closed. The output of the AND gate is high only if both inputs are high.

3. **OR Gate** The OR gate operates similarly to two switches that are wired together in parallel (Figure 12–41). If switch A or B is closed, the light turns on. A high signal to either input results in a high output.

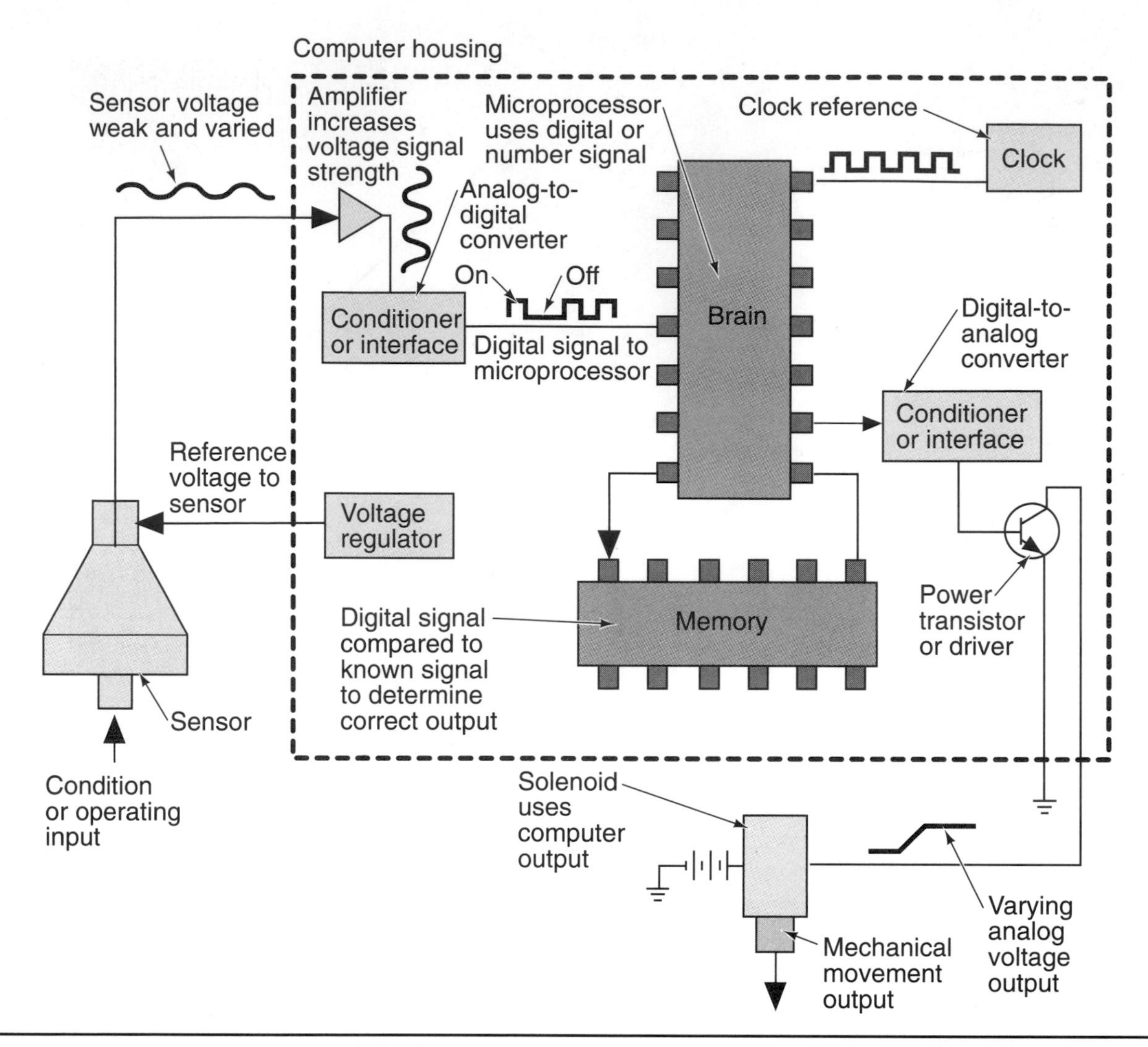

FIGURE 12–38 Computers in automobiles have the ability to process analog sensor information and generate digital signals that control various output circuits within specific parameters.

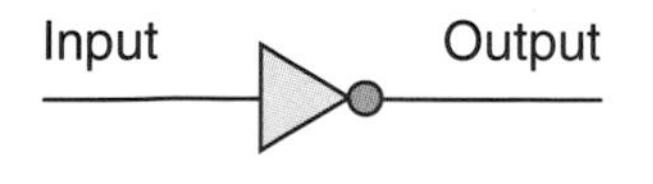

Truth table	
Input	Output
0	1
1	0

FIGURE 12–39 The NOT gate symbol and truth table, which shows the NOT gate inverting the input signal. The small circle on the symbol represents an inverting output.

4. **NAND and NOR Gates** A NAND (or NOR) gate is the equivalent of an AND (or OR) gate with an inverted output signal (Figure 12–42).

5. **XOR Gate** An exclusive OR gate is combination of NAND and NOR gates that produces a high output signal only if the inputs are different (Figure 12–43).

The application of the five gate circuits can be combined to form almost any type of logic circuit needed. Figure 12–44 shows an example of a power sunroof circuit combining multiple gate circuits. The input and output characteristics of each gate circuit can help the technician determine the output functions of the overall circuit.

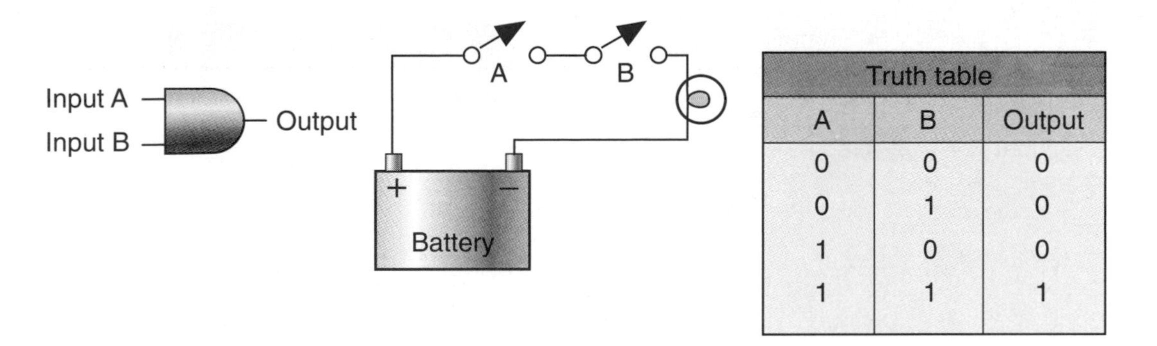

Truth table		
A	B	Output
0	0	0
0	1	0
1	0	0
1	1	1

FIGURE 12–40 The AND gate symbol and truth table.

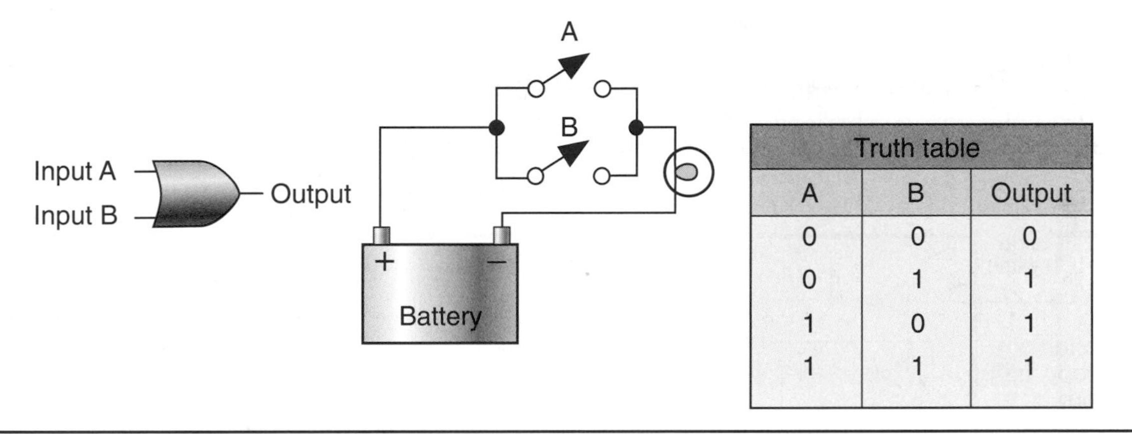

Truth table		
A	B	Output
0	0	0
0	1	1
1	0	1
1	1	1

FIGURE 12–41 The OR gate symbol and truth table.

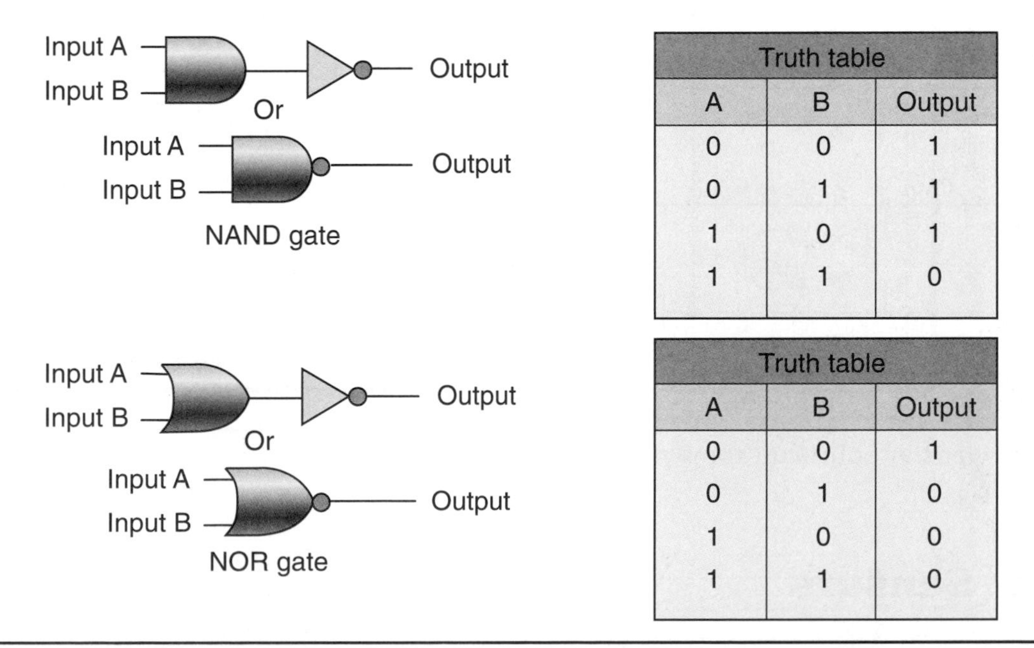

Truth table		
A	B	Output
0	0	1
0	1	1
1	0	1
1	1	0

Truth table		
A	B	Output
0	0	1
0	1	0
1	0	0
1	1	0

FIGURE 12–42 The NAND and NOR gate symbols and truth tables. The small circles represent an inverted output on any logic gate symbol.

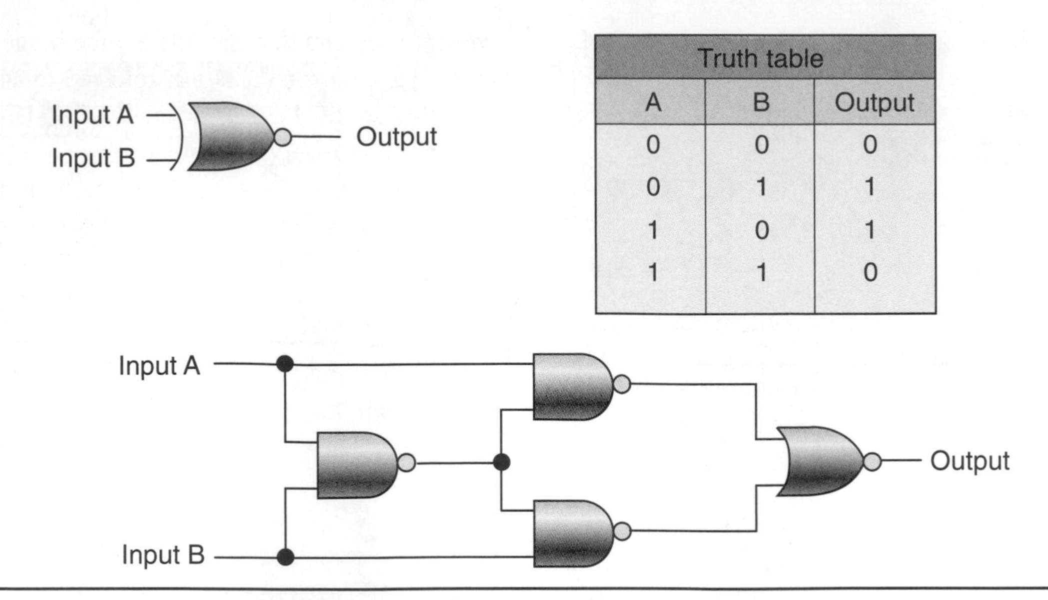

Truth table		
A	B	Output
0	0	0
0	1	1
1	0	1
1	1	0

FIGURE 12–43 The XOR gate symbol and truth table.

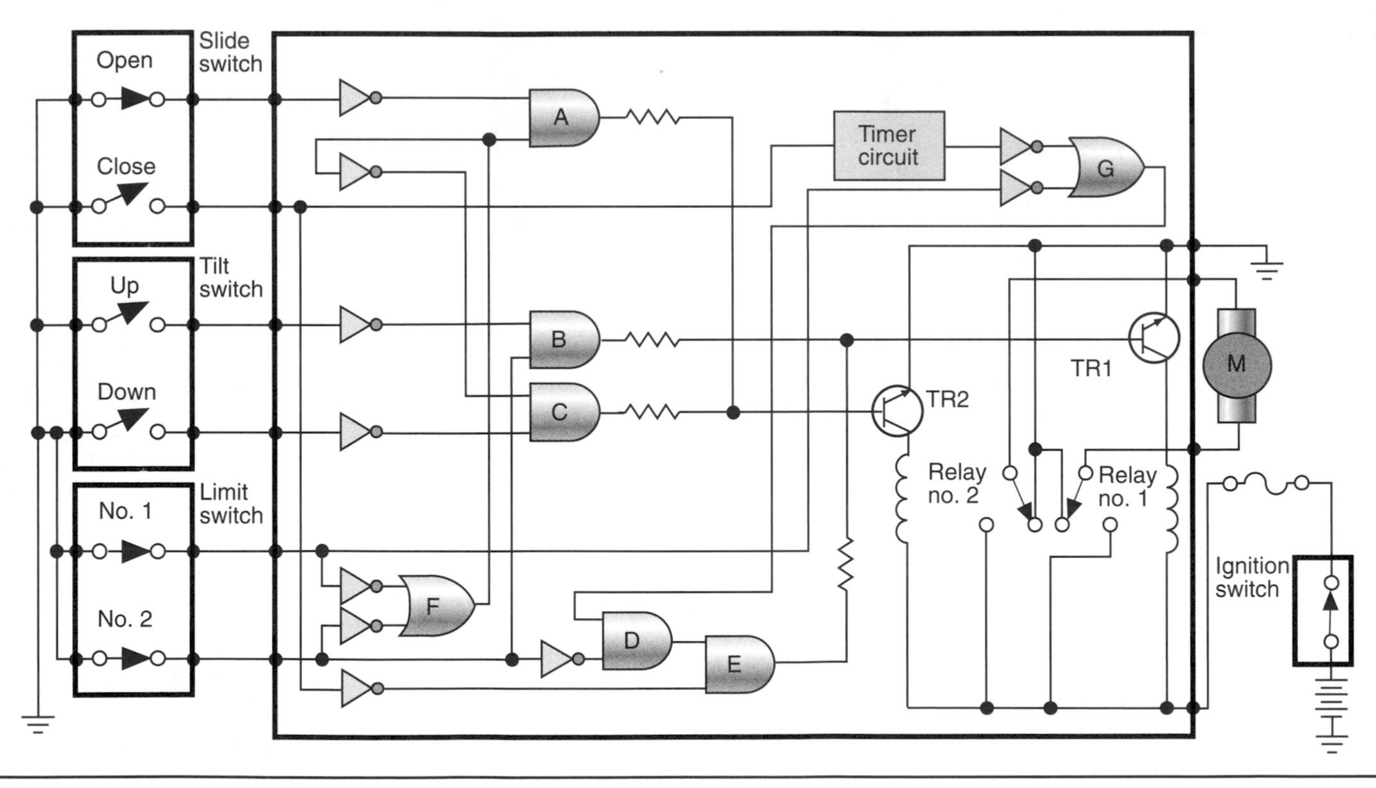

FIGURE 12–44 Power sunroof circuit combining multiple gate circuits.

Computer Sensors

Operation of an automotive computer requires several input and output sensors and control actuators (Figure 12–45). An **input sensor** transmits information to the control computer on coolant temperature, air inlet temperature, manifold pressure, barometric pressure, throttle position, vehicle speed, and many other conditions. After gathering, processing, and interpreting the information from the input sensors, the control computer provides output

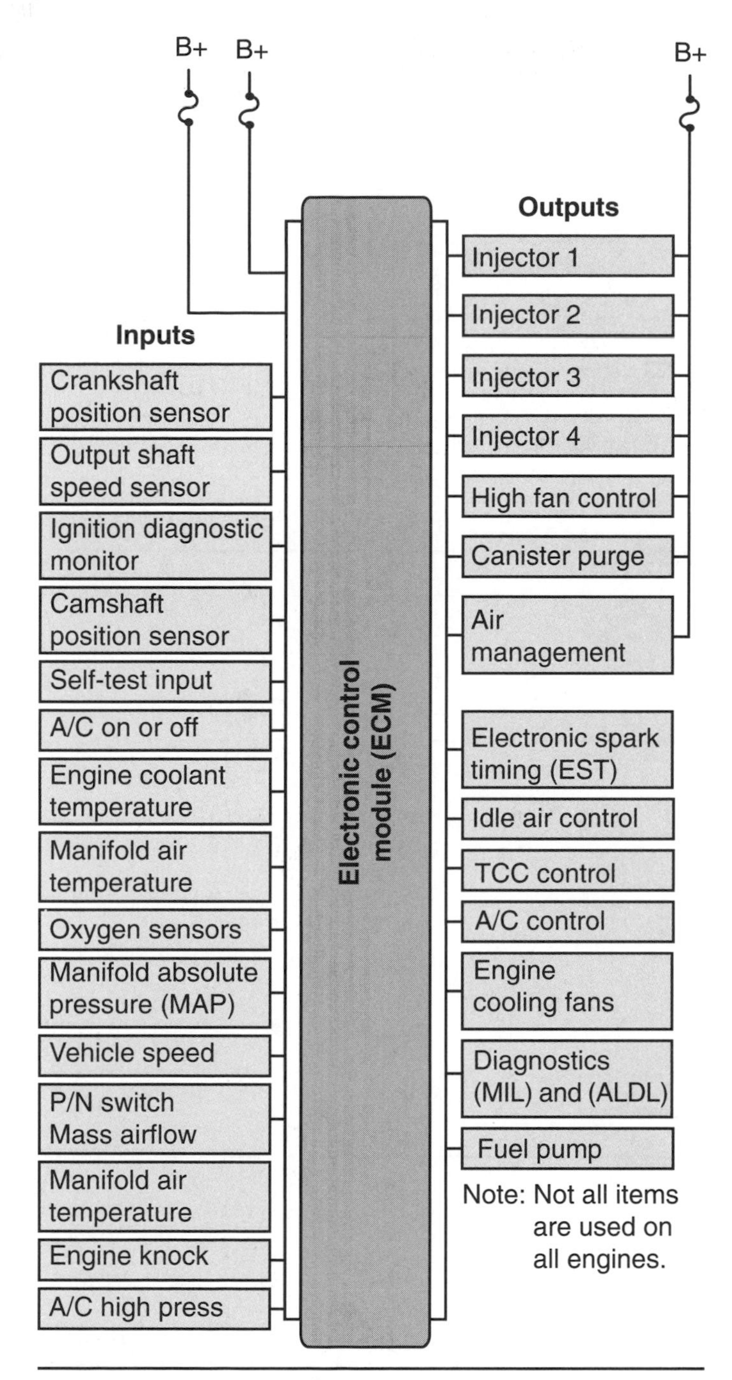

FIGURE 12–45 Typical computer-controlled inputs and outputs.

commands to actuators and controls, such as electronic spark timing, air conditioning, emissions devices, and transmission shifting.

Input Sensors

Input sensors are divided into two general categories: sensor *switches* that produce on or off information, and *variable* sensors that provide a wide range of input readings. Input sensor voltage is referred to as the **reference voltage.** This voltage is normally 5 or 12 volts, depending on the sensor and circuit.

In the first category, two common types of input sensor switches are used in circuits. The first is a switch-to-power or **pull-up switch** (Figure 12–46). This sensor switch circuit uses an *external* power source to the switch and a ground circuit *internal* to the control computer. When the signal switch is closed, a *high* reference signal from the external source voltage is sent to the computer. When the switch opens the reference signal is *low.* A pull-up switch can be found on an air conditioning (A/C) system to signal the control computer that A/C is requested and idle speed should be increased to compensate for the additional engine load.

The second type of input sensor switch is a switch-to-ground or **pull-down switch** (Figure 12–47). This sensor switch circuit uses an *internal* power source from the computer to the switch and an *external* ground circuit from the switch. When the signal switch is closed, a *low* reference signal is sensed in the computer. When the switch opens, the reference signal is *high.* A pull-down switch can be found in a park/neutral transmission switch (Figure 12–48). When the transmission is shifted into gear from park or neutral, the idle speed is increased to compensate for the additional engine load.

In the second category, a variable sensor provides voltage input signals that vary with the control device. Examples of these sensors are oxygen sensors, engine coolant sensor, and throttle position sensor. This type of sensor can be a rheostat or thermistor with one or two wires, or potentiometers that contain three wires.

Oxygen Sensors (O_2S) A typical example of a single-wire variable sensor is an **oxygen sensor (O_2S)** (Figure 12–49). This sensor uses one wire to carry the 450-millivolt reference voltage from the engine control module (ECM) through the sensor body to ground. The sensor body is constructed of a ceramic **zirconium dioxide** element (Figure 12–50) that reacts with the oxygen content in the exhaust gases. When the exhaust gas with low oxygen content strikes the O_2 sensor, there is a high oxygen content inside the sensor element from the atmosphere. This O_2 difference produces up to 1 volt for a rich mixture, and 0.3 volt or less for a lean mixture (Figure 12–51).

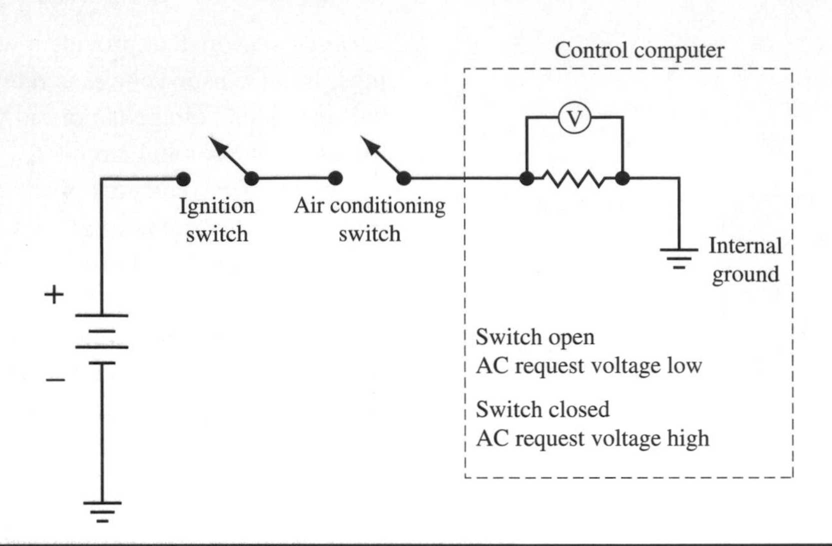

FIGURE 12–46 Typical pull-up switch sensor circuit.

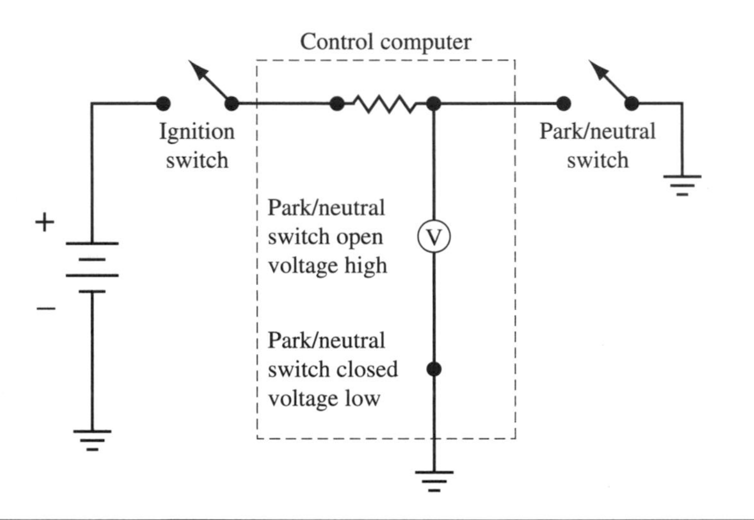

FIGURE 12–47 Typical pull-down switch sensor circuit.

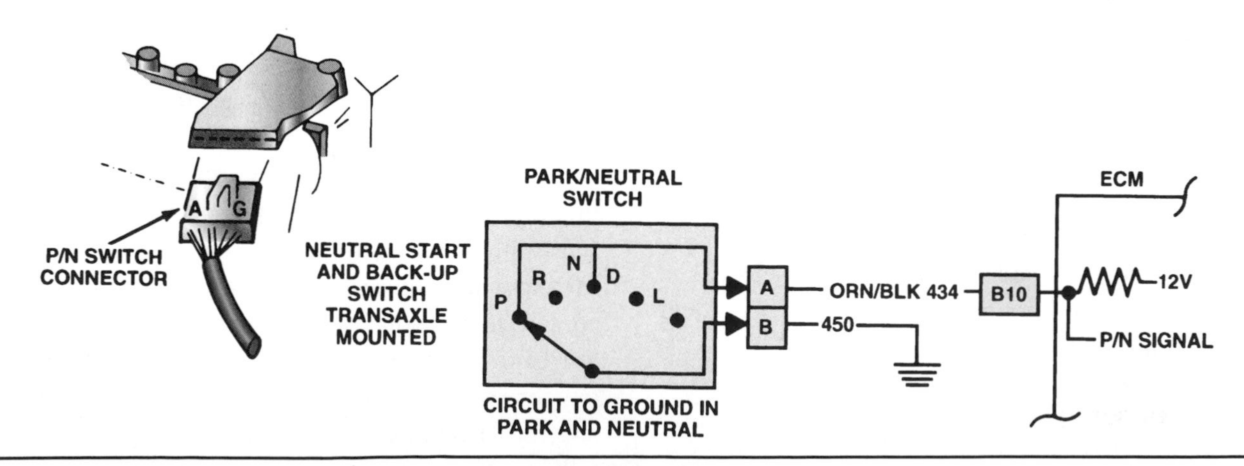

FIGURE 12–48 Typical pull-down park/neutral transmission switch.

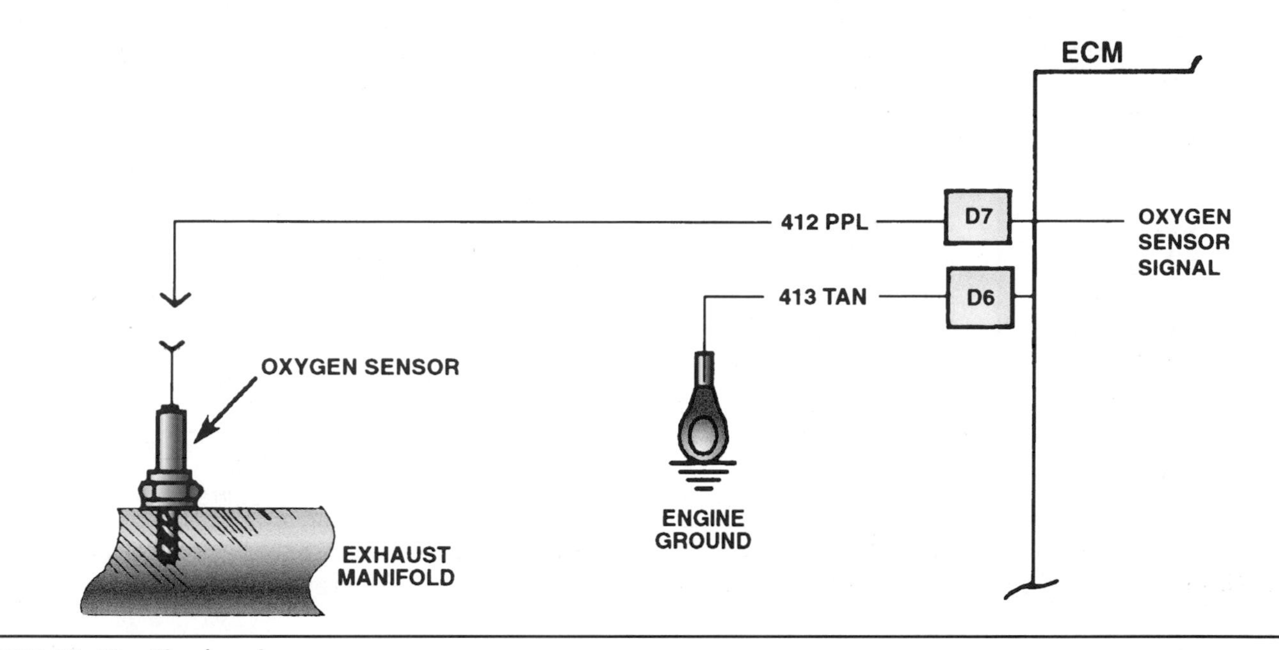

FIGURE 12–49 Single-wire oxygen sensor.

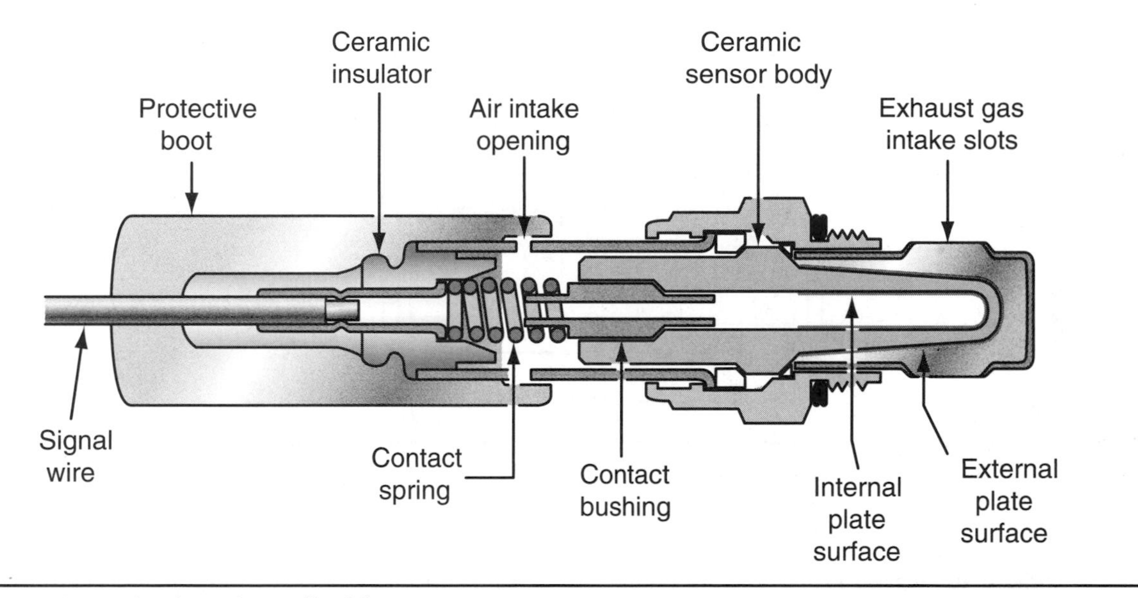

FIGURE 12–50 Ceramic zirconium dioxide oxygen sensor.

The atmosphere contains approximately 21% oxygen. The ideal air/fuel mixture is 14.7:1, which is referred to as a **stoichiometric** mixture. This mixture indicates that, for every 14.7 pounds of air entering the engine, the fuel system supplies 1 pound of fuel.

During normal engine operation the O_2 sensor provides an output voltage to the ECM that continually fluctuates in the closed-loop mode from a *maximum* of 0.9 volt (900 millivolts) to a minimum of 0.1 volt (100 millivolts), with a preprogrammed voltage value in the middle called a **set-point** that equates to the desired air/fuel ratio of 14.7:1. The voltage values below the set-point are considered lean mixtures, and voltage values above the set-point are considered rich mixtures (Figure 12–52). When observed on a scope the variable voltage would look like the sample in Figure 12–53.

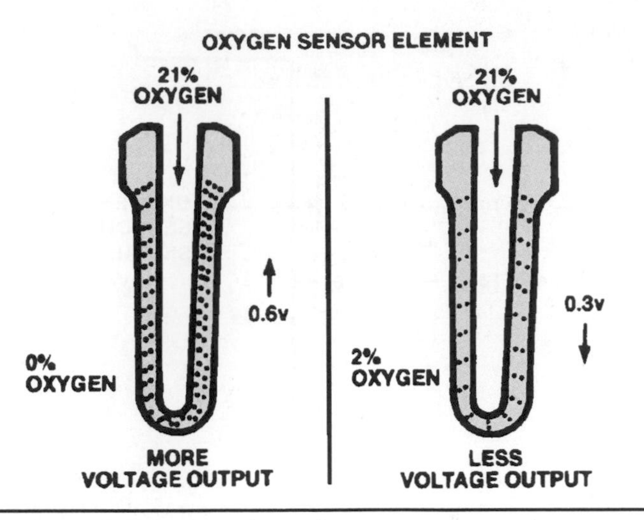

FIGURE 12–51 Reaction of oxygen content in exhaust gas compared to atmosphere.

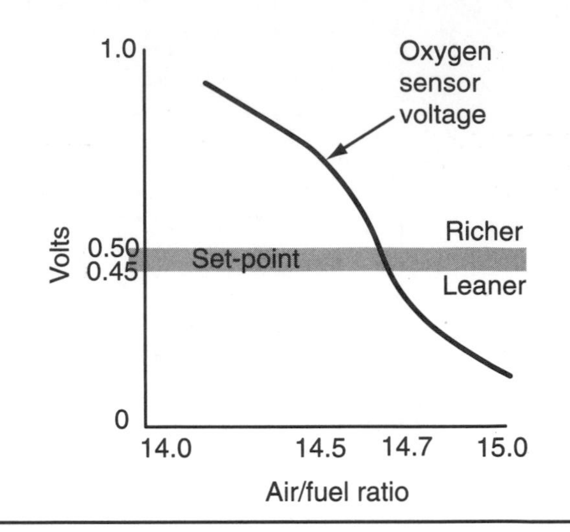

FIGURE 12–52 Oxygen sensor output voltage comparison to air/fuel ratio.

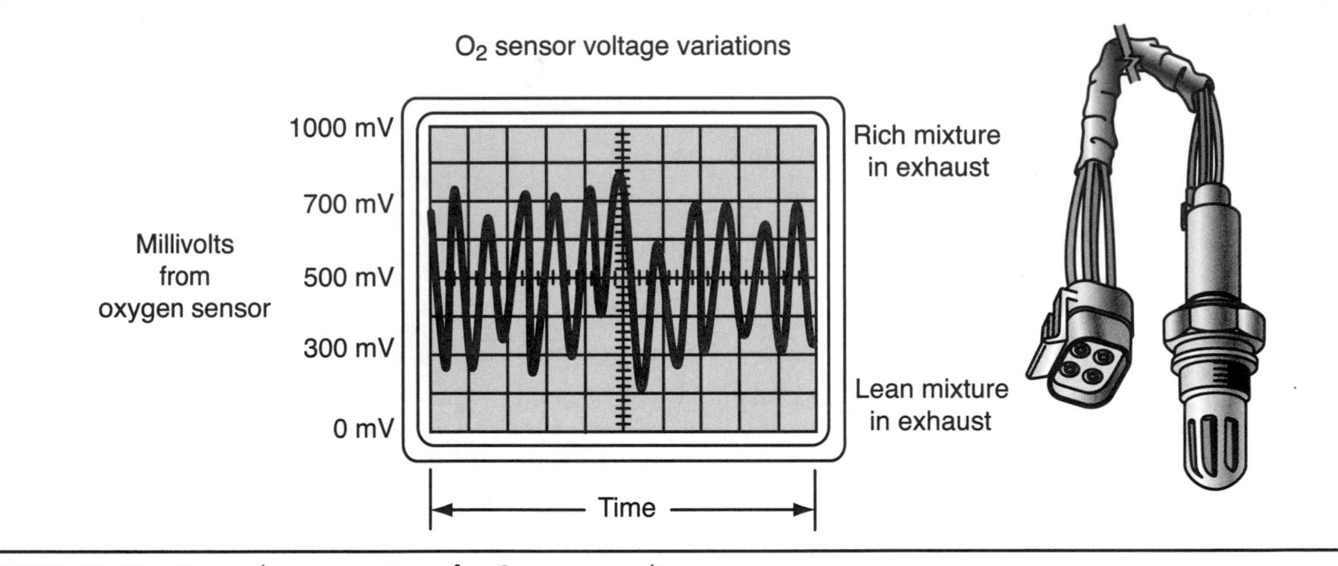

FIGURE 12–53 Normal scope pattern for O_2 sensor voltage.

Most late-model vehicles currently use O_2 sensors that incorporate a heating element (Figure 12–54), which preheats the O_2 sensor until the exhaust gas reaches normal operating temperature. The heating element also helps keep the O_2 sensor hot should the engine idle for extended periods and cool down. A heated O_2 sensor can be easily identified because there are three or more wires (a two-wire O_2 sensor normally consists of a signal wire and ground wire but no heater wire).

Engine Coolant Temperature Sensor The **engine coolant temperature (ECT)** sensor is a two-wire thermistor sensor (Figure 12–55) and is normally located in a threaded fitting in the intake manifold coolant passage. The ECT provides the ECM with engine coolant temperature information by providing a high-resistance, higher-voltage signal when cold, and low-resistance, lower-voltage reading at higher temperature (Figure 12–56).

Engine coolant temperature is a key input for the ECM to calculate several output operating functions. Some of these are air/fuel delivery, idle speed, cooling fan operation, emissions regulation, and ignition control.

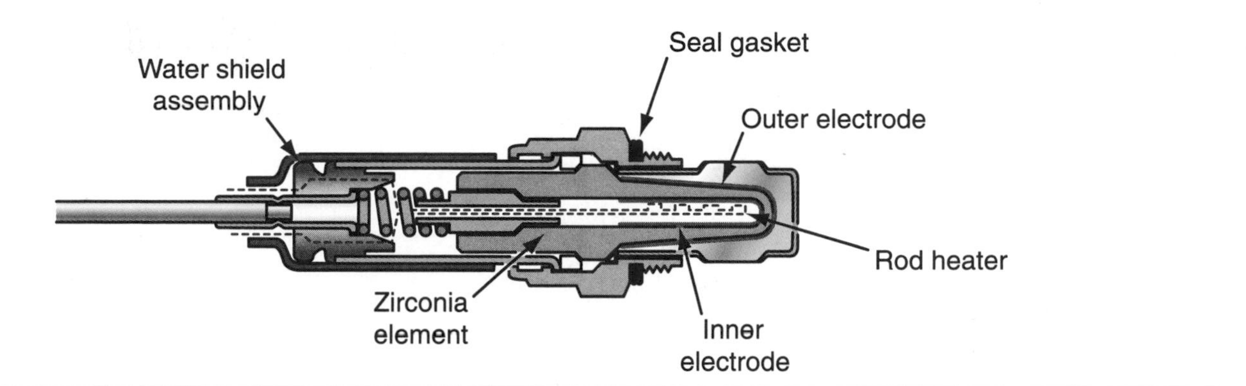

FIGURE 12–54 Typical heated oxygen sensor.

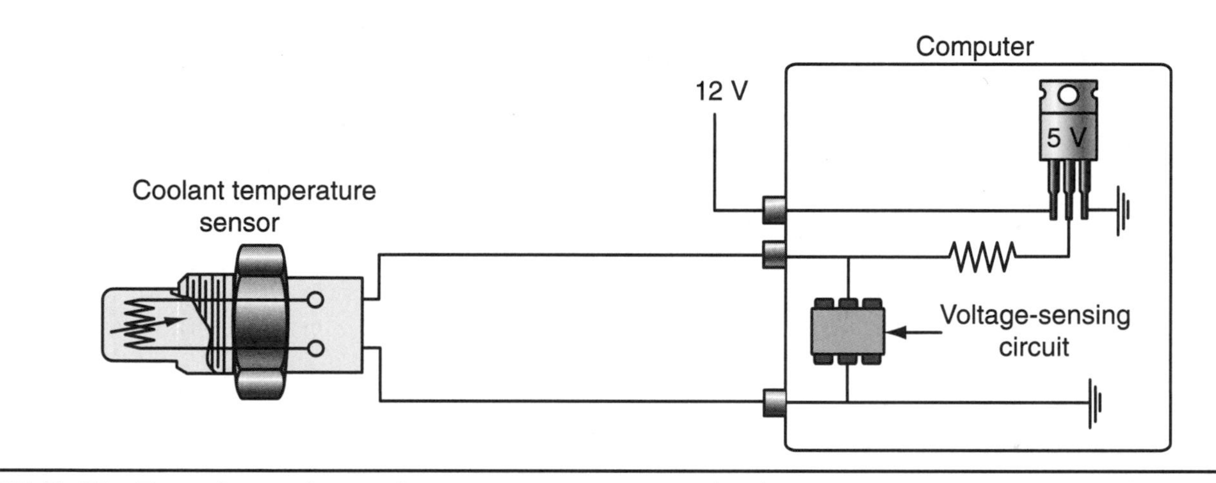

FIGURE 12–55 Two wire engine coolant temperature sensor circuit.

Throttle Position Sensor The **throttle position sensor (TP)** consists of a three-wire potentiometer connected to the throttle assembly and connected to the computer (Figure 12–57). The resistance of the circuit changes with throttle position. A typical TP sensor at idle position is approximately 1,000 ohms and 0.5 volt, and 4,000 ohms and 4.5 volts at wide open throttle (WOT) (Figure 12–58).

The voltage of the TP sensor should change gradually from idle to WOT. When the wiper arm on the resistive element has a worn spot (due to extended use in one position), a glitch can occur in smooth engine operation. A digital storage oscilloscope (DSO) can locate this condition (Figure 12–59) by monitoring the exact voltage measurement.

Output Devices

After gathering information from various input switches and sensors, the control computer provides electronic output commands through several mechanical motion devices, such as actuators, motors, relays, or solenoids. An example of output devices are relay-controlled circuits, idle air control (IAC) valves, and fuel injectors.

Relay Circuits Most relay-controlled circuits use electromechanical actuators that mechanically operate components. The control computer usually controls the ground circuit for the relay coil (Figure 12–60).

The output actuator can be cycled on and off rapidly to obtain the desired results, a process referred to as

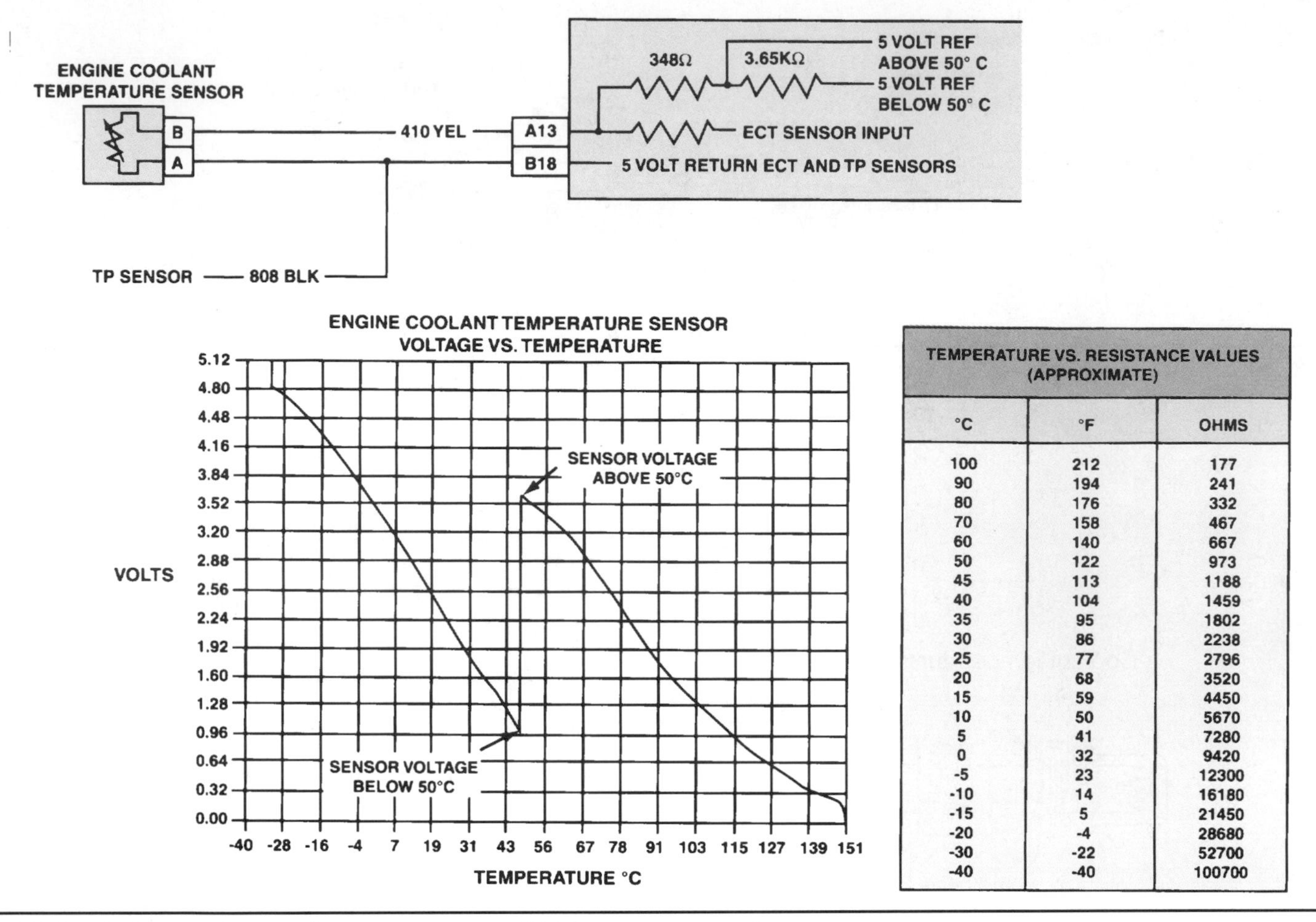

TEMPERATURE VS. RESISTANCE VALUES (APPROXIMATE)

°C	°F	OHMS
100	212	177
90	194	241
80	176	332
70	158	467
60	140	667
50	122	973
45	113	1188
40	104	1459
35	95	1802
30	86	2238
25	77	2796
20	68	3520
15	59	4450
10	50	5670
5	41	7280
0	32	9420
-5	23	12300
-10	14	16180
-15	5	21450
-20	-4	28680
-30	-22	52700
-40	-40	100700

FIGURE 12–56 Engine coolant temperature (ECT) sensor operating parameters.

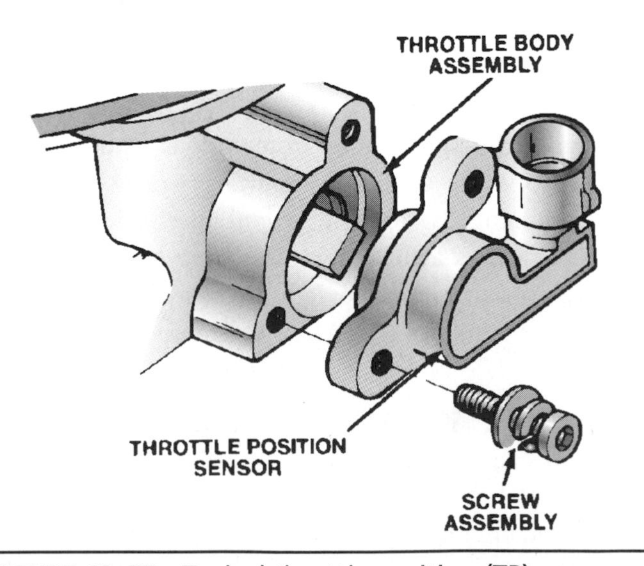

FIGURE 12–57 Typical throttle position (TP) sensor mounted to throttle body assembly.

pulse-width modulation. If this is used to control the illumination level of a digital instrument panel light cluster (Figure 12–61), the level of on time versus the off time dictates the light intensity.

Fuel Injector Controls A typical fuel injector control circuit controls the fuel that is pressurized. Each injector has an individual coil that is grounded by the control computer to complete the control circuit (Figure 12–62). The quantity of fuel injected depends on the amount of time the injector is energized, which is determined by engine temperature, load requirements, and other factors.

Idle Air Control Valve Many computer-controlled systems control engine idle speed with an **idle air control (IAC)** valve (Figure 12–63).

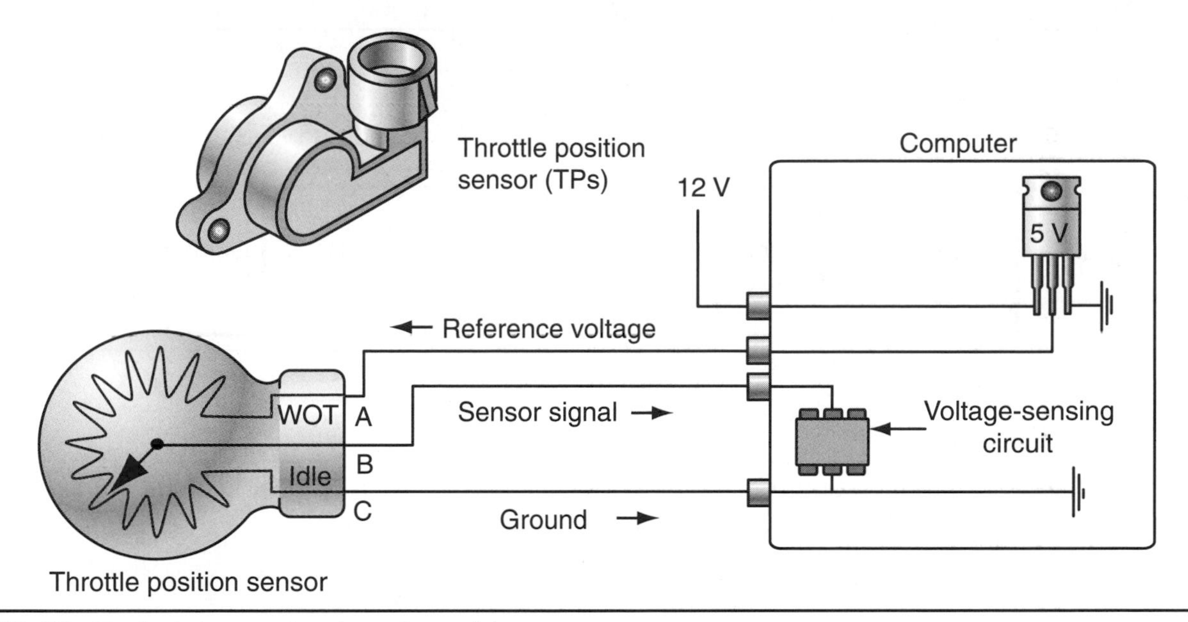

FIGURE 12–58 Typical three-wire throttle position sensor.

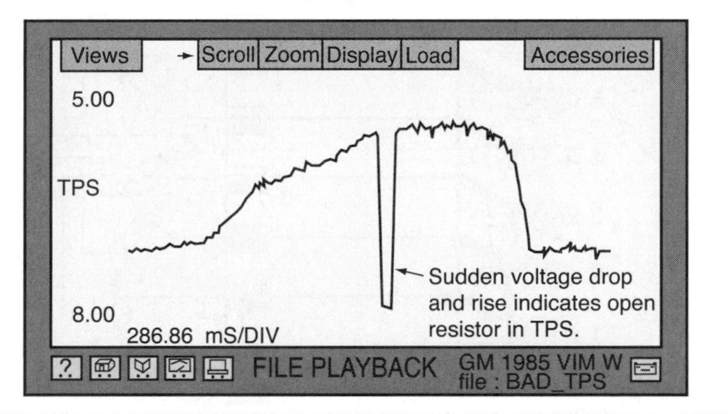

FIGURE 12–59 A throttle position (TP) sensor glitch is found using a digital storage oscilloscope (DSO).

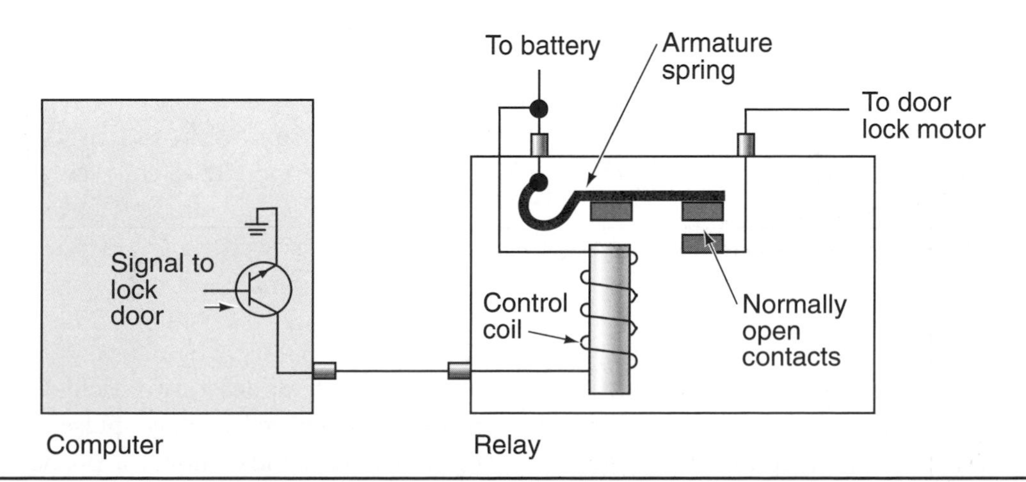

FIGURE 12–60 Computer output driver controls the relay coil ground.

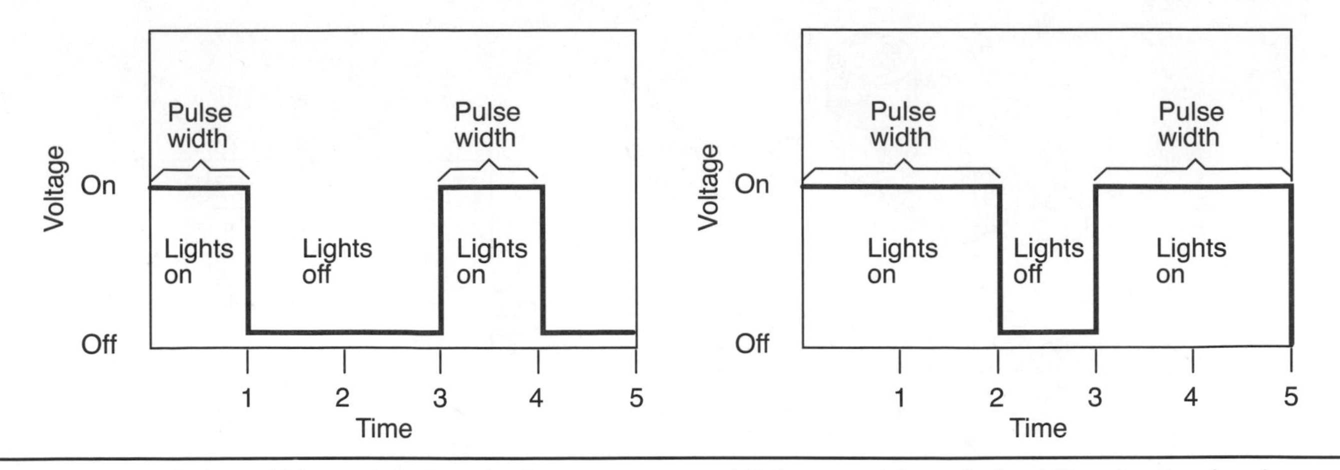

FIGURE 12–61 Pulse width modulating the instrument panel lights to achieve desired illumination level.

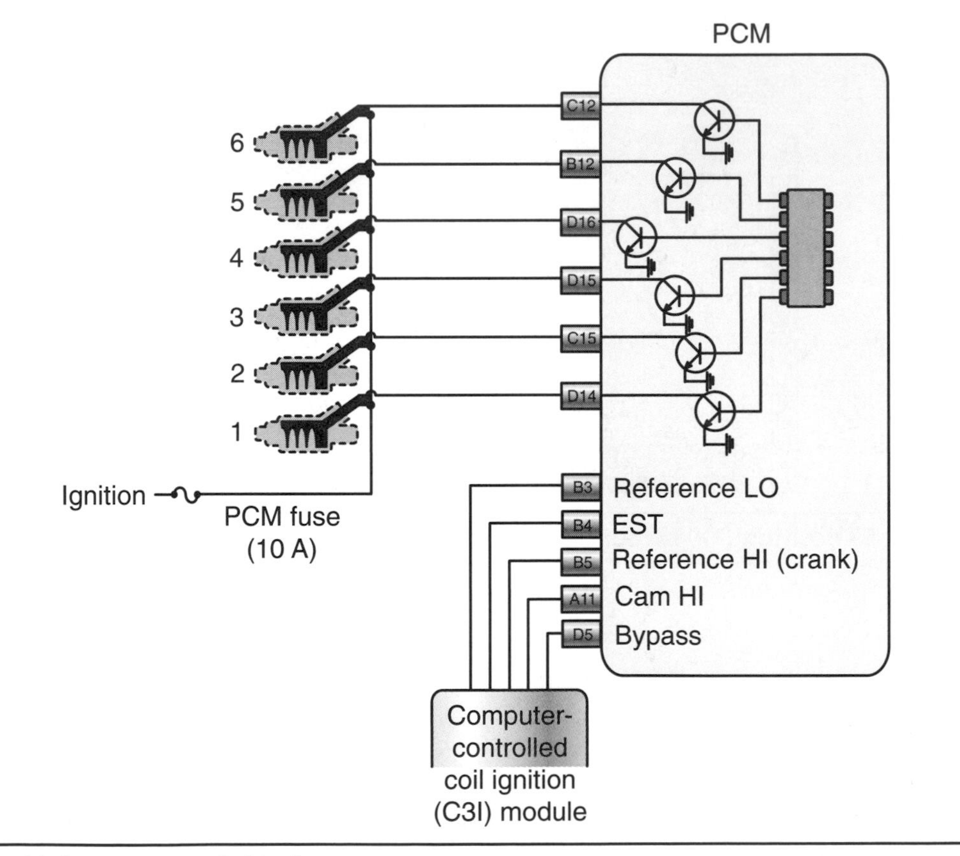

FIGURE 12–62 Fuel injector control circuit.

The engine control computer controls idle rpm by regulating the idle air passage on the throttle body (Figure 12–64). The IAC motor, also referred to as a stepper motor, contains a permanent magnet armature with at least two field coils. By applying properly sequenced voltage pulses to selected coils, the armature spindle moves a precise amount to control air flow.

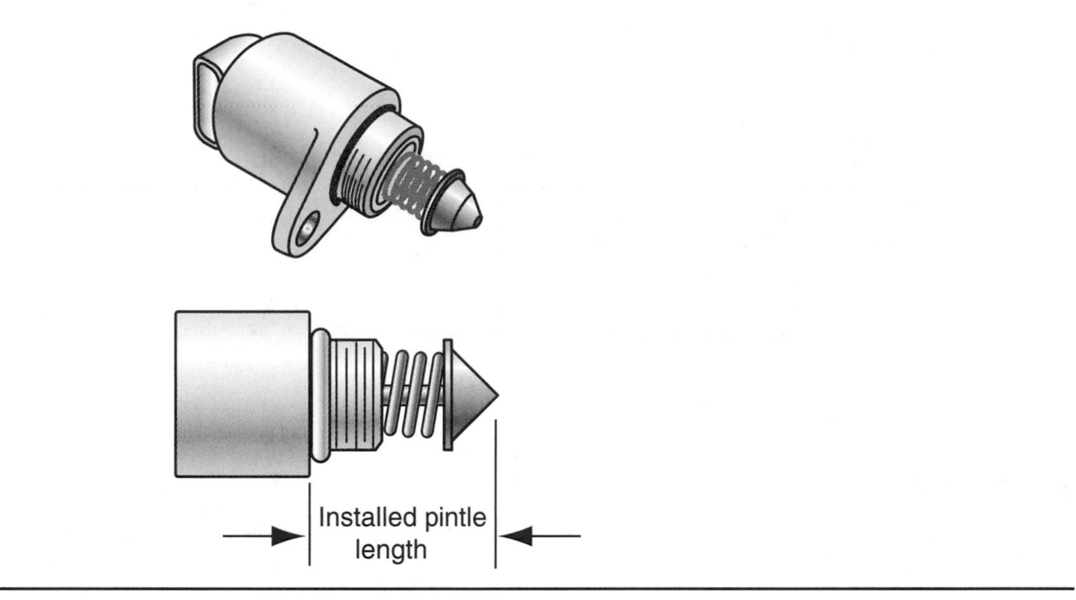

FIGURE 12–63 Idle air control (IAC) valve.

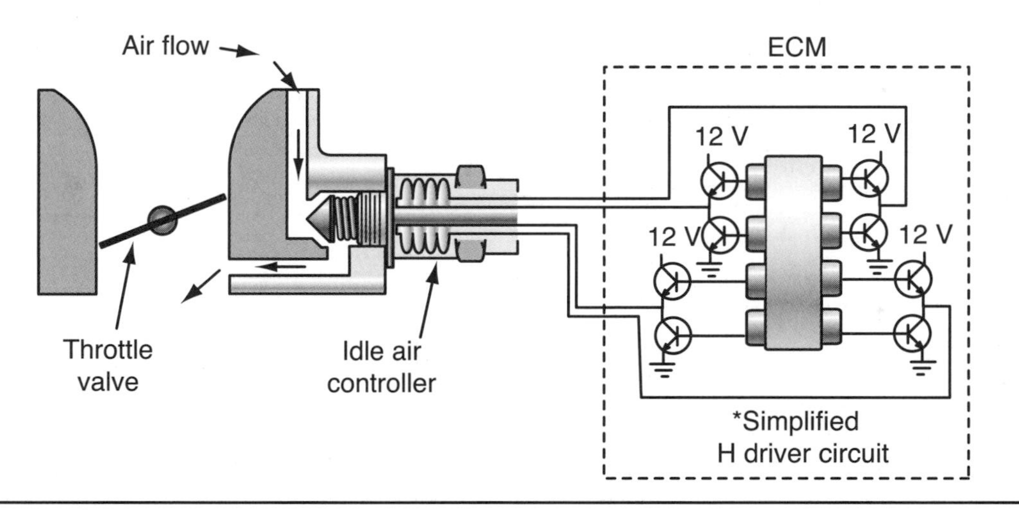

FIGURE 12–64 Typical idle air control (IAC) valve circuit.

Computer System Diagnostics

Depending on the vehicle, there are numerous computer-controlled input and output devices and systems. To perform effective troubleshooting requires the technician to retrieve fault codes stored in the computer memory. Always refer to specific manufacturer's procedures when retrieving computer memory codes. In most of the older computer-controlled vehicles, trouble codes can be retrieved by a blinking light on the dashboard (Figure 12–65) or a computer case that flashed a specific sequence of codes and subcodes for each trouble code stored (Figure 12–66).

The next means of accessing trouble codes is through a data link connector (DLC) located in the wiring harness under the dash, glove box area, or the engine compartment. Between the early 1980s and 1996, the DLC consisted of a 12-pin-or-less connector, depending on the vehicle manufacturer. Beginning in 1996, federal law required that all vehicles meet **OBD II** (on-board diagnostics, second generation) requirements (Figure 12–67). The

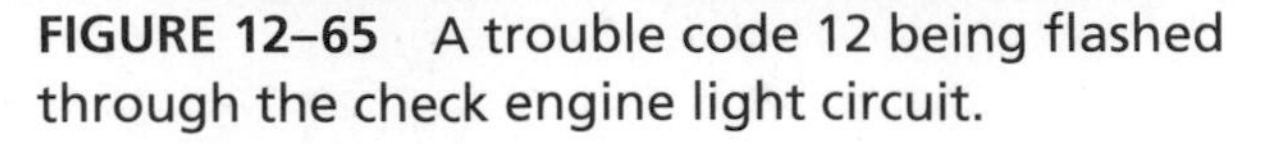
FIGURE 12–65 A trouble code 12 being flashed through the check engine light circuit.

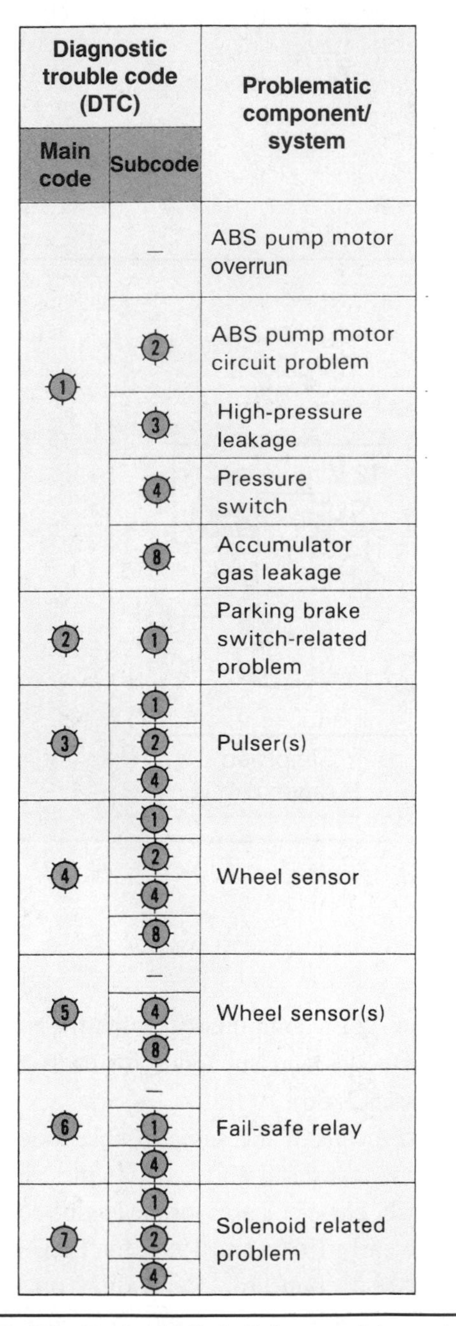

Diagnostic trouble code (DTC)		Problematic component/system
Main code	Subcode	
1	—	ABS pump motor overrun
	2	ABS pump motor circuit problem
	3	High-pressure leakage
	4	Pressure switch
	8	Accumulator gas leakage
2	1	Parking brake switch-related problem
3	1	Pulser(s)
	2	
	4	
4	1	Wheel sensor
	2	
	4	
	8	
5	—	Wheel sensor(s)
	4	
	8	
6	—	Fail-safe relay
	1	
	4	
7	1	Solenoid related problem
	2	
	4	

FIGURE 12–66 Typical blink light trouble codes for an antilock brake system.

(Courtesy of American Honda Motor Company, Inc.)

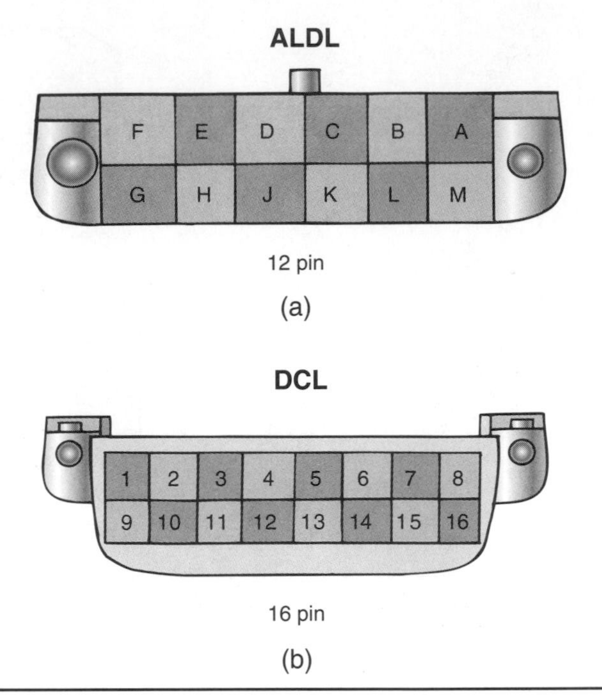

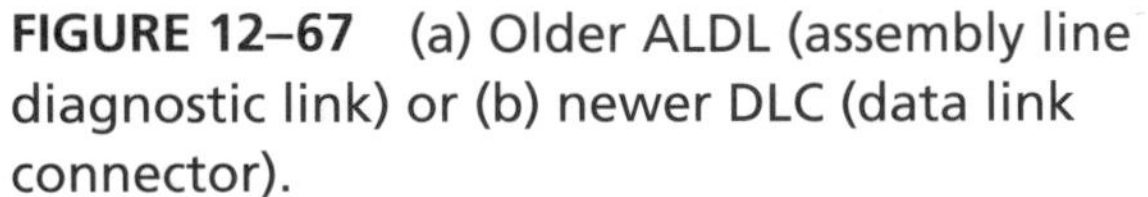
FIGURE 12–67 (a) Older ALDL (assembly line diagnostic link) or (b) newer DLC (data link connector).

DLC connector for OBD II is easily identified by the 16-pin universal connector common to all manufacturers' cars and light-duty trucks. In accordance with SAE Standard J1962, specific assignments are applied to each pin number on the connector (Figure 12–68).

A quick way to retrieve a computer code through a DLC is with a special scan tool (Figure 12–69). After the on-board computer identifies problems, the scan tool displays them as **diagnostic trouble codes (DTC).** OBD II and newer systems contain standardized diagnostic codes (Figure 12–70) that can give detailed descriptions of the faults detected by the engine control computer.

Safety

When repairing or handling computer system components, preventing static electricity is important. As more microprocessors are added to vehicles, the need to inform technicians about how to provide a static-free environment is even more important. To effectively work on electronic circuitry, the best precaution is to use a static grounding strap (Figure 12–71). Always refer to specific manufacturer procedures when handling computer-related components.

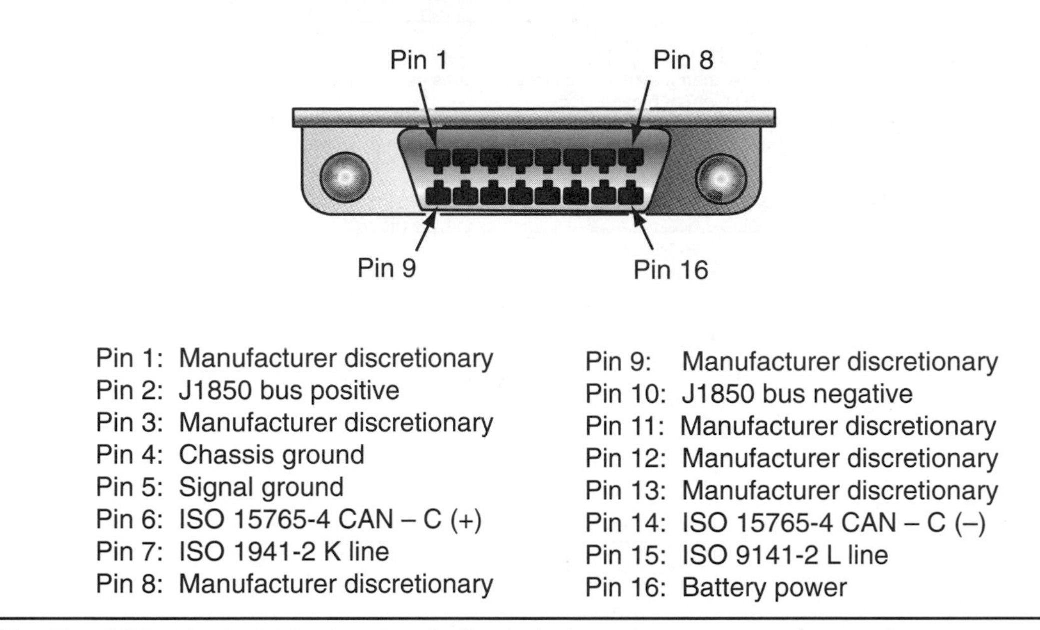

FIGURE 12–68 The OBD 16-pin DLC contains dedicated pin assignments and manufacturer-specific pins.

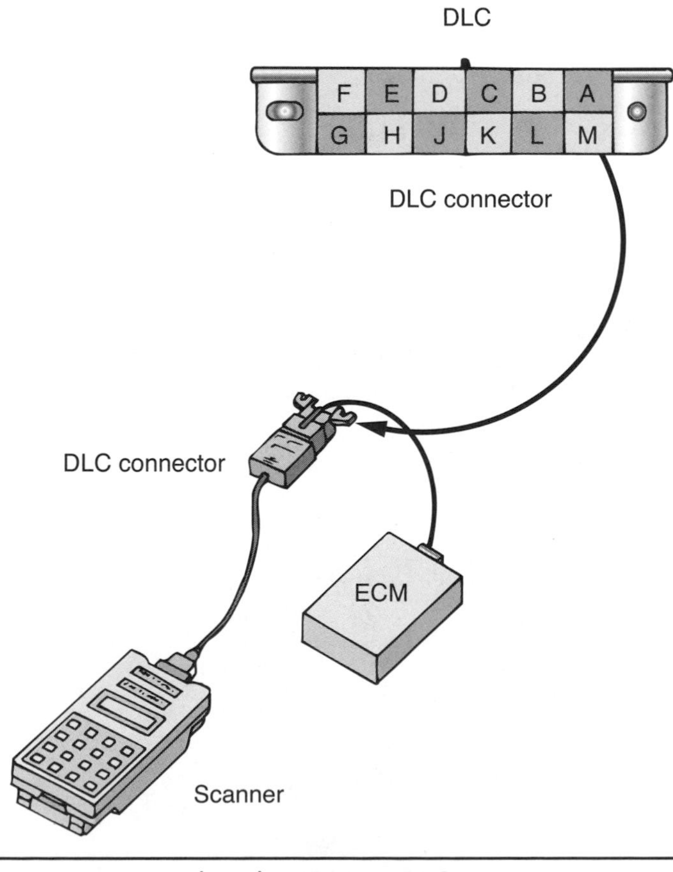

FIGURE 12–69 Handheld scanner connected to the ALDL or DLC.

• DTC No.
Indicates the diagnostic trouble code.

• Detection Item
Indicates the system of the problem or contents of the problem.

DTC chart (SAE controlled)

HINT: Parameters listed in the chart may not be exactly the same as your reading due to the type of instrument or other factors.

If a malfunction code is displayed during the OTC check in check mode, check the circuit for the code listed in the table below. For details of each code, turn to the page referred to under the "See page" for the respective "DTC No." in the DTC chart.

DTC No. (See page)	Detection Item	Trouble Area	MIL	Memory
PO100 (EG-244)	Mass air flow circuit malfunction	• Open or short in mass airflow meter circuit • Mass airflow meter • ECM		●
PO101 (EG-247)	Mass air flow circuit range/performance problem	• Mass airflow meter	●	●
PO110 (EG-248)	Intake air temp. circuit malfunction	• Open or short in intake air temp. sensor circuit • Intake air temp. sensor • ECM	●	●
PO115 (EG-251)	Engine coolant temp. circuit malfunction	• Open or short in engine coolant temp. sensor circuit • Engine coolant temp. sensor • ECM	●	●
		...olant temp. sensor		

• Page or Instructions.
Indicates the page where the inspection procedure for each circuit is to be found, or gives instructions for checking and repairs.

• Trouble Area
Indicates the suspect area of the problem.

FIGURE 12–70 Typical OBD II DTC chart.

(Courtesy of Toyota Motor Sales, U.S.A., Inc.)

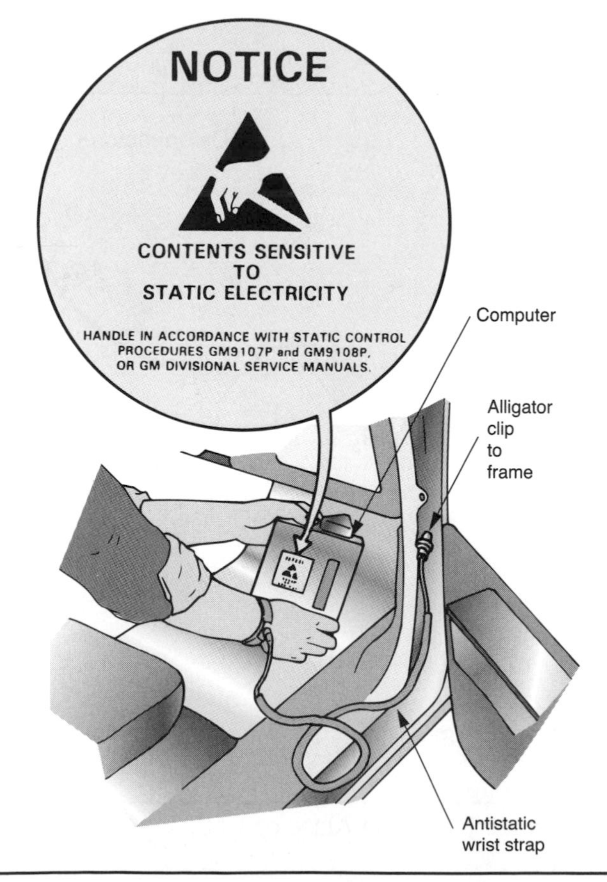

FIGURE 12–71 Typical static grounding strap used to ground the technician and prevent damage to sensitive computer components.

Summary

- A semiconductor is a stable element with exactly four valence electrons.
- The two most common semiconductor materials are silicon and germanium.
- A covalent bond is a bond between atoms that share valence electrons with neighboring atoms.
- An N-type semiconductor has loose or excessive electrons and a negative charge.
- A P-type semiconductor has positive charged material that carries a current.
- A diode is an electrical one-way check valve that allows current to flow in one direction only.
- A PN junction is the area where the anode and cathode material join.
- When a diode has the positive voltage on its anode side, it is forward biased.
- When a diode has the positive voltage on its cathode side, it is reverse biased.
- A clamping diode is used to control voltage surges or spikes in a circuit with a coil.
- An LED (light-emitting diode) is a diode that emits an electrical light energy when it is forward biased.
- A zener diode allows a reverse-biased voltage to flow at a designed voltage level.
- A transistor is an electronic switch that consists of three layers of semiconductor material joined together and designated as the emitter, collector, and base.
- An NPN transistor consists of two N-type materials and one P-type material.
- A PNP transistor consists of two P-type materials and one N-type material.
- An integrated circuit consists of many semiconductors contained on a single circuit.
- An analog voltage signal is usually produced by an input sensor that is continuously variable within a certain range.
- A digital voltage signal is a square wave signal that is either high or low.
- A binary code is the assignment of numeric values to digital signals.
- Each binary code numeric value equals 1 bit.
- Eight bits of information are equal to 1 byte.
- An A/D converter converts an analog signal to a digital signal to be used by a computer.
- A microprocessor is the decision-making chip in the computer that contains a large number of miniature transistor and diode switches.
- RAM (random access memory) is a temporary storage chip for use by the microprocessor to read, write, and erase information.
- A KAM is a keep alive memory that is lost if the source battery power is lost.

- ❑ ROM (read only memory) contains information that is permanently stored for use by the microprocessor.
- ❑ PROM (programmable read only memory) contains specific calibration programs that pertain to a model vehicle.
- ❑ EEPROM (electronic erasable programmable read only memory) can be updated only by the special manufacturer test equipment.
- ❑ A logic gate controls the output voltage signal, depending on the different combinations of input signals.
- ❑ A NOT gate reverses input and output data in a computer circuit.
- ❑ An AND gate has at least two inputs and only one output.
- ❑ An OR gate with a high signal to either of two inputs results in a high output.
- ❑ NAND and NOR gates are equivalent to AND and OR gates with an inverted output signal.
- ❑ An XOR gate is a combination of NAND and NOR gates that produces a high output signal only if the inputs are different.
- ❑ A computer input sensor transmits information to the control computer on various engine-related functions.
- ❑ Computer input sensors can provide either on/off information or variable information.
- ❑ A pull-up switch uses a ground circuit internal to the computer and an external power source.
- ❑ A pull-down switch uses a ground circuit external to the computer and an internal power source.
- ❑ An oxygen sensor reacts with the oxygen content in the exhaust gas to produce a voltage for monitoring by the computer.
- ❑ A stoichiometric mixture is the ideal air/fuel mixture of 14.7:1.
- ❑ An engine coolant temperature sensor is a two-wire thermistor sensor that provides a voltage signal to the computer depending on the coolant temperature.
- ❑ The throttle position sensor consists of a three-wire potentiometer that monitors the angle of the throttle plate.
- ❑ Output sensors are electromechanical motion devices controlled by the computer after receiving information from the input sensors.
- ❑ The on/off cycling of an output actuator is referred to as pulse-width modulation.
- ❑ The idle air control (IAC) valve controls engine idle speed by regulating the air flowing through the idle air passage on the throttle body.
- ❑ Older computer-controlled cars retrieved fault codes stored in the computer memory by means of a blinking light code.
- ❑ OBD II vehicles have a universal 16-pin data link connector (DLC) that a scan tool can hook up to for fault code retrieval.

Key Terms

amplified
analog-to-digital (A/D) converter
analog voltage signal
AND gate
base
binary code
bit
byte
clamping diode
collector
covalent bond
depletion zone
diagnostic trouble code (DTC)
digital voltage signal
diode
engine coolant temperature (ECT)
electronic erasable programmable read only memory (EEPROM)
emitter
field effect transistors (FET)
forward bias
idle air control (IAC)
input sensor
integrated circuit (IC)
keep alive memory (KAM)
light-emitting diode (LED)
logic gate
microprocessor
NAND gate
NOR gate
NOT gate
NPN transistor
N-type
OBD II
OR gate
oxygen sensor (O_2S)
powertrain control module (PCM)
P-type
PN junction
PNP transistor
programmable read only memory (PROM)
pull-down switch
pull-up switch
pulse-width modulation
random access memory (RAM)
reference voltage
reverse bias
read only memory (ROM)
set-point
semiconductor
square wave signal
stoichiometric
throttle position sensor (TP)
transistor
XOR gate
zener diode
zirconium dioxide

Review Questions

Short Answer Essays

1. Describe how to test a regular diode using a DMM equipped with a diode test feature.
2. Describe the differences between an NPN and a PNP transistor.
3. What is a stoichiometric fuel/air mixture?
4. Describe how the illumination level of digital dash lights is controlled by the rapid on/off cycling of the circuit.
5. Describe how to retrieve a stored computer fault code from a vehicle.

Fill in the Blanks

1. The AND gate is a logic gate that has ____________________ input and ____________________ output.
2. An engine coolant temperature sensor provides a ____________________ voltage signal when cold.
3. A throttle position sensor voltage is ____________________ at wide open throttle.
4. A transistor consists of three layers of semiconductor material joined together, called ____________________, ____________________, and ____________________.
5. The ____________________ is the area where the anode and cathode material join.

ASE-Review Questions

1. Technician A says the diode acts like a one-way check valve in an electrical circuit. Technician B says a diode allows current to flow in one direction only. Who is correct?
   A. A only
   B. B only
   C. Both A and B
   D. Neither A nor B

2. Technician A says a diode is forward biased if the positive voltage is connected to the cathode side of the diode and negative voltage is connected to the anode side. Technician B says a diode is reverse biased if the positive voltage is connected to the anode side of the diode and negative voltage is connected to the cathode side. Who is correct?
   A. A only
   B. B only
   C. Both A and B
   D. Neither A nor B

3. Voltage signals are being discussed. Technician A says a digital voltage signal is either high or low. Technician B says a digital voltage is variable within a certain range. Who is correct?
   A. A only
   B. B only
   C. Both A and B
   D. Neither A nor B

4. Technician A says a zener diode allows current to flow in one direction only. Technician B says a zener diode can be used to regulate circuit voltage. Who is correct?

   A. A only

   B. B only

   C. Both A and B

   D. Neither A nor B

5. Two technicians are discussing the operation of NPN transistors. Technician A says the transistor is on when a positive voltage is applied to the base. Technician B says when the transistor is on, the current flows from the collector to the emitter. Who is correct?

   A. A only

   B. B only

   C. Both A and B

   D. Neither A nor B

6. Technician A says a computer pull-up switch uses an external power source to the switch and a ground circuit that is internal to the control computer. Technician B says that, when a computer pull-up switch is closed, a high reference signal is sent to the computer. Who is correct?

   A. A only

   B. B only

   C. Both A and B

   D. Neither A nor B

7. Technician A says a PNP transistor is on when voltage is applied to the base. Technician B says current flows from the emitter to the collector when the transistor is on. Who is correct?

   A. A only

   B. B only

   C. Both A and B

   D. Neither A nor B

8. Technician A says a computer pull-down source uses an external power switch and an external ground circuit. Technician B says that, when a computer pull-down switch opens, the reference signal is low. Who is correct?

   A. A only

   B. B only

   C. Both A and B

   D. Neither A nor B

9. Technician A says RAM is a temporary storage area for information used by the microprocessor. Technician B says ROM is permanently stored information used by the microprocessor. Who is correct?
   A. A only
   B. B only
   C. Both A and B
   D. Neither A nor B

10. Two technicians are discussing logic gates in computer circuits. Technician A says an AND gate's output is high only if both inputs are high. Technician B says a NOT gate reverses the input and output signals. Who is correct?
    A. A only
    B. B only
    C. Both A and B
    D. Neither A nor B

11. Technician A says an oxygen sensor measures the amount of fuel the engine requires depending on the vehicle load. Technician B says the O_2 sensor produces an output voltage between 0.1 and 0.9 volt that is sent to the ECM. Who is correct?
    A. A only
    B. B only
    C. Both A and B
    D. Neither A nor B

12. Two technicians are discussing the basic operation of the engine coolant temperature sensor. Technician A says the ECT is a two-wire thermistor that provides the ECM with the engine coolant temperature information. Technician B says the ECT is a key input the ECM needs to calculate various output operating functions.
    A. A only
    B. B only
    C. Both A and B
    D. Neither A nor B

13. The IAC valve is used in many engine computer systems to control engine idle speed. Technician A says this is accomplished by regulating the air flowing through the idle air passage on the throttle body. Technician B says the IAC is a high-speed motor with a single field coil. Who is correct?
    A. A only
    B. B only
    C. Both A and B
    D. Neither A nor B

14. Technician A says an OBD II computer-controlled vehicle uses a standard 12-pin DLC connector. Technician B says the OBD II DLC connector is different for each vehicle manufacturer. Who is correct?

    A. A only
    B. B only
    C. Both A and B
    D. Neither A nor B

15. Technician A says the TP sensor consists of a three-wire potentiometer connected to the throttle assembly. Technician B says the TP sensor voltage changes gradually from idle to WOT. Who is correct?

    A. A only
    B. B only
    C. Both A and B
    D. Neither A nor B

13 Accessory Circuits and Basic Instrumentation

Introduction

Several accessory circuits are common in vehicles. Some of them are necessary safety accessories, like the horn, windshield wipers and washers, and the supplemental inflatable restraint (SIR) system. Others are accessory circuits that provide additional safety and comfort, like the climate control system, rear window defogger, power windows, power seats, power door locks, and power mirrors. Each of these circuits has design features and components that are common in most vehicles, helping the technician to diagnose circuits that do not work.

Objectives

When you complete this chapter you should be able to:

- ❑ Explain the operation and basis diagnosis of various horn circuits.
- ❑ Explain the operation and basic diagnosis of the blower fan control circuit.
- ❑ Explain the operation and basic diagnosis of wiper and washer circuits.
- ❑ Explain the principles of operation and basic diagnosis of power windows, power door locks, power seat circuits, power mirrors, and electrochromic mirrors.
- ❑ Explain the operation and basic troubleshooting procedures for window defogger circuits.
- ❑ Explain the operation and basic diagnosis of conventional instrument gauges, warning lights, and sending units.
- ❑ Explain the basic operation of supplemental inflatable restraint (SIR) components.

Accessory Circuits

Horn Circuits

Two types of horn circuits are typically used in vehicles.

Horn Circuit with No Relay The first type does *not* use a control relay to complete the circuit (Figure 13–1). The current to operate the horn(s) *must* go through the horn switch to complete the circuit from the battery to the horn(s). With this design, the horn switch will fail prematurely if the current to the horn(s) is beyond designed limits.

Horn Circuit with Relay The second horn circuit design—and the most common—uses a horn control relay to complete the circuit (Figure 13–2). The relay

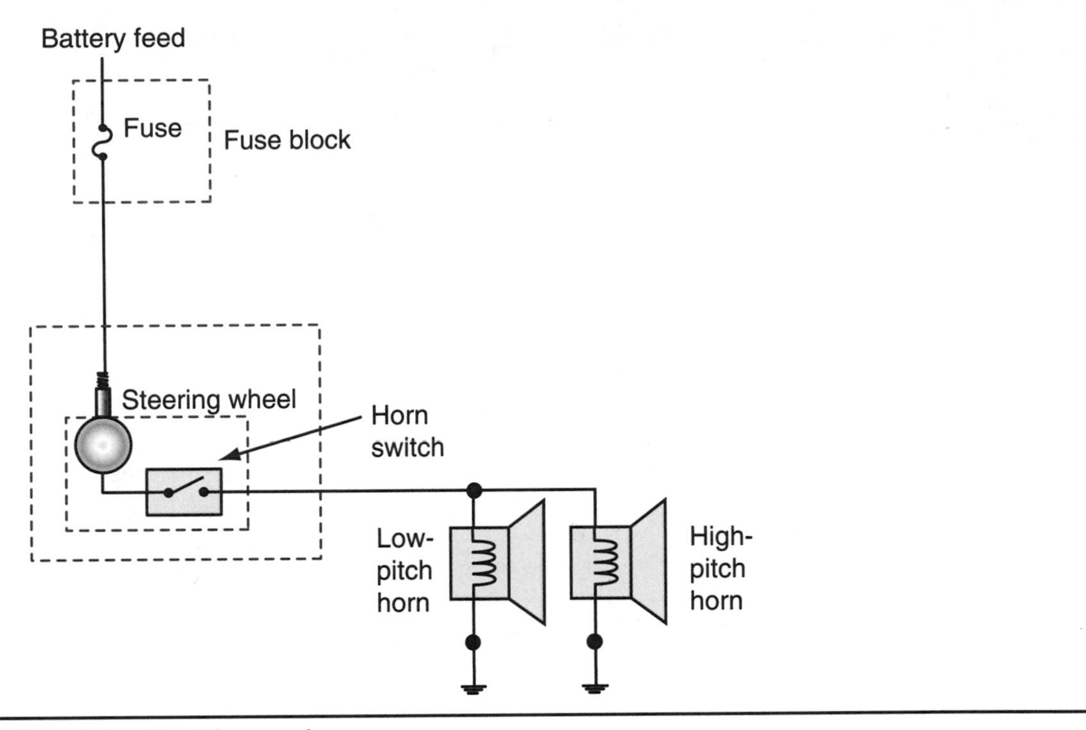

FIGURE 13–1 Simple two-horn circuit without relay.

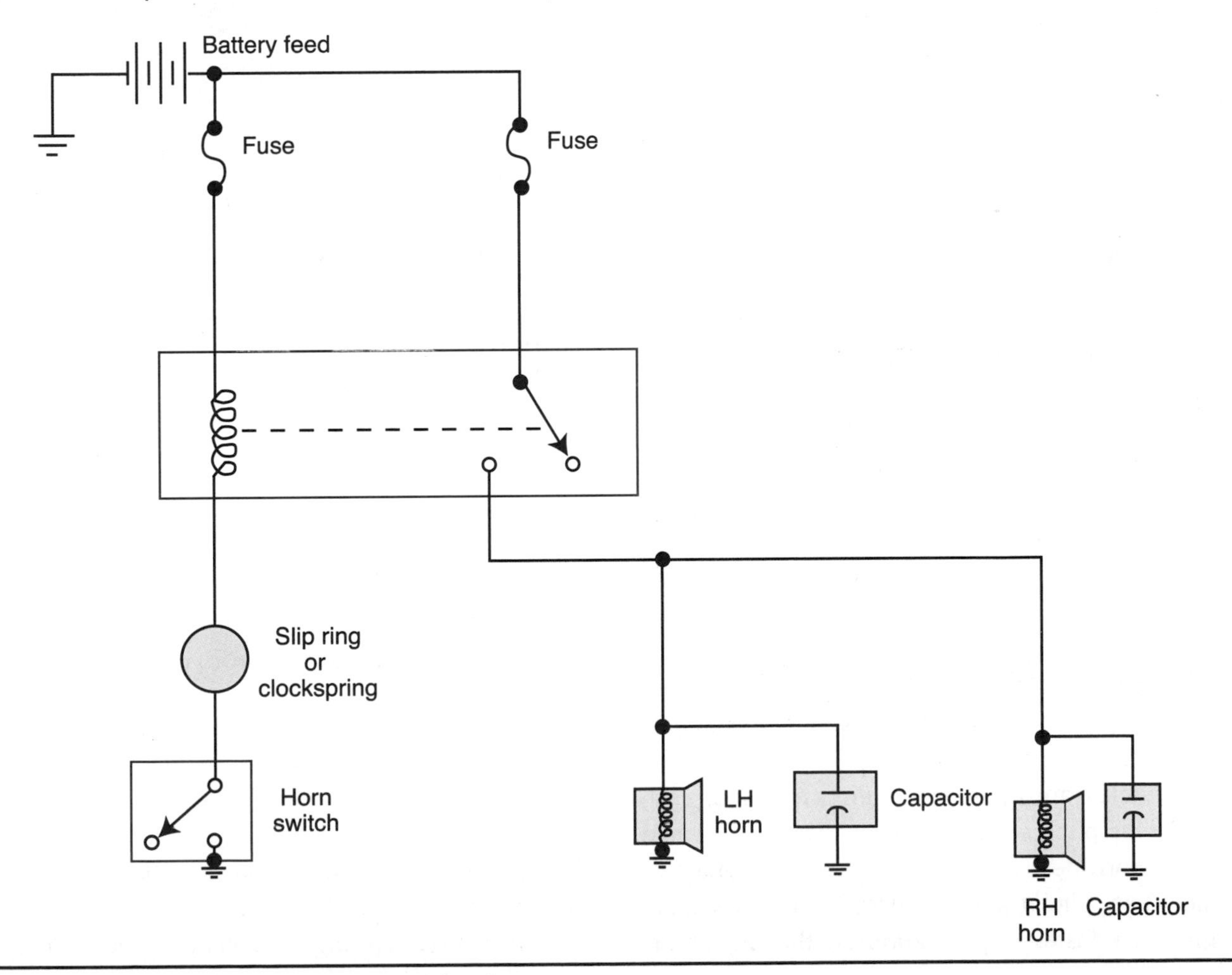

FIGURE 13–2 Typical two-horn circuit with relay.

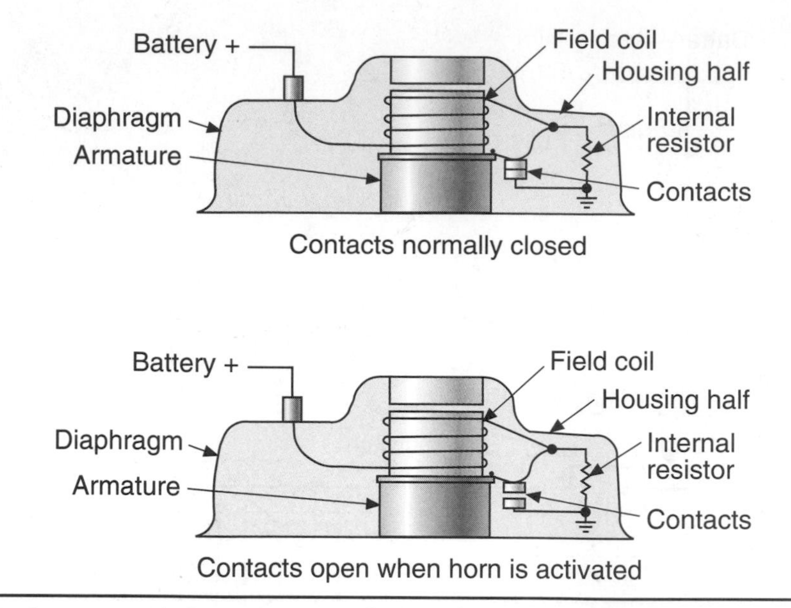

FIGURE 13–3 Typical electromagnetic horn construction and operation.

electrically isolates the horn switch from the horns and allows a higher amperage to go to the horn assemblies. The horn relay contacts will energize the horns when the horn switch completes the horn relay coil circuit.

Horn Assemblies Electric horns, operating on an electromagnetic principle, vibrate a steel diaphragm to produce an audible noise. Figure 13–3 is a typical horn with an electromagnet, flexible diaphragm, movable armature, field coil, and a set of normally closed contact points. The internal resistor allows a weak magnetic field to remain after the contact points open, thus reducing the time required to rebuild the magnetic field after the points close again.

To produce an audible noise, current flows through the *closed* contact points to the field coil, producing a magnetic field, which attracts the armature that is attached to the diaphragm. Movement of the armature will *open* the contact points and return the diaphragm to its original rest position. This event occurs several times per second, causing the diaphragm to vibrate. As the diaphragm vibrates, a column of air in the horn also vibrates to produce the audible noise.

Most vehicles are equipped with two horns, each with a different pitch from the other. The design and shape of the horn determine the frequency and tone of the sound. The pitch is controlled by the vibrating diaphragm. The faster the vibration is, the higher the pitch will be. Although by design a horn's pitch can be adjusted by turning the spring tension screw on the horn case (Figure 13–4), in reality the older the horn is, the greater the chance that this screw will break off due to rust or seizure of the adjusting screw. As a result, most horns are replaced when the horn changes pitch.

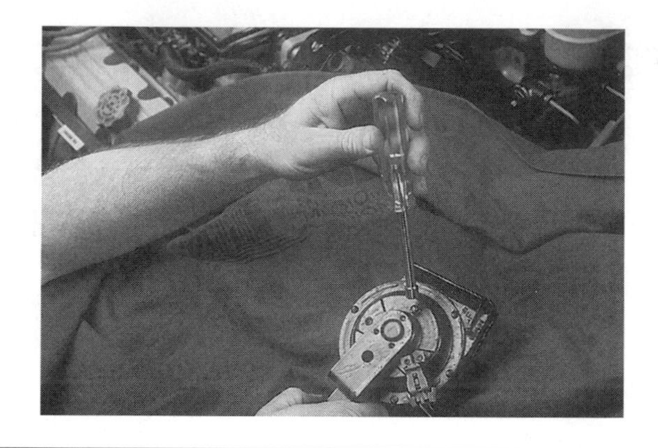

FIGURE 13–4 Changing the horn's pitch with an adjusting screw.

Horn Switch The horn switch in most vehicles is installed in the steering wheel (Figure 13–5), where a slip ring and spring-loaded sliding contact provide electrical continuity for the horn switch in all steering wheel positions. The horn switch may also be integrated as part of the air bag assembly (Figure 13–6). In this configuration, a **clock spring** transfers the electrical signals to the air bag assembly and to any other steering wheel mounted switches, such as the horn, radio controls, and cruise control.

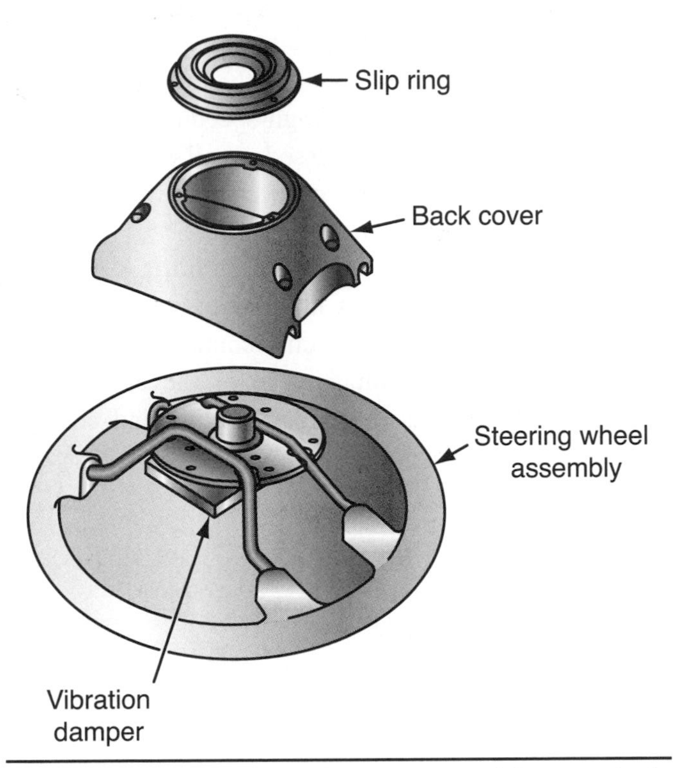

FIGURE 13–5 Location of slip ring used for a typical horn switch.

FIGURE 13–7 Typical air bag safety precaution symbol.

CAUTION *Remember to follow all manufacturer precautions and safe disabling techniques when working on vehicles with air bags or electronic controls in the steering wheel.*

Several vehicle manufacturers use a precaution symbol (Figure 13–7) in service publications to alert the service technician to any air bag safety precautions or steps to follow.

Blower Control Circuits

The blower control circuit is a good example of a basic series circuit that consists of a blower switch, stepped

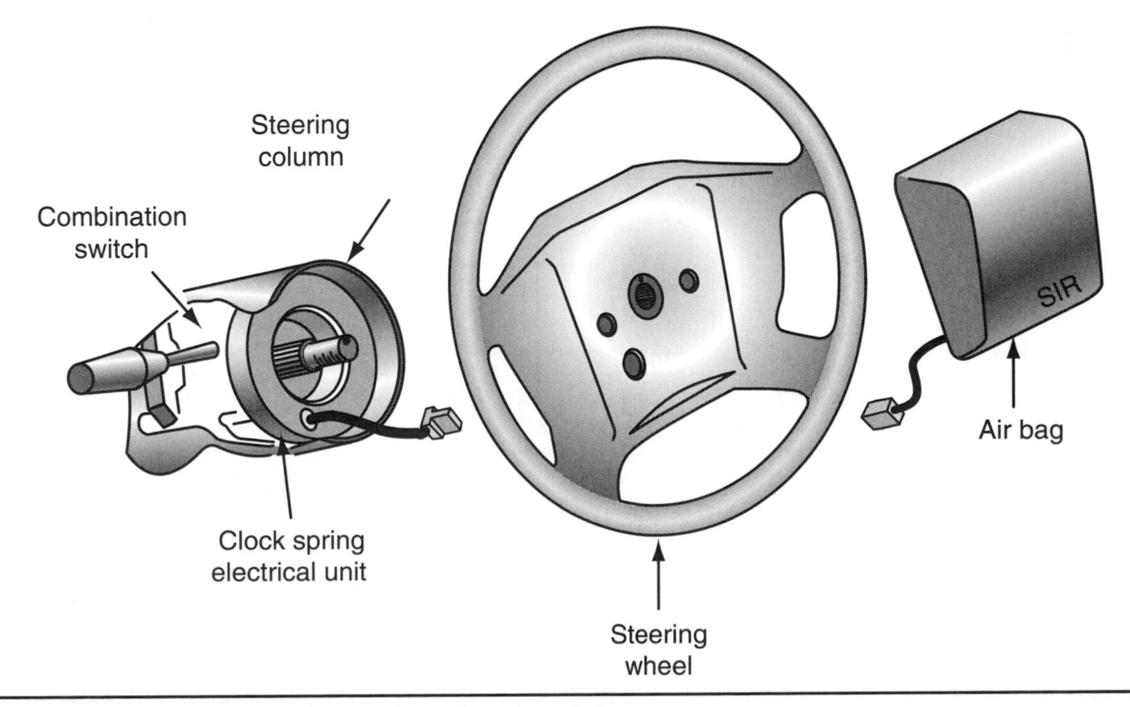

FIGURE 13–6 Typical horn switch assembly within the air bag system.

resistor block, blower motor, and in some circuits a high-speed relay in a series line. More expensive vehicles with automatic temperature controls often use solid-state electronics to control blower speed.

Blower Circuits Without Relay Control Two types of blower circuits do *not* use a relay for the blower's high-speed setting. Figure 13–8 is an example of a four-speed blower motor circuit with the blower switch and the resistor block located before the blower motor and the ground side of the circuit after the motor. In this circuit, when the air conditioner/heater switch is turned to the AC, Heat, or Def (defroster) settings, battery voltage is available to the right side of the resistor block when set for low-speed blower operation. The battery voltage goes through the three resistors in *series,* causing a voltage drop to occur before reaching the blower motor. (Remember Ohm's law states that as resistance goes up, amperage flow goes down.) When the blower switch is turned to the M1 speed position, resistor 3 is bypassed. This lowers the series resistance, allowing more current to the blower motor. As the blower switch is moved to the other positions, more resistors are bypassed. In the high-speed position *all* the resistors in the block are bypassed, so all the circuit current goes through the high-speed contact in the blower switch to the blower motor.

Figure 13–9 is another blower circuit that does *not* use a relay for the blower's high-speed setting. Unlike the previous circuit, this circuit connects the blower motor to the battery voltage side of the circuit. The resistor block controls the circuit ground and changes the circuit's resistance the same way as in the previous circuit.

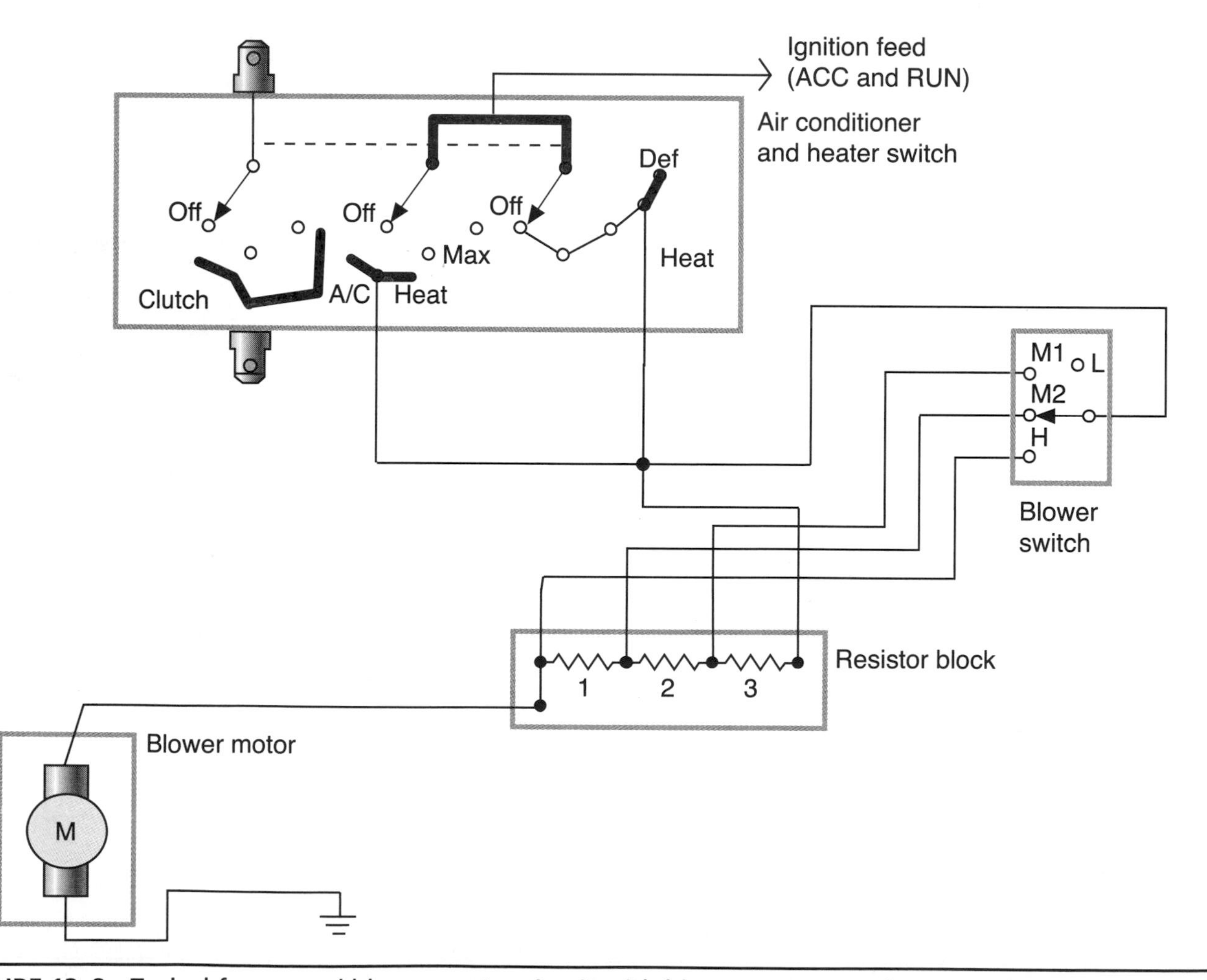

FIGURE 13–8 Typical four speed blower motor circuit with blower motor on the groundside.

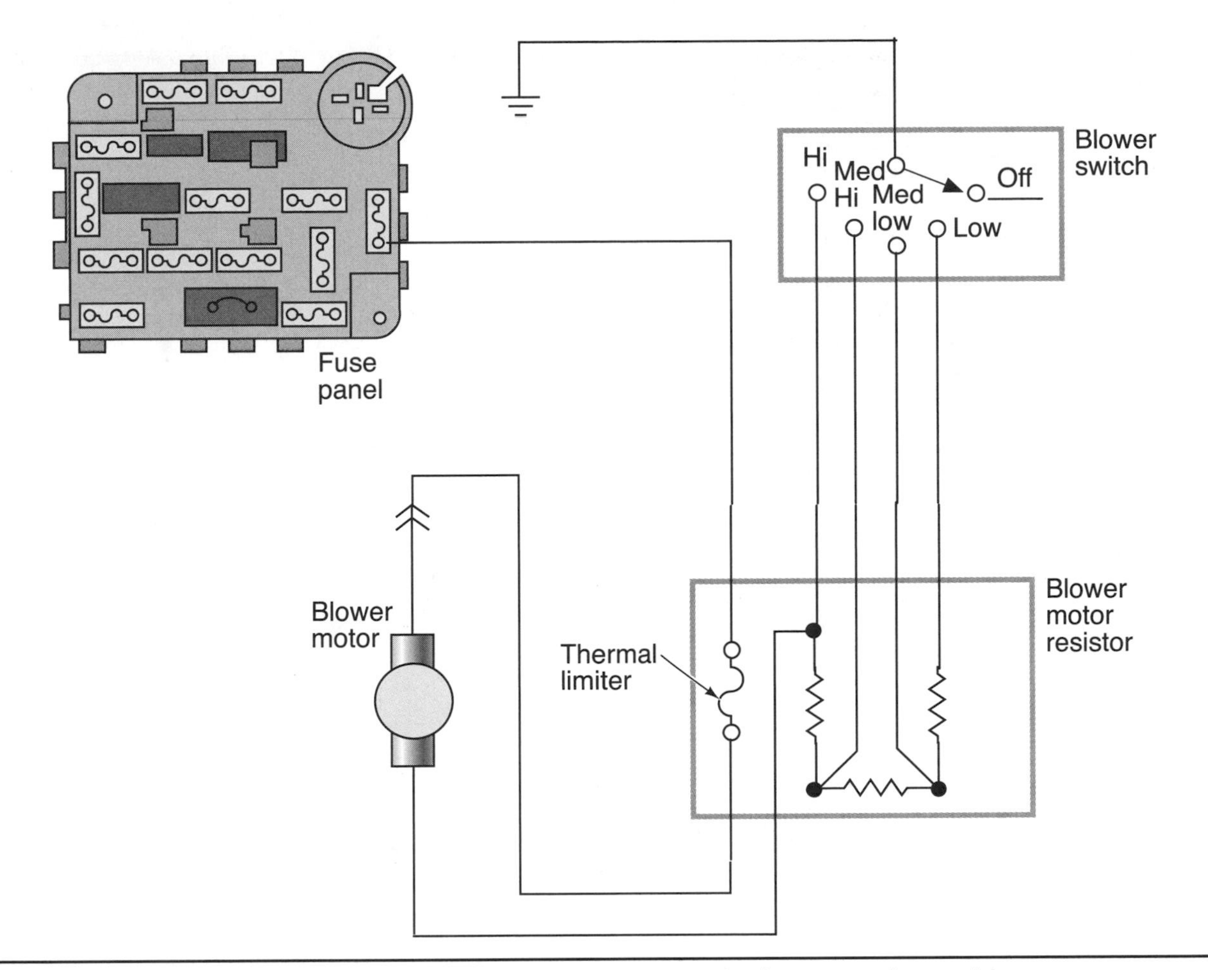

FIGURE 13–9 Typical blower motor circuit with blower motor on the battery voltage side.

Blower Circuits with Relay Control Because of increased demands on the blower motor in some vehicles, the high-speed operation of the motor is controlled through a relay assembly (Figure 13–10). Using the relay helps extend the life of the blower switch by enabling the higher current flow needed for the high-speed position to come from a separate circuit. Each of the blower Lo, M1, and M2 switch positions controls the amount of resistance total in the resistor block, as in the previous two circuits.

Resistor Block Testing Two methods are used to determine whether a blower resistor block is defective. The first method is to check for voltage drops across each resistor with a voltmeter with the blower switch in the low position. (Figure 13–11). Because the resistors are in series, if one resistor is burned out, the circuit shows source voltage up to the open in the circuit.

The second method to test a blower resistor block is to remove the assembly from the air plenum housing and, after a visual check (Figure 13–12), check each resistor with an ohmmeter. A reading of OL indicates an open circuit.

Electronic Blower Speed Controls Vehicles with electronic **climate control panel (CCP)** systems can be manually controlled by the driver, or, if the Auto function is selected, they automatically provide the desired cabin temperature and other select operating modes, such as A/C, heat, and vent (Figure 13–13).

The heater blower speed in this system is controlled by a power module (Figure 13–14) that amplifies computer command signals from the body control module (BCM). When a temperature selection is set by the driver through the CCP, the BCM monitors temperature inputs from the passenger compartment, ambient air,

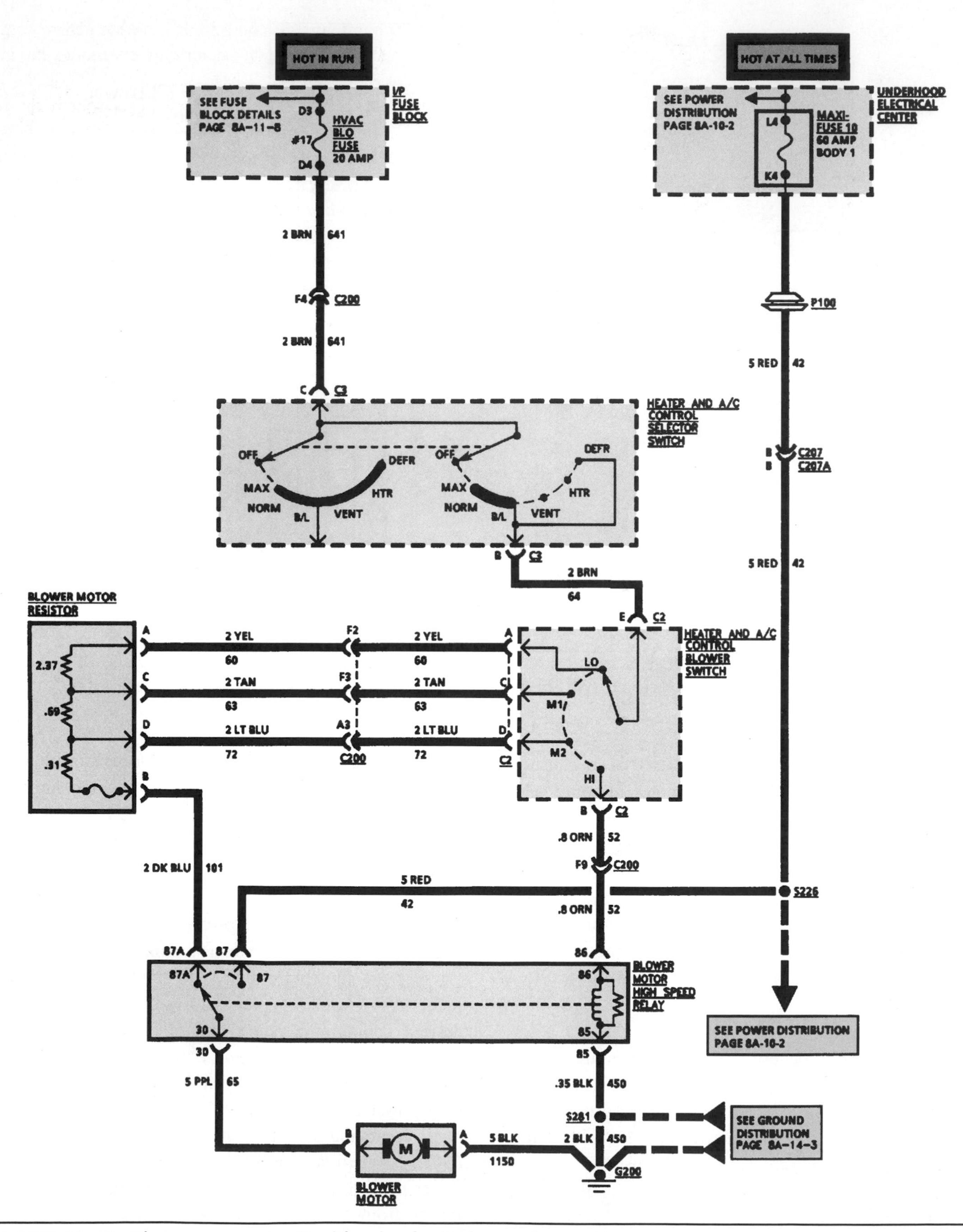

FIGURE 13–10 Blower motor circuit with relay for high-speed operation.

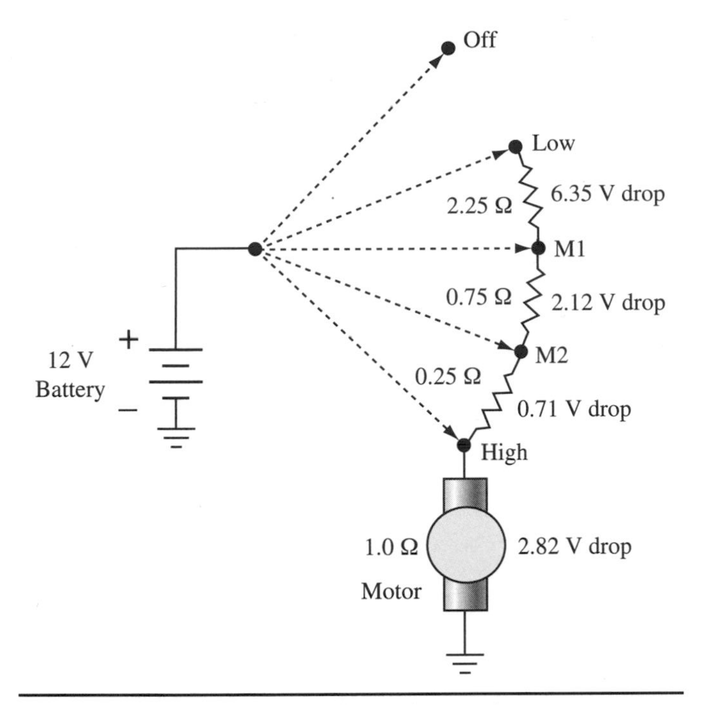

FIGURE 13–11 Checking for voltage drops across a blower resistor block.

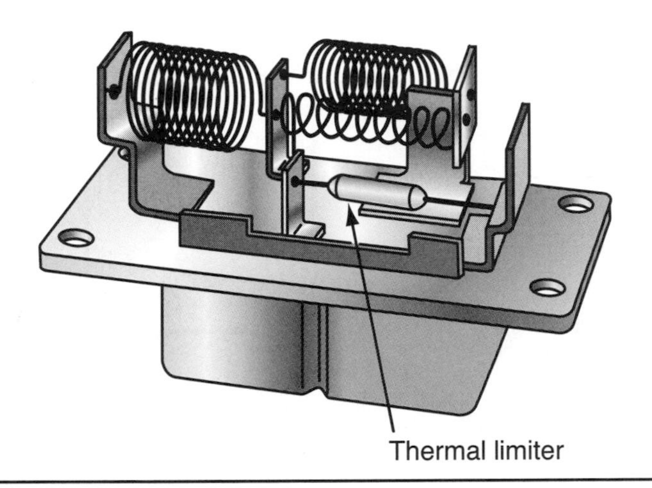

FIGURE 13–12 Blower motor resistor block.

engine coolant, and air conditioning system before commanding the fan blower speed, air delivery mode, and air blend door position.

In the electronic climate controlled schematic in Figure 13–15, the CCP is used to set the desired temperature, and the signal is sent to the BCM through data circuit line V. The BCM interprets the data and then transmits the information to the power module through circuit wire A9. The power module amplifies the voltage signal and controls the blower voltage through the blower motor driver circuit A.

Wiper and Washer Circuits

Several types of front and rear window wiper and washer systems are used by vehicle manufacturers today. Almost all have main components consisting of a wiper motor with park switch, speed control switch, and washer pump unit. Typical wiper systems can be divided into three design types: nondepressed park wiper circuit, depressed park wiper circuit, and interval wiper system.

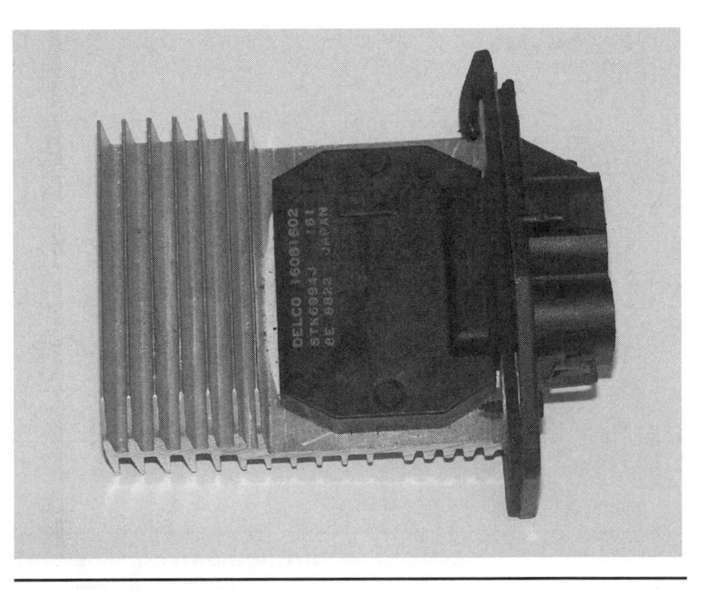

FIGURE 13–14 Blower circuit power module.

FIGURE 13–13 Typical electronic climate control panel (CCP).

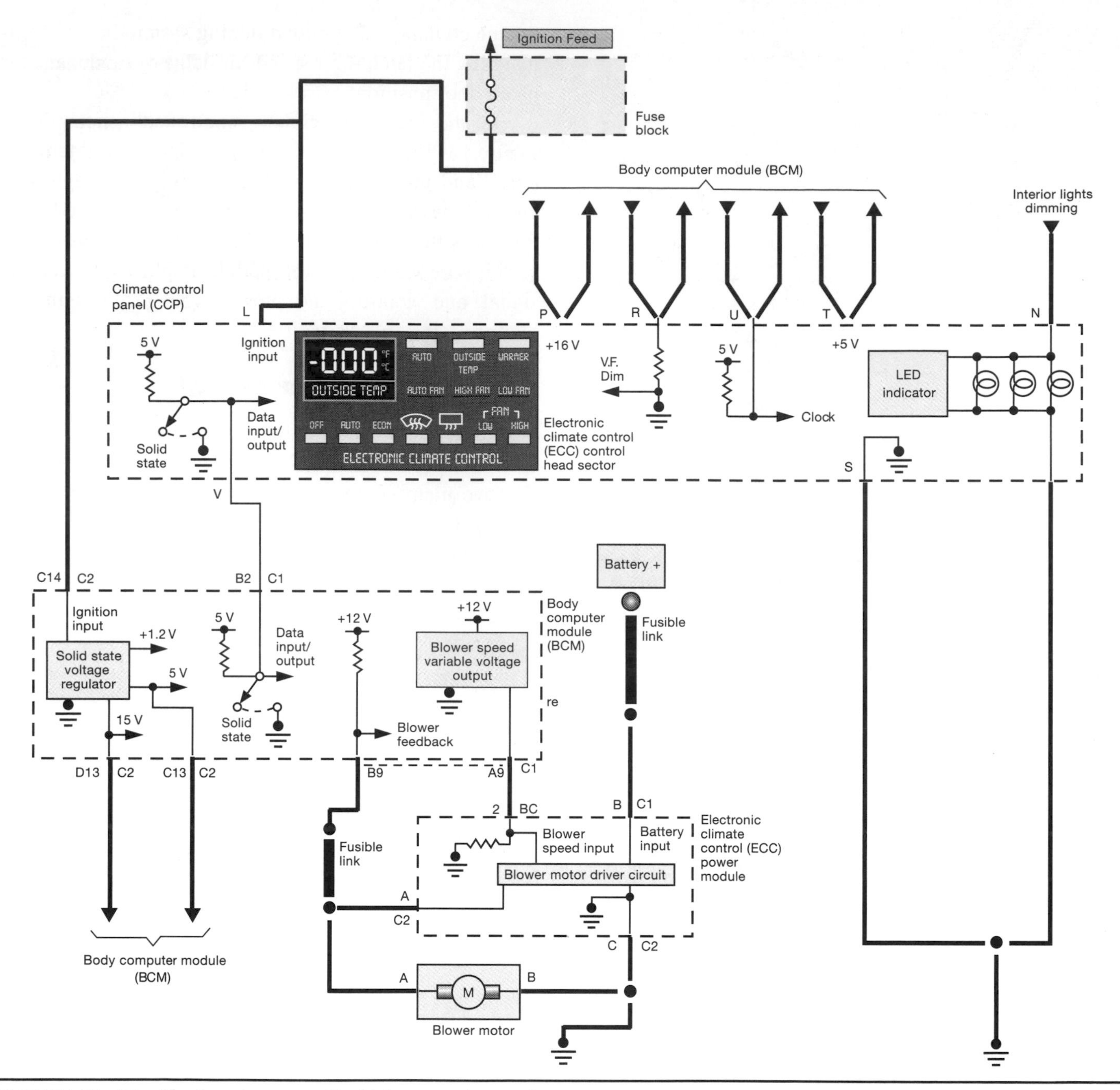

FIGURE 13–15 Electronic controlled blower schematic.

Nondepressed Park Wiper Circuit The first design, a **nondepressed park** wiper circuit, parks the wiper blades at the end of their normal stroke at the edge of the lower windshield molding. In a typical two-speed nondepressed wiper circuit (Figure 13–16), ignition feed voltage is available through an internal circuit breaker to the wiper control switch, wiper motor park switch, and the washer switch. Inside the two-speed wiper motor are three armature brushes (Figure 13–17) used for speed control. The low-speed brush and common brush oppose each other, and the high-speed brush is centered between them or is offset. When the wiper switch is moved to the low position (Figure 13–18), current moves through the wiper switch to the low-speed brush and then to ground. As the wiper switch is moved to the high position (Figure 13–19), the current is redirected to the high-speed

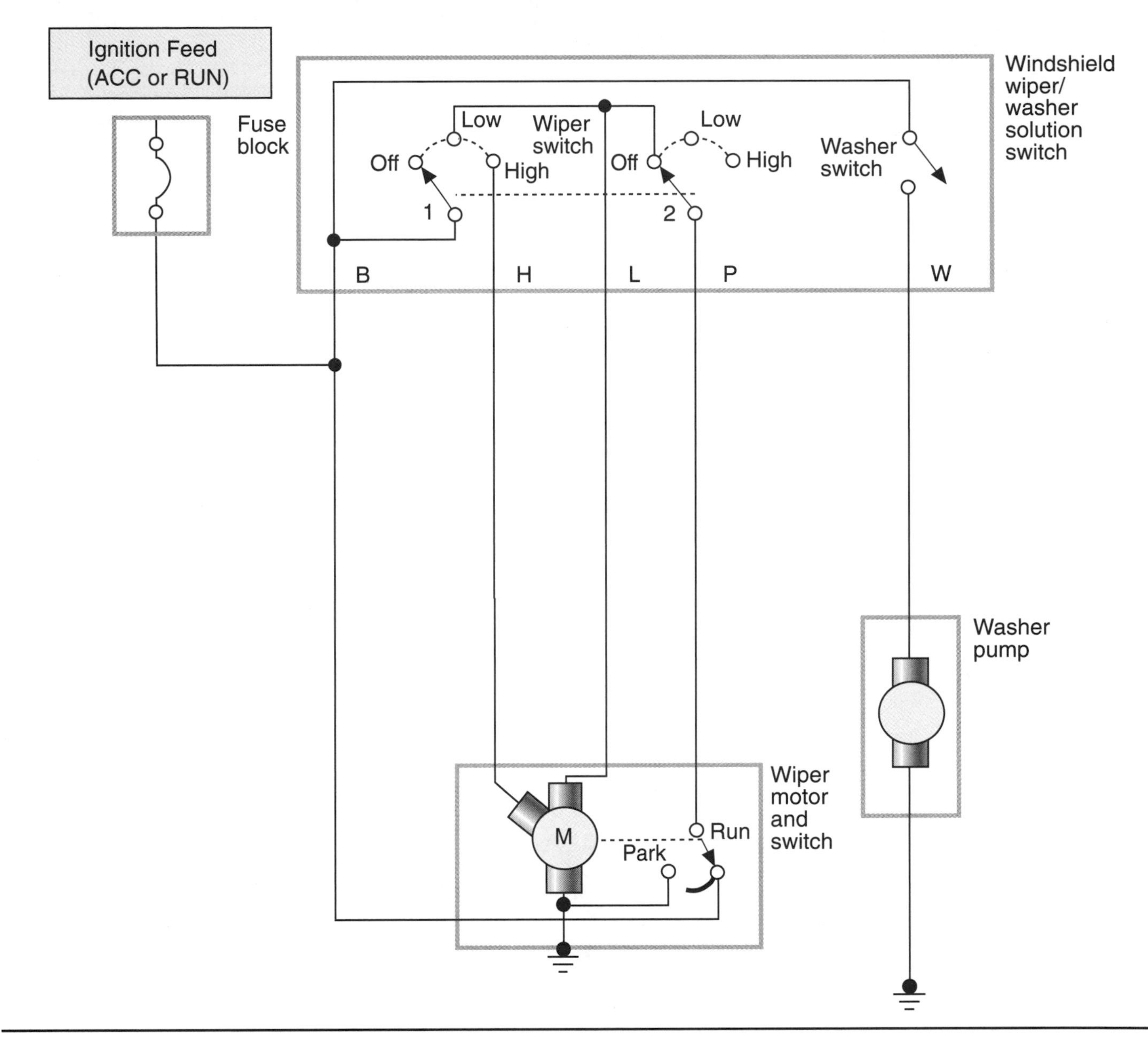

FIGURE 13–16 Typical nondepressed two-speed park wiper circuit.

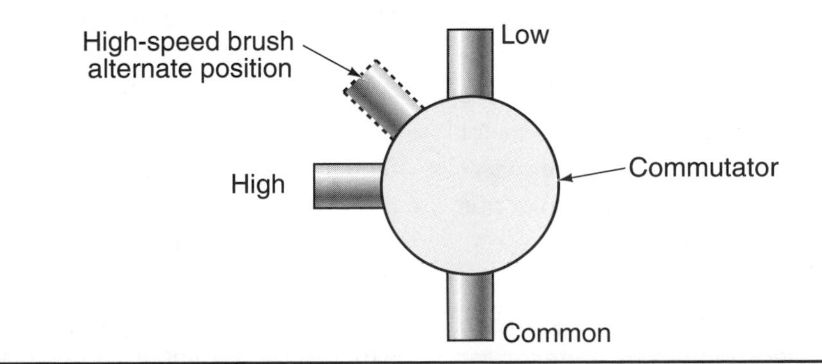

FIGURE 13–17 Typical armature brush arrangement for a two-speed wiper motor.

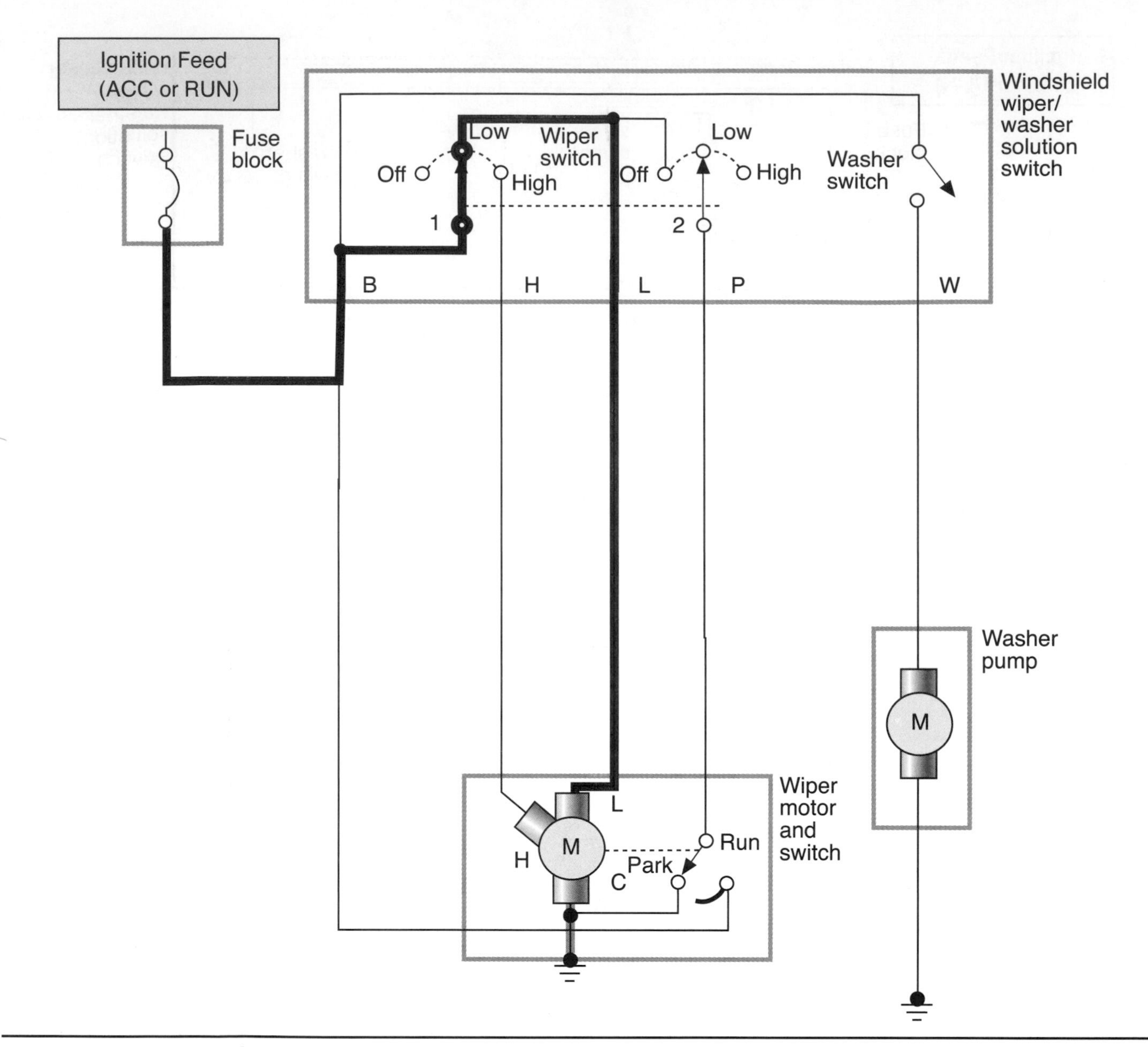

FIGURE 13–18 Flow of current through circuit in low position.

brush and then to ground. When the wiper switch is turned to the off position (Figure 13–20), battery voltage is still available to the wiper motor armature through the integral park contact switch located in the wiper motor assembly. When the contact points open, the wiper blades should be at the end of their normal stroke (the edge of the lower windshield molding).

CAUTION *Wiper arm levers are very strong when operating. Keep fingers clear when testing the wiper circuit.*

Depressed Park Wiper Circuit The second wiper circuit design is the **depressed park** wiper circuit (Figure 13–21). In vehicles where styling is important, the circuit parks the wiper blades *off* the glass and past the end of their normal stroke—usually below the edge of the lower windshield molding. In this circuit, a second set of contact points are used along with the park switch. When the wiper switch is turned to the off position, the additional set of park contacts reverse the rotation of the wiper motor for about 15 degrees after the wiper blades have reached the lower edge of the windshield molding.

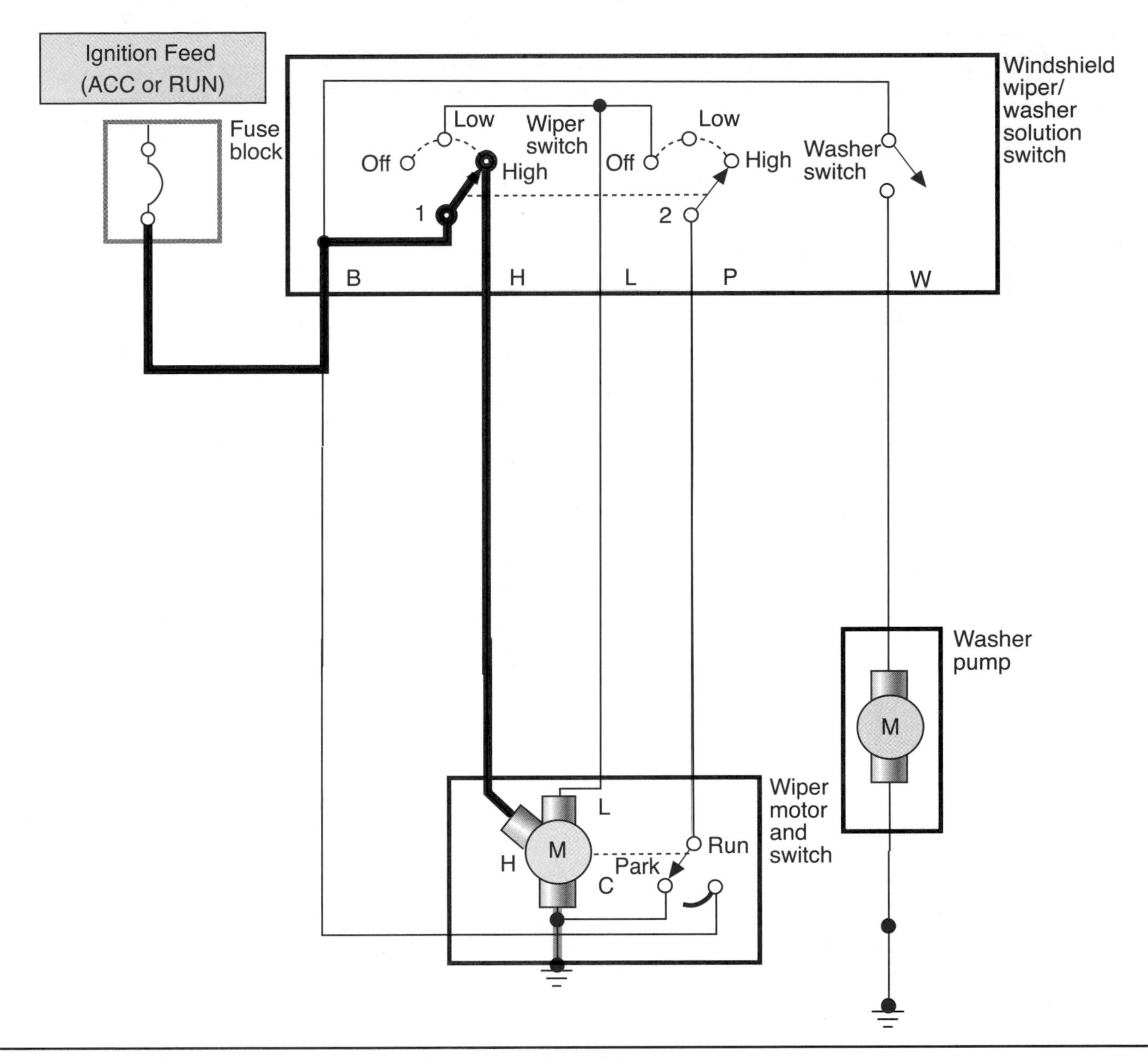

FIGURE 13–19 Flow of current through circuit in high position.

Current flows through the internal circuit breaker, through the park switch contact A, and to wiper switch contact 1. Current then flows through the low-speed brush to the motor armature. The ground path is directed to the common brush through wiper switch contact 3, to park switch contact B, and then to ground. Park switch A opens when the wipers reach the depressed park position.

Faulty park switches with contacts that are *open* all the time cause the wipers to stop whenever the wiper switch is in the off position. To get the wipers to the desired park position, the wiper switch must be turned on and off several times. A faulty park switch with contacts stuck or welded closed causes the motor to operate continually, until the wiper motor is disconnected or the ignition switch is turned off.

Interval Wiper System The third design of a wiper circuit is an **interval** wiper system, also known as an intermittent wiper system. The wipers are operated in time intervals depending on the delay setting requested by the driver. Figure 13–22 is an illustration of an interval

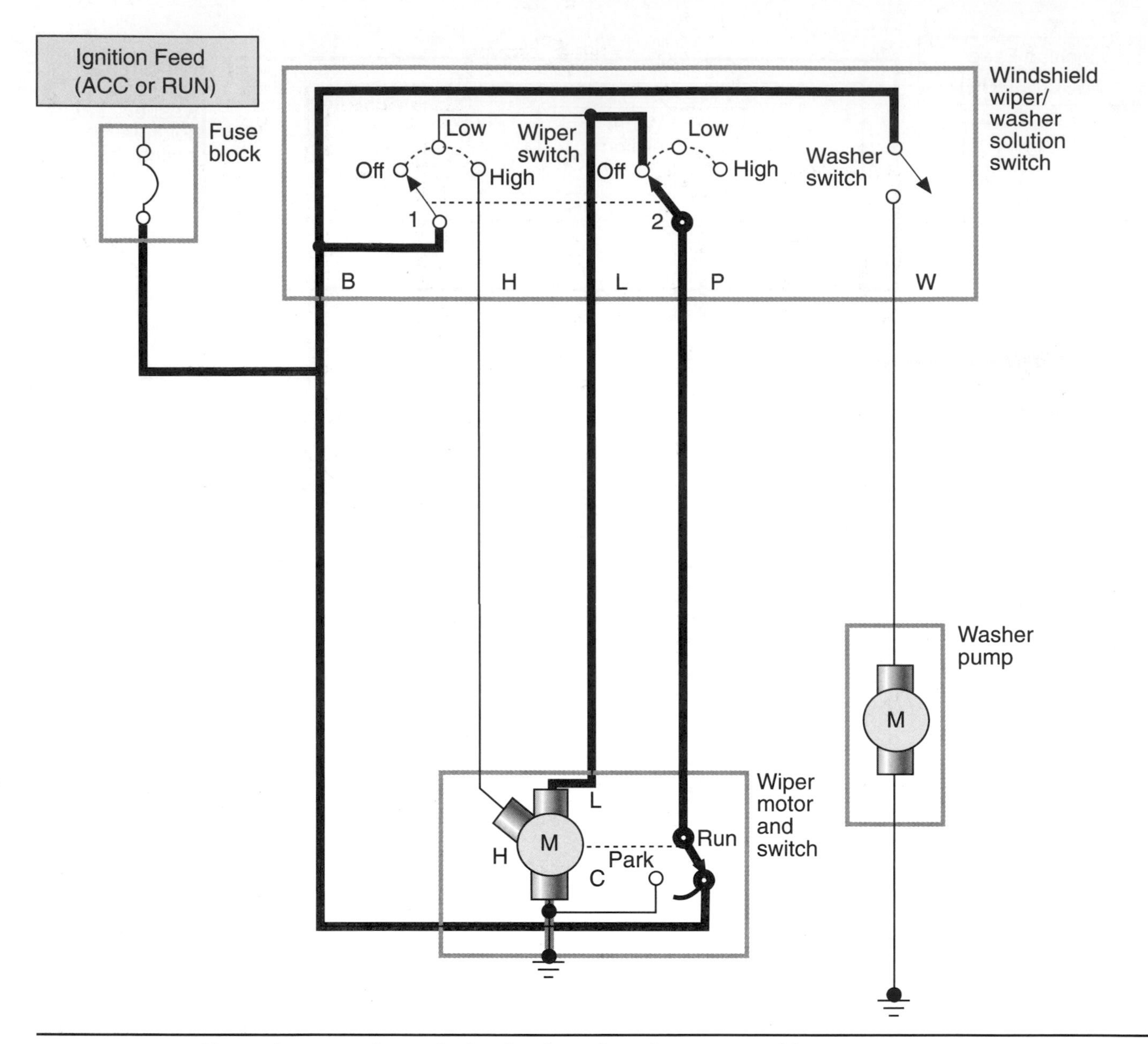

FIGURE 13–20 Flow of current through circuit when the wipers are parking.

wiper system activated in the intermittent wiper mode. The internal timer in the wiper module triggers the electronic switch to activate the governor relay contacts, which complete the circuit to the low-speed brush. The wiper interval is determined by the resistance setting of the interval control knob. The higher the resistance setting is, the longer the delay time is that is needed to saturate the capacitor triggering the electronic switch.

Washer Pump System The typical washer pump system works in conjunction with the wiper motor by automatically activating the wiper motor's low-speed setting (Figure 13–23). The washer switch is usually a spring-loaded, normally open switch that applies battery voltage to the washer pump when pushed closed. The washer motor is normally located in the fluid reservoir under the hood (Figure 13–24).

Window Defogger Circuits

The use of electric window defogger/defroster systems in vehicles has increased over the past several years. The

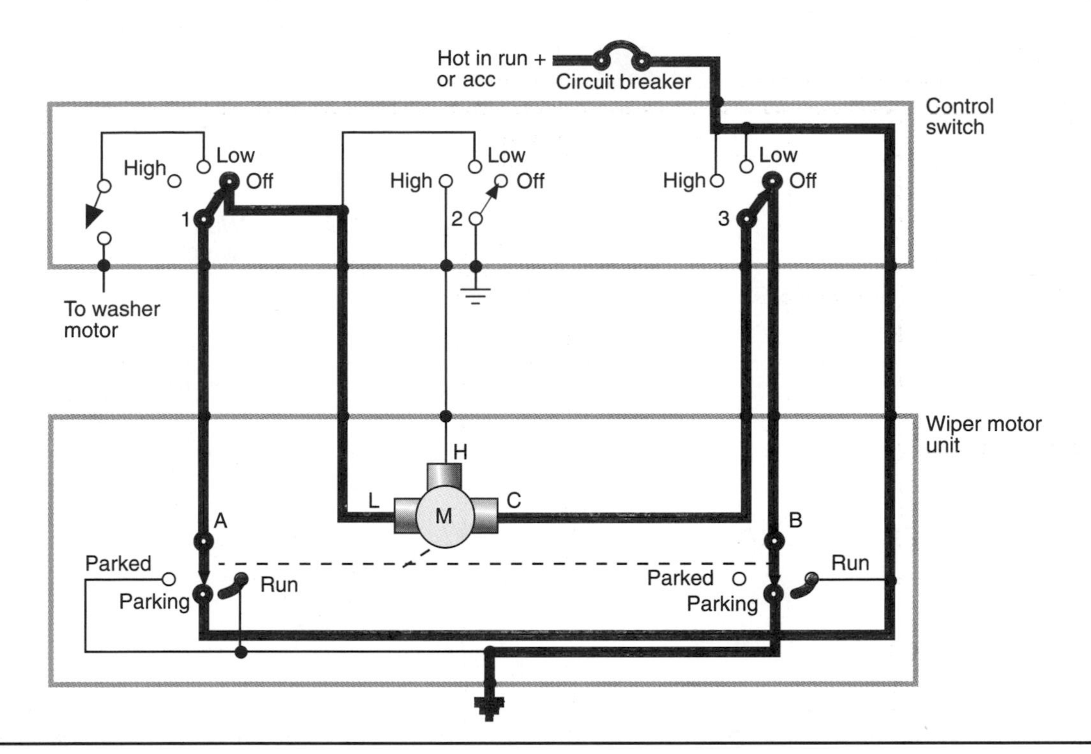

FIGURE 13–21 Typical depressed park wiper system with a second set of park contact points.

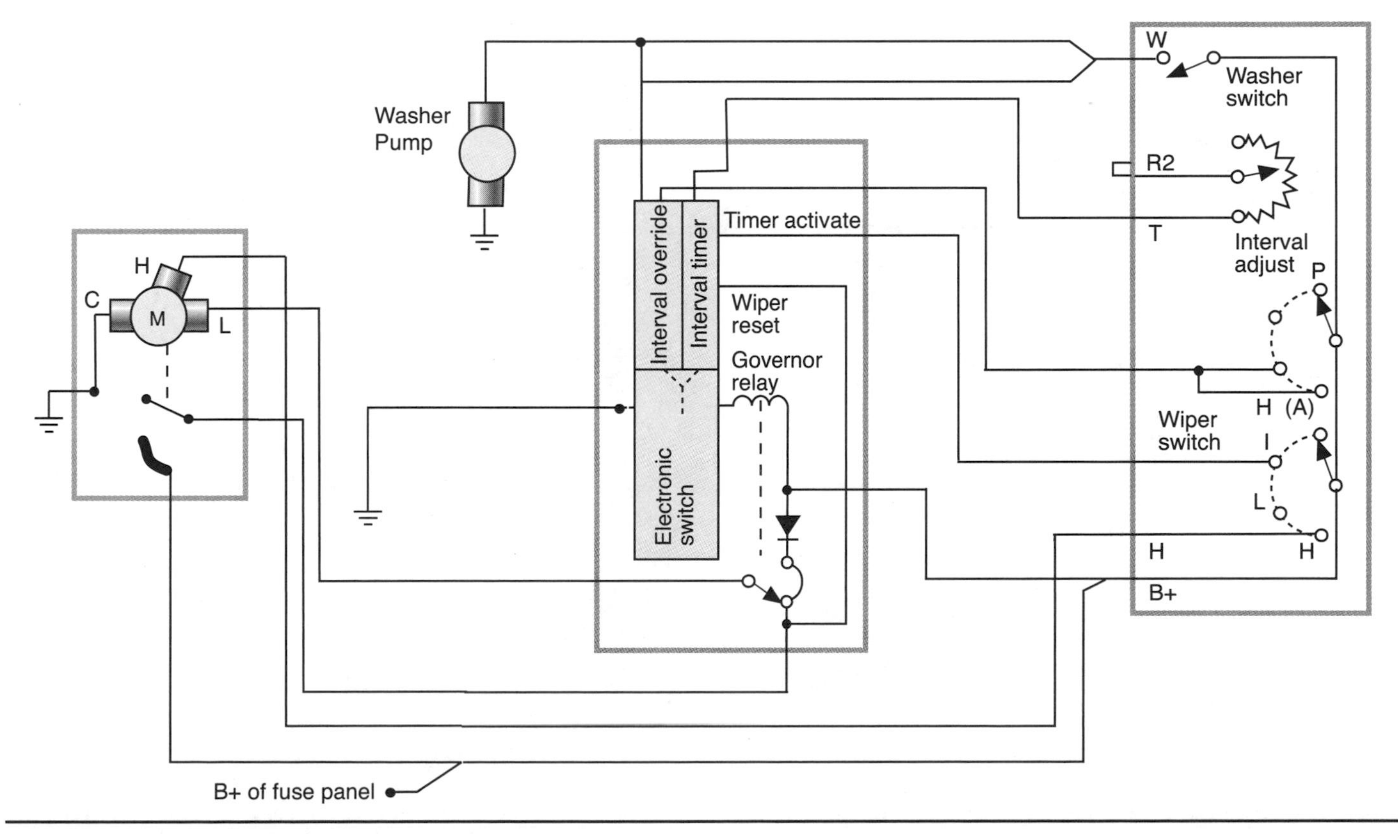

FIGURE 13–22 A typical interval wiper system.

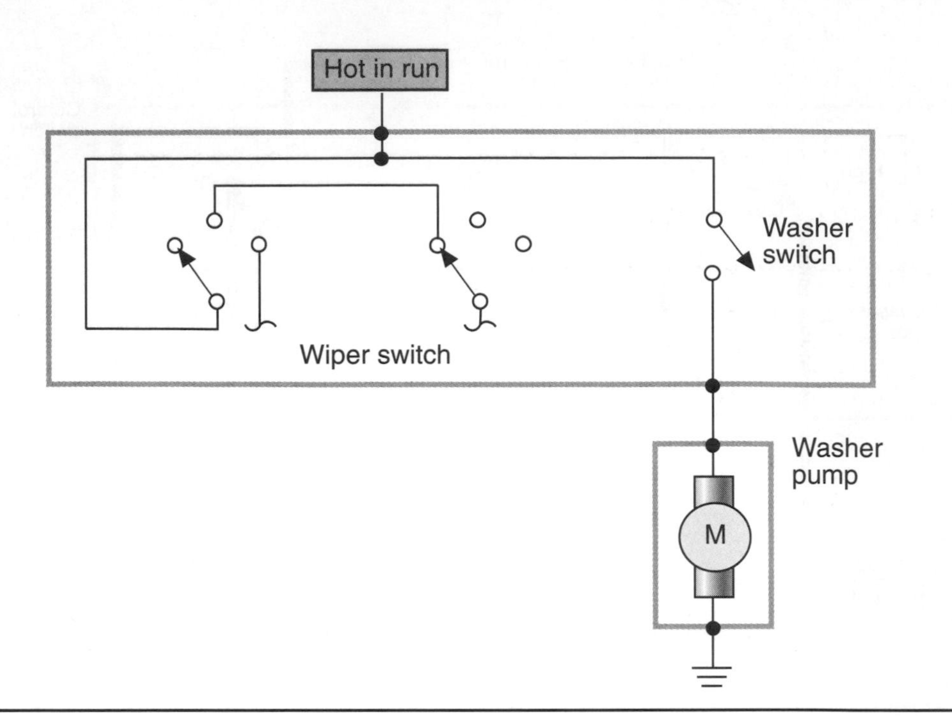

FIGURE 13–23 The typical washer pump works in conjunction with the wiper's low-speed setting.

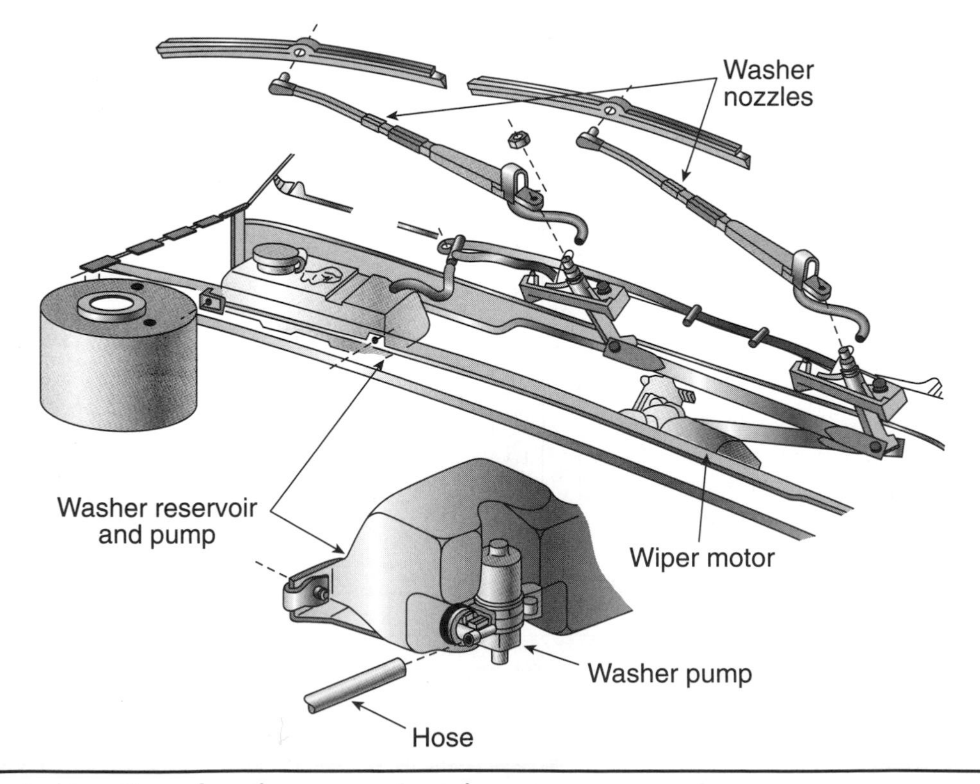

FIGURE 13–24 Typical location of washer reservoir and pump.

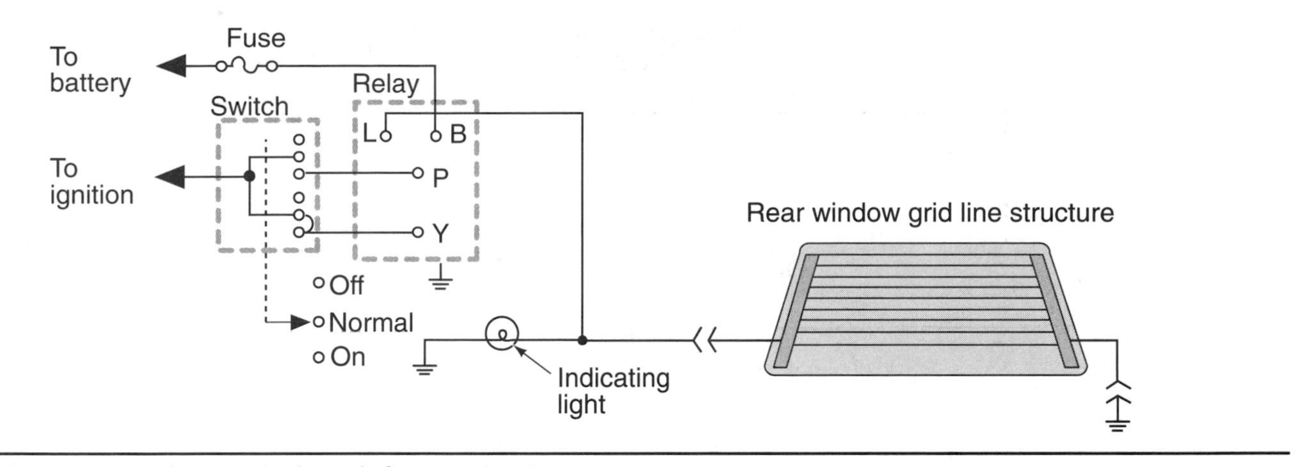

FIGURE 13–25 Typical rear window defogger circuit.

main components in a basic defogger/defroster system are the control switch, relay, indicator light, and the heating grids on the window. Some systems also incorporate a solid-state timer to shut off the defogger after a certain time. Figure 13–25 is an example of a simple rear window defogger circuit. When the switch is turned on, the relay applies battery voltage at the L terminal, which lights the indicator lamp and energizes the window heating grid. Most vehicles have window heating grid lines connected in parallel, allowing the heating grids to operate even if one grid is open. The heating grid can draw 20 amps or more; so using a relay is necessary.

Figure 13–26 is an example of a window defogger circuit that uses a solid-state timer to control the window heating grid. The timer opens the defogger control switch after the first 10 minutes of initial operation and every 5 minutes for succeeding operations.

The most common customer complaint about a window defogger system is that only a portion of the heating grids are working. This is normally due to broken grids wires in the window. Some breaks in the grid lines are easy to visually detect, but some breaks are too small to see. A few methods are commonly used by technicians to find a break in a grid line. The first method is using a test light to check the grid wire for a break (Figure 13–27). Because the grid wire is a resistive element that heats up as current is passed through it, the brilliance of the test light varies depending on where on the grid line the test light is applied.

If a grid line is broken, the test light has full brilliance up to the break in the grid and no light after the break (Figure 13–28). Remember that the window defogger circuit could be on a timer that shuts off the circuit after a set time (usually around 10 minutes).

CAUTION *Be careful when testing a heating grid wire with a test light. Probing the grid wire too hard with the test light tip might create a break in the grid wire by scratching it.*

Another method to check for breaks in a heating grid wire is to use a digital infrared tester (Figure 13–29), which is either an adapter unit for a DVOM or a standalone unit with its own digital readout numbers. To use an infrared tester, perform the following simple steps.

1. Turn on the rear window heating grid.
2. Wait approximately 2 minutes before checking the temperature of grid wires.
3. From the *outside* of the window, point the infrared tester at each grid wire no more than one-half inch away. Measure the temperature of wire.
4. Results: If the wire has a break in it, the temperature will be close to room temperature. If the wire is working, the infrared tester will show the heated temperature being produced (normally between 85° and 110°).

An infrared tester is a handy tool to help the technician quickly diagnose broken grid wires from outside the vehicle.

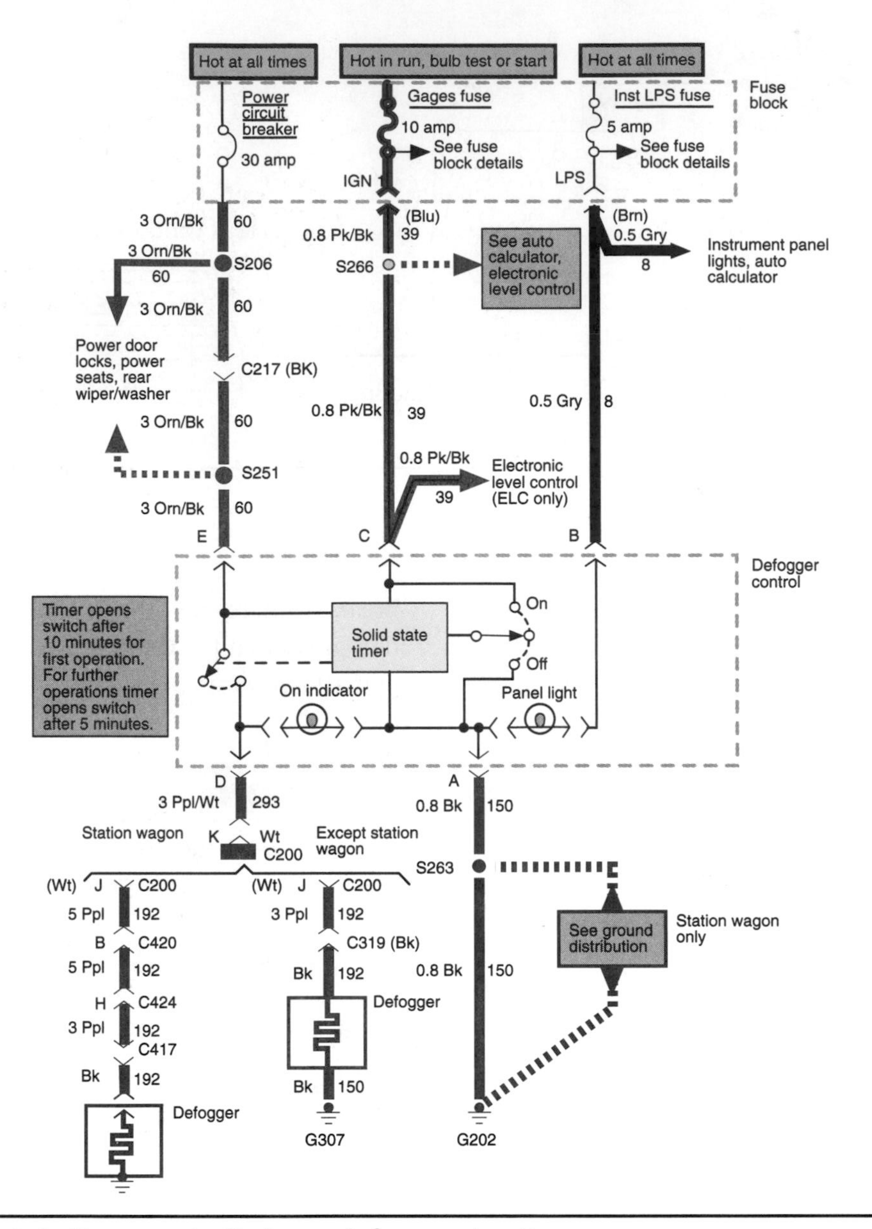

FIGURE 13–26 Solid-state timer-controlled rear defogger circuit.

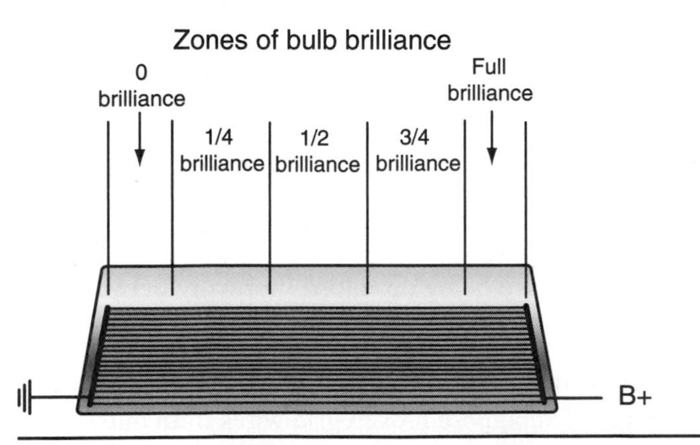

FIGURE 13–27 Test light brilliance while probing a window grid wire.

Defogger Grid Wire Repairs In the event a grid wire has a break, a special repair kit can be used to mend the break in the wire. The procedures listed in Photo Sequence 2 is typical for this repair.

Power Window Circuits

Several types of power window circuits are found in today's vehicles, depending on the vehicle body style and amount of installed options. The primary components of a power window circuit are a master control switch (normally on the driver's side) with a disable or

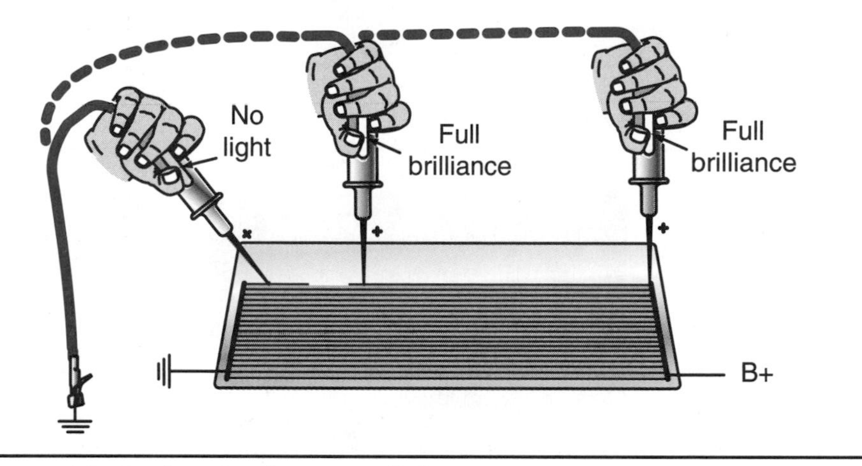

FIGURE 13–28 Finding a grid wire break using a test light.

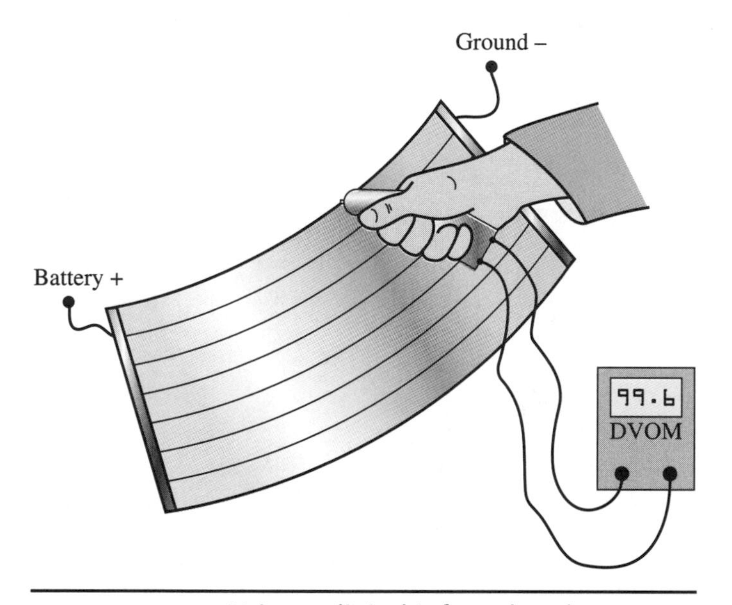

FIGURE 13–29 Using a digital infrared probe to check for broken heating grid wires.

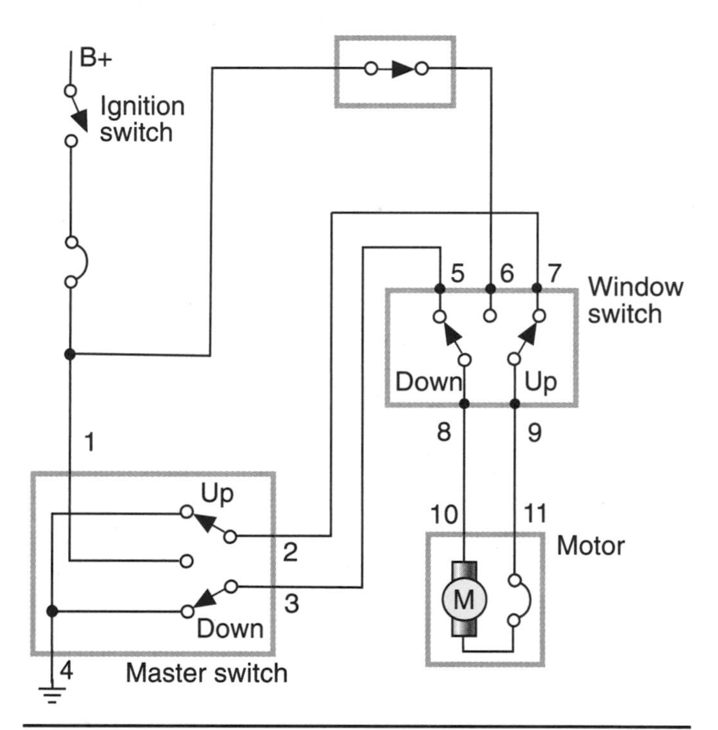

FIGURE 13–30 Simplified power window circuit.

lock-out switch, individual window control switches, and individual window control motors and sometimes rack-and-pinion gears.

Circuit Controls A simplified power window circuit is shown in Figure 13–30. When the ignition switch is closed, the window motor may be operated by either the window switch or the master control switch. In circuits with PM (permanent magnet) motors, the motor is insulated with the master switch, providing the path to ground. Also, a PTC (positive temperature coefficient) circuit breaker is installed internally to the motor to protect the electrical circuit from excessive current load.

In Figure 13–31, the left rear window motor is controlled by the master switch in the up position. When the left rear window switch is used to control the window motor, battery power is supplied from the circuit breaker directly to the LR window switch.

CAUTION *Power window circuits are very strong when operating. Keep fingers clear when testing or working on circuit components.*

PHOTO SEQUENCE 2

Repairing Defogger Grid Wires

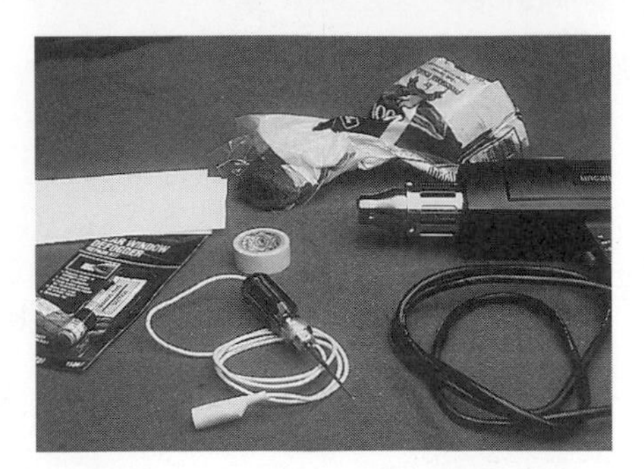

P13–1 Tools required to perform this task include masking tape, repair kit, 500°F heat gun, test light, steel wool, alcohol, and a clean cloth.

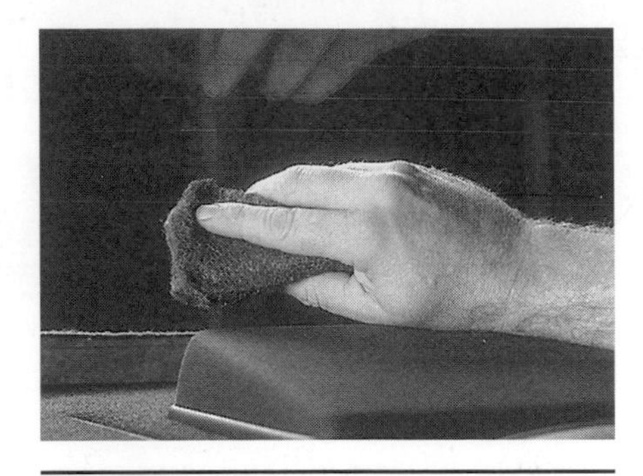

P13–2 Clean the grid line area to be repaired by buffing with steel wool. Wipe clean with a cloth dampened with alcohol. Clean an area about 1/4 inch (6 mm) on each side of the break.

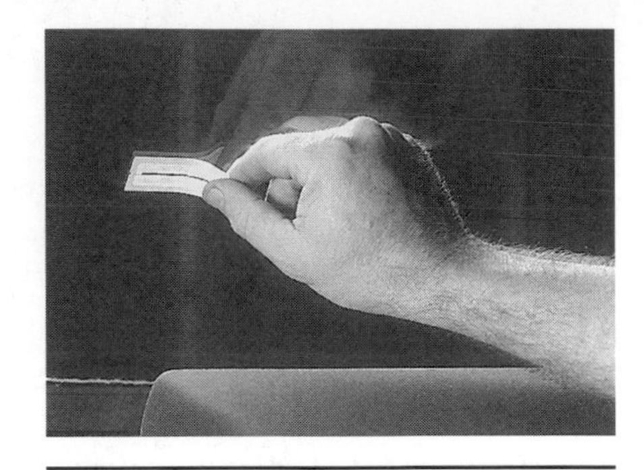

P13–3 Position a piece of tape above and below the grid. The tape is used to control the width of the repair.

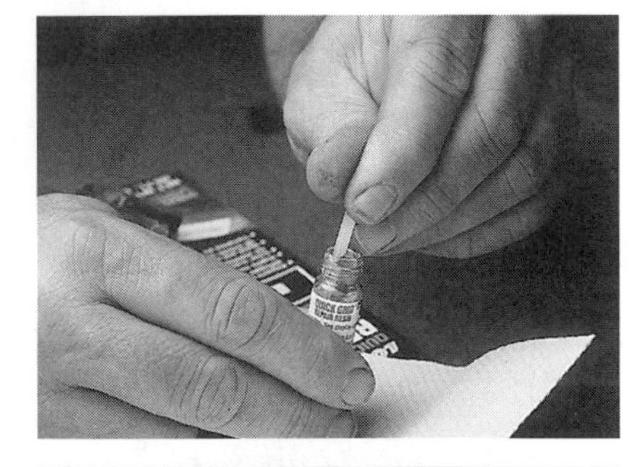

P13–4 Mix the hardener and silver plastic thoroughly. If the hardener has crystallized, immerse the packet in hot water.

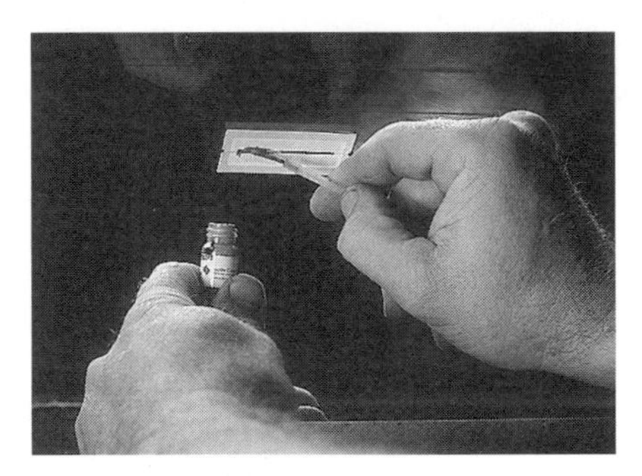

P13–5 Apply the grid repair material to the repair area using a small stick.

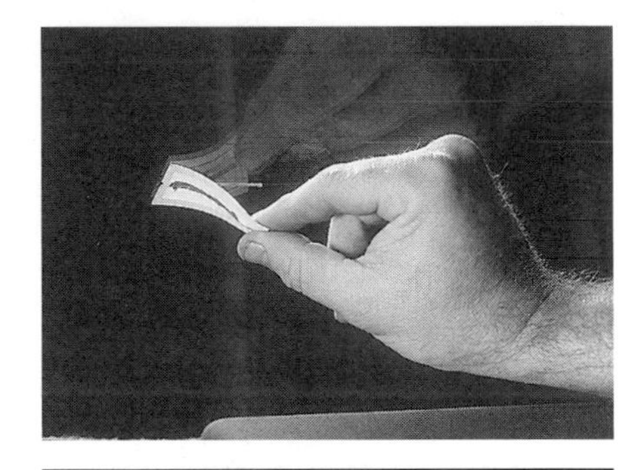

P13–6 Remove the tape.

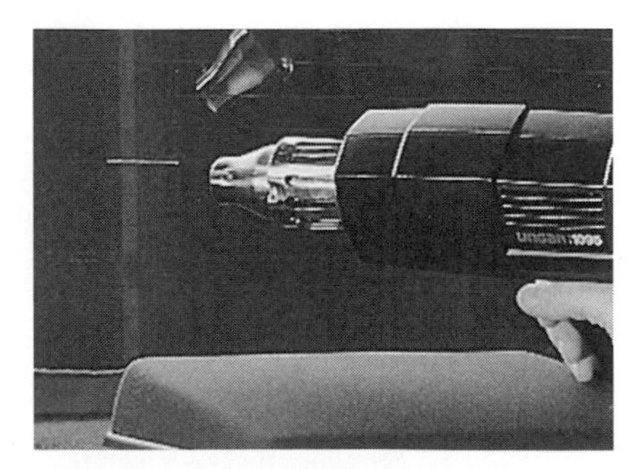

P13–7 Apply heat to the repair area for 2 minutes. Hold the heat gun 1 inch (25 mm) from the repair.

P13–8 Inspect the repair. If it is discolored, apply a coat of tincture of iodine to the repair. Allow to dry for 30 seconds; then wipe off the excess with a cloth.

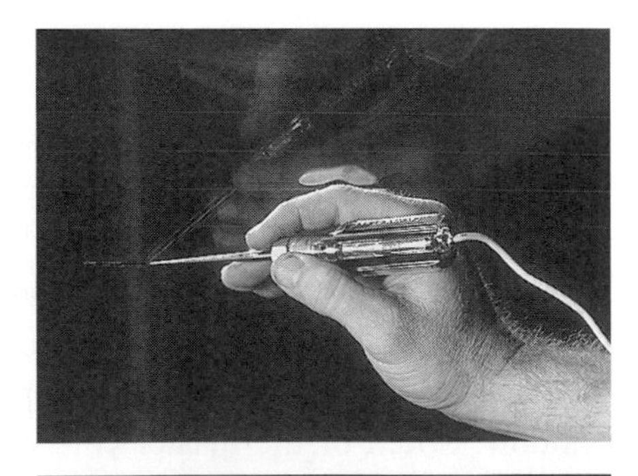

P13–9 Test the repair with a test light. Note: It takes 24 hours for the repair to fully cure.

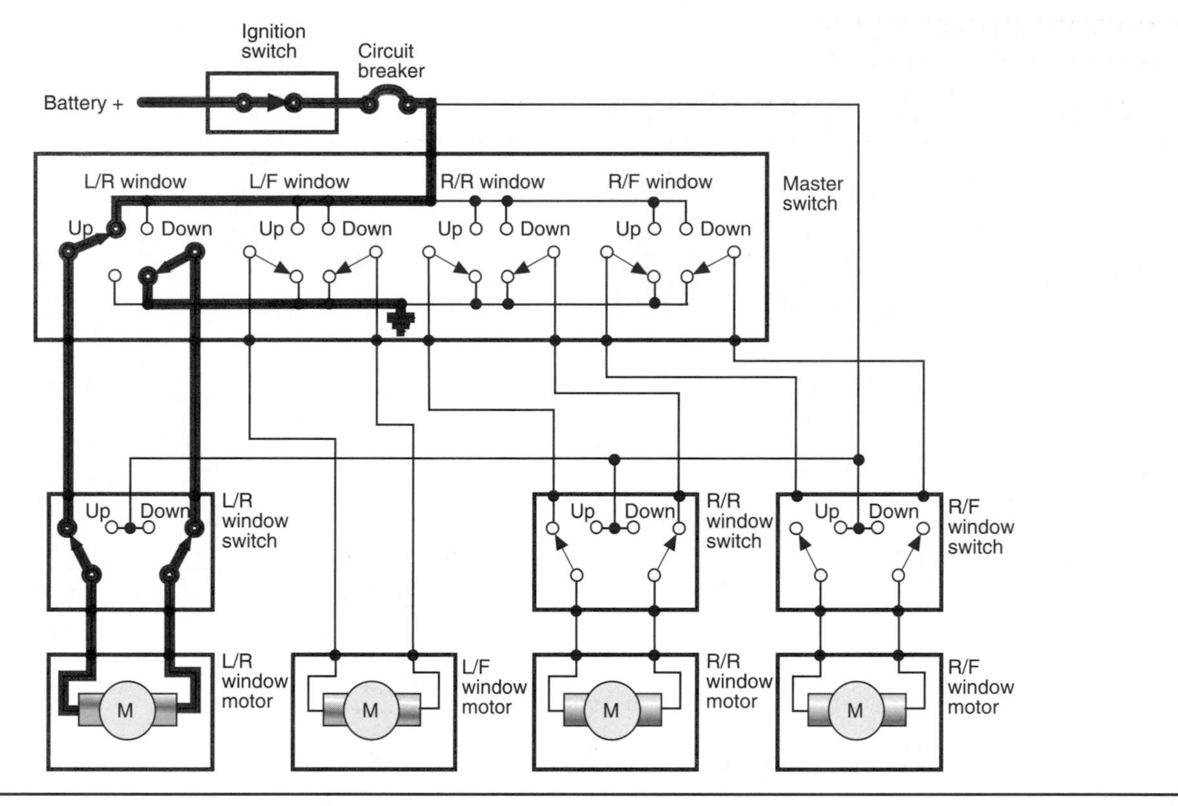

FIGURE 13–31 Master switch controls current to L/R window motor circuit.

Testing a master switch in a power window circuit can be accomplished with a test light (Figure 13–32). Connect the test light between connections 1 and 2 of the master switch. The test light should go on with the master switch in the off position, but should go out when the master switch is in the up or down position. The window switch can be checked in the same manner.

Power Window Motors Two types of power window motors are used to raise and lower a window. The first is a direct drive motor that uses a pinion gear meshed with a sector gear to raise the window regulator (Figure 13–33). When the window lowers, a spiral spring attached to the regulator is wound up. As the window raises, the spiral spring helps the motor raise the window.

The second type of power window motor uses a rack-and-pinion gear set (Figure 13–34). The rack is a flexible strip of gear teeth with one end attached to the window and the other end in a channel. The motor operates the window regulator through the rack assembly.

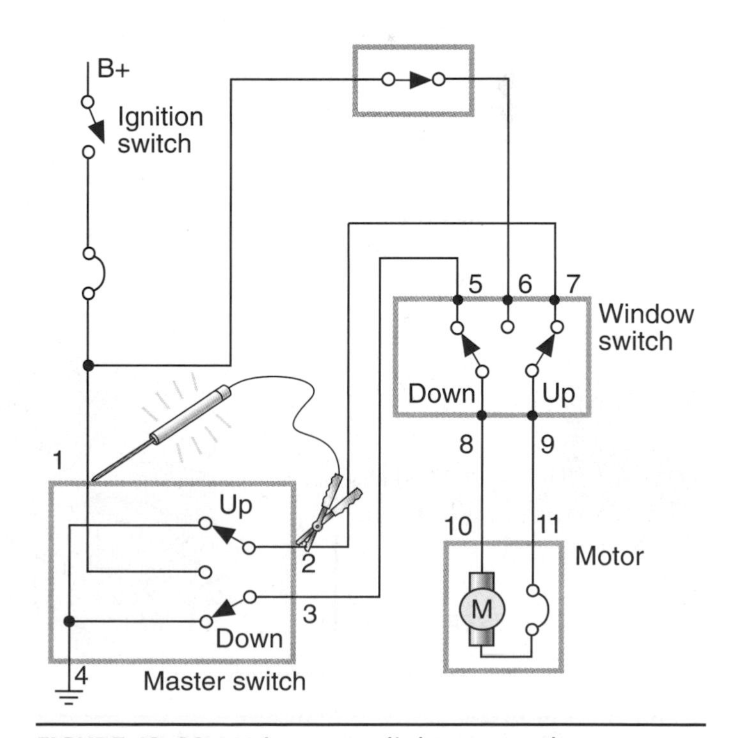

FIGURE 13–32 Using a test light to test the operation of the power window master switch.

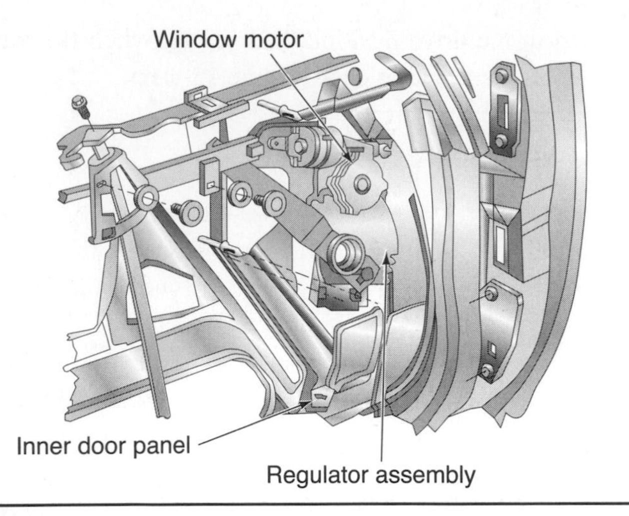

FIGURE 13–33 Direct drive motor and window regulator assembly.

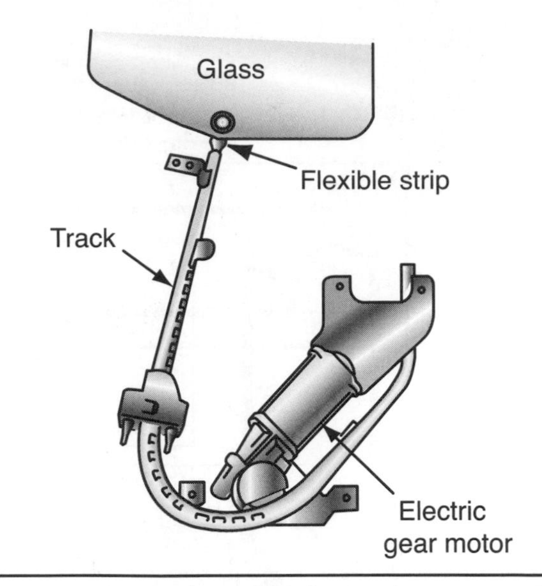

FIGURE 13–34 Rack-and-pinion-style power window motor and window regulator assembly.

Power Seat Circuits

The power seat circuit allows the driver and, in some vehicles, the passenger to adjust the seat for individual comfort (Figure 13–35). Power seats are classified into three basic configurations: The *two-way* type moves the seat forward and backward; the *four-way* moves the seat forward, backward, up, and down; and the *six-way* performs the same functions as the four-way but also tilts front and rear.

In Figure 13–36, a reversible permanent magnet motor pack called a **trimotor** controls the function of all six seat positions. Figure 13–37 is an example of a six-way power seat circuit that controls current flow direction through each motor depending on the individual function switches.

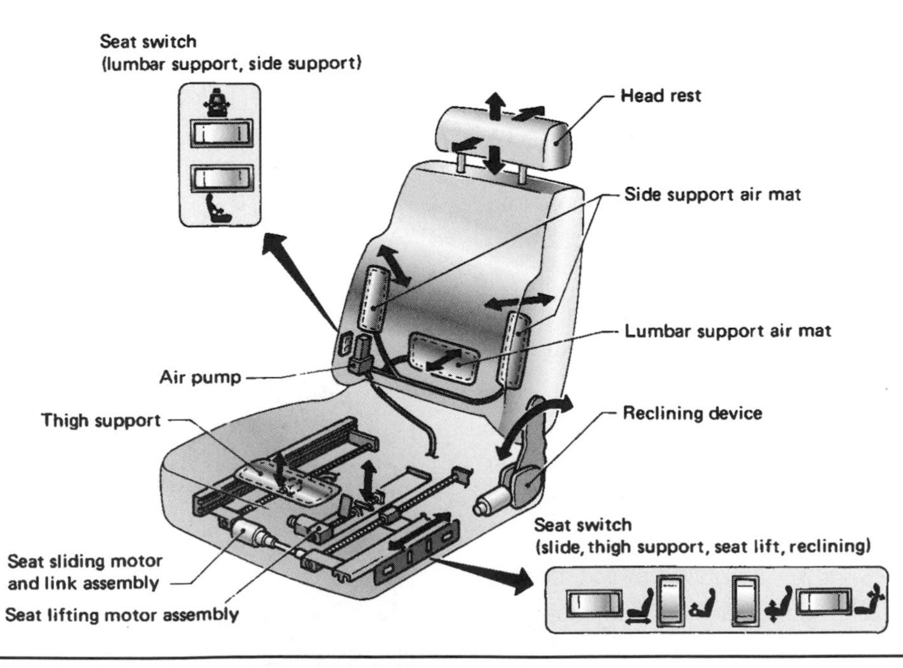

FIGURE 13–35 Typical six-way power seat components.

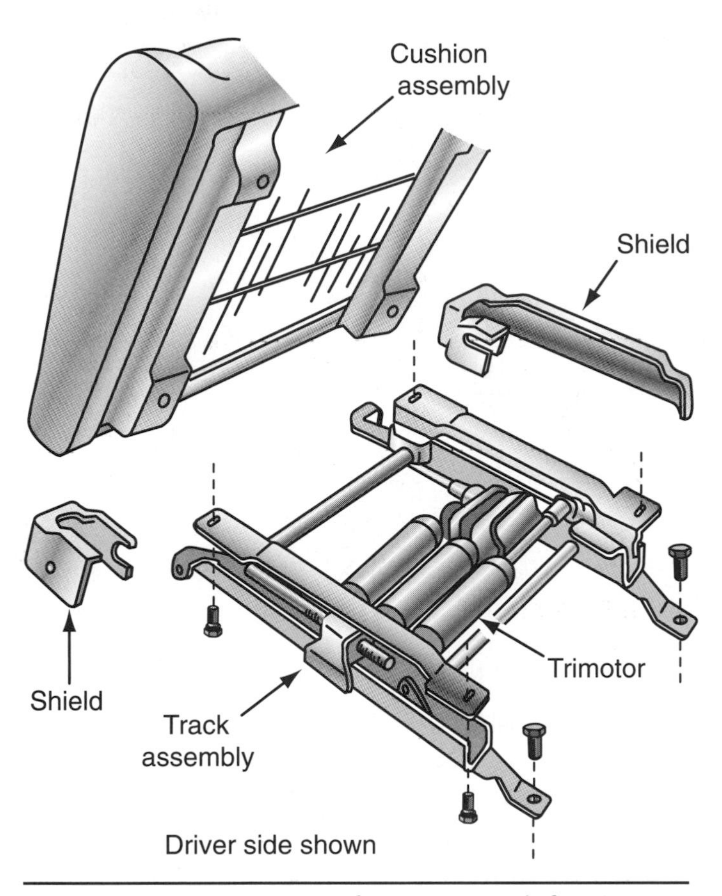

FIGURE 13–36 Location of trimotor pack for six-way adjustable seat.

In addition to power seats, several luxury vehicles have climate-controlled seats with heating elements located between the seat covering and the seat cushion (Figure 13–38). When the heated seat module (HSM) receives an on command over the data bus connection from the cabin compartment node (CCN), it provides pulse-width–modulated current to control the heated seat elements at the designed temperature setting.

Power Door Lock Circuits

Modern door locks have small electric motors that operate the door lock mechanisms. In Figure 13–39, a door lock circuit uses a double relay to control door lock motors. When a door lock switch is moved to the lock or unlock position, the relay energizes and sends battery voltage to the door lock motors or actuator solenoids (Figure 13–40).

Some vehicles are also equipped with door locks that activate automatically when the transmission is placed in the drive position, and unlock when the transmission is returned to park (Figure 13–41).

Power Mirror Circuits

Remote-Controlled Power Mirror A typical power mirror circuit controls the position of the outside mirrors through a position switch and individual motors for up/down and right/left functions (Figure 13–42). In Figure 13–43, fused voltage is supplied to the power mirror switch for use by the individual direction switches and the mirror select switch.

Electrochromic Mirrors More advanced power mirror circuits control glare intensity automatically through an electrochromatic process (Figure 13–44). The mirror material is constructed of a microthin layer of electrochromic material sandwiched between two conductive glass plates. Figure 13–45 is an example of an **electrochromic rear-view mirror.** Two internal photo cell sensors measure the light intensity on both sides of the mirror and adjust the electrochromic material to prevent glare.

Conventional Instrumentation

Instrument panel gauges and warning lights in today's vehicles are designed to closely monitor the proper function of several vehicle operating systems and provide feedback to the driver (Figure 13–46). This section looks at the basic operation of the analog instrument gauge cluster (Figure 13–47) and associated input sending units or sensors that control the gauges or warning lights. Some of the more complex instrument panel clusters consist of digital and liquid crystal displays (Figure 13–48) that obtain a variety of inputs from vehicle modules and analog sensors.

Analog Gauges

Two basic types of analog gauge designs are used: the thermal gauge and the electromagnetic gauge. The first type is a **thermal gauge** (Figure 13–49), also known as a bimetallic gauge. Used for many years but not normally

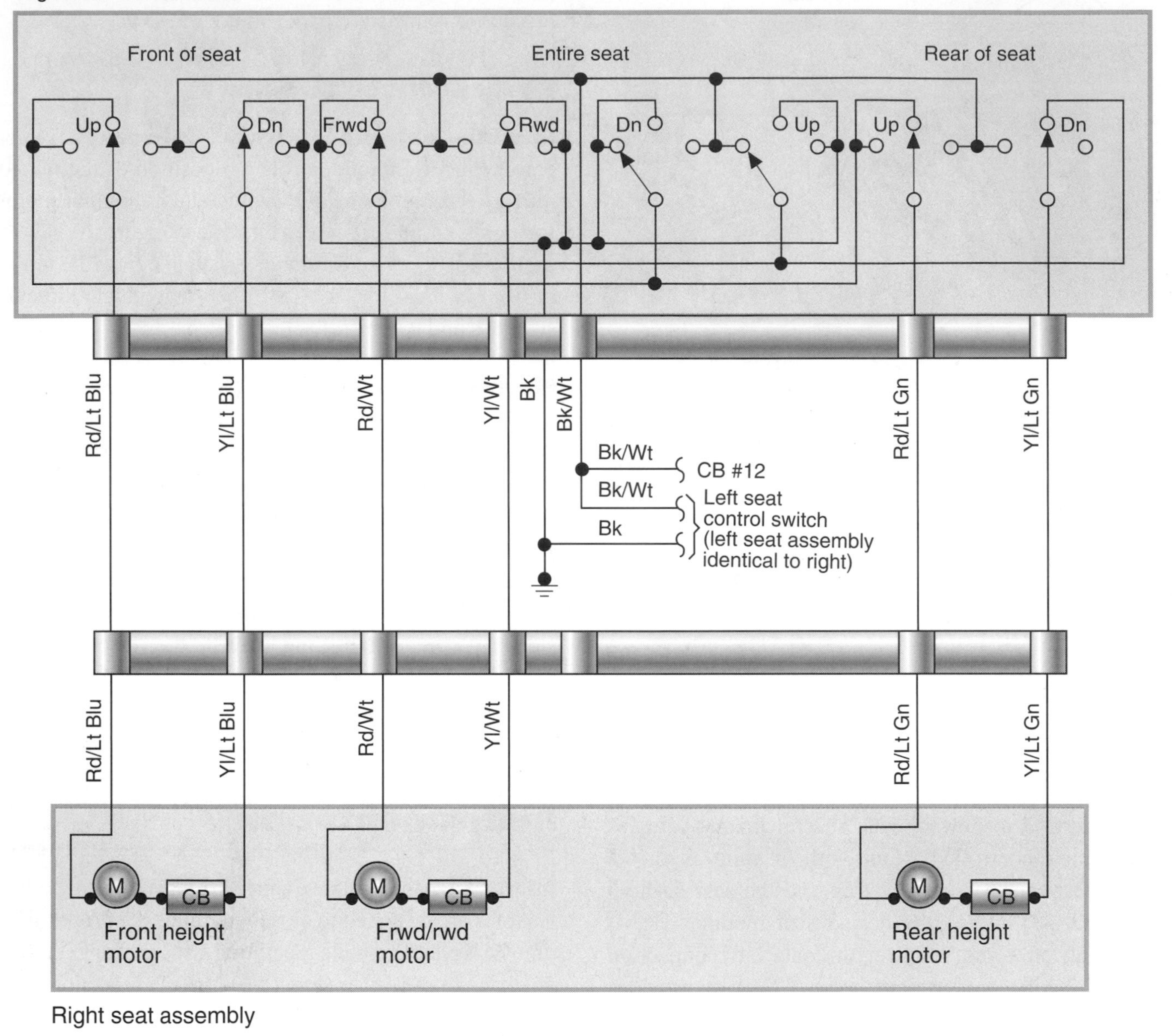

FIGURE 13–37 Typical six-way power seat circuit.

in today's vehicles, this gauge operates when controlled current flows through a heating coil in the gauge. The heat generated causes the bimetallic strip to bend in proportion to the amount of heat produced. Figure 13–50 is an example of a simple thermal fuel gauge circuit. The fuel level in the tank determines the variable resistance of the sending unit, which in turn controls the amount of current flowing through the gauge heating coil. The **instrument voltage regulator (IVR)** maintains a constant voltage to the gauge under all battery load and charging conditions. In Figure 13–51, a typical IVR consists of a bimetallic arm, heating coil, and a set of contact points. When current flows through the contact points, they open and close with the heating and cooling of the bimetal arm, producing a stable output voltage of up to 10 volts, depending on vehicle manufacturer.

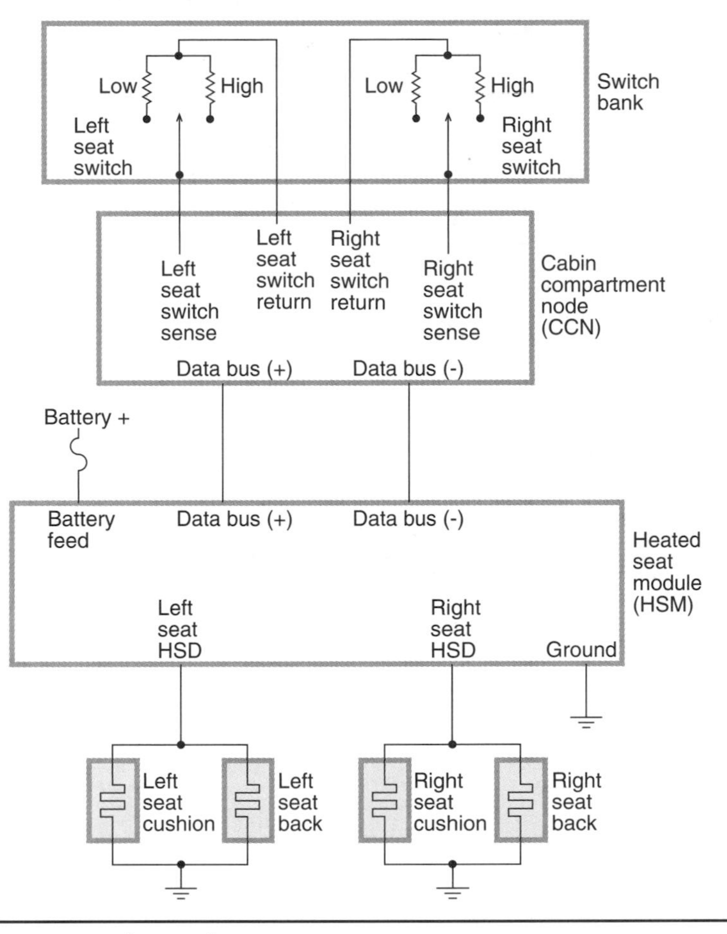

FIGURE 13–38 Heated seat system schematic.

The second type of analog gauge design, the type most modern vehicles use, is an electromagnetic gauge that produces needle movement by the magnetic force generated from current moving through small coils instead of heat. Four basic types of electromagnetic gauges are used: magnetic d'Arsonval gauge, three-coil (bobbin) gauge, two-coil (balancing coil) gauge, and the air-core gauge. Figure 13–52 is a **d'Arsonval gauge,** in which a coil of wire is wrapped around the base of the pointer needle to form an armature. When current flows through the coil of wire, a magnetic field is produced in the armature that opposes the permanent magnet and causes the pointer needle to swing. The second type of electromagnetic gauge is a **three-coil gauge** (Figure 13–53), also called a bobbin gauge. In this gauge, the current flowing through three electromagnetic coils determines the location of the pointer needle that is attached to a permanent magnet. In Figures 13–54 and 13–55, the flow of current through the circuit is determined by the resistance of the sending unit.

The third type of electromagnetic gauge is the **two-coil gauge** (Figure 13–56), also called a balancing coil gauge. This gauge uses the magnetic fields of the two coils to cause a magnetic attraction and repulsion of the needle in the gauge, depending on the resistance in the sending unit.

The last type of electromagnetic gauge, and probably the most common in modern vehicles, is the **air-core gauge** (Figure 13–57). This gauge is similar in operation to a two-coil gauge, except that the air-core gauge contains a permanent magnet attached to the pointer needle. The two windings do not have a core (thus the name air-core), and instead the permanent magnet is placed inside both windings. When current flows through the windings according to the sending unit resistance, the needle aligns itself with the stronger of the two magnetic fields.

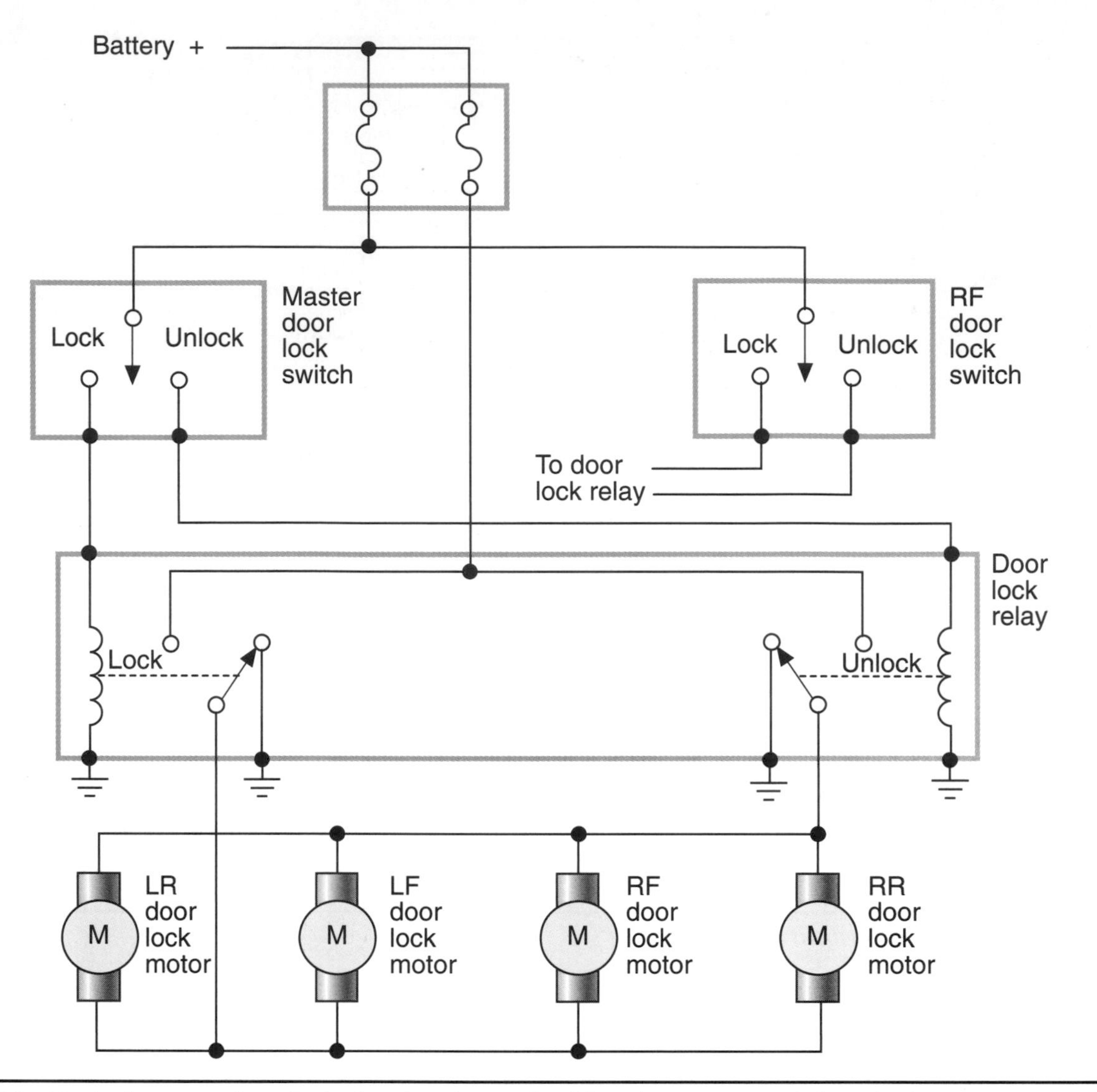

FIGURE 13–39 Typical power door lock circuit using a double relay to control door lock motors.

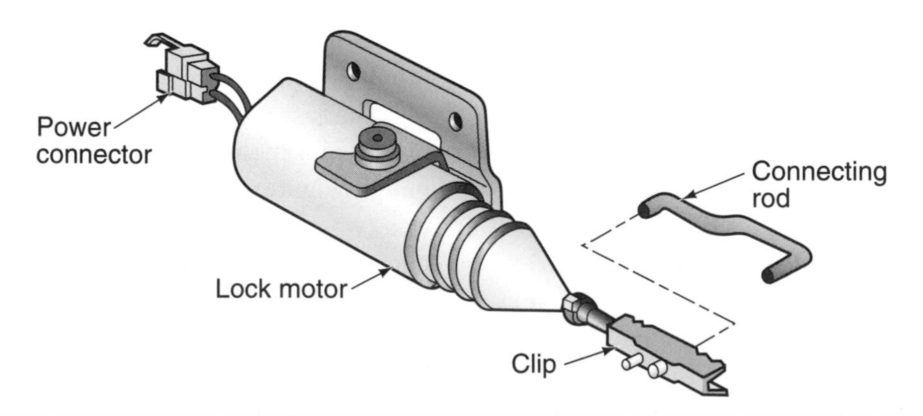

FIGURE 13–40 Typical permanent magnet power door lock motor assembly.

(Courtesy of DaimlerChrysler Corporation)

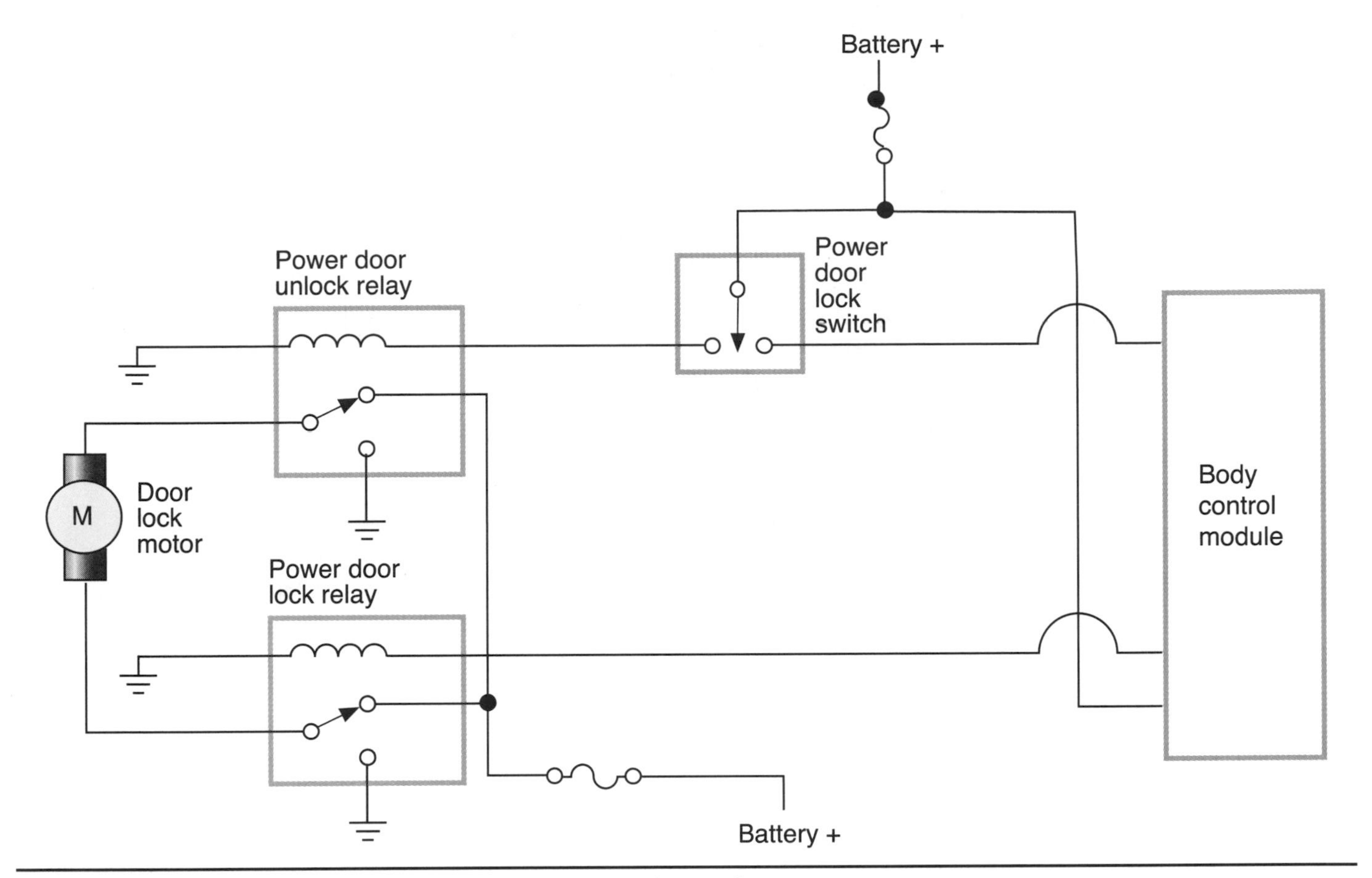

FIGURE 13–41 Typical automatic door lock circuit utilizing the BCM.

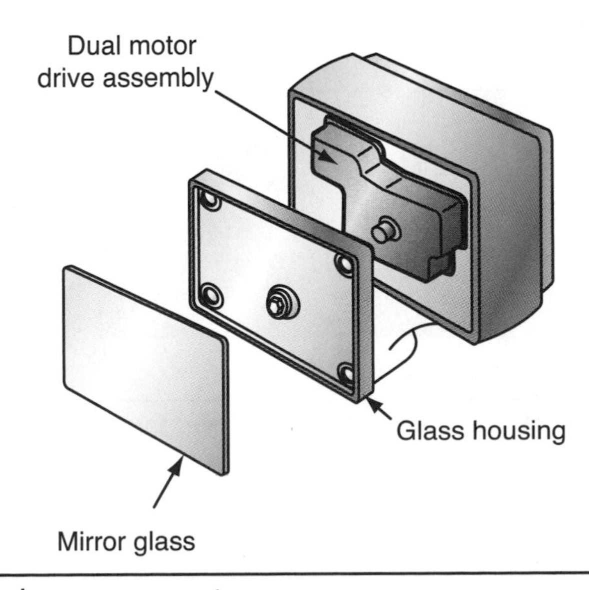

FIGURE 13–42 Remote power mirror components.

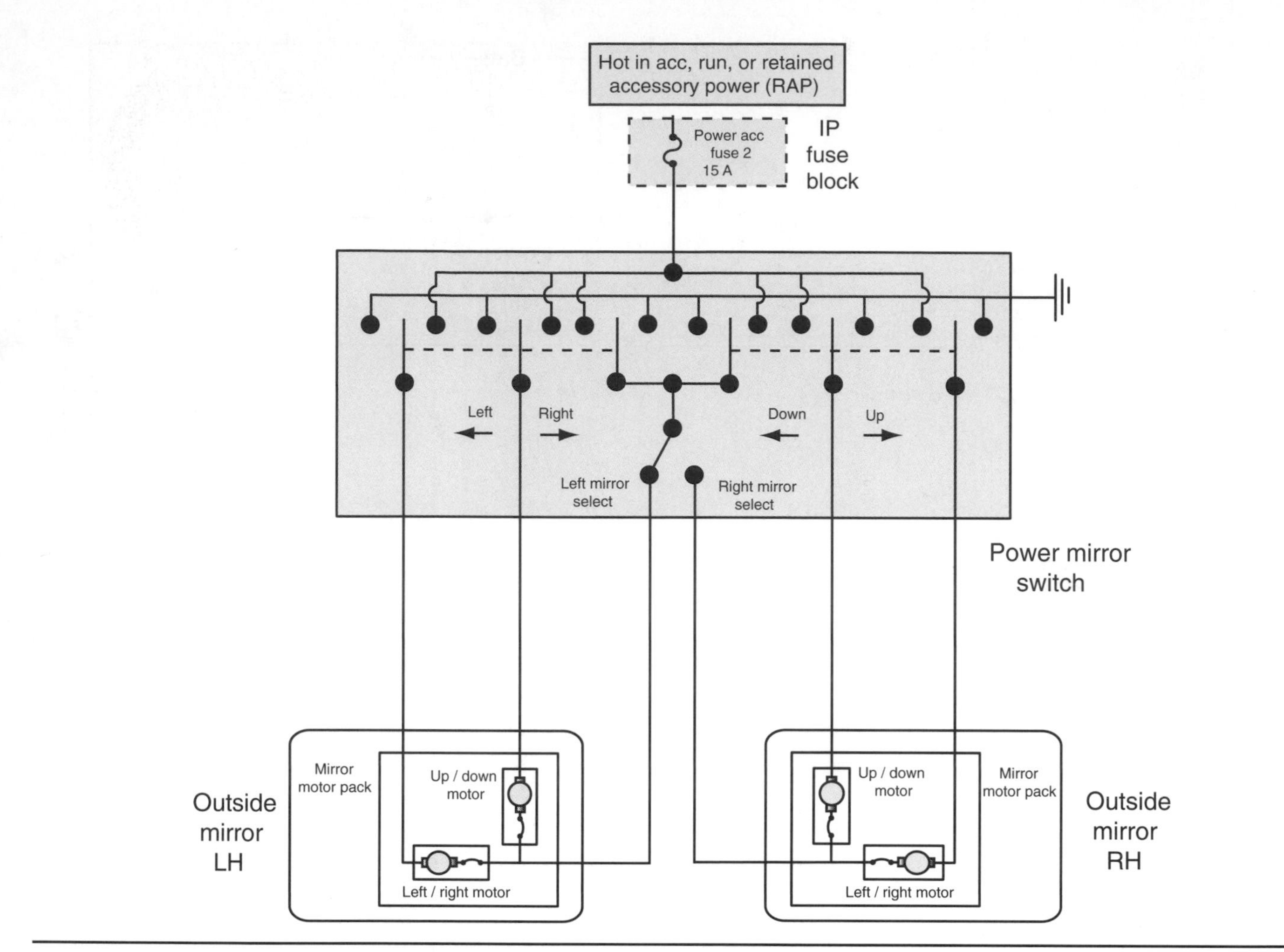

FIGURE 13–43 Typical power mirror circuit.

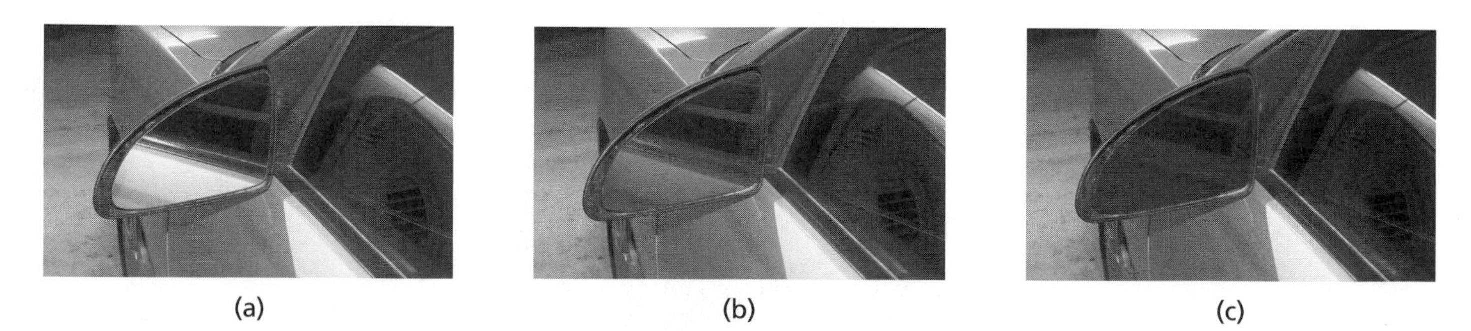

FIGURE 13–44 Operating characteristics of electrochromic mirror: (a) daytime, (b) mild glare, and (c) high glare.

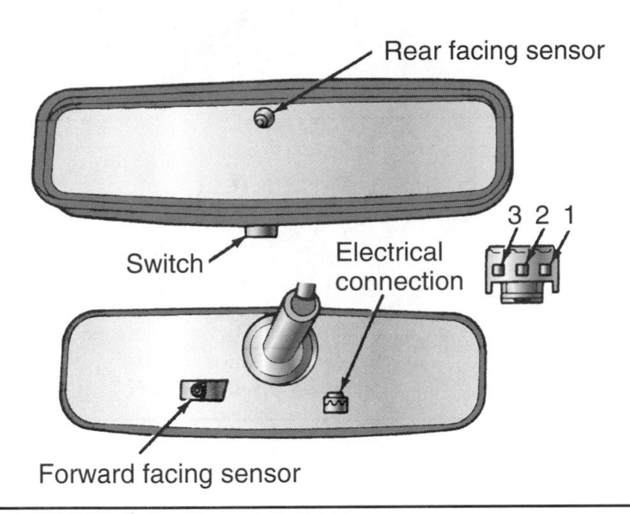

FIGURE 13–45 A typical automatic day/night electrochromic rear view mirror.

(Courtesy of DaimlerChrysler Corporation)

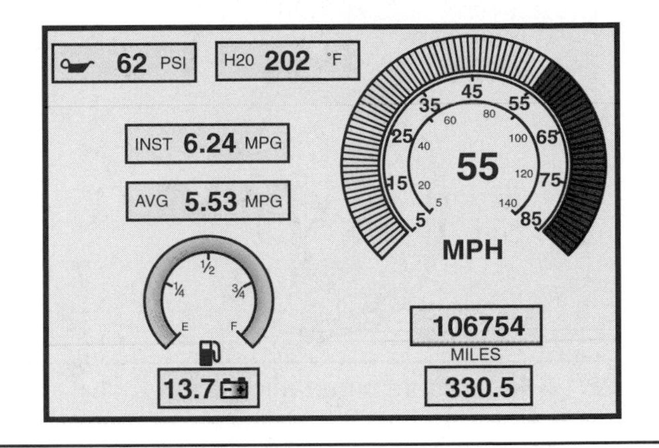

FIGURE 13–48 Typical digital instrument cluster.

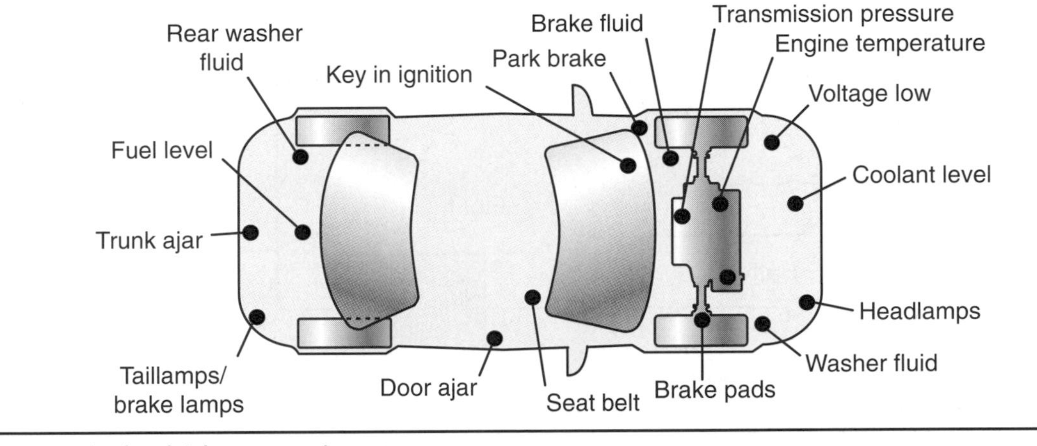

FIGURE 13–46 Typical vehicle system inputs.

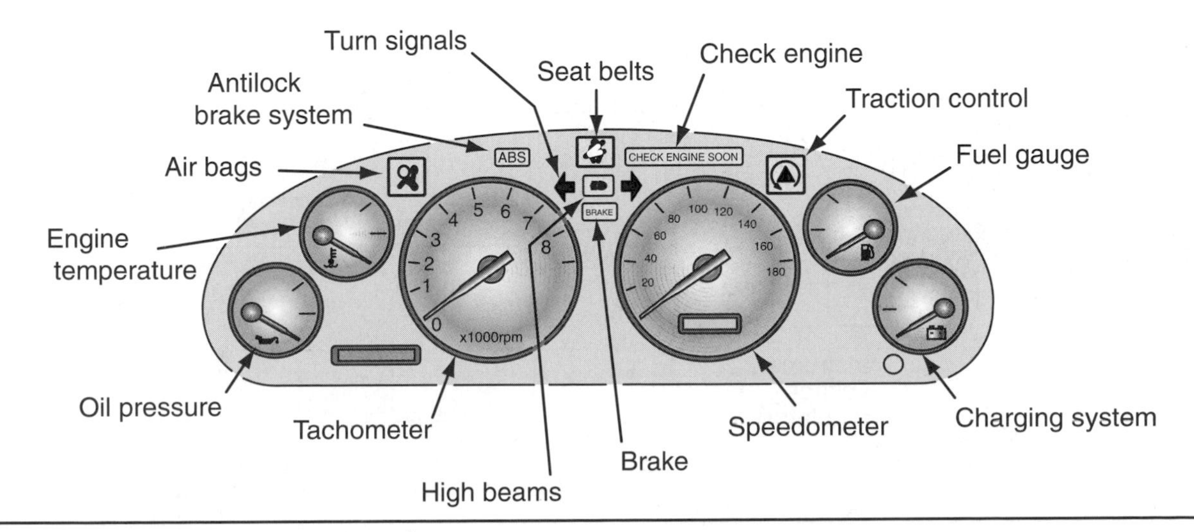

FIGURE 13–47 Analog instrument panel.

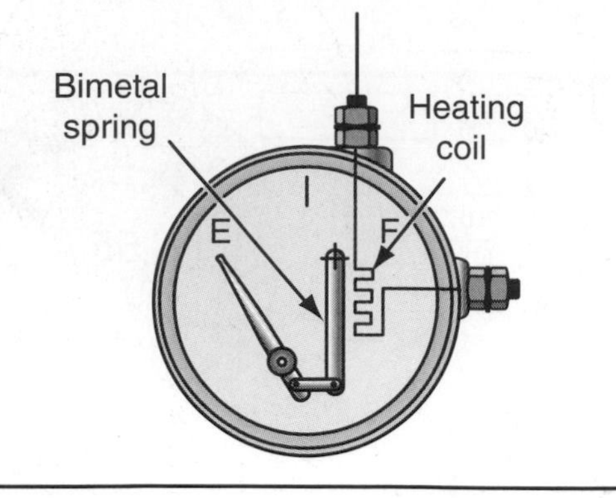

FIGURE 13–49 Simple bimetallic gauge construction.

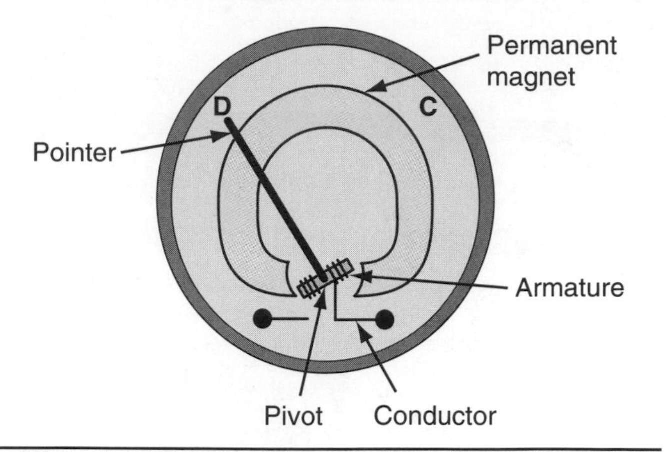

FIGURE 13–52 Simple d'Arsonval gauge assembly.

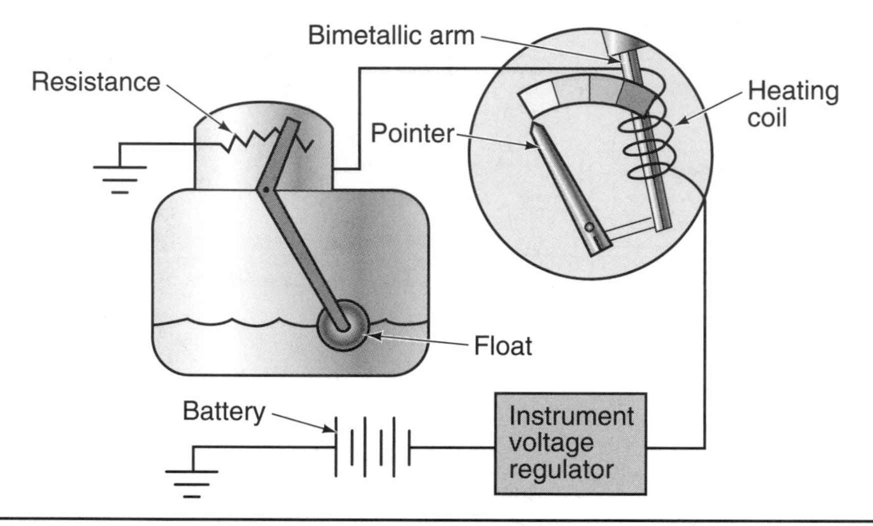

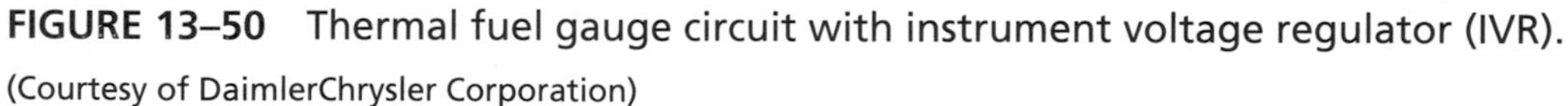

FIGURE 13–50 Thermal fuel gauge circuit with instrument voltage regulator (IVR).

(Courtesy of DaimlerChrysler Corporation)

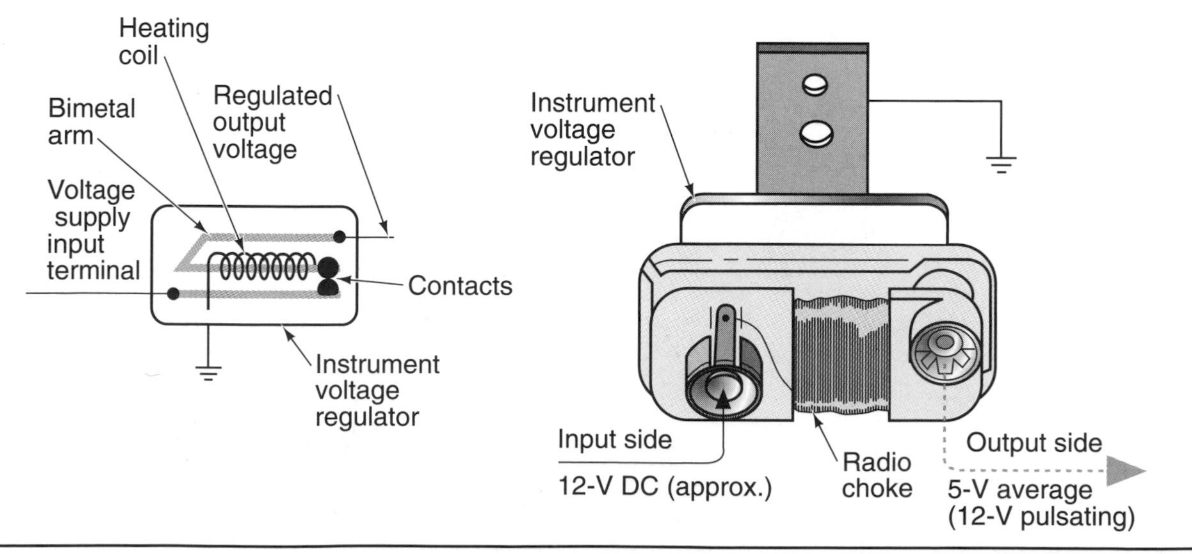

FIGURE 13–51 Typical instrument voltage regulator (IVR).

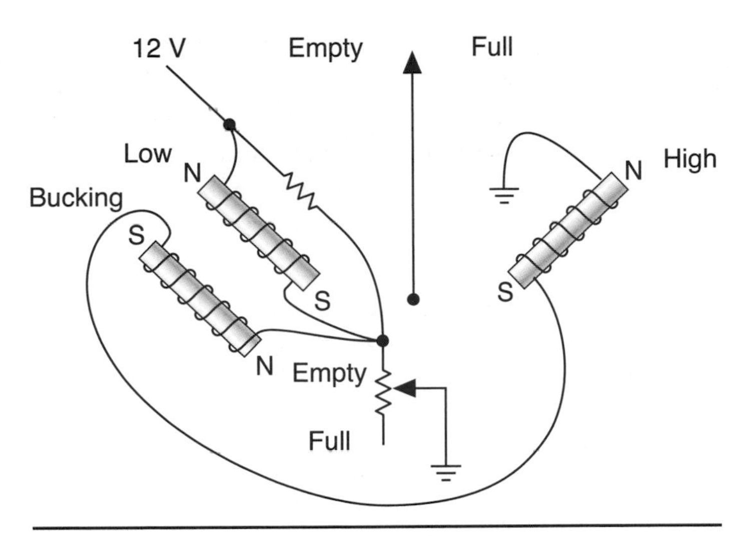

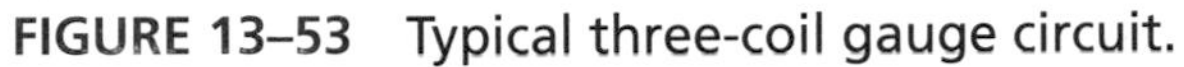
FIGURE 13–53 Typical three-coil gauge circuit.

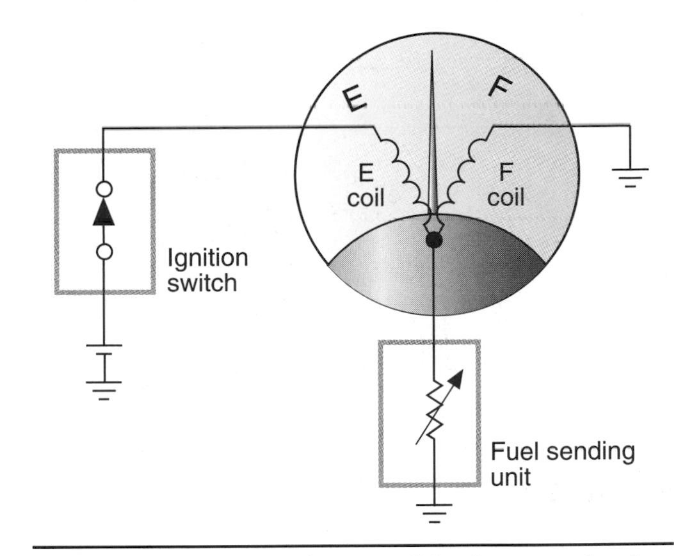

FIGURE 13–56 Typical two-coil fuel gauge circuit.

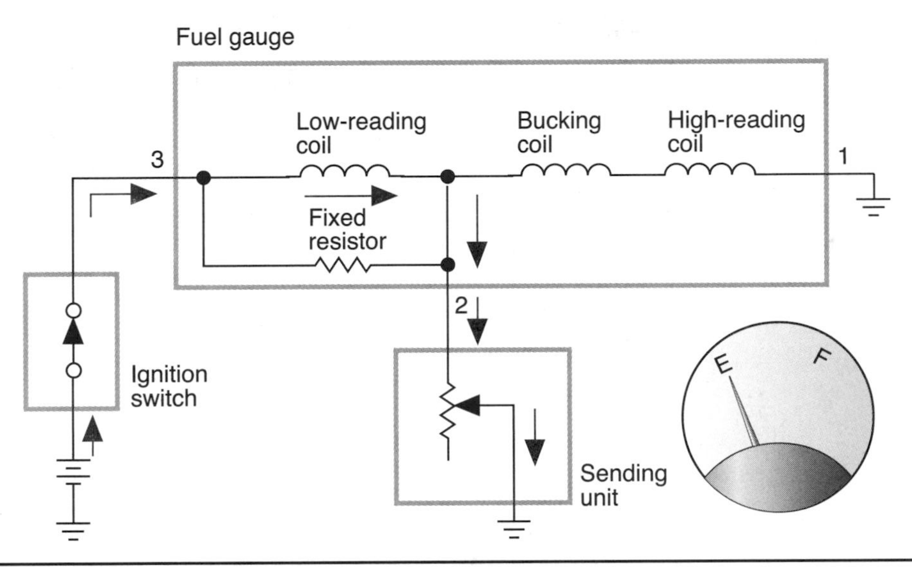

FIGURE 13–54 Flow of current through three-coil fuel gauge circuit when sending unit resistance is low.

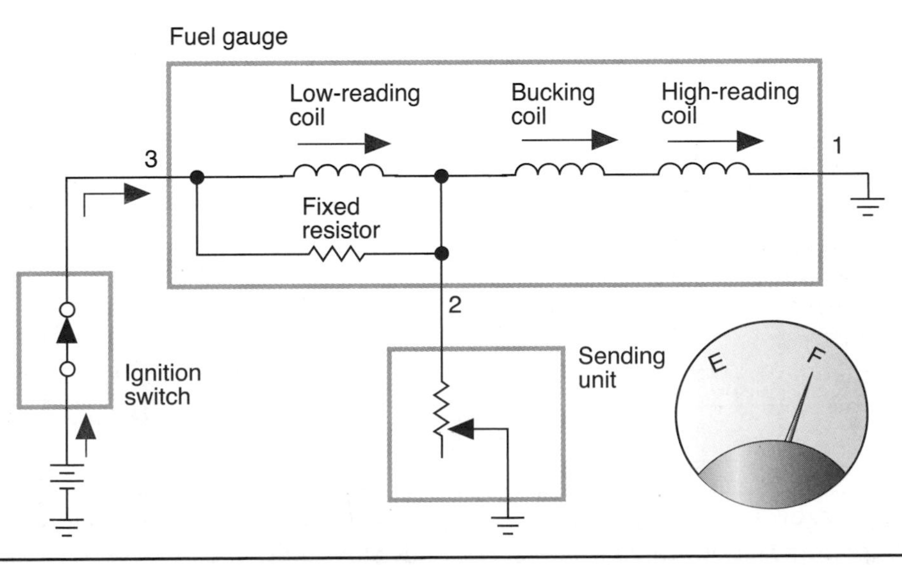

FIGURE 13–55 Flow of current through three-coil fuel gauge circuit when sending unit resistance is high.

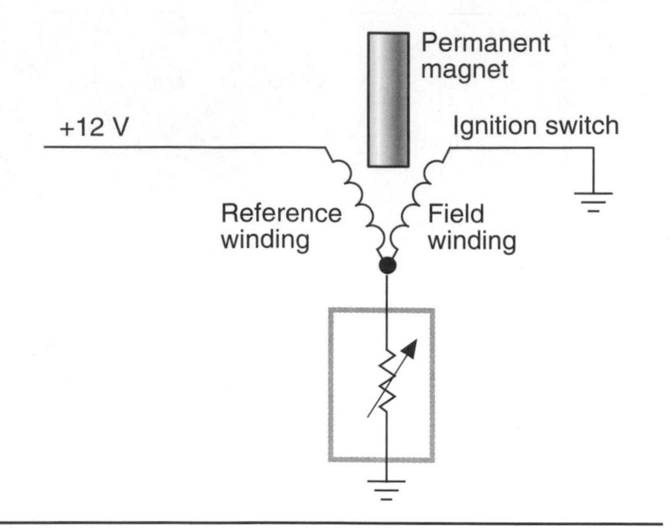

FIGURE 13–57 Basic air core fuel gauge circuit.

Gauge Sending Units

For both analog and digital gauge circuits, the sending unit is a vital part of the circuit design. In a typical oil pressure circuit, the oil pressure sending unit used for a gauge (Figure 13–58) consists of a piezoresistor that is moved against a contact arm when oil pressure is exerted on the flexible diaphragm.

The temperature gauge sending unit in Figure 13–59 is typical for a coolant temperature circuit. Current is sent from the temperature gauge to the sending unit terminal, through the thermistor (variable resistor), and then grounded to the engine. The resistance value of the thermistor depends on the temperature of the engine. As engine temperature rises, the thermistor resistance *decreases,* and the current flow through the gauge circuit *increases,* which raises the pointer needle toward the hot range.

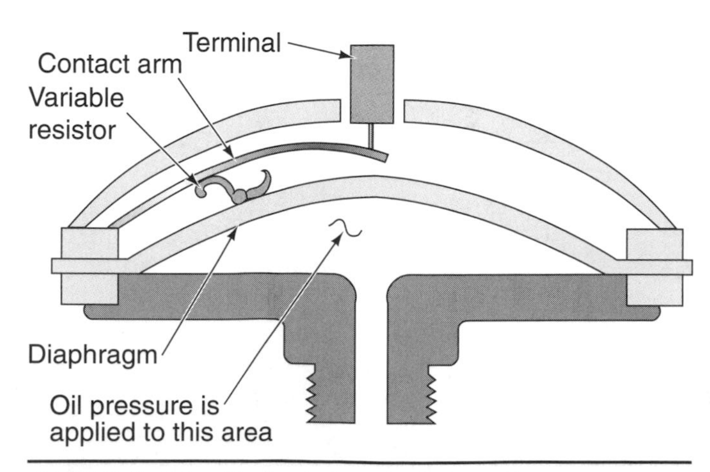

FIGURE 13–58 Typical piezoresistive oil pressure sensor used in an oil pressure gauge circuit.

The fuel gauge sending unit is typically a variable resistor, manually controlled by an arm and float that depends on the fuel level (Figure 13–60). Resistance of

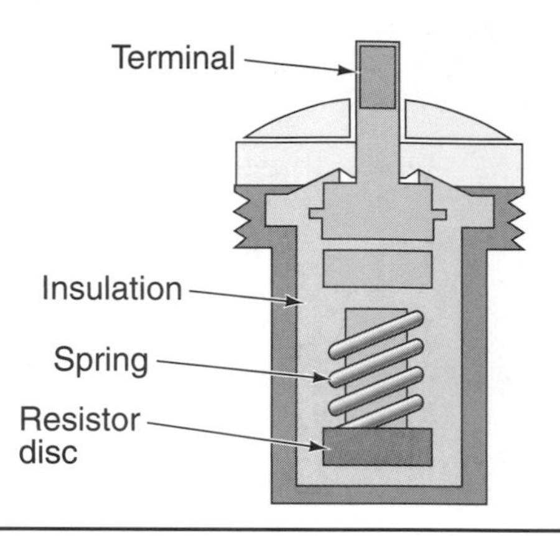

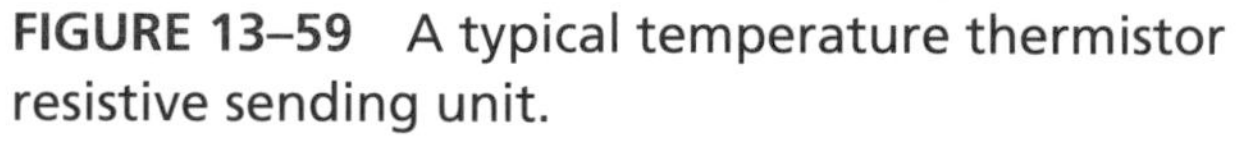

FIGURE 13–59 A typical temperature thermistor resistive sending unit.

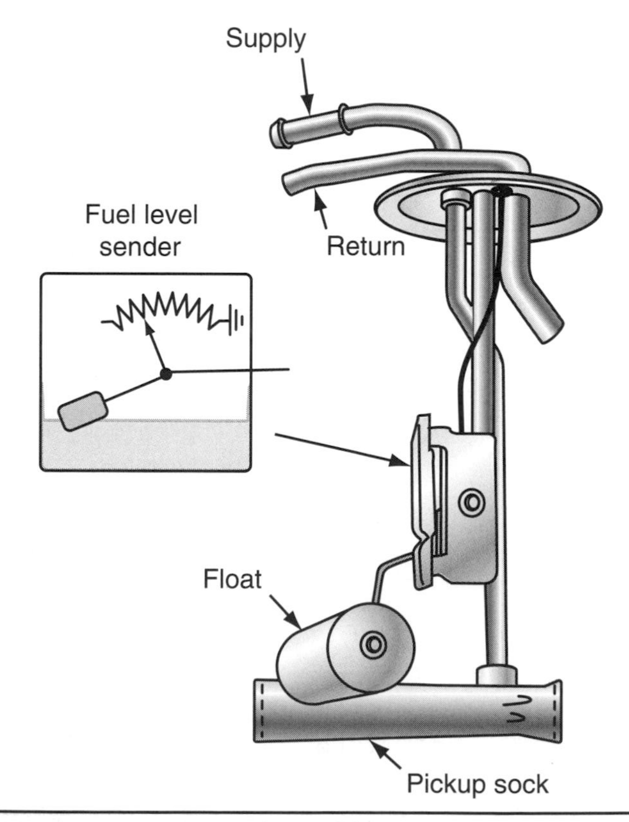

FIGURE 13–60 Typical variable resistance fuel gauge sending unit.

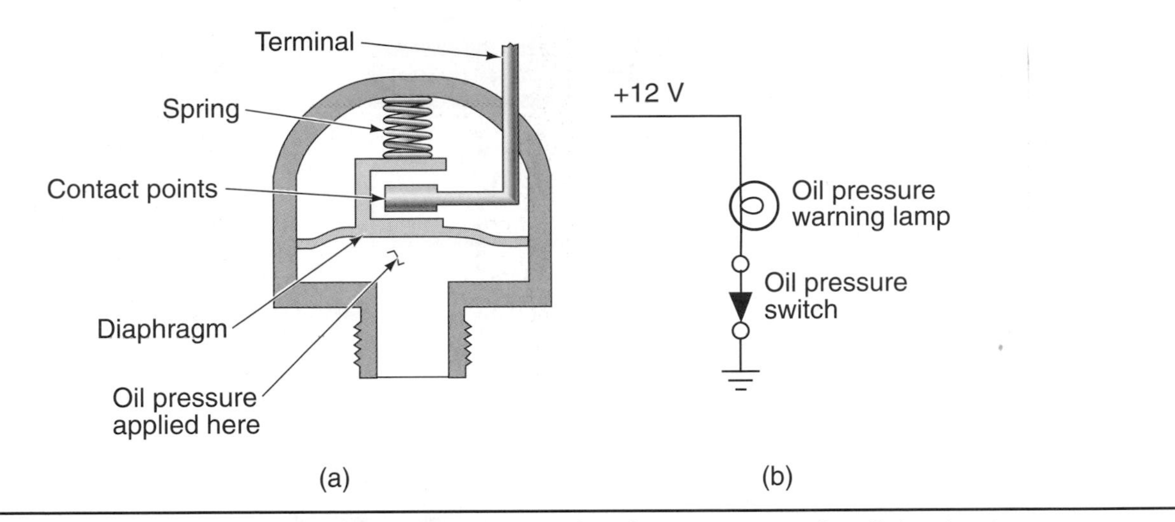

FIGURE 13–61 (a) Oil pressure light sending unit and (b) simple oil pressure warning light circuit.

the sending unit *increases* as the fuel level *decreases* and causes the fuel pointer needle to move toward empty.

When a warning lamp is used in place of a gauge, the sending unit is different than those used for gauges. The sending unit for a warning lamp circuit is either open or closed to complete the circuit to ground (Figure 13–61). When the engine is off or when there is no oil pressure, the contact points in Figure 13–61 are closed to allow a complete circuit for the warning lamp to light. When oil pressure is above the minimum specified pressure (usually 3–5 psi), the diaphragm moves the contacts apart and the warning lamp goes out.

Figure 13–62 is an example of a sending unit and simple circuit for a temperature warning light. The contact points in the temperature sending unit are *normally open* during normal engine temperatures. When a predetermined temperature level has been exceeded, the contacts close and complete the warning light circuit to ground.

Some instrument panels use a combination of gauges and warning lights (Figure 13–63). In the instrument cluster are individual sending units for each gauge or fluid level indicator, even though they share the same common power source line.

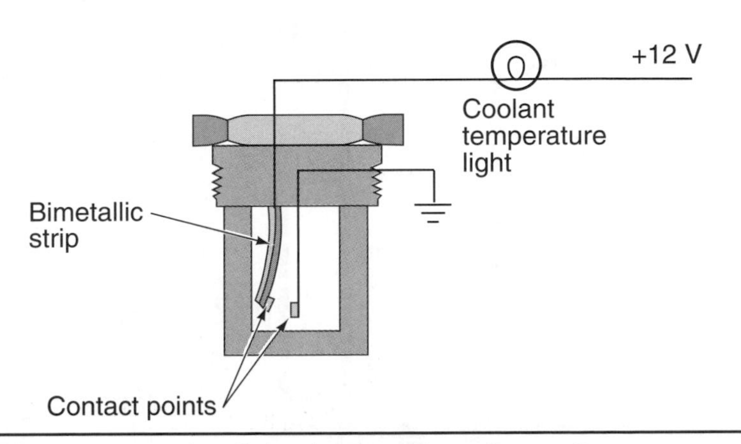

FIGURE 13–62 Typical temperature warning light circuit and sending unit.

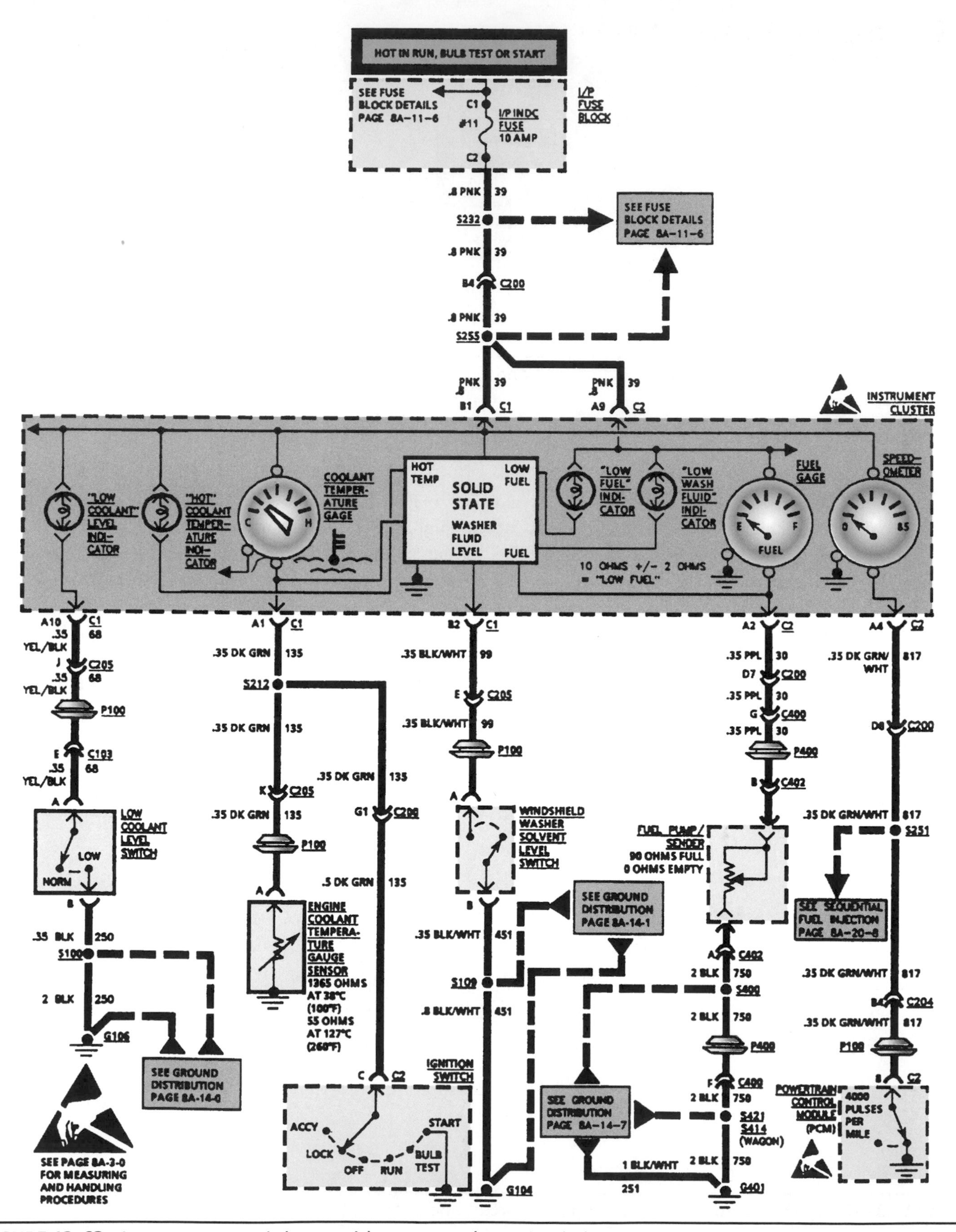

FIGURE 13–63 Instrument panel cluster with gauge and warning lights.

Air Bag Systems

A lot of technological improvements have gone into vehicle air bag or **supplemental inflatable restraint (SIR)** systems since they were first introduced over two decades ago. A first-generation simple air bag system (Figure 13–64) consisted of a single air bag module in the steering wheel, clock spring, control module, and various impact sensors. Newer vehicles can have as many as eight air bags (front, side, knee, side curtain) or more to protect all the vehicle occupants (Figure 13–65).

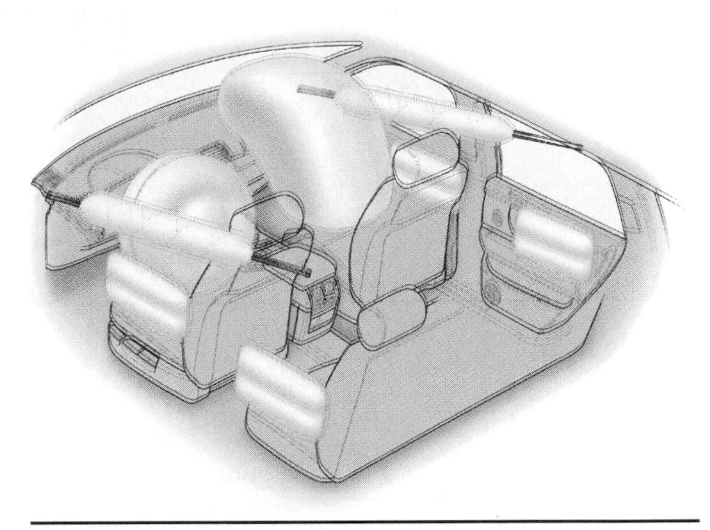

FIGURE 13–65 A total of eight air bags are standard on this vehicle.
(Courtesy of BMW of North America, Incorporated)

Air Bag Module The air bag module contains the inflatable porous nylon air bag assembly and inflator ignitor, which are packaged in a single unit (Figure 13–66). When a vehicle crash is indicated, the ignitor spark ignites a canister of zirconic potassium perchlorate (ZPP) that in turn ignites the inflator propellant

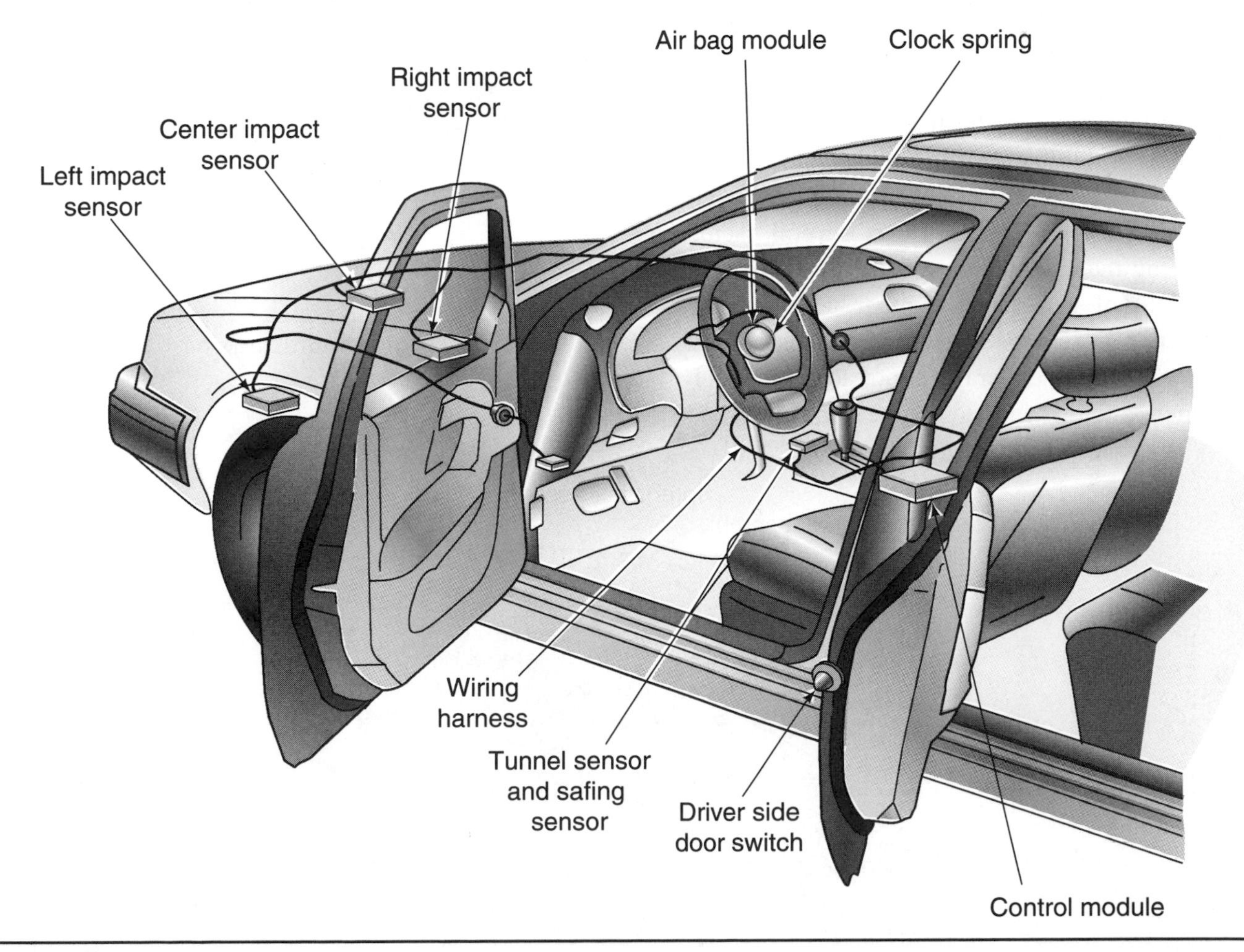

FIGURE 13–64 A first-generation single air bag system.

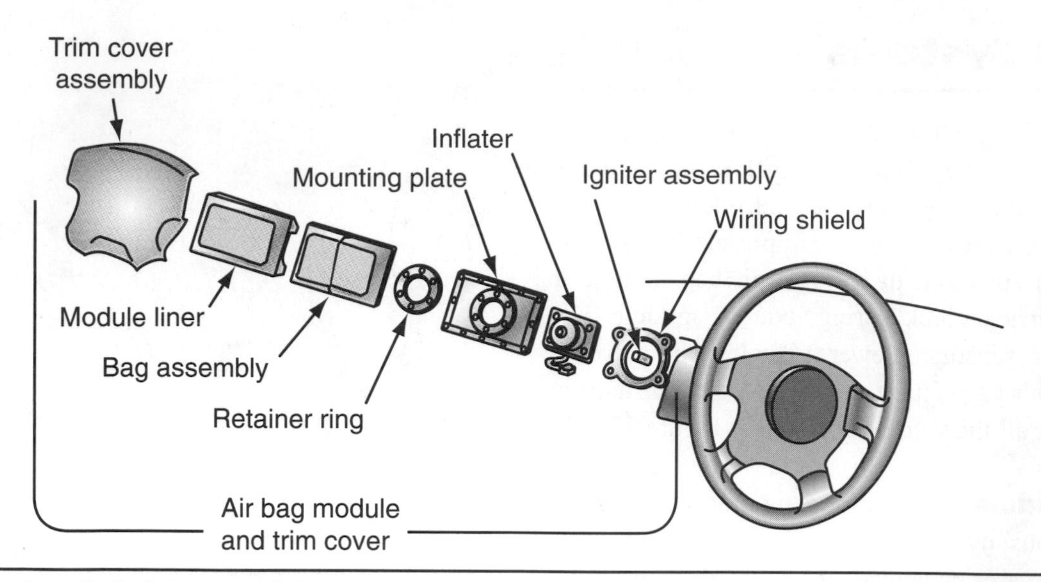

FIGURE 13–66 Detailed air bag module components.

charge containing sodium azide and copper oxide (Figure 13–67). As the inflator propellant burns, it rapidly produces nitrogen gas that inflates the air bag in milliseconds.

In the new generation of air bags (also called smart restraint systems or third-generation air bags), the deployment of the air bags can be multistage (Figure 13–68). The smart air bag system determines several factors before deploying the one or more of the air bag squibs; size and weight of occupants, number of occupants, position of driver in relation to steering wheel, and the severity of the crash. If the crash is not so severe, only one deployment squib is ignited by the control system.

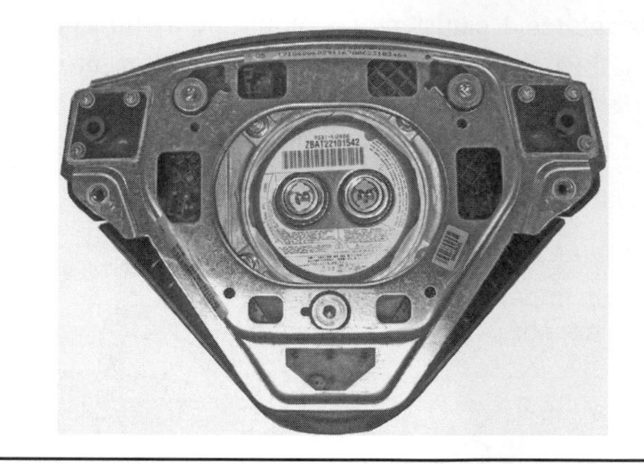

FIGURE 13–68 Typical multistage air bag with two deployment squibs.
(Courtesy of Tim Gilles)

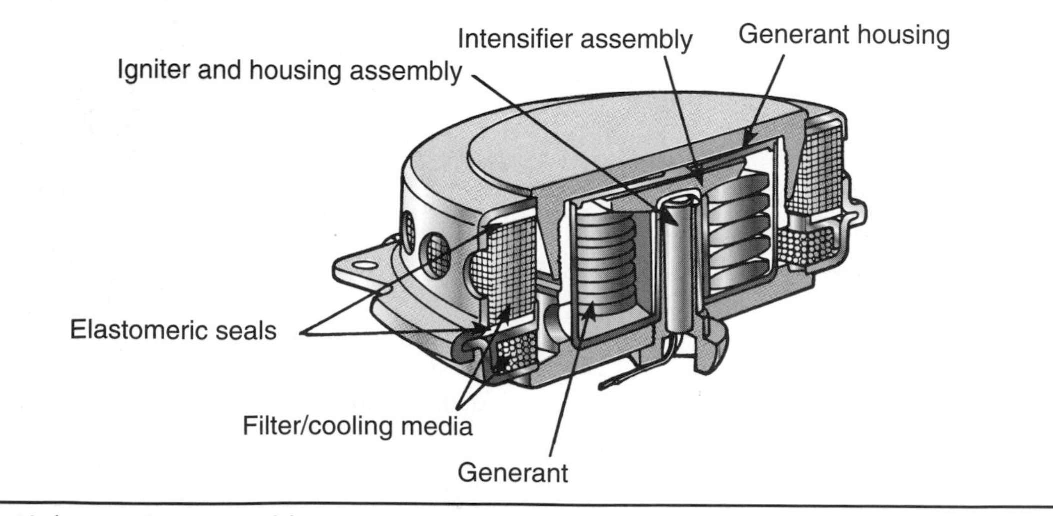

FIGURE 13–67 Air bag igniter assembly.

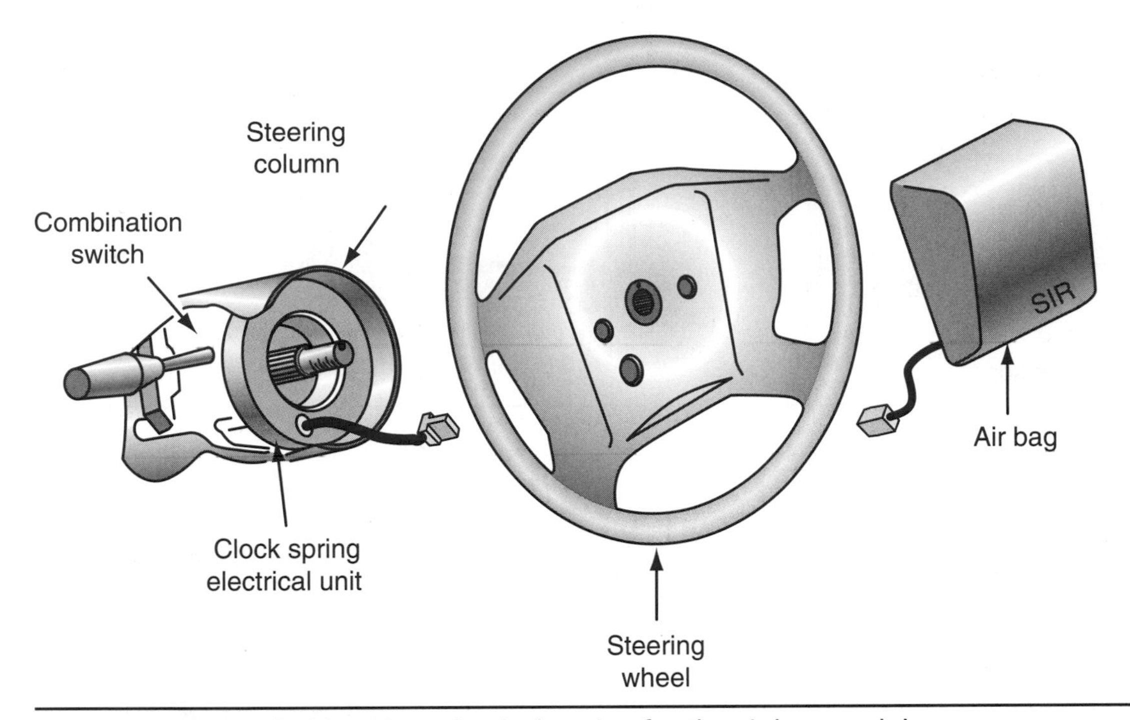

FIGURE 13–69 Typical location of a clock spring for the air bag module.

In more severe crashes, both squibs may fire at the same time or in sequence for greater occupant protection.

Clock Spring The clock spring provides electrical continuity to the air bag module through all steering wheel positions (Figure 13–69). One end of the electrical wires connects to the ribbon spring, and the other end connects to the air bag module. The ribbon spring winds and unwinds as the steering wheel is turned, but only for a certain distance. If the steering column is disconnected for service, the column must be locked in place to prevent accidental overwinding of the clock spring.

Impact Sensors Depending on the SIR system, there may be one to five impact sensors. Most SIR systems with more than one sensor require at least two impact sensors to close to deploy the air bag module. Three common impact sensors are used in SIR systems: mass-type, roller-type, and accelerometers.

A mass-type impact sensor (Figure 13–70) uses a gold-plated normally open switch and a gold-plated sensing mass ball that moves in a highly polished cylinder. In a frontal collision exceeding 10 to 15 miles per hour or more, the sensing ball moves forward enough to close the switch contacts.

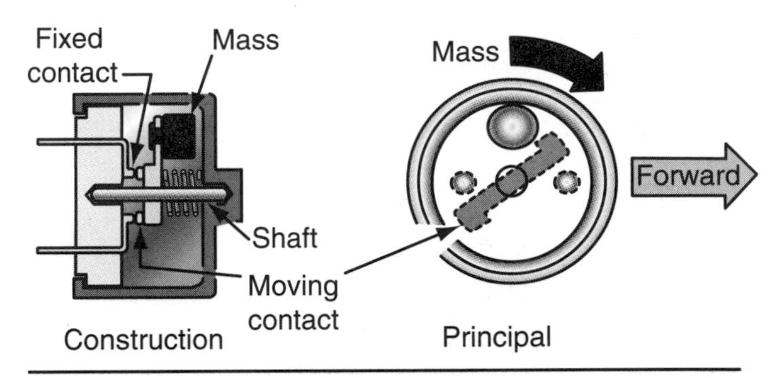

FIGURE 13–70 Mass-type impact sensor.

A roller-type impact sensor uses a spring-loaded roller mass mounted on a ramp (Figure 13–71). In a frontal collision exceeding 10 to 15 miles per hour or more, the roller moves up the ramp and strikes the spring contact to complete the circuit.

The third type of impact sensor is an **accelerometer,** (Figure 13–72) which contains a piezoelectric element that distorts in a frontal collision. The voltage generated by the distorted piezoelectric element is processed by the air bag computer, which deploys the air bag.

Hybrid Inflator Module All passenger vehicles produced after 1995 are required to have a passenger

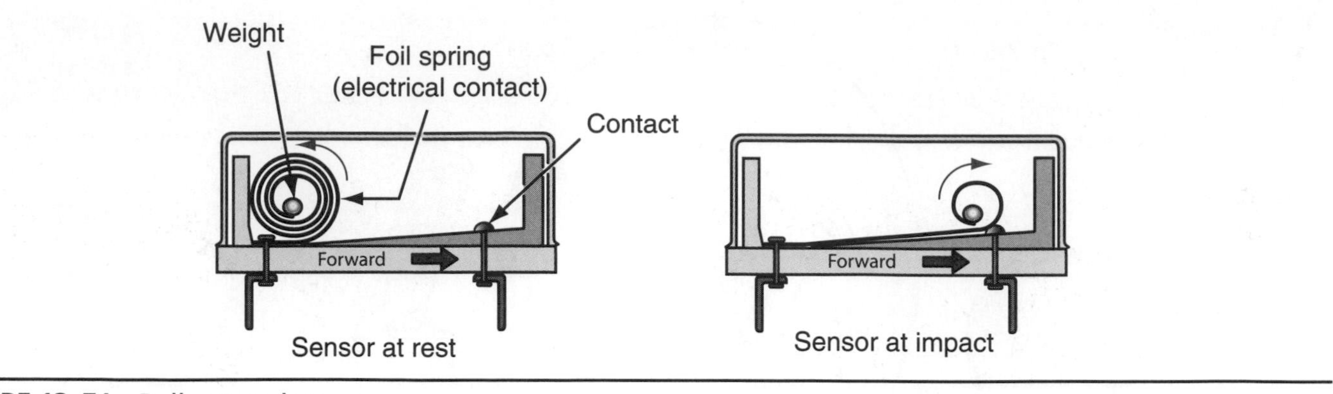

FIGURE 13–71 Roller-type impact sensor.

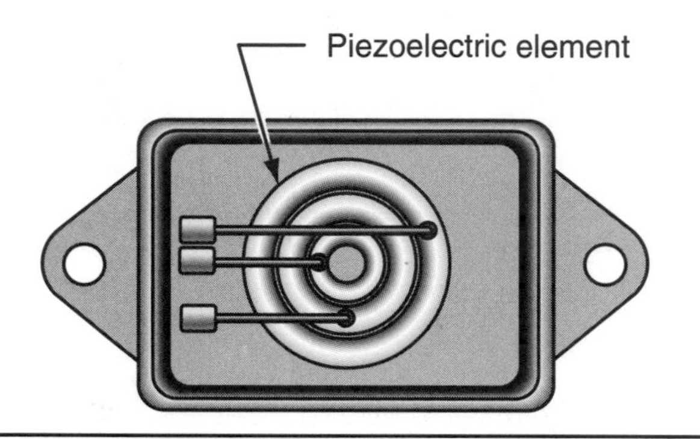

FIGURE 13–72 Accelerometer-type impact sensor.

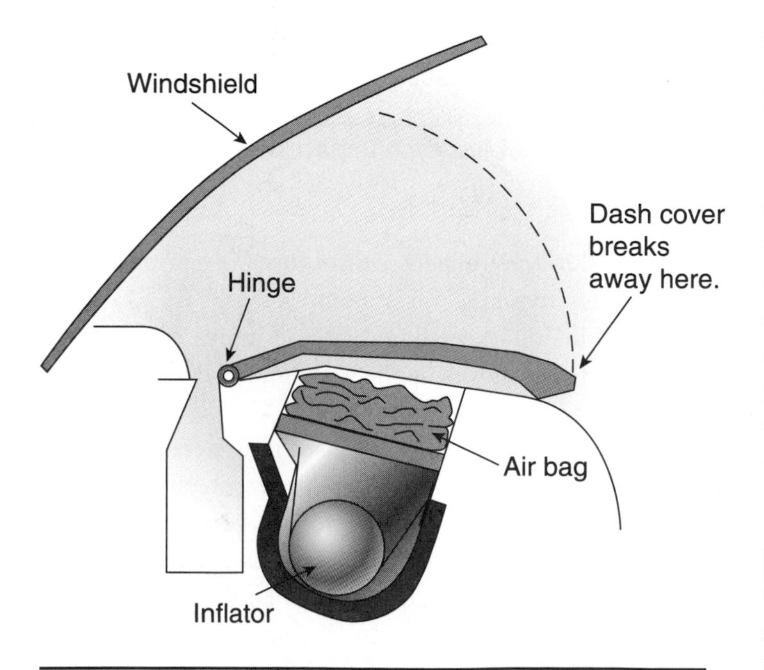

FIGURE 13–73 Components of a passenger side air bag.

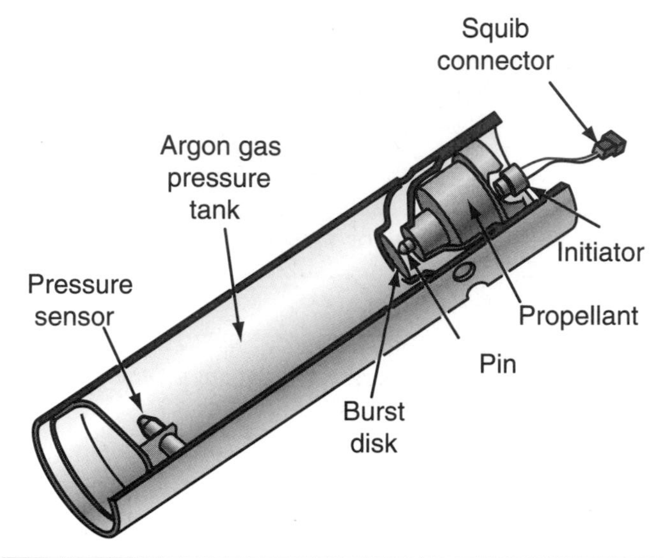

FIGURE 13–74 Hybrid inflator module with argon gas pressure chamber.

side air bag in addition to the driver's side (Figure 13–73), called a **hybrid inflator module.** Because of the distance between the passenger and dash panel compared to the driver/steering wheel, a larger air bag is needed. Some manufacturers use a container of pressurized argon gas (Figure 13–74) that fills the air bag when the initiator explodes and pierces the propellant container (Figure 13–75).

CAUTION *Because of the explosive force of passenger side air bags, never place a rearward facing child safety seat in the front seat. Serious injury or death may result if the air bag is deployed. Although the practice is not encouraged, at times small children need to ride in the front*

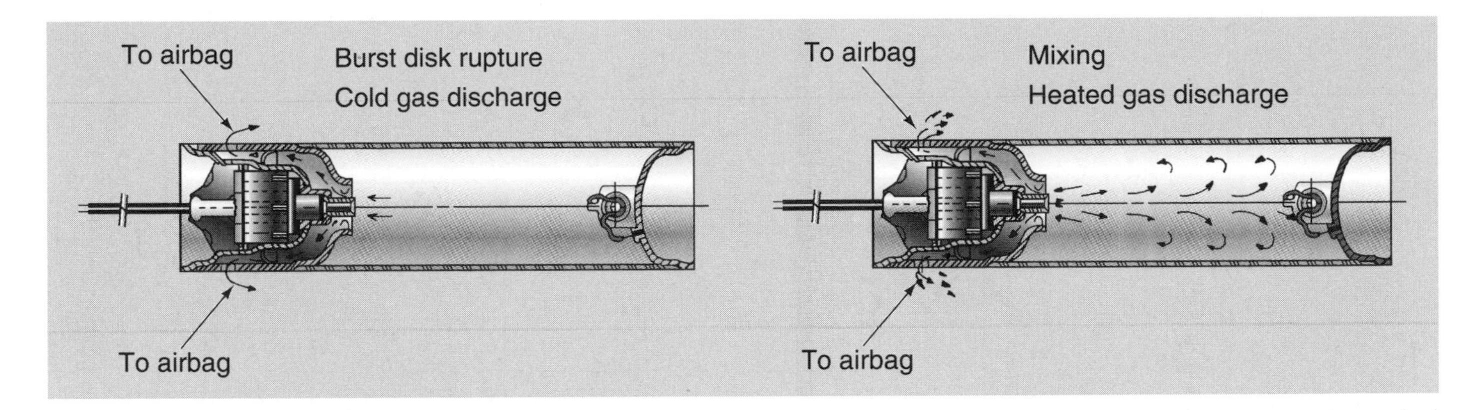

FIGURE 13–75 After initiator explodes and pierces the propellant container, argon gas rapidly fills the air bag.

FIGURE 13–76 A typical air bag deactivation switch.

passenger seat. Beginning in 2006, all new vehicles must be built with a system that allows the passenger side air bag to be deenergized manually with a deactivation switch (Figure 13–76). In addition, an ***occupant detection system*** *(pressure sensor) is installed in the passenger seat to automatically disable the passenger side air bag when an infant, child, or small adult is sitting in the seat (Figure 13–77).*

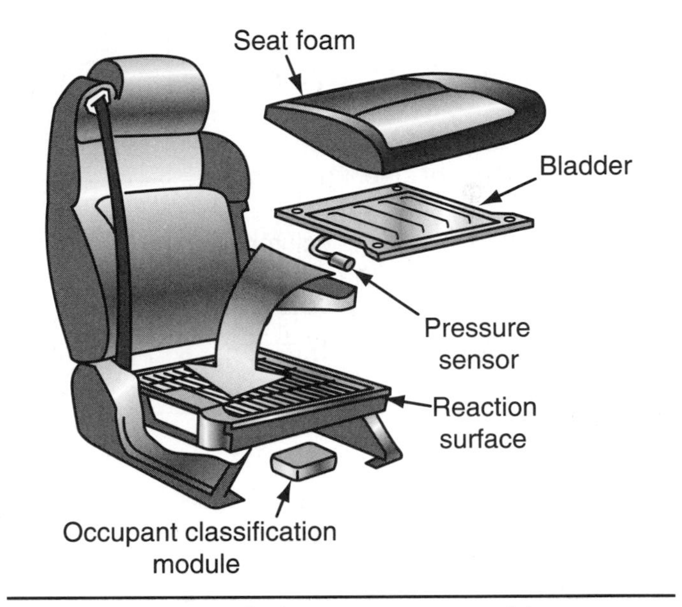

FIGURE 13–77 Typical passenger seat with pressure sensor.

Basic Air Bag Electrical System Figure 13–78 is a simple air bag wiring diagram that requires at least two sensor switches to close before the air bag deploys. Using more than one impact switch helps prevent the inadvertent deployment of an air bag in a minor accident.

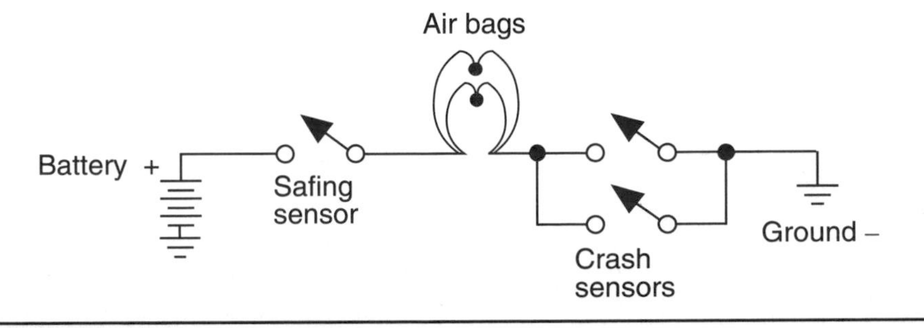

FIGURE 13–78 Simple air bag sensor circuit diagram.

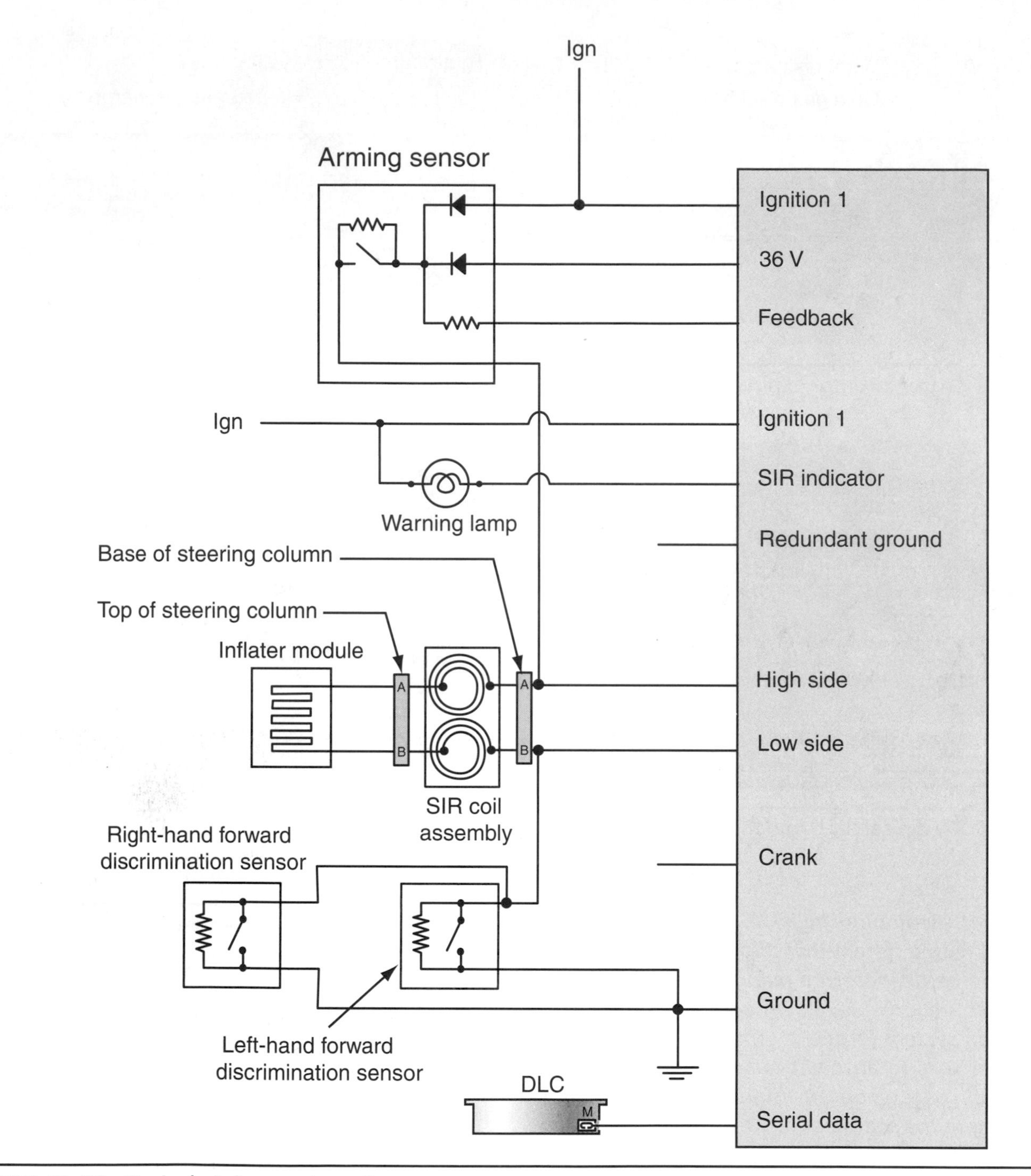

FIGURE 13–79 Typical air bag system diagram.

Figure 13–79 is a typical example of a more complex air bag wiring diagram with fault code detecting capabilities built into the system. Depending on the complexity of the vehicle air bag system and the number of components involved, always refer to specific vehicle manufacturer service information before diagnosing or repairing system components.

Air Bag Safety When you are performing air bag service, read all air bag caution labels on the vehicle (Figure 13–80) and observe the following safety precautions:

1. Avoid static electricity. Use an appropriate static grounding strap when handling air bag components.

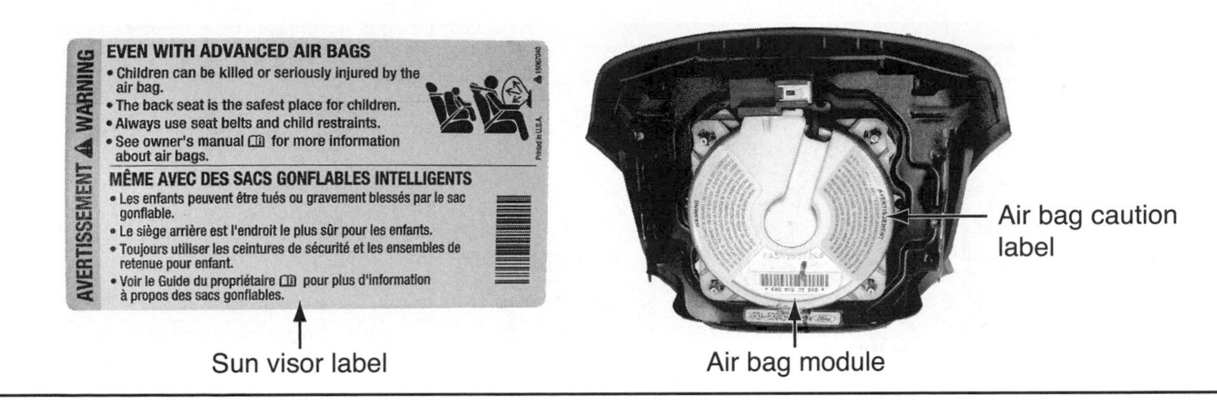

FIGURE 13–80 Typical air bag caution labels located on the back of sun visor and back of air bag module.
(Courtesy of Tim Gilles)

2. Always carry the air bag module facing away from you.
3. Store the air bag module on a clean bench facing upward and in a secure area.
4. Never store an air bag module in a high heat area.
5. Read all manufacturer service information before servicing an air bag system.

CASE STUDIES

CASE 1

This case study is typical of a horn circuit problem that technicians often need to diagnose. Knowledge of basic horn circuits and simple test procedures help you perform the diagnosis quickly and efficiently.

Customer Complaint

A customer brings her vehicle into the shop with a horn system problem. The customer replaced the horn relay, but the problem remains. Figure 13–81 is a diagram of the horn circuit for this vehicle.

Known Information

- ❑ The vehicle operating voltage = 14 volts.
- ❑ All circuit fuses are OK.
- ❑ The horn relay has been replaced.

Circuit Analysis

Answer the following questions on a *separate* sheet of paper.

1. With the information known, what is the *most* likely cause of the inoperative horn circuit?

 ____.

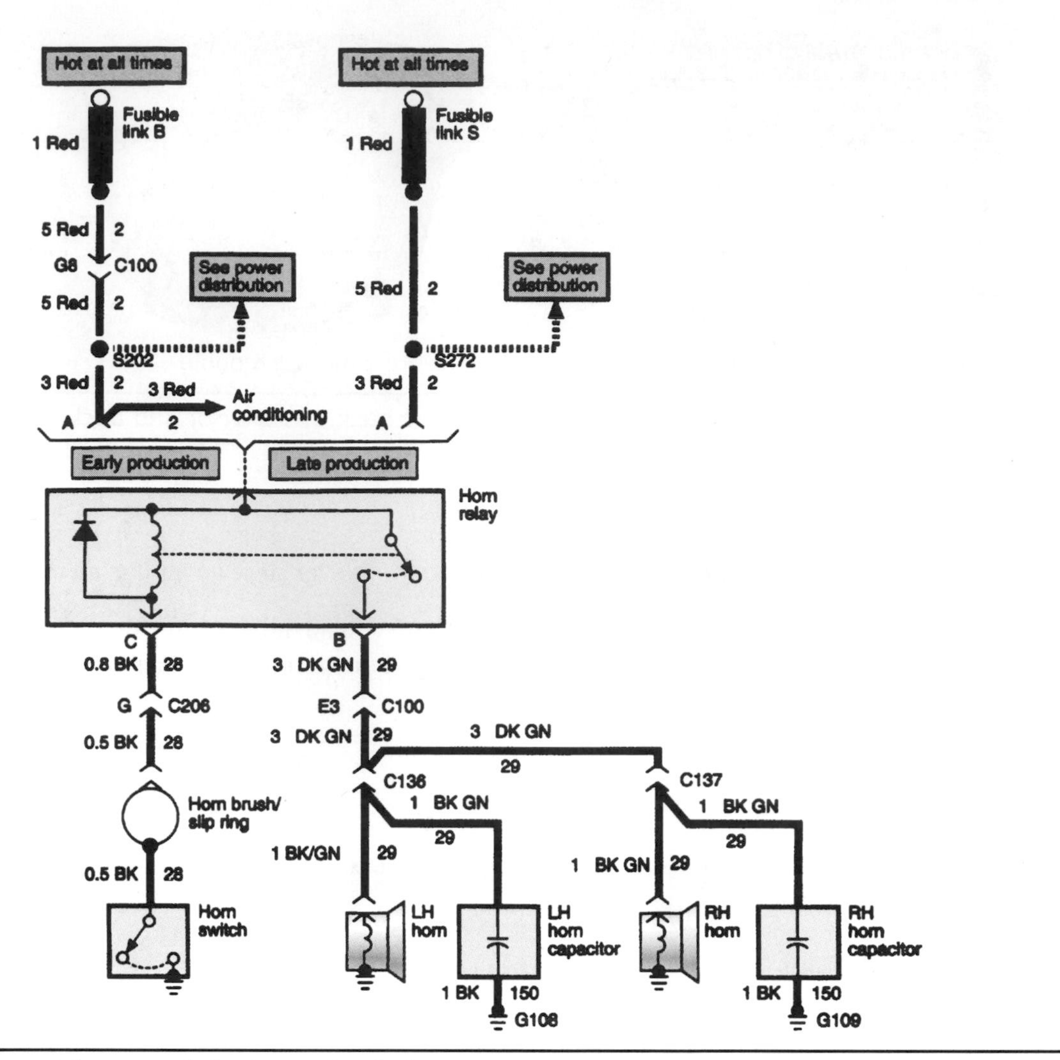

FIGURE 13–81 Diagram of the horn circuit for Case 1.

2. What diagnostic steps and tools would you use to find the suspected problem in this horn circuit?

____.

3. If you disconnected the C206 connector between the horn relay and the horn brush slip ring, what *should* happen in the circuit if the relay side of the C206 connector is grounded? ______________

____.

4. If the circuit is working properly, what should the voltage reading be at the C100 connector from the horn relay to the horns? ____.

5. How could you verify the circuit problem using a jumper wire?

____.

6. What is the best way to repair this system?

____.

7. What diagnostic steps were helpful when troubleshooting this system or explaining it to the customer? ____

____.

8. What is your analysis of this study? ____

____.

CASE 2

This case study is typical of a window defogger circuit problem that technicians often need to diagnose. Knowledge of defogger circuit components and wiring diagrams help you perform the diagnosis quickly and efficiently.

Customer Complaint

The rear window defogger does not work when the switch is turned on, but the indicator light illuminates. The relay was replaced by the customer, but the problem remains. Figure 13–82 is a diagram of the window defogger circuit.

Known Information

- ❑ The vehicle operating voltage = 14 volts.
- ❑ The defogger relay was replaced.
- ❑ The system fuse is OK.

Circuit Analysis

Answer the following questions on a *separate* sheet of paper.

1. With the information known, what is the *most* likely cause of the inoperative defogger circuit?

____.

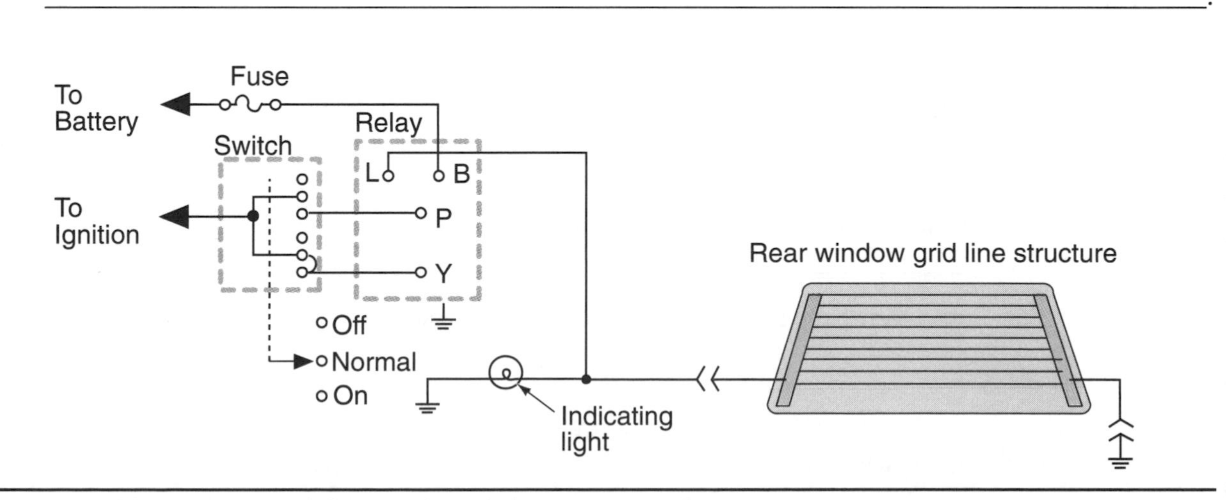

FIGURE 13–82 Diagram of the window defogger circuit for Case 2.

2. If the indicator light is working, what does that tell you about the voltage of the wire leading to the connector at the rear window grid assembly? ______________________________.
3. What diagnostic steps and tools would you use to find the suspected problem in this window defogger circuit? ______________________________.
4. If the voltage at the positive bus bar side of the rear window grid is operating voltage, what is the *next most* likely cause of the circuit problem? ______________________________.
5. How could you verify the circuit problem using a jumper wire? ______________________________.
6. What diagnostic steps were helpful when troubleshooting this system or explaining it to the customer? ______________________________

   ______________________________.
7. What is your analysis of this case study? ______________________________

   ______________________________.

CASE 3

This case study (see Figure 13–83) is typical of an instrument gauge problem. Knowledge of gauge designs, wiring diagrams, and simple test procedures can help you diagnose the gauge circuit safely and efficiently.

Customer Complaint

The customer complains that the instrument cluster fuel gauge reads empty all the time, regardless of the amount of fuel in the tank. The customer wants the system repaired.

Known Information

- ❑ The vehicle operating voltage = 14 volts.
- ❑ The circuit fuse is OK.

Circuit Analysis

Answer the following questions on a *separate* sheet of paper.

1. With the information known, what is the *most* likely cause of the inoperative fuel gauge circuit?

   ______________________________.
2. What diagnostic steps and tools would you use to find the suspected problem in this fuel gauge circuit? ______________________________.
3. Looking at the wiring diagram, what should the voltage supplied to the fuel gauge be? ______________

   ______________________________.
4. If you disconnected the C402 connector at the fuel gauge sending unit and grounded the gauge side of the connector with a jumper wire, what should the gauge needle show? ______________

   ______________________________.

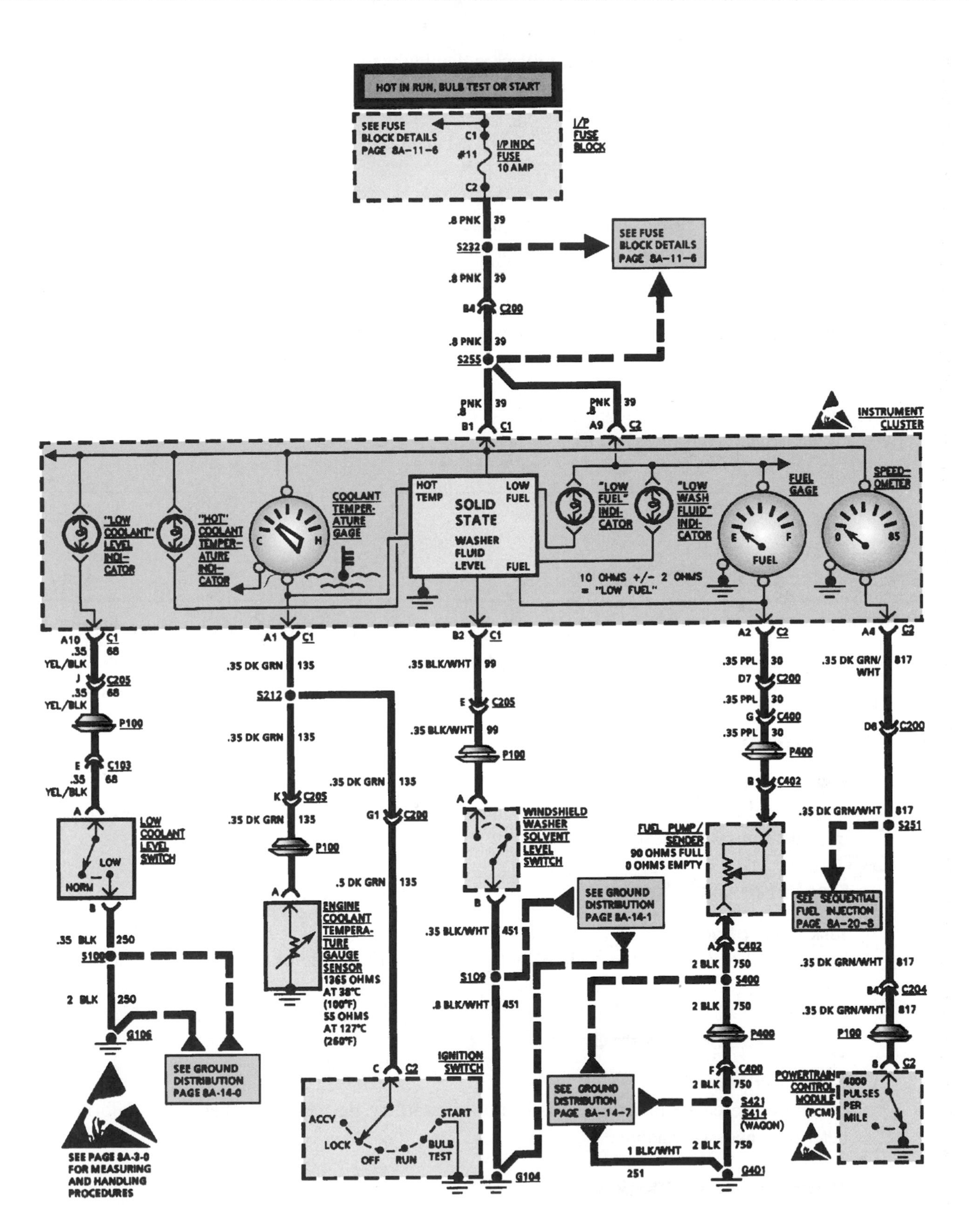

FIGURE 13–83 Diagram of the instrument cluster for Case 3.

5. For the fuel gauge needle to read full, what should the total fuel gauge circuit resistance be?

 ______________________________.

6. How is the fuel gauge circuit resistance varied to change the gauge reading needle? ______________

 ______________________________.

7. If the technician suspected a bad fuel gauge sending unit, what could he or she use to simulate the action of the sending unit to check for proper fuel gauge operation?______________

 ______________________________.

8. What diagnostic steps were helpful when troubleshooting this system or explaining it to the customer? ______________

 ______________________________.

9. What is your analysis of this case study? ______________

 ______________________________.

Hands-On Vehicle Tests

The following ten hands-on vehicle checks are included in the NATEF (National Automotive Technicians Education Foundation) Task List. Some tasks are grouped together due to their close association with each other. Complete your answers to the following questions on a *separate* sheet of paper.

Performance Task 1

Task Description

Inspect and test gauges and gauge sending units for cause of intermittent, high, low, or no gauge readings; determine necessary action.

Performance Task 2

Task Description

Inspect and test connectors, wires, and printed circuit boards of gauge circuits; determine necessary action.

Performance Task 3

Task Description

Diagnose the cause of incorrect operation of warning devices and other driver information systems; determine necessary action.

Performance Task 4

Task Description

Inspect and test sensors, connectors, and wires of electronic (digital) instrument circuits; determine necessary action.

Task Objectives

- ❑ Obtain a vehicle that can be used for these tasks. What model and year of vehicle are you using for these tasks? ____________________.
- ❑ Locate a wiring diagram and associated component locator references that show the instrument cluster gauges, warning lights, associated sending units, wiring, and connectors. List them in the following chart.

Gauge or warning light	Sending unit location

- ❑ Does the circuit contain an instrument voltage regulator (IVR)? ____________. How do you check an IVR for proper operation? ____________________.
- ❑ If a gauge or warning light is not reading or operating properly, what are some simple diagnostic procedures used to determine which circuit component is faulty?

Component	Suspected fault	Test procedure

- ❑ Which tools and equipment would be useful to inspect and test circuit wires and connectors? ______ ____________________.
- ❑ What diagnostic steps are the most helpful when diagnosing a gauge or warning light circuit problem? ____________________.

❑ What is your analysis of using wiring diagrams to help diagnose instrument cluster gauge and warning light circuits? ____

____.

Task Summary

After performing the preceding NATEF tasks, what can you determine will be helpful in knowing how to check instrument cluster gauges and warning lights for proper operation? ____________________

____

____.

Performance Task 5

Task Description

Diagnose incorrect horn operation; perform necessary action.

Task Objectives

❑ Obtain a vehicle that can be used for this task. What model and year of vehicle are you using for this task? ____.

❑ Locate a wiring diagram and associated component locators that show the horn circuit, fuses, and various components in the circuit. List them in the following chart.

Circuit component	Location

❑ If the horn relay is suspected faulty, what test procedures can you use to verify that the relay is faulty?

____.

❑ If the horn button switch or slip ring is suspected faulty, what test procedures can you use to verify that the horn button switch or slip ring is faulty? ____________________________

____.

❑ What voltage should be to the horns in the circuit? ____________________________________.

❑ To test the horn(s) for proper operation, how could you use a jumper wire to check their normal function? ____.

- ❑ What diagnostic steps are *most* useful when troubleshooting this system? ______________________________.
- ❑ What is your analysis of using wiring diagrams to help diagnose a horn circuit? ______________________________.

Task Summary

After performing the preceding NATEF tasks, what can you determine will be helpful in knowing how to check a horn circuit for proper operation? ______________________________

____

____

______________________________.

Performance Task 6

Task Description

Diagnose incorrect wiper operation; diagnose wiper speed control and park problems; perform necessary action.

Performance Task 7

Task Description

Diagnose incorrect windshield washer operation; perform necessary action.

Task Objectives

- ❑ Obtain a vehicle that can be used for this task. What model and year of vehicle are you using for this task? ______________________________.
- ❑ Locate a wiring diagram that shows the wiper/washer circuit, fuses, and associated components.
- ❑ Using the wiring diagram(s), identify areas in the circuit such as connectors, wires, fuses, and components that can cause a possible problem.

Identified area	Symptom	Cause	Test procedure

- ❑ ***Caution:*** Remember safety, and keep fingers clear of levers.

- ❑ Perform the diagnostic steps and procedures identified in the previous question. Does this help isolate the possible circuit problem(s)?____.
- ❑ Can the windshield washer operate independent of the wiper motor low speed selection? __________.
- ❑ What diagnostic steps are helpful when troubleshooting a wiper/washer system? ________________

  ____

  ____.
- ❑ What is your analysis of using wiring diagrams to help diagnose wiper/washer circuit problems?

  ____

  ____.

Task Summary

After performing the preceding two NATEF tasks, what can you determine will be helpful in knowing how to check wiper and washer circuits for proper operation? ______________________________

____.

Performance Task 8

Task Description

Diagnose incorrect operation of motor-driven accessory circuits; determine necessary action.

Task Objectives

- ❑ Obtain a vehicle with an electric window circuit that an be used for this task. What model and year of vehicle are you using for this task?____.
- ❑ Locate a wiring diagram for the vehicle that shows the electric door window circuit.
- ❑ Using the wiring diagram(s), identify areas in the circuit such as connectors, wires, fuses, and components that can cause a possible problem.

Identified area	Symptom	Cause	Test procedure

- ❑ Perform the diagnostic steps and procedures identified in the previous question. Does this help identify the circuit problem(s)?____.
- ❑ What is the circuit voltage supplied to the window motor assembly? ____________________.

❑ What is your analysis of using wiring diagrams to help diagnose a motor driven accessory circuit problem?____.

Task Summary

After performing the preceding NATEF tasks, what can you determine will be helpful in knowing how to check a motor driven accessory circuit for proper operation? ____.

Performance Task 9

Task Description

Diagnose incorrect heated glass, mirror, or seat operation; determine necessary action.

Task Objectives

❑ Obtain a vehicle that can be used for this task. What model and year of vehicle are you using for this task? ____.

❑ Find a wiring diagram that shows the rear window heated glass circuit for this vehicle.

❑ Does this heated glass circuit follow the characteristics of a parallel circuit? __________________.

❑ Identify the following components and list their location on the vehicle.

1. Rear window grid:____.
2. Control switch: ____.
3. Relay (if applicable):____.
4. Circuit fuse and fuse number: ____.
5. Circuit ground(s): ____.
6. Circuit indicator light:____.
7. Circuit timer assembly (if applicable): ____.
8. Other circuit components:____.

Circuit Checks

❑ Check the operation of the indicator light. If operating, proceed to the rear window heating grid. If indicator is not operating, check to see if the rear window grid feels warm. If not, is the circuit fuse OK?____.

❑ Check for circuit source voltage at rear window heater grid.

DVOM voltage reading: ____.

❑ What is the voltage from the positive end of the heating grid to the negative end? ________________.

- ❑ If you leave the DVOM hooked to the negative side and slowly move the positive lead along a grid line, what happens to the voltage readings? ______________________________.

  Why? ______________________________.
- ❑ Why are rear window grid lines wired in parallel? ______________________________.
- ❑ Remove the circuit fuse and install a DVOM in series in its place. What is the amperage of the circuit? ______________________________.
- ❑ What happens to the total circuit amps as the rear heating grid starts to heat up? ______________.

Task Summary

After performing the preceding NATEF task, what can you determine that will be helpful in diagnosing a rear window heating grid circuit? ______________________________

____

______________________________.

Performance Task 10

Task Description

Diagnose incorrect electric door lock operation; determine necessary action.

Task Objectives

- ❑ Obtain a vehicle with an electric door lock or hatch/trunk circuit that an be used for this task. What model and year of vehicle are you using for this task? ______________________________

  ______________________________.
- ❑ Locate a wiring diagram for the vehicle that shows the door lock or trunk/hatch circuit.
- ❑ Using the wiring diagram(s), identify areas in the circuit such as connectors, wires, fuses, and components that can cause a possible problem.

Identified area	Symptom	Cause	Test procedure

- ❑ Perform the diagnostic steps and procedures identified in the previous question. Does this help identify the circuit problem(s)? ______________________________.
- ❑ What is the circuit voltage supplied to the door lock solenoid assembly? ______________.

- ❑ What is the circuit voltage supplied to the trunk/hatch solenoid assembly?____________________.
- ❑ What is your analysis of using wiring diagrams to help diagnose door lock or trunk/hatch circuit problems?____

  ____.

Task Summary

After performing the preceding NATEF task, what can you determine will be helpful in knowing how to check door locks or trunk/hatch circuits for proper operation? ________________________________

____

____.

Summary

- ❑ Most horn circuits use a relay to separate the high-amperage horns from the control switch.
- ❑ A typical horn assembly uses electromagnetism to vibrate a flexible diaphragm and contact points.
- ❑ Horn circuits with two horns normally use a different pitch for each horn.
- ❑ Most horn switches installed in a steering wheel use a slip ring and spring-loaded sliding contact to complete the connection to ground, when needed.
- ❑ The horn switch may be integrated as part of the air bag module assembly.
- ❑ A typical blower control circuit uses a series resistor block to control the blower motor speed.
- ❑ Many blower circuits use a relay to control the high-speed blower motor speed.
- ❑ In an electronic climate control system, the heater blower speed is controlled by a power module.
- ❑ A nondepressed park wiper circuit parks the wiper blades at the end of their normal stroke by the edge of the lower windshield molding.
- ❑ A depressed park wiper circuit parks the wiper blades past the end of their normal stroke, below the edge of the lower windshield molding.
- ❑ An interval, or intermittent, wiper system operates in time intervals depending on the delay setting requested by the driver.
- ❑ Most wiper washer pumps operate in conjunction with the low-speed wiper motor setting.
- ❑ Typical rear window defogger grid wires are parallel to each other.
- ❑ Rear window defogger circuits with solid-state timers open the control switch after the first 10 minutes of operation and every 5 minutes on succeeding operations.

- ❑ Direct drive power window motors use a pinion gear meshed with a sector gear to raise the window regulator. Remember safety: Keep fingers clear.
- ❑ Rack-and-pinion window motors use a rack with a flexible strip of gear teeth operating a window regulator.
- ❑ There are three basic configurations of power seat circuits: two-way, four-way, and six-way.
- ❑ A typical power mirror circuit consist of two motors for each mirror: one for up/down and the other for left/right function.
- ❑ Two types of analog gauge designs are used today: the thermal gauge and the electromagnetic gauge.
- ❑ An instrument voltage regulator (IVR) maintains a constant voltage to the gauge circuit under all battery and charging system conditions.
- ❑ There are four basic types of electromagnetic gauges: the d'Arsonval gauge, the three-coil bobbin gauge, the two-coil balancing gauge, and the air-core gauge.
- ❑ A temperature gauge sending unit is typically a thermistor resistance device that measures coolant temperature.
- ❑ A fuel gauge sending unit is typically a variable resistor that is moved manually by the fuel level in the fuel tank.
- ❑ Warning lamp sending units are normally open or closed depending on the circuit design.
- ❑ The air bag module contains an inflatable porous nylon air bag assembly and an inflator ignitor.
- ❑ The clock spring in an air bag system provides electrical continuity to the air bag module through all steering wheel positions.
- ❑ In the third-generation air bag systems, the deployment of the air bag can be multistage depending on the severity of the crash.
- ❑ In most air bag systems, at least two impact sensors must close to deploy the air bag module.
- ❑ Beginning in 2006, all new vehicles must be built with an occupant detection system that allows the disabling of the passenger side air bag.

Key Terms

air-core gauge
accelerometer
climate control panel (CCP)
clock spring
d'Arsonval gauge
depressed park
electrochromic rearview mirror
hybrid inflator module
instrument voltage regulator (IVR)
interval
nondepressed park
occupant detection system
supplemental inflatable restraint (SIR)
thermal gauge
three-coil gauge
two-coil gauge
trimotor

Review Questions

Short Answer Essays

1. Describe how a blower motor circuit uses a resistor block assembly to control the low-, medium-, and high-speed operation.
2. Describe how a technician would repair breaks in the grid wire of a rear window defogger.
3. What is the purpose of an IVR in an analog gauge circuit?
4. Describe how a driver's side air bag deploys on a frontal collision impact.
5. What are some simple safety rules when handling air bag systems?

Fill in the Blanks

1. In a horn circuit with a relay, the ________________________ competes the circuit to ground from the relay coil.
2. In most vehicles with rear window defogger circuits, the grid lines are hooked in ________________________ to each other.
3. A reversible permanent magnet motor pack called a ________________________ controls the function of a ________________________-way power seat circuit.
4. When the engine coolant temperature increases, the temperature gauge sending unit thermistor resistance ____________________, and current flow to the gauge circuit ____________________.
5. In an air bag module the ignitor spark ignites a canister of ________________________ to ignite the inflator propellant.

ASE-Style Review Questions

1. Technician A says that *all* horn circuits have a relay to prevent excess voltage from flowing through the horn button switch. Technician B says the horn brush/slip ring completes the current path to ground in a typical horn circuit. Who is correct?
   A. A only
   B. B only
   C. Both A and B
   D. Neither A nor B

2. Technician A says the pitch of a horn is caused by a vibrating diaphragm. Technician B says that in a dual horn system, the pitch is normally the same for both horns. Who is correct?
   A. A only
   B. B only
   C. Both A and B
   D. Neither A nor B

3. Technician A says the resistor block in a blower motor circuit is wired in series between the control switch and the blower motor. Technician B says the blower fan speed is controlled by the resistor block and blower switch. Who is correct?
   A. A only
   B. B only
   C. Both A and B
   D. Neither A nor B

4. Two technicians are discussing a blower circuit with a relay. Technician A says the relay controls high-speed fan operation. Technician B says the blower's low-speed is controlled by the relay. Who is correct?
   A. A only
   B. B only
   C. Both A and B
   D. Neither A nor B

5. Technician A says a nondepressed park wiper circuit parks the wiper blades at the end of their normal stroke. Technician B says a depressed park wiper circuit parks the wiper blades below the edge of the lower windshield molding. Who is correct?
   A. A only
   B. B only
   C. Both A and B
   D. Neither A nor B

6. Two technicians are discussing the air bag occupant detection system. Technician A says this system can be manually deactivated with a key switch. Technician B says there is a pressure switch in the passenger seat to detect the presence of an infant or small child. Who is correct?
   A. A only
   B. B only
   C. Both A and B
   D. Neither A nor B

7. Technician A says an interval wiper system operates in time intervals depending on the delay setting requested. Technician B says an intermittent wiper system uses a spring-loaded pressure switch to time the activation of the wiper motor. Who is correct?
   A. A only
   B. B only
   C. Both A and B
   D. Neither A nor B

8. The following rear window defogger circuit is being discussed. Technician A says the indicator light illuminates even if the relay is defective. Technician B says the defogger switch controls current through the relay coil. Who is correct?

   A. A only

   B. B only

   C. Both A and B

   D. Neither A nor B

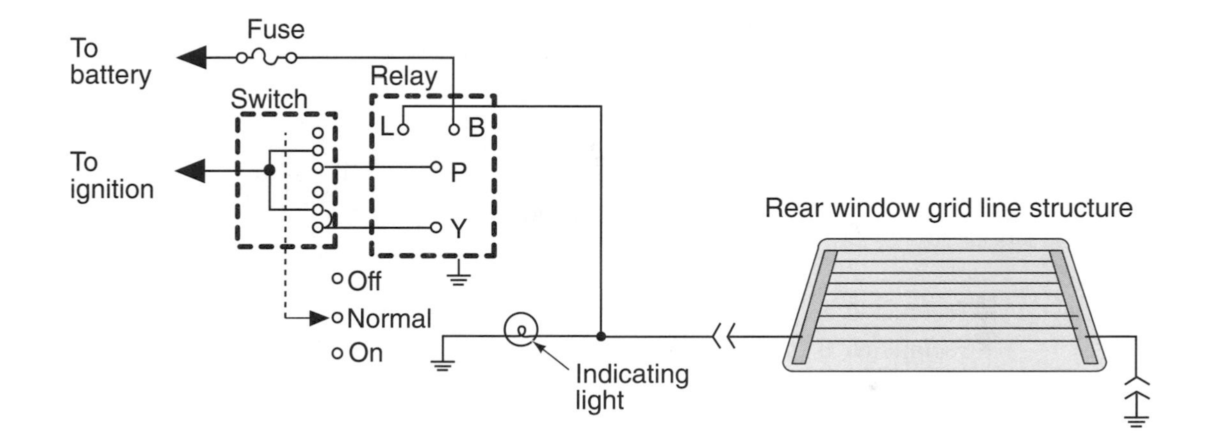

9. Technician A says the master control switch for power windows has total control of all windows in the system. Technician B says the rotation of the power window motor depends on which direction the control switch is moved. Who is correct?

   A. A only

   B. B only

   C. Both A and B

   D. Neither A nor B

10. The testing of the following power window circuit is being discussed. Technician A says the test light across the switch should light when the window switch is closed in either direction. Technician B says the test light should light when the master switch is closed in either direction. Who is correct?

    A. A only

    B. B only

    C. Both A and B

    D. Neither A nor B

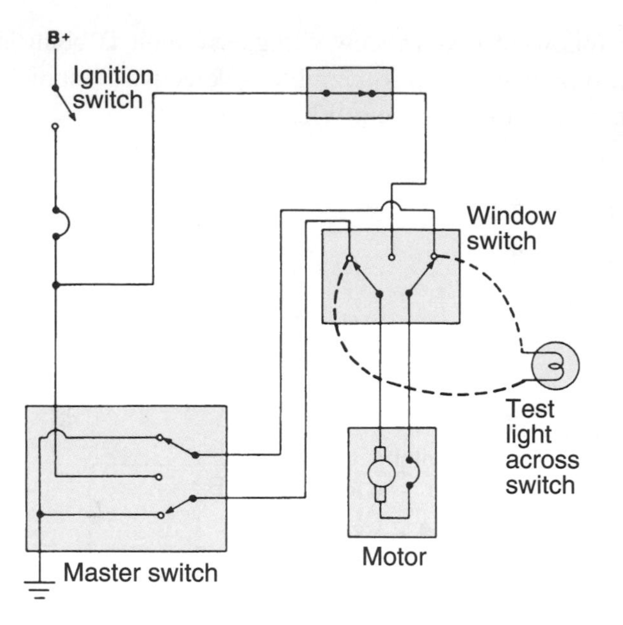

11. Technician A says that all power window motors are direct drive with a pinion and sector gear. Technician B says a flexible rack with gear teeth is used to attach the window regulator in all power window systems. Who is correct?
    A. A only
    B. B only
    C. Both A and B
    D. Neither A nor B

12. Two technicians are discussing the fuel gauge system on the next page. Technician A says that electromagnetism operates the gauge needle. Technician B says the resistance of the sending unit depends on the fuel level. Who is correct?
    A. A only
    B. B only
    C. Both A and B
    D. Neither A nor B

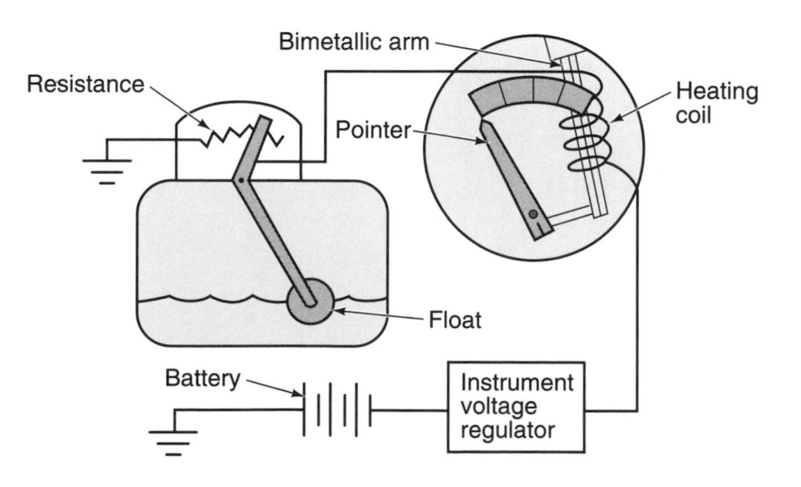

13. Technician A says an IVR can be used on bimetallic gauges. Technician B says the IVR opens if the alternator overcharges the battery. Who is correct?
    A. A only
    B. B only
    C. Both A and B
    D. Neither A nor B

14. Technician A says a sending unit for an oil pressure gauge can also be used for an oil pressure warning light system. Technician B says the sending unit for a temperature warning light is normally open during normal engine temperatures. Who is correct?
    A. A only
    B. B only
    C. Both A and B
    D. Neither A nor B

15. Technician A says the air bag module is electrically connected in the steering column through a clock spring assembly. Technician B says all air bag systems use a roller-type impact sensor to arm the air bag module. Who is correct?
    A. A only
    B. B only
    C. Both A and B
    D. Neither A nor B

14 Batteries and Testing

Introduction

The battery is often referred to as the heart of a vehicle's electrical system. It is an electrochemical device that changes chemical energy to electrical energy by the chemical reaction of two dissimilar plates immersed in an electrolyte solution. The design and testing of the battery has come a long way since Alessandro Volta invented the voltaic pile in the late 1700s. Increased demands from the vehicle electrical system, as well as new innovations in electric vehicles or hybrid electric vehicles, have contributed to recent advancements in battery technology. New construction case designs, plate material and grid improvements, and different electrolyte combinations all contribute to the batteries in use today.

Objectives

When you complete this chapter you should be able to:

- ❑ Describe the main components of a conventional battery.
- ❑ Describe the construction differences and features between conventional, maintenance free, deep cycle, and gel-cell batteries.
- ❑ Explain the chemical reaction that takes place during the discharge cycle and the charging cycle in the battery.
- ❑ Explain the different methods used to rate batteries.
- ❑ Explain the safety cautions required when working around or with batteries.
- ❑ Demonstrate how to test a battery's state of charge using a hydrometer and a voltmeter.
- ❑ Explain the procedures and test equipment used to load test a battery for proper capacity.
- ❑ Demonstrate how to properly use a battery charger and the appropriate load settings.
- ❑ Explain the cautions and procedures used to jump-start a vehicle safely.
- ❑ Demonstrate proper battery service procedures.

Battery Types and Construction

Several types of batteries are in use today, ranging from devices such as flashlights, watches, and calculators, to automobiles and trucks. The simplest form of battery is a single **voltaic cell** that uses two unlike metals and an alkali, salt, or acid solution to produce a voltage (Figure 14–1). In this example, a nickel and a penny are two unlike metals separated by a paper towel moistened with a person's saliva. The saliva contains sufficient alkali or acids that act as a suitable electrolyte solution to produce a small voltage. An **electrolyte** is a chemical compound of alkali, salt, or acid solution that is capable of conducting electrical energy.

The amount of voltage produced is determined by the materials used in the battery cell. Figure 14–2 is a table of metals used to produce various types of batteries. The chemical reactions that take place between the metals depends on their ability to receive or transfer free electrons.

The table of metals is divided into the two common types of available cells. The first is a **primary cell,** which is defined as a cell that *cannot be recharged* to its original voltage condition. The chemical reaction that takes place while voltage is produced erodes away one or both of the cell plates. In Figure 14–3, a carbon-zinc primary cell design is used in flashlight batteries that produces approximately 1.5 volts each. A carbon rod is used as a positive electrode with a zinc container acting as the negative electrode. The electrolyte mixture is a manganese dioxide paste that permits the battery to be used in any position without spillage.

The second common type of battery cell is a **secondary cell,** which is a cell that *can be recharged* once its original stored energy has been depleted. The lead-acid cell of an automotive battery is a perfect example of a secondary cell battery (Figure 14–4). The negative plate is made of pure lead (Pb), and the positive plate is made of lead dioxide (PbO_2). The electrolyte contains a diluted mixture of **sulfuric acid** (H_2SO_4), a conductive and reactive mixture consisting of 64 percent water and 36 percent sulfuric acid, by weight.

Conventional Battery Construction

The battery in a vehicle is a **conventional battery,** which is a lead-acid battery that uses a lead antimony mixture in the cell plates. The battery is constructed of several important parts. The first is the grid plate made of a **lead antimony** that contains up to a 10% mixture of lead and lead alloys to strengthen the plate grid frame (Figure 14–5). The grid plate has vertical and horizontal bars that hold the paste that forms the positive and negative plates.

The active material in the positive plate is a reddish-brown **lead peroxide** (PbO_2) paste. The negative plate material is a grayish **sponge lead** (Pb) paste. In each battery cell there are up to 13 plates arranged alternately (negative, positive, negative, positive, etc.) (Figure 14–6). Plate are divided by **separators** made of a microporous material that prevents contact between each positive and negative plate. When a group of positive plates and

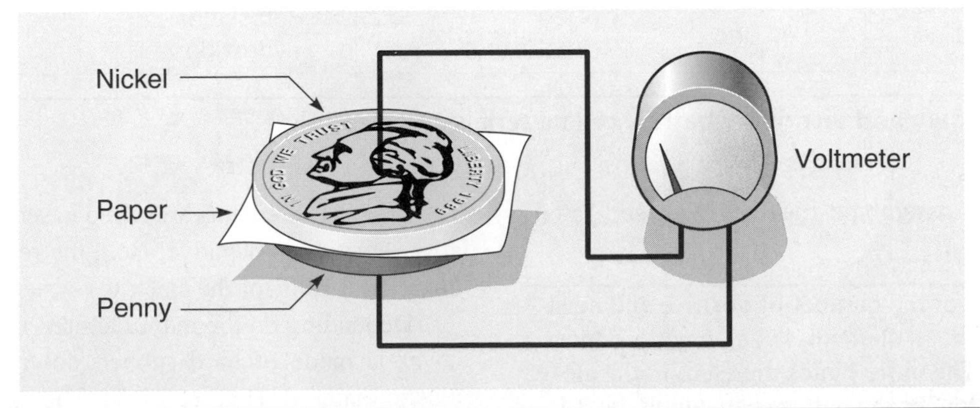

FIGURE 14–1 A simple voltaic cell.

Primary Cell Battery				
Cell type	Negative plate	Positive plate	Electrolyte	Volts per cell (V)
Alkaline	Zinc	Manganese dioxide	Potassium hydroxide	1.5
Carbon-zinc (Leclanche)	Zinc	Carbon Manganese dioxide	Ammonium chloride	1.5
Edison-Lalande	Zinc	Copper oxide	Sodium hydroxide	0.8
Mercury	Zinc	Mercuric oxide	Potassium hydroxide	1.35
Silver-zinc	Zinc	Silver oxide	Potassium hydroxide	1.6
Zinc-air	Zinc	Oxygen	Potassium hydroxide	1.4

Secondary Cell Battery				
Cell type	Negative plate	Positive plate	Electrolyte	Volts per cell (V)
Lead-acid	Lead	Lead dioxide	Diluted sulfuric acid	2.2
Lithium-ion	Carbon or graphite	Lithium metallic oxide	Lithium salt and organic solvents	3.6
Nickel-cadmium	Cadmium	Nickel hydroxide	Potassium hydroxide	1.2
Nickel-iron (Edison)	Iron	Nickel oxide	Potassium hydroxide	1.4
Nickel-metal hydride	Metal hydride	Nickel oxyhydroxide	Potassium hydroxide	1.2
Silver-cadmium	Cadmium	Silver oxide	Potassium hydroxide	1.1
Silver-zinc	Zinc	Silver oxide	Potassium hydroxide	1.5

FIGURE 14–2 Primary and secondary battery cell materials and electrolyte.

negative plates are assembled together with separators, they form an **element.**

Note: Regardless of the number of positive and negative plates in an element, the element produces 2.1 volts. The more plates there are in the element, the longer the battery can supply its 2.1 volts per element.

When six elements are wired in series (positive to negative, positive to negative, etc.), the resulting battery produces 12.6 volts for the electrical system (Figure 14–7).

Depending on the manufacturer, the battery case and cover is made of hard rubber, polypropylene, or other plastic materials (Figure 14–7). The case must:

1. Separate and stabilize the elements.

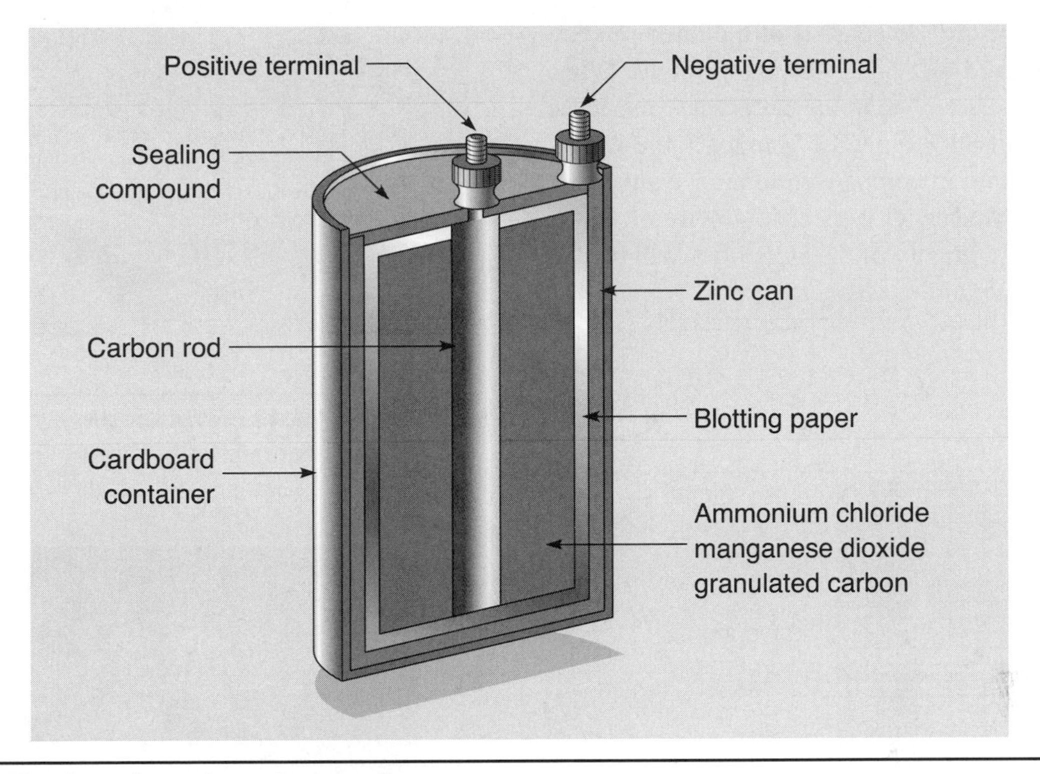

FIGURE 14–3 Simple carbon-zinc primary cell.

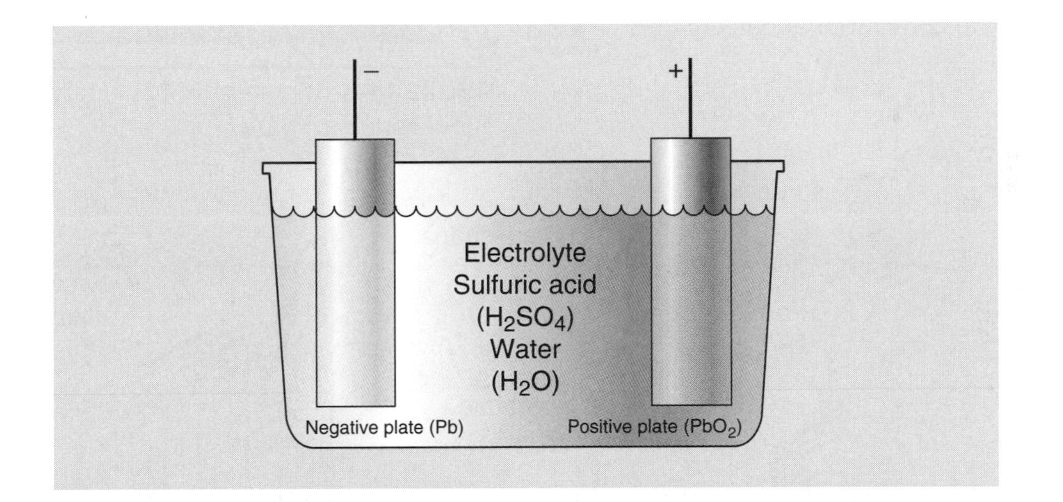

FIGURE 14–4 Simple secondary call lead acid battery.

2. Withstand temperature extremes without cracking.
3. Resist vibration from the vehicle under normal operating conditions.
4. Resist the absorption of the electrolyte acid mixture.
5. Provide raised supports in the bottom of the case for sediment and contaminants to fall.
6. Provide vents in the cover for release of any hydrogen gas generated by the charging or discharging process. (Non-maintenance-free batteries also provide access to the cells for testing or filling the electrolyte [Figure 14–8].)
7. Provide a stable platform and seal for the external positive and negative terminals (Figure 14–9).

The battery electrolyte consists of a diluted mixture of 36% sulfuric acid and 64% water, by weight. When fully charged, the electrolyte in the battery has a specific gravity of 1.265 at 80°F. **Specific gravity** is the weight of a volume of liquid in comparison to the weight of the same volume of water with a specific gravity of 1.000. Testing the specific gravity of the electrolyte will be discussed in depth later in this chapter.

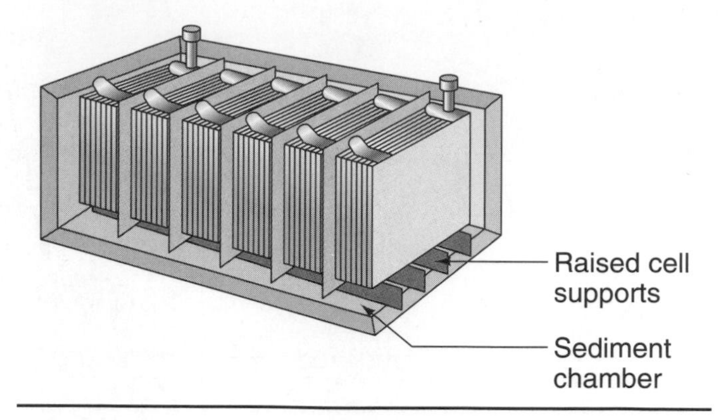

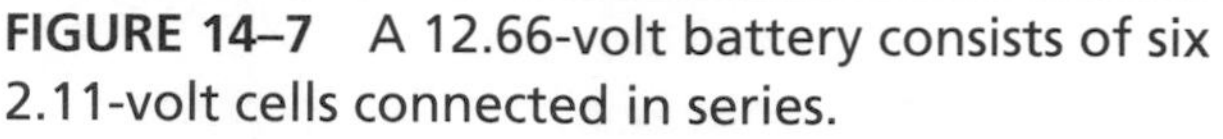

FIGURE 14–7 A 12.66-volt battery consists of six 2.11-volt cells connected in series.

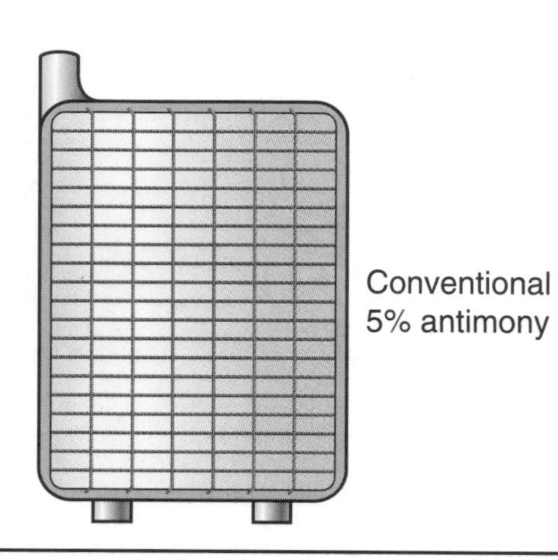

FIGURE 14–5 Typical conventional battery grid plate.

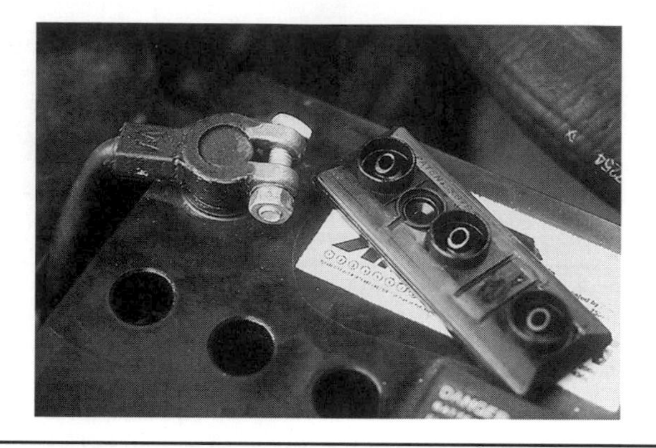

FIGURE 14–8 Conventional battery vent covers to allow escape of gases.

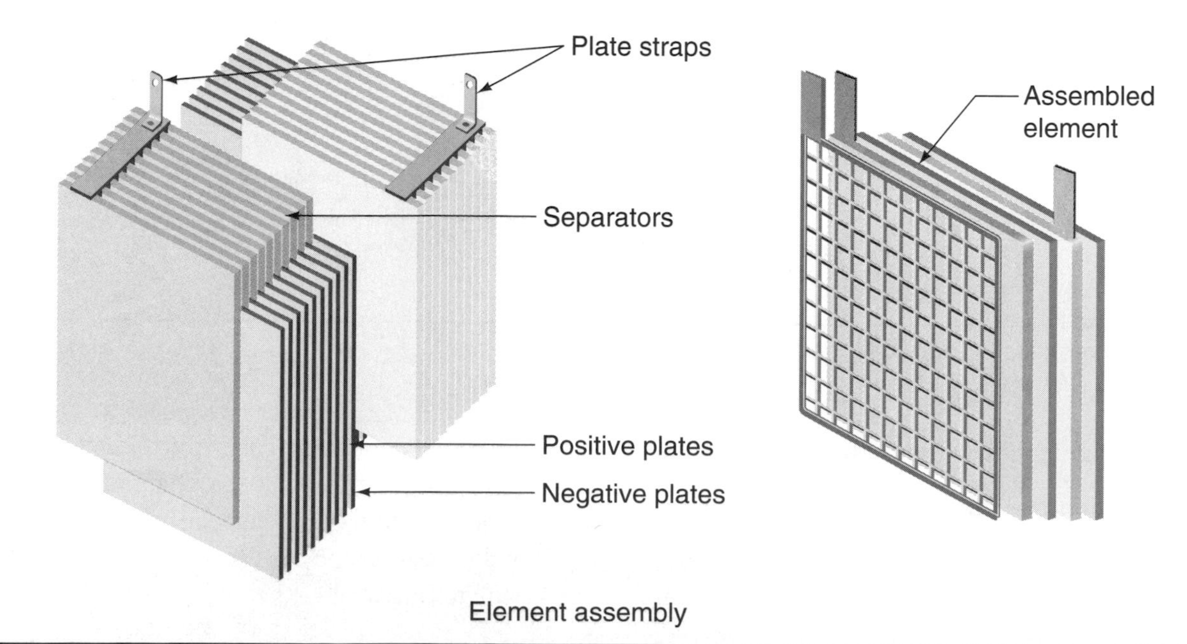

FIGURE 14–6 A battery cell, or element with positive and negative plated divided with separators.

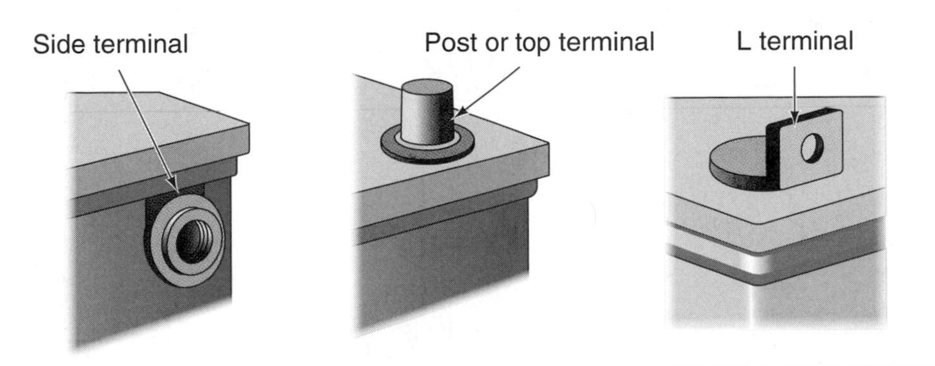

FIGURE 14–9 The most common types of battery terminals.

CAUTION *The electrolyte is a very corrosive acid, and it can cause severe injury to the eyes or skin (Figure 14–10). Always wear safety glasses and protective gloves and clothing when working around the battery electrolyte. A solution of baking soda and water can be used to neutralize the acid if it comes in contact with skin. If the acid is splashed in the eyes, flush them immediately with a large quantity of cool water, preferably from a emergency eyewash station (Figure 14–11), and then seek medical attention.*

The chemical production of electricity in the battery occurs when an electrical circuit is completed from positive to negative (Figure 14–12) and the discharge process begins. The oxygen (O_2) from the positive plate enters the electrolyte, and at the same time, the sulfuric acid (SO_4) from the electrolyte combines with the Pb on the positive plate to form lead sulfate ($PbSO_4$). And the SO_4 from the electrolyte combines with the Pb on the negative plate to form lead sulfate ($PbSO_4$).

In a discharged battery, the electrolyte mixture is only slightly heavier than the specific gravity of water. The following table indicates the specific gravity of a battery at 80°F.

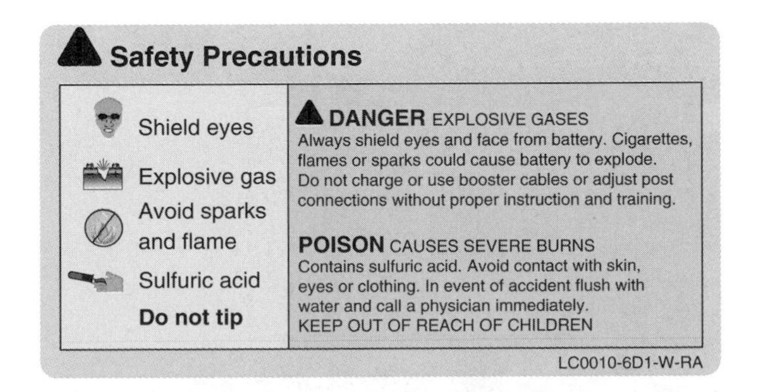

FIGURE 14–10 Safety and caution sticker located on battery cover.

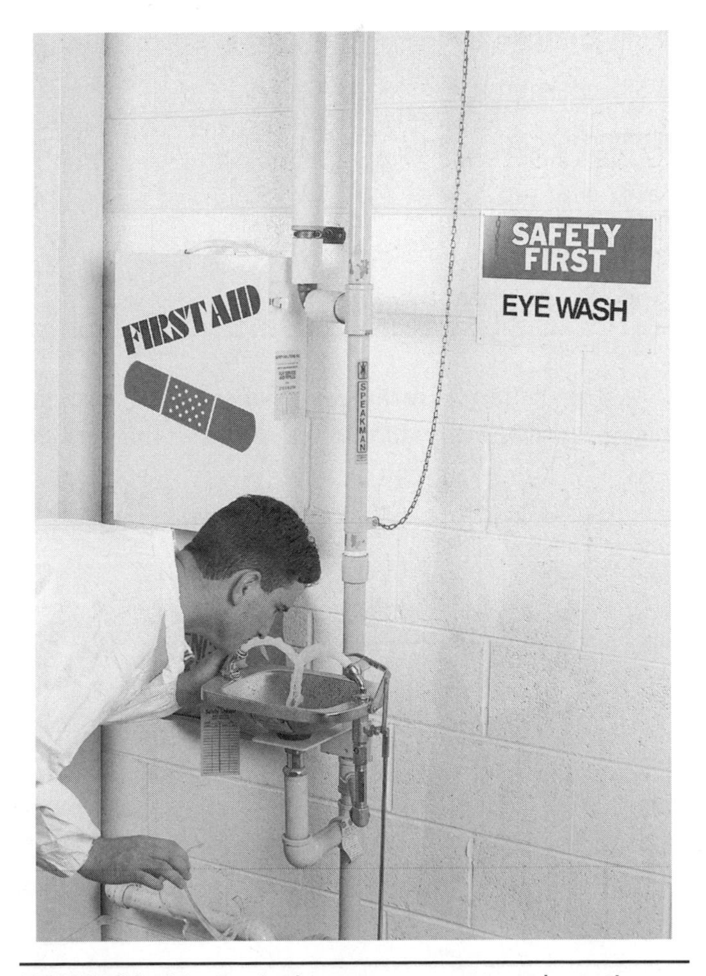

FIGURE 14–11 Typical emergency eyewash station.

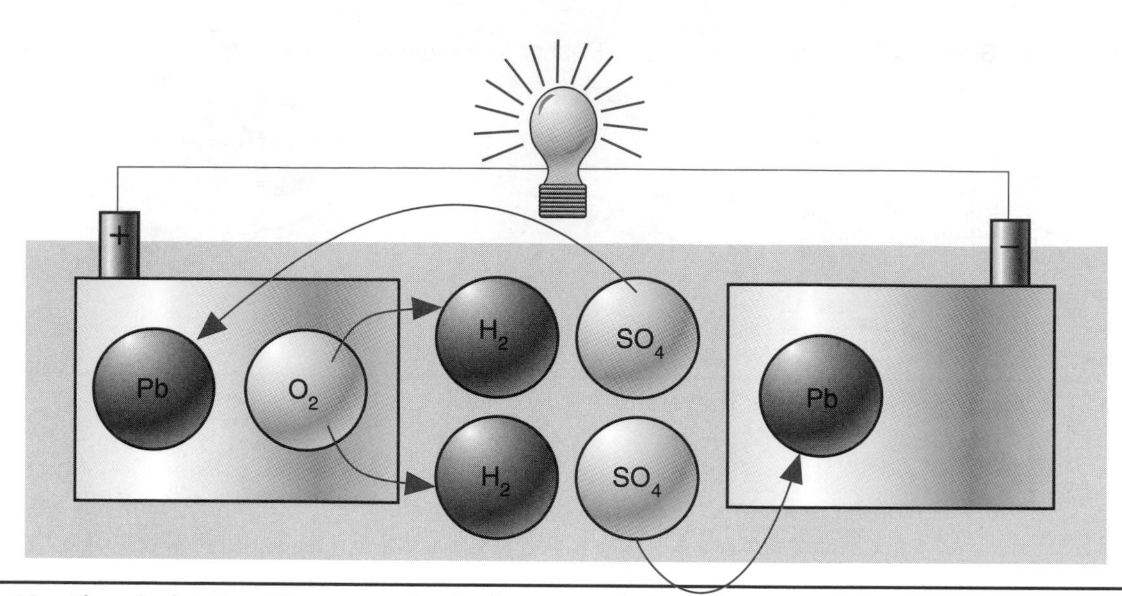

FIGURE 14–12 Chemical action that occurs in the battery cell during the discharge cycle.

Battery charge	Specific gravity
Fully charged	1.265
75% charged	1.225
50% charged	1.190
25% charged	1.155
Discharged	1.120 or less
Specific gravity at 80°F	

As the battery is recharged by the vehicle charging system or externally through a battery charger, the chemical action that took place during the discharge cycle now reverses itself to restore the battery to its original condition of PbO_2 in the positive plate, Pb in the negative plate, and an H_2SO_4 electrolyte mixture of 1.265 specific gravity (Figure 14–13).

Maintenance-Free Battery Construction

A **maintenance-free** battery is a lead-acid battery that uses a *lead calcium* mixture in the cell plates. Changing the grid plate material and design (Figure 14–14) reduces the amount of gasing of the electrolyte, which is prevalent with conventional batteries. **Gasing,** or **electrolysis,**

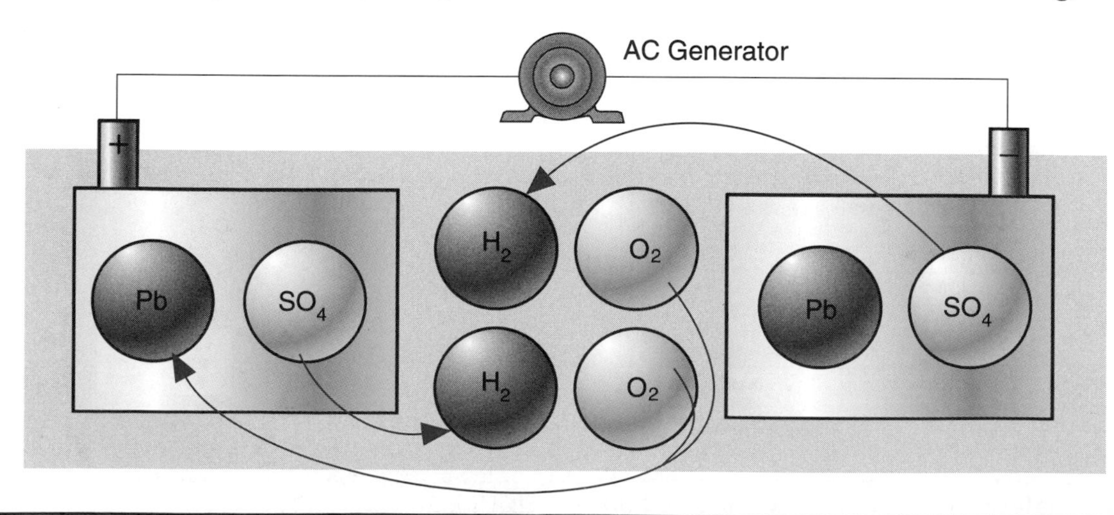

FIGURE 14–13 Chemical action that occurs in the battery during the charge cycle.

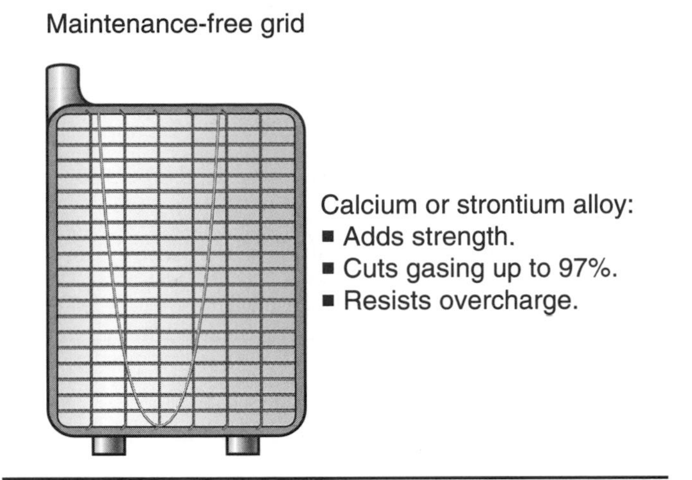

FIGURE 14–14 Typical maintenance-free battery grid plate.

in a battery occurs when the battery water is converted into explosive hydrogen and oxygen gas. The **lead calcium** grid contains calcium or strontium to reduce gasing and **self-discharge.**

Because heat and gasing are the main reasons for the vaporization of the electrolyte, using lead calcium grids almost eliminates electrolyte loss or the need to periodically add distilled water to the battery. The battery cover is sealed to prevent electrolyte or vapor loss and has only a small vent to relieve pressure during excess charging conditions (Figure 14–15).

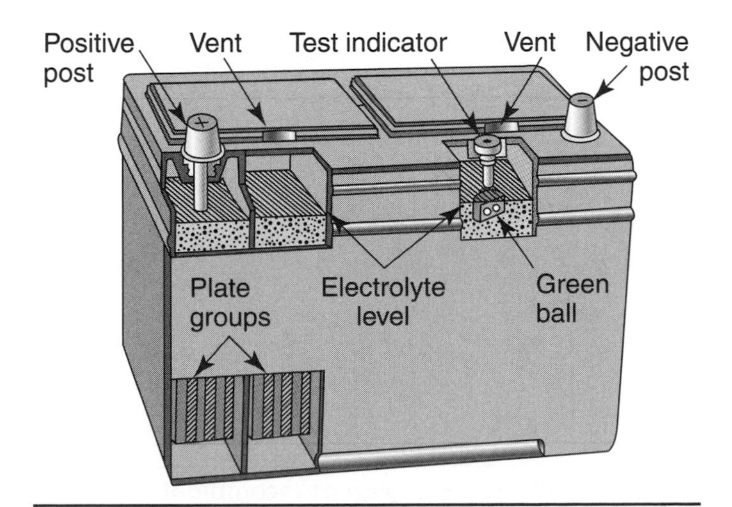

FIGURE 14–15 Typical construction of a maintenance-free battery.

Compared to conventional batteries, some of the advantages of maintenance-free batteries are:

- Larger electrolyte reserve due to a thin polypropylene or plastic case, and a reserve area above the plate cells.
- Longer shelf life of up to two years depending on the manufacturer.
- Higher cold cranking ampere rating.
- Little or no maintenance to battery terminal connections.

Compared to conventional batteries, there are some important *disadvantages* of maintenance-free batteries.

- Shorter life expectancy.
- Quick discharge from unwanted parasitic loads.
- Difficulty recharging if allowed to completely discharge.
- Lack of reserve capacity, especially in cold environments.
- **Grid growth** when the battery is exposed to high temperatures, a condition, that shorts out the plates when little metallic fingers grow through the separators to the plates. (To prevent grid growth and extend battery life, several manufacturers equip the vehicle's battery with an insulated heat shield [Figure 14–16]. The heat shield is normally made of plastic and should be reused when replacing the battery.)

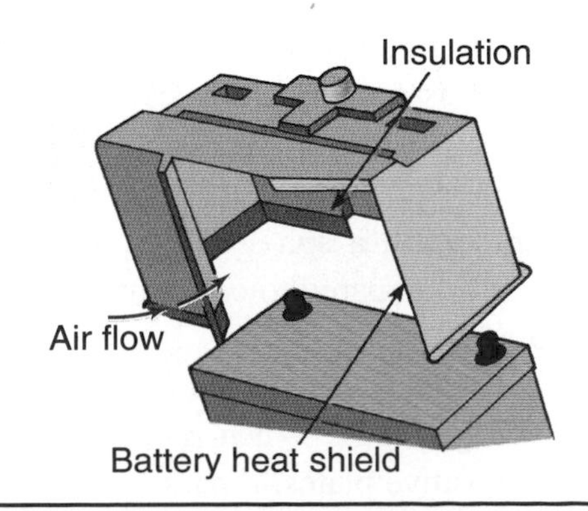

FIGURE 14–16 Typical insulated heat shield to protect battery from engine compartment heat.

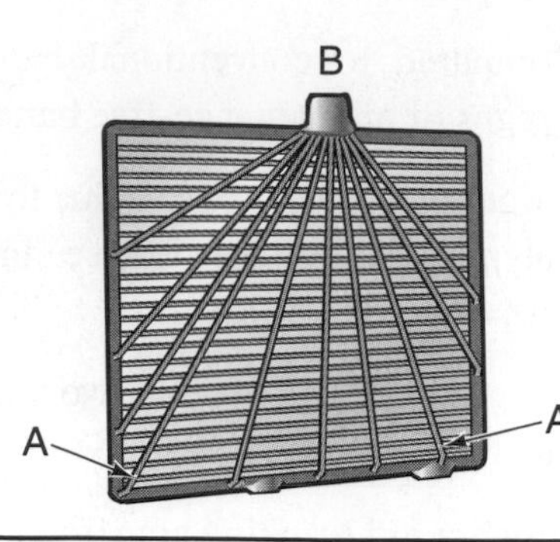

FIGURE 14–17 A typical deep cycle battery grid construction.

Deep Cycle Battery Construction

A **deep cycle** battery is a lead-acid battery that uses approximately 2.75% antimony on the positive plate and a calcium alloy on the negative plate. This type of battery is also known as a **hybrid** battery, a name that refers to the battery type, not necessarily to the batteries used in hybrid electric vehicles. A hybrid battery combines the best grid properties of conventional batteries and maintenance-free batteries, allowing it to withstand several deep cycles and still retain the original reserve capacity. The grid construction of a deep cycle battery provides more current at a faster rate because the vertical and horizontal grid bars are arranged in a radial design extending from a lug near the top center of the grid (Figure 14–17). Locating the lug near the center of the grid reduces resistance and provides a shorter path for current to follow to reach the lug.

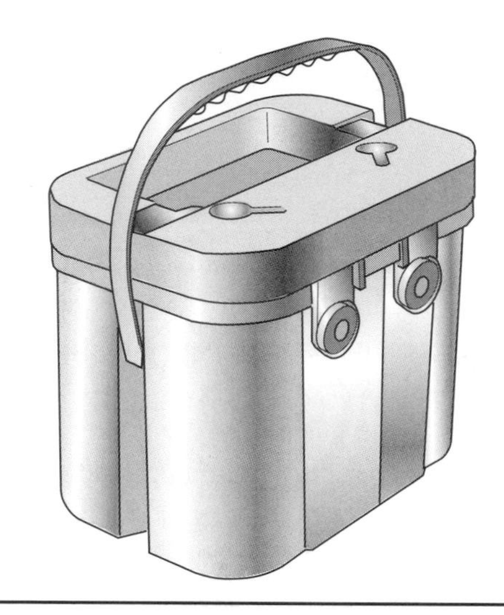

FIGURE 14–18 A typical sealed recombination, or gel-cell, battery.

Recombination Battery Construction

A **recombination battery,** or **gel-cell battery** as it is sometimes called, is a completely sealed lead-acid battery that uses electrolyte in a gel form (Figure 14–18). Some manufacturers of this battery design use cell plates that are constructed in a spiral design, which provides more surface area compared to conventional battery cell plates (Figure 14–19).

In a gel-cell battery, little or no gasing takes place to release hydrogen gas. When oxygen is internally released at the positive plates, it passes through the separators and is forced to recombine with the negative

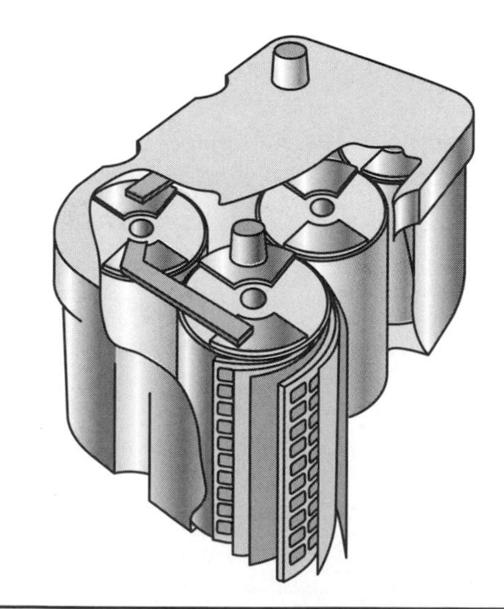

FIGURE 14–19 Spiral design of recombination battery cells.

plates. This process virtually eliminates gasing in the battery. Because of these design differences, follow manufacturer's instructions any time you are testing or charging the battery. Excessive current drains or charging rates might overheat the cells and cause damage.

Some of the advantages of a gel-cell battery are that it:

- Has no liquid electrolyte mixture to spill out, even if the case is cracked or mounted upside down. This feature makes the battery very suitable for off-road and 4-wheel drive vehicles.
- Can be deep cycled several times without damage to the reserve capacity.
- Has a completely sealed case that makes the battery totally corrosion- and maintenance-free.
- Outlasts conventional or maintenance-free batteries by up to four times.
- Provides more cell plate surface area with its battery spiral cell design, compared to flat cell plates in conventional batteries.
- Is available with ratings of over 800 cold cranking amperes.

A disadvantage of a gel-cell battery is the initial expense—normally two to three times the cost of a conventional or maintenance free battery.

New Technology Batteries

The recent surge of electric vehicles and hybrid electric vehicles (HEV) in the market today has produced a variety of new battery designs and applications. Two of the most common are the nickel-cadmium (NiCd) and the nickel metal hydride (NiMH). Both of these batteries are popular in HEVs because of their small size-to-energy ratio and their ability to tolerate deep cycling compared to standard lead-acid batteries (refer to Figure 14–2 for cell material comparison). Figure 14–20 is an example of a NiMH battery pack module for an HEV. This battery pack contains relays, solenoids, and battery packs wired in series to produce over 200 volts of DC current. Depending on the vehicle, the battery pack assembly units can weigh from 50 to 150 pounds. The future of lithium ion (Li-ion) as a preferred vehicle battery is still in question. However, the Li-ion battery cell produces 3.6 volts per cell compared to 1.2 volts for NiCd and NiMH, making the future for this battery in HEVs very promising.

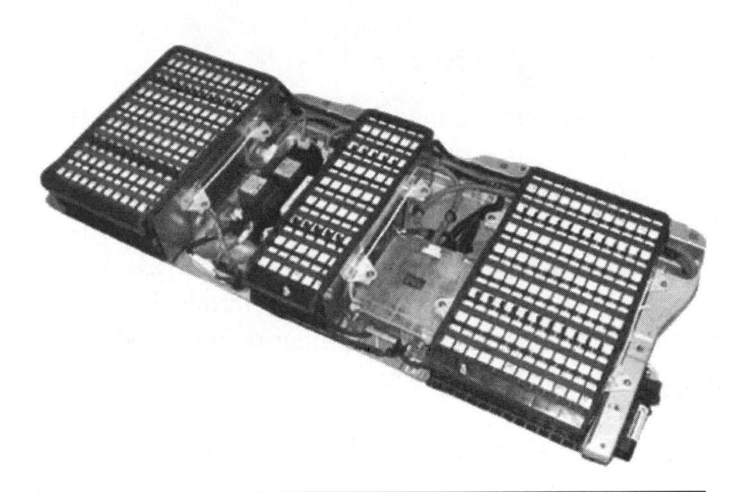

FIGURE 14–20 An HEV battery module pack assembly.

(Courtesy of Toyota Motor Sales, USA Inc.)

Battery Load Ratings

The Battery Council International (BCI) has established five capacity ratings that determine the current capacity of a battery. The current capacity is an indication of the battery's ability to provide high-amperage cranking power to the starter and provide reserve power to the electrical system.

Reserve Capacity Rating

The first battery rating is known as **reserve capacity (RC).** It is the amount of time in minutes a fully charged battery can be discharged at 25 amperes while maintaining a minimum battery voltage of 10.5 volts at 80°F in the event of a charging system failure. For example, a reserve capacity rating of 120 minutes indicates that in the event of a charging system failure, the battery could be discharged at a rate of 25 amperes for two hours (60 minutes × 2 = 120 minutes) at 80°F.

Amp-Hour Rating

The second battery rating, the **amp-hour rating (AH),** is the amount of current that a fully charged battery can produce over a 20-hour period at 80°F before the terminal voltage reaches a minimum of 10.5 volts. For example, if a battery can be discharged at a rate of 5 amperes for 20 hours before the terminal voltage reaches 10.5 volts, then the amp-hour rating is 100 A/H. This is computed as follows:

$$5 \text{ amps} \times 20 \text{ hours} = 100 \text{ A/H}$$

Most automotive manufacturers use original equipment batteries with an A/H rating of 50 to 120 A/H. A higher A/H rating does not necessarily indicate increased starter cranking capacity, but it does improve the ability to provide more circuit current over the 20-hour period.

Cold Cranking Ampere Rating

The third and most common battery rating is the **cold cranking ampere (CCA)** rating. It is the ability of the battery to provide an amperage load for 30 seconds at 0°F, without the terminal voltage dropping below 7.2 volts. The CCA rating on an original equipment battery can range between 300 CCA to slightly over 1000 CCA, depending on the vehicle.

If a battery has an A/H rating, its CCA is determined by multiplying the A/H by 5.25. For example, a 100 A/H battery has a CCA rating of 525 CCA.

$$100 \text{ A/H} \times 5.25 = 525 \text{ CCA}$$

To convert the battery's CCA rating to A/H, simply divide the CCA by 5.25.

$$525 \text{ CCA} \div 5.25 = 100 \text{ A/H}$$

Cranking Amps Rating

The fourth battery rating is the **cranking amps (CA)** rating, which is the battery's ability to deliver a cranking current at 32°F. A battery with a CA rating of 800 may confuse a technician who may assume it is a CCA rating number. To convert CA at 32°F to CCA at 0°F, divide CA by 1.25.

For example, an 800 CA rated battery is the same as a 640 CCA rated battery.

$$800 \text{ CA} \div 1.25 = 640 \text{ CCA}$$

Watt-Hour Rating

The final battery rating is the **watt-hour (WH)** rating, which is determined by how many watt-hours of mechanical energy a battery produces at 0°F The watt-hour rating is calculated by multiplying a battery's amp-hour rating by the battery's open circuit voltage. A 120 A/H rated battery of 12.6 volts has a watt-hour (W/H) rating of 1512 W/H.

$$120 \text{ A/H} \times 12.6 \text{ volts} = 1{,}512 \text{ W/H}$$

Battery Size and Capacity

When a battery needs to be replaced, several factors should be considered to find the best replacement match possible for the vehicle.

- Engine size and type
- Vehicle electrical options
- Climate conditions (Extreme heat or cold affects a battery's performance. The requirement for electrical energy to crank the engine increases as the temperature decreases [Figure 14–21].)
- Vehicle starting requirements, such as frequent starts and stops
- Size and weight limitations (If the battery is too high, it could possibly short out the terminal when the hood is closed. The BCI battery group numbers listed in Figure 14–22 are the most commonly used and indicate the proper dimensional size and characteristics of the battery.)

Temperature	Percentage of Cranking Power (%)
80°F (26.7°C)	100
32°F (0°C)	65
0°F (–17.8°C)	40

FIGURE 14–21 A battery's cranking power is affected by temperature.

Grp. Size	Vlt.	Cold cranking power—amps for 30 secs. at 0°F*	Number of months warranted	Size of battery container in inches (incl. terminals)		
				Lgth.	Wd.	Ht.
17HF	6	400	24	7 1/4	6 3/4	9
21	12	450	60	8	6 3/4	8 1/2
22F	12	430	60	9	6 7/8	8 1/8
	12	380	55	9	6 7/8	8 1/8
	12	330	40	9	6 7/8	8 1/8
22NF	12	330	24	9 1/2	5 1/2	8 7/8
24	12	525	60	10 1/4	6 7/8	8 5/8
	12	450	55	10 1/4	6 7/8	8 5/8
	12	410	48	10 1/4	6 7/8	8 5/8
	12	380	40	10 1/4	6 7/8	8 5/8
	12	325	36	10 1/4	6 7/8	8 5/8
	12	290	30	10 1/4	6 7/8	8 5/8
24F	12	525	60	10 1/4	6 7/8	8 5/8
	12	450	55	10 1/4	6 7/8	8 5/8
	12	410	48	10 1/4	6 7/8	8 5/8
	12	380	40	10 1/4	6 7/8	8 5/8
	12	325	36	10 1/4	6 7/8	8 5/8
	12	290	30	10 1/4	6 7/8	8 5/8
27	12	560	60	12	6 7/8	8 5/8
27F	12	560	60	12	6 7/8	9
41	12	525	60	11 9/16	6 13/16	6 15/16
42	12	450	60	9 5/8	6 7/8	6 3/4
	12	340	40	9 5/8	6 7/8	6 3/4
45	12	420	60	9 1/2	5 1/2	8 7/8
46	12	460	60	10 1/4	6 7/8	8 5/8
48	12	440	60	12	6 7/8	7 1/2
49	12	600	60	14 1/2	6 7/8	7 1/2
56	12	450	60	10	6	8 3/8
	12	380	48	10	6	8 3/8
58	12	425	60	9 1/4	7 1/4	6 7/8
71	12	450	60	8	7 1/4	8 1/2
	12	395	55	8	7 1/4	8 1/2
	12	330	36	8	7 1/4	8 1/2
72	12	490	60	9	7 1/4	8 1/4
	12	380	48	9	7 1/4	8 1/4
74	12	585	60	10 1/4	7 1/4	8 3/4
	12	525	60	10 1/4	7 1/4	8 3/4
	12	505	60	10 1/4	7 1/4	8 3/4
	12	450	55	10 1/4	7 1/4	8 3/4
	12	410	48	10 1/4	7 1/4	8 3/4
	12	380	40	10 1/4	7 1/4	8 3/4
	12	325	36	10 1/4	7 1/4	8 3/4

*Meets or exceeds Battery Council International rating standards.

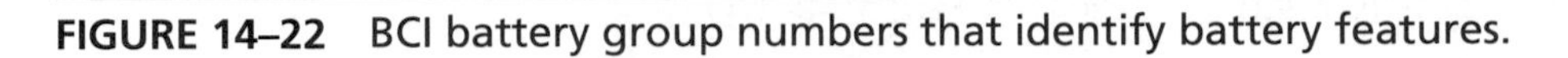

FIGURE 14–22 BCI battery group numbers that identify battery features.

Battery Testing

To properly check a discharged or weak battery for possible replacement, seven specific checks can be made to ensure that the battery is the source of the problem, not the result of another problem (such as a parasitic drain, discussed extensively in Chapter 10). Knowledge of how batteries operate, using quality test equipment, and following specific testing procedures all help isolate a battery problem.

1. Battery Terminal Voltage Drop Test

The first testing procedure to perform on a battery is a battery terminal voltage drop test, which tests the positive and negative terminal connections for a voltage drop. Figure 14–23 shows using a voltmeter connected from the battery post to the terminal connector while the starter is being cranked (ignition disabled to prevent starting). Voltage drops exceeding 0.1 volts indicate unwanted resistance at the terminal connection.

If a voltage drop exists, the battery terminals need to be removed (Figure 14–24) and cleaned. To prevent damage on a top post battery, use a terminal puller to remove the terminal from the post. Remember *always* to remove the *negative* terminal first to prevent the chance of the positive post contacting a ground through the wrench. Use a terminal cleaning tool or small wire brush to clean the terminals and posts (Figure 14–25).

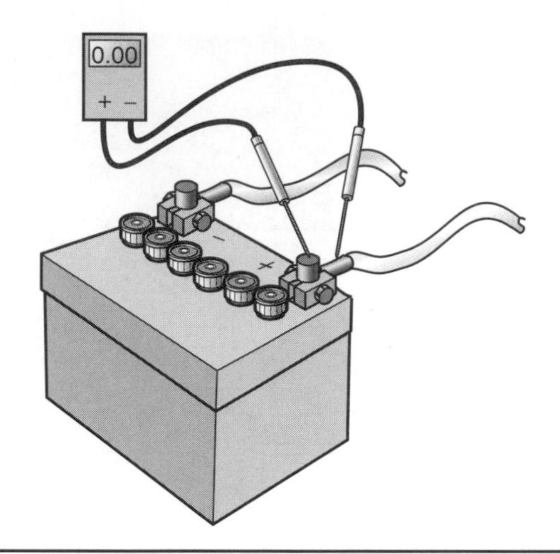

FIGURE 14–23 Performing a battery terminal voltage drop test.

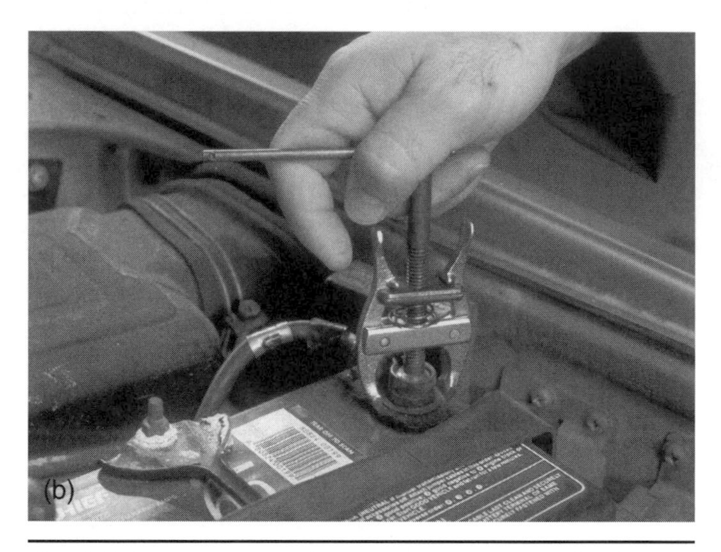

FIGURE 14–24 Removing a battery terminal using a battery terminal puller.

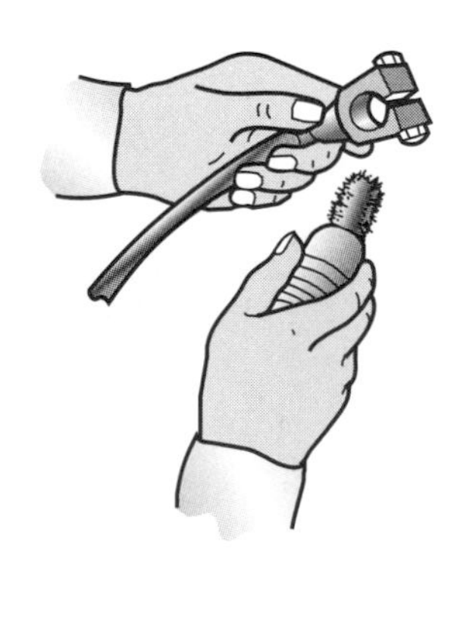

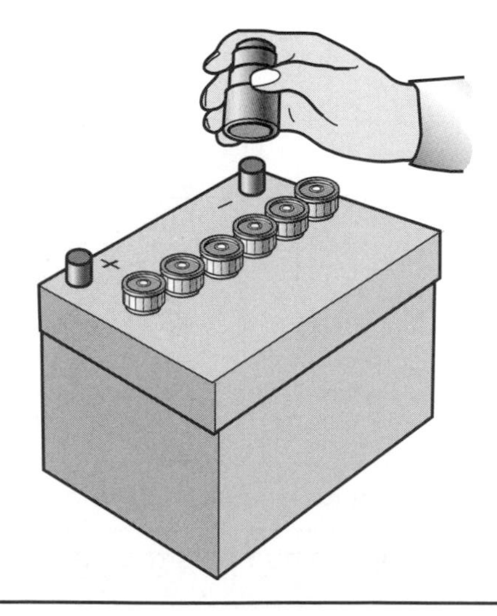

FIGURE 14–25 Using a terminal cleaning tool to clean the battery posts and terminals.

2. Battery Case Leakage Test

When the top of the battery case becomes wet or dirty, a small unwanted current draw can exist between the positive and negative posts through the dirty top. To find the unwanted current draw, perform a battery case leakage test (Figure 14–26) by connecting the negative lead of a voltmeter to the battery negative post and touching the positive meter probe to the case top in various locations. A *clean* battery should have *no* voltage reading between the posts and the battery case.

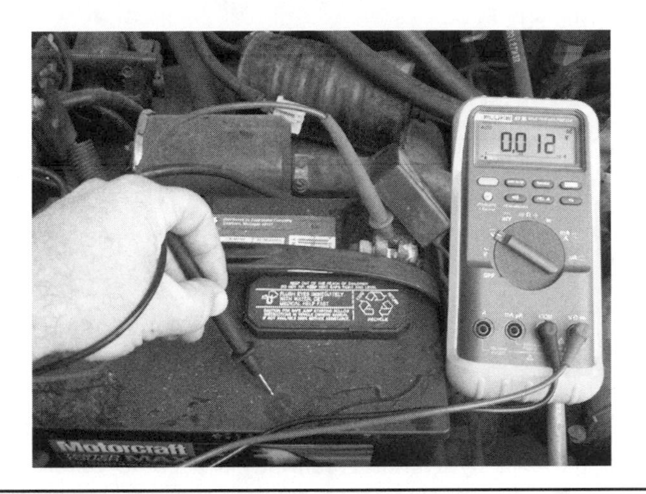

FIGURE 14–26 Performing a battery voltage leakage test on a batter cover.

3. Battery State-of-Charge Test

The state-of-charge test is done by using a **hydrometer** that measures the specific gravity content of the sulfuric acid mixture. When an electrolyte mixture is drawn into the hydrometer (Figure 14–27), the float rises to the specific gravity of the electrolyte solution (Figure 14–28). Because the battery may be tested at various temperatures, using a hydrometer that is temperature corrected provides a true specific gravity reading (Figure 14–29). For every 10°F above 80°F, add 0.004 points to the specific gravity reading. For every 10°F below 80°F, subtract 0.004 points from the specific gravity reading.

A fully charged battery has a specific gravity of 1.265 to 1.280 depending on the battery. As the battery becomes discharged, the water content of the electrolyte increases, lowering the specific gravity reading. A

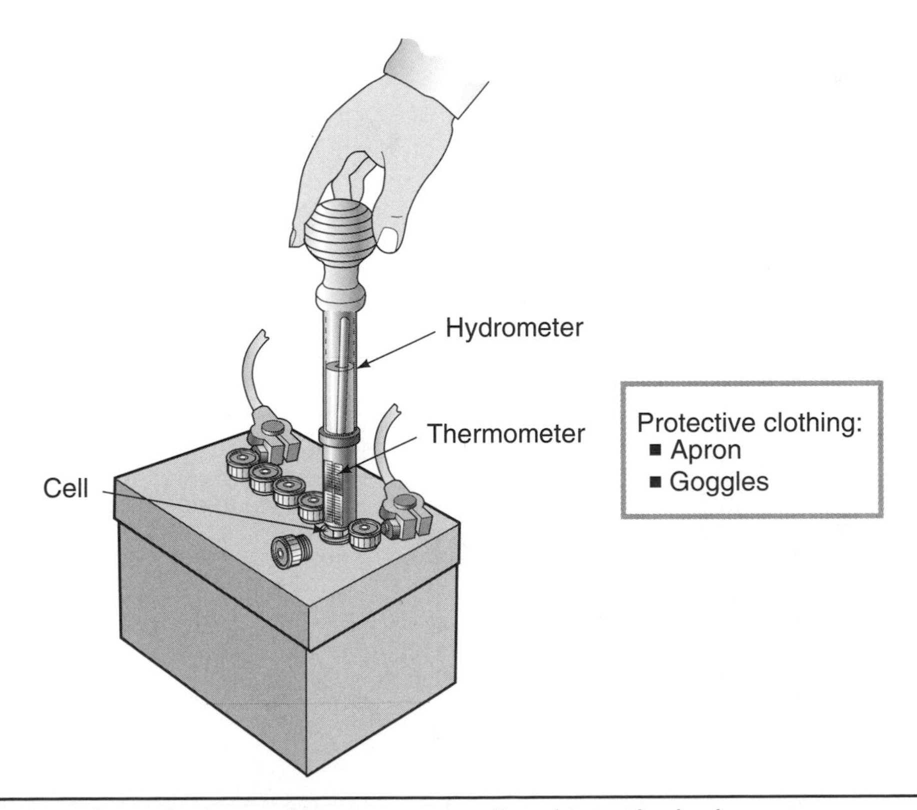

FIGURE 14–27 Drawing electrolyte out of the battery's cell and into the hydrometer.
(Courtesy of DaimlerChrysler Corporation)

FIGURE 14–28 Where the top of the electrolyte intersects the float is the specific gravity reading for the cell: (a) Float indicates a low reading, and (b) float indicates a high reading.

defective battery can be determined by recording the readings of each cell (Figure 14–30). If the specific gravity readings between the highest and lowest cells vary by 0.050 or more, consider the battery defective.

Note: Make sure distilled water was not recently added to the battery electrolyte, which can give a false specific gravity reading.

On some batteries, a built-in hydrometer is used as a quick reference indicator of a battery's state of charge (Figure 14–31). This indicator *only* checks the specific gravity in one cell. When a green dot appears on the indicator, the specific gravity charge is 65% or higher; below 65%, the dot appears dark. A clear dot indicates that the electrolyte level is low and the battery needs to be replaced.

Another method to determine the specific gravity reading is with an **optical refractometer** (Figure 14–32). A few drops of battery electrolyte are put on the refractometer lens, and, when looking through the lens window, the technician determines, by the reflected light, precise specific gravity readings (Figure 14–33).

4. Battery Open Circuit Voltage Test

When a battery, such as a maintenance-free battery, has a sealed top, a hydrometer cannot be used to check the state of charge. In this case, a battery open circuit voltage test can be used to substitute for a hydrometer test. To perform this test accurately, isolate (disconnect) the battery from the vehicle. Then use a voltmeter (Figure 14–34) to measure the voltage across the battery terminals. For an accurate reading, the battery temperature should be between 60°F and 100°F and stabilized for at least 10 minutes with no load applied. If the battery has been recently

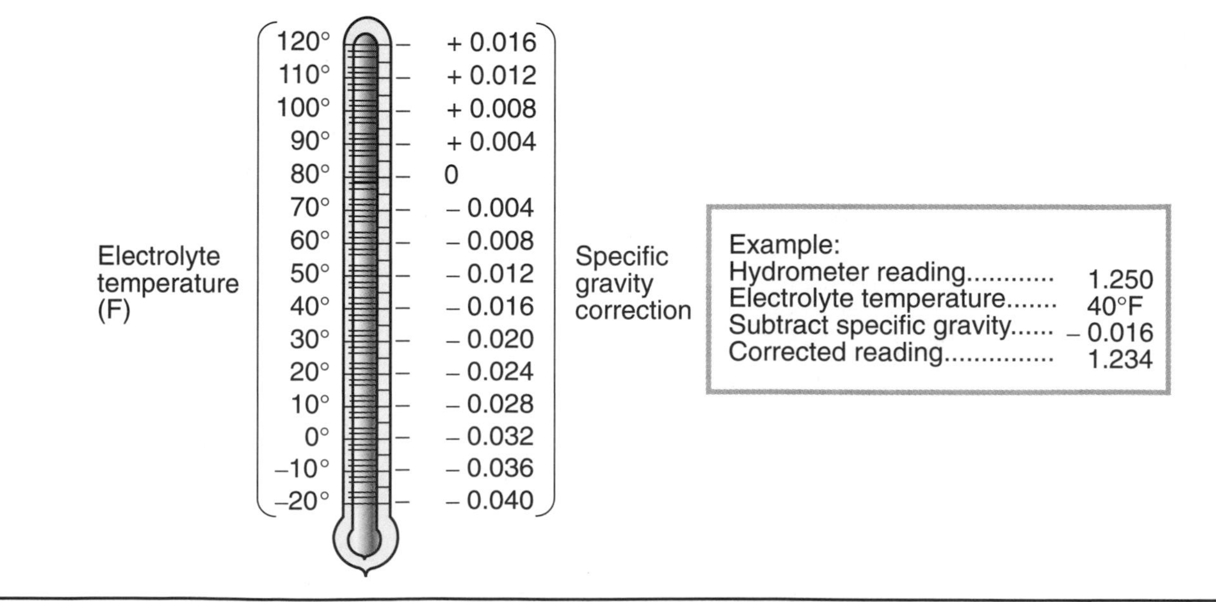

FIGURE 14–29 Temperature correcting the specific gravity reading.

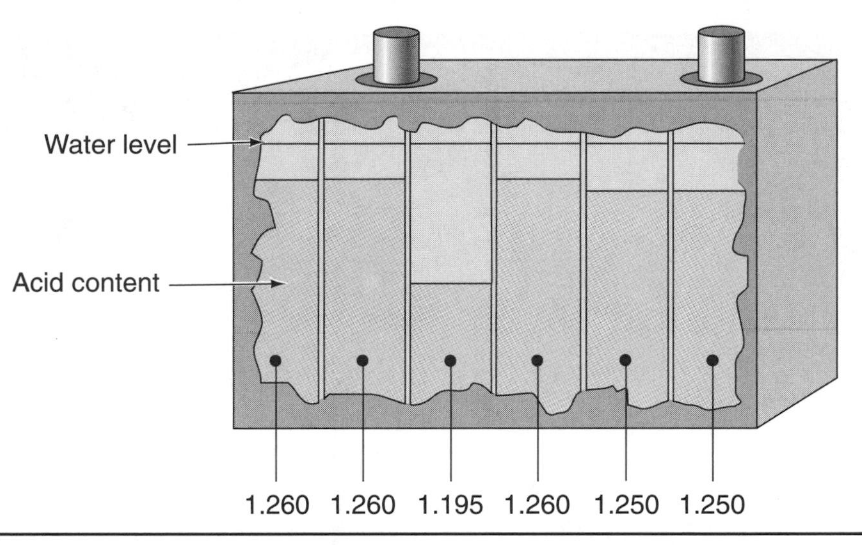

FIGURE 14–30 The specific gravity of a defective cell compared to fully charged cells.

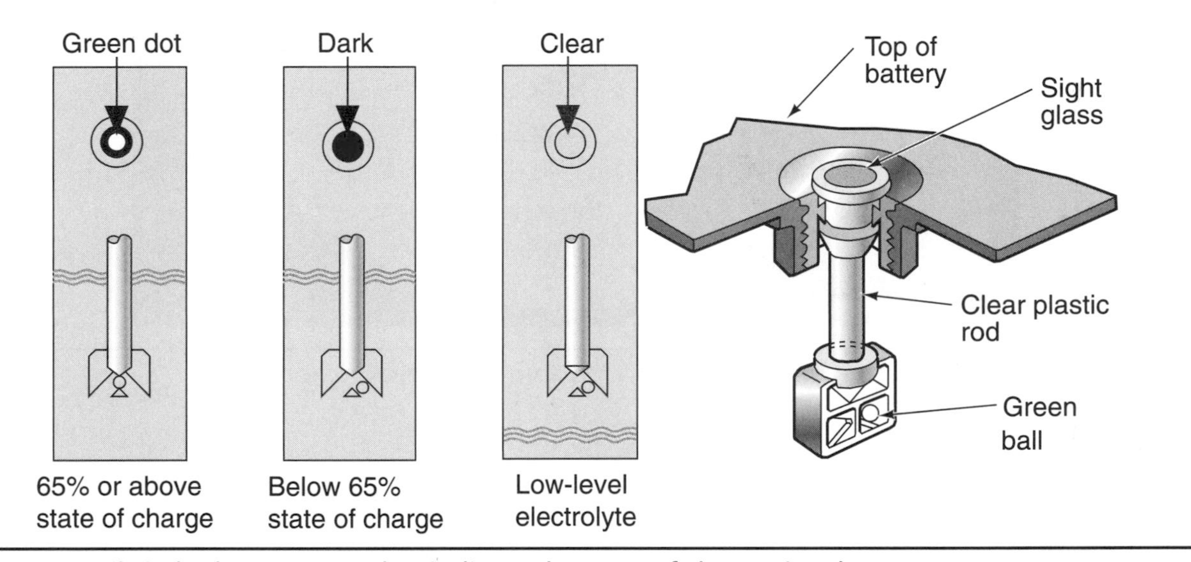

FIGURE 14–31 Built-in hydrometer used to indicate the state of change in a battery.

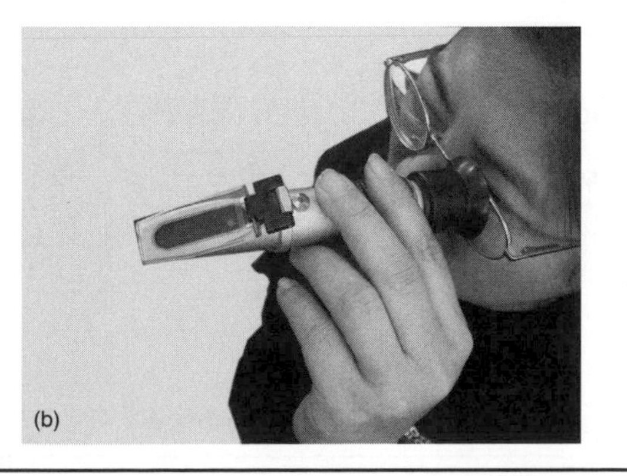

FIGURE 14–32 Using an optical refractometer to determine specific gravity.

(Courtesy of Tim Gilles)

charged, the surface charge must be removed to obtain an accurate open circuit voltage reading. To remove the surface charge, apply a 20-ampere load (high beams and heater blower) for 2 minutes, then shut off the circuits. Let the battery stabilize for 5 minutes before testing the open circuit voltage.

The relationship between the open circuit voltage and specific gravity is illustrated in Table 14–1. As you can see by the results in the table, the voltage difference between a fully charged battery and a discharged battery is only 0.7 volts. Using a quality digital voltmeter with a DC accuracy of ±0.1% is best to get an accurate reading.

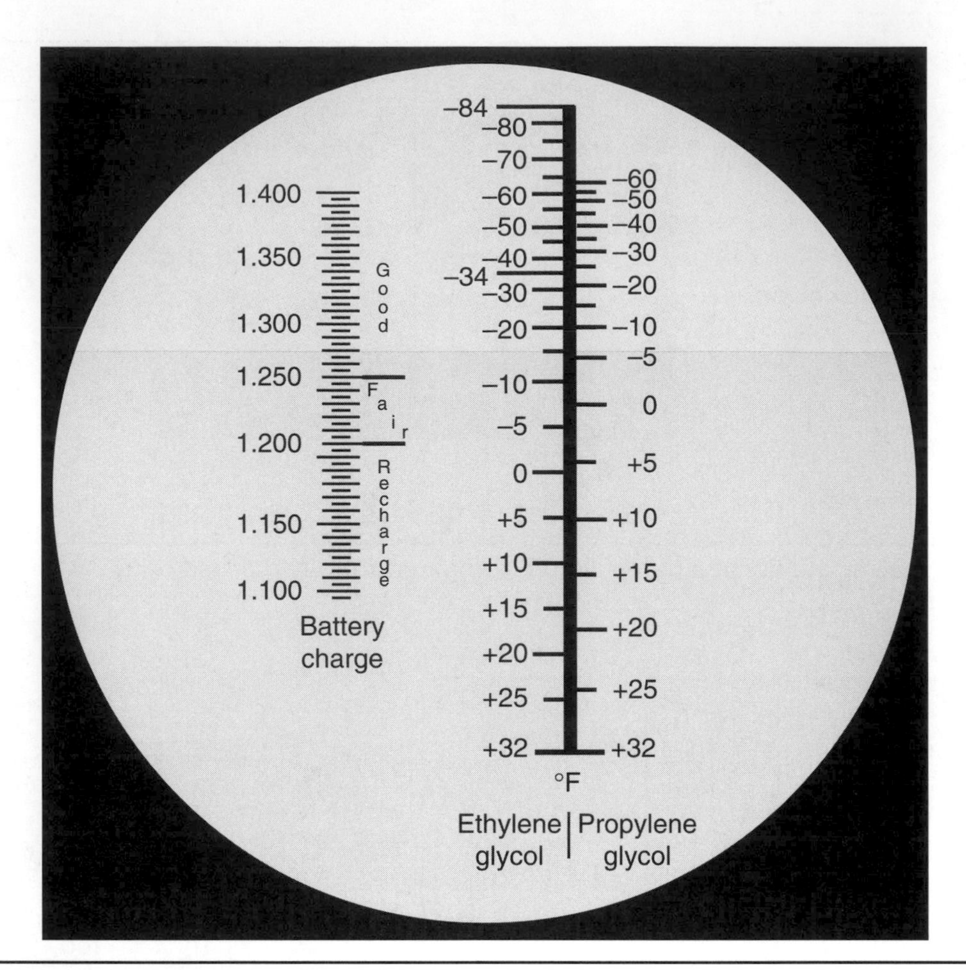

FIGURE 14–33 Viewing the battery specific gravity with an optical refractometer.

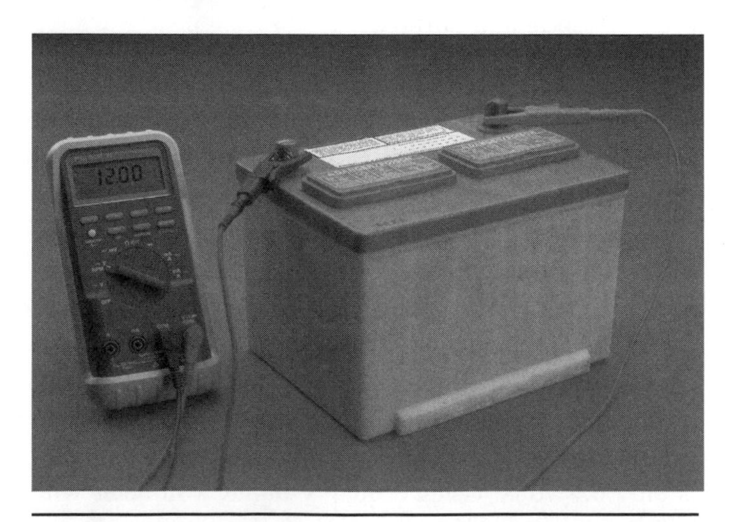

FIGURE 14–34 Measuring the open circuit voltage across battery terminals.

TABLE 14–1

Open Circuit Voltage (V)	State of Charge (%)	Specific Gravity
12.6 or greater	100	1.260
12.4	75	1.225
12.2	50	1.190
12.0	25	1.155
11.9	Discharged	1.100

5. Battery Capacity Load Test

The battery capacity load test is the best one to use to check a battery's condition. For this test to be accurate and realistic, the battery must be at least 75% charged and have passed the hydrometer state-of-charge test or the voltmeter open circuit test. The battery needs to supply high amperage to the starting system and still have sufficient voltage to operate the engine computer control systems. Depending on the load test equipment used, certain steps should be followed. If a carbon pile load tester is used (Figure 14–35), follow these general:

- Connect the tester cables following the manufacturer's instructions—normally to the positive and negative posts with the inductive amperage clamp placed around the negative load tester lead (not the negative battery lead).
- Determine the load test specifications for the battery. It is either one of the following.
  1. Battery A/H × 3 = load
  2. Battery CCA ÷ 2 = load
- Apply the calculated load for *15 seconds* while observing the voltmeter. The reading should be 9.6 volts or higher at 70°F. Table 14–2 calculates the minimum load test voltage in relation to the battery temperature. Batteries that fail this test should be charged and retested. If the battery fails the load test for a *second time,* replace it.

TABLE 14–2 Battery Load Test Temperature Chart

Minimum Voltage	At °F	At °C
9.6	70+	21+
9.5	60	16
9.4	50	10
9.3	40	4
9.1	30	−1
8.9	20	−7
8.7	10	−12
8.5	0	−18

Figure 14–36 shows an example of an automated battery load tester. One brand is called an ARBST (alternator/regulator/battery/starter tester). This tester performs a load test when the following steps are performed:

- Connect the tester's red lead to the positive battery post and the tester's black lead to the negative post. The tester's gray inductive clamp is *not* used to test batteries, only to test starters and alternators.

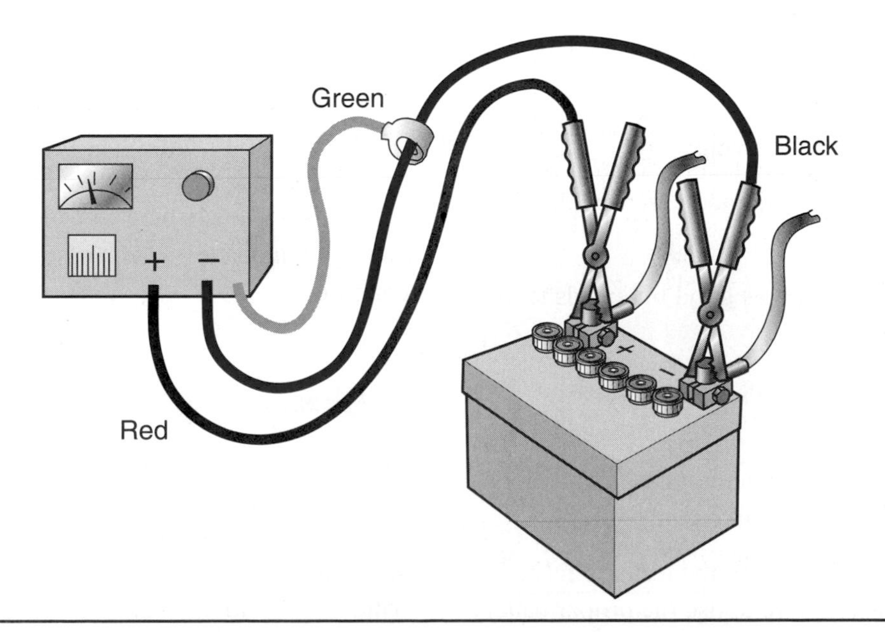

FIGURE 14–35 Connecting a carbon pile load tester to a battery.

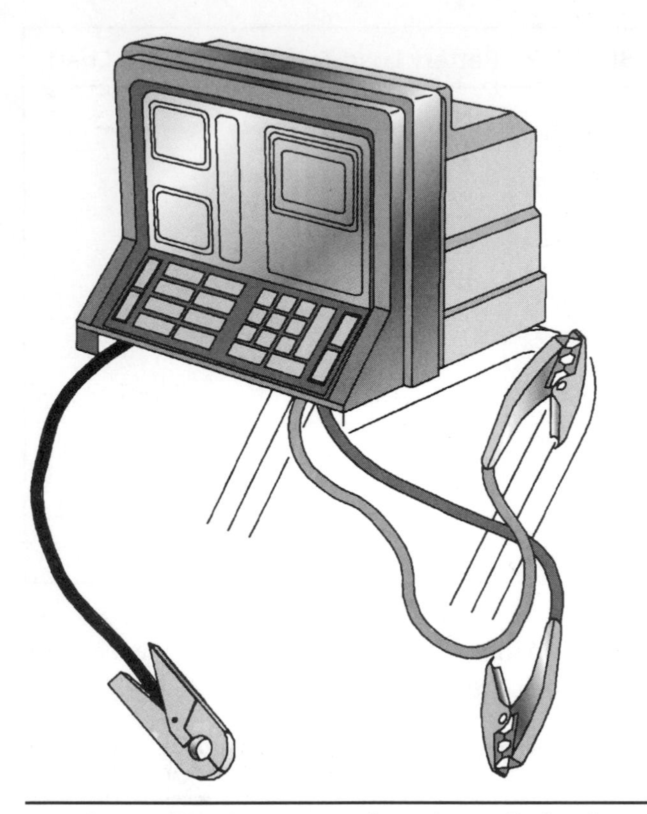

FIGURE 14–36 An automatic carbon pile load tester.

- Push the battery test button on the ARBST, and input the *actual* CCA of the battery (not one-half of CCA as when using a carbon pile tester).
- Push the enter button on the tester. The load test counts down through a 50-second test sequence and then records the results.
- Test results provide the following:

  Good battery—battery is charged and capable of performing as designed.

  Good, low charge—battery is good but needs to be recharged.

  Bad battery—battery will not hold a charge or perform as designed.

  Charge and retest—tester cannot determine battery condition—charge battery and retest.
- If the wrong CCA is input into the tester, a false reading is given. Make sure to use the proper battery CCA. Do not assume the CCA by the battery's size!

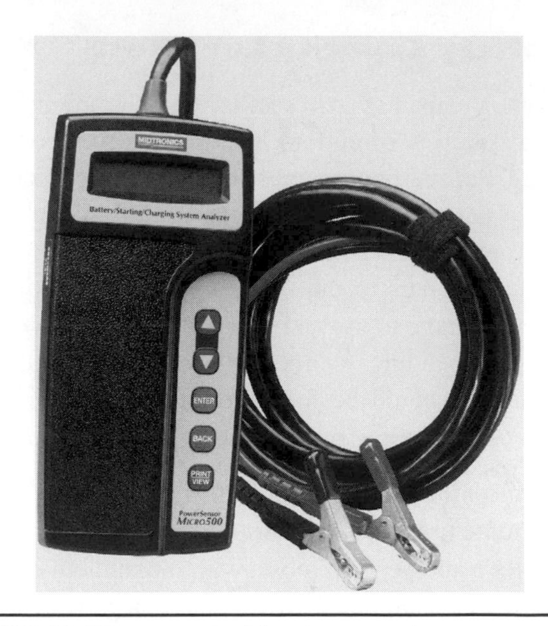

FIGURE 14–37 A typical battery conductance or capacitance tester.

6. Battery Conductance (Capacitance) Test

Some manufacturers recommend that a **conductance test,** or **capacitance test,** be performed on a battery instead of the using the carbon pile battery load tester. In a battery conductance test, a special tester (Figure 14–37) sends a low-frequency AC signal through the battery plates and then measures a percentage of the AC current response. This measurement is a calculation of the cell plate surface available for chemical reaction, and it has been shown to be highly accurate in detecting shorted or open cells, as well as other cell defects such as sulfation. A typical tester display indicates the results as a good battery, bad battery, recharge and test again, or poor condition.

7. Battery 3-Minute Charge Test

The battery 3-minute charge test is performed only when a battery fails a capacity load test. As a battery ages, deterioration of the cell plates occurs, which is called **sulfation.** The sulfation interferes with the ability of the battery cells to deliver a current and accept a charge (Figure 14–38).

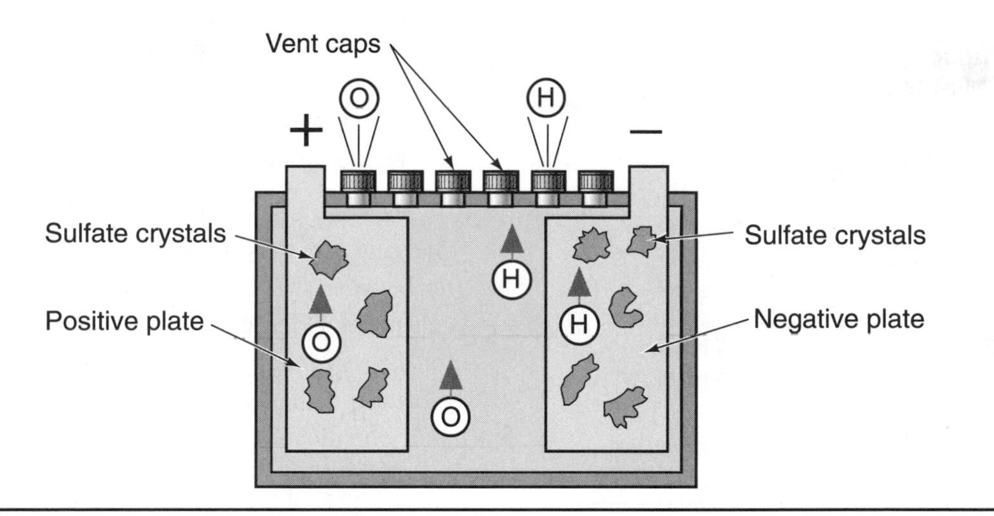

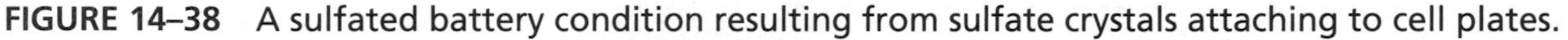

FIGURE 14–38 A sulfated battery condition resulting from sulfate crystals attaching to cell plates.

CAUTION *Some battery manufacturers, such as Delco, do not recommend the 3-minute charge test. Differences in plate design, size, and material used in the plates can all affect the results of the test. In these cases, the battery conductance tester may be a suitable alternative to check for cell plate sulfation.*

When a battery fails the load test, the 3-minute charge test is a reasonably accurate method to check for battery sulfation. To conduct the test:

- Remove the battery ground cable to isolate the vehicles electrical system from the battery.
- Connect the battery charger to the battery, observing polarity.
- Use a battery charger that can provide a continuous 40 amperes.
- Maintain a 40-ampere charge for 3 minutes while observing the voltmeter connected to the battery.
- If the voltage reading is less than 15.5 volts, then the battery is not sulfated. If the voltmeter reading is over 15.5 volts, the battery is sulfated. Some maintenance-free batteries use 16.5 volts as the division line between a good and a bad battery. Always refer to the specific battery manufacturer's guidelines before performing this test.

Battery Charging Procedures and Safety

In the event a battery needs to be charged with an external battery charger, several safety precautions and procedures need to be followed.

- Determine the condition of the battery that needs a charge. A battery that is severely sulfated can lead to excessive gasing and heat build-up of the plates during fast charging. If a battery is extremely discharged and remains in this state of (dis)charge for an extended period of time, it may be permanently damaged.
- Battery temperatures that exceed 125°F during charging can lead to a thermal runaway, where the internal heat will melt down the plates and create a short or an explosion.
- Disconnect the negative battery cable to isolate the vehicle's electrical system. If the charging voltage exceeds 15 volts during charging, the vehicle computers could be damaged if they were still connected to the battery.
- *Always* ensure that the battery charger is *off* when connecting the charger leads to the battery. Hydrogen and oxygen mixtures that may be in the

Battery Capacity (Reserve Minutes)	Slow Charge
80 min or less	10 hr @ 5 A 5 hr @ 10 A
Above 80 to 125 min	15 hr @ 5 A 7.5 hr @ 10 A
Above 125 to 170 min	20 hr @ 5 A 10 hr @ 10 A
Above 170 to 250 min	30 hr @ 5 A 15 hr @ 10 A

FIGURE 14–39 Determining the slow charge rate depending on the reserve capacity.
(Courtesy of Battery Council International)

immediate area of the discharged battery could explode if a spark is created.

- Determine the charge rate required to charge the battery. Slow charging a battery restores the battery grid plates to the fully charged state and minimizes the chance of overheating or overcharging the cells. Figure 14–39 determines the rate of slow charge according to the reserve capacity of the battery.
- If adequate charging time is not available, fast charging uses a high current over a short time to boost the battery enough for starting or to operate the electrical system. Figure 14–40 determines the rate of fast charge. A battery on a fast charge should be monitored at all times to prevent overheating.

The difference between slow charging and fast charging a battery affects the amount of chemical transfer between the electrolyte and the positive and negative cell plates (Figure 14–41). The longer the time allowed for battery charging at a slow constant current, the greater the chance is for the cell plates to make a complete chemical exchange.

CAUTION *The vehicle's alternator should* not *be used to charge a dead battery that has been jump-started. The extra load on the alternator, in addition to the vehicle electrical system, could overload the alternator.*

The percentage of charge of a battery can affect the temperature at which the battery will freeze (Table 14–3). Even though a frozen battery can be completely thawed and then recharged, the stress on the cell plates significantly shortens the performance and life of the battery. If there is physical damage to the battery case (if it is bloated), charging the battery should not take place.

If a battery needs to be charged or stored on a concrete floor, use an insulating mat or clean dry block of wood to insulate the battery. When a battery is stored on a

Battery High-Rate Charge Time Schedule					
Specific gravity reading	Charge rate amperes	Battery capacity—ampere hours			
		45	55	70	85
Above 1.225	5	★	★	★	★
1.200–1.225	35	30 min	35 min	45 min	55 min
1.175–1.200	35	40 min	50 min	60 min	75 min
1.150–1.175	35	50 min	65 min	80 min	105 min
1.125–1.150	35	65 min	80 min	100 min	125 min

★ Charge at 5-A rate until specific gravity reaches 1.250 @ 80°F.

FIGURE 14–40 Determining the fast charge rate depending on the state of change and capacity rate.

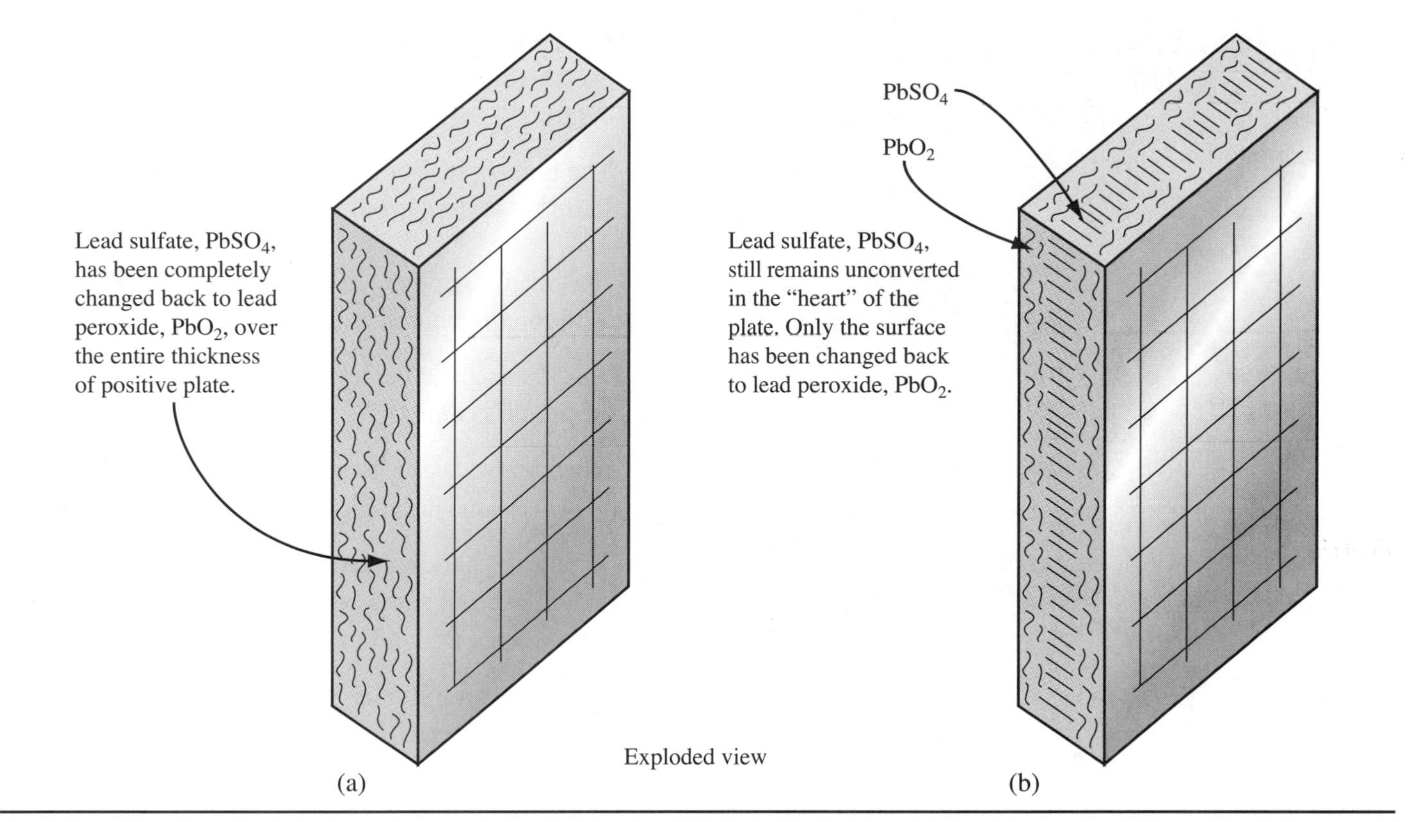

FIGURE 14–41 (a) Effects of a slow change on a positive plate and (b) effects of a fast change on a positive plate.

TABLE 14–3 Battery Freezing Temperatures According to the Specific Gravity of the Electrolyte

% Charge	Specific Gravity	Freezing Temp.
100%	1.265	−91°F
50%	1.200	−16°F
25%	1.150	+5°F
	1.100	+19°F
Water	1.000	+32°F

cold floor, there is a difference in temperature inside the cells between the relatively cold bottom section and the upper section, which is at room temperature. A difference in temperature within the same cell causes a self-discharge to occur, even while the battery is being charged. Self-discharge in batteries is a well-known fact in the submarine industry, which uses warm seawater cycled in tubes around their batteries to keep all the sections in each cell at the same temperature. This is also why the high-voltage battery packs in hybrid electric vehicles have designed cooling systems (Figure 14–42) to keep the batteries at an optimum operating temperature.

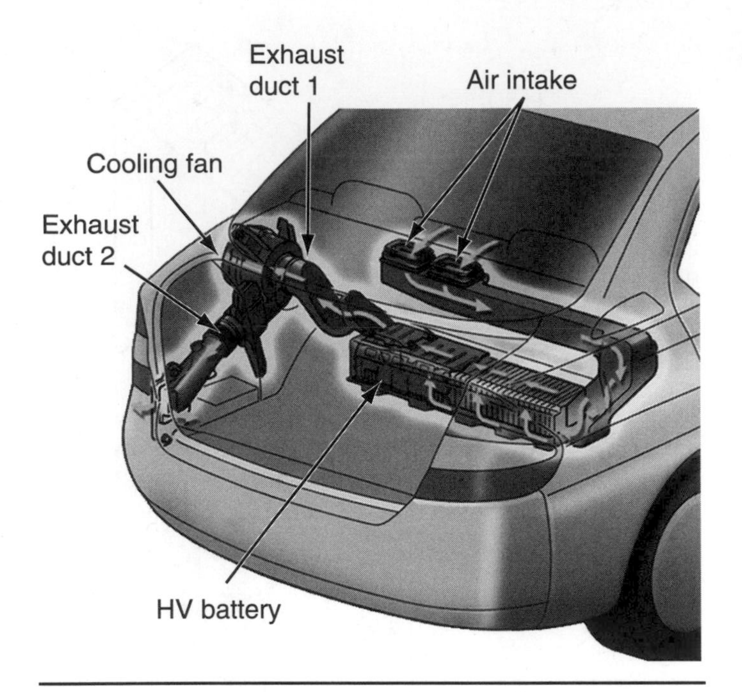

FIGURE 14–42 A battery pack air cooling system on a hybrid electric vehicle (HEV).

(Courtesy of Toyota Motor Sales, U.S.A., Inc.)

Battery Jump-Starting

Various vehicle manufacturers have sufficient differences in recommended jump-starting procedures and precautions; jump-starting procedures for one manufacturer's vehicle might turn out to be unsafe and costly for another manufacturer's vehicle. If one of the following three conditions exists, the disabled battery should *not* be jump-started.

1. Measure the battery terminal voltage on the disabled vehicle. A voltage reading of approximately 10.55 volts could indicate that the battery has a *shorted cell* (12.66 volts total − 2.11 volts of one cell = 10.55 volts) and should *not* be jump-started. A battery with one shorted cell could put an excessive load on the booster battery as it tries to compensate for the bad battery.
2. A voltage reading of 0.55 volts could indicate that the battery has a possible *open circuit* in a cell. This could quickly raise the alternator output on the disabled vehicle to 30 volts or more as the jumper cables are removed. With an *open* battery, the alternator has nothing but itself to use as a reference, and the excess voltage spike could damage vehicle computer controls.
3. The built-in hydrometer shows clear or light yellow. In this case, replace the battery.

The following generic safety precautions should be observed for all vehicles:

- Observe specific vehicle service manual(s) for any special procedures.
- Wear safety glasses. Explosive hydrogen gas may be present around the vehicle needing to be jump-started.
- Engage the parking brake on both vehicles and put both in neutral; some vehicles have been known to jump out of park and into reverse!
- Make sure the vehicles *do not* touch. High current should run directly from the battery to the starter through the battery cables. If the vehicles touch, current could flow through the vehicle body through small low-current body ground straps to the starter and burn up the straps.
- Turn off the ignition switch and *all* accessories on both vehicles before connecting jumper cables. Some manufacturers recommend turning on the heater blower motor on the vehicle with the dead battery to absorb any potentially damaging voltage spikes.
- Do not attempt to use the booster vehicle as a *battery charger.* Doing so could overload the alternator in the booster vehicle. Some vehicle manufacturers are very specific about not using this commonly practiced procedure, because it could severely overheat the alternator and cause premature failure to the windings or bearings (grease melts out of the bearing when it overheats).
- Obtain the ignition keys from the vehicle being jump-started. Having the keys *ensures* you have control of the whole jump-starting process. An overzealous driver trying to start the vehicle before you say it is ready could prove costly to both vehicles.
- Do *not* use more than 16 volts to jump start a vehicle equipped with an engine control computer.

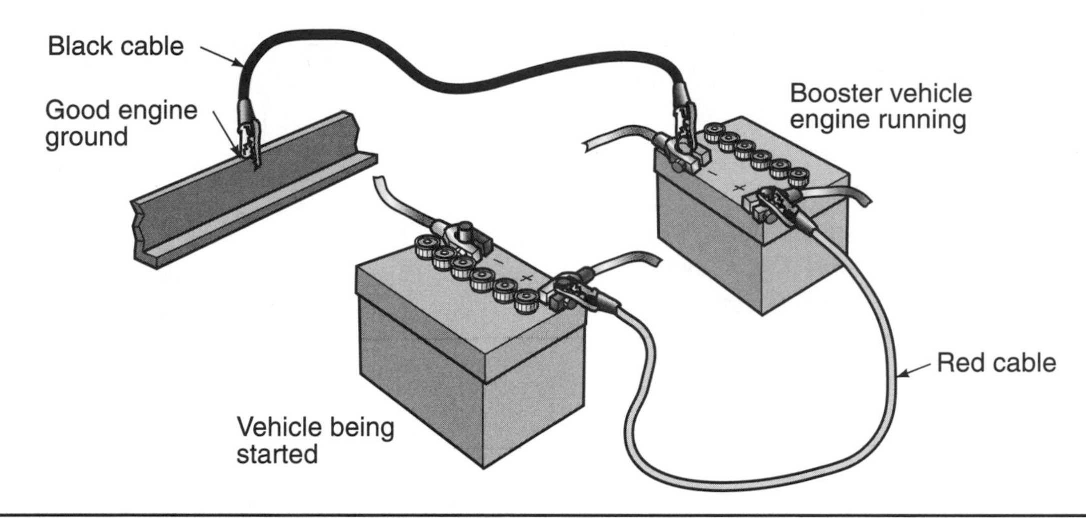

FIGURE 14–43 Connecting jumper cables between the booster vehicle and the vehicle being started.

The excess voltage may damage the electronic components. Most vehicles built during the last decade fall into this category.

- Do not disconnect jumper cables while the vehicle being started is in the cranking mode. A voltage spike may be generated that could damage electronic components on the booster vehicle.
- To hook up the jumper cables (Figure 14–43), attach one end of the positive jumper cable to the disabled battery's positive terminal and the other positive jumper cable to the booster battery's positive terminal. Attach one end of the negative jumper cable to the booster battery's negative terminal and the other negative jumper cable to a good engine ground on the disabled vehicle at least 3 feet away from the battery for safety.

The following jump-start procedures are recommended on electronically controlled vehicles only as a last resort if a proper battery charger is not available.

- Follow any safety precautions that apply from the previous list.
- Shut off the booster vehicle and hook up the jumper cables (Figure 14–43).
- Attempt to start the disabled vehicle. If the vehicle does not crank over sufficiently, stop cranking the engine. Start the booster vehicle and raise the idle to approximately 1,200–1,500 revolutions per minute for up to 5 minutes to assist in boosting the disabled battery. Attempt to start the disabled vehicle. If it still cranks slowly, *discontinue jump-starting.* Further running of booster vehicle could overload the alternator or cause a voltage spike in the system.

Battery Service Procedures

For premium battery efficiency, a quick visual inspection should be taken to inspect the condition of the battery case for cracking or bloating, the terminal connections and cables for corrosion, and the overall cleanliness of the battery. Corroded battery terminals can reduce the efficiency of a battery by creating an unwanted resistance to current flow for the vehicle's electrical system (Figure 14–44). It can also limit the

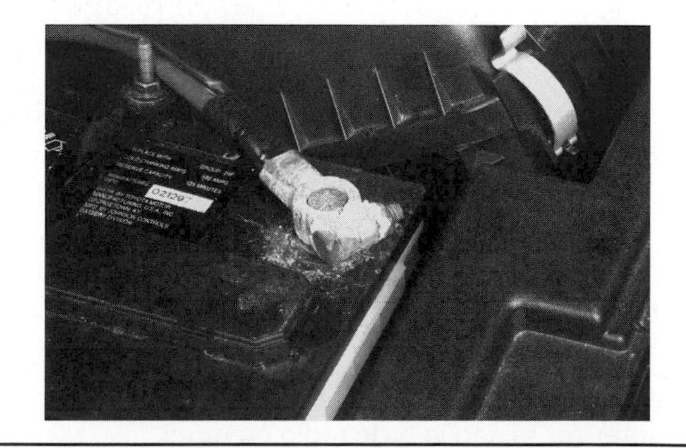

FIGURE 14–44 A typical corroded battery terminal.

current used to recharge the battery, possibly resulting in lead sulfate crystallizing on the battery plates over an extended period of time.

When battery servicing is necessary and the terminals have to be disconnected, it is important to save this memory on most vehicles with electronic memory functions in the radio or power train control computers. If the electronic memory is lost in some modern vehicles, it may take several drive cycles before the computer memory learns transmission shift points and engine idle and load conditions. It is not uncommon for these conditions to exist for up to 20 miles or more of driving time.

Electronic Memory Saving

Beginning in the early 1970s, most vehicles were introduced with radios having electronic tuning. A simple way to keep the memory alive in these earlier electronic radios is the use of a 9-volt memory saver adapter plugged into the cigar lighter receptacle (Figure 14–45). To prevent the 12-volt vehicle battery from back feeding into the 9-volt battery when they are both in parallel to the electrical circuit, a diode is placed in series with the adapter plug's positive lead.

Another 9-volt memory saver adapter (Figure 14–46) can be built with common electronic supply store parts. It uses a green LED light to indicate that the vehicle's electrical system is operating off the vehicle battery. When the battery's negative cable is removed, the memory saver automatically shifts to the 9-volt backup battery, and the red LED light goes on. Either of these memory saver tools is meant to save only the radio memory while the vehicle battery is changed. If a door light or other higher-amperage requirement is applied to the circuit, the 9-volt battery is drained quickly.

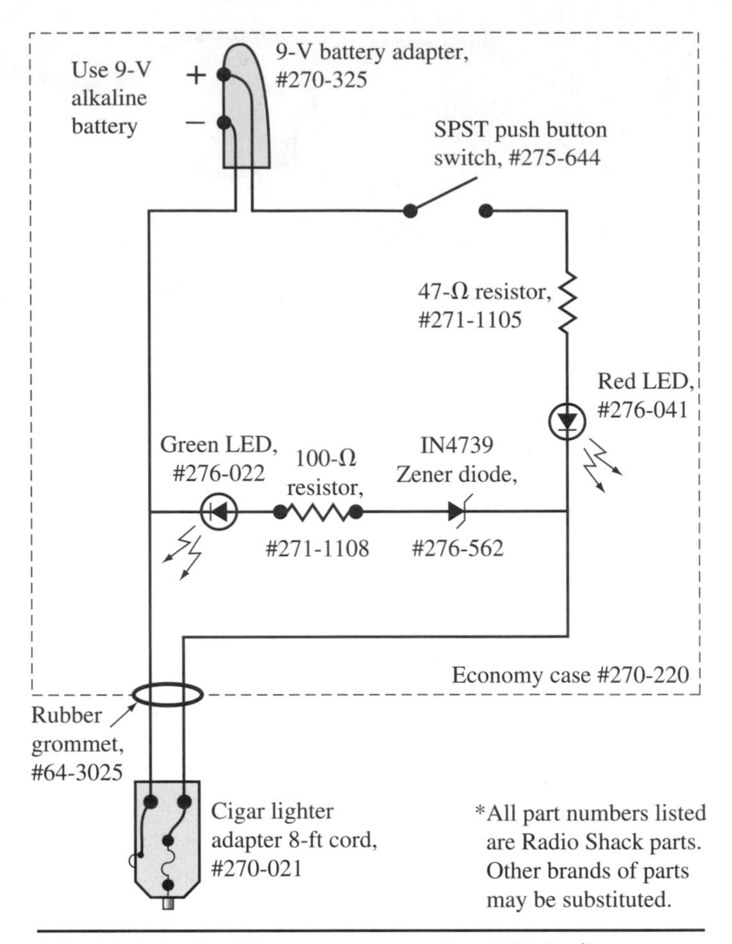

FIGURE 14–46 Memory saver using LED indicator lights, 9-volt battery, and cigar lighter adapter.

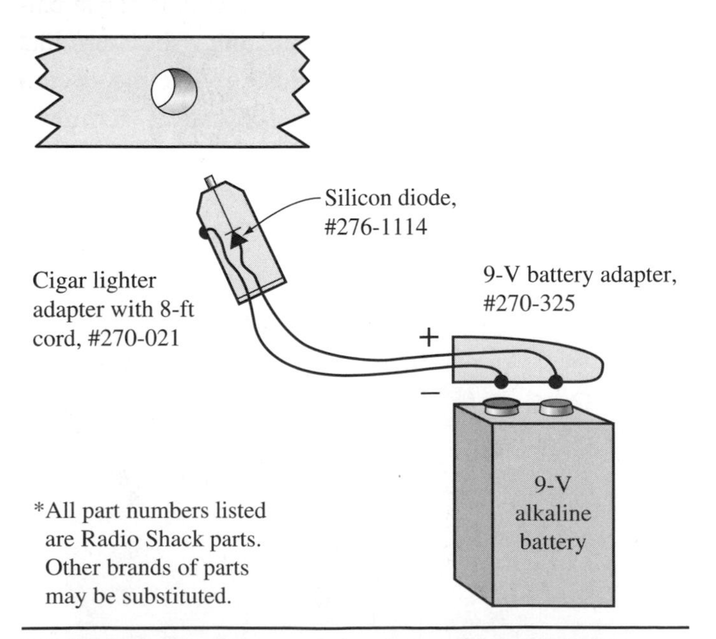

FIGURE 14–45 Nine-volt battery and cigar lighter adapter for saving vehicle radio memory.

If there is a chance that the vehicle door may be opened, some technicians use a 12-volt auxiliary battery in parallel with the vehicle battery, either directly to the positive and negative battery cables or through the cigar lighter plug if it is energized when the ignition key is in the off position (Figure 14–47).

On newer vehicles with computer controls or complex radio programming functions, the memory functions may be stored in a nonvolatile RAM memory, and

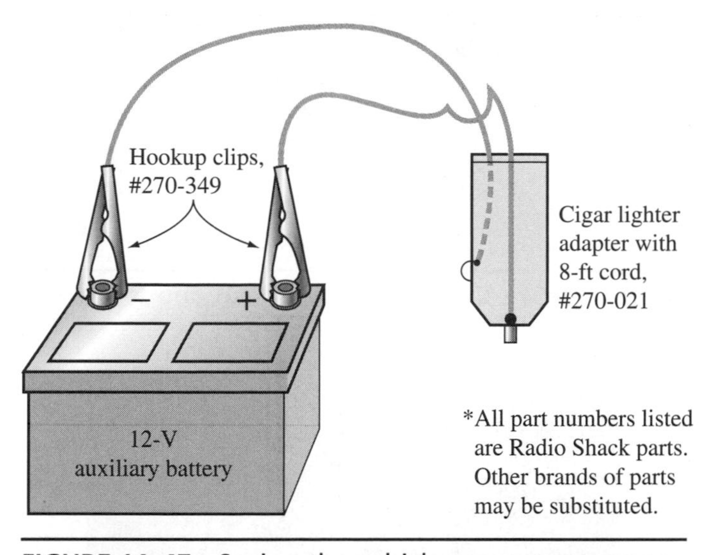

FIGURE 14–47 Saving the vehicle computer memory with a 12-volt auxiliary battery and cigar lighter adapter.

a memory-saving device is not needed when changing the vehicle battery. Always refer to vehicle manufacturer's directions when unsure of battery testing, service, and replacement procedures.

Battery Removal and Cleaning

The best way to clean a dirty battery is to remove it from the vehicle. Most batteries in modern vehicles are tightly packed into the engine compartment area. Attempting to clean a battery in the vehicle could spread the corrosive acids onto other engine or electrical components.

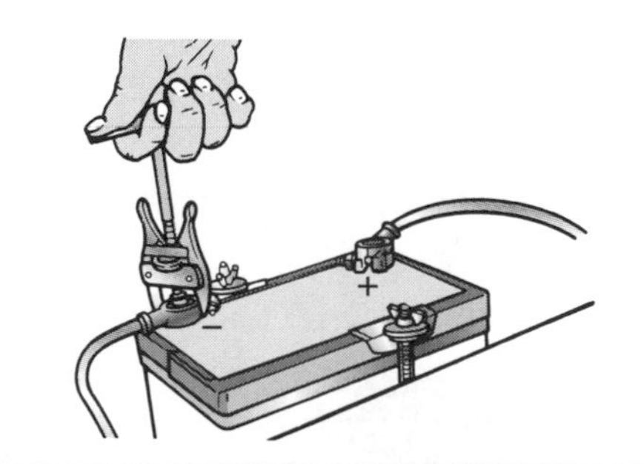

FIGURE 14–48 To prevent post damage, a terminal puller is used.

Begin by removing the battery negative terminal using a battery terminal puller (Figure 14–48); then remove the battery and carry it using a suitable lifting strap (Figure 14–49).

Apply a solution of baking soda and warm water (Figure 14–50) to the battery case to neutralize any accumulated corrosion on the battery case; then wipe the battery case dry with paper towels or an old rag. Do *not* use cotton shop towels from a laundry service to dry a battery because residual acids eat away the shop towel and eliminate it from the counted inventory. The terminals and posts should be cleaned with a cleaning tool or wire brush (Figure 14–51).

When the battery is replaced into the vehicle, make sure the hold-down connectors are secured. A loose battery could short out on the hood or tip over. Install the *positive* connector first, then the negative connector.

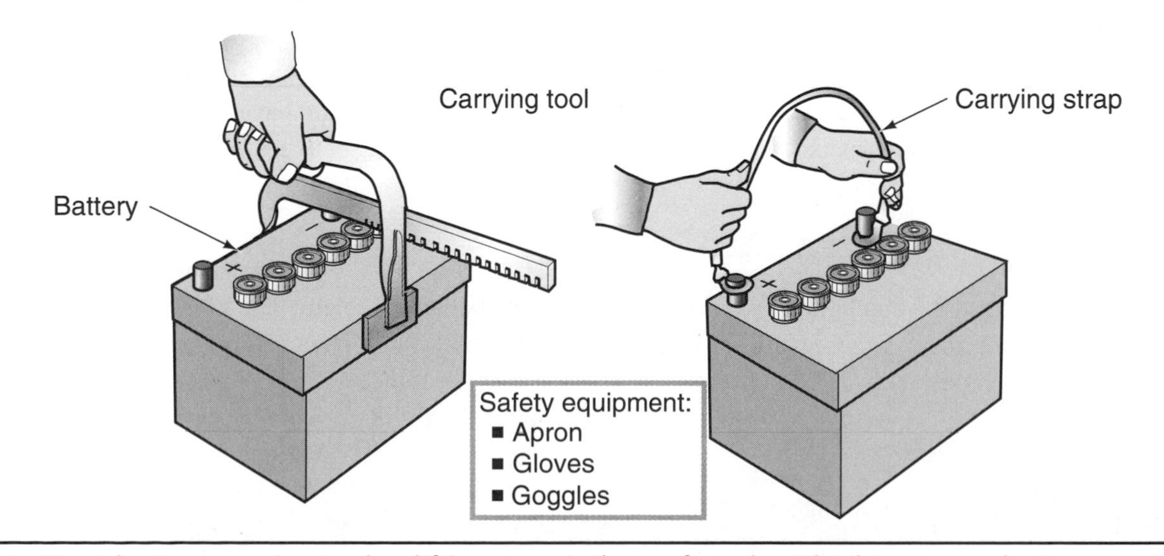

FIGURE 14–49 A battery carrying tool or lifting strap is the preferred method to move a battery.

FIGURE 14–50 A solution of baking soda and water will neutralize the battery acid corrosion.
(Courtesy of DaimlerChrysler Corporation)

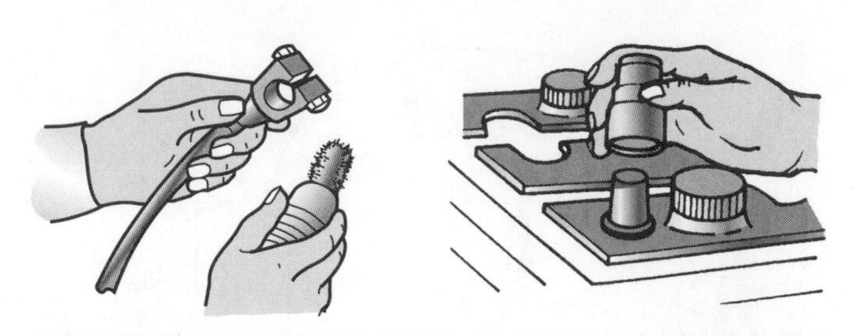

FIGURE 14–51 Cleaning battery posts and terminals with a battery cleaning tool.

FIGURE 14–52 Using corrosion protective pads to prevent terminal corrosion.

Battery terminal clamps should be sprayed with a protective coating to prevent corrosion. Grease or petroleum jelly can also be used to seal the terminals, but it also attracts dirt or other contaminants over time. Corrosion protective pads can also be used under the terminals to prevent corrosion (Figure 14–52).

CASE STUDIES

CASE 1

This case study is typical of a battery-related problem that technicians often have to diagnose. Knowledge of how batteries are constructed and of simple test procedures will help in performing the diagnosis quickly and efficiently.

Customer Complaint

A customer brings his vehicle into the shop with a possible battery problem. The vehicle cranks slowly in the morning when the weather is cold or under heavy load conditions. The customer changed the alternator, but the problem remains.

Known Information

- ❑ The vehicle operating voltage = 14 volts.
- ❑ The battery is a conventional design.

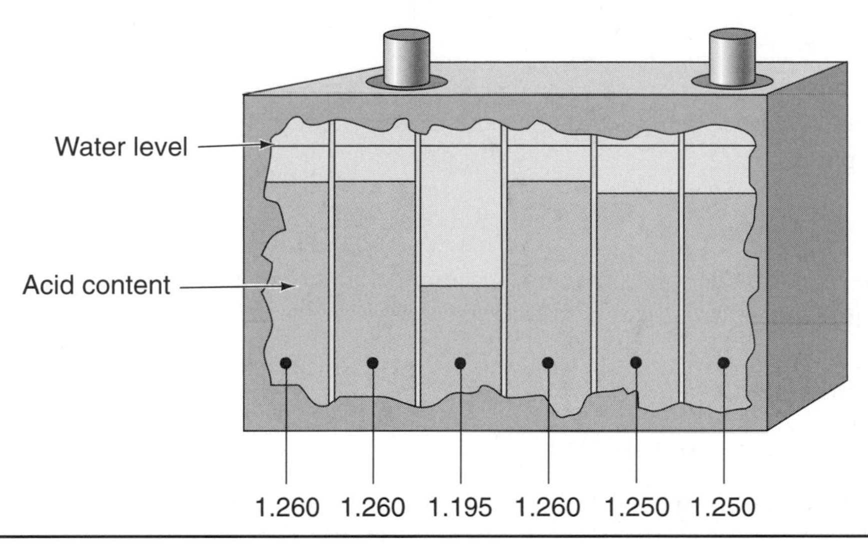

FIGURE 14–53 Hydrometer readings.

❑ The battery is four years old with a CCA rating of 550 CCA.

❑ The recorded battery hydrometer readings (Figure 14–53).

Circuit Analysis

Answer the following questions on a *separate* sheet of paper.

1. With the information known, what is the *most* likely cause of the battery performance during cold cranking conditions? ________________________.
2. To properly check a battery for performance under all load conditions, what are the proper steps to take in determining its condition?

   a. ________________________.
   b. ________________________.
   c. ________________________.
   d. ________________________.
   e. ________________________.
   f. ________________________.
3. What diagnostic steps were helpful when troubleshooting this system or explaining it to the customer?________________________

   ________________________.
4. What is your analysis of this case study? ________________________

   ____

   ________________________.

CASE 2

This case study concerns jump-starting a vehicle. Knowledge of jump-start procedures and safety precautions are helpful.

Customer Complaint

The battery in a vehicle was cranking successfully only an hour before the problem. The customer requests a "jump-start."

Known Information

- ❑ The battery is only two years old.
- ❑ A hydrometer check on the battery shows the specific gravity of all cells are approximately 1.265.
- ❑ A DMM is used to measure the open circuit voltage of the battery, which measured 10.55 volts.

Circuit Analysis

Answer the following questions on a *separate* sheet of paper.

1. With the known battery information, what is the most likely cause of the battery problem?

   ____.

2. Can the battery be safely jump-started any time a no-crank or slow-crank condition exists? ________.
3. What are some of the precautions that should be followed *before* attempting to jump-start any vehicle? ____.
4. What diagnostic information is helpful to know before attempting to jump-start any vehicle? ______

   ____.

5. What is your analysis of this case study? ____

   ____.

Hands-On Vehicle Tests

The following nine hands-on vehicle checks are included in the NATEF (National Automotive Technicians Education Foundation) Task List. Some tasks are grouped together due to their close association with each other. Complete your answers to the questions on a *separate* sheet of paper.

Performance Task 1

Task Description

Inspect, clean, fill, and replace the battery.

Performance Task 2

Task Description

Inspect and clean the battery cables, connectors, clamps, and hold-downs; repair or replace them as needed.

Task Objectives

❑ Obtain a vehicle that can be used for these tasks. What model and year of vehicle are you using for these tasks? ______________________________.

❑ Describe the general appearance of the battery.______________________________

______________________________.

❑ Inspect and record the condition of the following battery components:

Component	Condition	Repairs needed
Battery case		
Battery top		
Battery support tray		
Battery hold-down		
Battery posts		
Battery cables and clamps		
Surrounding area of battery		

❑ What tools and equipment would be useful in cleaning the battery and associated components?

______________________________.

❑ What are the necessary safety precautions that should be observed when working with or near a battery?______________________________

______________________________.

❑ If the battery needs to be filled, what type of water should be used and how full should it be filled?

______________________________.

❑ What procedures are most helpful when inspecting, filling, and cleaning a battery, and associated cables, connectors, and hold-downs?______________________________

______________________________.

Task Summary

After performing the preceding NATEF tasks, what can you determine will be helpful in knowing how to check instrument cluster gauges and warning lights for proper operation?______________________________

____

______________________________.

Performance Task 3

Task Description

Perform battery state-of-charge test; determine necessary action.

Performance Task 4 (not a NATEF task)

Task Description

Perform battery open circuit voltage test; determine necessary action.

Performance Task 5 (NATEF # VI-B-2)

Task Description

Perform battery capacity test (or conductance test); confirm proper battery capacity for vehicle application; determine necessary action.

Performance Task 6 (not a NATEF task)

Task Description

Perform a battery 3-minute charge test; determine needed repairs.

Task Objectives

- ❑ Obtain a vehicle that can be used for these tasks. What model and year of vehicle are you using for these tasks? ______________________________.
- ❑ Battery type: ______________ Battery rating: ______________
  Engine (cubic inches): ______________
- ❑ Is the battery of sufficient size and rating for the vehicle? ______________.
- ❑ Perform the following battery state-of-charge test.

  Hydrometer readings (if cells are accessible):

Cell	1	2	3	4	5	6

- ❑ Test results:
  1. All cells have a specific gravity of at least 1.265—continue with load test.
  2. All cells have an equally low specific gravity below 1.265—charge and retest battery.
  3. Cell specific gravity readings have a point variation of 0.050 between the highest and lowest cell—discontinue testing and replace battery.
- ❑ Perform the following battery open circuit voltage test:
  1. Load the battery with 20–25 amperes for 1 minute to remove any surface charge present on cell plates; turn on high-beam headlights and heater blower to high.

2. Remove any loads from the battery, and let the battery stabilize for 2 minutes.
3. Measure the open circuit voltage across the battery terminals with a quality DMM.

❑ Test results:
Refer to the chart below to convert the open circuit voltage to the state-of-charge or the specific gravity.

Open circuit voltage (V)	State-of-charge (%)	Specific gravity
12.6 or greater	100	1.260
12.4	75	1.225
12.2	50	1.190
12.0	25	1.155
11.9	Discharged	1.100

1. If the battery's open circuit voltage indicates a state of charge less than 100%, charge and retest the battery.
2. If the battery's open circuit voltage indicates a fully charged battery, continue with the load test.

❑ Perform the following battery capacity load test:

1. Determine which tester to use to perform the load test. Some testers perform the procedure more automatically than others.
2. Following are the instructions for operating a battery tester with a *manual controlled* carbon pile for load testing the battery.
   - Set tester for the high-amperage start position.
   - Determine the load rating of the battery. Calculate the load to apply with the carbon pile: CCA $\div$ 2 = load, or A/H $\times$ 3 = load.
   - Hook up large tester clamps to the battery's positive and negative terminals. The green inductive pickup clamp should go around the negative tester lead (not the battery negative lead).
   - Apply the calculated load to the carbon pile for 15 seconds while watching the voltmeter reading. Turn off the load after 15 seconds.
   - Determine the load test results.
     a. Good battery—reading remained above 9.6 volts at 70°F during the load test. Battery is satisfactory.
     b. Bad battery—battery voltage dropped below 9.6 volts at 70°F during the load test. Battery performance failed. Verify the condition of cell plates by performing a 3-minute charge test. Some manufacturers of maintenance-free batteries recommend that this test not be performed due to the differences in construction, materials, and size of batteries. Refer to specific manufacturer's instructions for testing the battery before proceeding.

3. Following are the instructions for operating a battery tester with an *automatic* load testing feature, such as the ARBST (alternator/regulator/battery/starter tester):
   - Hook up the large tester clamps to the battery's positive and negative terminals.
   - Push the battery test button, and enter the CCA rating from the battery.
   - Push enter and then wait for 50 seconds for test results.
   - Determine the load test results by reading the digital tester readout.
     a. *Good battery*—battery is charged and capable of performing as designed.
     b. *Good, low charge*—battery is good but needs to be recharged.
     c. *Bad battery*—battery will not hold a charge or perform as designed.
     d. *Charge and retest*—tester cannot determine battery condition. Charge battery and retest.

❑ Perform the following battery 3-minute charge test:

1. Disconnect the battery ground cable to isolate the vehicle electrical system from higher-than-normal voltages that might exist during this test.
2. Connect a battery charger to the battery, observing polarity and all safety precautions.
3. Connect a DMM across the battery terminals to monitor the battery voltage while the 3-minute charge is taking place.
4. Turn on battery charger to 40 amperes for 3 minutes. To get an accurate test, a battery charger that automatically reduces the charging amperage may not be suitable for this test.
5. Observe the voltmeter reading after 3 minutes.

❑ Three-minute charge test results:

1. Good battery—voltmeter reading of *less than* 15.5 volts (16.5 volts for some maintenance-free batteries). Battery is *not* sulfated; recharge battery and perform the capacity load test again.
2. Bad battery—voltmeter reading more than 15.5 volts (16.5 volts for some maintenance-free batteries). Battery is sulfated and not able to receive a charge. Battery should be replaced.

Task Summary

After performing the preceding NATEF tasks, what can you determine that will be helpful in knowing how to check the battery state-of-charge, open circuit voltage, capacity load, and 3-minute charge tests?

____

____

____.

Performance Task 7

Task Description

Perform slow/fast battery charge.

Task Objectives

- ☐ Obtain a vehicle that can be used for this task. What model and year of vehicle are you using for this task? ______.
- ☐ What are the safety precautions necessary when slow/fast charging a battery? ______ ______.
- ☐ What is the open circuit voltage of the battery needing charged? ______.
- ☐ How do you hook up the battery charger to the battery? ______ ______.
- ☐ What would be the charging rate for this battery with a slow charge? ______ ______.
- ☐ What would be the charging rate for this battery with a fast charge? ______ ______.
- ☐ How can you determine that the battery is fully charged? ______ ______.

Task Summary

After performing the preceding NATEF task, what can you determine will be helpful in knowing how to slow charge and fast charge a battery? ______ ______ ______.

Performance Task 8

Task Description

Start a vehicle using jumper cables and a battery or auxiliary power supply.

Task Objectives

- ☐ Obtain two vehicles that can be used for this task. What models and years of vehicles are you using for this task? ______.
- ☐ What are the safety precautions necessary when jump-starting a battery? ______ ______.
- ☐ What is the open circuit voltage of the battery needing jump-started? ______.
- ☐ What special precautions are necessary in reference to the open circuit voltage of the battery needing jump-starting? ______.
- ☐ What special precautions are necessary when either, or both, the booster vehicle and the dead battery are computer controlled.

❑ List the procedures you used to perform the jump start on this vehicle:

a. ______________ f. ______________

b. ______________ g. ______________

c. ______________ h. ______________

d. ______________ i. ______________

e. ______________ j. ______________

❑ What steps are most helpful when jump-starting a vehicle? ______________

______________.

Task Summary

After performing the preceding NATEF task, what can you determine will be helpful in knowing how to jump-start a battery? ______________

____

______________.

Performance Task 9

Task Description

Maintain or restore electronic memory functions.

Task Objectives

❑ Obtain a vehicle that can be used for this task. What model and year of vehicle are you using for this task? ______________.

❑ What type of memory-saving tool are you using to perform this task? ______________

______________.

❑ What precautions are necessary when using a memory saver tool while changing a battery?

______________.

❑ List the steps taken to properly use a memory-saving tool to save the vehicle's memory circuits:

1. ______________.
2. ______________.
3. ______________.
4. ______________.
5. ______________.

❑ What is your analysis of using a memory-saving tool to save the vehicle's memory circuits? ________

____.

Task Summary

After performing the preceding NATEF task, what can you determine will be helpful in knowing how to use a memory-saving tool? ____

____

____.

Summary

❑ The simplest form of battery is a voltaic cell, with two unlike metals and an alkali, salt, or an acid solution to produce a voltage.

❑ An electrolyte is a chemical compound of alkali, salt, or acid solution that is capable of conducting electrical energy.

❑ A primary cell cannot be recharged to its original voltage condition.

❑ A secondary cell can be recharged to its original stored energy.

❑ The negative plate in a secondary cell is made of pure lead (Pb).

❑ The positive plate in a secondary cell is made of lead dioxide (PbO_2).

❑ A diluted mixture of sulfuric acid contains 64% water and 36% sulfuric acid, by weight.

❑ The basic grid plate in a conventional battery is made of a lead antimony mixture.

❑ The active material in a positive plate is a reddish-brown paste called lead peroxide.

❑ Battery plates are separated by microporous materials called separators.

❑ Each cell in a lead acid battery produces 2.11 volts, regardless of the number of positive and negative plates used in each cell.

❑ A fully charged battery will produce 12.6 volts.

❑ A battery's specific gravity is 1.265 at 80°F when fully charged.

❑ A maintenance-free battery uses calcium in the grid plate material to reduce gasing and self-discharge.

❑ A deep cycle battery is a lead acid battery that uses approximately 2.75% antimony on the positive plate and a calcium alloy on the negative plate.

❑ A recombination battery is a completely sealed battery that uses the electrolyte mixture in a gel form.

- ❑ The battery cell in a recombination battery is a spiral, which provides more cell plate surface area than a standard battery cell plate.
- ❑ The nickel-cadmium (NiCd) and nickel metal hydride (NiMH) battery designs are two popular batteries used in hybrid electric vehicles (HEV) because of their small size-to-energy ratio and their ability to tolerate deep cycling, compared to lead-acid batteries.
- ❑ The reserve capacity (RC) of a battery is the amount of time in minutes a fully charged battery can be discharged at 25 amperes while maintaining a minimum battery voltage of 10.5 volts at 80°F in the event of a charging system failure.
- ❑ The amp-hour (A/H) rating of a battery is the amount of current that a fully charged battery can produce over a 20-hour period at 80°F before terminal voltage reaches a minimum of 10.5 volts.
- ❑ The cold cranking amperes (CCA) rating is the ability of the battery to provide an amperage load for 30 seconds at 0°F without the terminal voltage dropping below 7.2 volts.
- ❑ A battery's size is classified by BCI group numbers.
- ❑ A battery case leakage test determines whether there is a small current draw between the battery posts through a dirty or contaminated top.
- ❑ A battery hydrometer measures the specific gravity content of the sulfuric acid mixture.
- ❑ A fully charged battery has a specific gravity of 1.265–1.280.
- ❑ The specific gravity of a battery can be checked with an optical refractometer.
- ❑ The battery open circuit voltage test can be used to test the battery's state of charge when a hydrometer cannot be used.
- ❑ A battery capacity load test will check the ability of the battery to supply sufficient cranking amperage to the starter.
- ❑ A conductance test on a battery sends a low-frequency AC signal through the battery plates, then measures a percentage of the AC current response.
- ❑ The 3-minute charge test tests some batteries for sulfation on the cell plate surfaces.
- ❑ Self-discharge in a battery can occur when the internal temperature between the upper and lower portions of the cell plates are at different temperatures.
- ❑ A battery with an open circuit voltage of 10.55 volts should not be jump-started because one cell could be shorted.
- ❑ A battery with an open circuit voltage of 0.55 volts could indicate a battery with an open circuit, and it should not be jump-started.
- ❑ A memory-saving device is not needed on several newer cars to save the radio and computer memories, because the memory functions are stored in nonvolatile RAM memory.

Key Terms

amp-hour rating (AH)
capacitance test
cold cranking amperes (CCA)
conductance test
conventional battery
cranking amps (CA)
deep cycle
electrolysis
electrolyte
element
gasing
gel-cell battery
grid growth
hybrid
hydrometer
lead antimony
lead calcium
lead peroxide
maintenance-free
optical refractometer
primary cell
reserve capacity (RC)
recombination battery
secondary cell
self-discharge
separator
specific gravity
sponge lead
sulfation
sulfuric acid
voltaic cell
watt-hour (WH)

Review Questions

Short Answer Essays

1. Describe the difference between a primary cell battery and a secondary cell battery.
2. What causes the glass bubble to float in a battery hydrometer?
3. Describe how to perform a battery capacity load test using the cold cranking rating method.
4. Describe how would you perform a battery case leakage test.
5. Describe the procedures and safety precautions when charging a battery.

Fill in the Blanks

1. A fully charged battery will have an electrolyte reading of ________________.
2. When a battery has an open circuit voltage of 12.2 volts, the state of charge is ________________%.
3. During a battery 3-minute charge test, if the battery voltage exceeds ________________, then the battery is considered sulfated.
4. When battery service is necessary, the ________________ cable is disconnected first and then reconnected ________________.
5. A battery with a CA rating of 600 has a CCA rating of ________________.

ASE-Style Review Questions

1. As a lead-acid battery is being discharged, the positive plate material is converted to:
   A. Sulfuric acid.
   B. Lead sulfate.
   C. Sponge lead.
   D. None of the above.

2. Technician A says the lead-acid battery contains an electrolyte with a mixture of water and sulfuric acid. Technician B says the electrolyte in a lead-acid battery is capable of conducting electrical energy. Who is correct?
   A. A only
   B. B only
   C. Both A and B
   D. Neither A nor B

3. Technician A says a primary cell battery can be recharged to its original condition. Technician B says a secondary cell battery cannot be recharged once its original stored energy has been depleted. Who is correct?
   A. A only
   B. B only
   C. Both A and B
   D. Neither A nor B

4. Technician A says that a conventional battery has a grid construction made from a mixture of lead and lead antimony. Technician B says the negative grid in a conventional battery has lead peroxide paste pressed into it to form the negative plate. Who is correct?
   A. A only
   B. B only
   C. Both A and B
   D. Neither A nor B

5. A 500 CCA battery is being load tested with a manual carbon pile load tester. Technician A says the correct load to apply to the load tester is one-third of the rated A/H for 15 seconds. Technician B says the correct load to apply to the tester is 50% of the rated CCA for 15 seconds. Who is correct?
   A. A only
   B. B only
   C. Both A and B
   D. Neither A nor B

6. Technician A says the specific gravity of a fully charged battery is 1.265 at 80°F. Technician B says the specific gravity of a battery is measured with a hydrometer. Who is correct?
   A. A only
   B. B only
   C. Both A and B
   D. Neither A nor B

7. Technician A says a maintenance-free battery produces less gasing than a conventional battery. Technician B says a maintenance-free battery uses a grid frame made with a lead calcium mixture. Who is correct?
   A. A only
   B. B only
   C. Both A and B
   D. Neither A nor B

8. Technician A says a negative battery cable is always removed first when removing a battery. Technician B says a negative battery cable should always be put on first when replacing a battery. Who is correct?
   A. A only
   B. B only
   C. Both A and B
   D. Neither A nor B

9. Two technician are discussing the temperature correction readings on a hydrometer. Technician A says 0.004 should be added to the reading for every 10° above 80°F. Technician B says 0.004 should be subtracted from the reading for every 10° below 80°F. Who is correct?
   A. A only
   B. B only
   C. Both A and B
   D. Neither A nor B

10. Technician A says the battery must be stabilized before performing a battery open circuit voltage test. Technician B says a voltage reading of 12.0 volts is acceptable for a battery open circuit voltage test. Who is correct?
   A. A only
   B. B only
   C. Both A and B
   D. Neither A nor B

11. Two technicians are discussing the results of a 3-minute charge test. Technician A says a reading greater than 15.5 volts on a conventional battery indicates a sulfated battery. Technician B says a load of 40 amperes should be applied to the battery to get an accurate reading. Who is correct?
    A. A only
    B. B only
    C. Both A and B
    D. Neither A nor B

12. Technician A says a deep cycle battery uses a gel material for the electrolyte mixture. Technician B says a deep cycle battery grid plate has the vertical and horizontal grid bars in a radial design expanding from a center lug for faster amperage flow. Who is correct?
    A. A only
    B. B only
    C. Both A and B
    D. Neither A nor B

13. Technician A says a reserve capacity load rating indicates the amount of time in minutes a battery be discharged at 25 amperes while maintaining a minimum battery voltage of 10.5 volts. Technician B says a battery with an RC capacity of 120 minutes will sustain 25-ampere load for 2 hours. Who is correct?
    A. A only
    B. B only
    C. Both A and B
    D. Neither A nor B

14. Technician A says a baking soda and water mixture can be used to neutralize battery surface corrosion. Technician B says baking soda can be used in the battery cells to refresh the sulfuric acid mixture. Who is correct?
    A. A only
    B. B only
    C. Both A and B
    D. Neither A nor B

15. Technician A says a CCA rating on a battery determines the ability to provide an amperage load for thirty seconds at 0°F, with the terminal voltage dropping no lower than 7.2 volts. Technician B says the CCA can be divided be 5.25 to get the approximate A/H rating. Who is correct?
    A. A only
    B. B only
    C. Both A and B
    D. Neither A nor B

16. Technician A says the battery capacity rating is determined by the battery case size. Technician B says the battery case is always made of a hard rubber material. Who is correct?

    A. A only

    B. B only

    C. Both A and B

    D. Neither A nor B

17. Technician A says a voltage terminal drop test of 0.50 volts maximum is satisfactory. Technician B says the battery terminal voltage drop test is performed by removing the battery cable and hooking the voltmeter in series between the post and the terminal. Who is correct?

    A. A only

    B. B only

    C. Both A and B

    D. Neither A nor B

18. Technician A says battery temperature over 125°F helps the battery charge faster. Technician B says the battery negative cable should be disconnected before charging to isolate the vehicle's electrical system from voltage conditions above 15.5 volts. Who is correct?

    A. A only

    B. B only

    C. Both A and B

    D. Neither A nor B

19. Technician A says a battery will self-discharge if left on a cold floor. Technician B says a battery will self-discharge if the internal temperature of the cell plates vary. Who is correct?

    A. A only

    B. B only

    C. Both A and B

    D. Neither A nor B

20. Technician A says a 9-volt battery and cigar plug adapter can be used to save the radio memory while the vehicle battery is being changed. Technician B says an auxiliary 12-volt battery hooked in parallel to the vehicle battery is preferred for computer-controlled vehicles to save memory when changing the vehicle battery. Who is correct?

    A. A only

    B. B only

    C. Both A and B

    D. Neither A nor B

15 Starting System Theory

Introduction

The starter circuit in modern vehicles is a high-amperage circuit that combines electrical and mechanical components to crank the engine for starting. The electrical energy produced by the battery is converted to mechanical energy through the interaction of magnetic fields in the starter motor. The typical small internal combustion engine of today requires 90 to 150 amperes to turn the engine over at 400 rpm or more to be effective. In addition to standard starter circuits, the recent advancement in hybrid electric vehicle (HEV) technology has created several new starter system designs that assist the HEV in its ability to start and stop the engine/motor unit under a variety of operating conditions.

Objectives

When you complete this chapter you should be able to:

- ❑ Describe the main components of the starter motor system and their purpose.
- ❑ Explain the principal theory of magnetism in a DC starter motor.
- ❑ Explain the differences and operation of various solenoid control switches.
- ❑ Explain the differences between starter drive mechanisms.
- ❑ Describe the main features and operation of different types of starter motors.
- ❑ Explain the differences and operation of various starter safety switches.
- ❑ Perform a systematic diagnostic check on a starter circuit.
- ❑ Perform system and individual tests on the starting system components.
- ❑ Remove and reinstall a starter motor.
- ❑ Disassemble, clean, inspect, provide necessary repairs, and reassemble a starter motor.
- ❑ Explain how a typical starter system works in a hybrid electric vehicle.

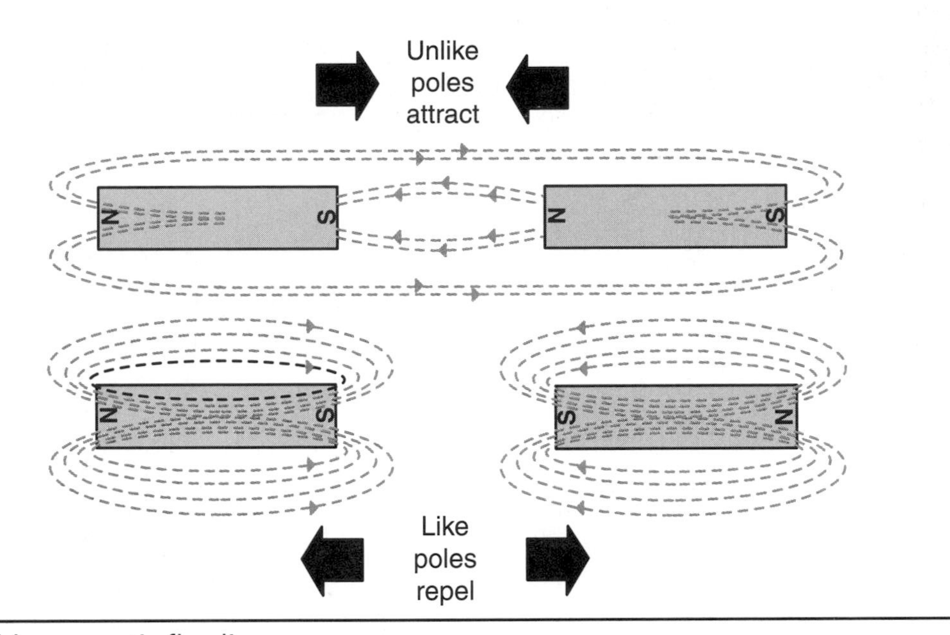

FIGURE 15–1 Invisible magnetic flux lines.

Magnetism and Motor Principles

Learning about the operation of a DC starter motor begins with a thorough understanding of magnetism and the interaction of magnetic fields in the starter as the electrical energy from the battery is converted into mechanical energy to crank the engine. If you were able to see the magnetic fields surrounding a permanent magnet (Figure 15–1), you would find that *like* poles *repel* each other and *unlike* poles *attract* each other. Magnetic repulsion and attraction form the basis of the electromagnetic motor.

When current flows through a conductor in a complete circuit, a weak magnetic field is produced around the conductor (Figure 15–2). The concentration of magnetic lines of force is called **flux density.** If a current-

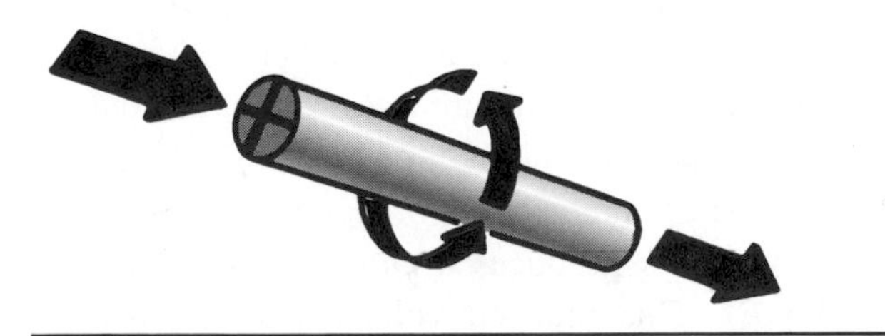

FIGURE 15–2 A weak magnetic field surrounds a conductor that has a current flowing through it.

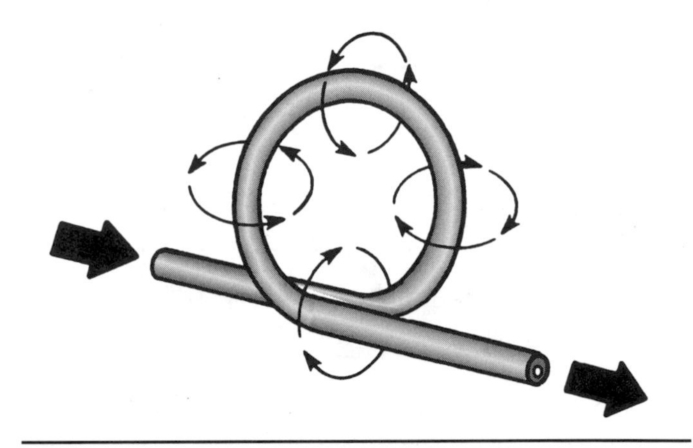

FIGURE 15–3 The flux density and magnetic field increase when a conductor is looped in a circle.

carrying conductor is looped in a circle (Figure 15–3), where the conductor is placed side by side, then the flux density and magnetic field both increase (Figure 15–4).

When a current-carrying conductor is placed the magnetic field of a horseshoe magnet (Figure 15–5), the conductor moves toward the weaker magnetic field on the outside of the magnet (Figure 15–6).

A Simple Electromagnetic Motor

A simple electromagnetic starter motor consists of several important components that form a motor strong enough to crank an engine (Figure 15–7).

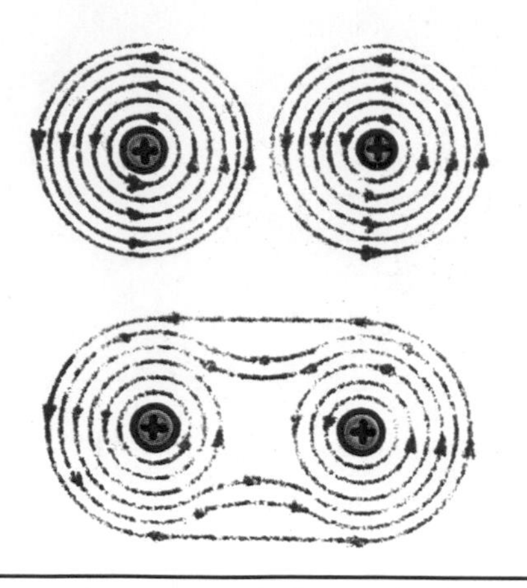

FIGURE 15–4 The magnetic lines of a force join when the current-carrying conductors are placed close together.

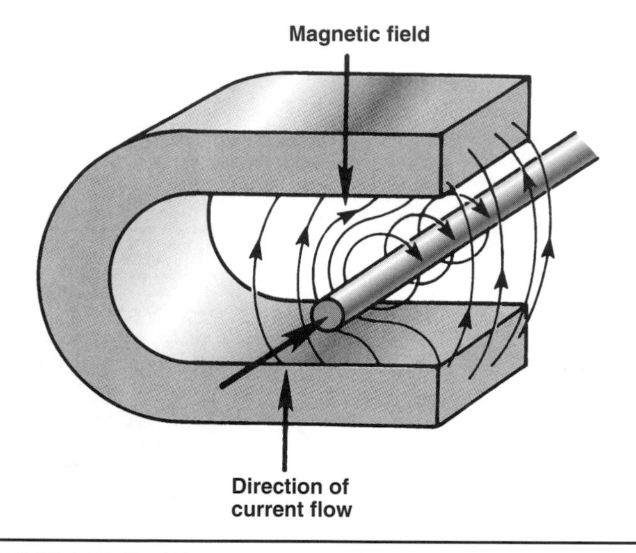

FIGURE 15–5 Placing magnetic current-carrying conductor in the magnetic field of a horseshoe magnet.

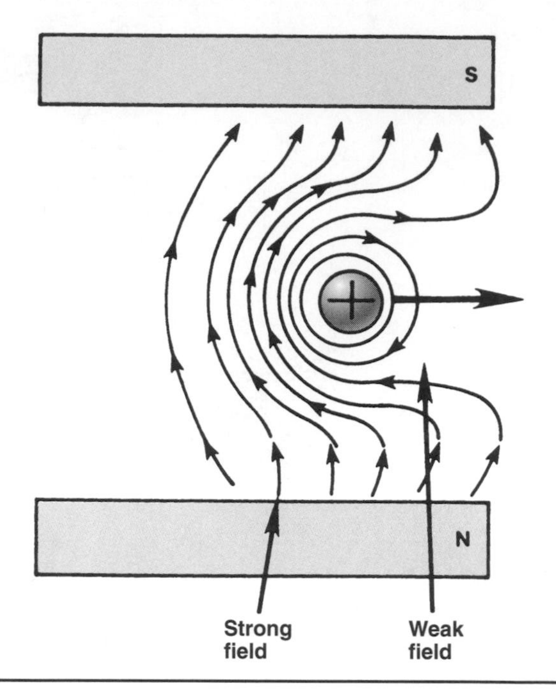

FIGURE 15–6 The conductor movement in a magnetic field tends to move toward the weaker outside magnetic field.

The first important component in a starter motor consists of the **field coils,** which are constructed of heavy copper wire wrapped in a coil around an iron core to form an electromagnet (Figure 15–8). The soft iron core in the field coil is called the **pole shoe,** which helps to concentrate and direct the lines of force in the field coil.

The second important component is the **armature** assembly that rotates within the stationary field coils. It consists of several heavy loops of copper wire wrapped around, and insulated from, a laminated iron core (Figure 15–9). At one end of the armature is the **commutator,** a series of split ring copper segment bars that form a sliding contact with a set of electrically conductive brushes. The **brushes** are made of a mixture of carbon and copper that conducts high amperage through the armature windings.

When current is applied to the field coils and the armature windings, both produce magnetic flux lines (Figure 15–10). The current flowing though the conductor interacts with the north and south pole magnetic field coils and causes a rotational force of the armature in the direction of the weaker magnetic field. To keep the armature spinning, the split ring copper commutator attached to the end of the armature keeps the current flowing in one direction until the current is turned off. The commutator, in fact, is what converts the alternating current produced in the armature into a direct current.

When the battery supply voltage pushes current through the starter motor, electromagnetic induction in the starter windings produces a **counter electromotive force (CEMF).** This induced voltage acts against the supply battery voltage and opposes the flow of current through the starter motor. There is a relationship between

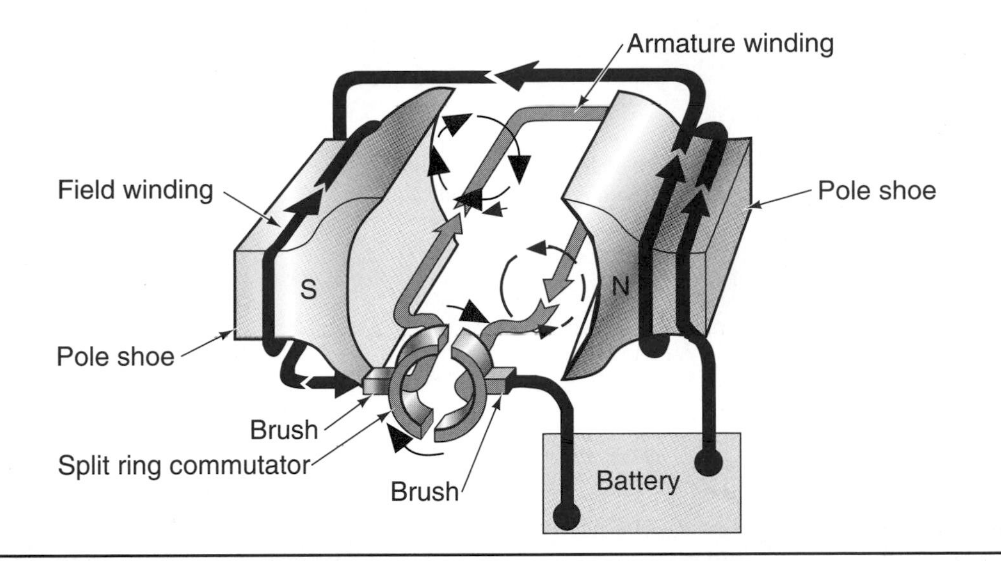

FIGURE 15–7 A simple electromagnetic starter motor.

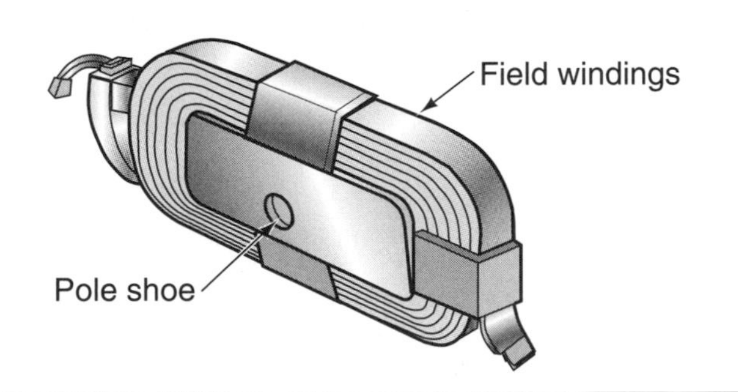

FIGURE 15–8 Heavy copper wire wrapped around a soft iron core called a pole shoe.

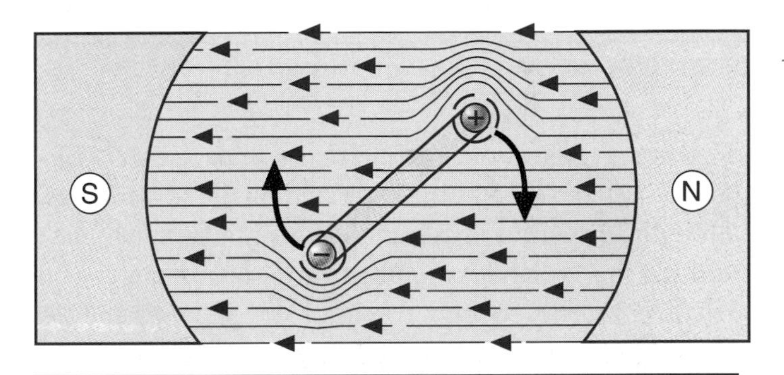

FIGURE 15–10 Rotation of the conductor is in the direction of the weaker magnetic field.

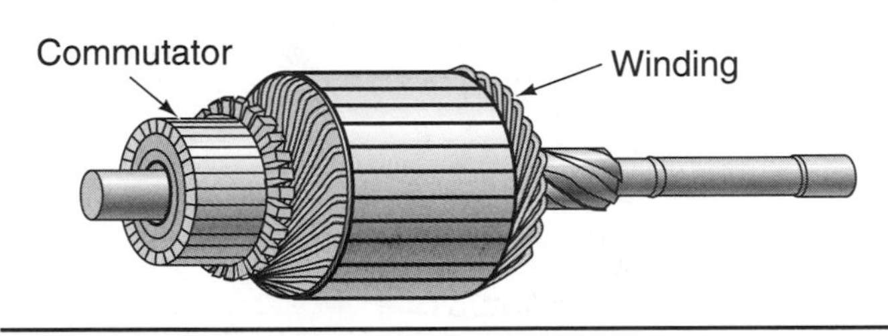

FIGURE 15–9 Starter armature assembly.

CEMF, motor speed, and current draw (Figure 15–11). As the starter speed and CEMF increases, the current or torque decreases.

The designed operating characteristics of the starter depend on how the field coils are wound inside the starter housing. There are four common basic types of starters, each using a different type of winding.

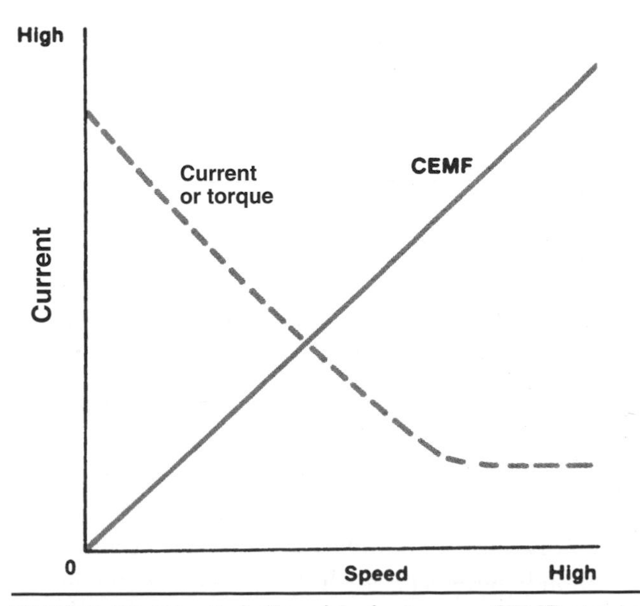

FIGURE 15–11 Relationship between CEMF, starter motor speed, and current draw.

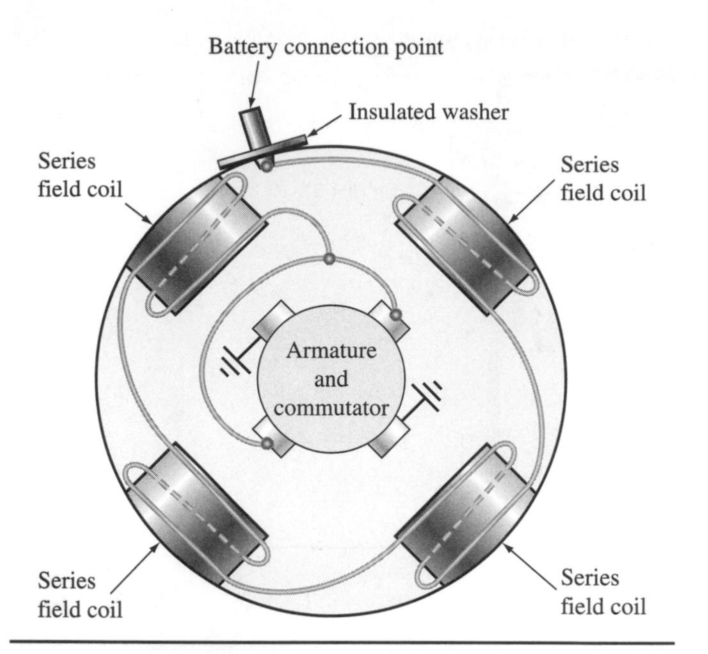

FIGURE 15–12 A series wound starter motor.

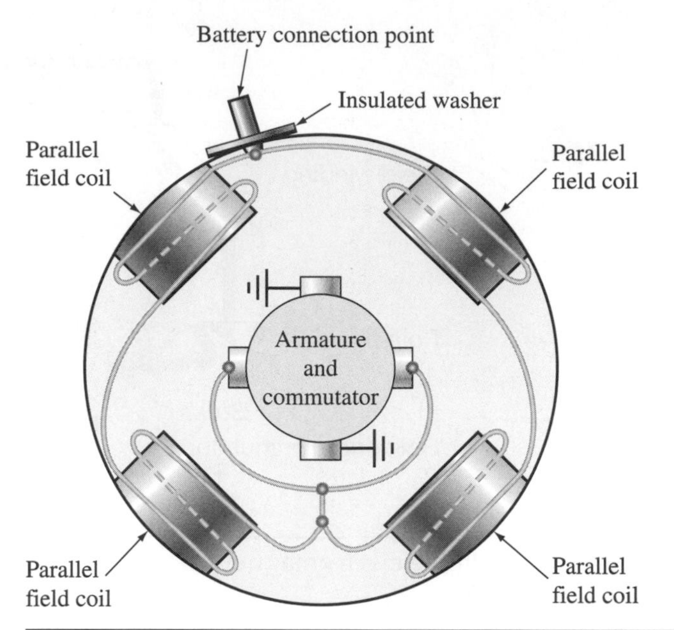

FIGURE 15–13 A parallel wound, or shunt wound, starter motor.

In the first type, a **series wound** starter (Figure 15–12), the current flows through the field windings, through the positive brushes to the commutator, and then through the armature windings to the commutator and negative brushes before reaching negative ground. A series wound motor develops its maximum torque output during initial starter cranking. As the speed increases, the torque output of the starter motor decreases because of the CEMF produced in the field windings.

The second type of starter is a **parallel wound** starter (Figure 15–13), also known as a **shunt wound** starter. The current flows in parallel through the field windings, through the positive brushes and commutator, to the armature windings, then to the commutator and negative brushes to negative ground. A parallel wound motor does not produce a high cranking torque and spins at a slower cranking speed. The CEMF produced in the field windings does not decrease the field coil strength, even as the starter motor speed increases.

In the third type of starter motor, a **compound wound,** or series/parallel wound starter (Figure 15–14), most of the field coils are connected in series to the armature, and a single field coil is connected to negative ground. This starter combines the best properties of a series wound starter and a parallel wound starter into a starter with good starting torque and a constant operating speed. The field coil that is connected to negative ground is called the shunt coil, and limits the speed of the starter motor.

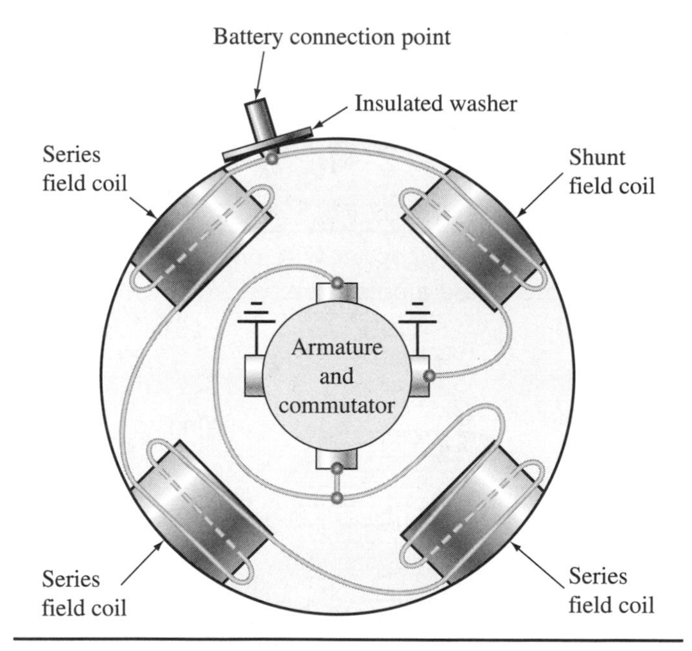

FIGURE 15–14 A compound wound starter motor.

The fourth and most popular type is a **permanent magnet starter** (Figure 15–15). It uses permanent magnets in place of the electromagnet field coils. The permanent magnets are extra high strength in comparison to normal permanent magnet motors for other applications.

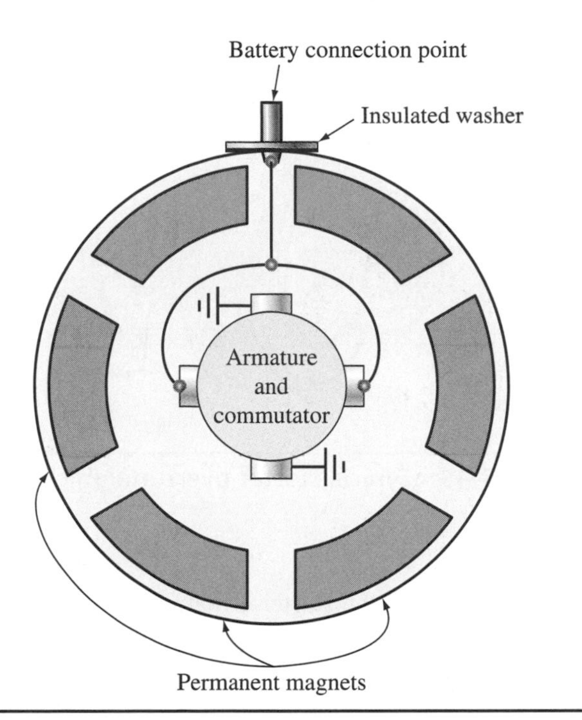

FIGURE 15–15 A permanent magnet starter motor.

Using permanent magnets instead of field coils reduces the weight of the starter and allows for a significantly smaller motor housing.

Starting Circuit Components

A vehicle starting system is a combination of electrical and mechanical parts that operate together to crank the engine fast enough for starting. The typical starting system consists of the following main components (Figure 15–16): battery, battery cables, starter motor, ignition switch, harness wiring, starter relay or solenoid, pinion drive, flywheel ring gear, and a neutral safety switch.

Starter Pinion Drives

Two types of starter pinion drives are designed to mesh with the flywheel ring gear (Figure 15–17): the Bendix inertia drive and the overrunning clutch drive. The engagement of the drive pinion gear into the flywheel ring gear is critically timed to prevent grinding of the teeth of one or both gears. The drive pinion must be engaged into the ring gear *before* the starter motor spins.

Bendix Inertia Drive The *least* common type of starter drive pinion is the **Bendix inertia drive** (Figure 15–18), which was common on vehicles from

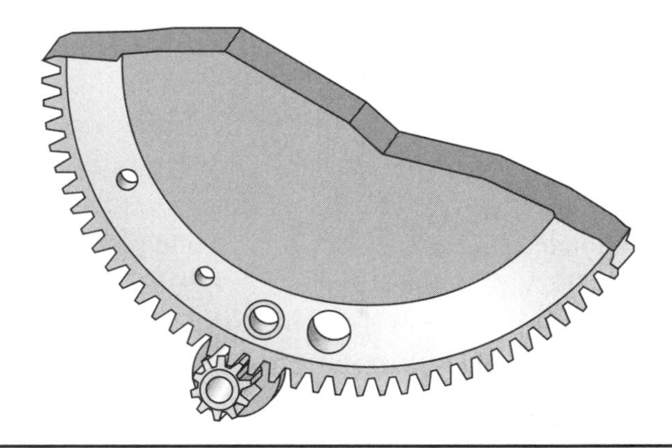

FIGURE 15–17 The starter drive pinion gear is engaged with the flywheel ring gear before the starter motor turns.

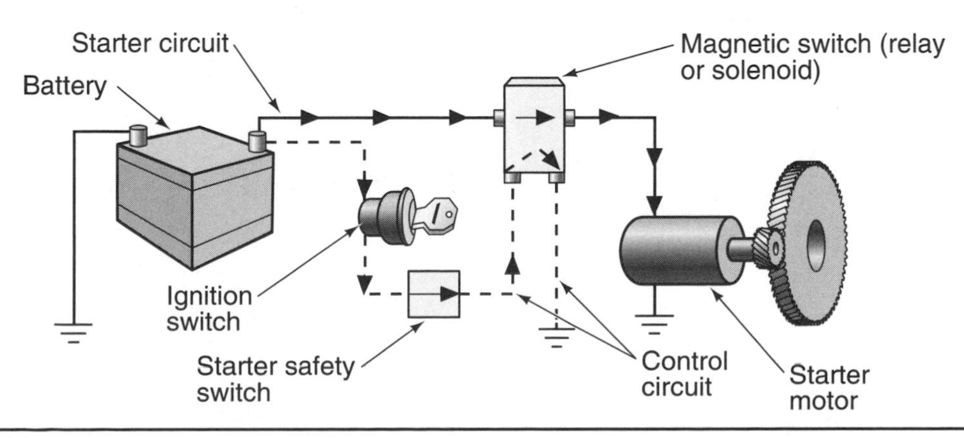

FIGURE 15–16 The major components of a starter motor system. The solid line indicates the high-amperage cranking circuit, and the dashed line indicates the starter control circuit.

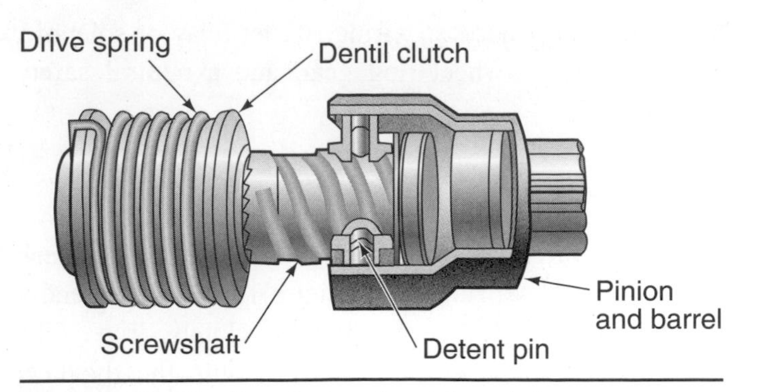

FIGURE 15–18 Bendix inertia drive pinion.

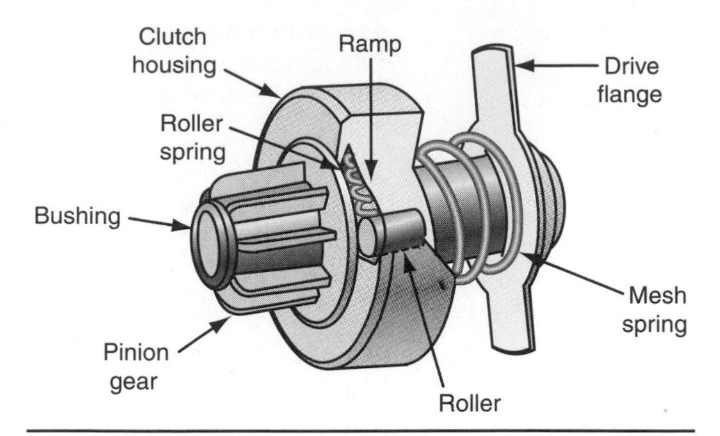

FIGURE 15–19 Typical starter overrunning clutch drive.

the 1950s and earlier but which is currently found only on farm or lawn equipment. This drive pinion depends on the inertia movement of the starter armature spinning to mesh the drive pinion forward into the ring gear teeth. When the engine starts, it is rotating faster than the starter, causing the pinion drive gear to back down its own shaft and disengage from the flywheel teeth.

Overrunning Clutch Drive The *most* common type of starter drive pinion is the **overrunning clutch drive** (Figure 15–19). This is a roller-type clutch wedges spring-loaded rollers between an inner and outer housing to engage the pinion teeth with the flywheel teeth. The inner clutch housing is splined to the starter armature shaft, and the pinion barrel contains the freely spinning pinion teeth. When the starter armature turns, the spring-loaded rollers are wedged tightly into the small ends of the tapered slots to transfer starter motor torque to the ring gear (Figure 15–20). When the engine starts and is rotating faster than the starter, the spring-loaded rollers unlock and allow the pinion gear to rotate freely around its shaft until the starter motor is deenergized.

Starter Relays and Solenoids

A starter motor can require a current draw of up to 250 amperes or more to crank the engine fast enough for starting. To control this high-amperage requirement, an

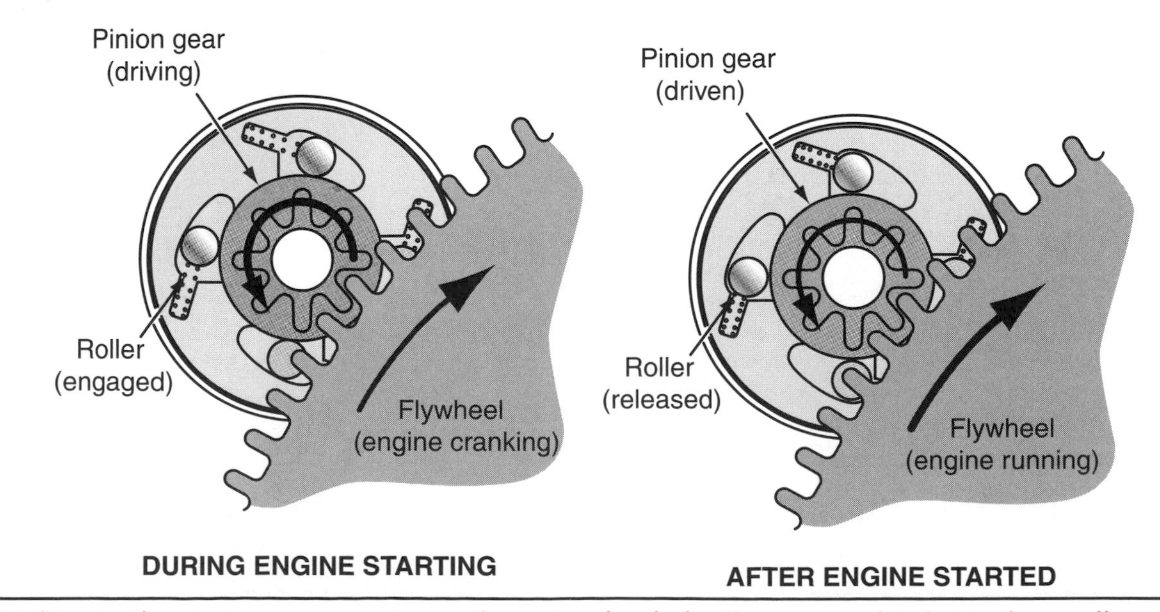

FIGURE 15–20 As the starter armature turns, the spring-loaded rollers are wedged into the smaller ends of the tempered slots.

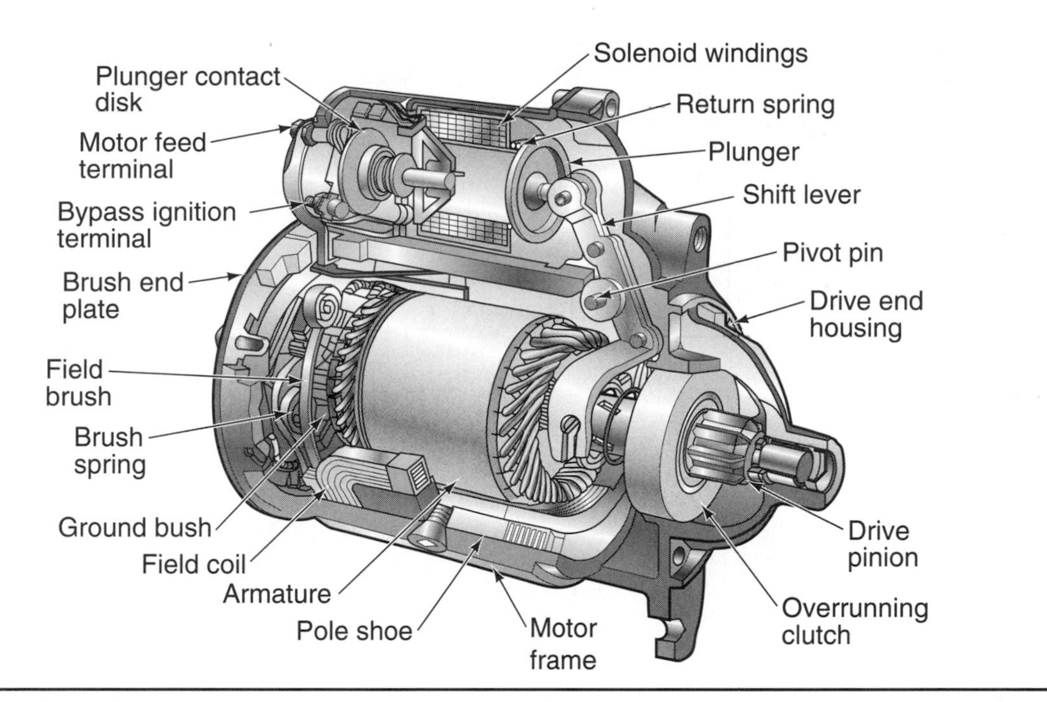

FIGURE 15–21 A solenoid-operated starter motor has the solenoid mounted directly to the motor.

electromagnetic switch is used. Two basic types of switches are used today to perform this function: the starter solenoid and the remote starter relay.

Starter Solenoid In a **starter solenoid**–actuated starter system, the solenoid is attached directly to the starter motor assembly (Figure 15–21) and controls two important functions sequenced in a specific order. The first function is to move the solenoid plunger, which is connected through a linkage to the pinion gear that meshes with the flywheel ring gear. The second function occurs after the teeth of the pinion are engaged into the flywheel; the solenoid plunger connects the battery B+ terminal to the motor M terminal, which cranks the starter motor (Figure 15–22). This sequence must take place to prevent the overrunning clutch from spinning before contacting the flywheel ring gear, thus damaging the teeth on one or both.

The schematic in Figure 15–22 consists of two separate solenoid windings: the **pull-in winding** and the **hold-in winding.** Both are energized when the battery voltage that is going through the ignition switch in *start* mode to the solenoid S terminal electromagnetically pulls in the solenoid plunger. When the copper disk at the end of the solenoid plunger contacts the B+ and the M terminals, the pull-in windings are de-energized because equal battery voltage is present on both sides of the pull-in windings. The hold-in winding continues to hold the copper disc that is connected between the B+ and the M terminal, allowing the starter motor to crank.

The R terminal is an **ignition bypass terminal** that supplies full battery voltage to the ignition coil during engine cranking conditions only. This bypass circuit normally applies to older vehicles that have point-style ignition systems, which use a ballast resistor or a resistance wire to control the ignition coil primary voltage to be less than full battery voltage. Modern electronic ignition systems normally do not have a bypass terminal because their ignition systems operate on full battery voltage.

Starter Relay In a **starter relay**–actuated starter system, the relay is normally mounted separately from the starter and controls the high current to the starter motor (Figure 15–23). This type of starter relay controls only the high current to the starter motor and does not control the pinion gear meshing into the flywheel teeth. When the ignition switch is in start mode, battery voltage is supplied to the S terminal (Figure 15–24) of the ignition switch, and current then flows through the relay coil. The relay coil pulls the plunger disk downward into

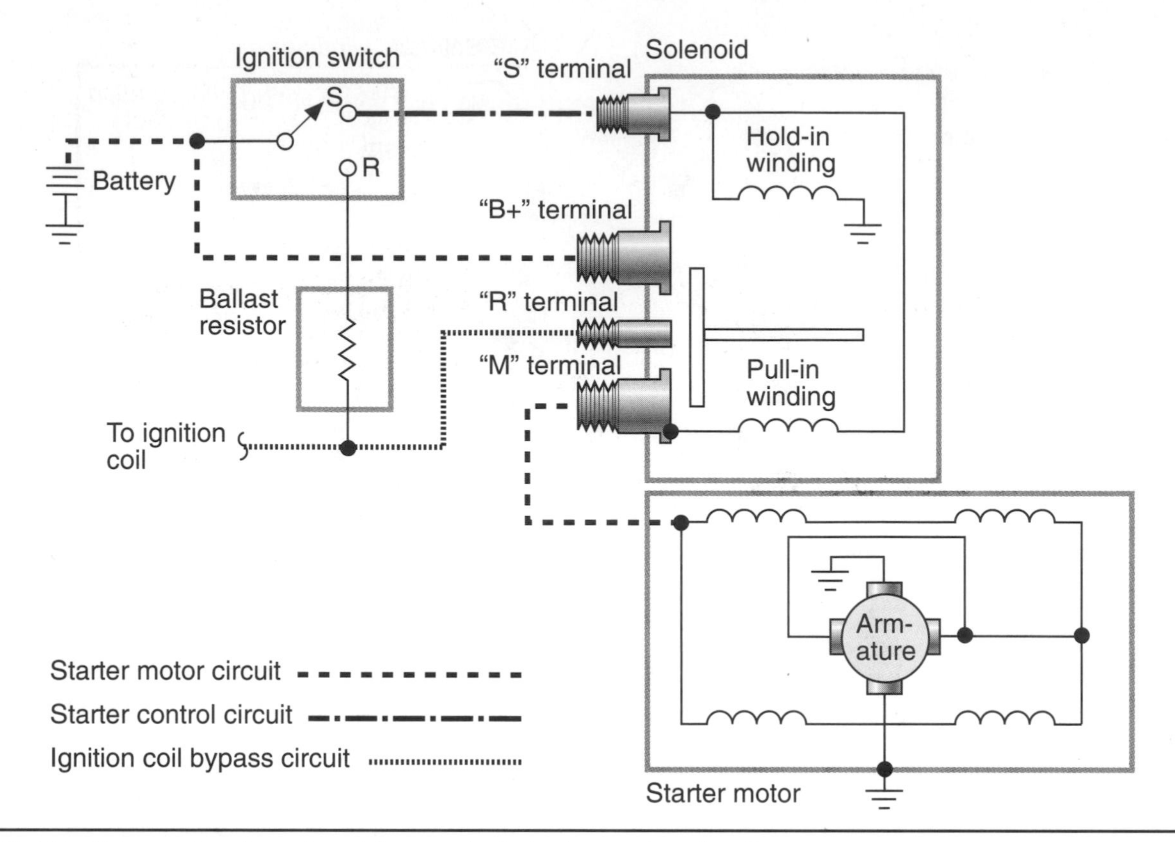

FIGURE 15–22 Schematic of a solenoid-operated starter motor circuit.

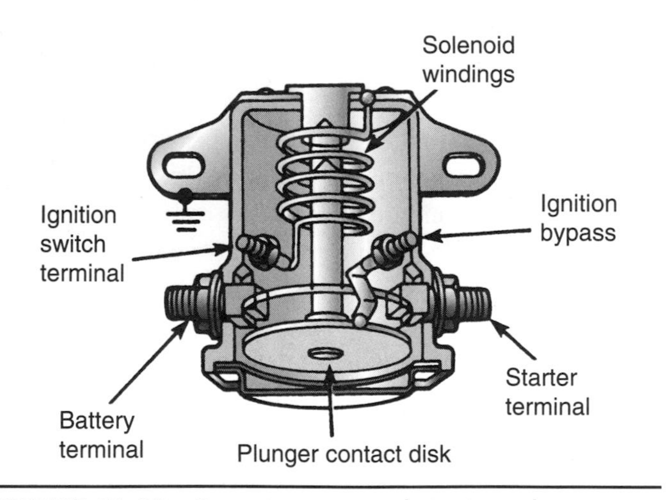

FIGURE 15–23 Remote mounted starter relay.

the battery and starter terminals (Figure 15–23) to supply battery current to the starter motor. The I terminal serves the same purpose as the ignition bypass R terminal in the starter solenoid.

Starter Motor Designs

Four basic starter motor designs that use the overrunning clutch pinion drive to crank the engine are used in today's vehicles, and some new technology starter motor designs are emerging in hybrid electric vehicle (HEV) applications. Depending on the engine application, some manufacturers use the same basic overrunning clutch starter, with only slight variations for several models of vehicles. Be sure the starter replacement is the specific one designed for the vehicle being serviced.

Direct Drive Starter The **direct drive starter** (Figure 15–25) design is probably the most commonly used in vehicles today. This starter design uses a solenoid directly attached to the starter motor to control the pinion gear movement and the high amperage required for the starter motor windings. When the solenoid is energized, it first engages the overrunning clutch pinion into the flywheel and then energizes the starter motor windings so that they will turn.

Positive Engagement Starter The **positive engagement starter** (Figure 15–26) was very popular on Ford vehicles in the past and for industrial engine applications. It uses an internal movable pole shoe that is connected through linkage to the pinion gear, which meshes

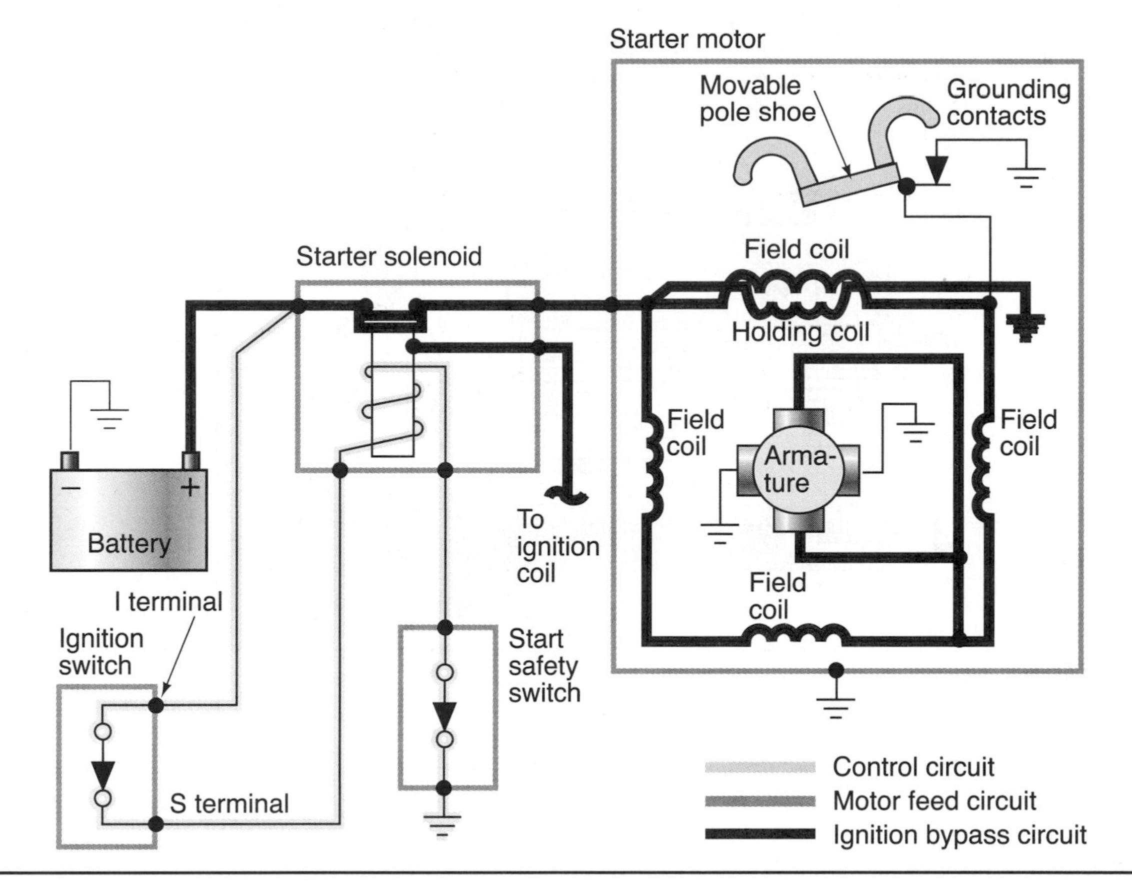

FIGURE 15–24 Current flow in a starter circuit with a remote mounted starter relay.

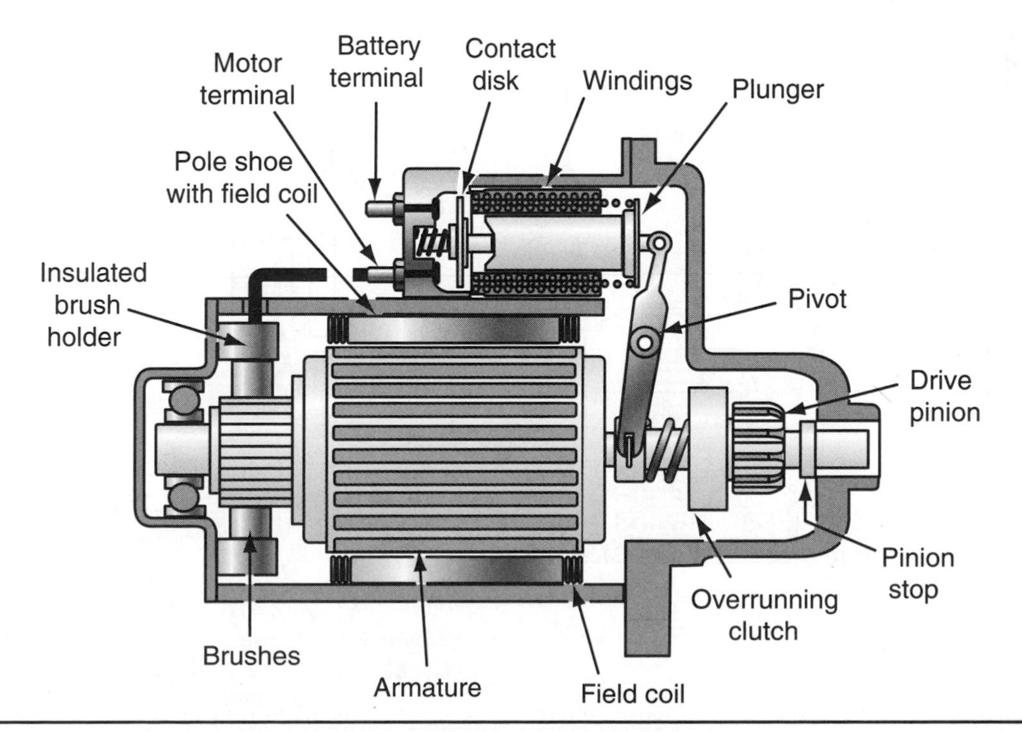

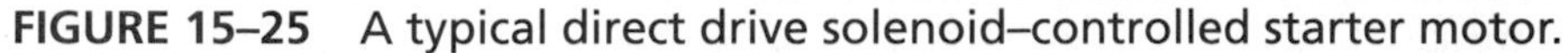

FIGURE 15–25 A typical direct drive solenoid–controlled starter motor.

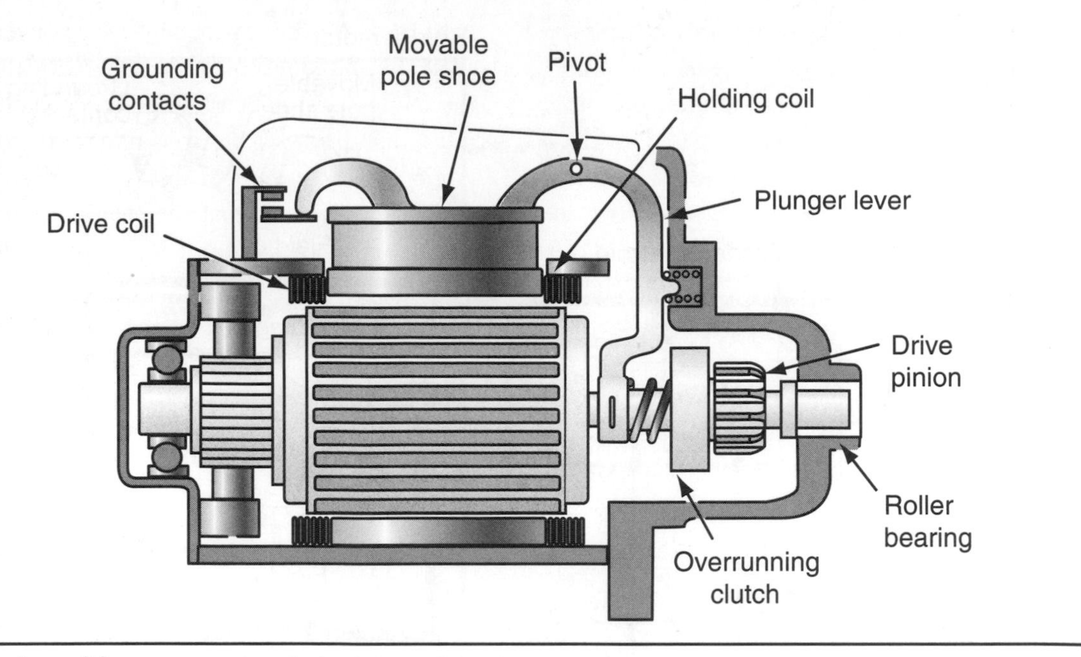

FIGURE 15–26 A positive engagement starter motor.

with the flywheel ring gear. This starter uses a separate starter relay (Figure 15–23) to control the high amperage from the battery to the drive coil. When current flows from the relay to the drive coil, electromagnetism pulls a movable pole shoe that is attached through a linkage to the overrunning clutch pinion. This sliding action engages the pinion gear into the flywheel teeth and then opens a set of normally closed grounding contacts, allowing current to flow to the starter motor windings.

Gear Reduction Starter The **gear reduction starter** (Figure 15–27) contains a small gear reduction

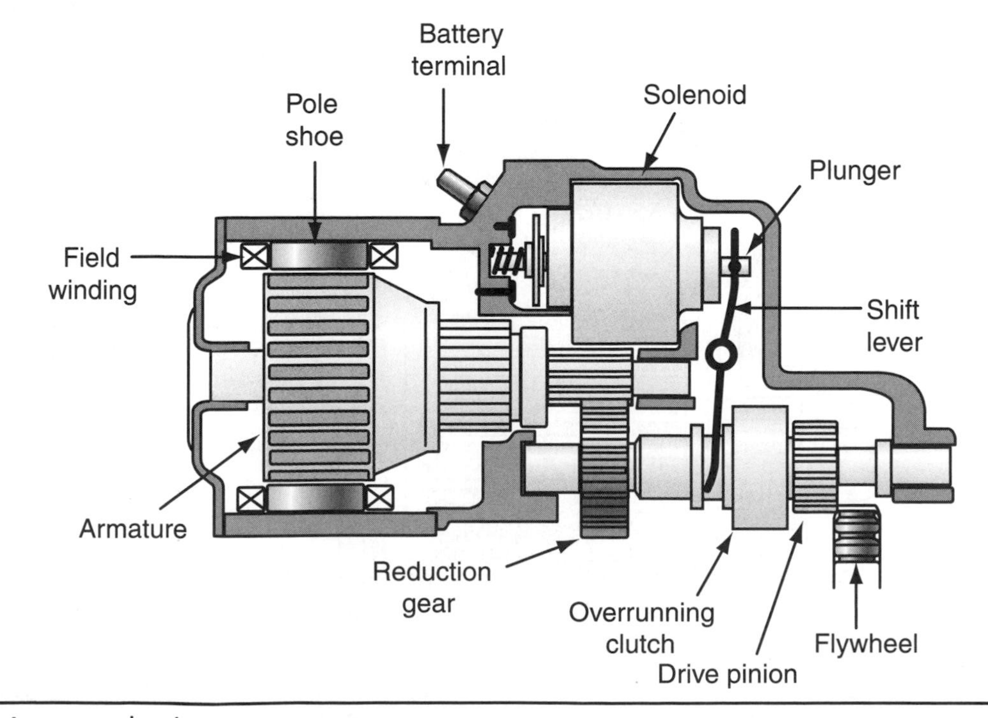

FIGURE 15–27 A gear reduction starter motor.

drive gear between the starter motor and the drive pinion. The design of the starter solenoid, armature assembly, and overrunning clutch is similar to those on the direct drive starter, with the exception of the starter's physical size and a gear reduction assembly driving the overrunning clutch pinion. The reduction ratio is normally between 2:1 and 3.5:1 depending on the application. The higher the gear ratio is, the greater the available starting torque is for the same-size starter housing. Many smaller vehicles use a smaller motor running at a higher operating speed to produce the necessary gear reduction torque.

Permanent Magnet Starters The **permanent magnet starter** (Figure 15–28), the newest design of starter, creates its magnetic field from four to six strong permanent magnets instead of from electromagnetic field coils. The starter amperage, previously supplied to the field coils and armature assembly, now supplies only the armature assembly through the brushes and commutator. Combined with a planetary gear reduction set (Figure 15–29) that produces a 4.5:1 reduction, the starter is physically smaller, overall current draw on the battery is reduced, and starter torque output is increased.

New Technology Starter Applications Over the past several years, hybrid electric vehicle technology has produced several new starter design applications. Because the major operating principle of an HEV is to increase fuel efficiency over conventional vehicles, the HEV has the ability to automatically start and stop the engine unit depending on the vehicle load and operating condition. The constant start/stop of the HEV engine would wear out a standard design starter, which spins the engine crankshaft at only around 250–400 rpm, compared to about 600–1,000 rpm needed for the seamless operation of an HEV.

Depending on the HEV engine/motor combination, one of the most recent starter designs to emerge is the **integrated starter/generator (ISG).** This starter is integrated with the generator to form a combination unit (Figure 15–30) that serves as a starter, a generator, and a power booster at certain engine load speeds. The ISG consists of a three-phase motor unit with a stator assembly attached to the inside the transmission/engine bell housing, along with the rotor attached to the engine crankshaft (Figure 15–31). When current flows through the stator assembly, it creates a magnetic field in the rotor. This field causes the rotor to spin at 600–1,000 rpm, turning the engine crankshaft sufficiently for a seamless start of the engine or to propel the vehicle forward at slow speeds.

Another application of an ISG is a belt-driven alternator/starter (Figure 15–32) that mounts in the same location as a typical engine-mounted alternator. When the engine shuts off at idle to conserve fuel, this system is designed only to restart the engine, not to propel the HEV down the road.

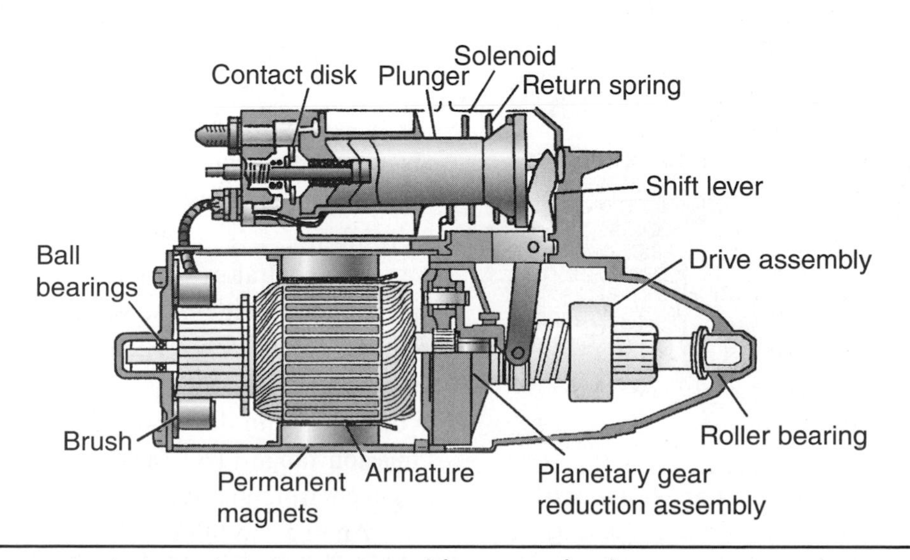

FIGURE 15–28 A permanent magnet starter motor with gear reduction.

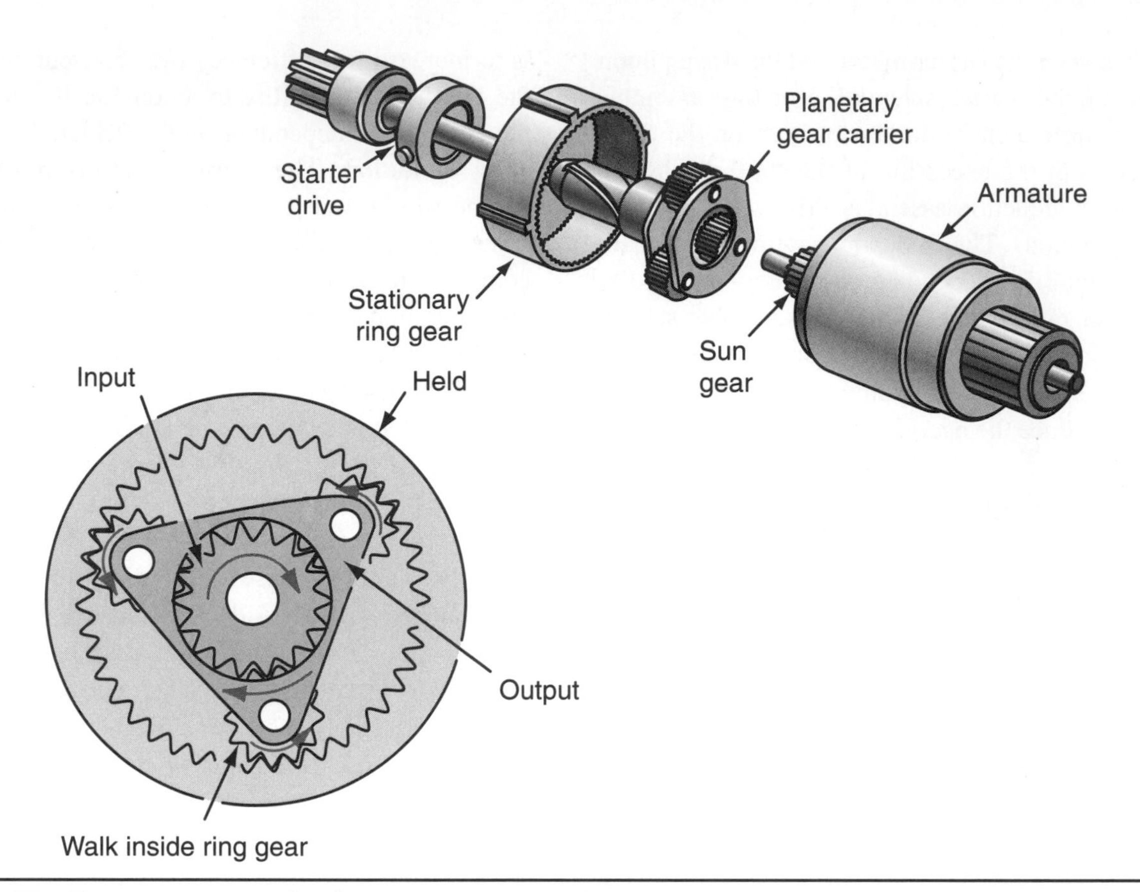

FIGURE 15–29 Planetary gear reduction components on a starter motor.

FIGURE 15–30 A cutaway view of an HEV electric-assist motor/generator used for starting the engine.

(Courtesy of American Honda Motor Co., Inc.)

Starting Safety Switch

The **starting safety switch,** also referred to as a **neutral safety switch (NSS),** prevents the starter from cranking in an automatic transmission vehicle unless the gear selector is in park or neutral or in a manual transmission vehicle only when the clutch is disengaged. Figure 15–33 is a starter schematic in an automatic transmission vehicle that has a safety switch attached to the gear selector assembly.

The neutral safety switch may be located in several places depending on the vehicle and manufacturer's preferences. The safety switch may be located on the side of the transmission housing (Figure 15–34). It may also be combined with the backup light switch and/or the transmission range (TR) sensor that tells the PCM (powertrain control module) the position of the gear shift lever.

Another location of the neutral safety switch on older automatic transmission vehicles is through a

FIGURE 15–31 Components of an integrated starter/generator assembly (ISG).

(Courtesy of BMW of North America, Incorporated)

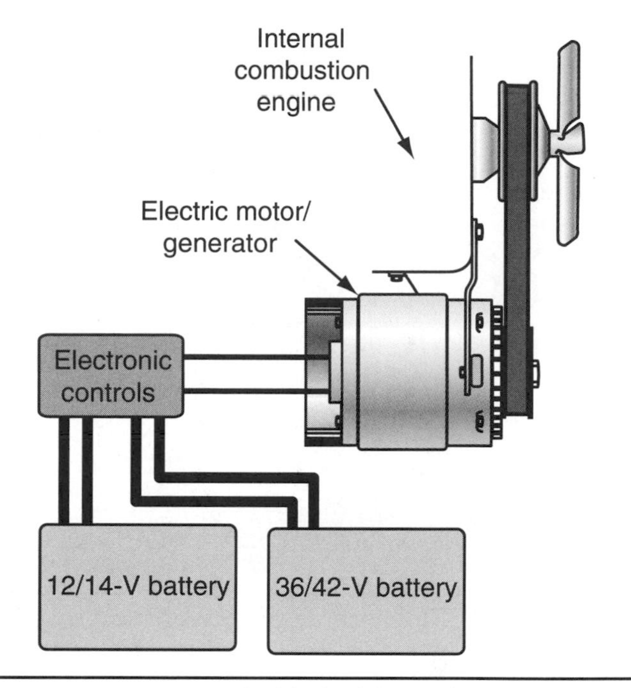

FIGURE 15–32 A typical belt-driven alternator/starter ISG system.

mechanical linkage that blocks the movement of the ignition switch lock cylinder unless the transmission is in park or neutral (Figure 15–35).

In a manual transmission vehicle, a popular location for the neutral safety switch is attached to the gear selector or to the clutch pedal assembly (Figure 15–36). The safety switch contact points are closed only when the clutch pedal is fully depressed.

Several modern vehicles use the powertrain control module (PCM) to control the starter system circuit

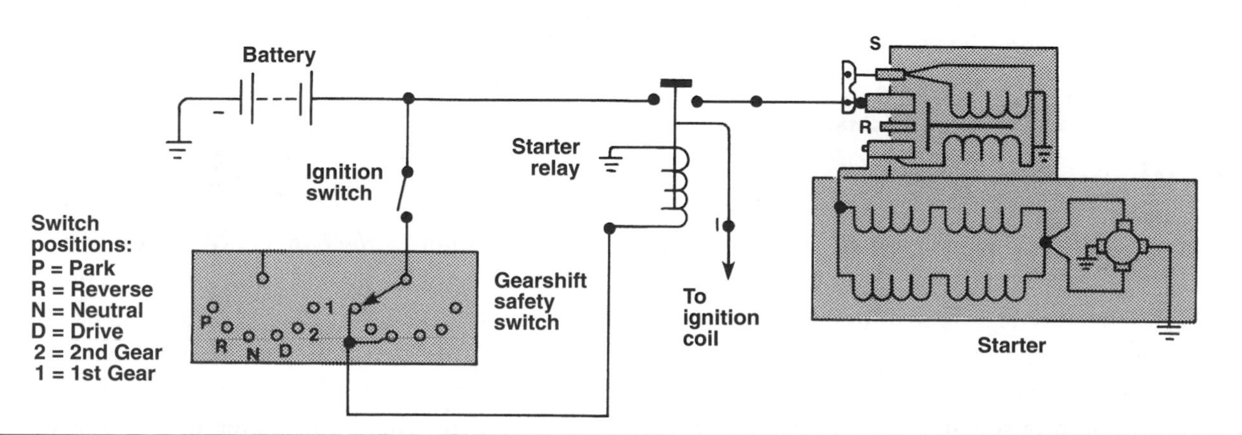

FIGURE 15–33 The neutral safety switch (NSS) must be closed to complete the starter control circuit.

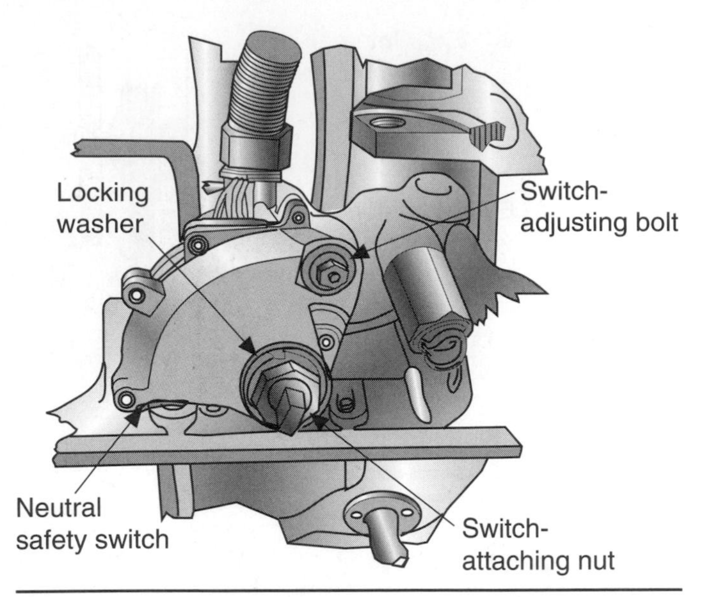

FIGURE 15–34 The neutral safety switch attached to a transmission.

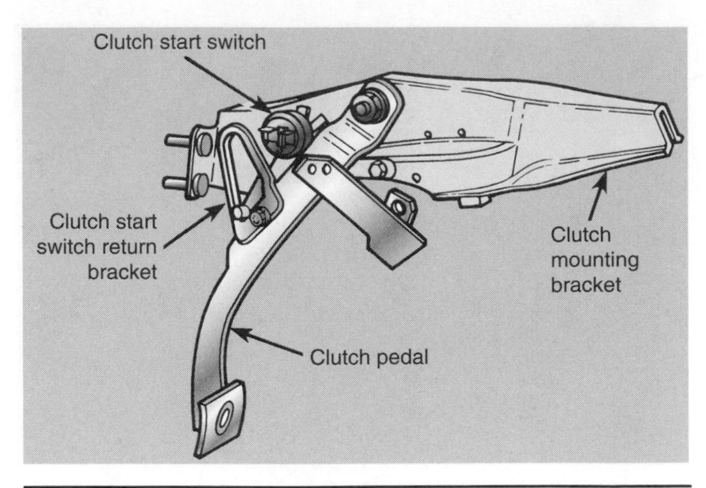

FIGURE 15–36 A starter safety switch located on the clutch pedal linkage for a manual transmission vehicle.

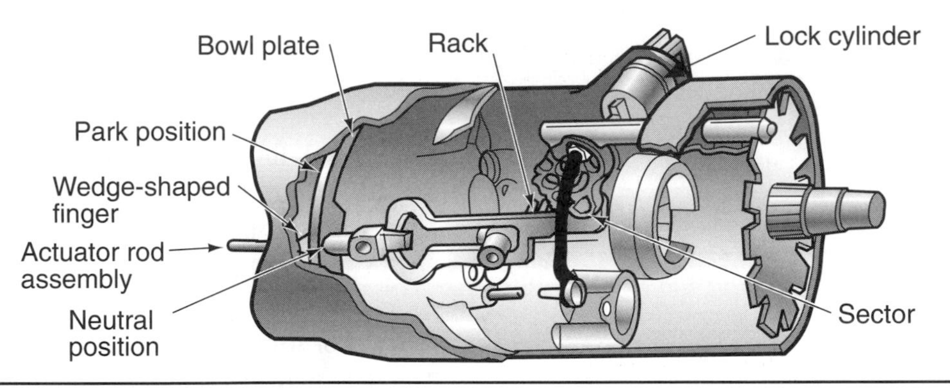

FIGURE 15–35 Mechanical linkage in the steering column prevents the ignition lock cylinder from turning when the transmission is in gear.

(Figure 15–37). Specifically, the PCM controls the ground circuit of the starter relay coil.

Basic Starter System Testing

The starting system in a vehicle normally provides years of trouble-free service, but it can produce aggravating experiences to the driver when it fails to operate as designed. To properly test and diagnose a starting system circuit, systematic troubleshooting procedures should be followed to prevent a misdiagnosis. This is important if you consider that over three-quarters of the starters returned for warranty replacement test satisfactorily at the rebuilding facility and that they were probably returned because of a bad battery; a bad, loose, or corroded cable or connection; a bad switch or relay; or a circuit wiring fault.

Before testing a starting system, make three important preliminary vehicle checks:

1. *Condition of the battery* A three-quarter or greater charged battery *must* be used to accurately test the starting system. When a partially charged battery is used, the starter may crank the engine more slowly than normal or fail to engage completely into the flywheel ring gear. Refer to the previous chapter about batteries and testing them to perform any battery tests.

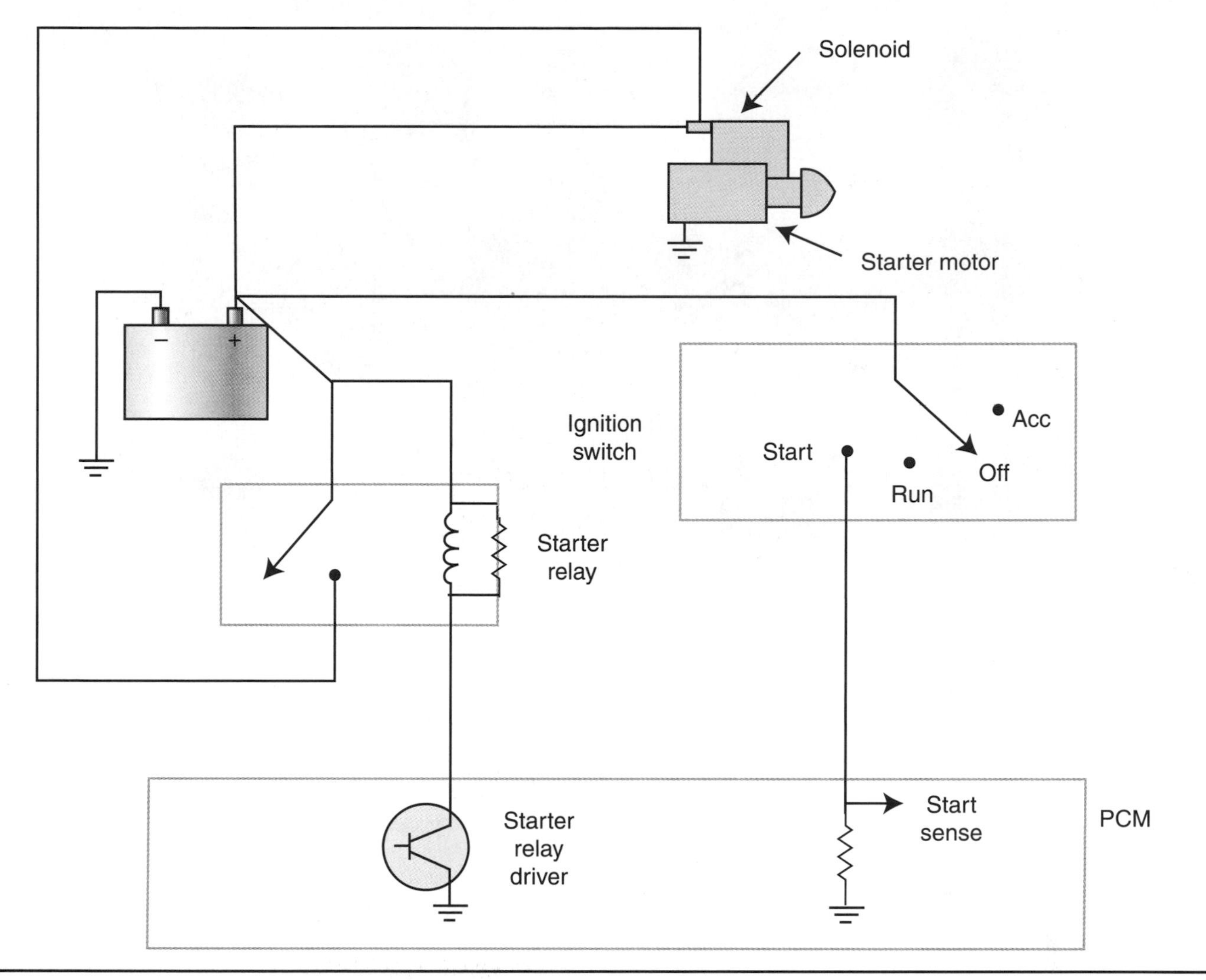

FIGURE 15–37 Typical starter circuit controlled by the PCM.

2. *Engine mechanical condition* Satisfactory condition of the engine internally should allow the starter motor to crank the engine at 200–400 rpm, depending on the vehicle and the engine. The following mechanical problems could effect cranking conditions:

   - Engine crankshaft bearing failed or is too tight, causing the crankshaft to seize.
   - Broken internal engine components, such as connecting rods and loose bearing caps, are wedged against the crankshaft and prevent rotation.
   - Hydrostatic lock in one or more cylinders. Liquid antifreeze cannot be compressed, and if allowed to fill the engine cylinder(s) through, say, a bad head gasket, it can prevent the piston from rising in the cylinder, creating a hydrostatic lock.

3. *Other conditions* The engine should be filled with a proper weight oil, as recommended by the vehicle manufacturer for specific weather conditions. If a heavier weight oil is used in combination with cold weather conditions, the cranking rpm required to start the engine could be below the designed acceptable minimum (Figure 15–38).

Engine mechanical conditions that create a no-start condition can be verified by rotating the engine using the crankshaft pulley nut. The engine should be rotated, without excessive resistance, for at least two rotations in a clockwise direction only, which prevents unthreading the crankshaft pulley nut.

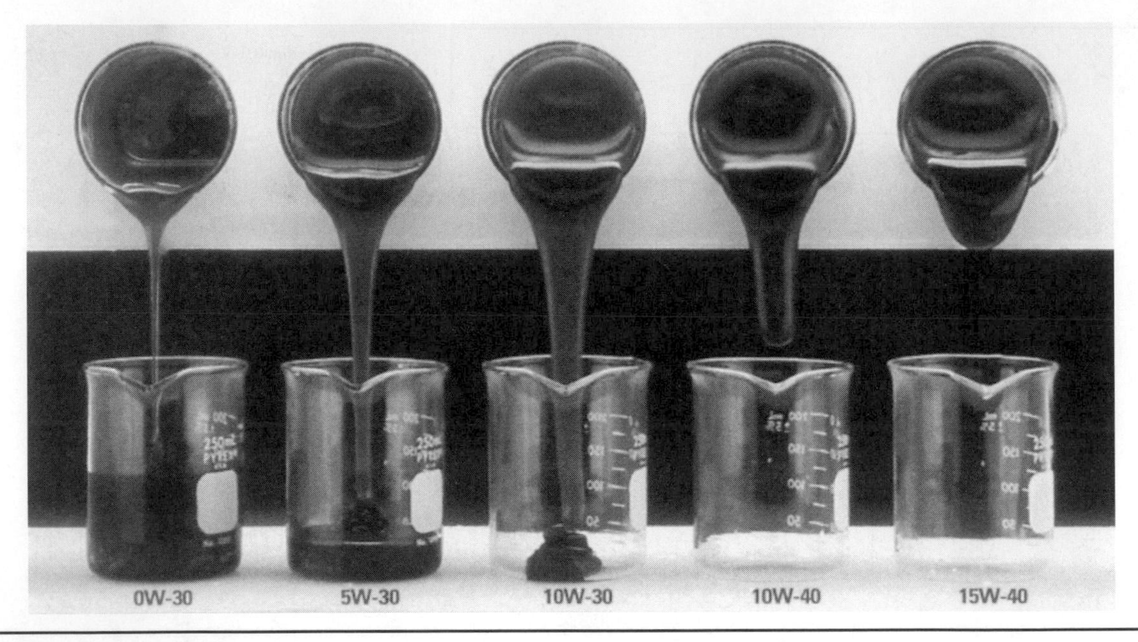

FIGURE 15–38 Engine oil viscosity comparison during cold weather conditions.
(Courtesy of Esso/Imperial Oil)

If these preliminary conditions check out satisfactorily, then the following six starter system tests should be performed:

1. Starter circuit isolation test
2. Starter current draw test
3. Starter circuit voltage drop test
4. Starter open circuit test
5. Starter bench testing, if necessary
6. Starter pinion gear clearance

1. Starter Circuit Isolation Test

If the starter does *not* crank the engine and the preliminary engine mechanical condition checks out satisfactory, then a quick isolation test can be performed on the starter system to determine whether the starter motor, solenoid, relay, or control switch circuit is at fault (Figure 15–39).

To perform a quick isolation test on a starting system, make sure the parking brake is set and the transmission is in neutral or park. Turn on the headlights, and turn the ignition switch to the start position while watching the headlight intensity. Depending on the results of this headlight intensity test, three conditions can be determined.

1. *The headlights go out.* This result is most likely caused by a poor battery terminal connection due to corrosion or looseness, which should be corrected before proceeding with any other tests.

2. *The headlights dim.* This can be caused by a discharged battery, engine mechanical condition, or internal starter motor damage. If the battery and engine mechanical condition check out satisfactorily during the preliminary tests, the problem is most likely internal to the starter: armature shorts, worn bearings, shorted field windings, or loose or dragging conditions. A starter in this condition draws excessive amperage from the battery during starter cranking.

3. *The headlight intensity remains the same.* This can be caused by an open in the circuit if the starter makes *no* sound. If, however, there is a heavy clicking noise, either the solenoid or the control circuit can be at fault. To verify this, connect a remote starter switch across the starter solenoid (Figure 15–40) to bypass the ignition control and safety circuits. If the starter motor cranks while using the remote starter switch, the control circuit is at fault. If the starter does not rotate and the lights do not go dim, then the solenoid or relay is at fault.

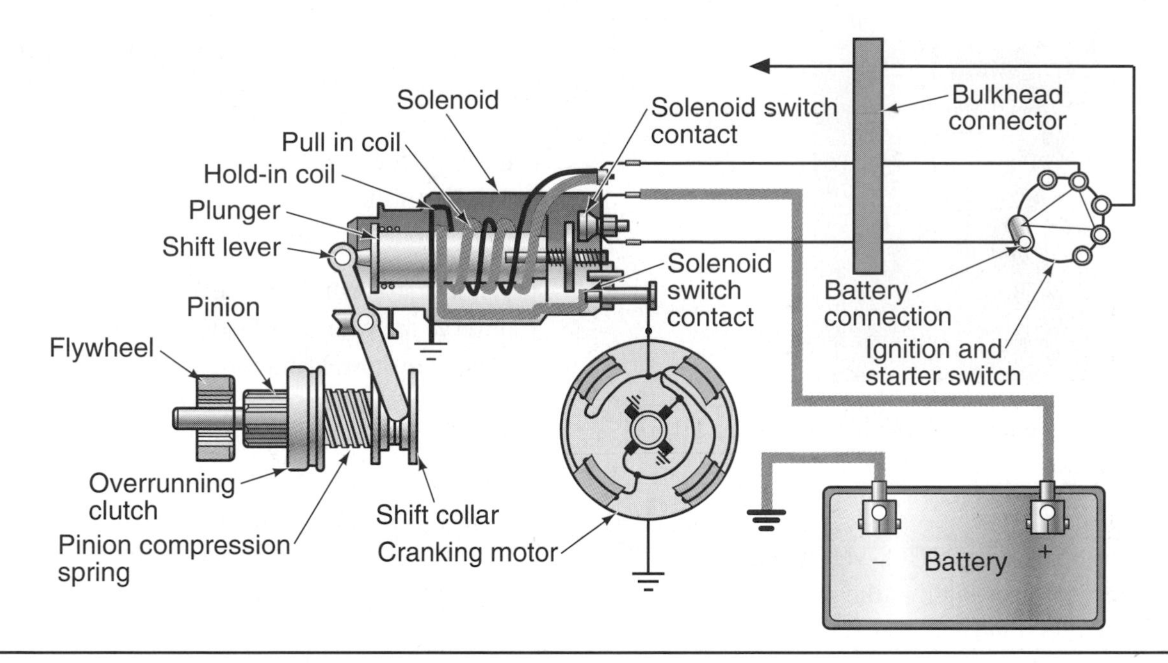

FIGURE 15–39 Potential connections and components that could cause a no-crank condition.

2. Starter Current Draw Test

If the starter *does* at least crank the engine, a current draw test should be performed with a battery/starter/charging system tester (Figure 15–41) that is capable of recording several hundred amperes.

The following steps are used with a VAT40 tester to perform a current draw test on a starter. The procedures for most other starter current draw testers are similar to these:

- Connect the large red lead to the battery positive post and the black tester lead to the battery negative

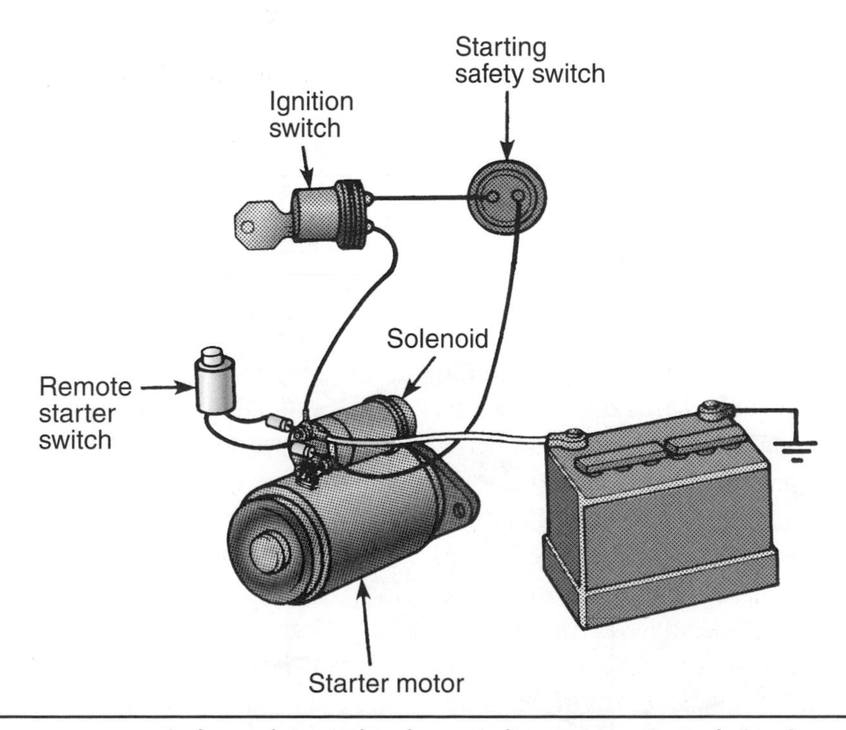

FIGURE 15–40 A remote starter switch can be used to bypass the starter control circuit.

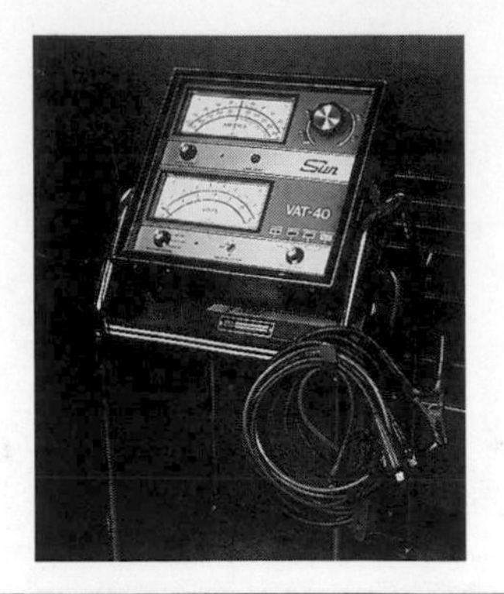

FIGURE 15–41 A typical carbon-pile battery/starter/current draw tester.

battery post. Select the INT 18 Volt meter scale. Place the test selector knob in the starting position (0–600 amperes); then zero the ammeter needle. Connect the green inductive clamp around the battery ground cable (Figure 15–42).

- Turn off all vehicle accessories so that the battery can provide its full amperage to the starter.
- Disable the engine from starting by referring to the preferred method recommended by the specific vehicle manufacturer. In some vehicles, especially in older carbureted engines and early electronic ignitions, the ignition system was disabled by grounding the coil secondary wire (Figure 15–43). On newer fuel injected electronic ignition vehicles, the preferred method of disabling the engine is to remove the PCM fuse, ECM fuse, or the ignition fuse (Figure 15–44), depending on the specific vehicle (some vehicles do not crank with the PCM or ECM fuse removed). Removing the fuse disconnects power to the electronic ignition module and stops the pulsing of the injectors, which can load up the engine and exhaust system with raw fuel.
- While cranking the engine for 10 to 15 seconds, observe the voltmeter reading and the starter cranking amperage draw. Minimum starter cranking voltage should remain at 10.5 volts or higher. Minimum starter cranking revolutions per minute should be at 200–400, depending on the engine. Typical starter amperage draw is dependent on the size of the engine, the number of cylinders, special modifications to the engine, and severe weather conditions. If manufacturer's specifications are not available, refer to the amperage chart in Table 15–1 as a guide to determine typical starter amperage draw.

TABLE 15–1 Typical Starter Current Draw Test Results

Engine Size	Amperage Draw Range (A)	Normal (A)
4 cylinder	75–150	125
6 cylinder	100–175	150
Small V-8	125–200	150
Large V-8	150–300	200

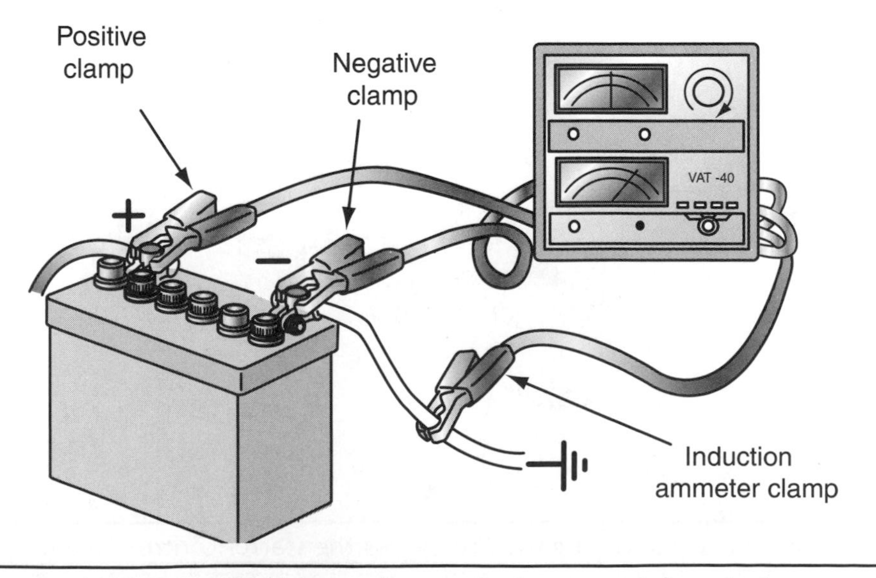

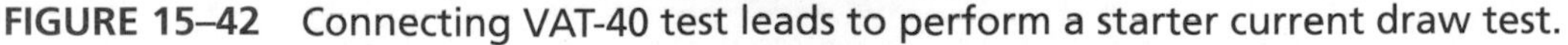

FIGURE 15–42 Connecting VAT-40 test leads to perform a starter current draw test.

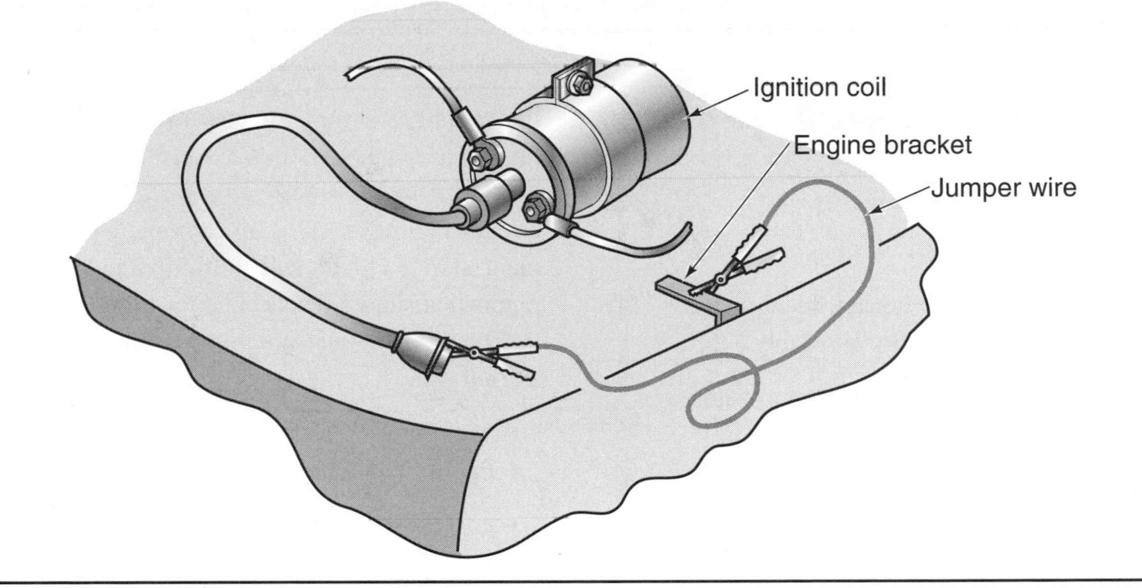

FIGURE 15–43 Disabling the ignition system on an older vehicle by grounding the secondary coil wire.

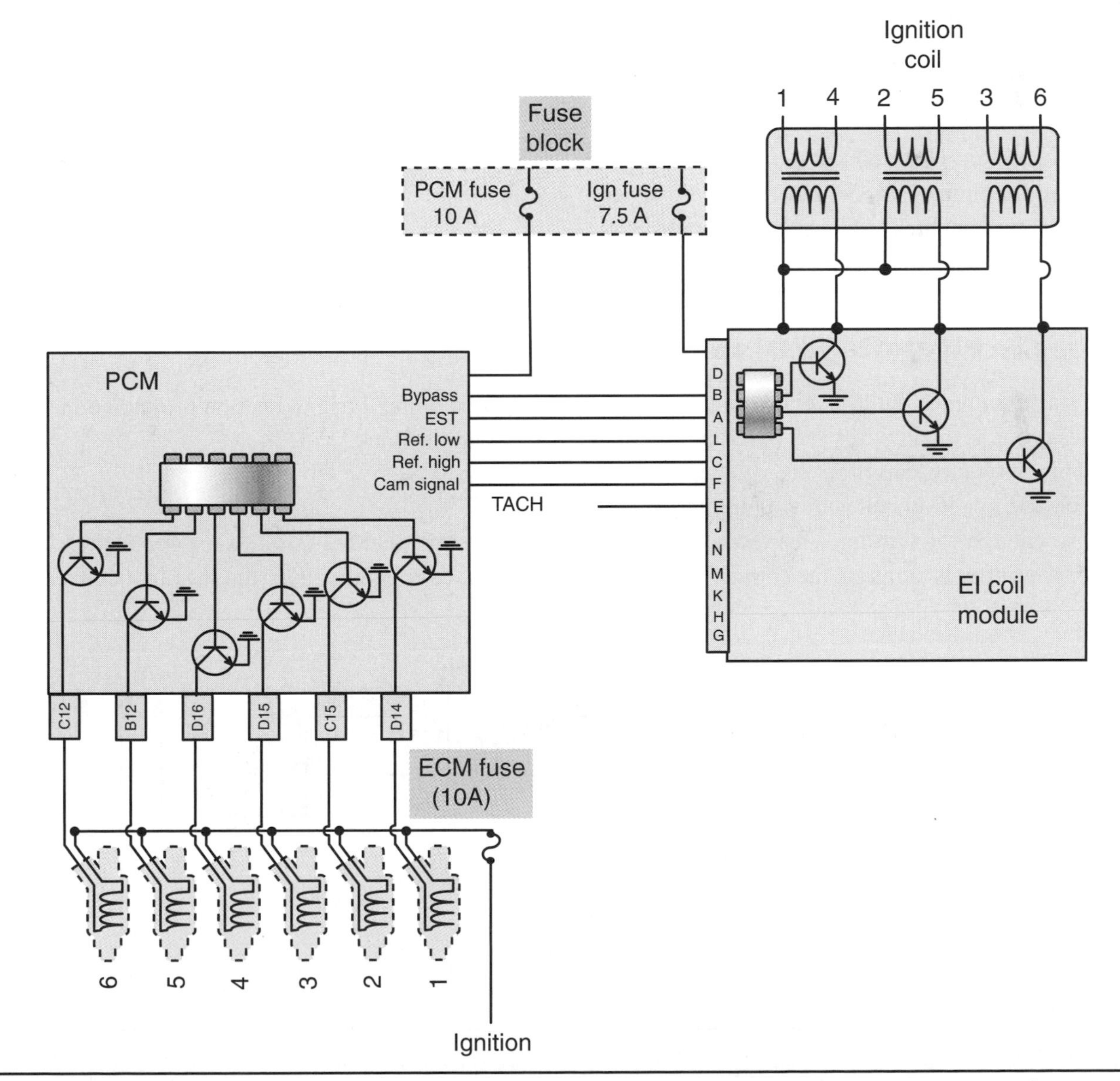

FIGURE 15–44 Disabling the ignition system and fuel injection system by removing the control fuses.

Results of Starter Current Draw Test If the starter's current draw exceeds the standard range for the engine size, this can be due to:

- A short in the starter motor windings.
- Spinning resistance in the starter due to worn bushing or dragging armature assembly.
- Broken, binding, or misaligned starter motor components.
- Engine mechanical problems.
- Excessively advanced ignition timing beyond specifications.

If the starter's current draw is below the standard range for the engine size, this can be due to:

- Excessively worn engine components, such as from a high-mileage engine.
- Voltage loss in the starter circuit due to bad cables, connections, or control solenoid or relay. (A voltage drop test on the insulated side and the ground side of the starter circuit should be performed.)

3. Starter Circuit Voltage Drop Test

In a normal starter motor circuit, several connections, terminals, cables, switches, and wires can develop unwanted resistance. This resistance can drop the voltage available to crank the engine to a level below the minimum required to crank it fast enough for starting. This test must be performed while the starter is cranking the engine. Remember that voltage is the pressure that pushes the amperage though a completed circuit. A voltage drop indicates that unwanted resistance is reducing the current flow in the circuit. Figure 15–45 is an example of a positive engagement starter circuit and the individual voltage drop test points that should be checked for unwanted resistance.

Each of the test points in Figure 15–45 is described as follows:

- Test points A to B to test the connection between the battery positive post and terminal conductor.
- Test points B to C to test the battery positive cable 1.
- Test points C to D to test the connection from the battery positive cable to the solenoid.
- Test points D to E to test the contact points inside the solenoid.
- Test points E to F to test the connection from the solenoid to the battery positive cable 2.
- Test points F to G to test the battery positive cable 2.
- Tests points G to H to test the connection from the battery positive cable 2 to the starter motor.
- Tests points H to I to test the actual voltage drop across the starter motor.
- Test points I to J to test the ground connection at the engine block.
- Test points J to K to test the battery negative cable.
- Test points K to L to test the connection between the battery negative post and terminal connector.

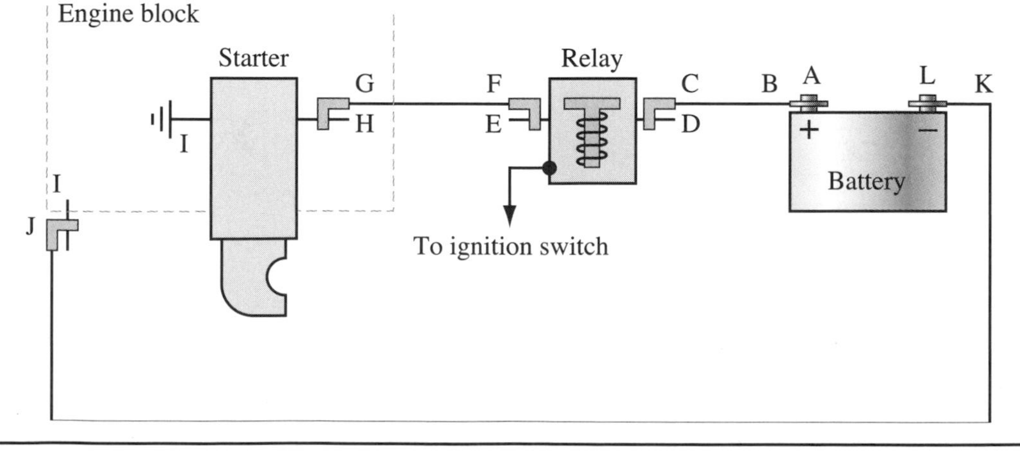

FIGURE 15–45 Starter circuit voltage drop test points.

If the amounts for allowed voltage drops are not available from the vehicle manufacturer, the following general specifications can be used as a general rule for circuit voltage drops.

Connection Point	Allowed Voltage Drop (V)
Ground connections	0.1
Switch contacts	0.2
System wiring	0.2
Terminal connections	0.0
Computer sensor connections	0.0 to 0.05

The starter circuit is designed to operate on the entire voltage supplied by the battery. A voltage drop in any one of these connections that exceeds the amount allowed can affect the proper operation of the starter motor.

4. Starter Open Circuit Test

When a starter system has a no-crank problem, a voltage drop test cannot be performed. The most common no-crank problem is caused by an open circuit in the starter or the control circuit. The simplest way to diagnose a no-crank problem is with a high-wattage test light. A custom-made high-wattage test light can be made from a #1156 bulb, a socket, and connection leads. Standard commercial test lights are insufficient for this test because they are 3 watts or less and need only 0.25 amp or less to glow brightly. To perform the test on a starter system with the solenoid mounted on the starter motor (Figure 15–46), connect the test light positive lead to the M terminal and the test light negative lead to a good engine ground. The test light should go on brightly when the ignition switch is turned to the start position *if* the control circuit is operating properly. If the test light does not come on, or if it goes on very dimly, there is an open or high resistance in the circuit somewhere between the M terminal and the battery. To find the open circuit, use the test light to backtrack though the system from the starter to the battery.

To test the ground side of the starter motor circuit, connect the test light positive lead to the starter M terminal and the test light negative lead to the starter housing. The test light should come on brightly when the ignition switch is turned to the start position, indicating a good ground. If the light comes on brightly, then suspect a bad starter motor.

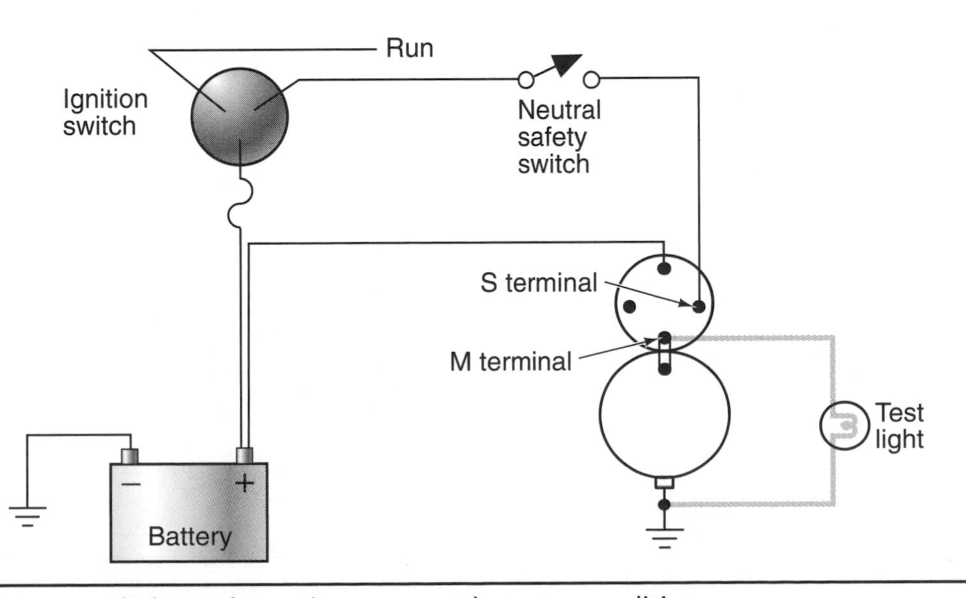

FIGURE 15–46 Using a test light to determine a no-crank starter condition.

5. Starter Bench Testing

If a starter has failed the previous tests, or if a new starter is going to be installed, a free running (no-load) bench test should be performed to determine:

- Starter free running revolutions per minute.
- Starter current draw.
- Operation of the overrunning clutch pinion drive.
- Operation of the solenoid or relay, if attached to the starter.

To perform a starter free running, no-load bench test, the starter should be fastened in a vice snugly, but not so tightly as to crack the frame or pole shoes (especially on a permanent magnet starter). Connect a starter load tester (VAT 40, for example, Figure 15–47) to load the battery until a voltage reading of 10 volts is obtained, thus preventing the starter from overspeeding the armature and causing damage to the armature windings. If available, a manual hand tachometer can be used to determine the starter motor revolutions per minute. Energize the starter solenoid with a remote starter switch, and observe the current draw. Typical results should be:

- Starter free running revolutions per minute equals 6,000–12,000.
- Starter current draw equals 60–100 amperes.

If the starter fails the free running bench test, there are four criteria to help determine the cause.

1. If the revolutions per minute are slower than specified and the current draw is excessive, then the cause can be worn starter bushings, a shorted or grounded armature, shorted field windings, a bent armature, or a combination of these.
2. If the revolutions per minute are lower than specified and the current draw is lower than normal, then

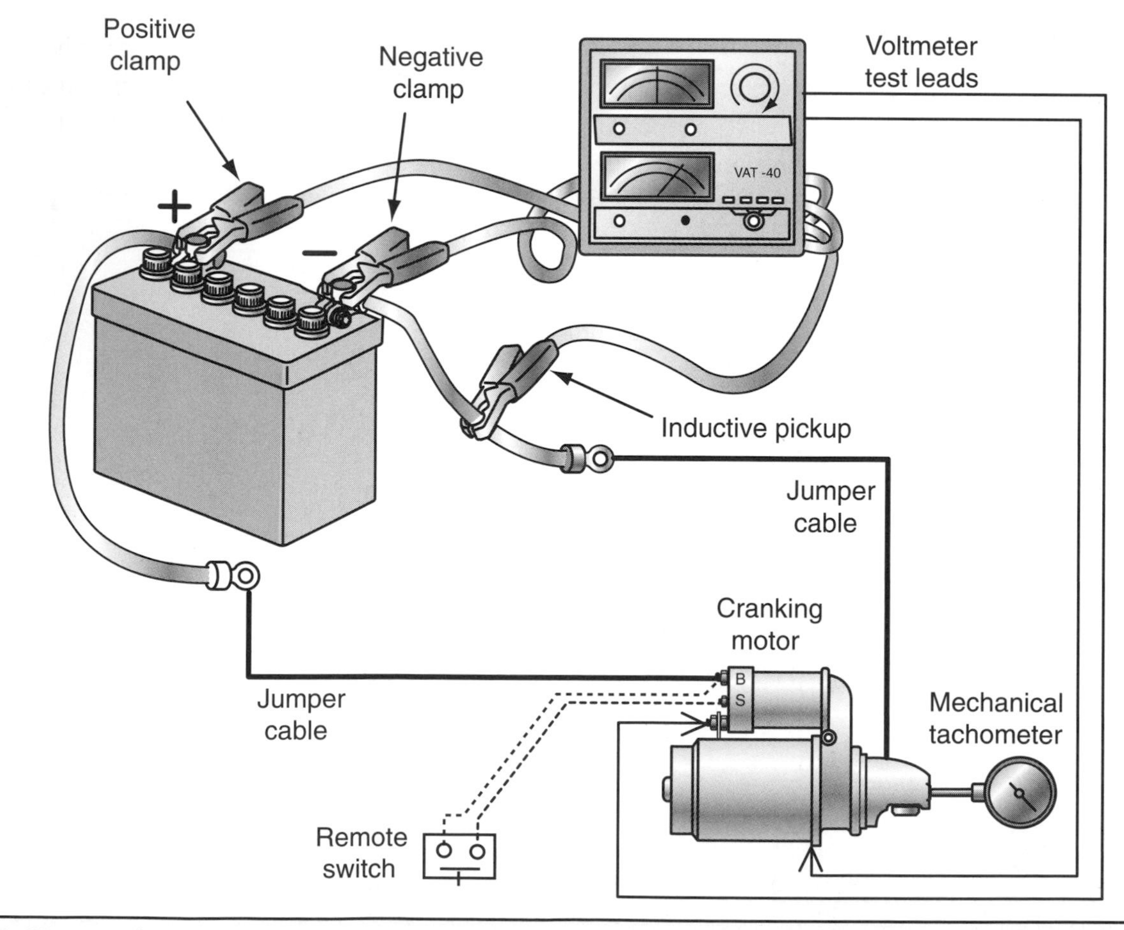

FIGURE 15–47 Bench testing a starter motor with a battery and a starter system tester.

excessive resistance is present in the starter. This condition is normally caused by a poor connection between the commutator and the brushes or by a poor connections in the starter.

3. If the revolutions per minute and current draw are both higher than normal, check for a shorted field winding in the starter motor.
4. If the starter did not rotate and there is no current draw, the cause could be open field windings, open armature coils, or broken brush springs or brushes.

The final starter bench test to perform is a load check on the overrunning clutch pinion. Using a piece of hard wood, such as an old hammer handle, push against the overrunning clutch gear while the starter is engaged. If the clutch is slipping, the hardwood handle stops the gear from turning.

Because some starters are difficult to replace, some manufacturers recommend that the replacement starter be bench tested to ensure proper operation before installation.

6. Starter Pinion Gear Clearance

On some older General Motors starter motors, there is a specified clearance between the overrunning clutch pinion gear and the flywheel ring gear. To check this clearance, insert a flat blade screwdriver into the access hole on the side of the pinion drive housing and pry the drive pinion into the engaged position (Figure 15–48). Use a piece of 0.020-inch diameter wire to check the clearance between the gears (Figure 15–49). If the clearance is excessive, the starter produces a high-pitched whine while the engine is being cranked. If the clearance

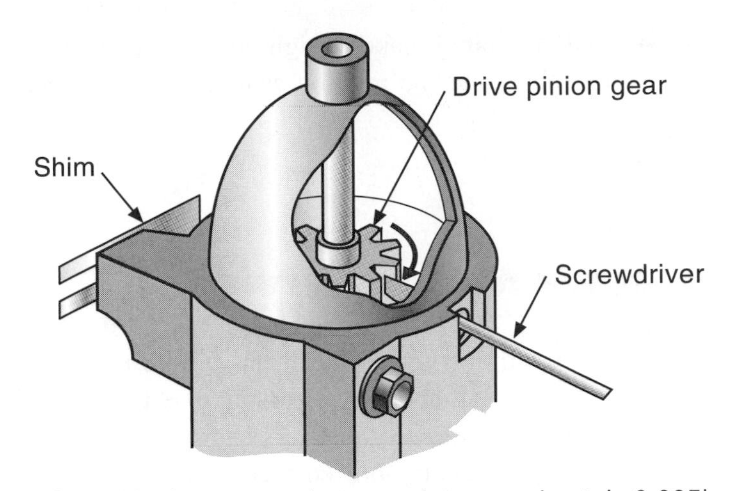

FIGURE 15–48 Prying the drive pinion into the engaged position to check clearance with ring gear.

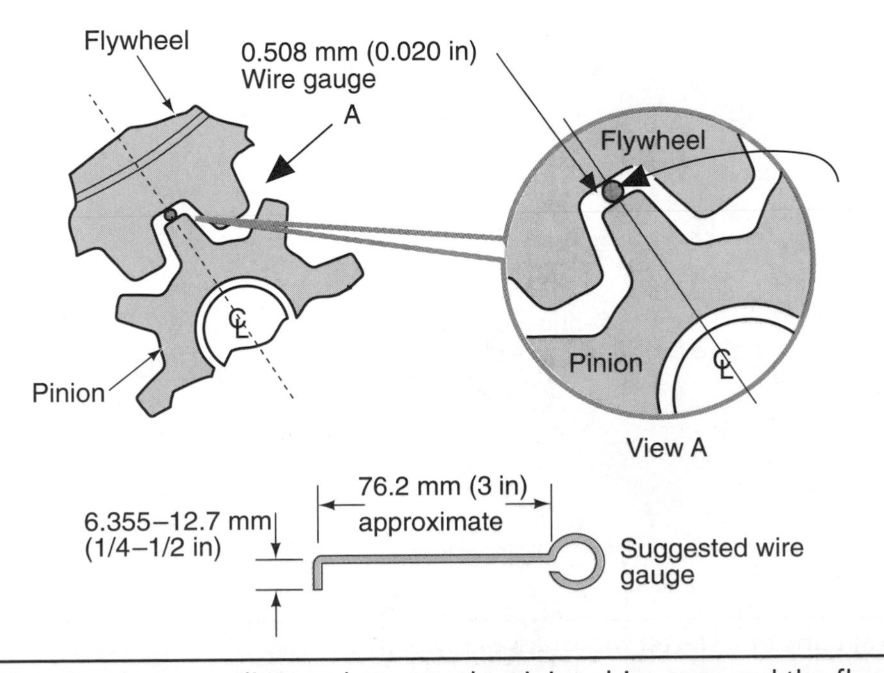

FIGURE 15–49 Checking the clearance distance between the pinion drive gear and the flywheel ring gear.

is too small, the starter makes a high-pitched whine after the engine starts and the ignition switch is returned to the run position. This is also the major cause of a broken pinion drive housing.

Starter Motor Service

Starter motor removal and replacement varies with vehicle manufacturer and brand. The servicing and overhaul of a starter motor are becoming less and less common, as repair shops find it more cost effective to replace rather than repair. For applications in the automotive and light truck world that still provide starter repair parts, the following general information can be used as a basic testing and repair guide. Figure 15–50 is typical of the parts list for a newer vehicle starter motor. Many of the parts listed can be replaced or rebuilt depending on:

1. What is preferred by the customer.
2. Shop warranty policies.
3. Rebuild equipment available in the shop.
4. Specific knowledge of the shop employees.
5. Cost of a new or remanufactured starter compared to an in-house rebuild.
6. The time required to repair or rebuild the starter.
7. Complexity of the starter needing service.

If the decision is made to rebuild the starter motor, then the manufacturer's service manual should be consulted on the specific starter motor disassembly procedures.

After the starter motor is disassembled, clean the parts with an electrical cleaning solution or denatured alcohol. *Do not use solvent or gasoline*. Then test. Begin

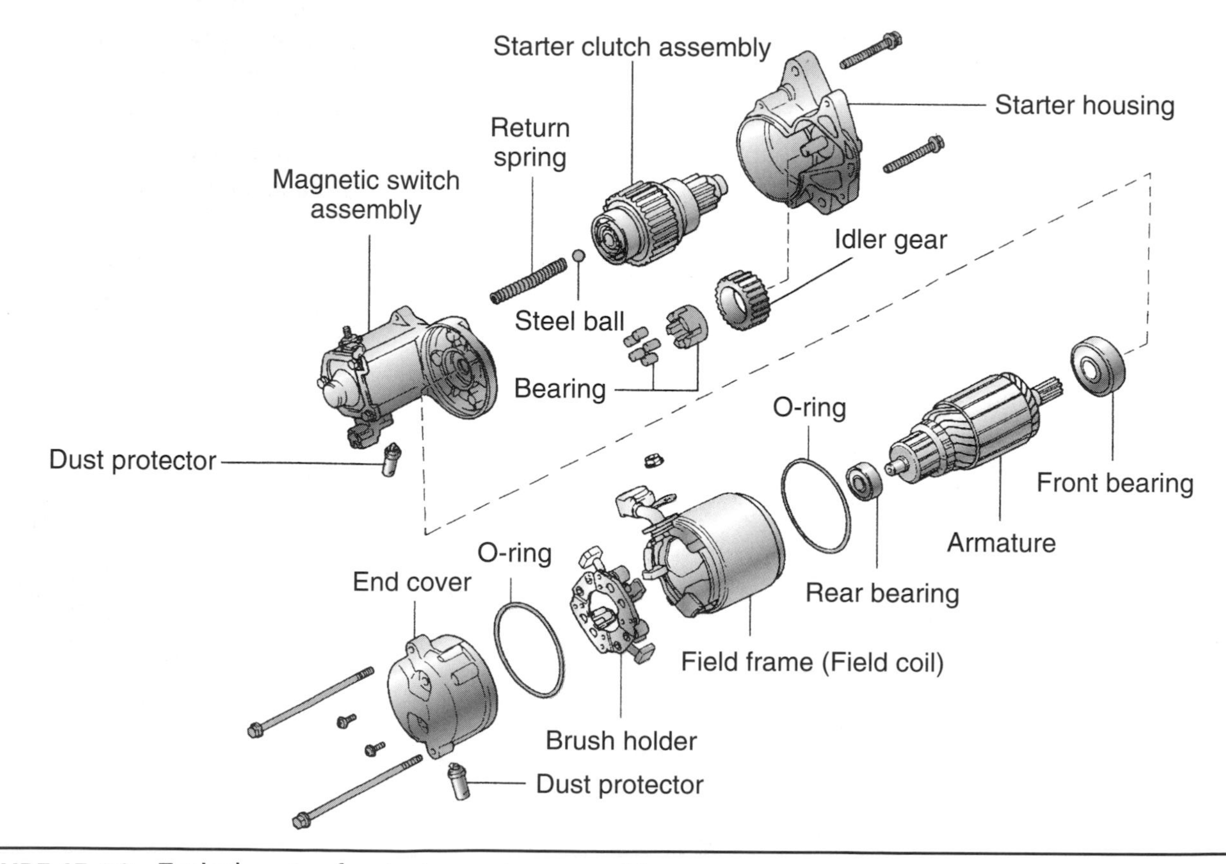

FIGURE 15–50 Typical parts of a starter motor assembly.

(Reprinted with permission)

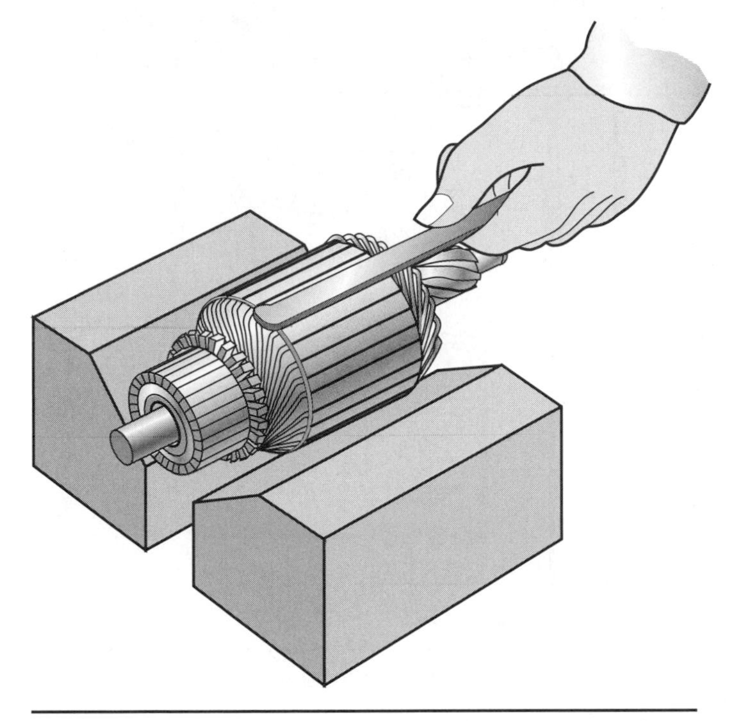

FIGURE 15–51 If a short exists in the armature, the hacksaw blade vibrates when over the shorted area.

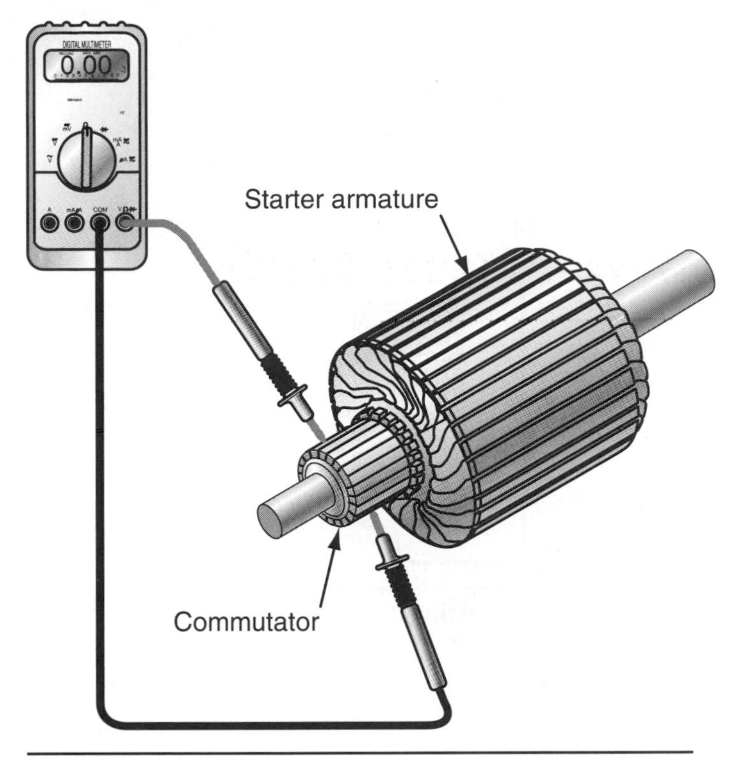

FIGURE 15–52 Using an ohmmeter to test an armature for open windings.

by testing the armature for shorts using an electrical armature growler. The armature is placed in the growler and slowly rotated with a steel hacksaw blade held parallel to the armature core (Figure 15–51). If the blade vibrates, the armature is shorted and should be replaced.

The armature can be tested for opens using an ohmmeter. Figure 15–52 shows connecting the ohmmeter test leads to any two commutator bar sections. There should be 0 ohms of resistance in a good armature. If there is resistance, the armature has to be replaced.

The armature can be tested for grounds using an ohmmeter. Figure 15–53 shows connecting the ohmmeter test leads between each commutator bar and the armature shaft. There should be an infinite reading (no continuity). If there is continuity, the armature has to be replaced.

The field coil windings can be tested for both opens and grounds using an ohmmeter. When the field coil windings are disconnected, they should show an infinite reading (OL) when measuring from the coil windings to the starter housing (Figure 15–54). To test the field coils for an open circuit, measure from each end of the field coil leads (Figure 15–55). The field coils should read

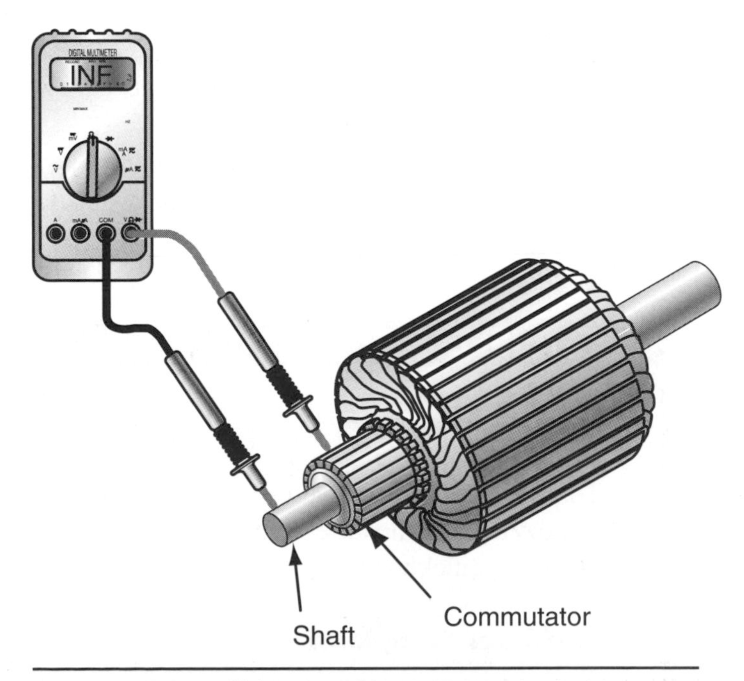

FIGURE 15–53 Using an ohmmeter to test an armature for shorted-to-ground windings.

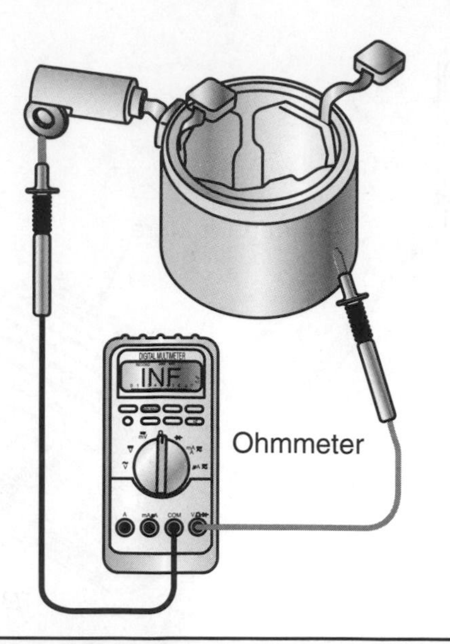

FIGURE 15–54 Testing the field coils for shorts-to-ground.

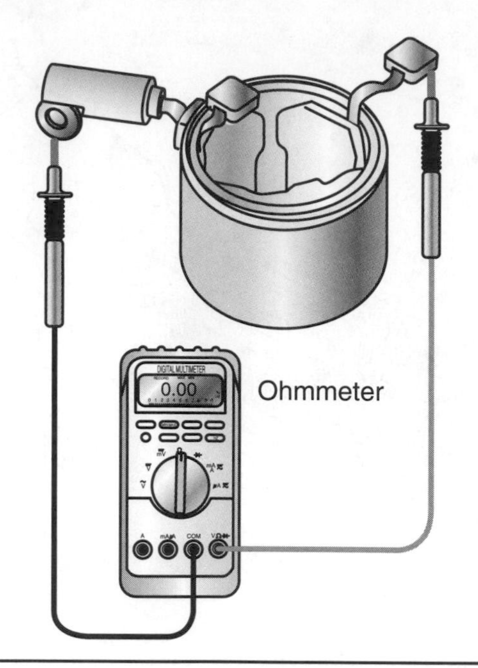

FIGURE 15–55 Testing the field coils for an open circuit.

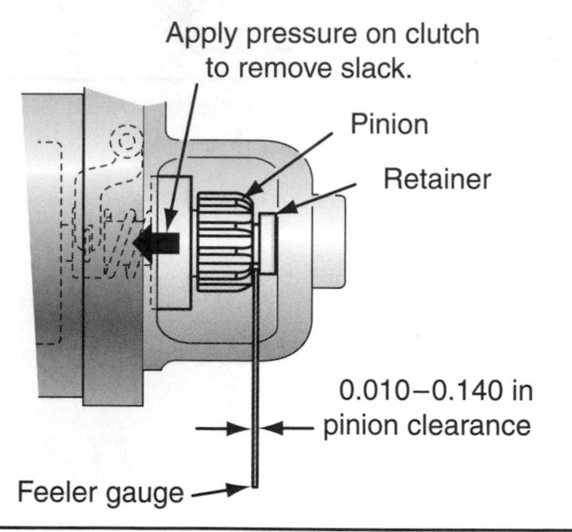

FIGURE 15–56 Checking the starter pinion gear-to-drive housing clearance.

0 ohms, whereas an infinite reading indicates an open circuit and that the coil is not serviceable.

In the reassembly of the starter, a few brands make it necessary to check the pinion gear clearance with a feeler gauge (Figure 15–56). Depending on the starter, specifications range from 0.010 to 0.140 inches. Excessive clearance indicates excessive wear in the shift lever or solenoid linkage.

Note: Remember to perform the free running bench test after starter repair or rebuild.

CASE STUDIES

CASE 1

This case study is typical of a starter circuit problem that technicians often have to diagnose. Knowledge of how individual starter circuit components work and simple test procedures will help you perform the diagnosis quickly and efficiently.

Customer Complaint

A vehicle is towed into the shop with a no-crank condition. The battery and starter were replaced, but the condition remains.

Known Information

- ❑ The new battery has an OCV of 12.66 volts.
- ❑ The starter circuit is a positive engagement design with a remote solenoid relay assembly.
- ❑ The starter motor was replaced.
- ❑ The starter does not crank the engine when ignition switch is turned to the start position.
- ❑ The engine is free when turned with a socket wrench attached to the front pulley nut.

Circuit Analysis

Answer the following questions on a *separate* sheet of paper:

1. With the information known, what is the most likely cause of the starter system problem?

   ____

   ____

2. What tests can you perform on the starter system to determine whether the problem is with the starter motor or the control circuit? ____

   ____.

3. What diagnostic procedures are helpful to know to determine the cause of the no-crank condition?

   ____.

4. What is your analysis of this case study? ____

   ____

   ____.

CASE 2

This case study is typical of a slow-crank condition that technicians often have to diagnose. Knowledge of the starter circuit being tested and the simple test procedures discussed in this chapter will help you perform the diagnosis quickly and efficiently.

Customer Complaint

A vehicle with a direct drive starter cranks slowly all the time. The customer replaced the starter motor and solenoid assembly, the battery, and the alternator, with no improvement in starter cranking speed. The customer is thinking of replacing the engine when the vehicle is towed into the shop for one final diagnosis.

Known Information

- ❑ The new battery has an OCV of 12.66 volts.
- ❑ The starter motor and solenoid assembly is new and has been bench tested as OK.

- ❑ The alternator has been changed (out of desperation).
- ❑ The battery positive and negative cables and terminals *look* OK, with only a slight amount of corrosion at the terminal ends.

Circuit Analysis

Answer the following questions on a *separate* sheet of paper:

1. With the formation known, what is the *most* likely cause of the slow-crank problem above?

   ____.

2. What tests can you perform on the starter system to determine whether the problem is with the starter motor, the control circuit, or the cables and connections? ____________________

   ____.

3. What diagnostic procedures are helpful to know to determine the cause of the slow-crank condition?

   ____.

4. What is your analysis of this case study? ____________________

   ____.

Hands-On Vehicle Tests

The following seven hands-on vehicle checks are included in either the NATEF (National Automotive Technician Education Foundation) Task List or as an extra recommended check. Some of the tasks are grouped together due to their close association with each other. Complete your answers to the following questions on a *separate* sheet of paper.

Performance Task 1

Task Description
Perform starter current draw test; determine necessary action.

Performance Task 2

Task Description
Perform starter circuit voltage drop tests; determine necessary action.

Performance Task 3

Task Description
Inspect and test starter relays and solenoids; determine necessary action.

Performance Task 4

Task Description
Inspect and test switches, connectors, and wires of starter control circuits; perform necessary action.

Task Objectives

❑ Obtain a vehicle that can be used for these tasks. What model and year of vehicle are you using for these tasks? ______________________________.

❑ Set the vehicle emergency brake and place the vehicle in neutral. Observe all standard shop safety precautions.

❑ Perform a visual inspection of the vehicle starter motor circuit. What type of starter circuit does the vehicle have?

Type of starter circuit: ______________________________.

List the main components of the circuit:

1. ______________ 5. ______________

2. ______________ 6. ______________

3. ______________ 7. ______________

4. ______________ 8. ______________

❑ What are the preliminary vehicle checks that should be made to any vehicle before performing tests on a starter circuit? ______________________________

______________________________.

❑ List the equipment you would use to perform diagnostic tests on a starter system.

Diagnostic equipment	Purpose of test equipment
1.	
2.	
3.	
4.	

❑ List the steps required to perform a current draw test on a starter motor using the equipment you have. ______________________________

____

______________________________.

❑ Perform a current draw test on the vehicle being tested. List the results:

Starter circuit voltage: ______________.

Starter amperage draw: ______________.

❑ What are the current draw specifications for the vehicle being tested? If this information is unavailable, what are the standard current draw limits for the size of the engine in the test vehicle?

______________________________.

❑ How do you test a starter relay and/or solenoid for proper operation? ______________________

____.

❑ What can you use to bypass a solenoid or relay that you suspect is faulty? __________________

____.

❑ Bypass the solenoid or relay. What happened to the starter motor? _________________________.

❑ How could this test help you when diagnosing circuit-related problems?_____________________

____.

____

❑ Perform a voltage drop test on all starter circuit components and record the results:

Component	Voltage drop found	Industry limits
1.		
2.		
3.		
4.		
5.		
6.		
7.		
8.		
9.		
10.		

❑ How would voltage drops in the circuit affect the performance of the starter motor?______________

____

____.

❑ What procedures are most helpful when inspecting and testing the starter motor circuit? __________

____.

Task Summary

After performing these NATEF tasks, what can you determine will be helpful in knowing how to check a starter motor system and individual components for proper operation? ______________________

____

____

____.

Performance Task 5

Task Description

Remove and install starter in a vehicle.

Performance Task 6 (Not an NATEF Task)

Task Description

Disassemble, clean, inspect, and test starter components; replace as needed.

Performance Task 7 (Not an NATEF Task)

Task Description

Perform starter bench tests; determine needed repairs.

Task Objectives

❑ Obtain a vehicle that can be used for these tasks. What model and year of vehicle are you using for these tasks? ______________________________.

❑ List the procedures for removing and replacing the starter motor according to the manufacturer of the vehicle being worked on. ______________________________

______________________________.

❑ What tools and equipment did you use to remove and replace the starter motor?______________________________

______________________________.

❑ List the steps that you used to disassemble, clean, and inspect the starter motor. ______________________________

______________________________.

❑ Perform individual starter motor component tests according to the manufacturer of the specific vehicle being worked on. Record the results:

Starter component	Test equipment used	Results of test
1.		
2.		
3.		
4.		
5.		
6.		

❑ What are the steps and precautions necessary when performing a free running bench test of a starter motor? ______________________________.

- ❑ Perform a free running bench test on the vehicle starter motor. Record the results.

  Starter RPM ______________________. Starter current draw ______________________.

- ❑ Does the starter meet standard industry specification for a free running bench test? ____________.

Task Summary

After performing the above tasks, what can you determine will be helpful in knowing how to check a starter motor system and individual components for proper operation? ______________________

____.

Summary

- ❑ The magnetic repulsion and attraction of electromagnets form the basis of a DC starter motor.
- ❑ Flux density is the concentration of the magnetic lines of force.
- ❑ Field coils are several turns of heavy copper wire wrapped around an iron core called a pole shoe.
- ❑ The commutator is a series of split ring copper segment bars at one end of the armature that provide the sliding contact to a set of electrically conductive carbon brushes.
- ❑ CEMF is the counter electromotive force that is induced within the starter windings when current flows through the windings.
- ❑ A series wound starter directs current through the starter field winding in a series line.
- ❑ A parallel wound, or shunt wound, starter directs current through the starter field windings in two parallel paths.
- ❑ A compound wound starter directs most current through the starter in a series line, but it has a separate field winding in parallel to control starter speed.
- ❑ A permanent magnet starter uses strong permanent magnets in place of electromagnets for the field coils.
- ❑ The typical starter system consists of a battery, battery cables, starter motor, ignition switch, harness wiring, starter relay or solenoid, pinion drive, flywheel ring gear, and a neutral safety switch.
- ❑ The Bendix inertia drive is an older style drive pinion that uses inertia speed to engage the pinion teeth into the flywheel ring gear.
- ❑ The overrunning clutch drive is the most common type of starter drive pinion. It uses a roller-type clutch that wedges spring-loaded rollers between inner and outer housings to engage the pinion teeth with the flywheel teeth.
- ❑ A starter relay, or solenoid, is a high-amperage switch that controls the battery amperage going to the starter motor.

- ❑ The solenoid pull-in and hold-in windings both energize to engage the overrunning clutch into the flywheel ring gear.
- ❑ The ignition bypass terminal supplies full battery voltage to the ignition coil only during engine cranking conditions.
- ❑ The direct drive starter is the most popular style of starter that uses a solenoid attached directly to the starter motor.
- ❑ The positive engagement starter uses a separate relay to control an internal movable pole shoe in the starter to engage the pinion gear.
- ❑ A gear reduction starter contains a small gear reduction drive gear between the starter motor and the drive pinion.
- ❑ A permanent magnet starter uses strong permanent magnets instead of electromagnets to form the field coils.
- ❑ A hybrid electric vehicle (HEV) may use an integrated starter/generator (ISG) that combines a starter, generator, and power booster in one unit.
- ❑ The starting safety switch, also called a neutral safety switch (NSS), is used to prevent the starter motor from cranking on an automatic transmission with the transmission in gear or on a manual transmission only if the clutch is pushed in.
- ❑ A starter circuit isolation test is a quick test that determines whether the starter motor, solenoid, or control circuit is at fault during a no-crank test.
- ❑ A starter current draw test tests the starter current draw during cranking conditions.
- ❑ A starter circuit voltage drop test checks for unwanted resistance in the circuit that creates a voltage drop condition, which could affect cranking voltage available to the starter motor.
- ❑ A starter open circuit test is performed on a starter with a no-crank complaint.
- ❑ Starter bench testing checks the starter motor revolutions per minute, current draw, and overrunning clutch operation during no-load conditions.

Key Terms

armature
Bendix inertia drive
brushes
commutator
compound wound
counter electromotive force (CEMF)
direct drive starter
field coil
flux density
gear reduction starter
hold-in winding
ignition bypass terminal
integrated starter/generator (ISG)
neutral safety switch (NSS)
overrunning clutch drive
parallel wound
permanent magnet starter
pole shoe
positive engagement starter
pull-in winding
series wound
shunt wound
starter relay
starter solenoid
starting safety switch

Review Questions

Short Answer Essays

1. When an engine starts, what causes the starter motor overrunning clutch to disengage itself for safety before the ignition key is returned to the run position?
2. Describe how the pull-in and the hold-in windings work in a typical starter solenoid.
3. What is the purpose of the neutral safety switch in a starter system?
4. Describe the procedures used to perform a starter current draw test using a VAT-40 tester?
5. What are the allowed voltage drops in a typical starter motor circuit?

Fill in the Blanks

1. The soft iron core in the field coil in the starter motor is called the ________________.
2. On a starter solenoid the R terminal is an ________________, that supplies battery voltage to the ________________ during engine cranking conditions.
3. A gear reduction starter contains a ________________ between the starter motor and the drive pinion.
4. To perform a starter system test, the battery must be ________________.
5. Normal starter current draw for a small V-8 engine should be ________________ amperes.

ASE-Style Review Questions

1. Technician A says the ignition coil is part of the starter system. Technician B says the ignition switch is part of the starter system. Who is correct?
   A. A only
   B. B only
   C. Both A and B
   D. Neither A nor B

2. Technician A says the commutator is a series of conducting segments located around one end of the armature. Technician B says the armature is the stationary component that consists of several turns of copper wire wrapped around a pole shoe. Who is correct?
   A. A only
   B. B only
   C. Both A and B
   D. Neither A nor B

3. Technician A says the starter solenoid moves a plunger to engage the overrunning clutch. Technician B says the two windings of the starter solenoid are called the pull-in and hold-in windings. Who is correct?

   A. A only

   B. B only

   C. Both A and B

   D. Neither A nor B

4. Technician A says the CEMF is present only in high-amperage starters. Technician B says CEMF is caused by excessive resistance in the starter motor field windings. Who is correct?

   A. A only

   B. B only

   C. Both A and B

   D. Neither A nor B

5. Technician A says the neutral safety switch controls only the starter circuit on all vehicles. Technician B says the neutral safety switch prevents the starter from cranking when the automatic transmission is in gear. Who is correct?

   A. A only

   B. B only

   C. Both A and B

   D. Neither A nor B

6. Two technicians are discussing the results of a starter current draw test on a 4-cylinder engine. Technician A says the minimum voltage should not drop below 9.6 volts. Technician B says the amperage draw of 125 amperes is considered satisfactory. Who is correct?

   A. A only

   B. B only

   C. Both A and B

   D. Neither A nor B

7. Technician A says a permanent magnet starter uses strong permanent magnets in place of electromagnetic field coils. Technician B says a permanent magnet starter uses a planetary gear set for maximum gear reduction in a small starter. Who is correct?

   A. A only

   B. B only

   C. Both A and B

   D. Neither A nor B

8. Technician A says the maximum allowed voltage drop for a battery cable is 0.5 volts. Technician B says a voltage drop at a connection should not exceed 0.3 volts per connection. Who is correct?
   A. A only
   B. B only
   C. Both A and B
   D. Neither A nor B

9. Technician A says if starter pinion gear clearance is excessive, the starter produces a high-pitched whine while cranking the engine. Technician B says if starter pinion gear clearance is excessive, the starter drive housing will break. Who is correct?
   A. A only
   B. B only
   C. Both A and B
   D. Neither A nor B

10. An engine cranks slowly. Technician A says a possible cause could be a low battery charge. Technician B says a possible cause could be incorrect ignition timing. Who is correct?
    A. A only
    B. B only
    C. Both A and B
    D. Neither A nor B

11. Technician A says the purpose of the starter relay is to complete the circuit from the battery to the starter motor. Technician B says the purpose of the starter solenoid is to complete the circuit from the battery to the starter motor. Who is correct?
    A. A only
    B. B only
    C. Both A and B
    D. Neither A nor B

12. A starter system has a no-crank condition. The headlights remain on with the same intensity when the ignition switch is turned to the start position. Technician A says this can be caused by a dead battery. Technician B says this can be caused by a faulty solenoid switch. Who is correct?
    A. A only
    B. B only
    C. Both A and B
    D. Neither A nor B

13. A vehicle is being tested for a slow-crank condition. A starter current draw test indicates 450 amperes. Technician A says this could be caused by a battery with a CCA rating too large for the system. Technician B says this could be caused by a pinion drive that slips when engaged with the flywheel ring gear. Who is correct?
    A. A only
    B. B only
    C. Both A and B
    D. Neither A nor B

14. Technician A says the R terminal on the starter solenoid is used for the reactor windings in the solenoid. Technician B says the pull-in windings become de-energized when the overrunning clutch engages into the flywheel ring gear. Who is correct?
    A. A only
    B. B only
    C. Both A and B
    D. Neither A nor B

15. A slow-crank condition exists in a starter system. Technician A says corroded conditions at the battery can be the cause. Technician B says excessive battery CCA capacity can be the cause. Who is correct?
    A. A only
    B. B only
    C. Both A and B
    D. Neither A nor B

16 Charging System Theory

Introduction

The purpose of the automotive charging system is to convert the mechanical energy of the engine into electrical energy to recharge the battery, when necessary, and to provide all the current needed to operate the vehicle electrical accessories while the engine is running. Accessory electrical demands on the vehicle have increased almost fivefold to 100 amperes or more over the past three decades. To keep up with this demand, charging systems have made significant design improvements since the DC (direct current) generator was discontinued in the mid-1960s and AC (alternating current) generators became the predominant charging device. Recent developments in hybrid electric vehicle (HEV) technology has created several new charging systems that assist the HEV in its ability provide the high voltages necessary to charge the battery systems and assist the engine/motor under a variety of operating conditions.

Objectives

When you complete this chapter you should be able to:

- ❑ Explain the purpose of the charging system in a vehicle.
- ❑ Explain the principles of magnetism in the operation of a charging system.
- ❑ Explain the differences between DC generators and AC generators in a charging system.
- ❑ Identify and explain the function of the major components of the charging system.
- ❑ Explain the function of regulation devices and circuits in charging systems.
- ❑ Explain the basic operation of an HEV integrated starter/generator (ISG).
- ❑ Perform charging system output tests to diagnose an undercharge, overcharge, or no-charge condition, and determine needed repairs.
- ❑ Perform tests on individual charging system components and determine needed repairs.
- ❑ Perform charging system voltage drop tests and determine needed repairs.

Charging System Principles

All vehicle charging systems use the principles of electromagnetic induction to generate the current necessary to charge the battery and operate the vehicle accessory circuits. The electromagnetic principle states that a voltage is produced if motion between a conductor and a magnetic field occurs (Figure 16–1). The amount of voltage produced in the conductor is dependent on:

1. The strength of the magnetic field.
2. The speed at which the conductor is cutting through the magnetic field.
3. The number of conductors passing through the magnetic field.

Direct Current (DC) Generators

Although they are not used in modern automobiles, knowledge of DC generator principles helps you understand the benefits of an AC generator. The DC generator (Figure 16–2) is similar to the DC starter motor (discussed in Chapter 15). Figure 16–3 is a simple DC motor that

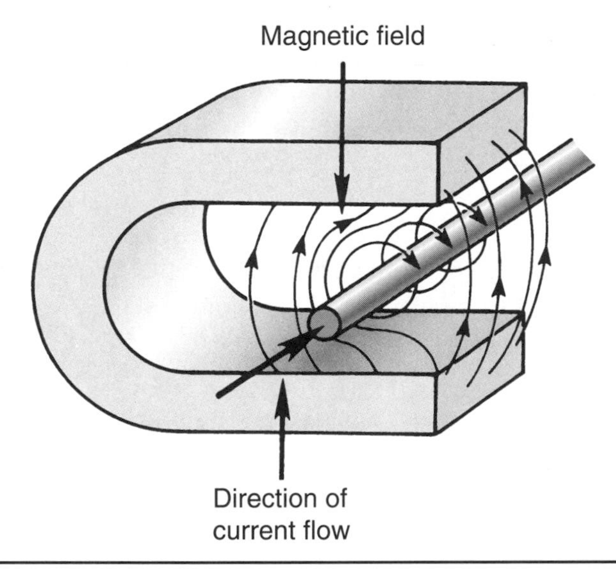

FIGURE 16–1 Conductor interacting in a magnetic field.

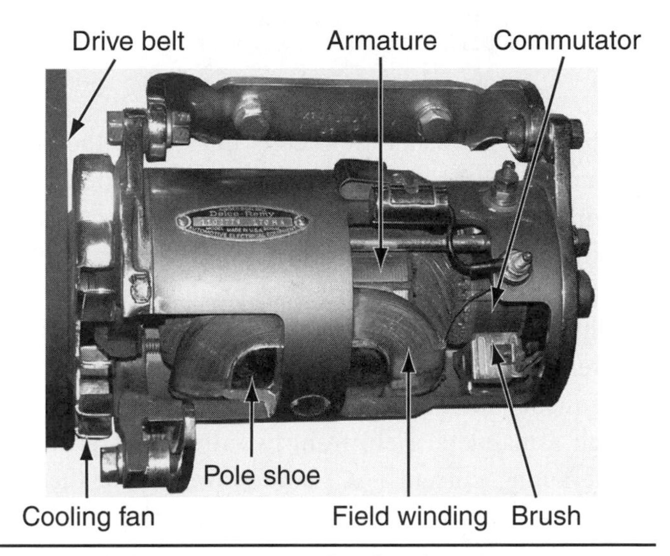

FIGURE 16–2 A cutaway of a simple DC generator. (Courtesy of Tim Gilles)

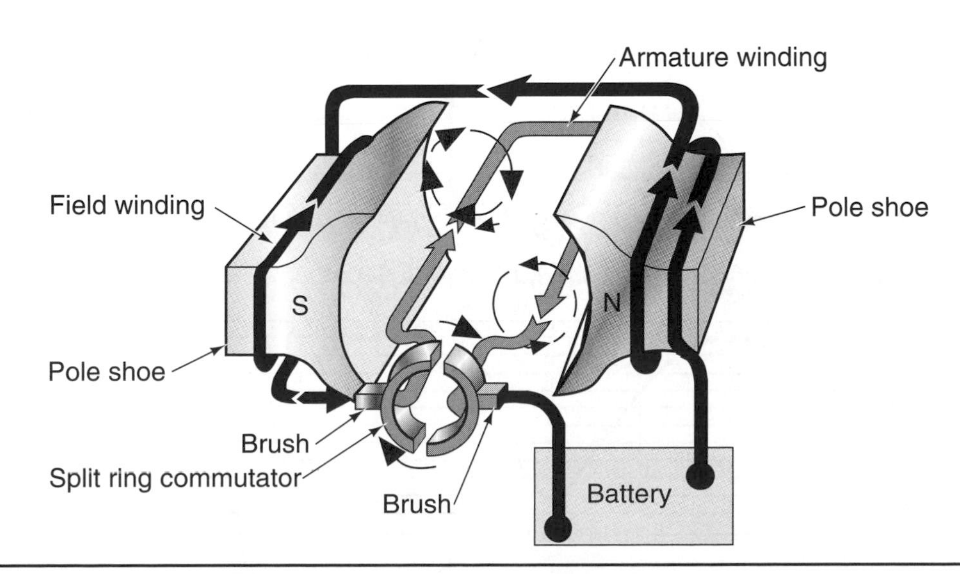

FIGURE 16–3 Simple DC motor with two stationary field coils and rotating armature output.

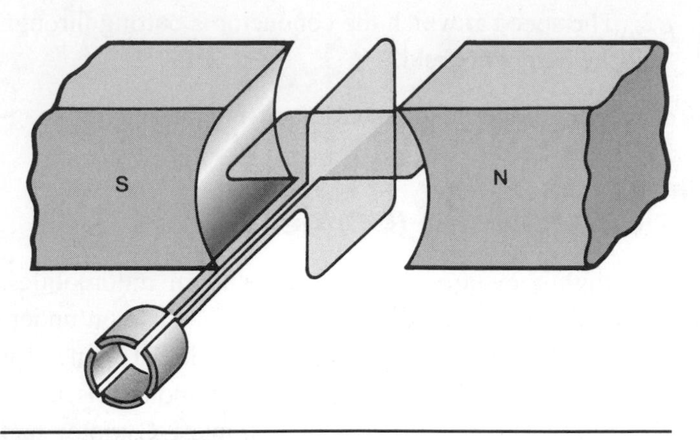

FIGURE 16–4 Strengthening the voltage output by increasing the number of conductors.

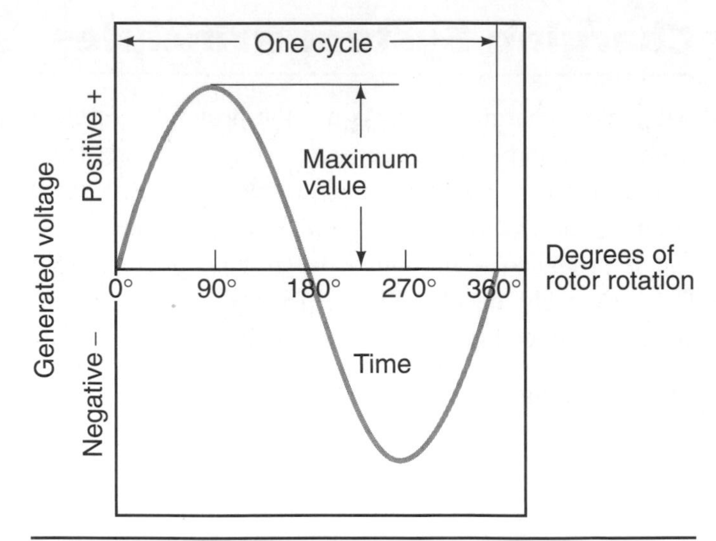

FIGURE 16–5 Sine wave produced in one revolution of a conductor in a magnetic field.

contains two stationary field coils that create the north and south magnetic poles. Output voltage is generated in the wire loop(s) of the armature as it rotates through the magnetic field. The more wire loops there are in the armature, the more voltage will be produced at the same revolutions per minute and strength of the magnetic field (Figure 16–4).

In theory the output voltage of a conductor passing through a magnetic field from two directions produces an alternating current (AC) **sine wave,** where there is positive and negative voltage generated during one revolution (Figure 16–5). In a DC generator the AC sinewave is changed to a pulsing DC sine wave by the armature commutator bars (Figure 16–6).

Alternating Current (AC) Generators

The AC generator (also called an alternator) is widely used in all modern vehicles because of its ability to

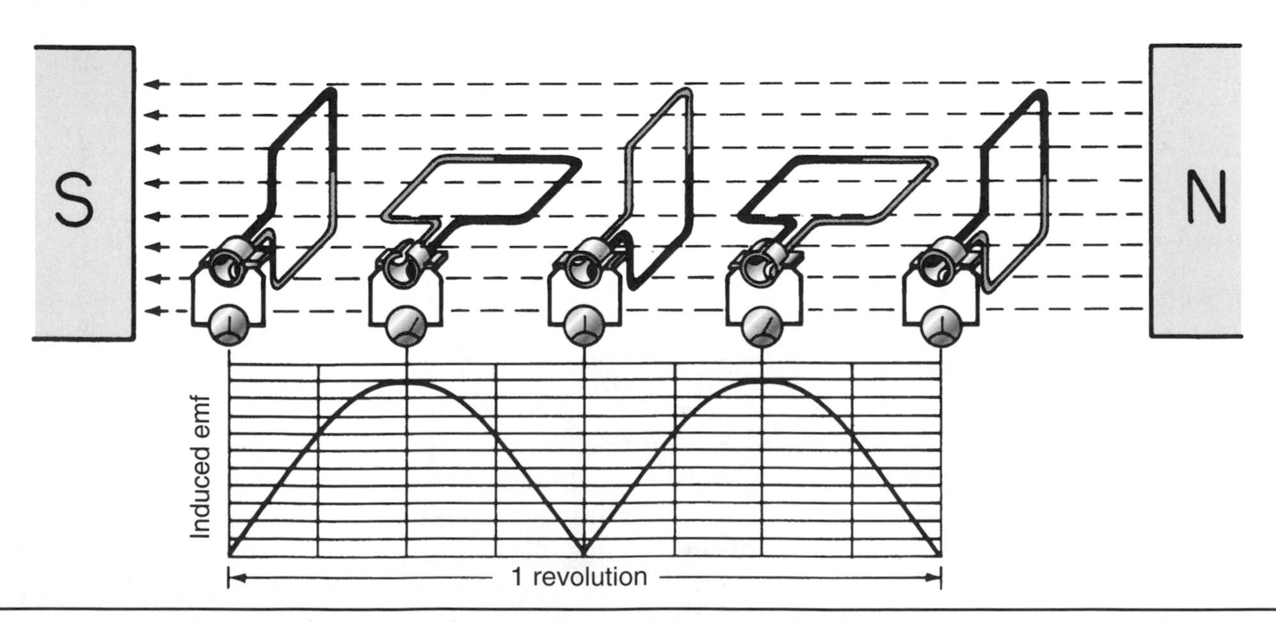

FIGURE 16–6 AC sine wave is converted to a pulsing DC sine wave by the armature commutator bars.

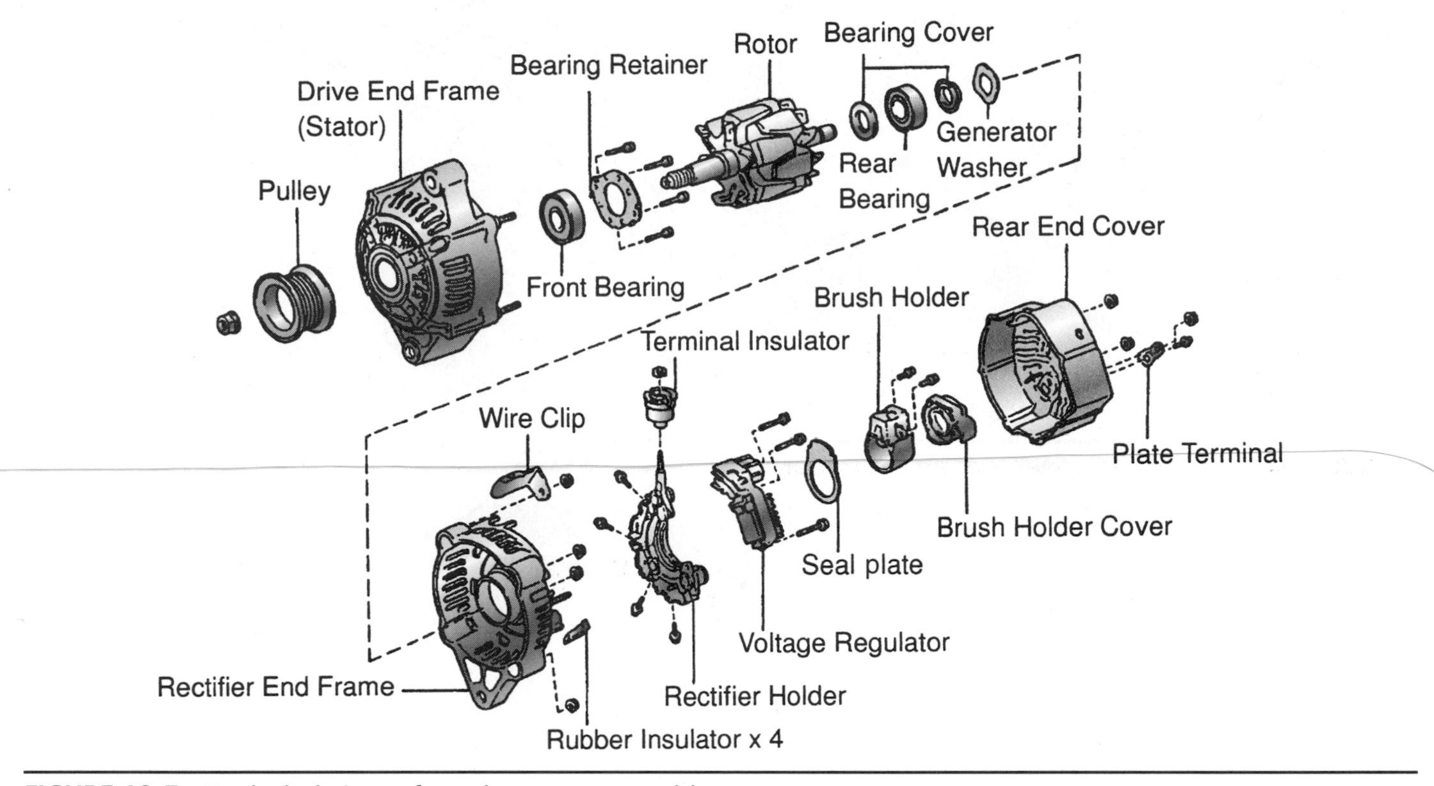

FIGURE 16–7 Exploded view of an alternator assembly.
(Reprinted with permission)

produce sufficient amounts of current at lower engine revolutions per minute. The current produced by a DC generator was limited to 40 amperes or less. The speed at which a conductor is cutting through the magnetic field is important when electrical accessory load demands are high, and the engine revolutions per minute is low. Figure 16–7 shows the components of a typical AC generator, which include:

- Front and rear housings (frames).
- Front and rear bearings.
- Cooling fan.
- Rotor assembly.
- Stator assembly.
- Rectifier assembly.
- Brush holder and brushes.
- Voltage regulator.

Charging System Components and Operation

Rotor Assembly

The **rotor** assembly is considered the heart of the AC generator. It is a rotating electromagnetic field with several north and south poles. The rotor is constructed of many turns of insulated copper wire wrapped around an iron core called a field coil. The field coil is surrounded on each end by soft metal pole pieces with fingers that interlace but do not touch each other (Figure 16–8). When current passes through the coil, the pole pieces become magnetized and take on north and south pole polarities (Figure 16–9). The poles alternate their north and south pole arrangement around the rotor coil, causing the magnetic flux lines to move in opposite directions between adjacent poles (Figure 16–10).

The rotor **slip rings** provide a current path connection from the insulated carbon brushes to the rotor

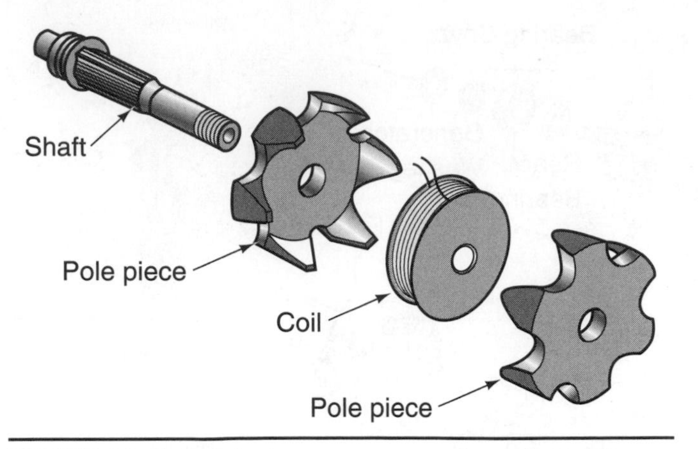

FIGURE 16–8 Typical components of an alternator rotor.

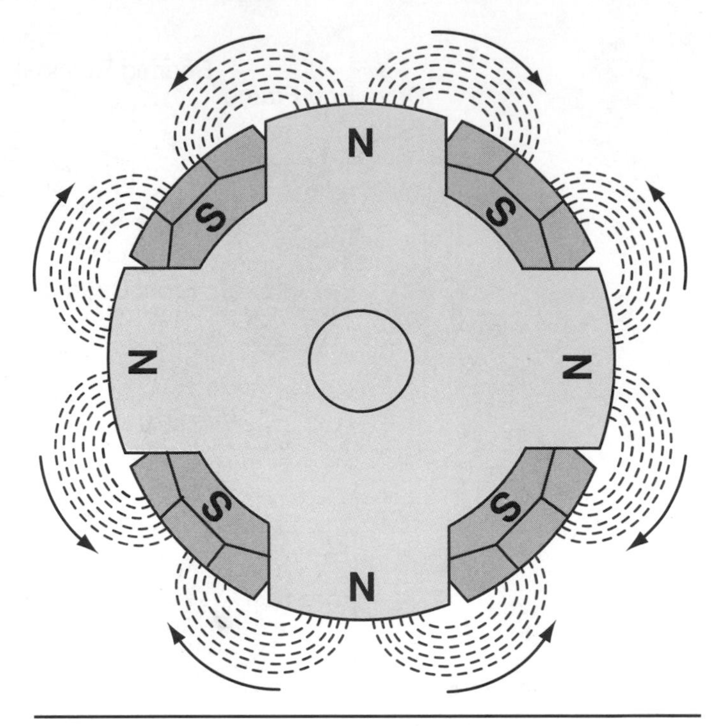

FIGURE 16–10 The magnetic flux lines move in opposite directions between the rotor poles.

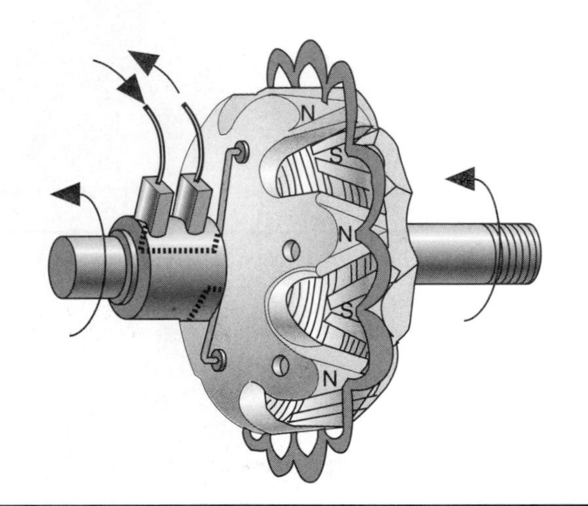

FIGURE 16–9 Multiple north and south poles of a rotor.

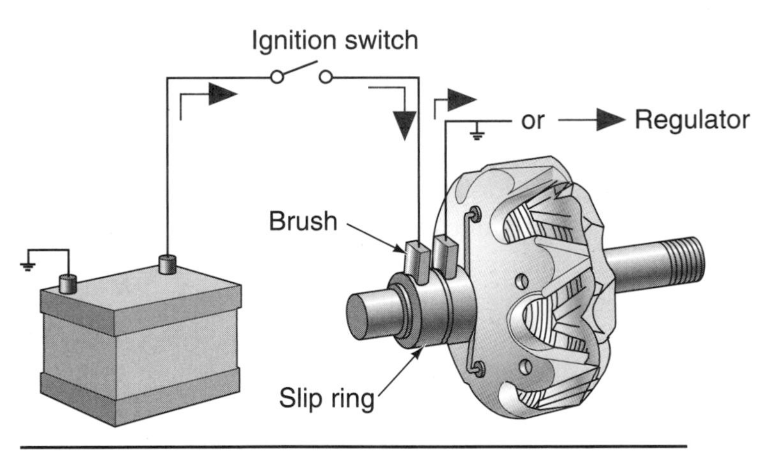

FIGURE 16–11 The slip rings and brushes provide an insulated current path to the rotor coil.

coil assembly (Figure 16–11). The slip rings are insulated from the rotor shaft and from each other. When the circuit is energized, regulated current flows through the first brush and slip ring to the field coil. Current then passes through the field coil and the second slip ring and brush to the ground in the alternator or regulator.

There are three basic checks to perform on a rotor that has been removed from the AC generator.

1. The first check is a complete visual inspection of the rotor field windings for overheating or discoloration. The slip rings should be smooth and flat with no grooves. A rotor with one of these conditions is unusable.
2. The second check is to measure the resistance between the slip rings (Figure 16–12) for an open condition. Always refer to the vehicle manufacturer service manual for correct specifications. Typical resistance values are between 2.4 and 6.0 ohms. If the resistance value is below the specified value, the rotor coil is shorted. If there is an infinite resistance (O/L), the rotor is open.

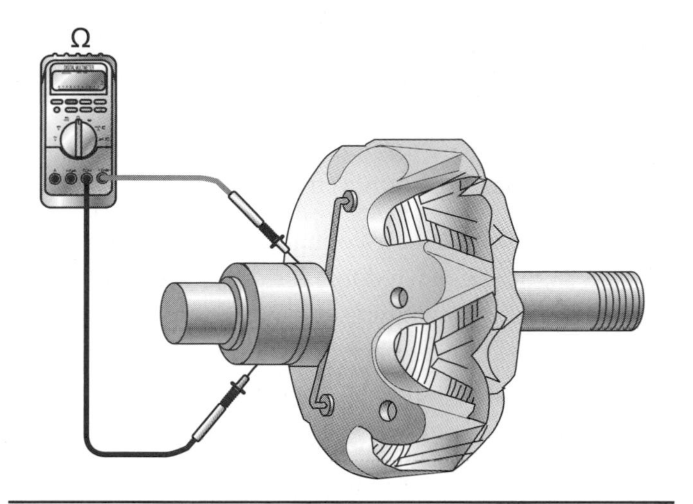

FIGURE 16–12 Testing the rotor field coil for an open or shorted condition.

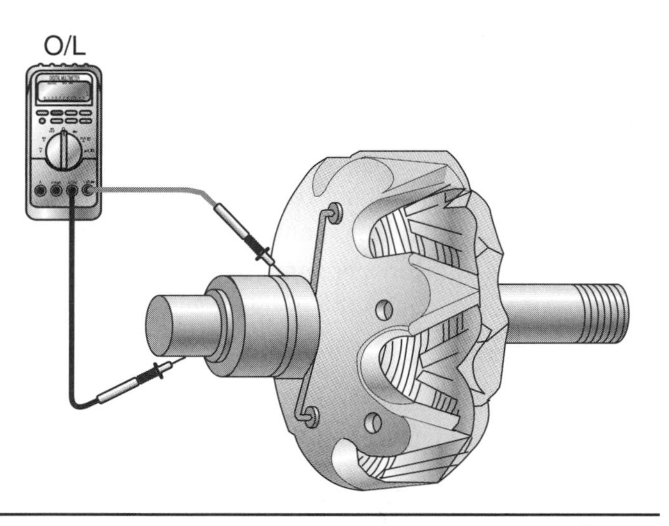

FIGURE 16–13 Testing the rotor slip rings and field coil for a short-to-ground with the rotor shaft.

3. The third check is to test the slip rings and field coil for a short to the rotor shaft (Figure 16–13). Using an ohmmeter from each slip ring and the rotor shaft is one way to check for this condition. The ohmmeter reading should be infinite (O/L) resistance. If the reading is low, the field coil or one of the slip rings is shorted, and the rotor is unusable.

Stator Assembly

The **stator** assembly in an AC generator is the stationary coil windings in which electricity is induced. It is located between the front and rear halves of the alternator shell

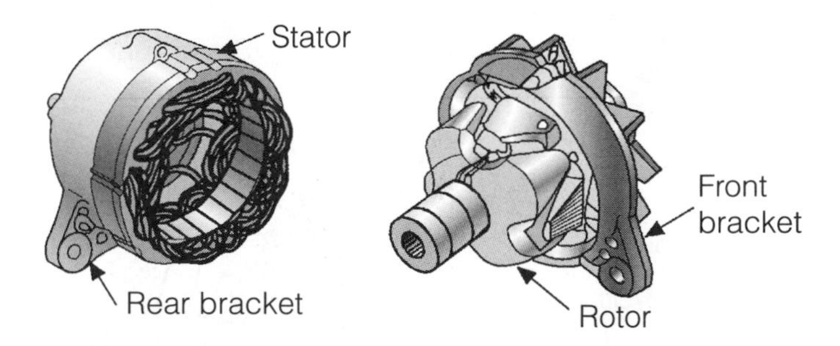

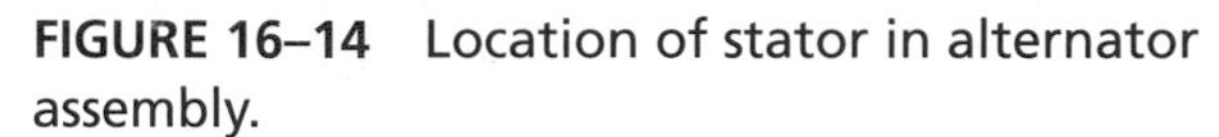

FIGURE 16–14 Location of stator in alternator assembly.

(Courtesy of DaimlerChrysler Corporation)

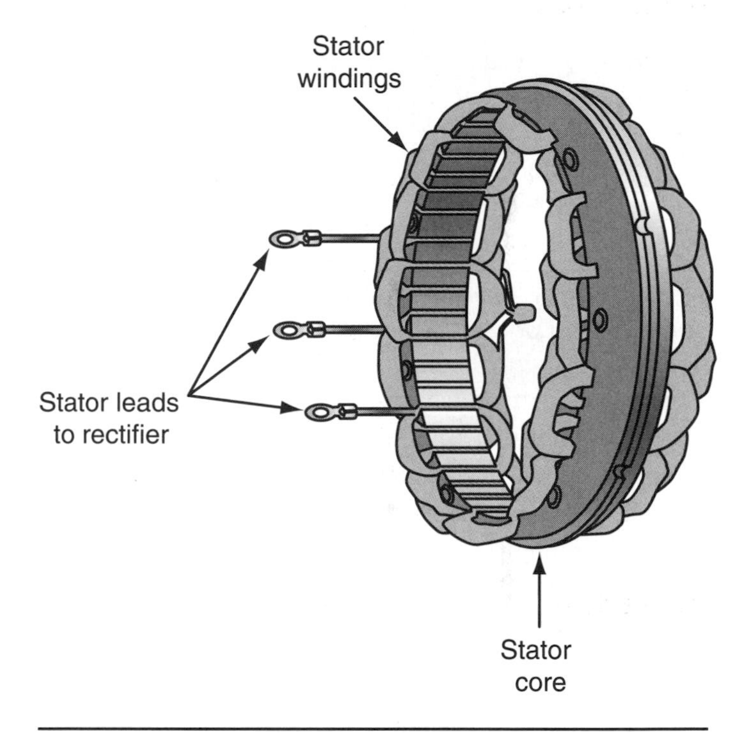

FIGURE 16–15 Components of a typical stator assembly.

(Figure 16–14). The stator consists of three sets of insulated windings (Figure 16–15), with coils equal to the number of north and south poles on the rotor. The windings are wrapped in insulated slots around a laminated circular soft iron frame and overlapped to produce the required phase angles. (Figure 16–16).

Depending on the electrical requirements of the vehicle, two basic configurations of stator winding connections are used on AC generators in modern vehicles. The first is a **wye wound** stator winding (Figure 16–17), which consists of a common junction and three output

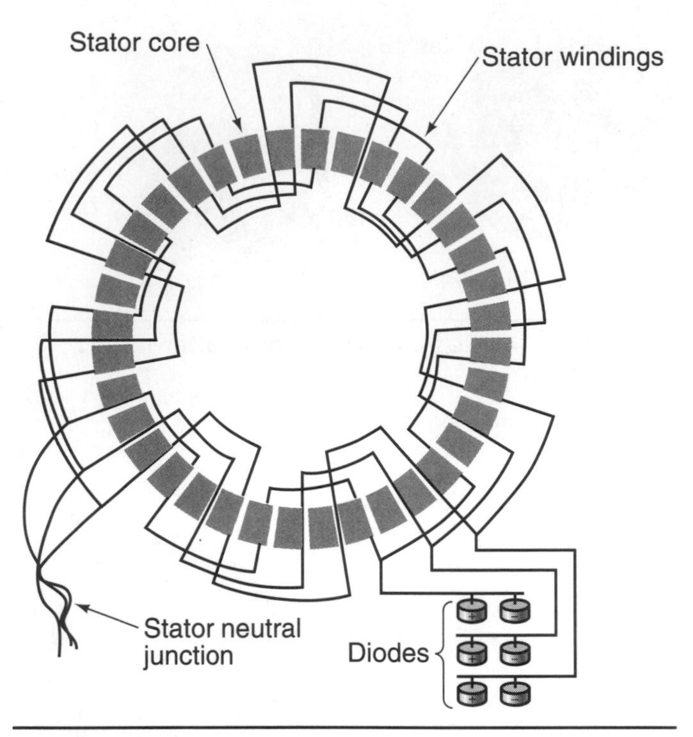

FIGURE 16–16 Overlapping stator windings produce the required phase angles.

connections that branch out in a Y pattern. A wye wound alternator is usually found in applications that do not require high-amperage output.

The second stator winding configuration is a **delta wound** winding (Figure 16–18), which consists of three windings connected in a triangle shape with three output leads. The delta connection is the most common style in use today because of its high-amperage output for its small size. The reason for the increased amperage output is explained in the diode section of this chapter.

There are three basic checks to perform on a stator winding that has been removed from the AC generator.

1. The first check is a complete visual inspection of the stator phase windings for overheating or discoloration. Phase windings that have blackened varnish insulation indicate that overheating has occurred, and the stator should be replaced.
2. Next, check for an open condition in any combination of the stator windings (Figure 16–19). Connect an ohmmeter to any two stator output leads—three combinations in all. The results should show a resistance of less than 1.0 ohm. If the ohmmeter reading is infinity (O/L) between any two leads, the stator has an open and must be replaced.
3. The third check is to test the stator windings for a short-to-ground (Figure 16–20) by connecting the ohmmeter to each stator output lead and the support frame. The ohmmeter should read infinity (O/L) on all three stator leads. For any reading less than infinity, the stator is shorted to ground and is unusable.

The rotor assembly is fitted inside the stator with a clearance of approximately 0.015 inch of air gap, which maximizes the magnetic force of the rotor field to all of the stator windings at the same time (Figure 16–21).

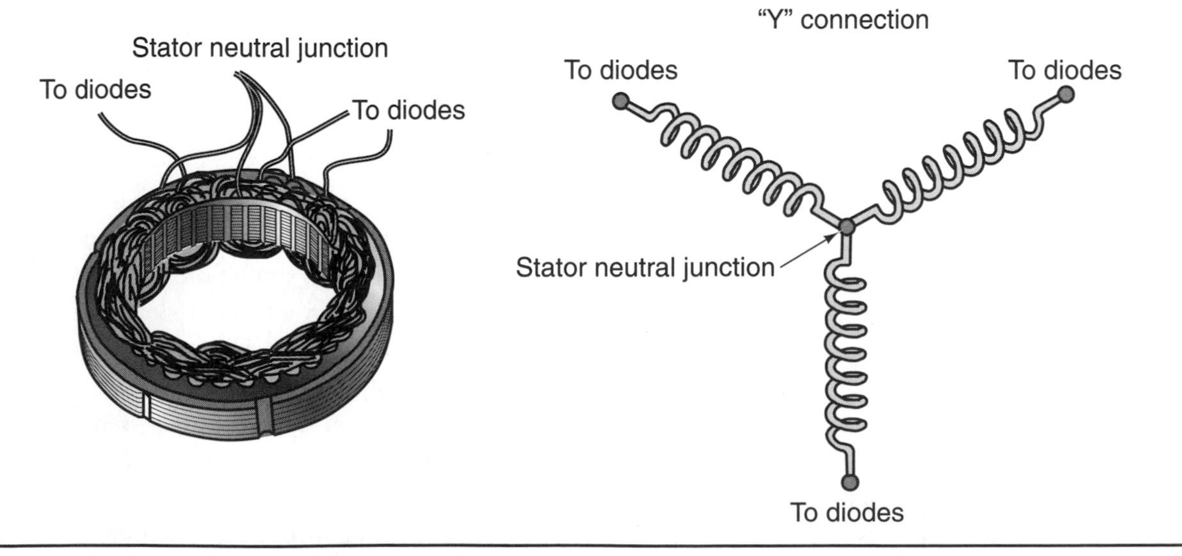

FIGURE 16–17 Typical wye wound stator windings.

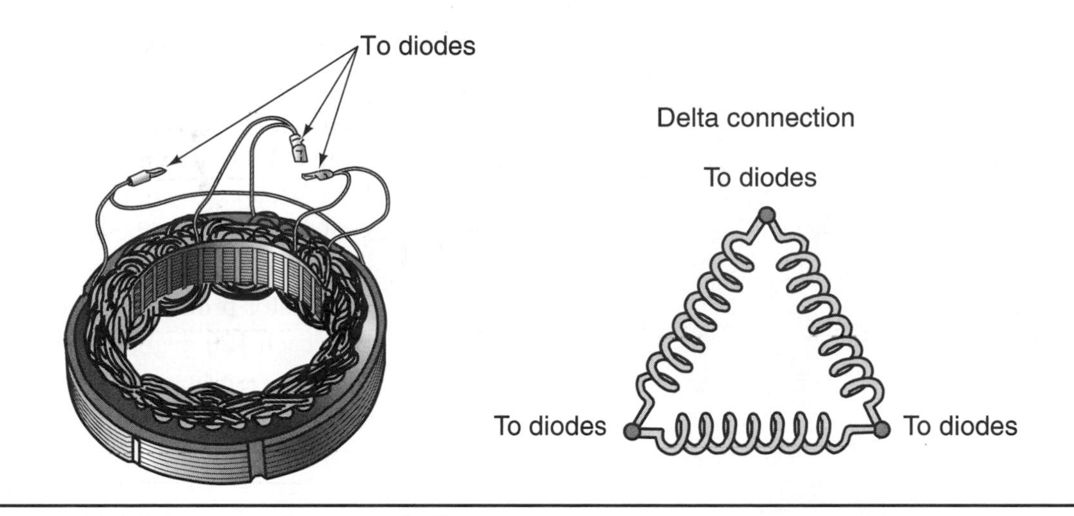

FIGURE 16–18 Typical delta wound stator windings.

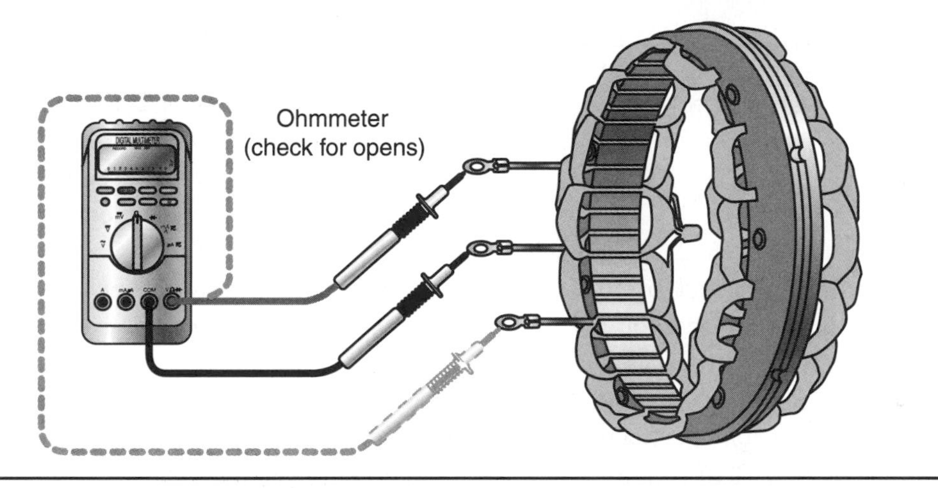

FIGURE 16–19 Testing the stator assembly for opens.

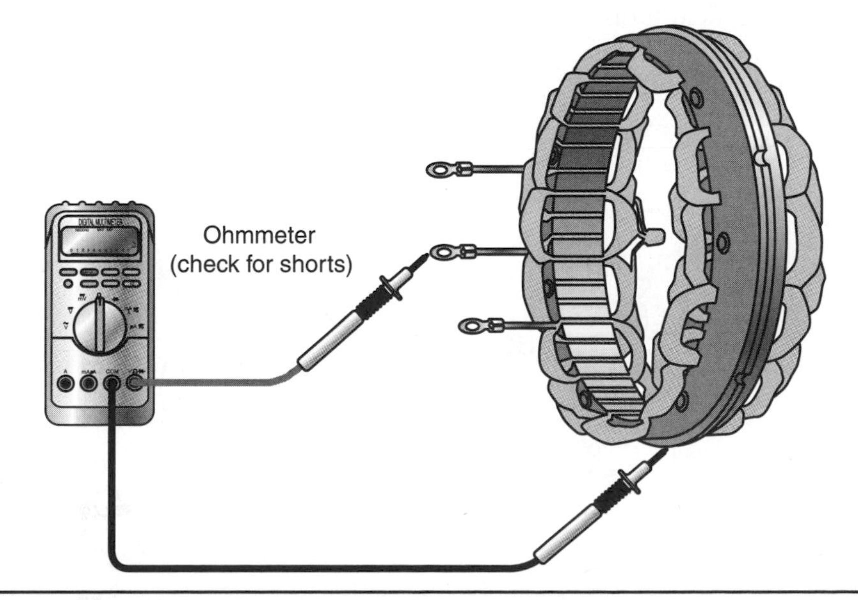

FIGURE 16–20 Testing the stator assembly for shorts-to-ground.

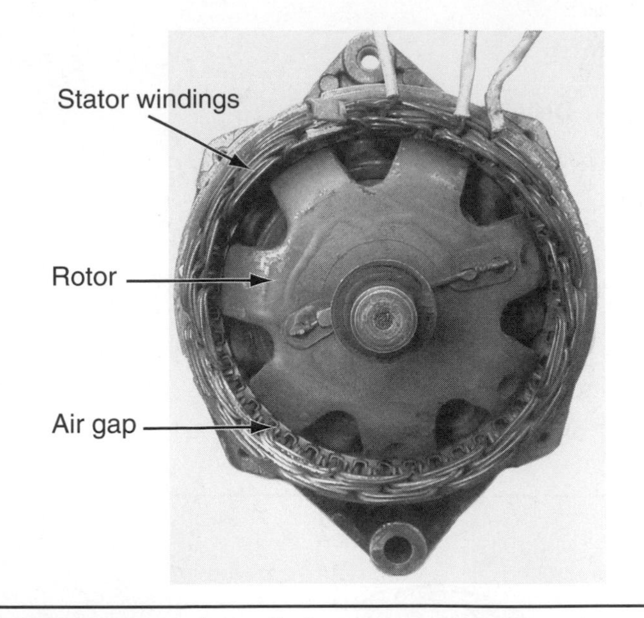

FIGURE 16–21 A small air gap between the rotor and stator maximizes the magnetic force.
(Courtesy of Tim Gilles)

In a simple alternator (Figure 16–22), a single wire is induced with an alternating current by the rotating north and south poles. This can be seen on an oscilloscope as an alternating current sine wave (Figure 16–23) that begins at (a) 0 volts, when the phase winding is parallel to the magnetic field. When the magnetic field rotates 90 degrees, the magnetic field is at a right angle to the phase winding and produces (b) its *maximum positive* voltage. When the magnetic field turns another 90 degrees, being parallel with the phase winding, the voltage returns to (c) 0 volts. A third 90-degree rotation results in the magnetic field being reversed at the top phase winding, producing its (d) *maximum negative* voltage. The last 90-degree rotation completes the 360-degree revolution and returns the voltage to (e) 0.

In a wye wound and a delta wound stator winding, each of the three groups of windings occupies one-third (120 degrees) of the circle. As the rotor, with its multiple north and south poles, rotates inside the stator windings, each loop of the stator produces a voltage with a different phase angle, each 120 degrees apart from the others (Figure 16–24). The resulting overlap of the three phases of each stator winding group produces a **three-phase voltage** output.

Alternating Current Rectification

A vehicle's electrical circuits and battery operate with direct current (DC). The AC current produced by the alternator needs to be converted to a DC voltage. This process is called **rectification.** In a DC generator, the split ring commutator is used to rectify the AC voltage to DC voltage. In an AC generator, the AC voltage is rectified by diodes that act as one-way check valves, allowing current to flow in one direction only. In a simple circuit, if AC voltage passes through a positively biased diode, the diode blocks the negative portion of the sine wave. The result, as seen on the oscilloscope pattern in Figure 16–25, would be a pulsing DC current, called **half-wave rectification** (Figure 16–25).

Rectifying the negative half of a simple AC sine wave requires four diodes (Figure 16–26). The direction of current flow through the phase windings determines the path of current flow through the diode circuit. The final result is current that flows through the load circuit

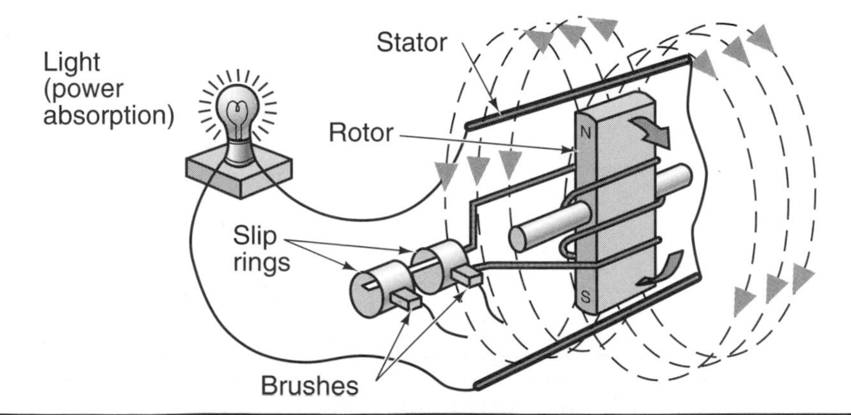

FIGURE 16–22 Simplified alternator showing electromagnetic induction.

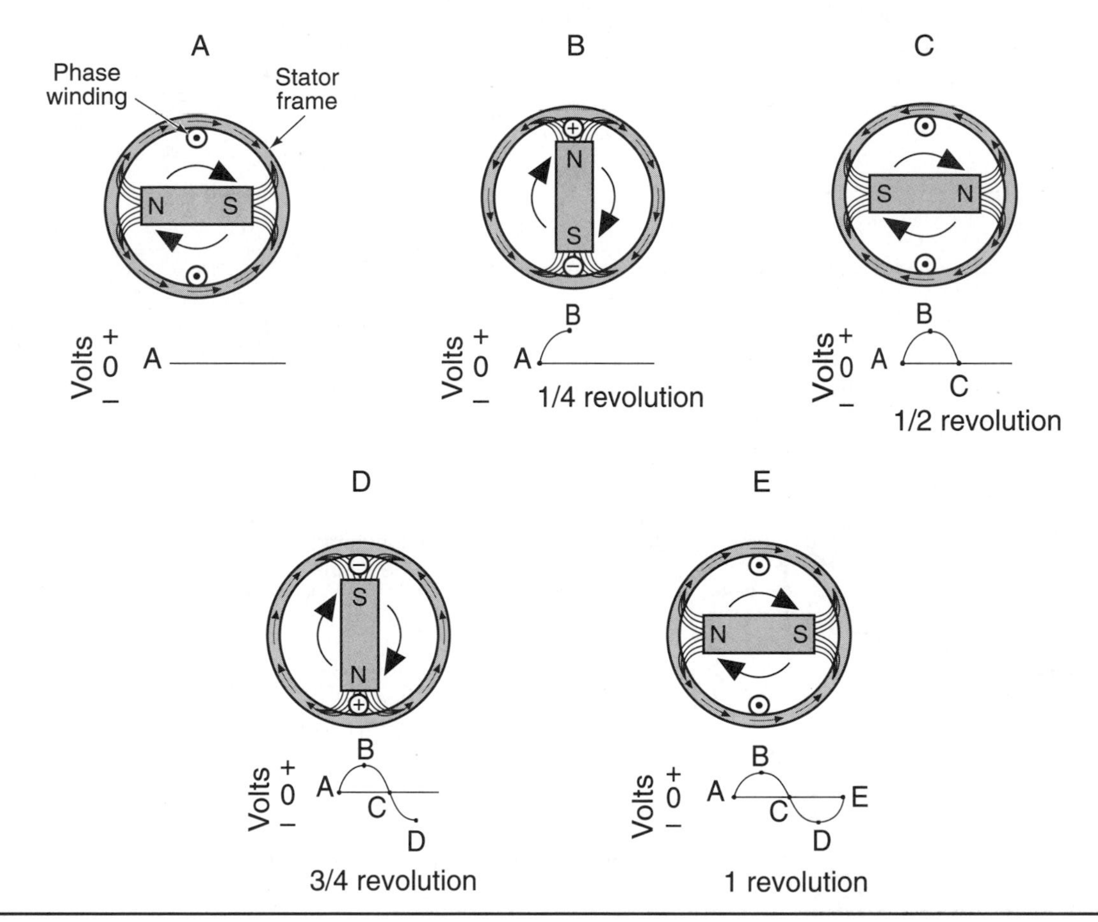

FIGURE 16–23 Producing a simple alternating current (AC) sine wave as the magnetic field is rotated.

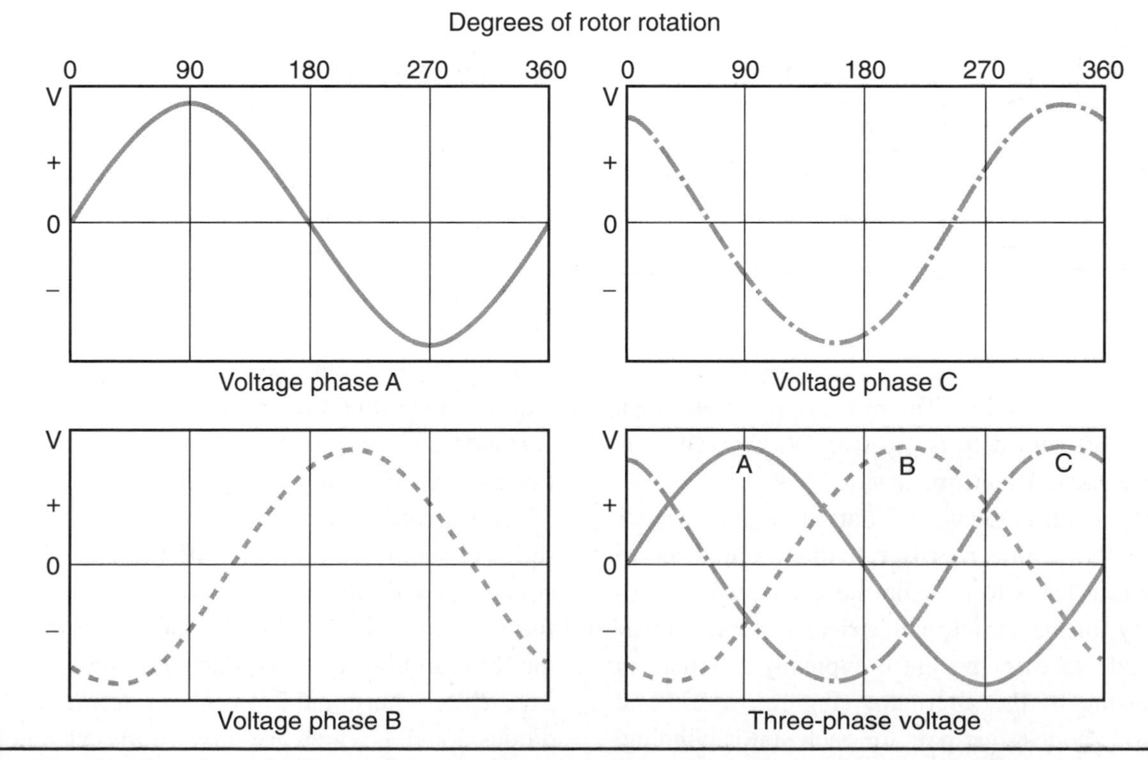

FIGURE 16–24 Overlapping of three stator windings creates a three-phase voltage output.

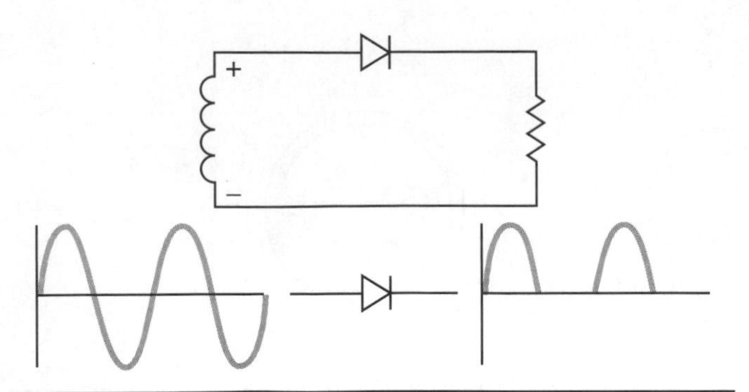

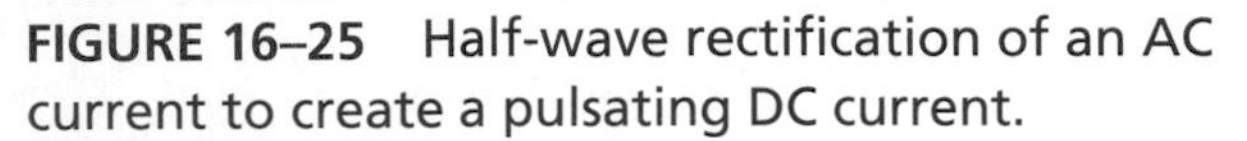

FIGURE 16–25 Half-wave rectification of an AC current to create a pulsating DC current.

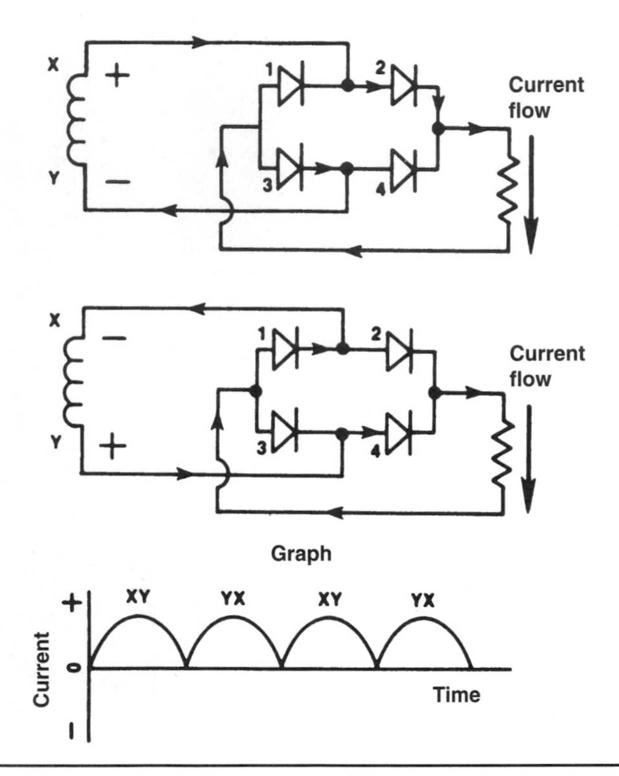

FIGURE 16–26 Full-wave rectification of a simple AC current requires four diodes.

and battery in one direction. The process of converting a complete AC waveform to a positive DC waveform is called **full-wave rectification.**

To rectify the three-phase AC output of an alternator to a DC output, a **diode rectifier bridge** is attached to the stator output leads to provide the constant DC voltage necessary for the vehicle's electrical system and battery. The diode rectifier bridge is typically mounted in the rear housing of the alternator (Figure 16–27) and consists of six diodes, one pair for each stator winding.

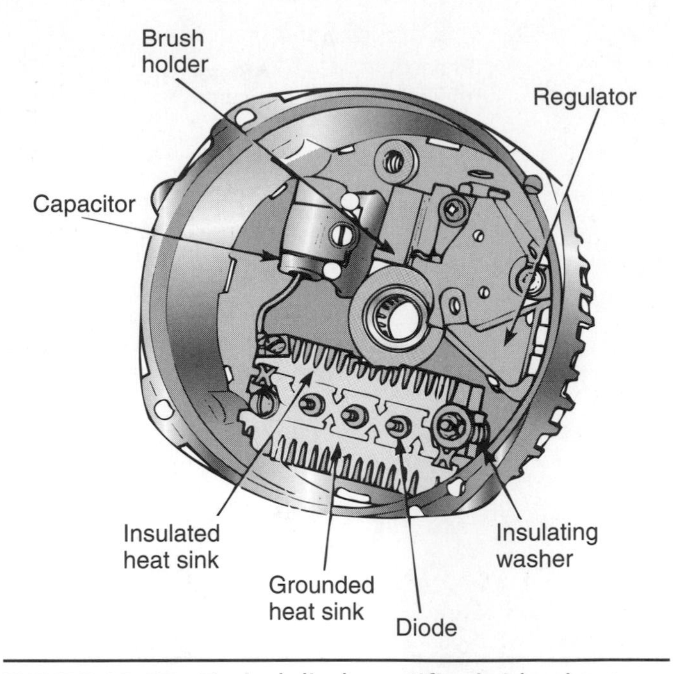

FIGURE 16–27 Typical diode rectifier bridge heat sink located in the rear housing of an alternator.
(Courtesy of Prestolite Electric Corp.)

Three of the diodes are positively biased and are mounted in a finned heat sink (Figure 16–28), and the three remaining diodes are negatively biased and either attached to the heat sink frame mounted inside the rear alternator housing or pressed directly into the rear alternator housing. The **heat sink** absorbs the heat produced by the diodes and dissipates it through the generator frame and by the air flowing through the alternator.

Testing a rectifier bridge consists of checking the six individual diodes for continuity in one direction only. On an ohmmeter suitable for diode testing, each diode shows infinite (O/L) in one direction and low resistance in the opposite direction (Figure 16–29).

Full-wave rectification of a wye wound stator requires a six-diode rectifier bridge, which redirects the current from the stator windings so that all the current flows in one direction through the battery (Figure 16–30). When the rotor turns, two stator windings are aligned in series to supply voltage, and the third winding is always connected to diodes that are reversed biased. Current can follow six possible paths, depending on the position of the rotor. In Figure 16–31, current is supplied to the positive battery terminal from stator terminal A through diodes 2 and 3, which are forward biased, and then back

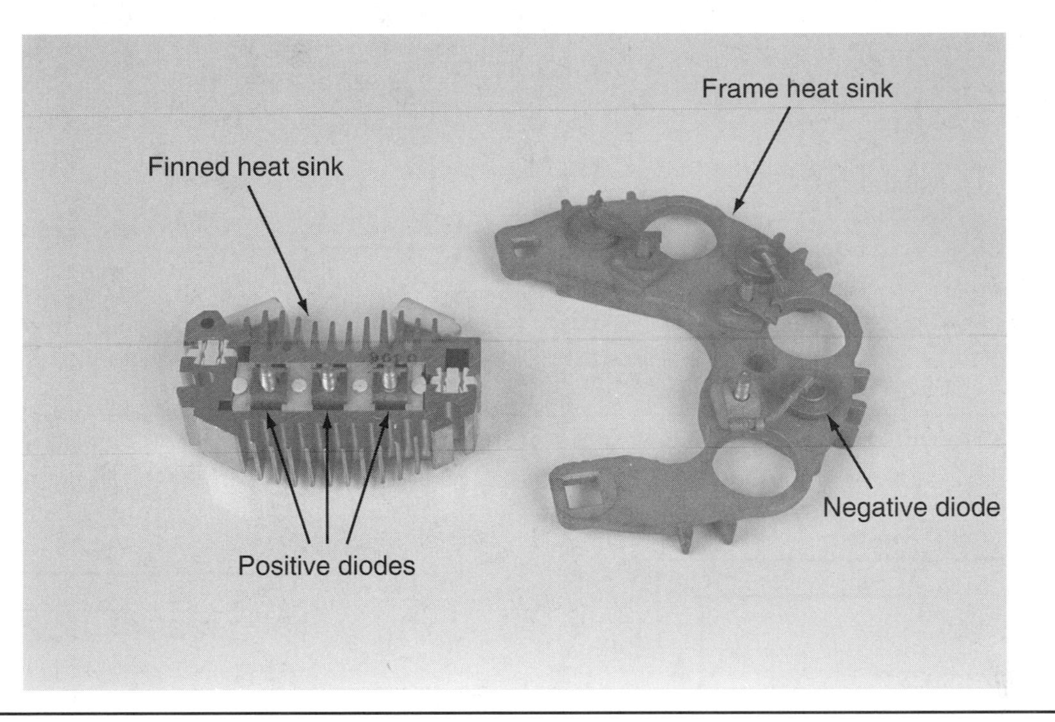

FIGURE 16–28 Typical rectifier assembly consisting of positive and negative diodes and heat sinks.

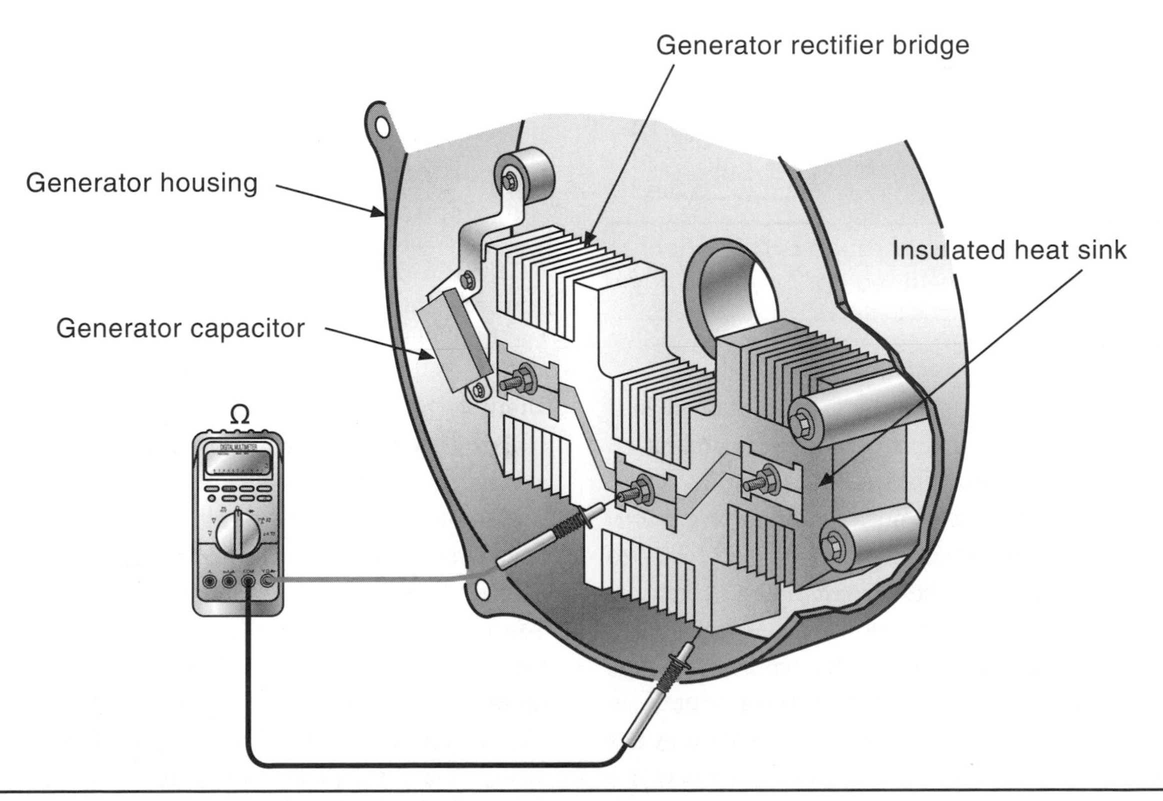

FIGURE 16–29 Testing the rectifier diodes with an ohmmeter.

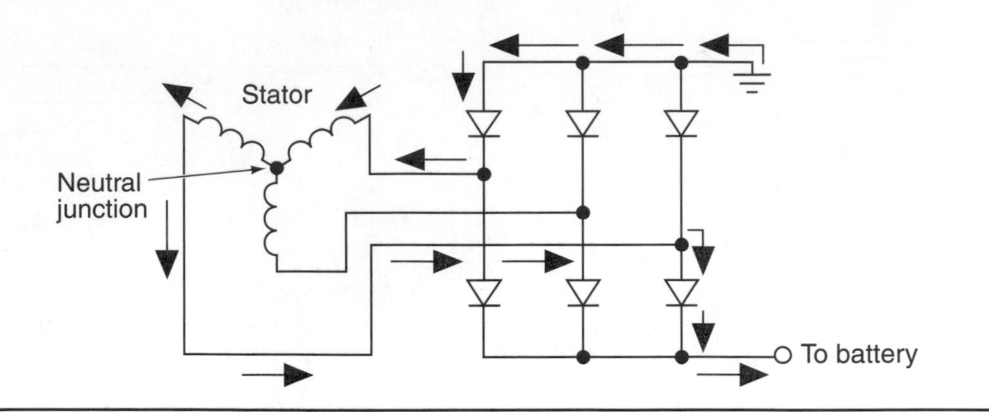

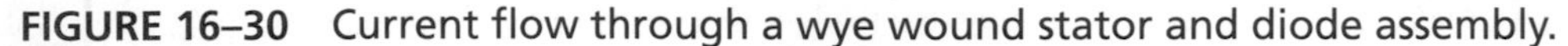

FIGURE 16–30 Current flow through a wye wound stator and diode assembly.

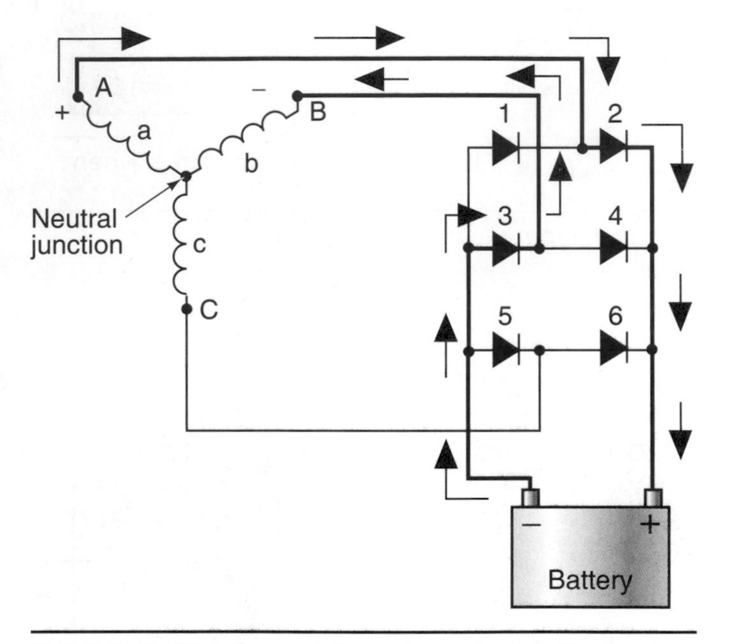

FIGURE 16–31 Current flow through circuit when stator terminals A and B are lined up.

FIGURE 16–32 Current flow through circuit when stator terminals A and C are lined up.

to stator terminal B to complete the circuit. Stator winding C does not produce current because it is connected to diodes that are reversed biased.

This process continues through the other five possible current paths before it repeats itself (Figure 16–32 to Figure 16–36). Notice that the current flow through the battery is always in the *same* direction.

Full-wave rectification in a delta wound stator is accomplished with the same diode rectifier bridge as in a wye wound stator, but all the stator outputs are in use at the same time (Figure 16–37). The three outputs of the delta stator are in parallel; so current flows from each winding all the time, thus increasing the alternator output current potential compared to the wye wound alternator.

Some older designs of alternators used a **diode trio,** which rectifies the current from the stator to create the needed magnetic field excitation in the rotor field coil. To excite the field coil, the voltage on the anode side of the diode trio (Figure 16–38) must be at least 0.6 volt more positive than the cathode side. When the ignition switch is turned on, the current load of the warning lamp acts as a small magnetizing current because it preexcites the field coil until the rotor turns (Figure 16–39). Each

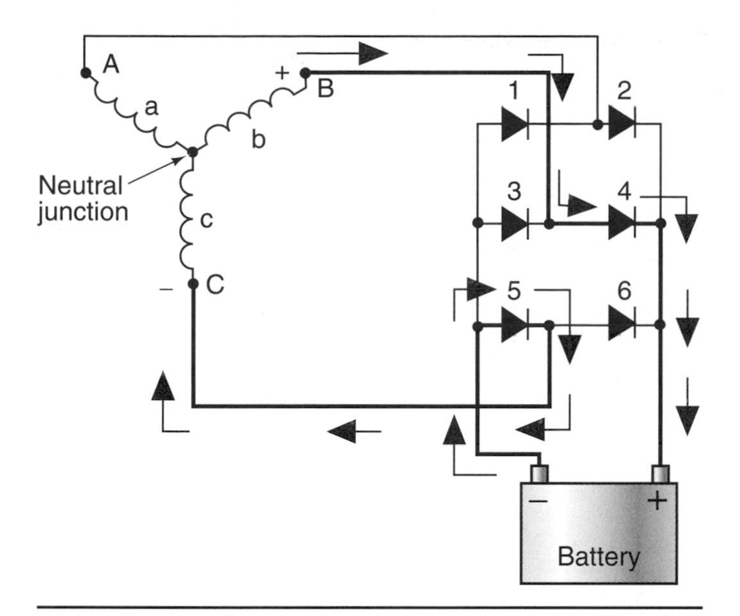

FIGURE 16–33 Current flow through circuit when stator terminals B and C are lined up.

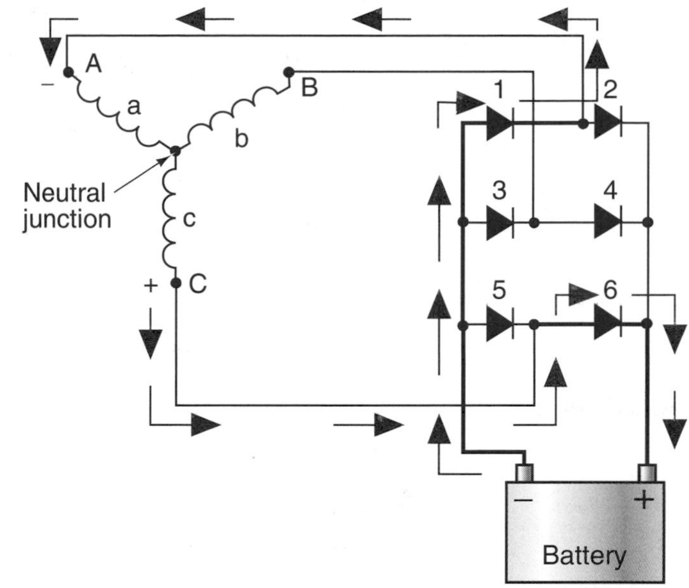

FIGURE 16–35 Current flow through circuit when stator terminals C and A are lined up.

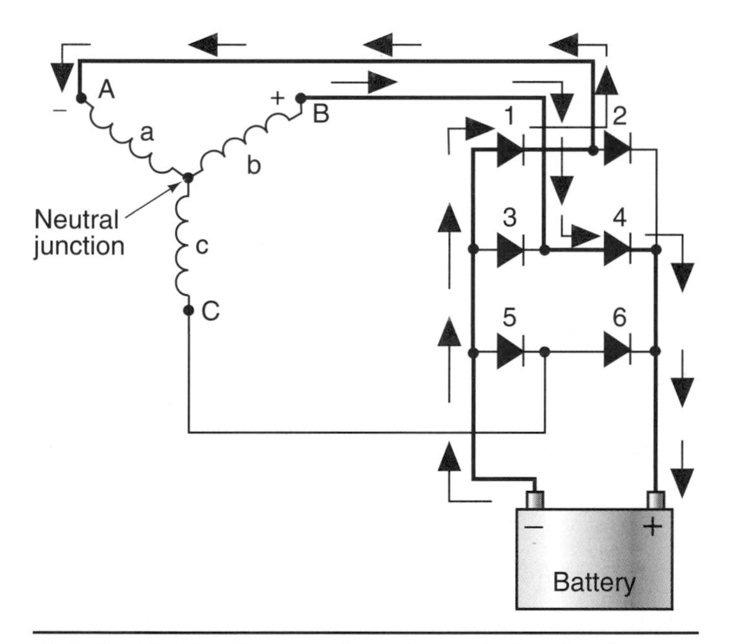

FIGURE 16–34 Current flow through circuit when stator terminals B and A are lined up.

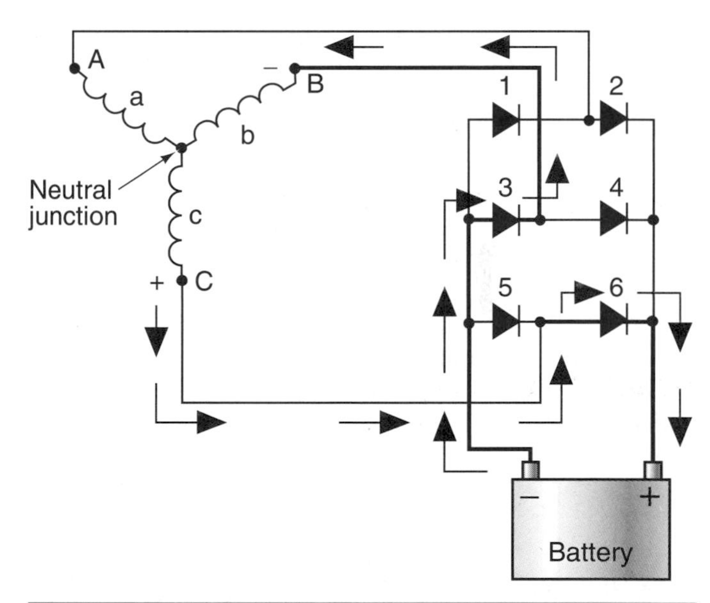

FIGURE 16–36 Current flow through circuit when stator terminals C and B are lined up.

stator output is connected to a diode in the diode trio, which rectifies the current entering the voltage regulator, in turn controling the strength of the magnetic field in the rotor.

Testing a diode trio for opens and shorts is similar to testing a diode rectifier bridge. Disconnect all wires to the diode trio and connect one lead of an ohmmeter to the single large connector and the other lead to each of the small diode connectors (Figure 16–40). Record the ohmmeter reading while it is connected to each small lead before repeating with the ohmmeter leads reversed. The ohmmeter should read less than 300 ohms in one

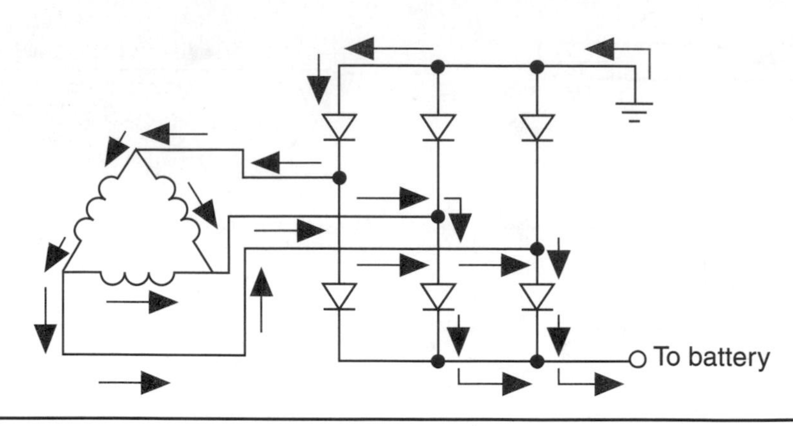

FIGURE 16–37 Current flow through a delta wound stator and diode assembly.

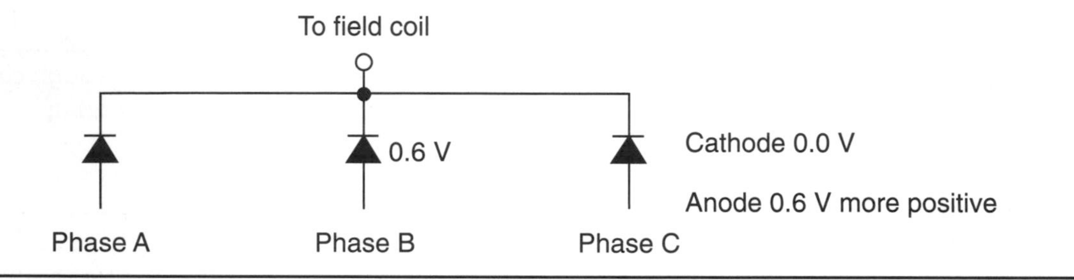

FIGURE 16–38 A typical diode trio connects the stator phase windings to the field coil to create field excitation.

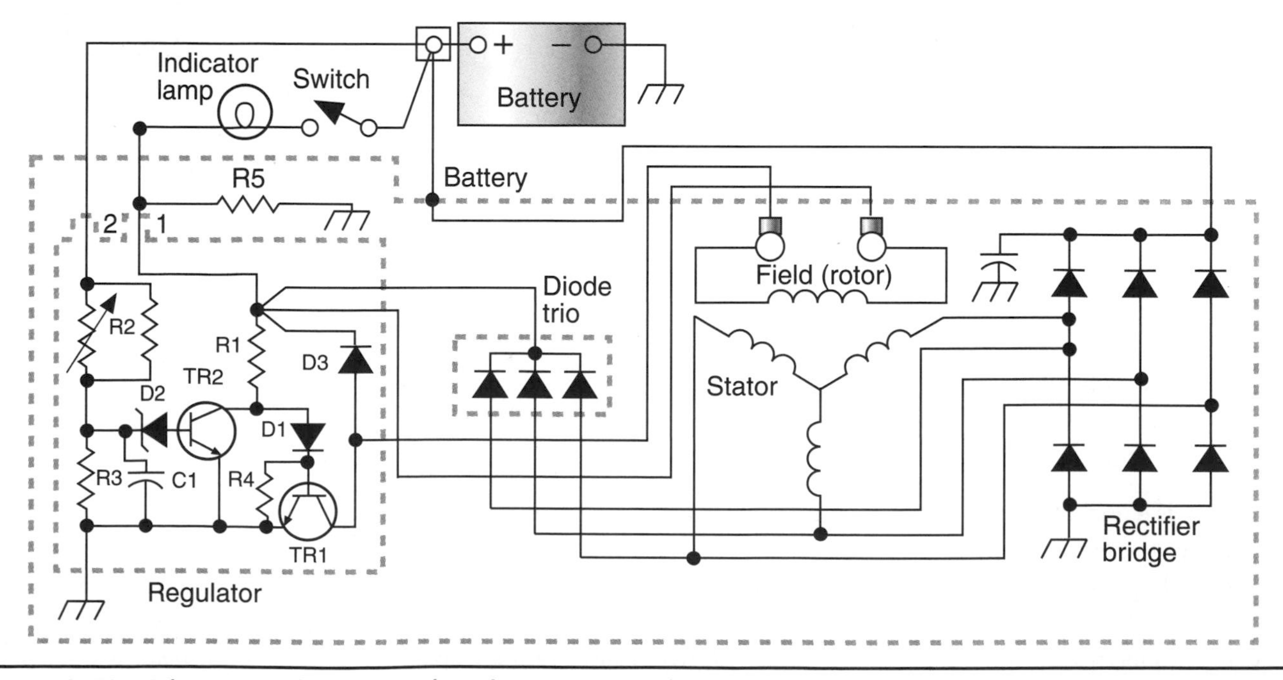

FIGURE 16–39 Schematic of an older charging system with a diode trio.

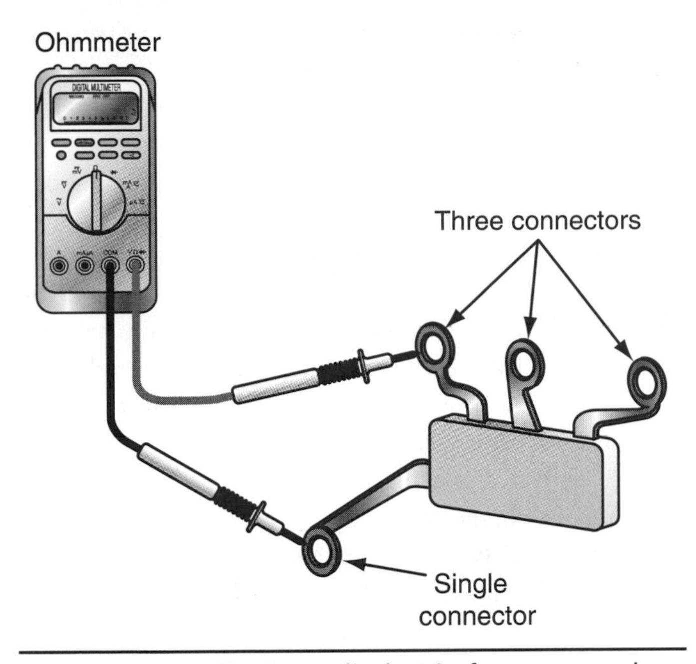

FIGURE 16–40 Testing a diode trio for opens and shorts with a suitable ohmmeter.

direction and O/L in the other direction. A diode trio that is not within specifications should be replaced. A digital multimeter (DMM) with a *diode check* function may also be used for this test.

Many newer alternators use a delta wound stator assembly; the field current is directly applied to the stator, eliminating the need for a diode trio. Depending on the charging load demands, the ground circuit for the field windings is pulsed output, controlled by the PCM at a fixed frequency of around 400 cycles per second (Figure 16–41).

Charging System Regulation

A vehicle's electrical system and battery must be protected from unstable or excessive charging system voltage. Computer controlled circuits are especially sensitive to voltage fluctuations that may develop as the alternator

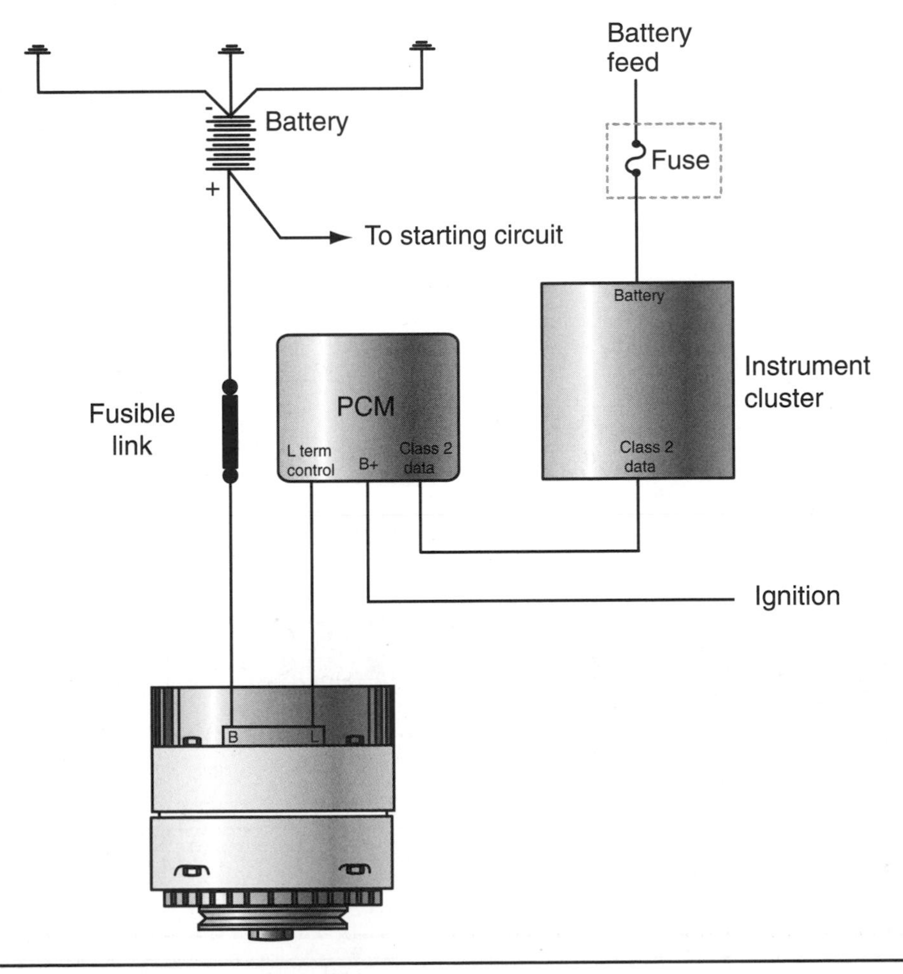

FIGURE 16–41 Controlling field current strength by the PCM.

adjusts for varying electrical load demands while keeping the battery fully charged. If the charging current is lower than the vehicle's electrical demands, the battery gradually runs down over time. If the charging current is higher than needed, the battery or electrical system components could be damaged.

As shown in Figure 16–42, the **voltage regulator** controls and stabilizes the charging system voltage by (a) controlling the resistance in series to the rotor field coil or (b) toggling the field circuit very rapidly between open and closed (Figure 16–42). The amount of regulated **field current** going to the rotor coil varies depending on

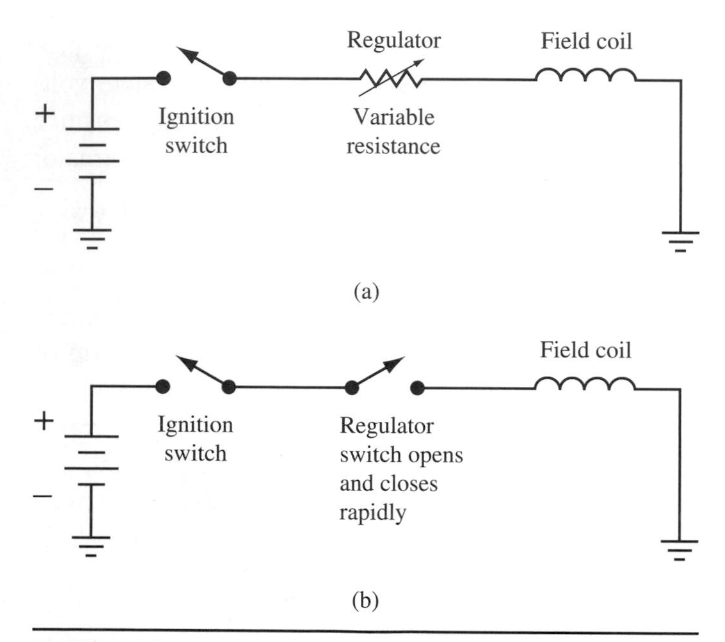

FIGURE 16–42 The regulator controls the rotor field by (a) controlling the resistance in the series to the rotor coil or (b) toggling the field circuit on and off very rapidly.

system voltage needs. Voltage regulators use a **sensing voltage** input from the battery, interpret the voltage, and then regulate the alternator output voltage between 13.5 and 14.5 volts.

In addition to the sensing voltage input, another input that affects regulation is temperature. Regulators are temperature compensated because ambient temperature influences the rate of charge that a battery can accept. Figure 16–43 shows the relationship between temperature and charge rate. Batteries are more reluctant to accept a charge at lower ambient temperatures; so the regulator compensates for this by increasing the system voltage until it is at a high enough level that the battery will accept a charge quickly.

Three basic types of field circuits are used in charging systems. Knowledge of each type helps you in diagnosing basic charging system problems.

1. The first type of field circuit is called the **A-circuit,** which has the regulator on the ground side of the field coil (Figure 16–44). It is also referred to as an *externally grounded field circuit.* The B+ used for the field coil is picked up from within the AC

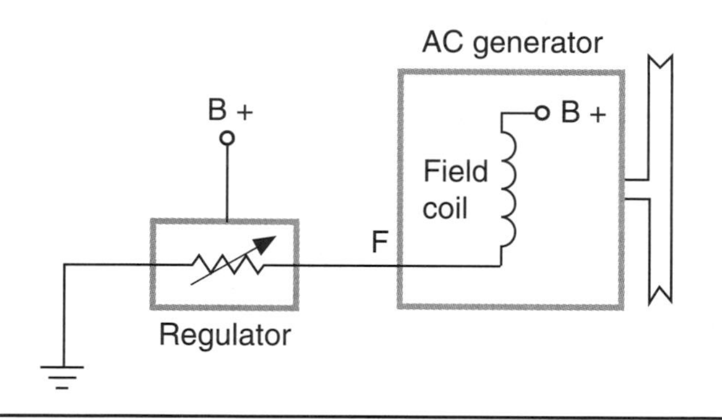

FIGURE 16–44 A simplified A-circuit style field circuit.

	Volts	
Temperature (°F)	Minimum	Maximum
20	14.3	15.3
80	13.8	14.4
140	13.3	14.0
Over 140	Less than 13.3	–

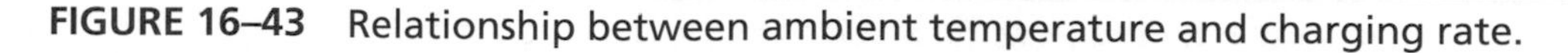

FIGURE 16–43 Relationship between ambient temperature and charging rate.

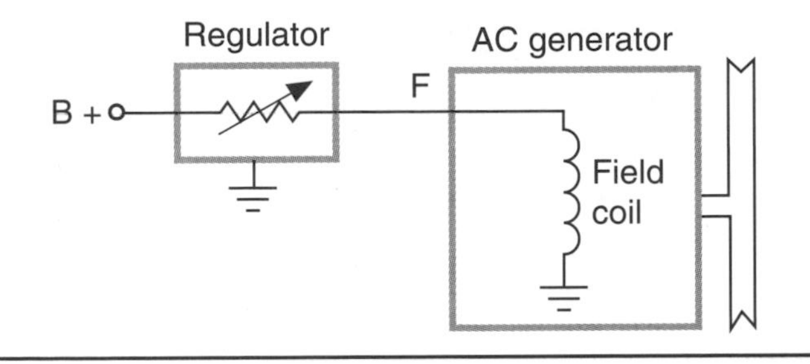

FIGURE 16–45 A simplified B-circuit style field circuit.

generator assembly. The regulator is located in series on the ground side and controls the field current by varying the current flow to ground.

2. The second type of field circuit, the **B-circuit,** has the regulator controlling the power input side of the field circuit (Figure 16–45). It is also referred to as an internally grounded circuit. The field coil completes the circuit by being grounded within the AC generator assembly.

3. The third type of field circuit is called an **isolated field circuit** because it picks up both B+ and ground externally. It can be easily identified by the *two* field wires attached to the outside of the alternator case. Isolated field circuit alternators can have the regulator located on A-circuit (ground) side or the B-circuit (positive) side (Figure 16–46).

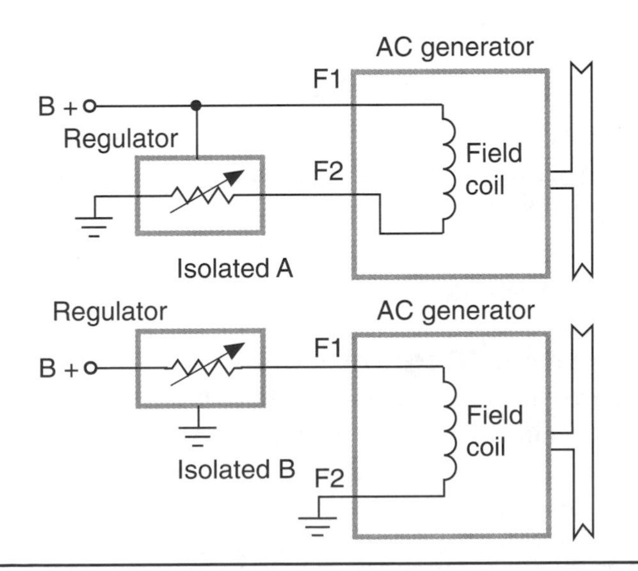

FIGURE 16–46 In an isolated field circuit for an AC generator, the regulator can be installed on either side of the rotor field coil.

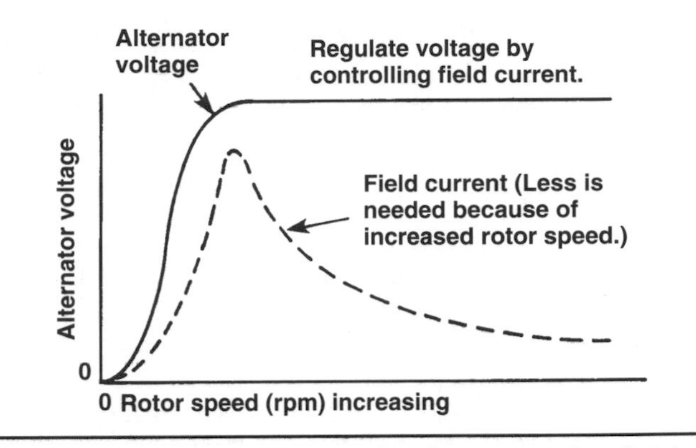

FIGURE 16–47 The relationship between the field current, rotor speed, and regulated voltage changes depending on the vehicle electrical loads.

Regardless of the type of field circuit, the alternator is designed to control the amount of current flowing through the field windings. There is a relationship between the field current, rotor speed, and the regulated voltage (Figure 16–47). As the rotor speed increases, the field current must be reduced to maintain a regulated voltage.

Electromechanical Regulators Very common on most older vehicle charging systems and on farm or industrial machinery, an external electromechanical regulator with two vibrating electromagnetically operated contact point relays is used to control the charging circuit (Figure 16–48). One relay coil is the field relay with two contact points and the other is the voltage regulator with three contact points (Figure 16–49).

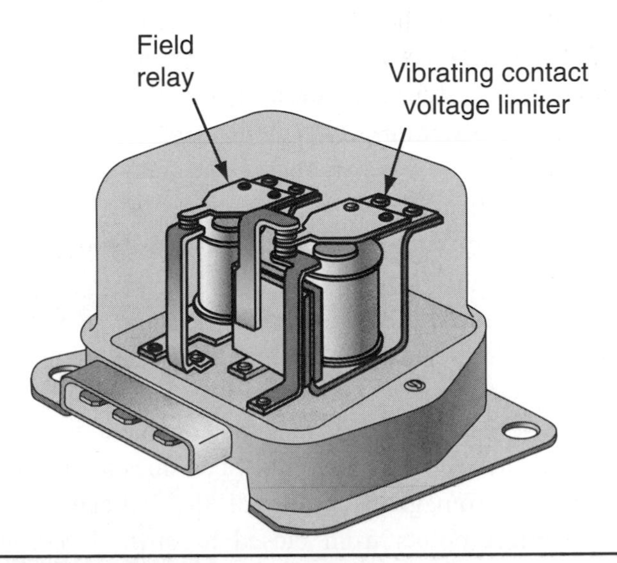

FIGURE 16–48 An external electromechanical contact point regulator.

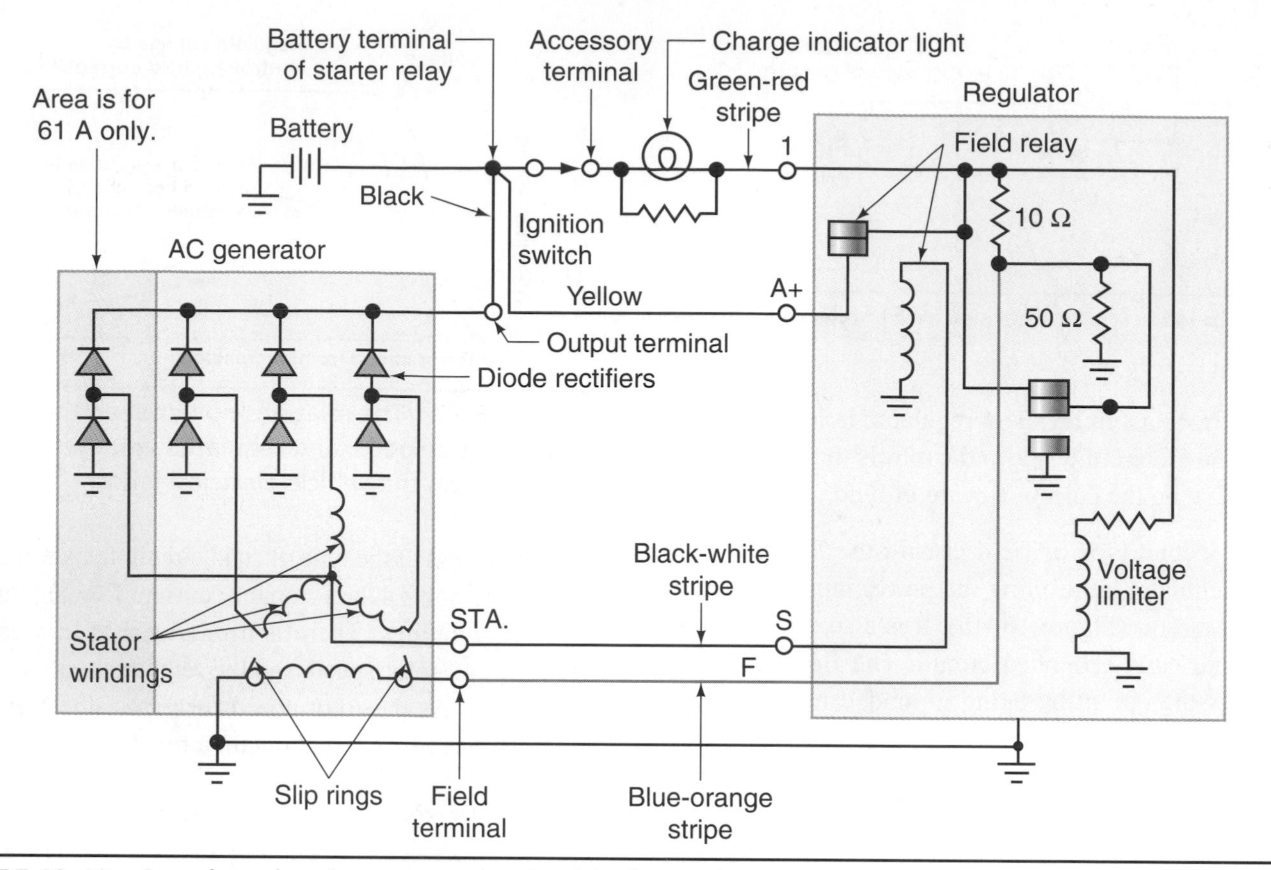

FIGURE 16–49 Complete charging system circuit with electrochemical regulator.

In Figure 16–49, when the ignition switch is closed, current flows through the resistor, illuminating the indicator lamp, then through the lower contacts shown in the regulator section, out the F terminal, through the alternator field coil, and then to ground. When the engine has started and the alternator rotor is turning, system voltage may be below 13.5 volts; however, some voltage is produced at the S terminal, which energizes the field relay. When the field relay is energized, the field relay points close magnetically, and the indicator bulb goes out because battery voltage is now present on both sides of the light circuit.

At the same time, the voltage regulator relay—also known as the **voltage limiter** relay—automatically determines the field circuit voltage for the required amount of charging. When battery voltage is below 13.5 volts, the voltage limiter relay coil does not produce enough electromagnetism to pull the voltage limiter relay contact points from closed to open. Maximum charging conditions are allowed when the voltage limiter relay points are closed. As the battery charges and its voltage rises, the increased voltage strengthens the electromagnetic field of the voltage limiter relay coil and causes the voltage limiter contact points to open, grounding the field circuit. The contact points constantly vibrate open and closed several times per second to maintain the system voltage output between 13.5 and 14.5 volts.

After the engine shuts off, there is no current flow from the STA terminal; so the field relay coil is deenergized. This opens the points and prevents the battery from discharging.

Electronic Regulators Modern vehicle charging systems use an electronic regulator to provide precise control of the field current. All of the control circuitry and components are located on a single silicon chip sealed in a plastic module normally located inside the alternator (Figure 16–50). The regulator has no moving parts, and it can cycle several thousand time per second, if necessary, for accurate control of the field circuit voltage.

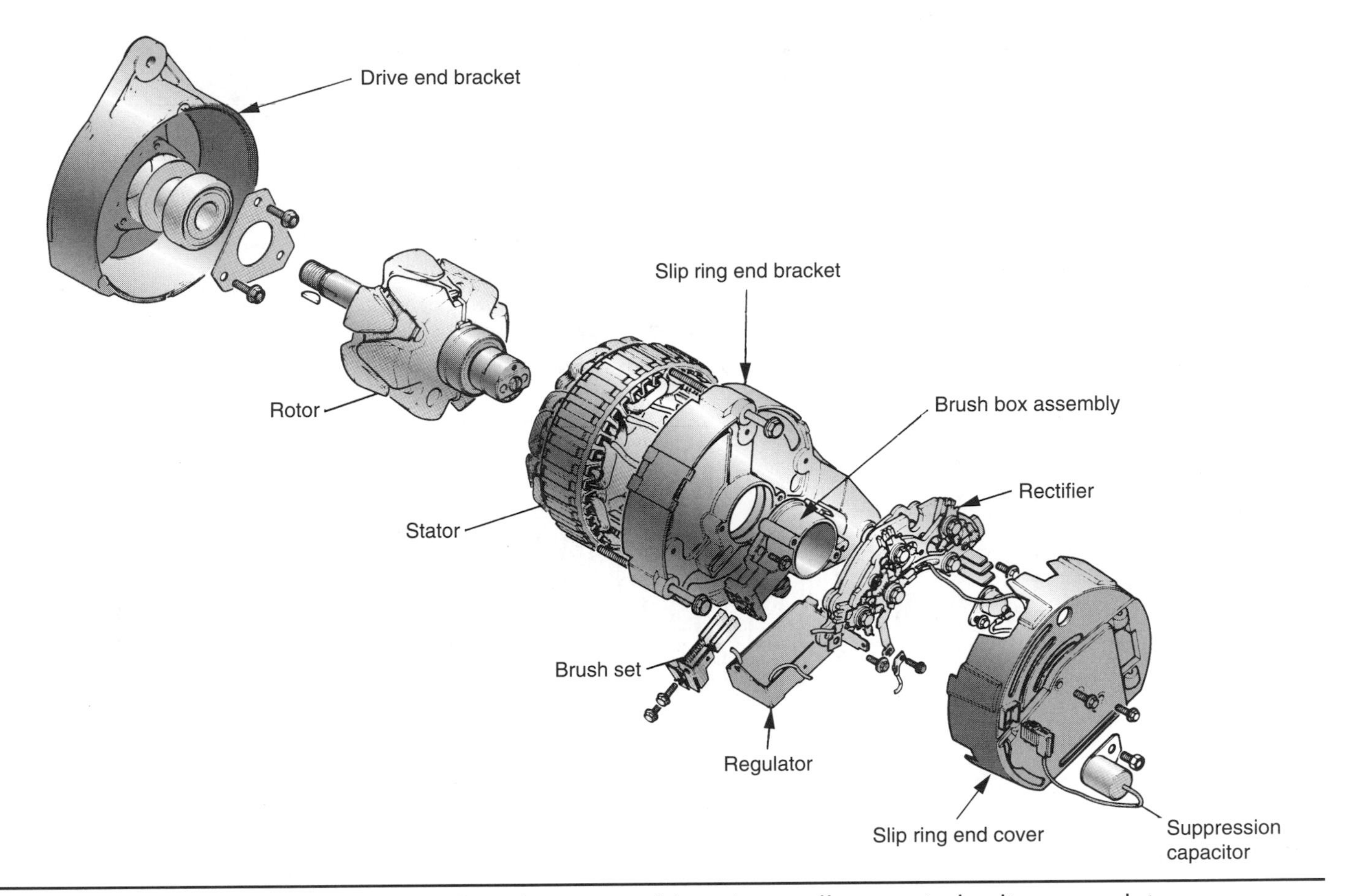

FIGURE 16–50 Component locations of an alternator with an internally mounted voltage regulator. (Courtesy of Lucas Aftermarket Operation)

The on/off cycling of the field circuit is an example of **pulse-width modulation.** The cycling modulation controls the alternator output by varying the amount of time the field coil is energized based on the vehicle's electrical demand. For example, if a 100-ampere alternator on a vehicle had an electrical system demand of 75 amperes, the regulator would energize the field coil 75% of the cycle (Figure 16–51). The period of time for each cycle does not change; only the amount of ON time in each cycle changes. The percentage of ON time relative to the total cycle time is known as the **duty cycle.** In part (b) of Figure 16–51, the pulse-width modulated output is continually varying depending on the electrical load. When averaged, this output provides a precise voltage input for an electrical device.

In an electronic regulator, one of the main components is a zener diode (Figure 16–52), which is similar to the diodes in other electronic circuits except that it has the ability to conduct current in the reverse direction without damage. The zener diode in an electronic regulator is designed to conduct in reverse once a predetermined voltage of 13.5–14.5 volts has been achieved. In Figure 16–53, the zener diode has B+ on the anode side—the side with the arrow—and no current flowing through the diode. When the upper design voltage limit (approximately 14.5 volts) is reached, the zener diode is said to break down, allowing a voltage signal to occur at the base of transistor 1. This turns on transistor 1 and switches off transistor 2, which is in control of the field current and alternator output. Transistor 2 modulates the field circuit on and off as many time per second as necessary to maintain the designed voltage of the system. The thermistor shown in the upper left of Figure 16–53 changes circuit resistance according to temperature changes, thus providing for the temperature-related voltage change necessary to keep the battery charged in cold weather conditions.

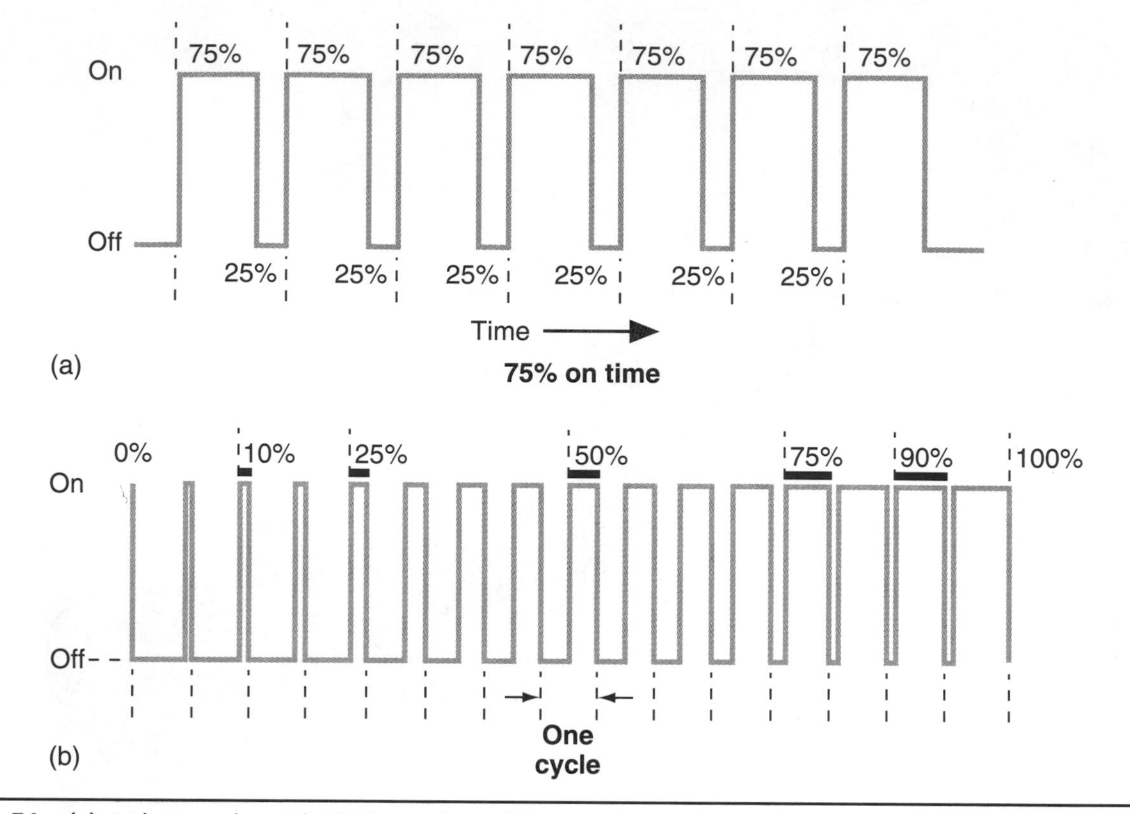

FIGURE 16–51 (a) A duty cycle with 75% on time; (b) a varying pulse-width modulated duty cycle.

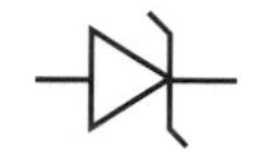

FIGURE 16–52 Zener diode symbol.

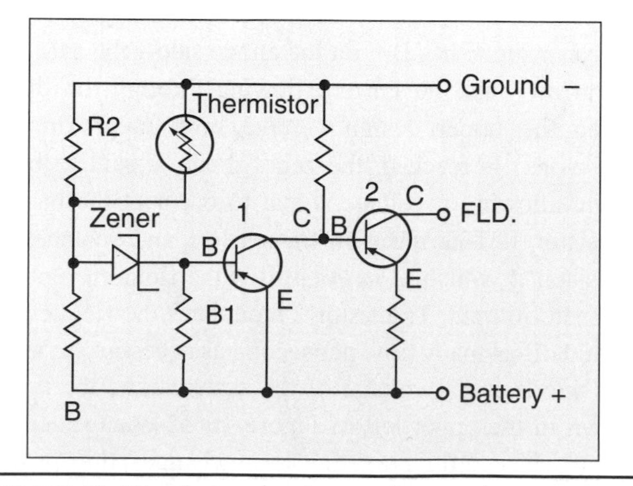

FIGURE 16–53 A simple electronic transistor regulator circuit using a zener diode.

An application of a zener diode in the electronic regulator of a charging system is shown in Figure 16–54. This particular system has only two input connections, located in the upper left of Figure 16–54. Input 1 receives B+ that comes through the ignition switch and through the indicator lamp and resistor that are in parallel. Input 2 is for connecting the sensing circuit in the regulator to the battery.

When the ignition switch is turned on, B+ is available at input 1 and at the field circuit through the common battery connection at R1. Transistor TR1 is on and conducts the field current from the field coil to ground, producing a weak magnetic field in the field coil. When the alternator begins to produce a current output, the diode trio conducts, and B+ is available for the field and for terminal 1 at the common battery connection at R1. This turns off the indicator lamp because there is B+ on both sides of the lamp.

As the alternator output reaches its designed charging voltage, sufficient voltage at the sensing circuit connected to input 2 is applied to the zener diode so that it breaks down and turns on transistor TR2, which turns off

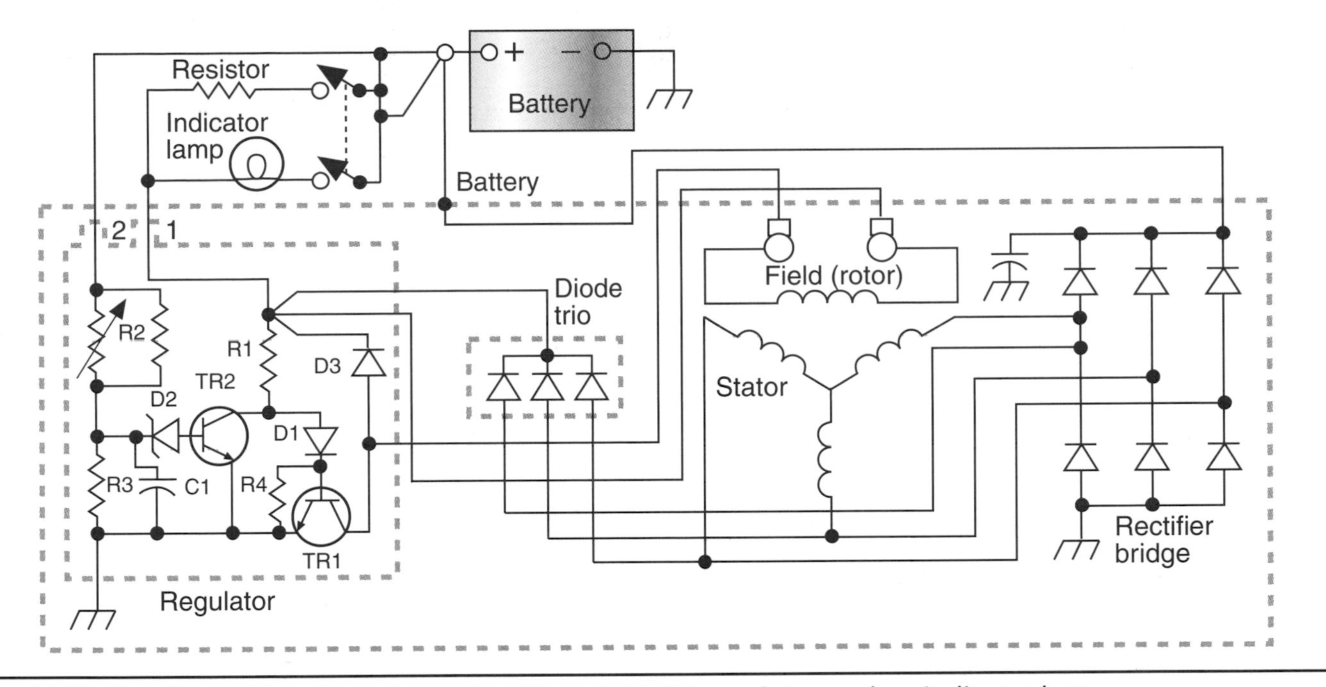

FIGURE 16–54 Charging system schematic with an internal regulator and an indicator lamp.

transistor TR1. This process continuously turns the switching transistor TR1 on and off as the voltage rises above and drops below the designed voltage.

Computer Controlled Regulators On many newer vehicles, the regulator function has been incorporated into the powertrain control module (PCM) (Figure 16–55). The operation of the PCM-controlled regulation is designed to pulse the field current ON and OFF at a fixed frequency of about 400 cycles per second. By pulse-width modulating the duty cycle, an average field current is produced to meet the designed alternator output. At high engine speeds with minimum system load requirements, the field circuit ON time may be a little as 10% (Figure 16–56). At low engine speeds with maximum system load requirements, the field circuit ON time may be 75% more.

A computer-controlled regulator has the ability to precisely manage the charging rate according to vehicle electrical requirements, ambient temperature, and various sensor inputs (Figure 16–57). As a result, some of the benefits of computer regulation are:

- A small, lightweight storage battery.
- Reduced magnetic drag on the alternator, resulting in increased engine horsepower.

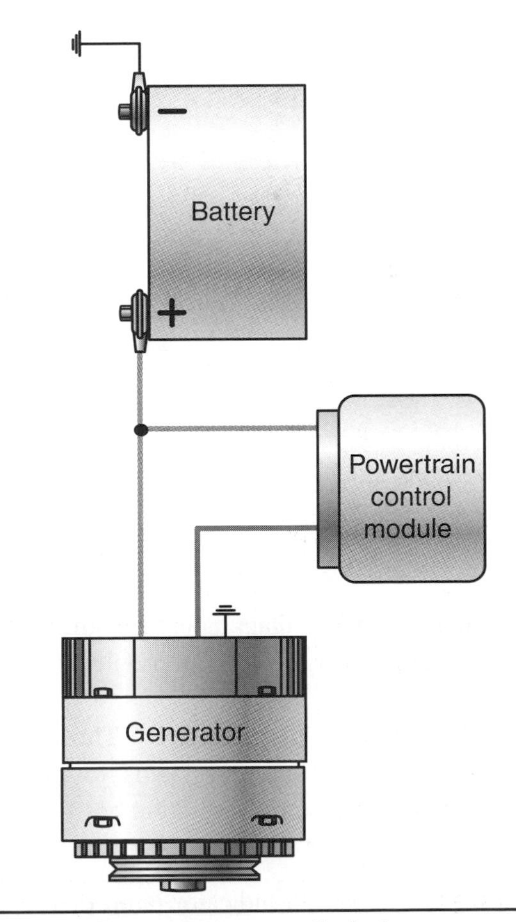

FIGURE 16–55 A PCM-controlled alternator system.

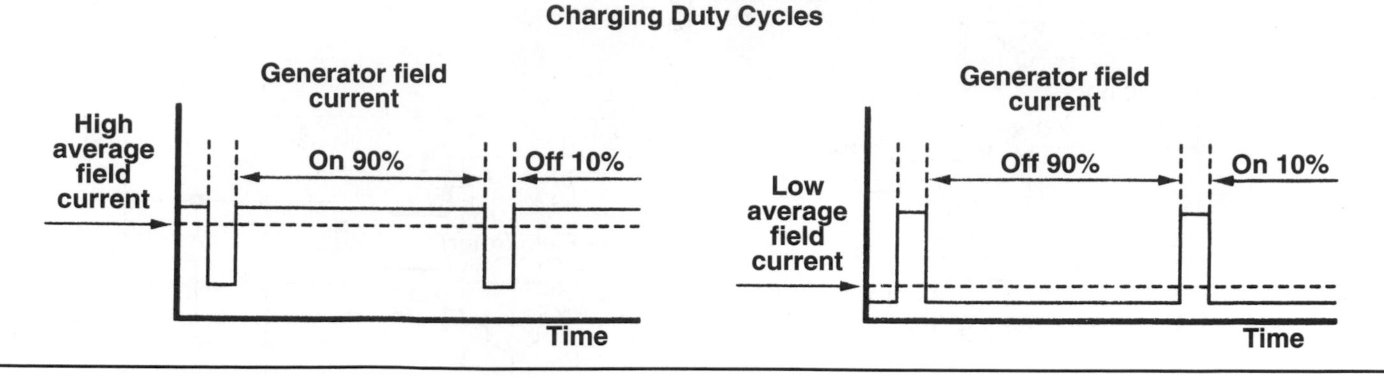

FIGURE 16–56 On/off duty cycling of the alternator.

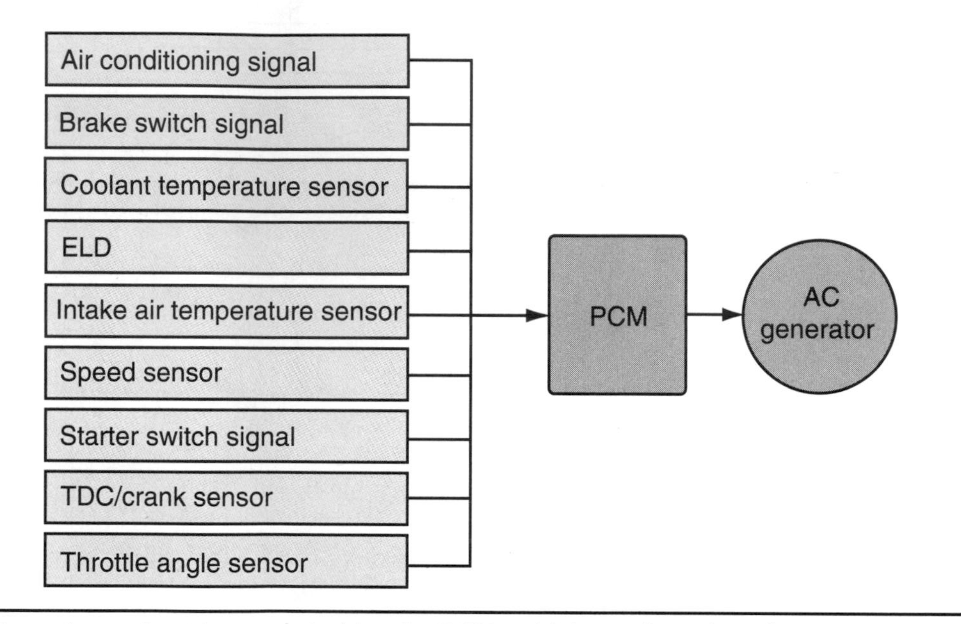

FIGURE 16–57 Alternator output is regulated by the PCM, which receives data from various input sensors.

- Increased gas mileage.
- Reduced rough idle speed problems caused by parasitic voltage loss.
- Use of the computer's diagnostic capabilities in troubleshooting charging system problems.

Charging System Indicators Three common methods are used in charging systems to inform the driver of the charging system's condition or of a potential problem.

1. The first method is an indicator lamp that operates, in most circuits, on the basis of opposing voltages (Figure 16–58). If the alternator output is less than battery voltage, there is an electrical potential difference in the lamp circuit and the lamp lights. If the output voltage is equal to the battery voltage, then the equal voltage potentials result in the lamp's not lighting (Figure 16–59). Figure 16–60 is an example of an electronic regulator that uses an indicator lamp to warn of a charging system problem. If there is no stator output through the diode trio, then the indicator lamp circuit is completed to ground through the rotor field and transistor TR1.
2. The second method is an ammeter wired in series between the alternator and battery (Figure 16–61).

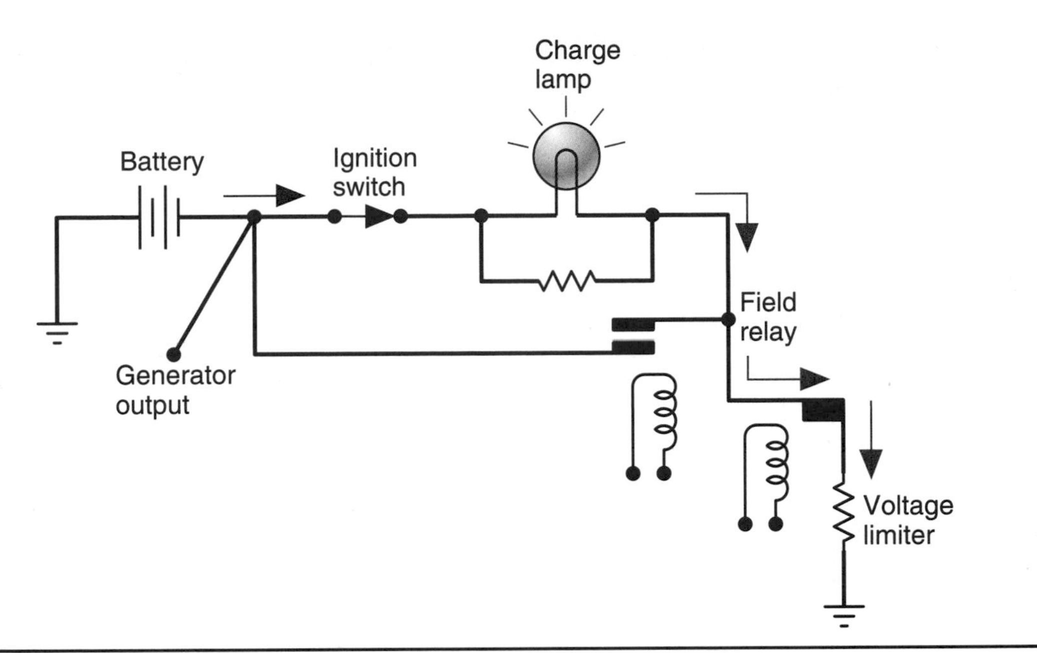

FIGURE 16–58 Electromechanical regulator indicator lamp is on when there is no alternator output.

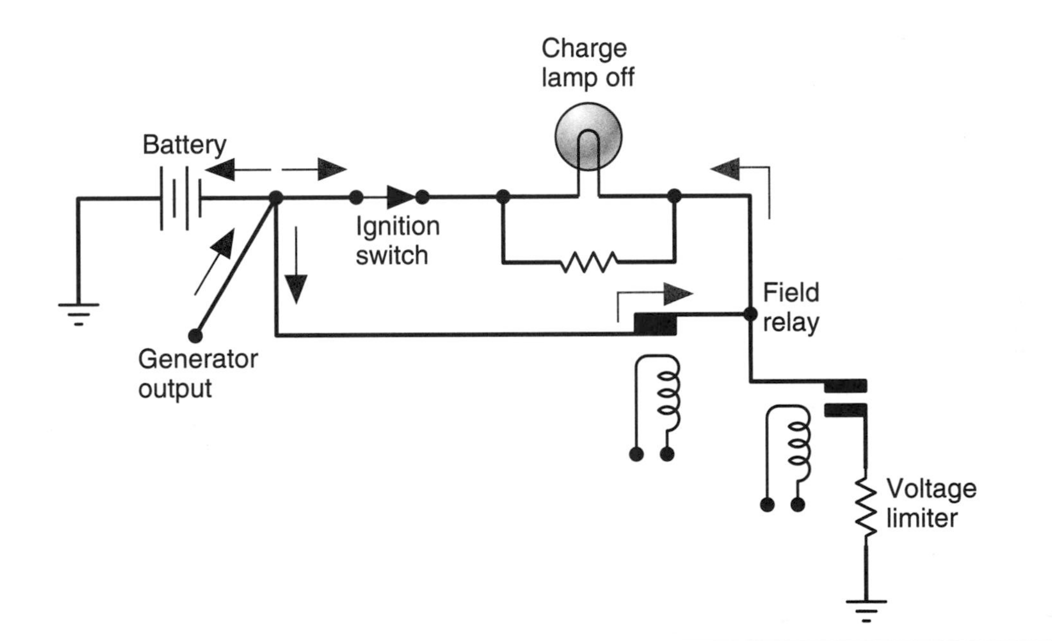

FIGURE 16–59 Electromechanical regulator indicator lamp is off when there is alternator output.

The ammeter is normally mounted in the dash and gives the driver more information compared to an indicator lamp (Figure 16–62). If the charging system is not generating sufficient current to keep up with vehicle electrical load demands, or if the charging system is inoperative, the needle points to discharge. After the engine initially starts or the electrical load demand is high, the gauge needle indicates a high rate of charge until designed system voltage is maintained.

3. The third method is a voltmeter connected in the battery electrical feed circuit to monitor system operation (Figure 16–63). With the ignition switch on

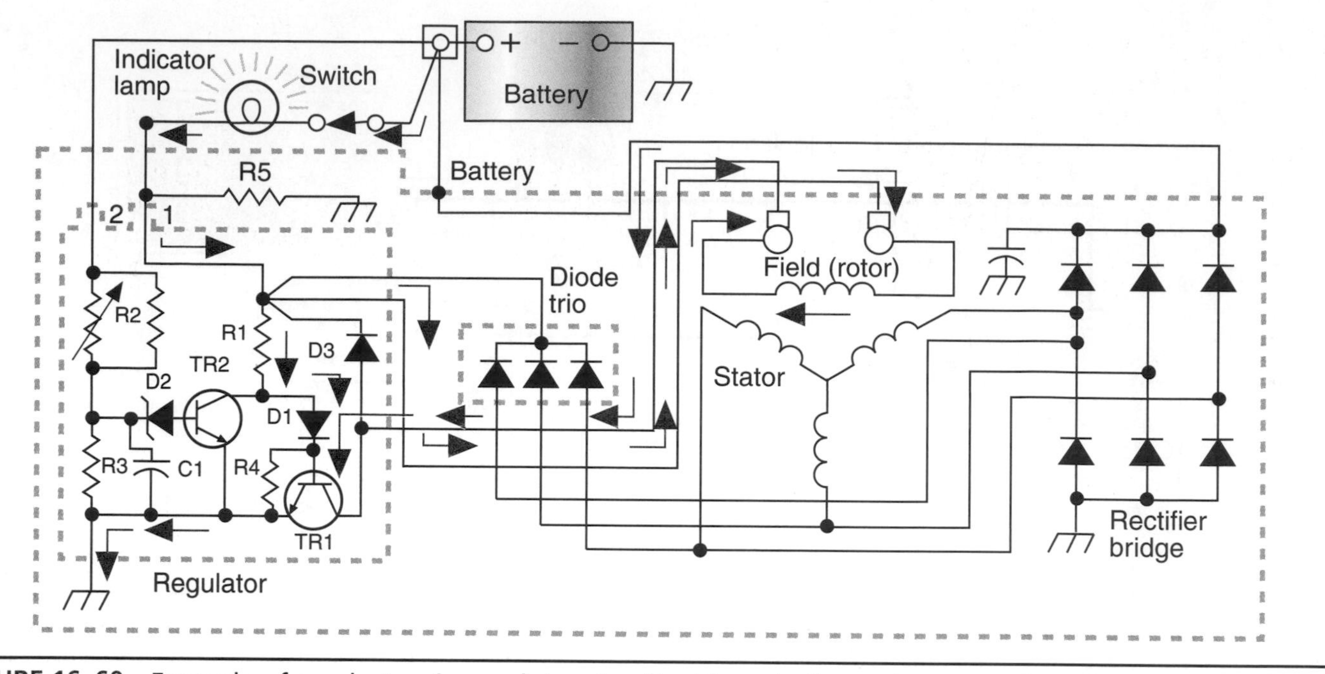

FIGURE 16–60 Example of an electronic regulator circuit with an indicator lamp on when there is no alternator output.

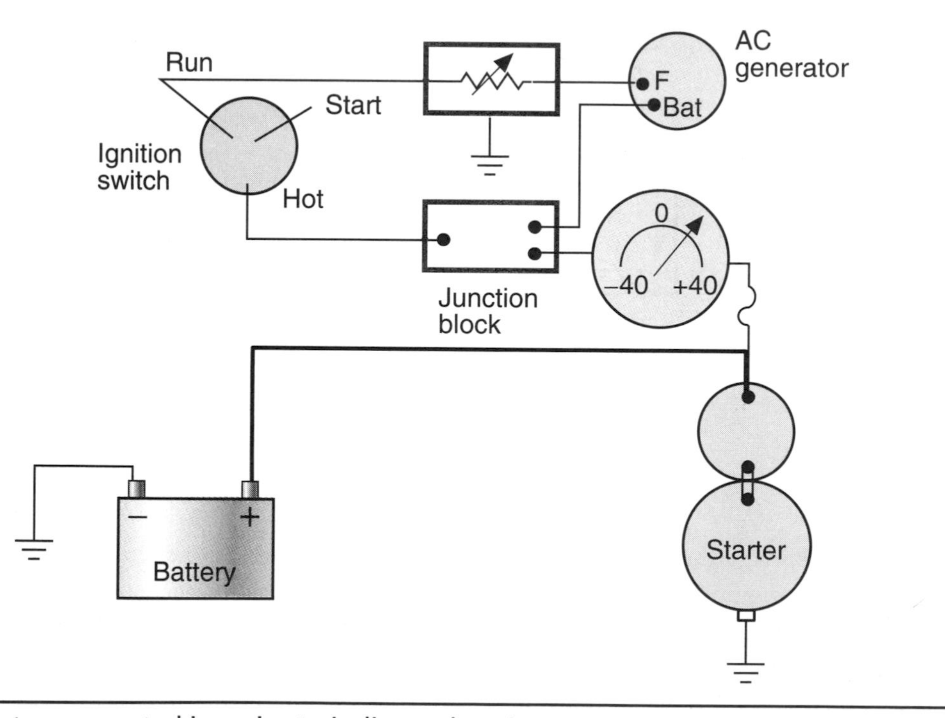

FIGURE 16–61 Ammeter connected in series to indicate charging system operation.

and the engine off, the voltmeter should show the voltage available to the electrical circuit from the battery—normally around 12.66 volts. When the engine is started, the voltmeter normally indicates a reading between 13.2 and 15.2 volts, depending on the vehicle. If the voltage level is below 13.2 volts, the charging system is not providing sufficient current for the electrical load demands. If the voltmeter indicates a voltage reading above 15.2–15.5 volts, depending on manufacturer, an overcharge condition

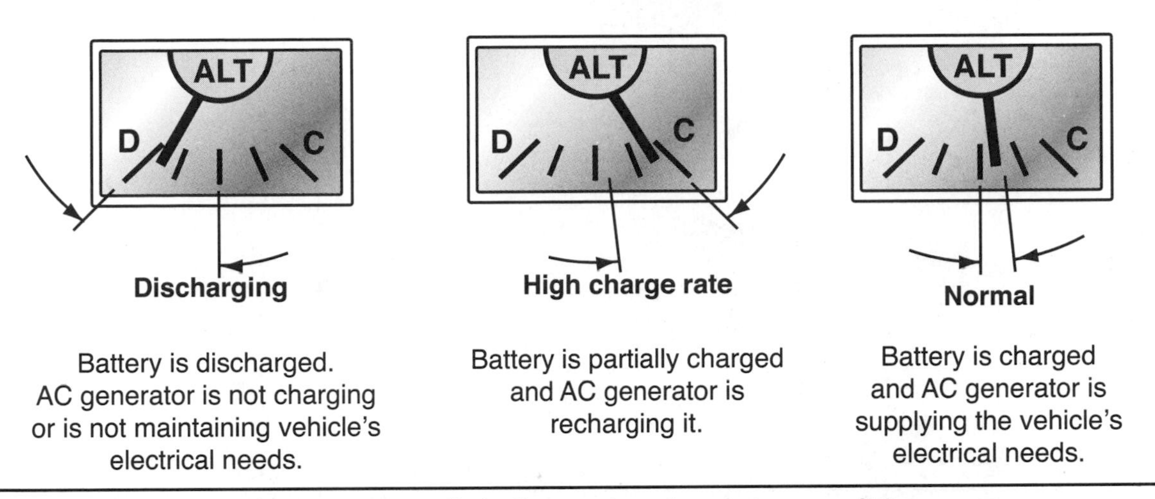

FIGURE 16–62 Dash-mounted ammeter needle indicates charging system conditions.

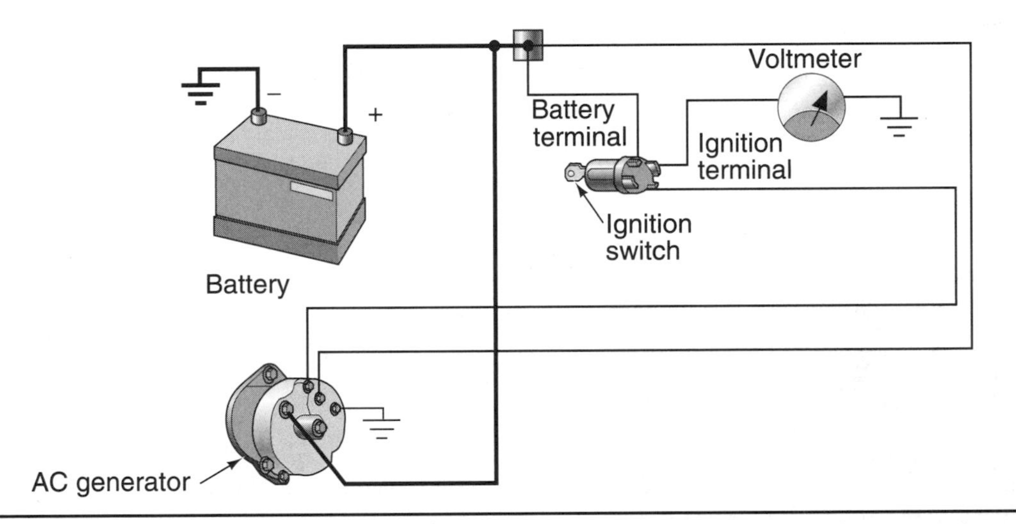

FIGURE 16–63 Voltmeter connected in parallel to the charging circuit to monitor the operation.

is occurring that will damage the battery and sensitive electrical circuit components.

New Alternator Technology

With the recent advancement in hybrid electric vehicle (HEV) technology, several new alternator/generator system designs are available that assist either the HEV in its ability to charge or the engine motor unit under a variety of operating conditions.

Depending on the HEV engine/motor combination, one of the most recent alternator designs to emerge is the integrated starter/generator (ISG), introduced in the last chapter because the starter unit is integrated into the generator to form a combination unit (Figure 16–64). The ISG consists of a three-phase motor unit with a stator assembly attached to the inside of the transmission/engine bell housing and a rotor attached to the engine crankshaft (Figure 16–65). Due to their size and windings, some ISGs can produce in excess of 200 amperes at low engine speeds and over 300 amperes maximum current at full speed. The ISG is designed to perform several functions: assist the engine during acceleration as a motor unit, seamlessly start the engine depending on

FIGURE 16–64 A cutaway view of an HEV electric-assist motor generator used to charge the engine.
(Courtesy of American Honda Motor Co., Inc.)

FIGURE 16–65 Components of an integrated starter/generator assembly (ISG).
(Courtesy of BMW of North America, Incorporated)

power needs, and serve as a generator to recharge the battery packs during deceleration or heavy load conditions.

Another application of an ISG is a belt-driven alternator/starter (Figure 16–66) that mounts in the same location as the typical engine-mounted alternator. In one design of this system, the crankshaft is fitted with an electromagnetic clutch that is engaged and driving the ISG belt when the engine is operating. When the engine shuts off at road intersections to conserve fuel, the electromagnetic clutch disengages from the crankshaft, and the ISG works as a motor to operate the electrical accessories.

Manufacturers continue to develop new charging systems to meet the increased electrical demands placed on them. In the early 1970s, the typical 12/14-volt electrical load on a vehicle was about 40 amperes or 500 watts. The typical alternator maximum output in those days was around 50 amperes or less. Thirty years later the electrical load has increased to over 2,000 watts (Figure 16–67), requiring an alternator with a capacity of

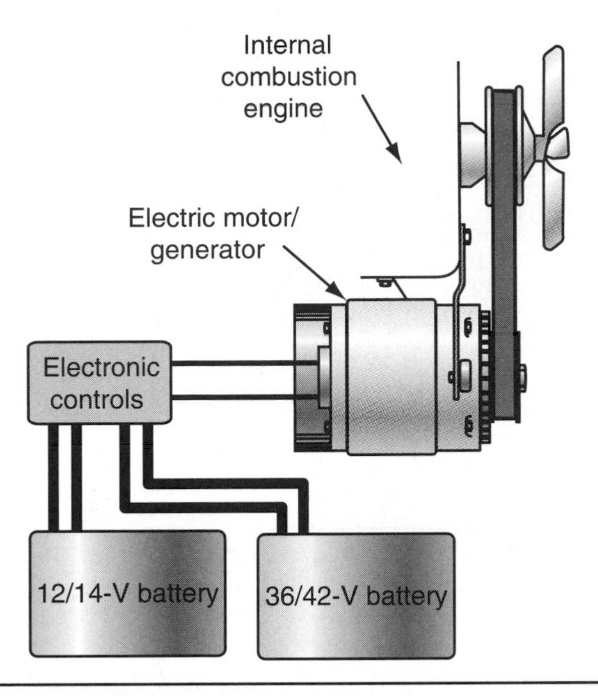

FIGURE 16–66 A typical belt-driven alternator/starter ISG system.

150 amperes or more just to keep up with the load. It has been estimated that future vehicle electrical loads could reach as high as 8,000 watts or more. According to Watt's law, 8,000 watts divided by 14 volts equals 571 amperes. This current level, of course, is not feasible using today's alternator technology; so manufacturers have been developing a variety of 42-volt system combinations (Figure 16–68) to power the electrical accessories that would perform better using a higher voltage, while keeping the typical 14-volt systems. If the charging system voltage is increased to 42 volts, using Watt's law, 8,000 watts divided by 14 volts would equal 190 amperes, thus decreasing the number of amperes the alternator has to produce.

Typical alternators over the years have been air cooled (Figure 16–69) to keep the diodes and windings cool. But as alternator output capacity has increased, so has the need to find better ways to extend alternator life.

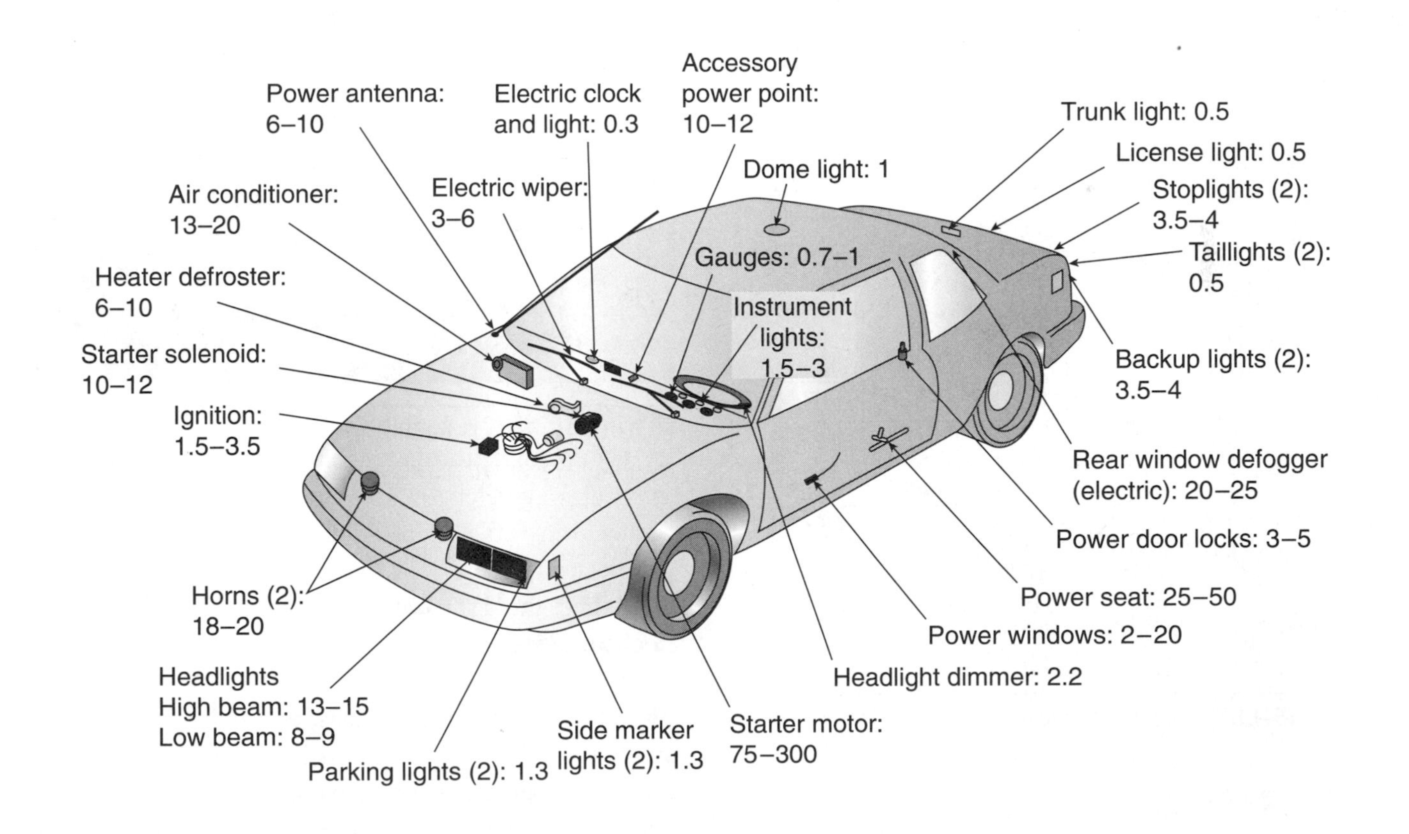

FIGURE 16–67 Typical current flow for vehicle accessories.

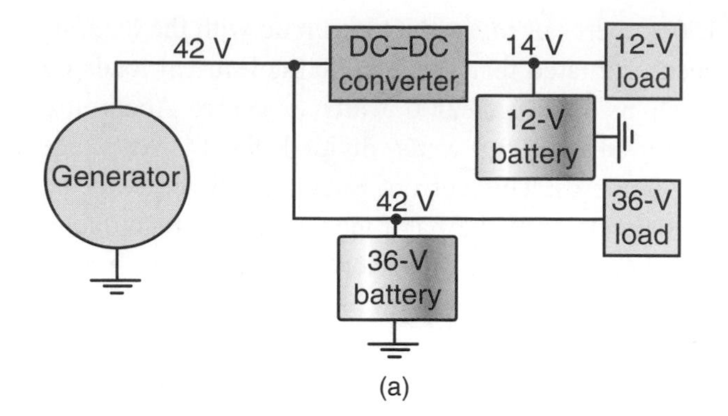

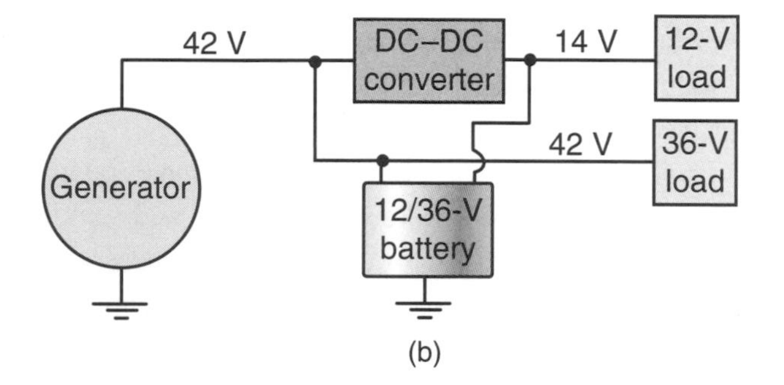

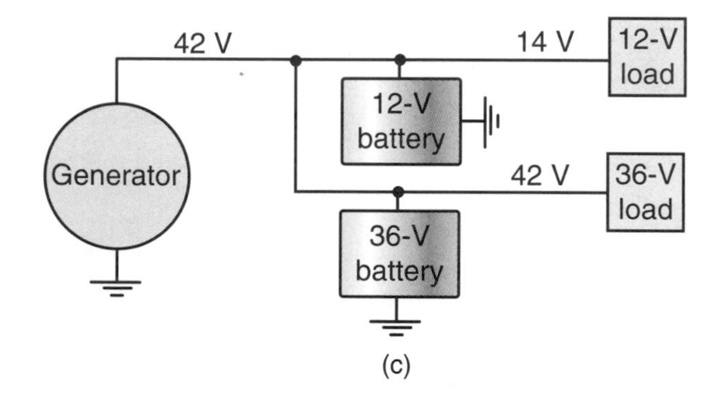

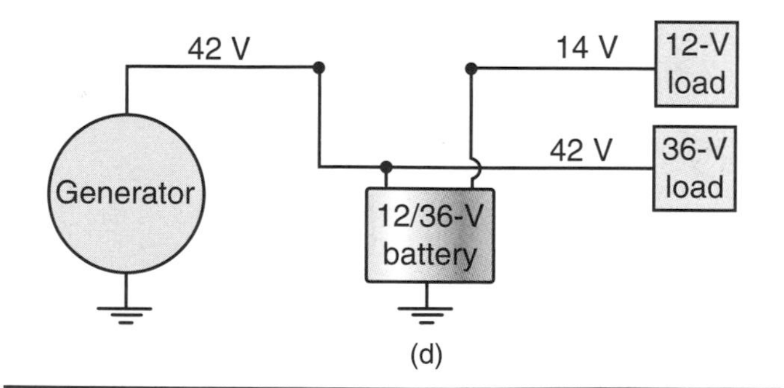

FIGURE 16–68 Various 42-volt electrical system layouts.

One such way is a liquid-cooled alternator (Figure 16–70). Using coolant rather than a fan is a very efficient method of cooling the diodes and windings, and it is much quieter than a fan.

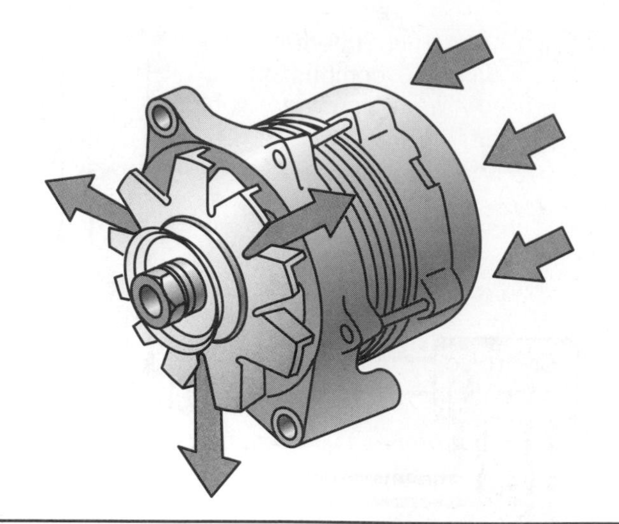

FIGURE 16–69 Air is drawn in from the rear of the alternator to keep the diodes and windings cool.

FIGURE 16–70 A liquid-cooled alternator. (Courtesy of BMW of North America Incorporated)

Basic Charging System Testing

The reliable testing, diagnosis, and repair of a charging system problem is determined by several factors: the type of charging system, knowledge of the circuit being checked, service manual information available, detailed preliminary checks of the system, the type of charging system test equipment available, the actual charging system tests, and the availability of repair parts.

CAUTION *Due to the complexity of modern charging systems, always refer to specific charging system diagnostic procedures to verify possible problems. Do not assume that general diagnostic tests that work for one charging system will work for a different system.*

Preliminary Charging System Checks

Follow this preliminary checklist before attempting to diagnose charging system problems:

- *Battery condition* The charging system is designed to maintain the charge in a battery, not to charge a dead battery. The state of charge and condition of the battery must be determined and corrected, if necessary, before performing charging system tests.
- *Condition of alternator drive belts* A drive belt that is loose, worn, glazed, damaged, or oil soaked (Figure 16–71) may cause the alternator belt to squeal when under a heavy charging condition. If the drive belt condition checks out satisfactorily, adjust the belt tension to the manufacturer's specification using a handheld tension gauge (Figures 16–72 and 16–73). On some vehicles with spring-loaded serpentine V-ribbed belts, the belt tension is determined by the belt length indicator marks stamped on the tensioner (Figure 16–74).

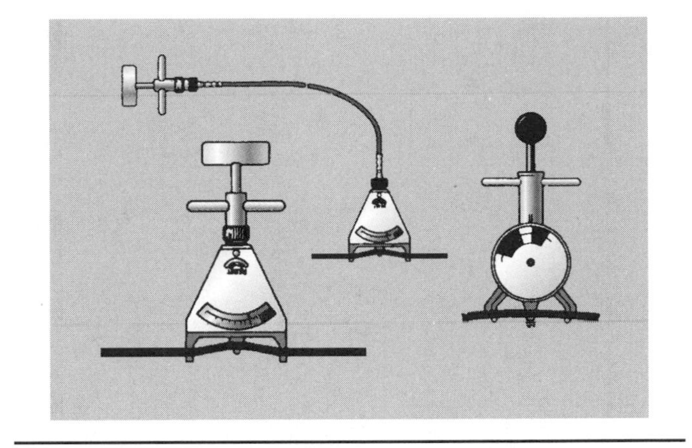

FIGURE 16–72 Checking belt tension with a belt tension gauge.

- *Condition of electrical cables and terminal connections* Loose, corroded, or disconnected electrical connections to the alternator, regulator, or battery could damage the charging system. The battery acts as a buffer to stabilize voltage spikes or

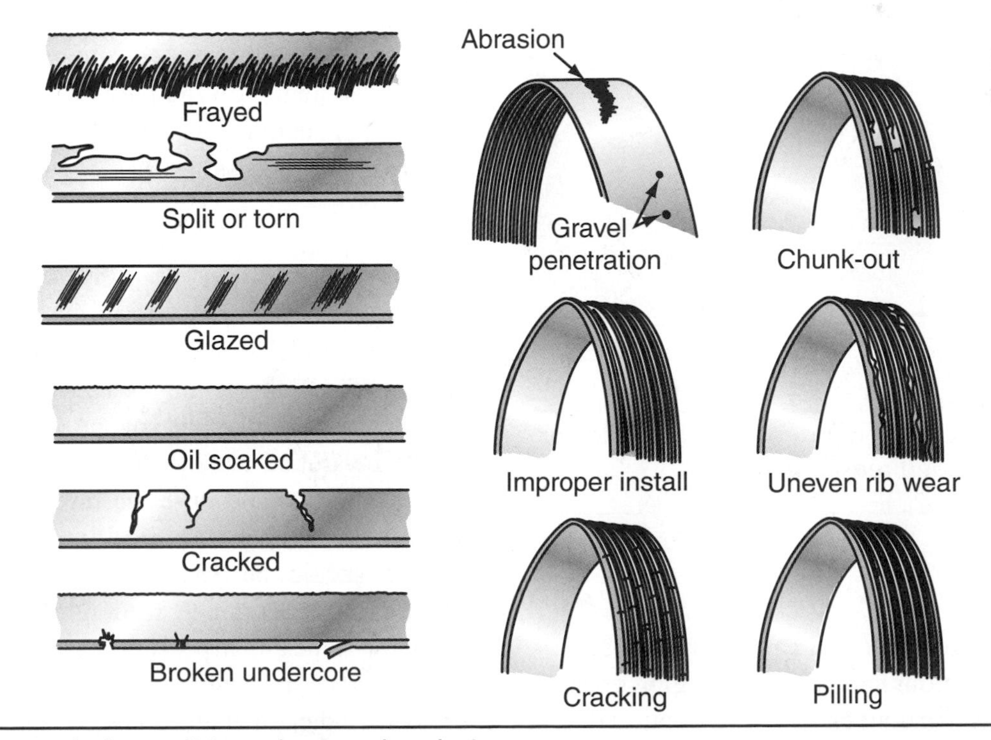

FIGURE 16–71 Drive belt conditions that need replacing.

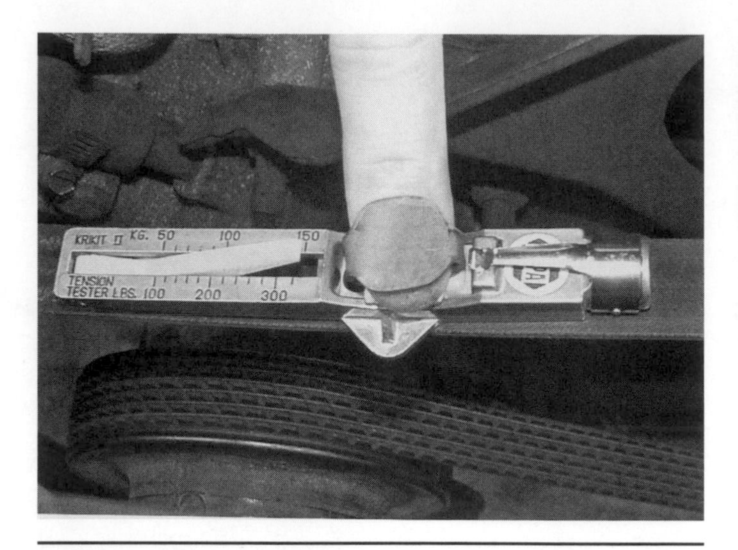

FIGURE 16–73 Checking belt tension with a pocket-sized click-type tension gauge.
(Courtesy of Tim Gilles)

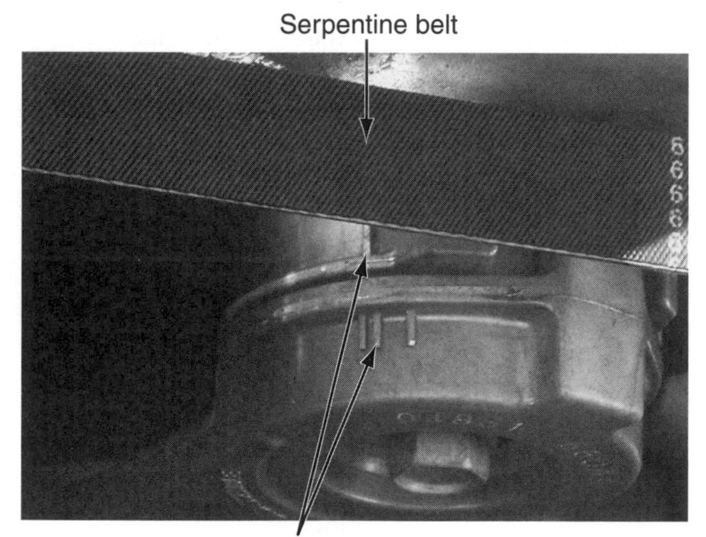

FIGURE 16–74 Belt tension determined by indicator marks on spring-loaded tensioner.
(Courtesy of Tim Gilles)

surge conditions in the electrical system. Loose or corroded connections can give the charging system regulator a false reading on the battery state of charge and cause an overcharge or undercharge condition.

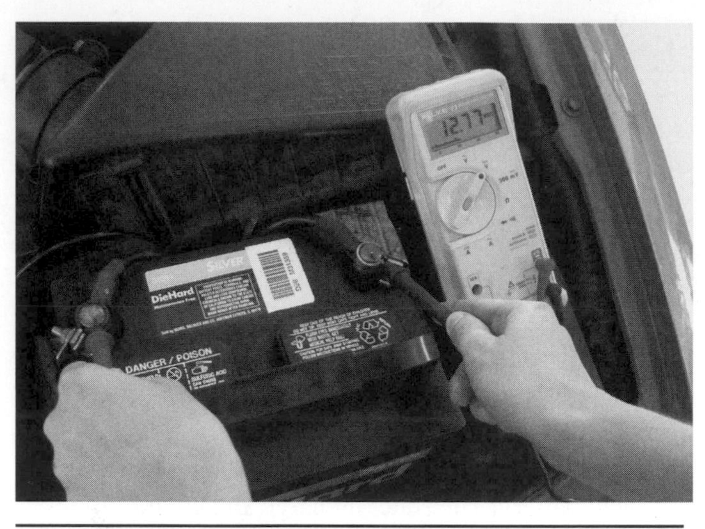

FIGURE 16–75 Performing an engine-off/engine-running charging system output test.
(Courtesy of Tim Gilles)

On-Car Charging System Tests

Vehicle manufacturers are very specific about the procedures used for performing charging system tests. Refer to service manual information to identify circuit component locations, wiring color codes, component specifications, and *actual* test procedures.

Five basic charging system checks may be used to determine alternator and regulator condition: battery and charging system OCV/output voltage test, charging system current output test, charging system voltage drop test, regulator bypass test (on older models), and diode pattern test.

Battery and Charging System OCV/Output Voltage Test To perform this test, connect a voltmeter in parallel to the battery cable connections (Figure 16–75). With the engine and all accessory systems off, note the open circuit voltage (OCV) reading, which should be around 12.66 volts on a fully charged battery. Start the engine and raise the revolutions per minute to about 1,500. With no electrical load, the charging system voltage should be 13.5 to 15.0 volts depending on specific manufacturer. A reading of less than 13.0 volts or below indicates a possible charging system problem that will drain the battery over time. A reading of 16 volts or more indicates that severe overcharge is occurring. If the unloaded charging system voltage is within specifications, test the voltage output under a load condition.

Operate the engine at 2,000–2,500 revolutions per minute, and turn on the headlamps, heater fan, rear window defogger, and other high-current accessories. The charging system output under a heavy electrical load condition should be at least 13.2 volts or more.

Charging System Current Output Check Using a professional starting/charging/battery carbon pile tester is the preferred method for performing this test. There are several good testers. Always refer to the tester's operator's manual when performing any tests. Figure 16–76 is an example of the tester connections for performing most charging system checks. The charging system must be loaded with the carbon pile to obtain the alternator current output at 12 volts. Typical testing procedures are as follows.

- Hook up the charging system test as shown in Figure 16–76 or for specific tester you are using. Some testers specify the inductive pickup lead be attached around the alternator output lead to obtain an accurate reading.
- Start the engine and operate at 1,500–2,000 revolutions per minute.
- Load the system with the carbon pile until the highest ammeter reading is obtained without dropping the voltmeter reading below 12.6 volts. The maximum current produced under these conditions should not be lower than 10% below the alternator specified output plus the ignition current draw. For example, if the ignition system draw is 8 amperes with the key on and engine off, and the alternator is rated at 100 amperes, then:

  $$10\% \times 100\ \text{A} + 8\ \text{A} = 18\ \text{A}$$

  Minimum alternator output should be no less than 82 amperes (100 amperes − 18 amperes). Output readings lower than specifications indicate that further tests are needed.

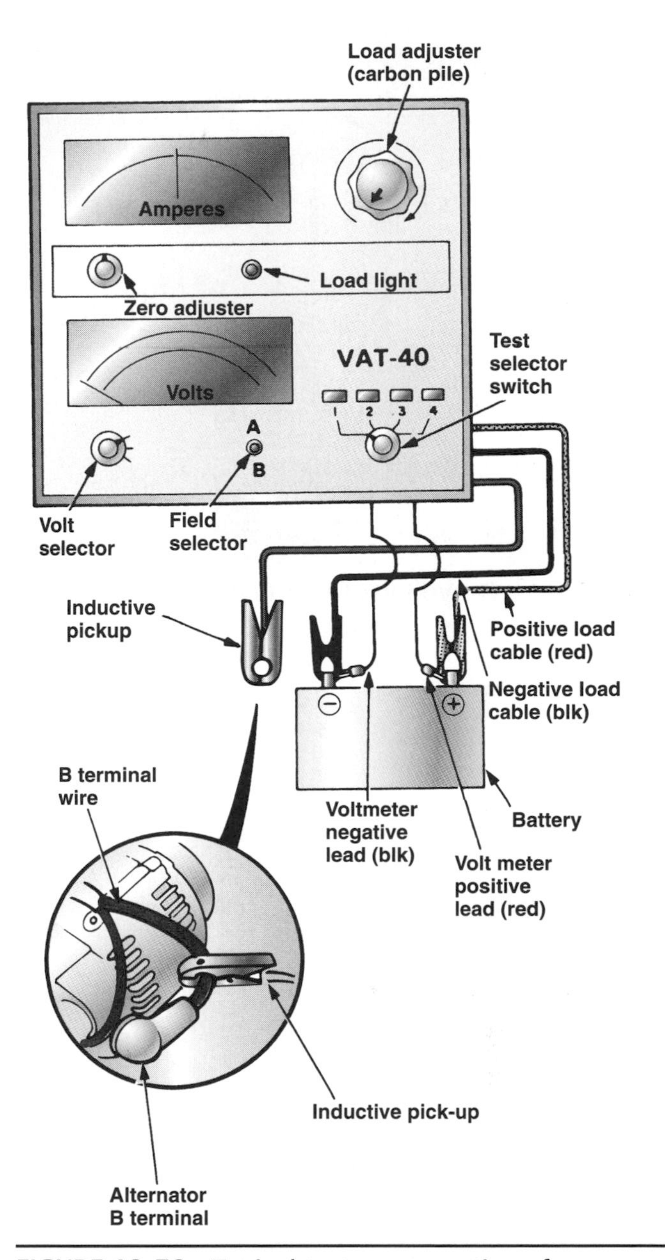

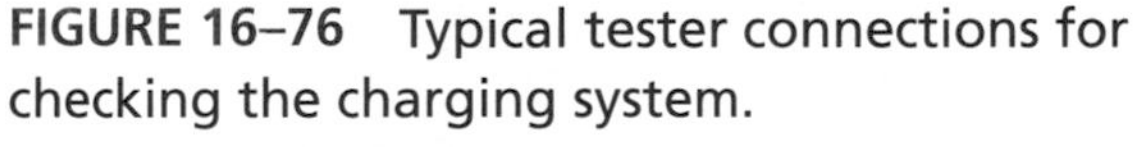
FIGURE 16–76 Typical tester connections for checking the charging system.
(Courtesy of American Honda Motor Company, Inc.)

Charging System Voltage Drop Test This test determines whether the battery, alternator, and regulator are all functioning at the same potential. Over a long period of time, resistance may build up in wiring and connections, affecting the ability of the alternator to meet the electrical load demand. The meter hookups in Figure 16–77 indicate voltage drops on the positive side and negative side of the charging circuit. After hooking up the meter(s), start the vehicle and load the system between 10–20 amperes. This can be accomplished by using a carbon pile on the volt/ampere tester or by turning the headlights on. The total voltage drop should not exceed 0.7 volts (0.5 volts positive side and 0.2 volts negative side). Any readings that exceed the maximum indicate resistance in the circuit wiring or connections.

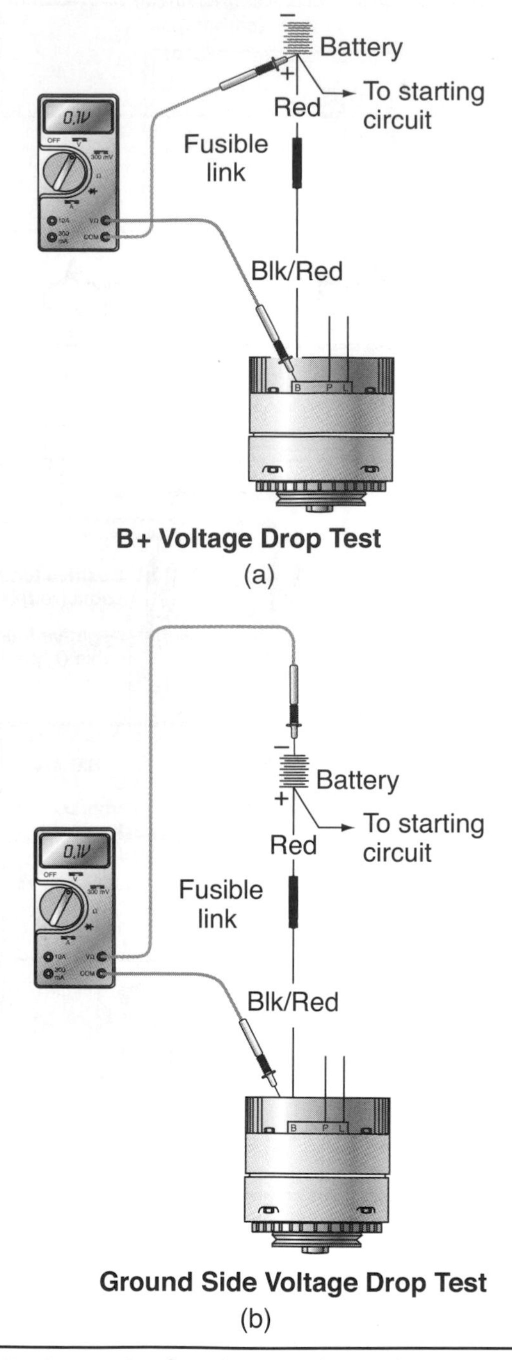

FIGURE 16–77 Performing a voltage drop test on the positive (a) and negative (b) sides of a charging system.

Regulator Bypass Test The fourth charging system check is *not normally used* on most modern vehicles, but it is a still a valid test on older alternator systems. This test is performed by fully fielding the rotor field windings in the alternator using battery supply voltage with the regulator bypassed.

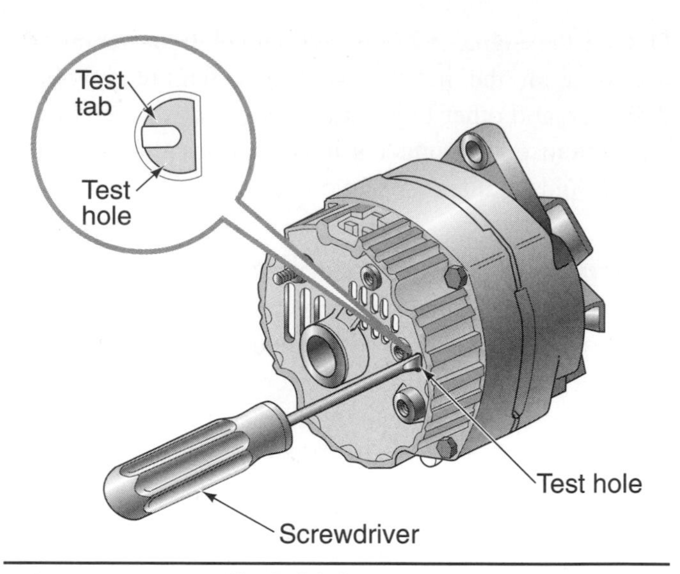

FIGURE 16–78 Fully field the GM10SI alternator by grounding the test tab through the D-hole.

CAUTION *For older vehicles, follow the model-specific procedures when performing a full field test. Not all alternators can be fully fielded. Check the manufacturer's procedures before attempting to perform this test.*

The full field test produces maximum alternator output at the available engine revolutions per minute. To protect electrical components and circuits, do not allow the voltage output to exceed 16 volts. Figure 16–78 is an example of a full field test on an older General Motors 10SI alternator with an internal regulator. Inserting a small screwdriver about one-half inch into the D-hole in the slip ring housing and grounding it to the alternator housing—for 10 seconds maximum—fully fields the system. If the current output is within specifications, the regulator is at fault.

Diode Pattern Test The diodes are responsible for transferring the alternating current (AC) from the stator into the direct current (DC) used by the electrical system.

An alternator may have an open diode, yet be close to its manufacturer's specifications for alternator current output. Failing to find a bad system diode may result in a repeat failure of new replacement components. One way to test for an open diode requires an oscilloscope to visually see the diode patterns under a *load* condition. Figure 16–79 shows four examples of diode test patterns

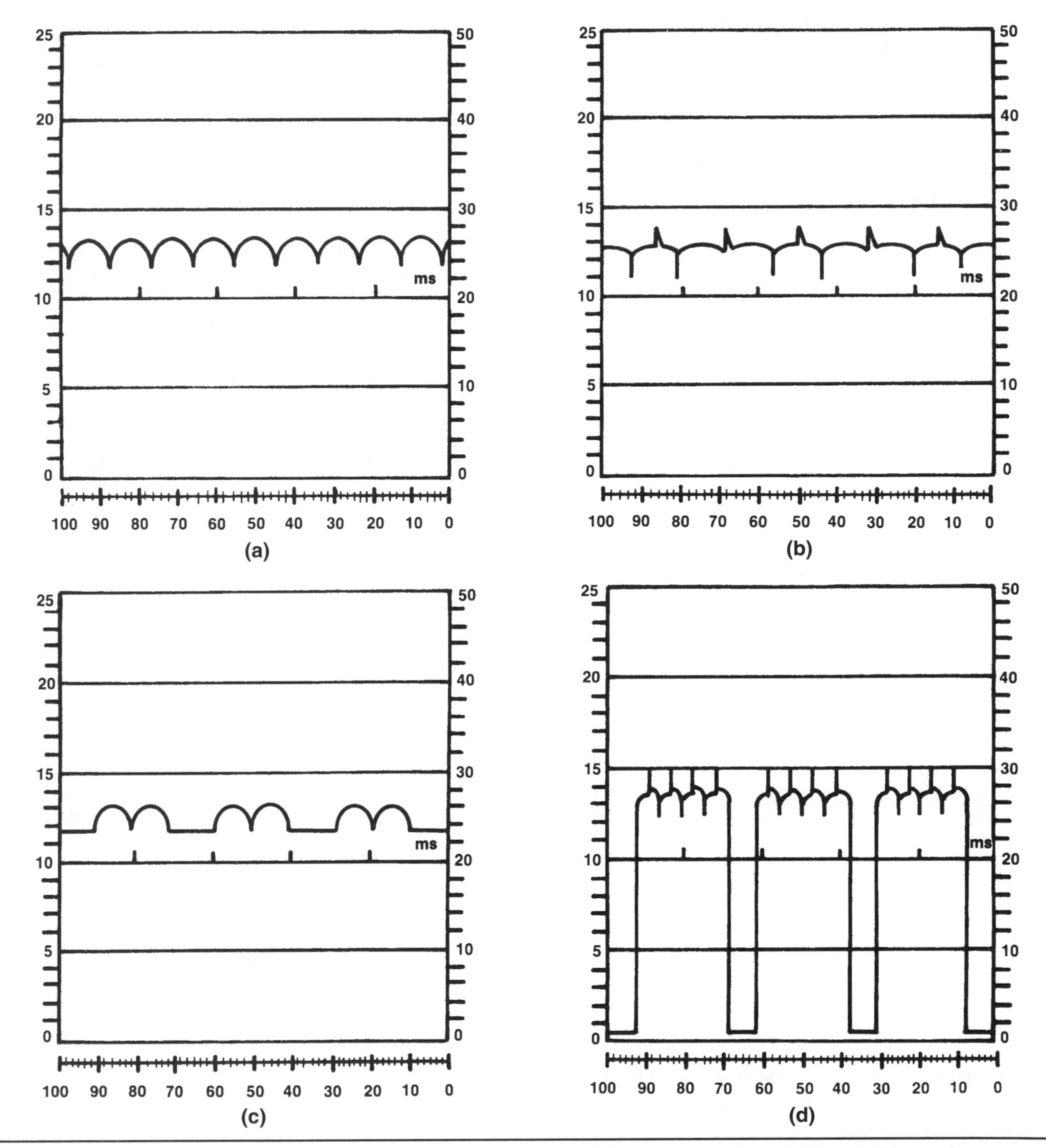

FIGURE 16–79 Typical alternator oscilloscope patterns: (a) good alternator under a full load; (b) good alternator under no load; (c) alternator with a shorted diode and/or stator winding under a full load; and (d) alternator with an open diode trio.

that illustrate comparisons between good diode patterns and ones that have problems.

Another way to quickly check an alternator for an open diode is to use the special diode test feature on a DMM (Figure 16–80). With the alternator electrically disconnected, one meter test probe is touched to the alternator housing and the other probe to the alternator output terminal. A meter reading of approximately 0.8 volts is satisfactory, and a meter reading of approximately 0.4 volts indicates a shorted diode. The measurements are repeated with the meter probes reversed. The meter will exhibit a continuous tone with two diodes shorted. To determine which diodes are faulty, the alternator must be dismantled and each diode checked individually.

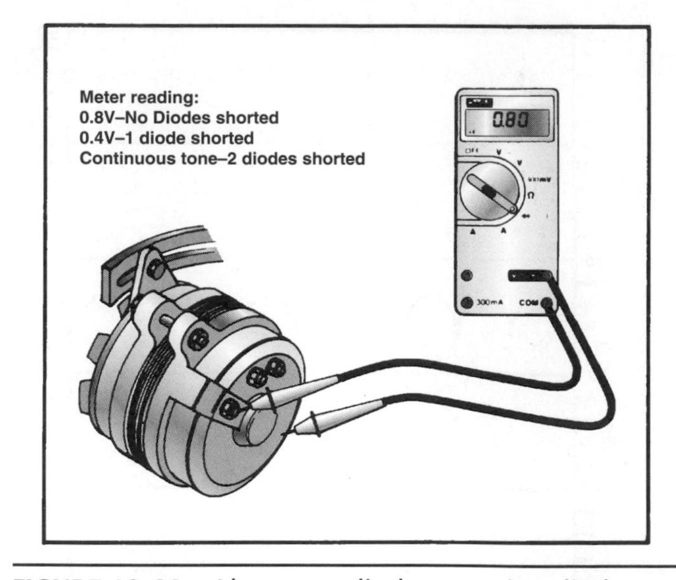

FIGURE 16–80 Alternator diode test using diode feature of a DMM.
(Courtesy of The Fluke Corporation)

After you diagnose a faulty alternator, the preferred method of almost all mainstream repair shops is simply to replace it with a new or rebuilt unit. This is mainly due to a combination of one or more of the following factors: the alternator is not rebuildable, the cost of a replacement alternator cheaper than that of an in-shop rebuild, the availability of repair parts, the level of repair/rebuild experience of the technician, the shop labor rate to rebuild it, and the possible extended downtime of the vehicle.

CASE STUDIES

CASE 1

This case study is typical of a charging circuit problem that technicians often need to diagnose. Knowledge of various charging systems, testing equipment, and simple test procedures will help you perform the diagnosis quickly and efficiently.

Customer Complaint

A vehicle is driven into the shop with a complaint that the battery runs dead if the vehicle is driven with many accessories on. The battery was replaced by the customer, but the condition returned after several hours of driving.

Known Information

- ❑ The new battery has an OCV of 12.2 volts.
- ❑ The charging system indicator lamp is off while the engine is running.
- ❑ Upon examination, the alternator belt appears loose and glazed.

Circuit Analysis

Answer the following questions on a *separate sheet* of paper:

1. With this information, what is the *most* likely cause of the charging system problem?____________
____.
2. What preliminary checks should you perform before proceeding with any charging system tests?
____.
3. After these checks and repairs are done, what diagnostic system tests would you perform to verify the condition of the charging system? ____
____
____.
4. What is your analysis of this case study? ____
____
____.

Hands-On Vehicle Tests

The following six hands-on vehicle checks are included in the NATEF (National Automotive Technician Education Foundation) Task List or as additional independent tasks. Some tasks are grouped together due to their close association with each other. Complete your answers to the following questions on a *separate* sheet of paper.

Performance Task 1

Task Description
Diagnose charging system for the cause of an undercharge, no-charge, or overcharge condition.

Performance Task 2

Task Description
Perform charging circuit voltage drop tests; determine necessary action.

Performance Task 3

Task Description
Perform charging circuit voltage output test; determine necessary action.

Task Objectives

❑ Obtain a vehicle that can be used for these tasks. What model and year of vehicle are you using for these tasks? ____.

❑ Set the vehicle emergency brake and place the vehicle in neutral. Observe all standard shop safety practices and precautions.

❑ Perform a preliminary charging system check on the vehicle and determine the repairs needed.

Preliminary check made	Repairs needed
1.	
2.	
3.	
4.	

❑ What diagnostic equipment are you going to use to test the condition of the charging system?

____.

❑ List the steps used to perform a voltage output test on a charging system. ____________

____

____.

❑ Perform a voltage output test on the charging system. Record the results:

Battery OCV test: ____.

Charging system output voltage at 1,500 revolutions per minute: ____________.

❑ List the steps used to perform a current output test on a charging system. ____________

____

____.

❑ Perform a current output test on a charging system. Record the results:

System specifications: ____.

Charging current output: ____.

❑ List the steps used to perform a charging circuit voltage drop test. ____________

____.

❑ When performing a charging circuit voltage drop test, what can you use to load the alternator when a carbon pile is not available? ____________

____.

❑ What is the maximum voltage drop reading allowed in the charging circuit? ____________.

❑ What was the charging circuit voltage drop in the positive side of the circuit? ____________.

❑ What was the charging circuit voltage drop in the negative side of the circuit? ____________.

❑ What procedures are most helpful when inspecting and testing the charging system circuit? ________

____.

Task Summary

After performing the preceding NATEF tasks, what can you determine will be helpful in knowing how to check a charging system and individual components for proper operation? ____________________

____

____.

Performance Task 4

Task Description

Remove, inspect, and install generator (alternator).

Performance Task 5 (Not a Current NATEF Task)

Task Description

Disassemble, clean, inspect, and test alternator components.

Performance Task 6

Task Description

Inspect, adjust, or replace generator (alternator) drive belts, pulleys, and tensioners; check pulley and belt alignment.

Task Description

❑ Obtain a vehicle that can be used for these tasks. What model and year of vehicle are you using for these tasks? ____.

❑ List the procedures for removing and replacing the alternator according to the manufacturer of the vehicle being worked on. ____

____

____.

❑ What tools and equipment did you use to remove and replace the alternator? ________________

____.

❑ List the steps that you used to disassemble, clean, and inspect the alternator. ________________

____.

❑ Perform individual alternator component tests according to the specific vehicle being worked on. Record the results:

Alternator component	Test equipment used	Results of test
1.		
2.		
3.		
4.		
5.		
6.		

❑ What procedures and special tools did you use to check the alternator belt tension after replacing the alternator? ______________________________.

❑ How did you check the pulleys, belts, and tensioners for alignment? ______________________________

______________________________.

Task Summary

After performing the preceding NATEF tasks, what can you determine will be helpful in knowing how to check a starter motor system and individual components for proper operation? ______________________________

____

______________________________.

Summary

❑ The amount of voltage produced in a conductor is dependent on three things: the strength of the magnetic field, the speed at which the conductor is cutting through the magnetic field, and the number of conductors passing through the magnetic field.

❑ The DC generator has two stationary field coils that create the north and south magnetic poles.

❑ Output voltage in a DC generator is produced in the wire-looped armature as it rotates through the magnetic field.

❑ In a DC generator, the AC voltage is changed to a DC voltage by the armature commutator bars.

❑ The rotor assembly in an AC generator—an alternator—contains several rotating north and south poles.

❑ The rotor slip rings provide a current path connection from the insulated carbon brushes to the rotor field coil assembly.

❑ The stator assembly in the AC generator is the stationary coil windings in which electricity is produced.

- ❑ The wye wound stator consists of a common junction and three output connections.
- ❑ The delta wound stator consists of three windings connected in a triangular shape with three output leads.
- ❑ Half-wave rectification of an AC sine wave converts the positive portion of the sine wave through a positively biased diode, while blocking the negative portion.
- ❑ Full-wave rectification of an AC sine wave is the process of converting the complete AC waveform to a positive DC waveform.
- ❑ A diode rectifier bridge consists of six diodes: three positive biased diodes mounted in a heat sink and three negative biased diodes mounted in the alternator rear housing.
- ❑ A diode trio rectifies current from the stator to create the magnetic field excitation in the rotor field coil.
- ❑ A voltage regulator controls and stabilizes the charging system voltage by controlling the resistance in series with the rotor field coil.
- ❑ The A-circuit regulator has field circuit control on the ground side of the field coil.
- ❑ The B-circuit regulator has field circuit control on the power input side of the field circuit.
- ❑ An isolated regulator field circuit picks up both B+ and ground externally.
- ❑ An electromechanical regulator uses two vibrating electromagnetically operated point contact relays to control the charging circuit.
- ❑ An electronic regulator controls the field circuit by modulating the on/off cycle of the field circuit, depending on the electrical circuit loads.
- ❑ When the field circuit in an electronic regulator modulates, it is called pulse width modulation.
- ❑ The zener diode in an electronic regulator is designed to conduct current in a reverse direction at a designed-in voltage level.
- ❑ Computer-controlled regulators incorporate a powertrain control module (PCM) to modulate the field circuit.
- ❑ A charging system indicator lamp operates on the principle of opposing voltage to control lamp function.
- ❑ An ammeter is sometimes wired in series to monitor the function of a charging system.
- ❑ A voltmeter is sometimes wired in parallel to the battery feed circuit to monitor the function of a charging system.
- ❑ Explain the basic operation of an HEV integrated starter/generator (ISG).
- ❑ One type of ISG consists of a three-phase motor with a stator assembly attached to the inside of the transmission/engine bell housing and a rotor attached to the engine crankshaft.
- ❑ Another application of an ISG is a belt-driven alternator/starter that mounts in the same location as the typical engine-mounted alternator.
- ❑ A voltage output test can be used to compare the battery OCV with the charging system voltage.

- ❑ A current output test uses a carbon pile tester to determine the alternator current output while the electrical system is under a load condition.
- ❑ A regulator bypass test, also referred to as a full field test, determines whether the alternator or regulator is faulty.
- ❑ Alternator diodes can be tested using an oscilloscope to visually see the diode pattern under a load condition.

Key Terms

A-circuit
B-circuit
delta wound
diode rectifier bridge
diode trio
duty cycle
field current
full-wave rectification
half-wave rectification
heat sink
isolated field circuit
pulse-width modulation
rectification
rotor
sensing voltage
sine wave
slip rings
stator
three-phase voltage
voltage limiter
voltage regulator
wye wound

Review Questions

Short Answer Essays

1. List the parts of a typical AC generator.
2. Describe the difference between a wye wound stator and a delta wound stator.
3. Describe how the AC current produced by the alternator is converted to DC current.
4. What is the purpose of the voltage regulator in a typical charging system?
5. Describe the purpose of a zener diode in a charging system electronic regulator.

Fill in the Blanks

1. The rotor ________________ provide a current path connection from the insulated carbon brushes to the rotor coil.
2. A wye wound stator winding has a common junction and ________________ output connections. The delta wound stator winding has the windings connected in a ________________ shape with ________________ output connections.
3. The process of converting a complete AC waveform to a positive DC waveform is referred to as ________________.
4. The on/off cycling of the field circuit is called ________________.
5. Normal charging system output under a heavy load condition should be at least ________________.

ASE-Style Review Questions

1. Two technicians are discussing the magnetic principles of charging systems. Technician A says the output is rotating in a DC generator. Technician B says the output is stationary in an AC generator. Who is correct?
   A. A only
   B. B only
   C. Both A and B
   D. Neither A nor B

2. Technician A says the battery has no effect on the charging system operation. Technician B says the charging system is used only to restore the electrical power to the battery that was used up during engine starting. Who is correct?
   A. A only
   B. B only
   C. Both A and B
   D. Neither A nor B

3. Technician A says the slip rings on an alternator rotor convert the AC voltage to a DC voltage. Technician B says the field coil on an alternator rotor must be insulated from the rotor shaft. Who is correct?
   A. A only
   B. B only
   C. Both A and B
   D. Neither A nor B

4. Technician A says a wye wound stator winding consists of three output connections and a common neutral junction. Technician B says a wye wound stator winding produces a higher amperage output than a delta wound stator winding. Who is correct?
   A. A only
   B. B only
   C. Both A and B
   D. Neither A nor B

5. Before adjusting the alternator drive belt tension, Technician A checks for proper pulley alignment. Technician B refers to the vehicle's service manual to determine the proper belt tension. Who is correct?
   A. A only
   B. B only
   C. Both A and B
   D. Neither A nor B

6. Technician A says the charging system regulator uses a sensing voltage input from the battery to interpret the alternator output voltage. Technician B says an A-circuit charging system regulator controls the field circuit on the power side of the circuit. Who is correct?
   A. A only
   B. B only
   C. Both A and B
   D. Neither A nor B

7. Technician A says a delta wound winding consists of three windings connected in a triangular shape with three output leads. Technician B says a delta wound winding provides a high-amperage output for its size. Who is correct?
   A. A only
   B. B only
   C. Both A and B
   D. Neither A nor B

8. The charging system voltage of a vehicle with an external regulator is 16.5 volts at 1,500 revolutions per minute. Technician A says the alternator's field circuit may have excessive resistance. Technician B says the regulator sensing circuit may have excessive resistance. Who is correct?
   A. A only
   B. B only
   C. Both A and B
   D. Neither A nor B

9. Technician A uses only a current output test to check alternator output. Technician B uses only a voltage output test to check alternator output. Who is correct?
   A. A only
   B. B only
   C. Both A and B
   D. Neither A nor B

10. Technician A says the positively biased diodes in a rectifier bridge are attached directly to the alternator rear housing. Technician B says the negatively biased diodes in a rectifier bridge are mounted in a heat sink. Who is correct?
    A. A only
    B. B only
    C. Both A and B
    D. Neither A nor B

11. Technician A says a diode trio rectifies current from the stator to create the magnetic field excitation in the field coil. Technician B says the diode trio attaches to the rotor slip rings. Who is correct?
    A. A only
    B. B only
    C. Both A and B
    D. Neither A nor B

12. Technician A says the integrated starter generator on a hybrid electric vehicle is located between the engine and transmission unit. Technician B says the ISG may be belt driven and is located in the same place as a standard alternator. Who is correct?
    A. A only
    B. B only
    C. Both A and B
    D. Neither A nor B

13. Technician A says the pulse-width modulation in an electronic voltage regulator is the on/off modulation of the field circuit. Technician B says the modulation in a electronic voltage regulator varies the amount of time the field coil is energized depending on the vehicle's electrical demand. Who is correct?
    A. A only
    B. B only
    C. Both A and B
    D. Neither A nor B

14. Technician A says a slipping belt can cause a squealing noise when the alternator is under a heavy load. Technician B says a slipping belt can become glazed over a period of time. Who is correct?
    A. A only
    B. B only
    C. Both A and B
    D. Neither A nor B

15. A charging system voltage drop test is being conducted. Technician A says the total voltage drop in the positive side of the circuit should not exceed 0.7 volts. Technician B says the total voltage drop on the negative side of the circuit should not exceed 0.5 volts. Who is correct?
    A. A only
    B. B only
    C. Both A and B
    D. Neither A nor B

17 Ignition System Theory

Introduction

Over the past several decades automotive ignition system designs have made a steady growth in complexity from the pre-1970s point-style distributor to the computer-controlled ignitions of today. Demand for more fuel-efficient and cleaner burning engines have been primary reasons for ignition system design changes. The need for maintenance-free ignition systems that perform to exact engine combustion efficiency standards is important to prevent tailpipe emissions from exceeding federal vehicle pollution standards. A tremendous number of ignition systems are in use today, even within a single vehicle manufacturer. This chapter provides a look at basic ignition system theory, components, diagnosis, and typical service needs. When diagnosing and servicing a specific ignition system, always refer to the appropriate manufacturer's service manual information.

Objectives

When you complete this chapter you should be able to:

- ❑ Describe the basic function of an ignition system.
- ❑ Name the components and their function in the primary side of an ignition system.
- ❑ Name the components and their function in the secondary side of an ignition system.
- ❑ Describe the operation of and basic tests on an ignition coil.
- ❑ Describe the functions of spark plugs and plug wires in an ignition system.
- ❑ Identify combustion chamber conditions by observing spark plug wear patterns.
- ❑ Describe the purpose and operation of advancing units in ignition systems.
- ❑ Describe the operation and characteristics of distributorless ignition systems.
- ❑ Describe the operation and characteristics of a typical waste spark ignition system.
- ❑ Describe the operation and characteristics of a coil-on-plug direct ignition system.
- ❑ Describe the basic purpose of an oscilloscope in checking an ignition system.

Ignition System Principles

The ignition system in an internal combustion engine is designed to supply:

1. Step-up voltage high enough to jump the spark plug gap.
2. Control of ignition timing based on engine speed and load.
3. Distribution of the spark to each cylinder based on firing order.

The combustion of the fuel/air mixture in the cylinder occurs in milliseconds depending on several design and operating factors. As a result, the combustion process must begin **before-top-dead-center (BTDC)** near the end of the compression stroke. This ensures that the fuel/air mixture burns completely and delivers its maximum compression push when the piston is at 10 to 23 degrees **after-top-dead-center (ATDC)** (Figure 17–1).

Calculating when the ignition spark should occur for a controlled burning of the fuel/air mixture depends on several design and operating factors. Some of these are engine load, intake air temperature, barometric pressure, humidity, cylinder compression, and the fuel/air mixture turbulence in the cylinder.

Much has been learned over the past two decades concerning the tendency of an engine to knock, or ping, under abnormal combustion conditions. Knowledge of these conditions is important because it may not be ignition-related problems affecting the engine operation.

Two typical combustion conditions may cause a knock or ping. The first is engine **detonation.** It can cause a knock to occur when unburned fuel/air mixture self-ignites and collides with the normal combustion flame front (Figure 17–2). Detonation that is allowed to continue can cause severe piston, cylinder head, or spark plug damage over time (Figure 17–3).

The second is **preignition,** a condition caused by an early ignition of the fuel/air mixture by a source other than the spark plug and ignition system (Figure 17–4). The most common causes of preignition are hot carbon deposits or overheating of sharp edge surfaces in the combustion chamber. In a preignition condition, the fuel/air mixture is ignited by a hot carbon deposit that creates a flame front that collides with the normal spark ignition flame front. A pinging or light tapping noise may result from an engine with this condition, but it may not be noticeable from a quiet passenger compartment. Figure 17–5 is an example of the destructive results of a preignition condition on a piston by using a spark plug with too high a heat range.

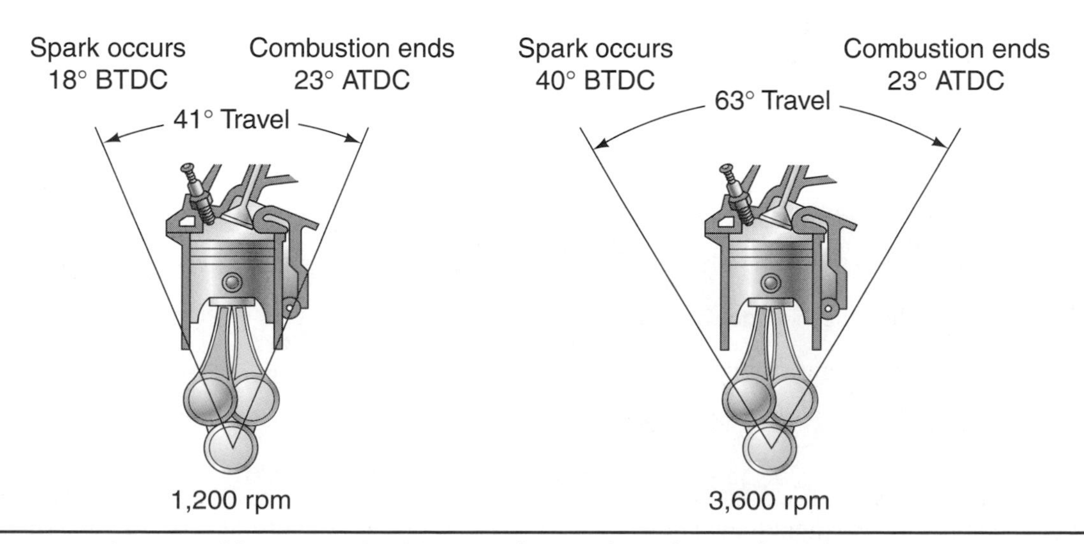

FIGURE 17–1 As engine speed increases, ignition spark must be advanced to obtain maximum pressure by 23 degrees ATDC.

(Courtesy of Federal-Mogul Corporation)

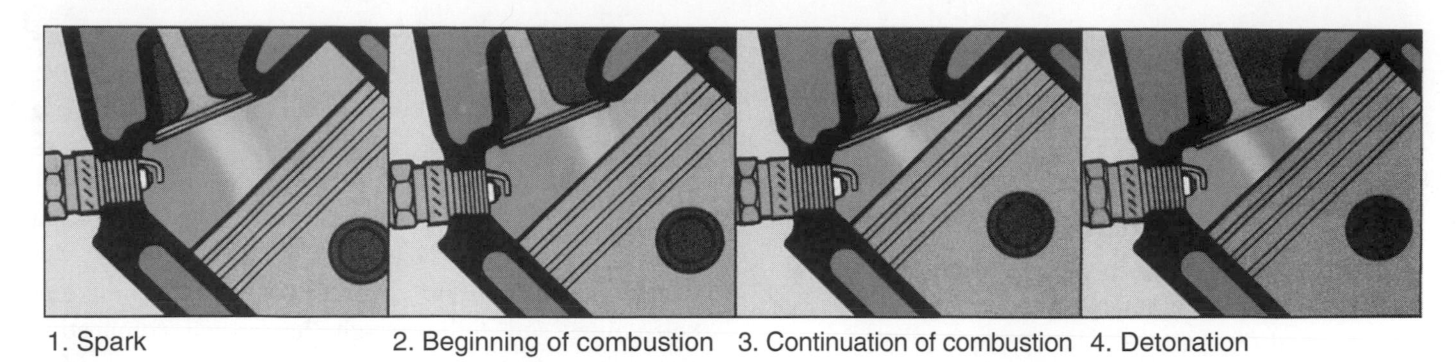

FIGURE 17–2 Detonation stages in combustion chamber.

(Courtesy of Champion Spark Plug Company)

FIGURE 17–3 The effects of detonation on a piston and spark plug.

(Courtesy of Champion Spark Plug Company)

FIGURE 17–5 The results of a preignition condition caused by using a spark plug with a heat range that is too hot.

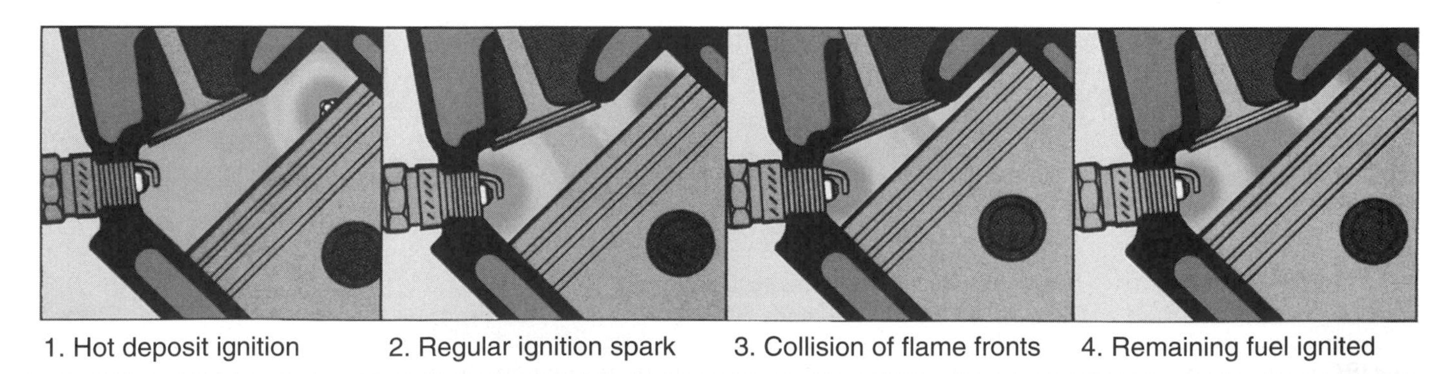

FIGURE 17–4 Preignition stages in a combustion chamber.

(Courtesy of Champion Spark Plug Company)

Remember: Detonation occurs *after* the start of normal spark ignition conditions, and preignition occurs *before* the start of normal spark ignition conditions.

Ignition System Components and Operation

All ignition systems consist of two interconnected electrical circuits and components designed to work together to produce the necessary high voltage to jump the spark plug gap. The two systems are a low-voltage **primary circuit** and a high-voltage **secondary circuit** (Figure 17–6).

Depending on the exact type of ignition system, the typical components of the low-voltage primary circuit in a distributor-type ignition system are:

- Battery.
- Ignition switch.
- Ballast resistor or resistance wire.
- Ignition coil primary windings.
- Ignition bypass circuit during starting mode.
- Points or electronic triggering device.
- Electronic control module or switching device.
- Ground connection.

The typical secondary circuit components in a distributor-type ignition system are:

- Ignition coil secondary windings.
- Coil wire and spark plug wires.
- Distributor cap and rotor.
- Spark plugs.
- Ground connection.

Ignition Coil

Regardless of the era in which the vehicle was built or the style of the ignition system, the heart of *any* ignition system is the ignition coil. The ignition coil can also be referred to as a **pulse transformer** that generates a high voltage when a magnetic field moves across a wire.

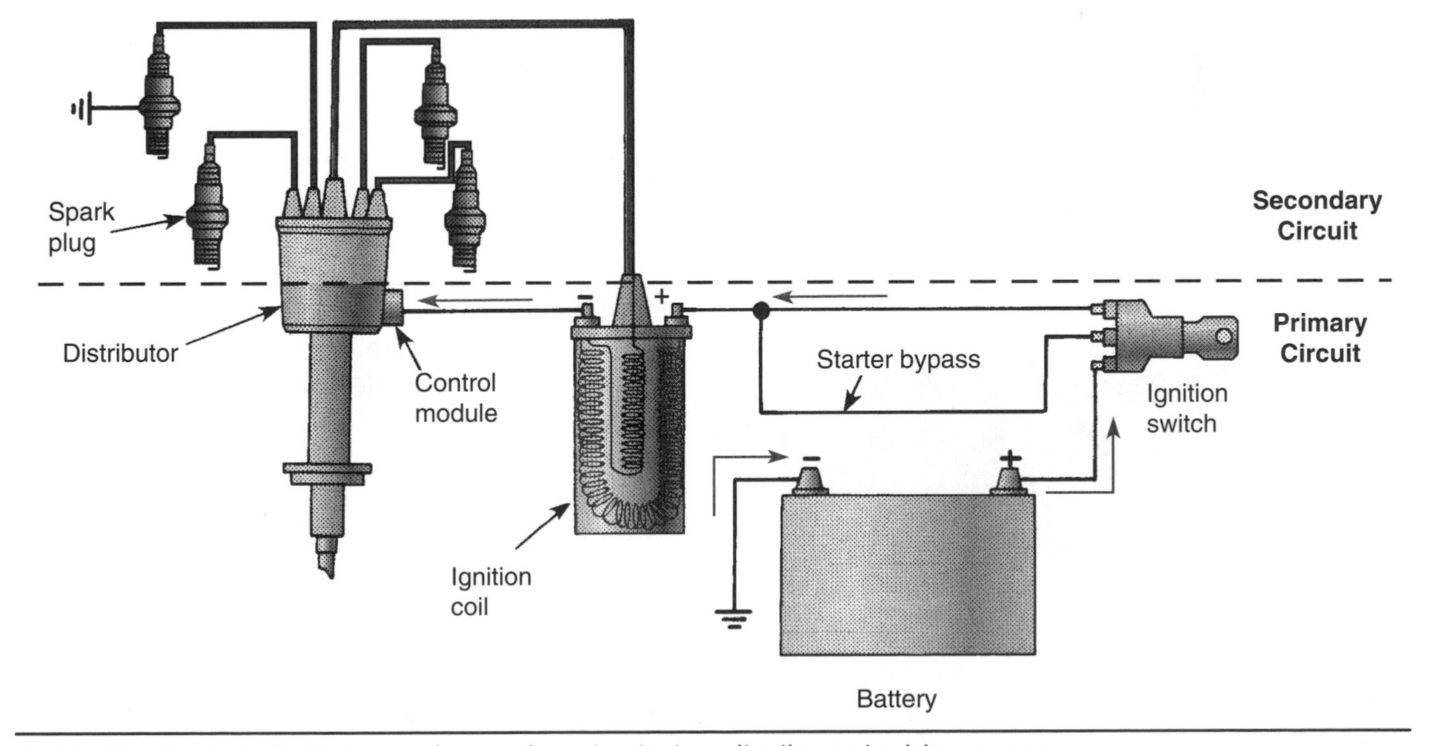

FIGURE 17–6 Typical primary and secondary circuits in a distributor ignition system.

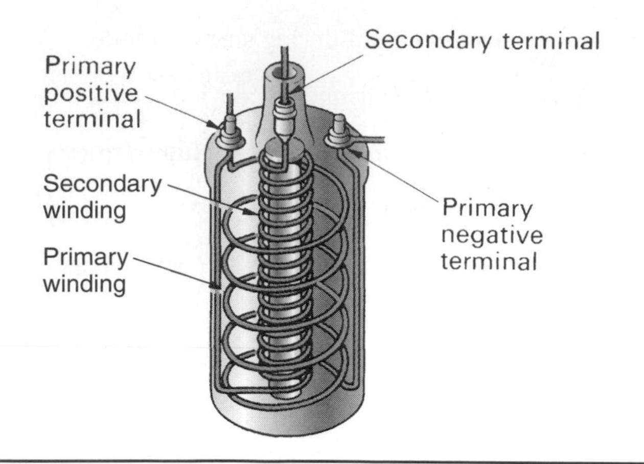

FIGURE 17–7 A typical old-style ignition coil.

The design function of the ignition coil is to transform 12 volts from the battery to a high-voltage potential capable of jumping across the spark plug gap. This voltage may reach 50,000 volts or more in some modern systems.

The construction of a typical ignition coil (Figure 17–7) begins with a couple hundred turns of 20 to 24 gauge copper wire connected between the coil positive and negative posts to form the primary coil windings. The secondary coil windings consist of 20,000 turns or more of very fine copper wire, connected normally between the coil positive or negative post and the coil center tower. Both sets of copper windings are coated with an insulating enamel for protection. The soft laminated iron core has a low **inductive reluctance** that freely expands or strengthens the magnetic field around it. The core and windings are immersed in a metal case that is filled with oil to dissipate heat generated from the windings. The top of the coil is hermetically sealed to prevent the oil from leaking out or moisture and condensation from entering.

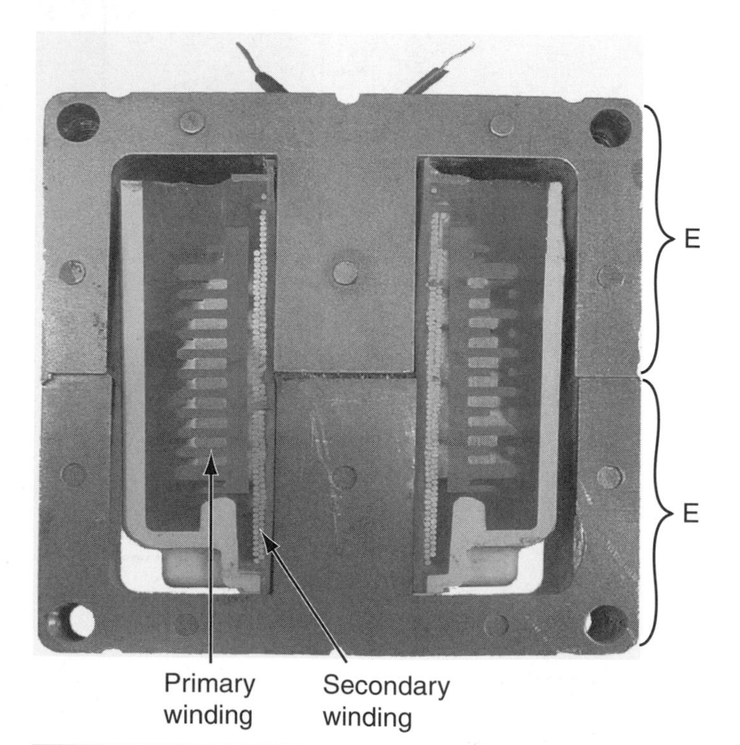

FIGURE 17–8 Typical air-cooled epoxy resin-sealed E coil.

(Courtesy of Tim Gilles)

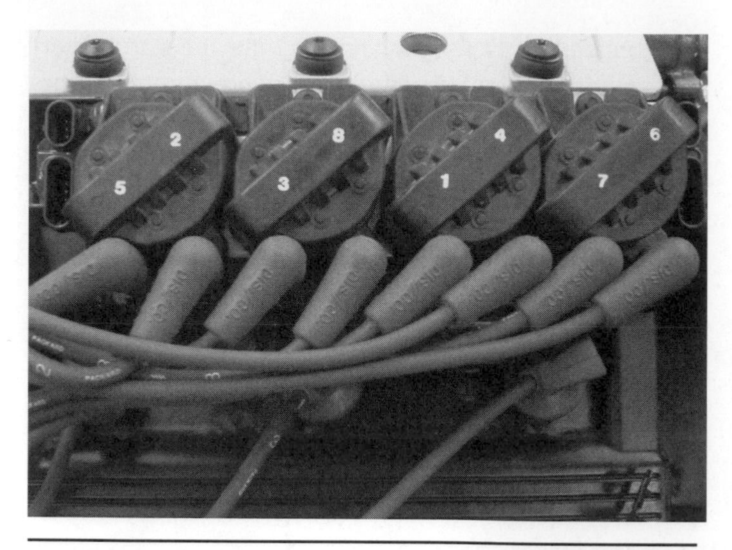

FIGURE 17–9 One coil pack for every 2 cylinders on an 8-cylinder distributor or less ignition system.

(Courtesy of Tim Gilles)

Another common ignition coil design used mostly in older electronic ignition systems is an air-cooled epoxy resin-sealed **E-coil** (Figure 17–8). This type of coil consists of a soft iron laminated core shaped like the letter E. The primary and secondary windings are wrapped around the middle core of the E-coil and are then sealed with epoxy resin.

On most modern vehicles, the coil assembly may be found in a variety of combinations depending on the ignition system design. Figure 17–9 is an example of a distributorless four-coil system on an eight-cylinder engine. Figure 17–10 is an example of an ignition system in which each spark plug has an individual coil (coil-over-plug or coil-on-plug).

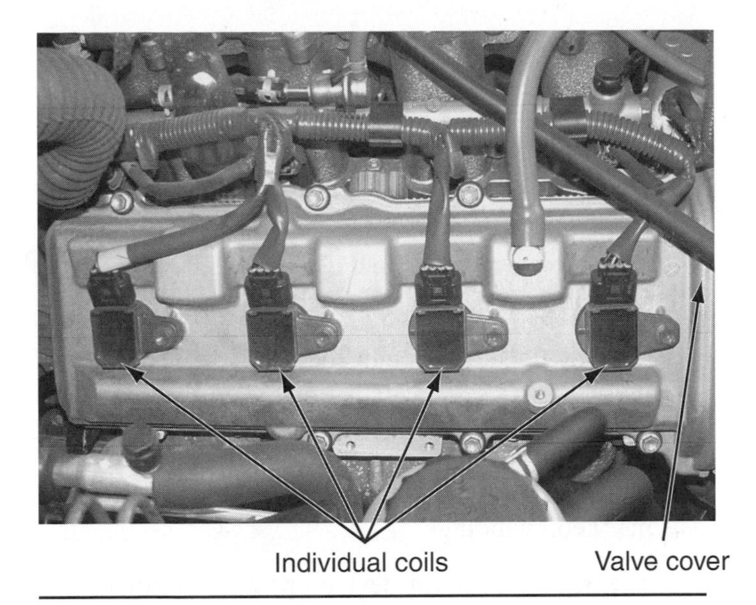

FIGURE 17–10 A typical ignition system where individual coils are mounted directly over, or on, the spark plugs.

(Courtesy of Tim Gilles)

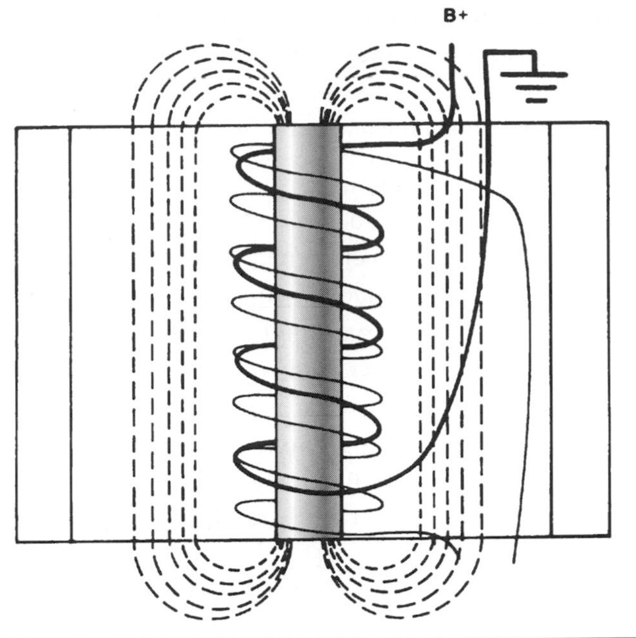

FIGURE 17–11 Magnetic field develops as current flows in the primary windings.

Regardless of the coil location, or the type of ignition system, the operation of an ignition coil involves three important principles of magnetism in the primary and secondary windings:

1. The first principle is called electromagnetism. We have discussed this principle before in previous chapters. Any time a current flows through a wire, magnetic lines of force create an electromagnet. The strength of the electromagnet depends on the number of wire loops and the amount of current flowing through these loops (Figure 17–11).
2. The second principle, **mutual induction,** occurs as the magnetic lines build up around the primary windings. The magnetic lines cut across the secondary windings and mutually induce a voltage (Figure 17–12).
3. **Self-induction** occurs when the primary winding induces a voltage onto itself. Self-induction creates a countervoltage in the primary winding called a counterelectromotive force (CEMF) that tries to resist the electromotive force (EMF) from the battery. The effects of CEMF on the primary windings allow time for the coil to become fully magnetized with voltage

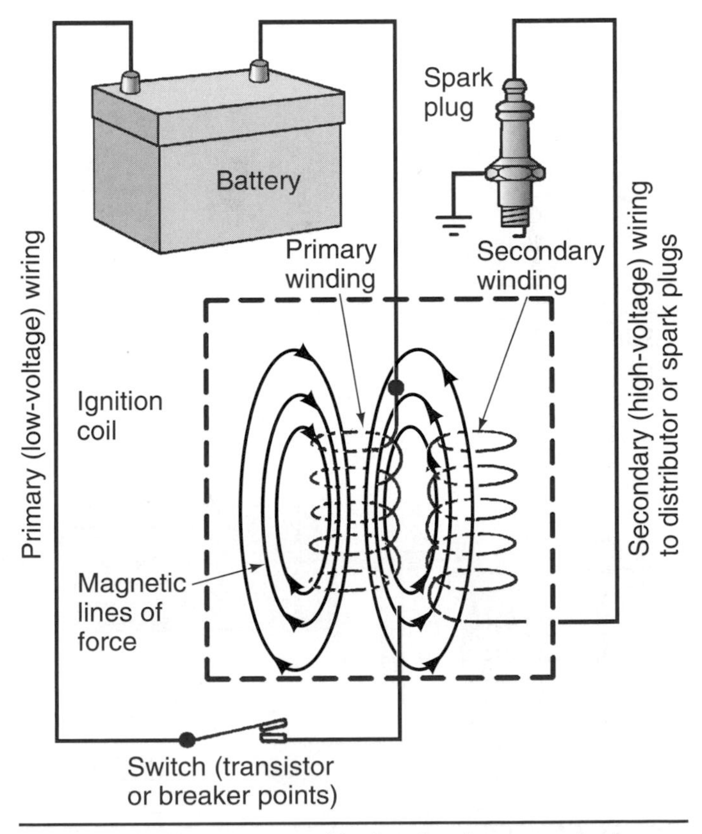

FIGURE 17–12 A mutual induction is created when magnetic lines of force from the primary windings cut across the secondary windings.

from the battery. The buildup of the magnetic field in the coil is called the **saturation** time. The longer the time is that current flows through the primary windings, the stronger the magnetic field will become.

When the primary circuit is suddenly opened by the primary control device, such as a transistor switch or breaker points, the magnetic field instantly collapses. This sudden collapse produces a very high voltage in the secondary windings that rises high enough to jump the spark plug gap. On all older point-style ignition systems, to protect the points from arcing and to speed up the collapse of the magnetic field, a **capacitor,** also known as a condenser, is wired in parallel to the circuit to absorb the surge of electrons induced by the collapsing magnetic field (Figures 17–13 and 17–14).

Capacitor

FIGURE 17–13 Current flow through primary circuit with points closed.

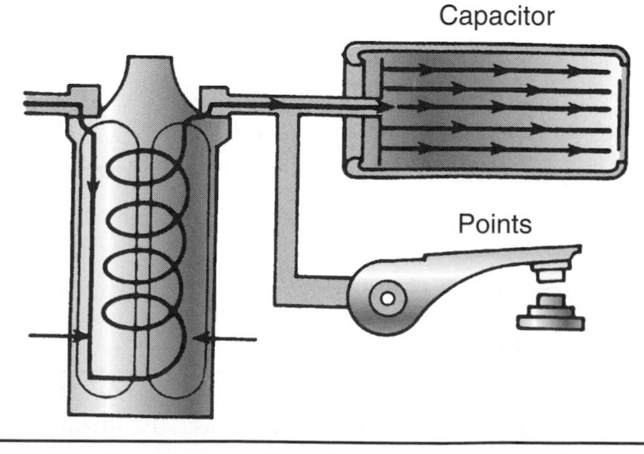

FIGURE 17–14 Capacitor absorbing surges of electrons as points open.

The steps of ignition coil operation can be summarized as follows:

1. Battery current flows through the primary windings when the primary control device closes.
2. Current flowing through primary windings creates an electromagnetic field that mutually induces a voltage onto the secondary windings.
3. Current flowing through the primary windings induces a CEMF on itself, called self-induction.
4. The coil's magnetic field becomes fully saturated as the battery EMF overcomes the CEMF in the primary coil windings.
5. The primary control device opens the primary circuit, stopping the current flow.
6. The magnetic field rapidly collapses, and the magnetic lines of force cut across the secondary windings.
7. The system capacitor absorbs the surges of electrons induced in the primary windings by the collapsing magnetic field.
8. The moving magnetic lines of force induce a voltage into the secondary windings that is high enough to jump the spark plug gap.
9. The high voltage—several thousand volts—jumps the spark plug gap and completes the circuit to ground.
10. The sequence repeats itself up to 200 times per second, depending on number of cylinders and engine revolutions per minute.

Ignition Coil Testing Three basic tests of the coil windings can be performed with an ohmmeter. The first test checks the resistance of the primary coil windings. With the key off and the wiring disconnected from the coil, connect the ohmmeter leads across the positive and negative terminals of the coil (Figure 17–15). Primary resistance usually ranges from 0.5 to 2.0 ohms maximum. Always refer to the vehicle's service manual for detailed specifications.

The second ignition coil test checks the resistance of the secondary windings. Again, with the key off and the wiring disconnected from the coil, connect the ohmmeter leads across the positive terminal and the coil's high-voltage

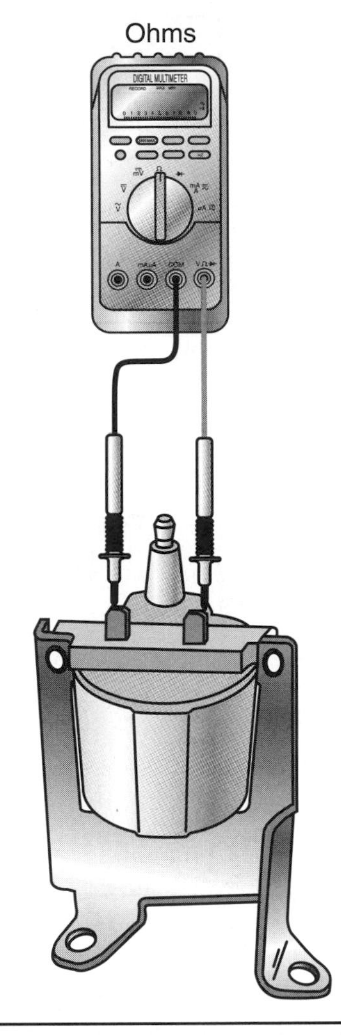

FIGURE 17–15 Testing the resistance of the coil primary windings using an ohmmeter.

center tower connection (Figure 17–16). Secondary resistance usually ranges from 5,000 to 20,000 ohms maximum. Always refer to the vehicle's service manual for actual specifications.

The third ignition coil test checks for grounding of the primary and secondary windings to the coil case. Three individual ground tests should be performed while the circuit wiring is disconnected from the coil. The reading in each test should indicate infinite (OL). A resistance reading indicates a shorted primary or secondary winding to the coil case.

1. Connect one ohmmeter lead to the coil case and the other to the primary positive terminal.
2. Connect one ohmmeter lead to the coil case and the other to the primary negative terminal.

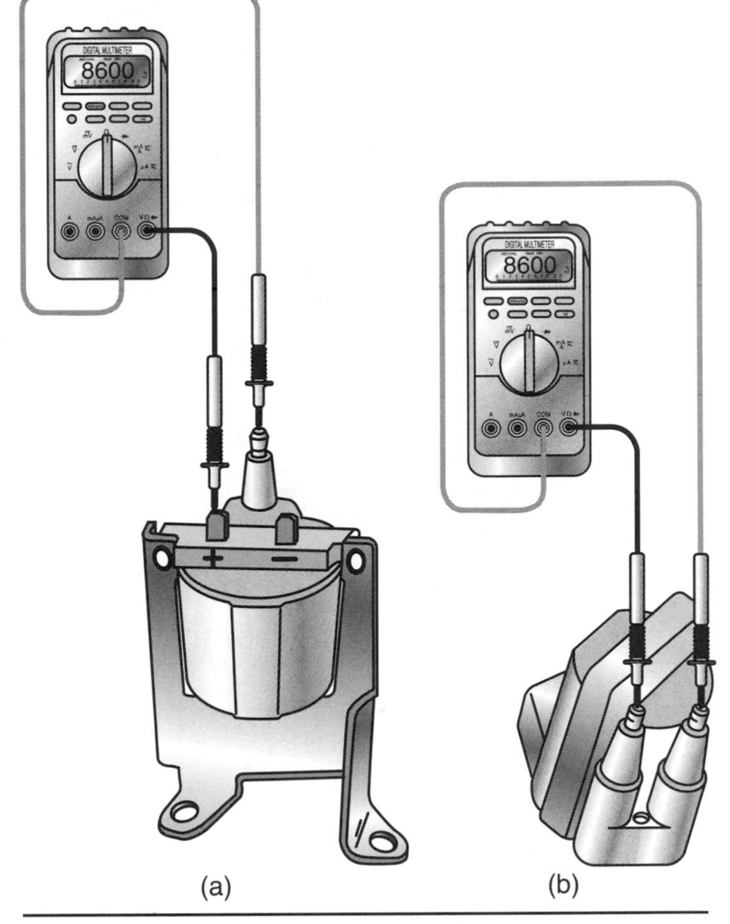

FIGURE 17–16 Testing the secondary winding resistance of two types of coils: (a) typical coil design and (b) waste spark coil.

3. Connect one ohmmeter lead to the coil case and the other to the secondary center post.

The ignition coil tests give a general view of the condition of the primary and secondary coil windings. These tests do not guarantee that the coil will work as designed. Some defects can be detected only while the coil is in operation and on an oscilloscope that checks for defective insulation on the windings and secondary tracking or arcing shorts to the case (Figure 17–17).

Primary System Control Switching Devices

There have been major advancements in ignition primary circuit switching devices over the past two decades. Even though there are tremendous variations among vehicle manufacturers, the central purpose has

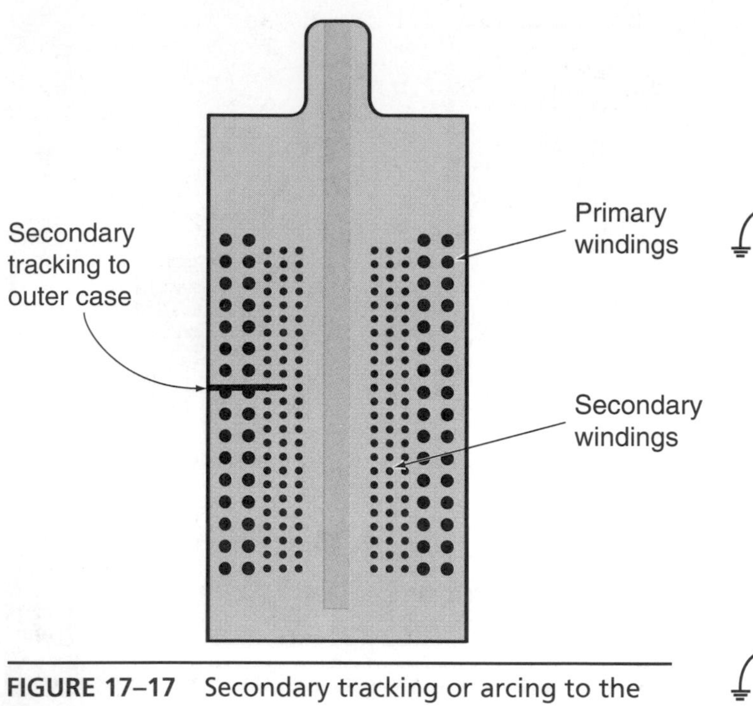

FIGURE 17–17 Secondary tracking or arcing to the coil area.

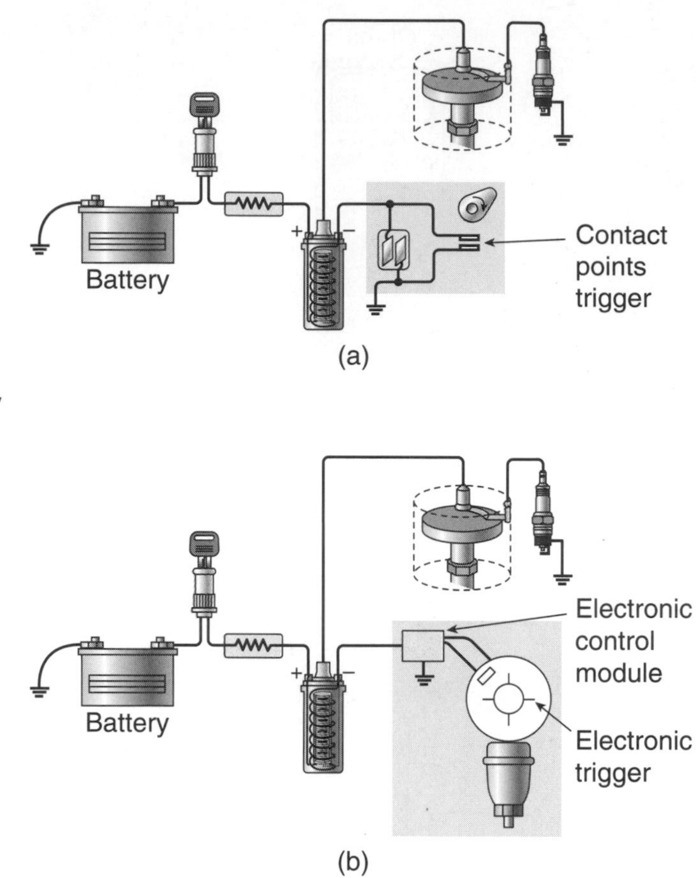

FIGURE 17–18 Basic comparison between an ignition system with (a) a mechanical trigger and (b) an ignition system with an electronic controlled trigger. (Courtesy of OTC/SPX Service Solutions)

never changed: to break the current flow in the coil primary windings at a precise time so that the saturated field may collapse. Figure 17–18 is a comparison between the older style of mechanical contact point–triggered ignition system and an older style of distributor ignition system with electronic trigger control. In this section, we briefly discuss the simple mechanical form of switching device that was used for decades and then the complex computer-controlled systems in use on newer vehicles.

Breaker Point Ignition For over 60 years, the **breaker point ignition** was the main method of controlling the primary system circuit. It consists of a mechanical contact point set and capacitor (condenser) mounted on a breaker plate inside the distributor (Figure 17–19). The contact point set consists of a fixed contact point that is grounded to the breaker plate base and a movable contact point connected to the negative terminal of the coil's primary winding. The movable contact points are connected to a rubbing block that rubs against the rotating distributor cam, causing the points to open and close a set distance.

When the contact points are closed, current flows into the primary circuit windings and coil saturation occurs. The amount of time that the contact points are

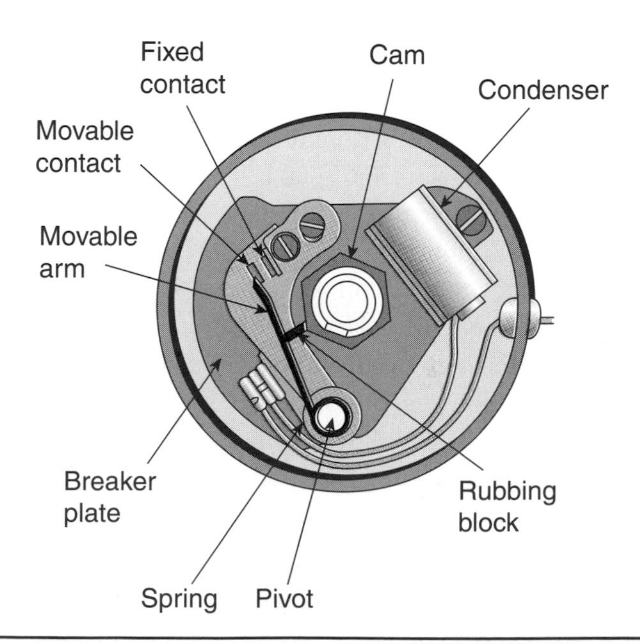

FIGURE 17–19 Typical older ignition system using breaker points as the primary switching device.

closed is called the **dwell,** or cam angle (Figure 17–20). Ignition dwell is adjusted to the manufacturer's specifications by adjusting the point gap. Moving the points away from the distributor cam increases the dwell; moving the points closer decreases the dwell.

Although common in pre-1980s vehicles and in farm or industrial equipment, high-maintenance breaker point ignition systems were abandoned and replaced by electronically controlled ignition systems. These systems helped decrease auto emissions and increase fuel efficiency.

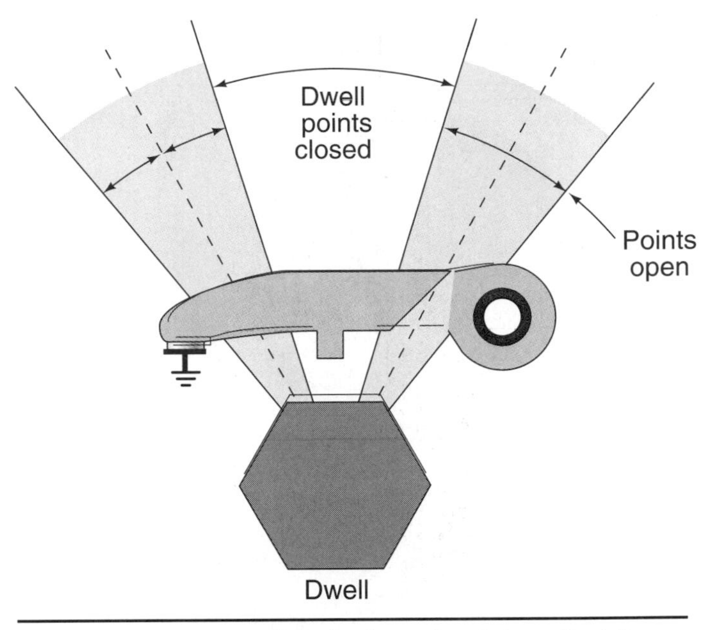

FIGURE 17–20 Dwell is the length of time that the ignition contact points are closed.

Solid-State Ignitions Several designs of electronically controlled primary ignition circuits are in use today. Breaker points have been replaced with electronic triggering devices that are connected to a **solid-state ignition** control module that precisely controls the on and off time of the primary circuit (Figure 17–21).

Four common electronic triggering systems are used by major vehicle manufacturers: magnetic pulse generator

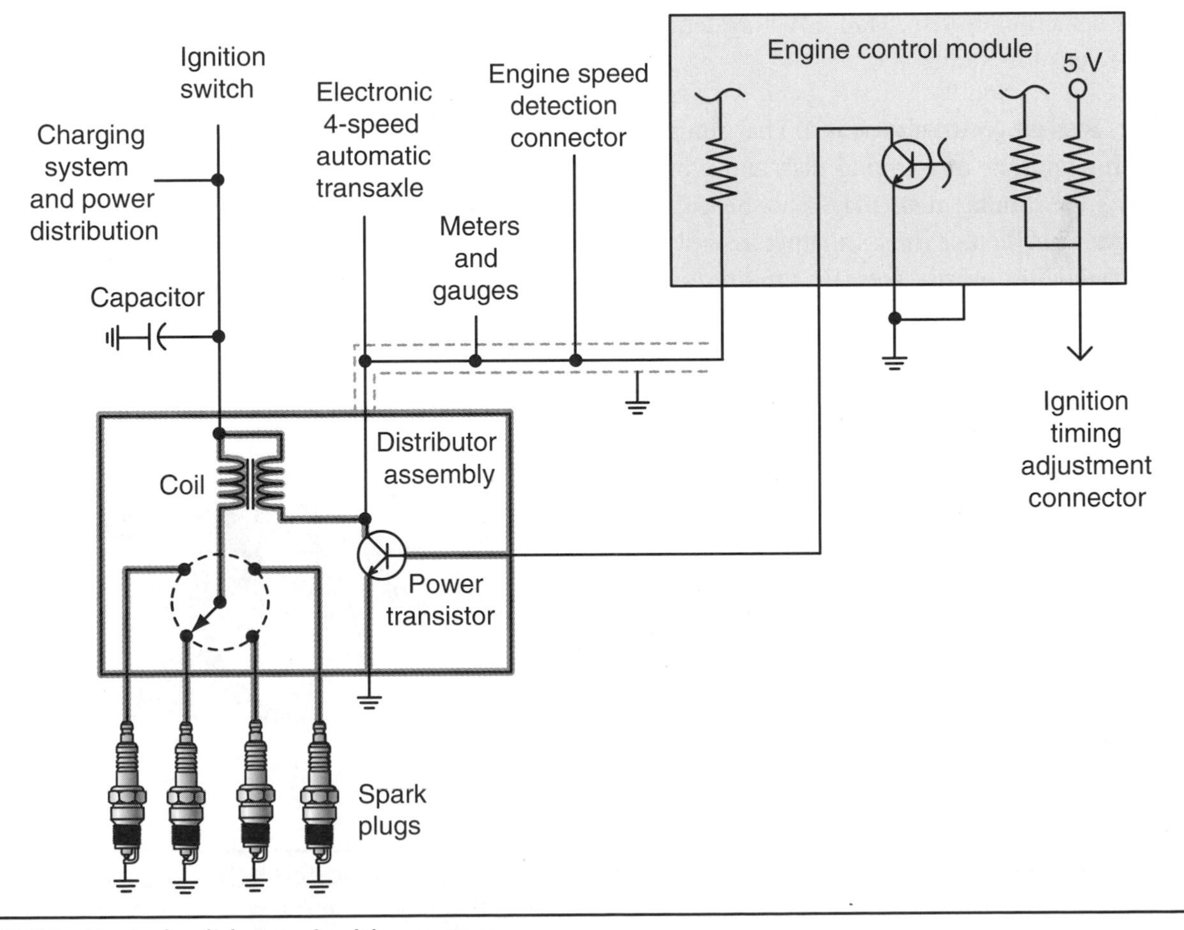

FIGURE 17–21 Typical solid state ignition system.

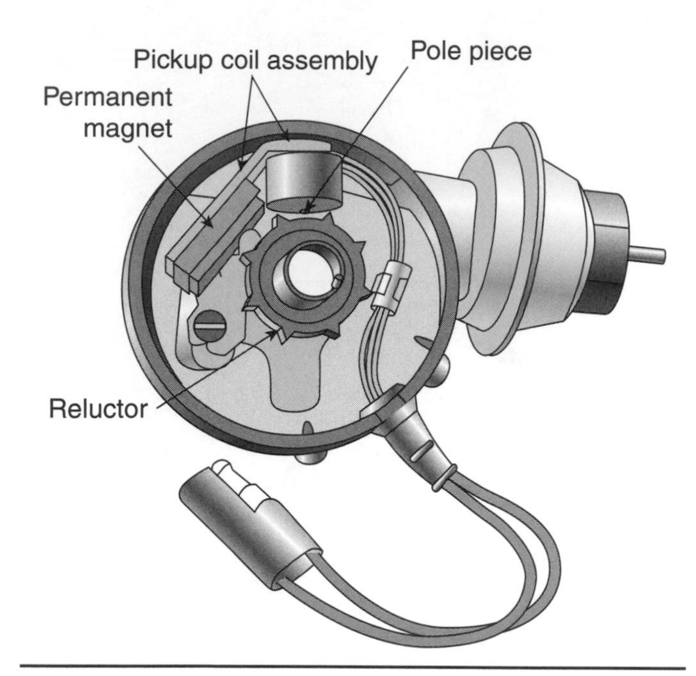

FIGURE 17–22 A magnetic pulse AC generator assembly located inside a distributor.

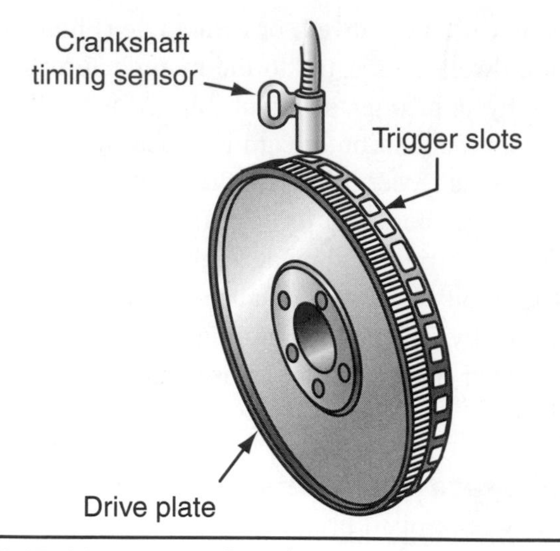

FIGURE 17–23 Magnetic pulse AC generator assembly positioned to sense flywheel rotation through drive plate trigger slots.

system, metal detection system, Hall effect system, and a photoelectric sensor system.

Magnetic Pulse Generator The **magnetic pulse generator** consists of a timing disk and a pickup coil assembly. The timing disk may also be called a reductor, trigger wheel, pulse ring, armature assembly, or timing core depending on the specific ignition system manufacturer. The timing disk may be mounted on the distributor shaft in the distributor (Figure 17–22), on the engine flywheel drive plate (Figure 17–23), on the crankshaft vibration damper (Figure 17–24), or near the center of the crankshaft (Figure 17–25).

The magnetic pulse generator or **PM generator** operates on the basis of electromagnetism. The magnetic field for the PM generator is provided by the **pickup coil** assembly and by the movement of the timing disk, or reductor, through the magnetic field. In Figure 17–26, the high points of the left reductor are out of alignment with

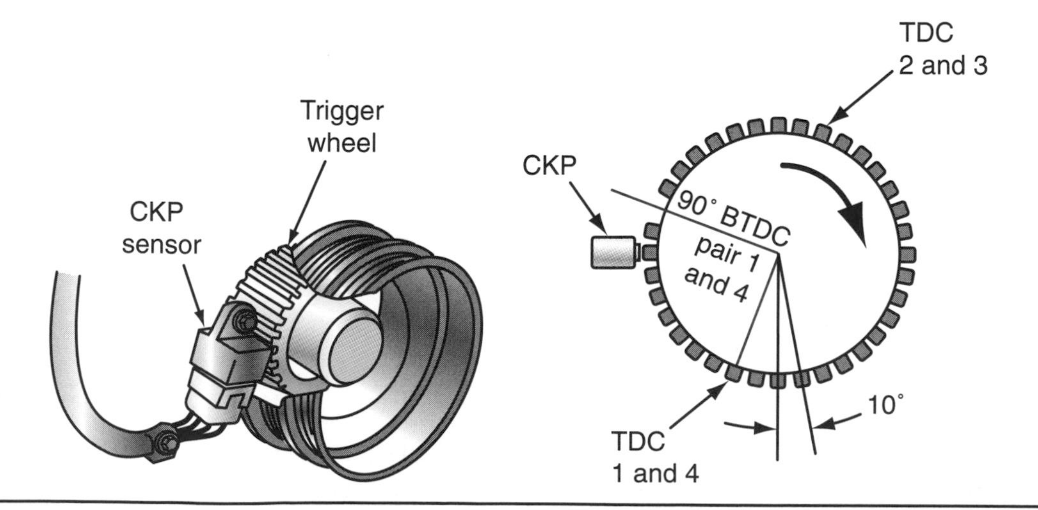

FIGURE 17–24 Magnetic pulse AC generator (crank position sensor, CKP) located behind crankshaft vibration damper.

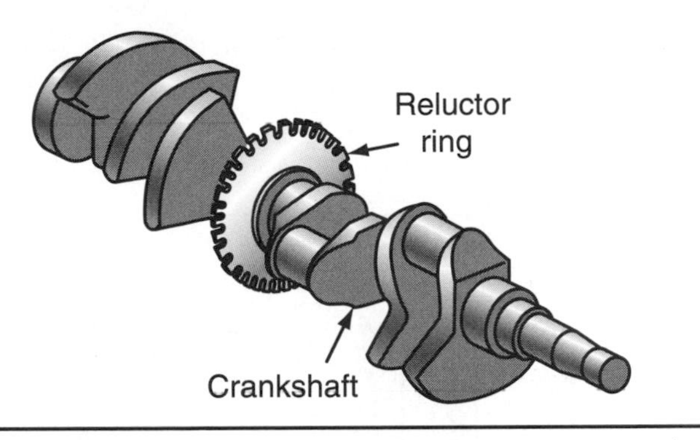

FIGURE 17–25 A midcrankshaft reluctor ring used with a magnetic pulse AC generator.

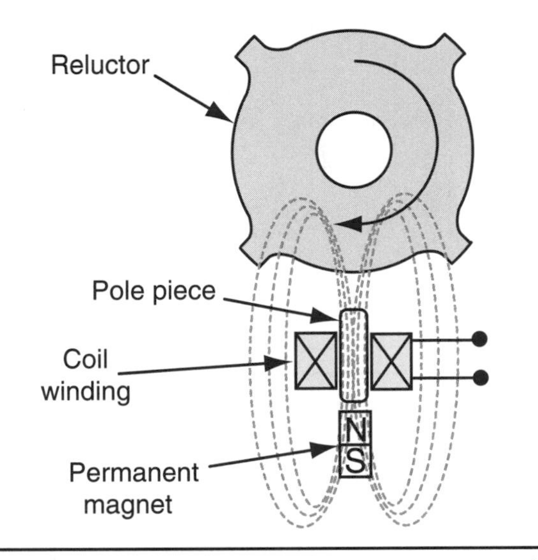

FIGURE 17–26 A weak magnetic field exists in the PM generator when the reluctor high points are out of alignment with the pickup cell.

the pickup coil, so a weak magnetic field and voltage exist in the pickup coil. When the reductor tooth is directly in line with the pickup coil (B in Figure 17–27), the magnetic field is not expanding or contracting, so voltage in the pickup coil at this precise moment drops to 0. The ignition module switching device reacts to the 0 voltage and turns off the ignition primary circuit current. When the reductor high point passes the pickup coil, the magnetic field expands again and produces another voltage signal. The increasing and decreasing of the voltage in the pickup coil, in relation to 0, trigger the ignition module to turn the primary circuit on and off. Figure 17–28 is an example of a typical ignition control module mounted externally on the distributor.

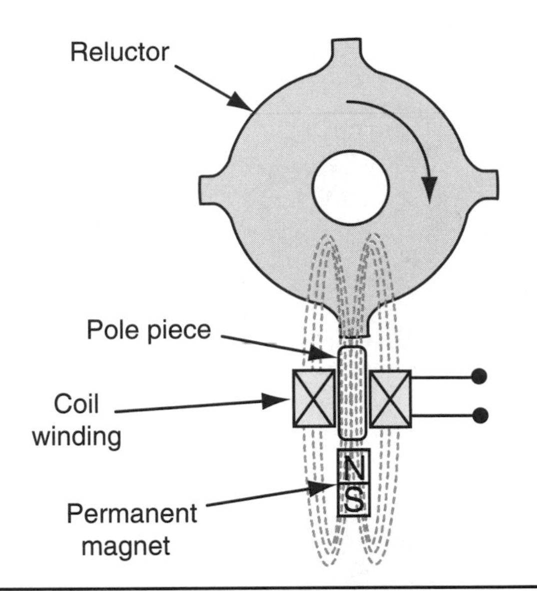

FIGURE 17–27 A strong magnetic field exists in the PM generator when the reluctor high points are aligned with the pickup coil.

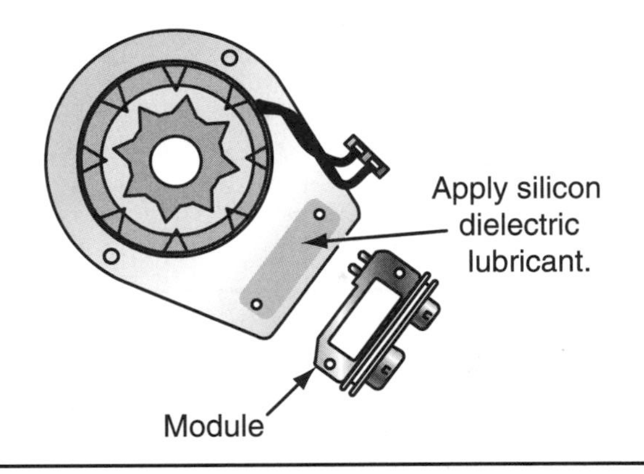

FIGURE 17–28 A typical ignition control module mounted on the side of the distributor.

The pickup coil of a PM generator can be checked for an open or shorted condition (Figure 17–29). With the ohmmeter 2 connected as shown between the two pickup coil leads, the pickup coil resistance can be checked. Most pickup coils have a resistance of between 150 and 900 ohms, but always refer to the manufacturer's specifications for allowed resistance values. If the pickup coil is open, the reading would be infinite (OL).

With ohmmeter 1 connected as shown between the distributor housing and one of the pickup coil leads, the pickup coil can be checked for a short-to-ground condition. The ohmmeter reading should be infinite (OL).

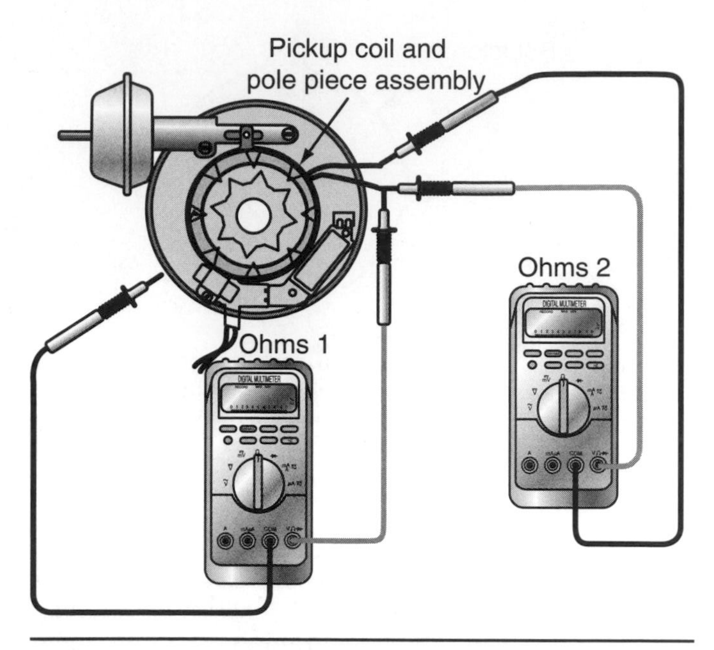

FIGURE 17–29 Ohmmeter 1 testing for short-to-ground and ohmmeter 2 testing the resistance of the pickup coil.

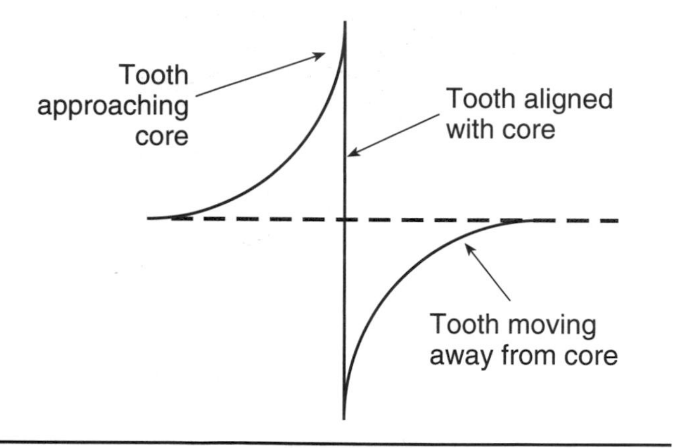

FIGURE 17–30 The modified AC pulse waveform produced by a PM generator.

The pickup coil operation can also be tested with an oscilloscope (Figure 17–30). With the scope leads hooked to the pickup coil leads, a pulsing AC waveform should appear on the screen while the distributor is turning. The trace is not a true sine wave, but it should have both positive and negative pulses. This test can also be performed with a voltmeter set on the AC scale.

Metal Detection Sensors On a few early electronic ignition systems, **a metal detection sensor** is used to control the ignition primary circuit (Figure 17–31).

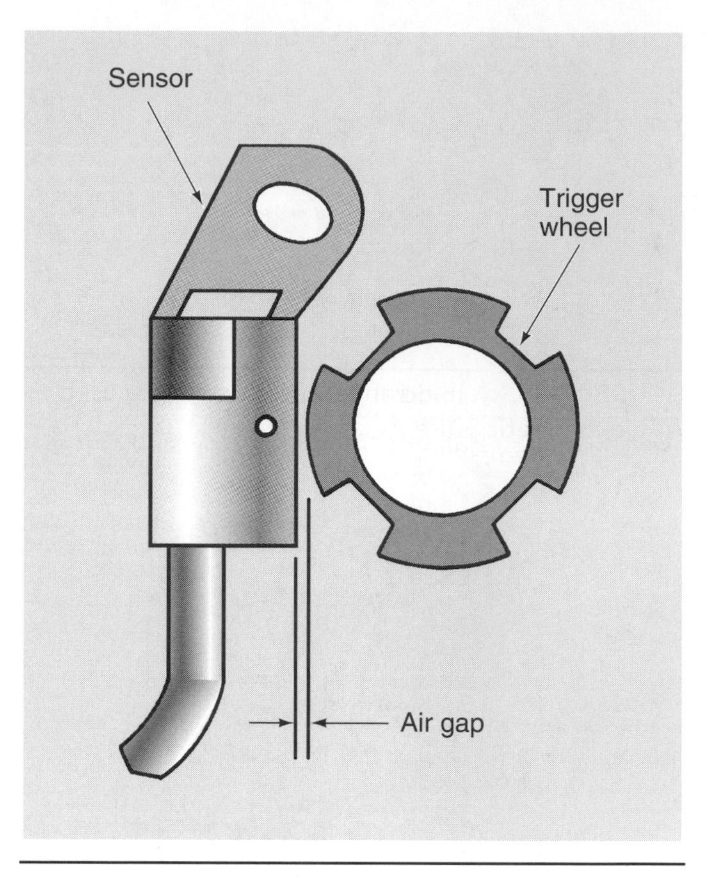

FIGURE 17–31 Typical metal detector sensor.

The metal detection sensor works similarly to a PM generator, but the pickup coil of a metal detection sensor does not have a permanent magnet. Instead, it is a weak electromagnet with a low-level current supplied to the pickup coil by the electronic control unit. The operation of the rest of the system is the same as the PM generator.

Hall Effect Sensor Since the early 1980s, the **Hall effect sensor** has worked well as a precise triggering mechanism for the primary ignition system. Unlike the PM generator or the metal detection sensor, the Hall effect sensor produces an accurate square wave voltage signal throughout the entire range of the engine's revolutions per minute. The Hall effect switching method is based on the principle that if a current is allowed to flow through a thin conducting material, and when that material is exposed to a magnetic field, another voltage is produced (Figure 17–32).

The heart of the Hall effect system is a thin semiconductor gallium arsenate crystal with one positive and one negative terminal that provides source current for the

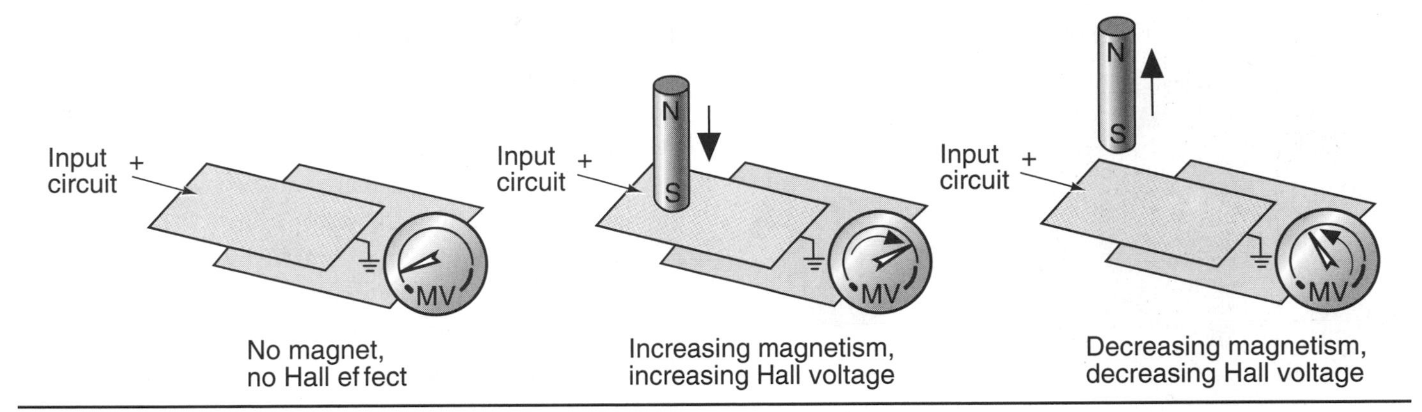

FIGURE 17–32 Hall effect principles of voltage induction.

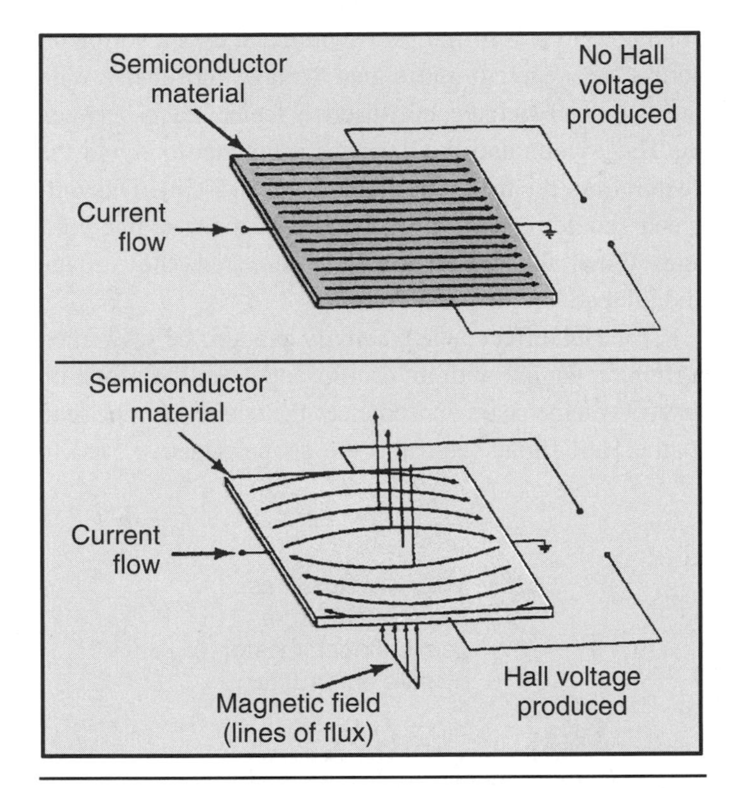

FIGURE 17–33 Hall effect semiconductor material producing a small voltage.

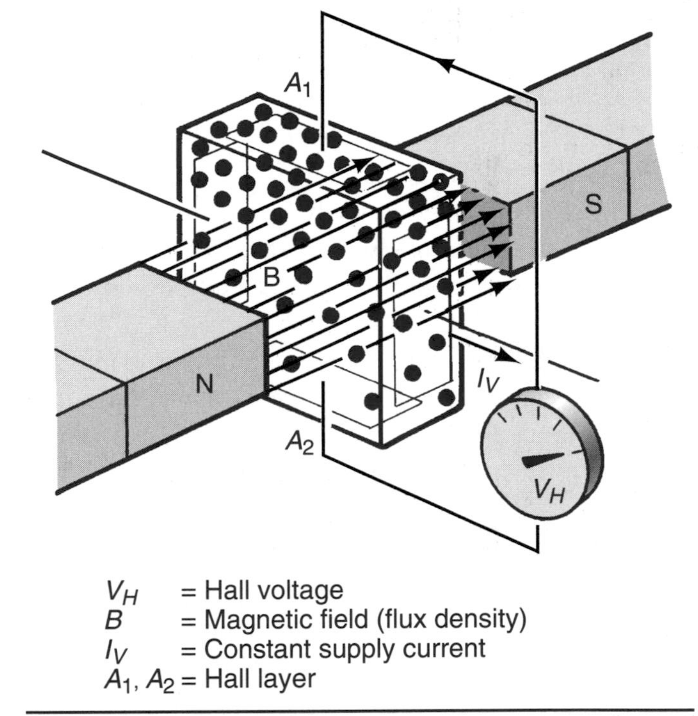

FIGURE 17–34 Action of a Hall effect switch.

Hall transformation, as well as two additional terminals that form the signal output from the Hall layer. Directly across from the semiconductor element is a small permanent magnet positioned so that its magnetic flux lines bisect the Hall crystal at right angles to the current flow (Figure 17–33).

When current is passed through the Hall layer, a voltage is produced perpendicularly to the direction of current flow and magnetic flux (Figure 17–34). The signal voltage produced is the direct result of the magnetic field's effect on the electron flow. As the magnetic lines of force collide with the electrons in the supply current, a weak voltage potential is produced through the crystal from end to end.

Any time the Hall switch is being exposed to a magnetic field, it is on and producing a very weak voltage signal. To shield the Hall switch from the magnetic field, a shutter wheel (Figure 17–35), consisting of a series of alternating windows and vanes, passes between the Hall layer and permanent magnet to create a magnetic shunt

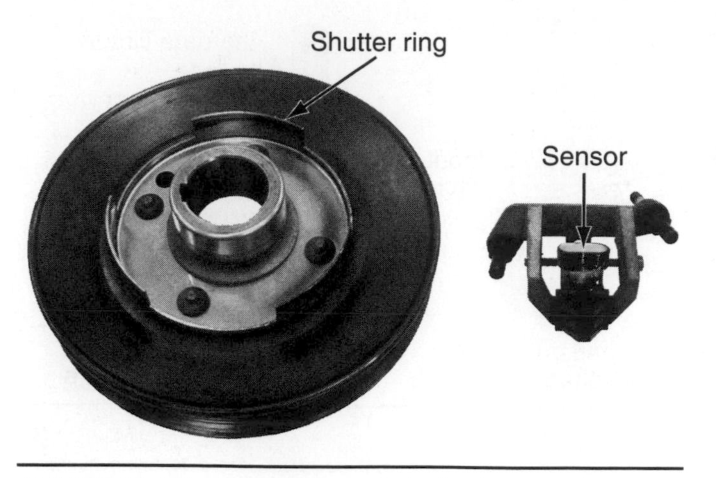

FIGURE 17–35 A typical crankshaft damper shutter ring or wheel for a Hall effect sensor.
(Courtesy of Tim Gilles)

that blocks the magnetic field (Figure 17–36). As a result, the magnetic field is concentrated in the shutter blade, and the electrons in the supply current are no longer disrupted and return to a normal state. This results in a low-voltage potential in the signal circuit portion of the Hall switch.

Before the Hall signal can be of any use, it must be modified (Figure 17–37). The weak voltage signal is first sent to an amplifier, which strengthens and inverts it so that the signal reads high when it is actually coming in low, and vice versa. Once it has been inverted, the signal goes through a pulse-shaping device called a **Schmitt trigger,** where it is cleaned up to be a square wave digital signal. The square wave signal is then fed to the base of a switching transistor to turn it on and off in response to the signals generated by the Hall switch assembly. The computer's voltage-sensing circuit senses the switching transistor being turned on and off, then determines the frequency of the signals and calculates engine speed.

Most Hall effect sensors can be tested by connecting a 12-volt battery across the positive and negative supply voltage terminals of the Hall switch and then a voltmeter across the negative and signal voltage terminals. With the voltmeter hooked up, insert a feeler blade between the Hall switch and the permanent magnet to shield the switch from the magnetic field (Figure 17–38). The voltmeter should read within 0.5 volt of battery source voltage. When the feeler gauge is removed, the voltage should read less than 0.5 volt.

The Hall effect switch activity can also be viewed on a running engine with an oscilloscope. Set the scope on an AC voltage scale and connect the scope positive lead to the Hall signal lead and the scope negative lead to

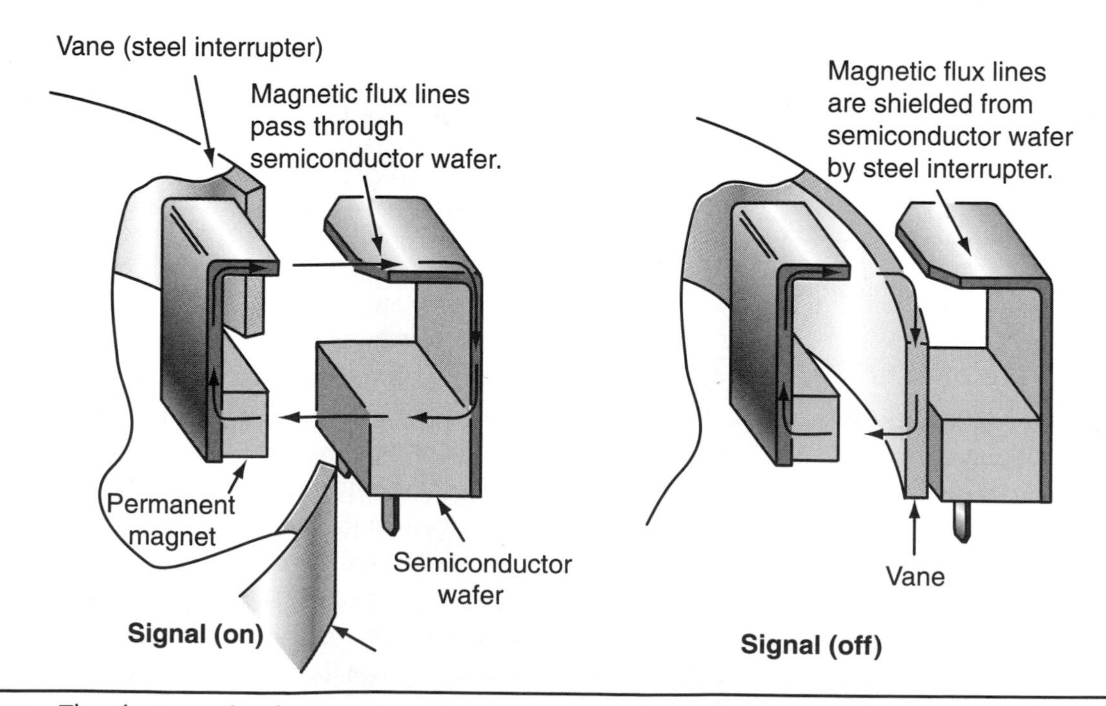

FIGURE 17–36 The shutter wheel or vane creates a magnetic shunt that blocks the magnetic flux lines.

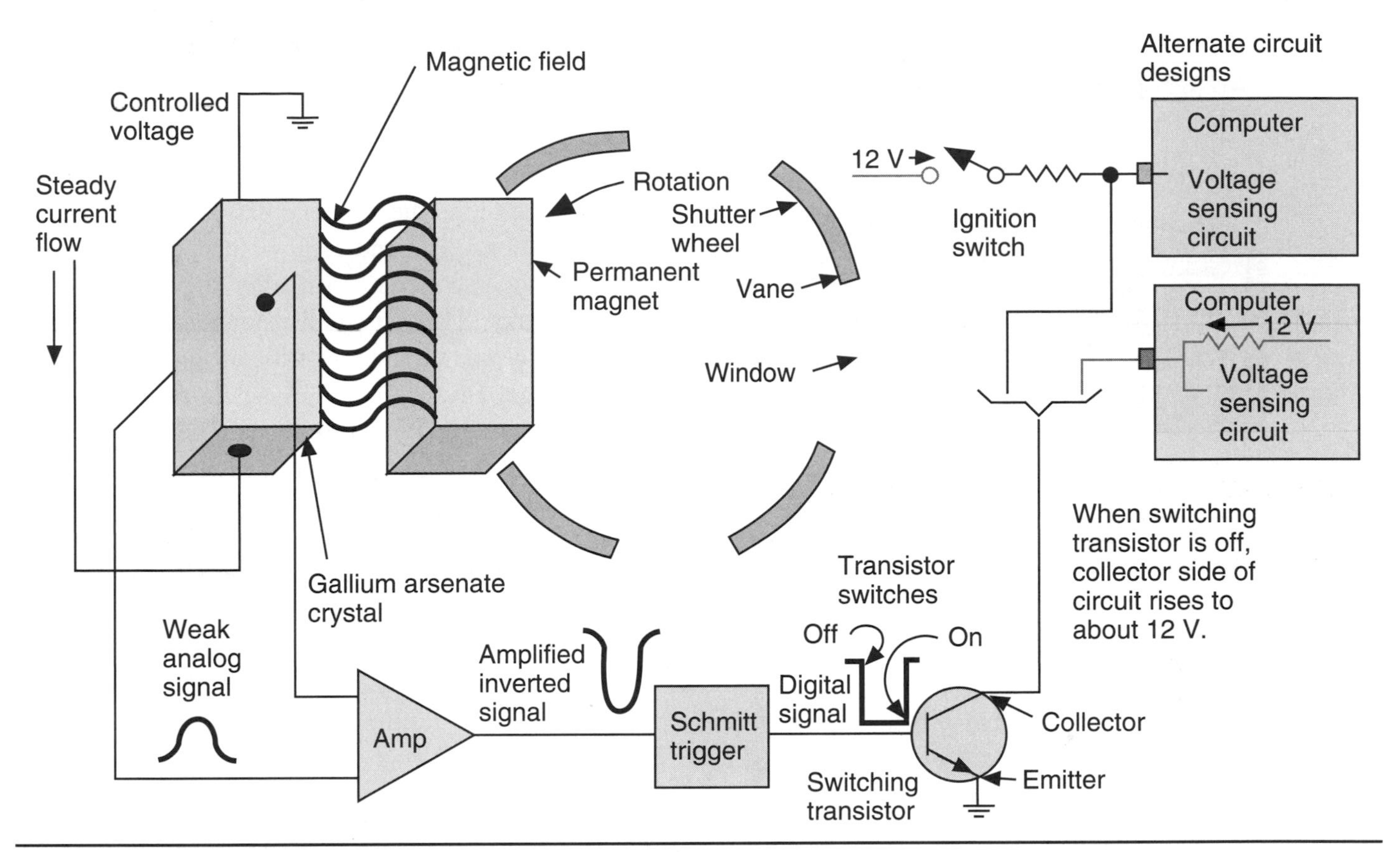

FIGURE 17–37 Typical Hall effect switch circuit.

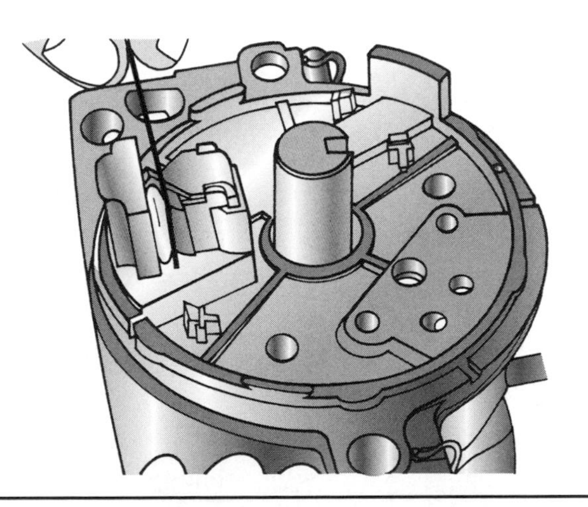

FIGURE 17–38 Testing a Hall effect sensor for proper operation with a steel feeler gauge.

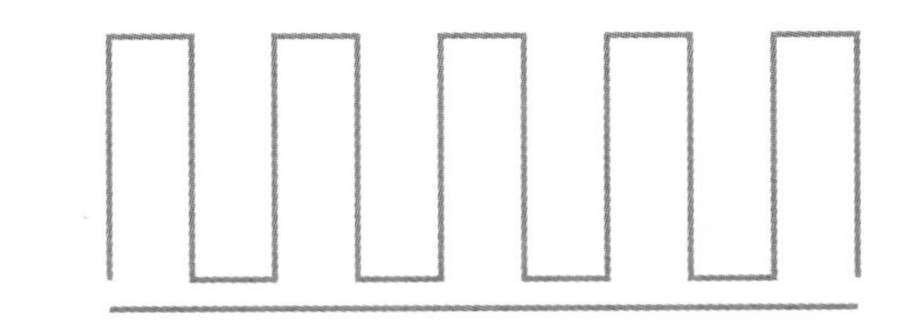

FIGURE 17–39 Square wave trace produced by a normal Hall effect switch.

ground or the ground terminal at the Hall switch connector. With the engine running, the scope pattern should be a square wave pattern ranging from approximately 0 to 12 volts (Figure 17–39). If the pattern is distorted or the range is out of specifications, the Hall switch should be replaced.

Photoelectric Sensor The last type of sensor used to control the ignition primary circuit is the **photoelectric sensor.** It consists of a light-emitting diode (LED), light-sensitive photo cells, and a slotted disk called a light beam interrupter (Figure 17–40). The slotted disk is attached to the distributor shaft and rotates between the LED and the photo cells. Light from the LED shines through the slotted disks to the photo cells, which translate the intermittent flashes of light into a voltage signal pulse similar to that of a Hall effect sensor. When the voltage signal occurs, the control unit turns the primary ignition system on until the voltage signal ceases from the photo cell.

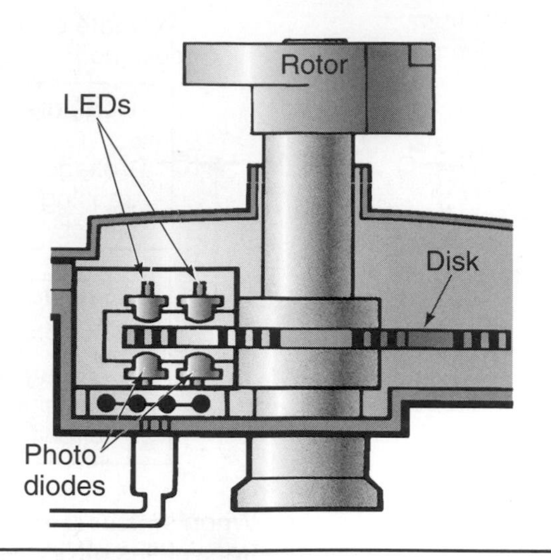

FIGURE 17–40 A distributor with a photoelectric sensor.
(Courtesy of DaimlerChrysler Corporation)

The photoelectric sensor system in Figure 17–41 uses two photo diode cells: one for the 360-degree camshaft signal and the other a 60-degree signal corresponding to the number of cylinders and crankshaft degrees. The slotted disc is for a six-cylinder engine with a cylinder firing every 120 degrees of crankshaft rotation.

Timing Advance Mechanisms

Early electronic ignition distributors changed the ignition timing mechanically, as breaker point distributors did. The need to change ignition timing is determined by the burn time of the fuel, the engine speed changes, and load fluctuations. The end result is that the peak combustion power must occur just after the piston passes top-dead-center (TDC). Two mechanical timing advance systems were used on older distributor ignition systems: the centrifugal advance and the vacuum advance. Both are covered only briefly in this chapter, because they are becoming increasingly rare in today's vehicle repair environment.

Centrifugal Advance The **centrifugal advance** mechanism consists of a set of weights and springs connected to the distributor shaft and armature plate assembly (Figure 17–42). During idle speeds, the springs keep the weights in their resting position and the armature and distributor shaft rotate as one assembly. When engine speed increases, centrifugal force causes the weights to move outward, allowing the armature assembly to move ahead of (advance) the distributor shaft rotation. The ignition's triggering device is mounted to the armature plate

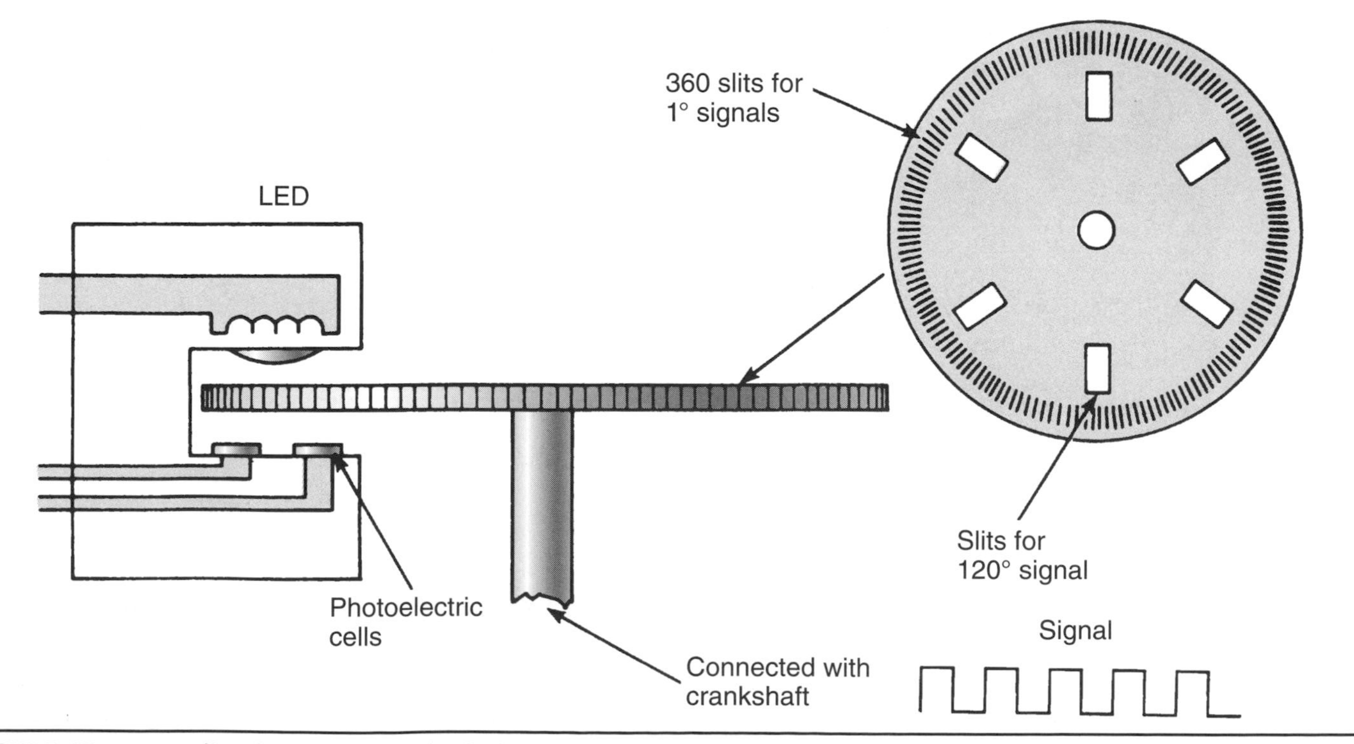

FIGURE 17–41 A distributor with a dual photoelectric sensor.

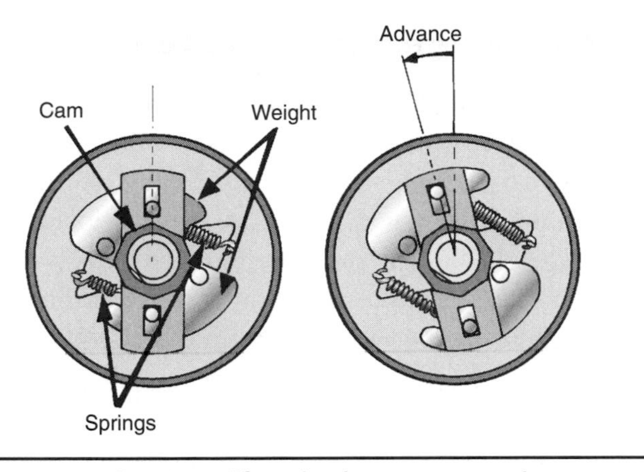

FIGURE 17–42 Centrifugal advance operation on an old-style distributor.

(Courtesy of Allied Signal Automotive Aftermarket/Autolite)

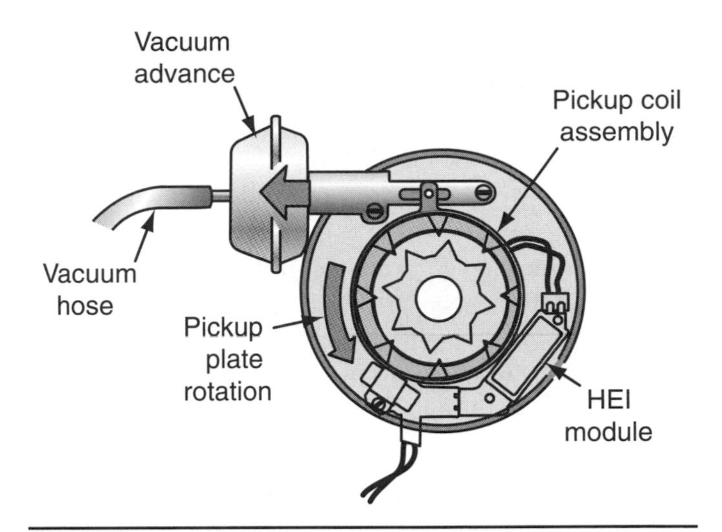

FIGURE 17–44 Typical action of a distributor vacuum advance mechanism.

assembly. The amount of centrifugal advance depends on engine speed and usually follows a designed power curve (Figure 17–43).

Vacuum Advance When the engine is at part-throttle operation, the **vacuum advance** mechanism advances the ignition timing according to the vacuum in the intake manifold. The burn rate of the fuel/air mixture at a part throttle and load is slower than at full throttle and heavy load. To ensure that peak combustion pressure occurs just after the piston passes TDC, the vacuum advance mechanism advances the plate in the distributor that holds the primary switching device in relation to the distributor shaft (Figure 17–44). The heart of this system is the spring-loaded diaphragm in the vacuum advance unit. It is connected to a movable plate in the distributor containing the primary circuit's triggering switch components. When vacuum is applied to one side of the diaphragm in the advance unit, atmospheric pressure pushes on the other side of it, causing the movable plate to turn and advance the ignition timing.

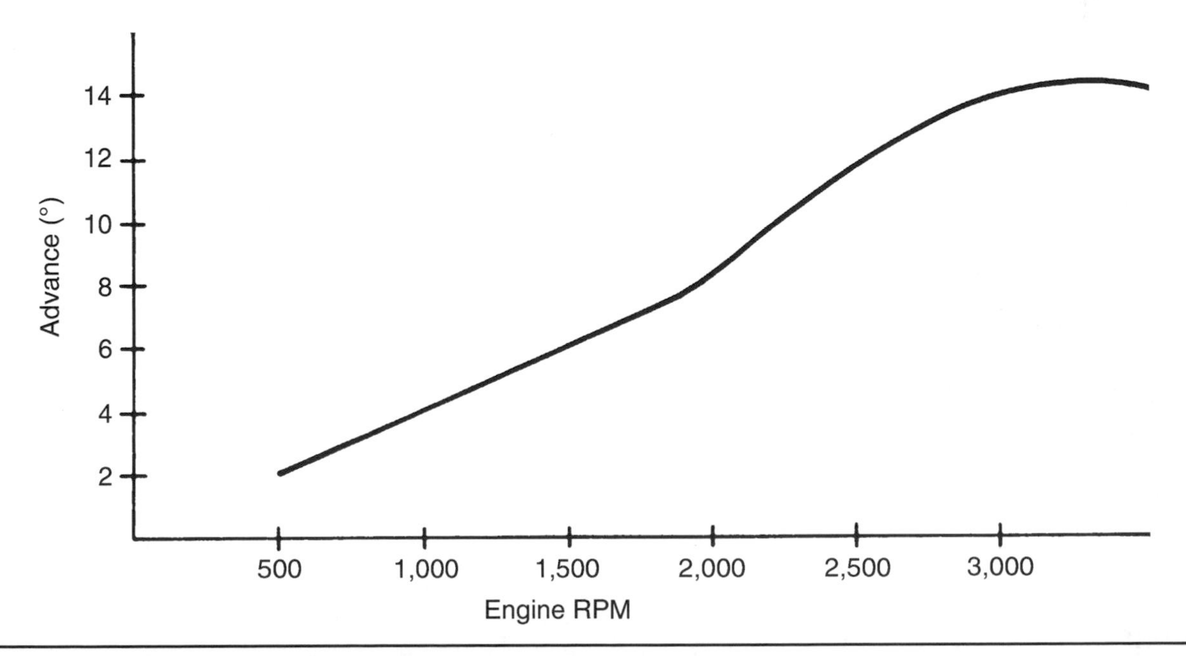

FIGURE 17–43 Engine speed affects centrifugal advance operation.

Electronically Controlled Timing Advance

In an effort to increase engine fuel efficiency and improve tailpipe emissions, the centrifugal and vacuum advance mechanisms have been replaced in most modern automobiles by computer-controlled systems, especially those produced in North America. One of the main reasons for the change is the inability of the mechanical or vacuum advance systems to recognize the individual timing needs of each cylinder.

With the elimination of the mechanical advance systems, the distributor's sole purpose is to generate the primary circuit switching signal and distribute the secondary voltage to the spark plugs. The ignition system in Figure 17–45 is an example of a **distributor ignition (DI)** system that has no centrifugal or vacuum advance units, only the primary switching function. Timing advance is controlled totally by an **electronic spark timing (EST)** function within the vehicle PCM. Some DI systems have even removed the primary switching mechanism from the distributor by using a crankshaft position sensor (CKP). The function of a distributor in this case is only to distribute the secondary voltage to the spark plugs.

Electronic Ignition Systems

In an effort to replace the ignition distributor with systems that have no moving parts, **electronic ignition (EI)** systems have emerged to electronically perform all the functions once performed by the distributor (Figure 17–46). In an EI system, the basic purpose is to provide the same ignition functions that a distributor has done in the past, but with no moving parts. The earlier EI systems were called distributorless ignition systems (DIS), but with the standardization of the SAE J1930 terminology, the term *electronic ignition (EI)* replaces all previous terms for distributorless ignition systems.

In an EI system the function of an ignition coil is the same as in previously discussed ignition systems. The coil transforms battery voltage to a high-voltage potential that jumps the spark plug gap. On most EI systems, there is one coil for every two spark plugs. This system is called the **waste spark** method of spark distribution. Each coil secondary winding is connected to a pair of spark plugs in the cylinders containing pistons that rise and fall together (Figure 17–47). When one cylinder is on the end of its compression stroke, the other is on the exhaust stroke. Both cylinders get spark simultaneously,

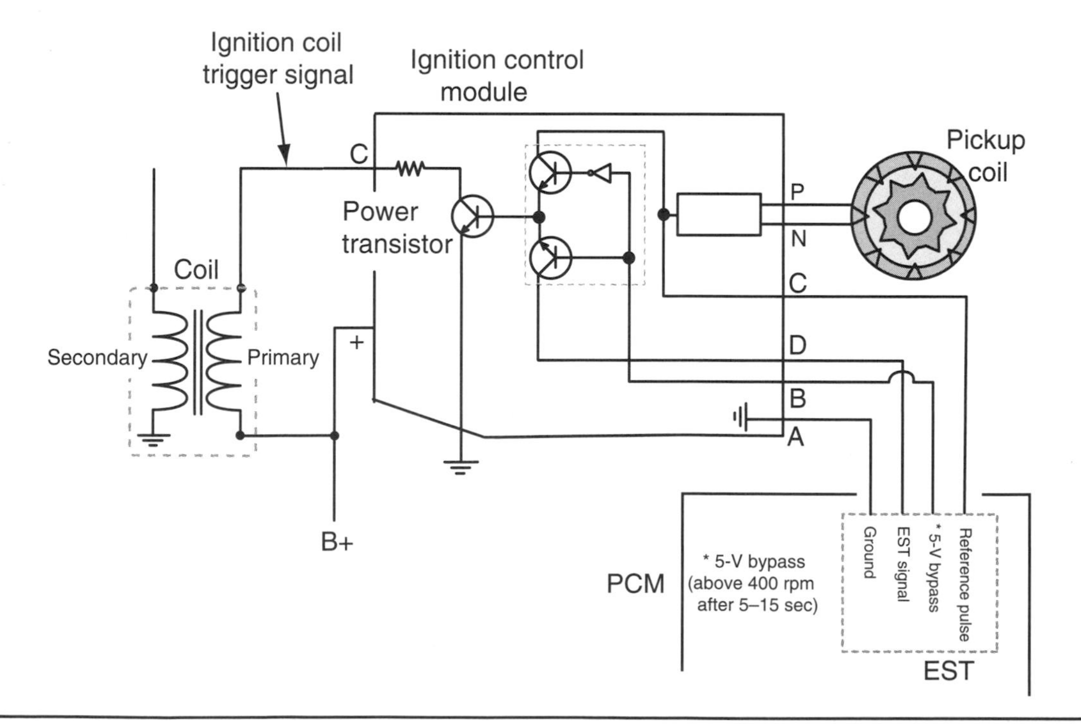

FIGURE 17–45 A typical electronic distributor ignition (DI) system using electronic spark timing (EST) to control ignition advance functions.

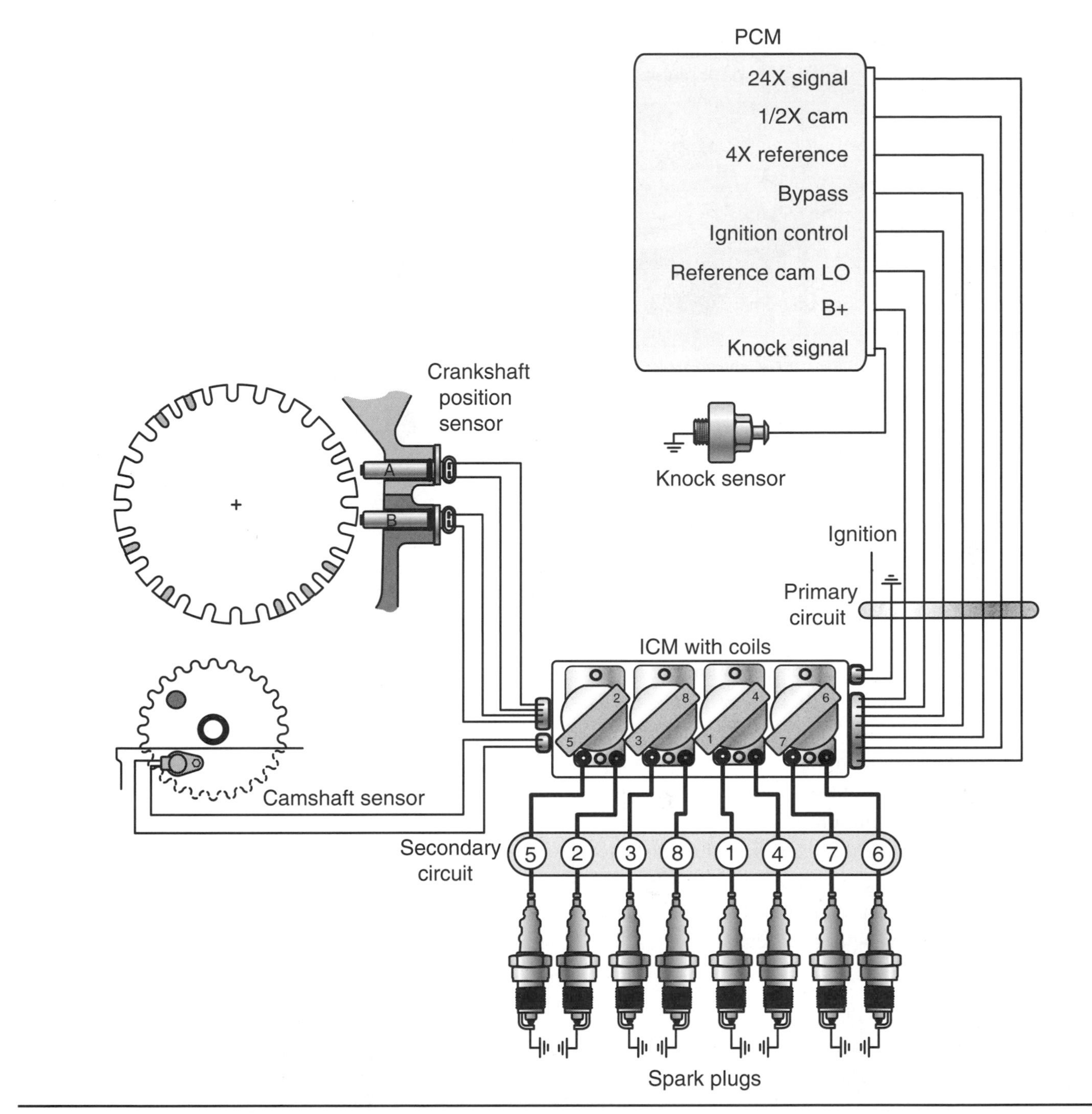

FIGURE 17–46 A typical electronic ignition (EI) system.

but only one spark generates power while the other is wasted out the exhaust.

When the induced voltage that cuts across the primary and secondary windings collapses, one spark plug fires in the normal direction from the center electrode to the outer electrode. The other spark plug fires in the reverse direction from the outer electrode to the center electrode to complete the series circuit. Each plug always fires in the same direction on both the compression and exhaust stroke (Figure 17–48).

Even though there is a reversed polarity in one of the spark plugs, the coil is able to produce a voltage as high as 100,000 volts to fire the spark plugs. There is very little resistance across the plug gap on the exhaust stroke,

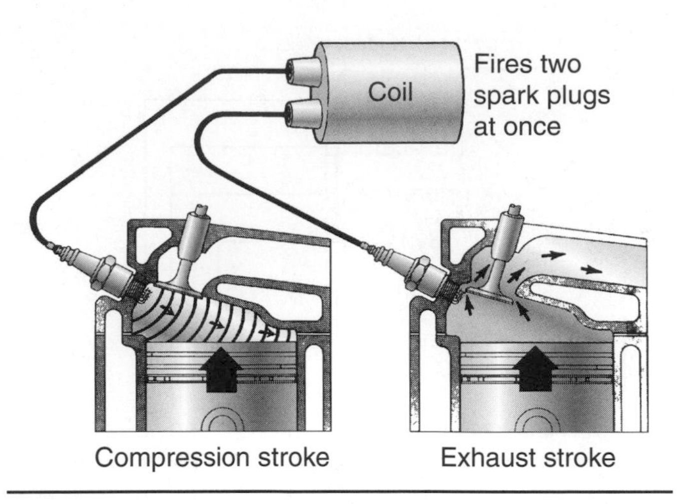

FIGURE 17–47 A typical waste spark ignition system. (Courtesy of Federal-Mogul Corporation)

so the plug requires very little voltage fire, thereby providing the plug on the end of the compression stroke with enough available voltage (Figure 17–49).

In a typical EI system (Figure 17–46), the ignition module uses crank/cam sensor data to control the timing of the primary circuit in the coils. The computer collects and processes this information to determine the ideal amount of spark advance for the operating conditions. The ignition module synchronizes the coils' firing sequence in relation to the crankshaft position and firing order of the engine. The ignition module is the component that takes the place of the distributor to distribute the high-voltage spark to the cylinders.

Figure 17–50 shows an example of a waste spark system called the **coil-over-plug** system, in which the coils are mounted directly over the spark plugs. This eliminates the need for plug wires between the coils and plugs, reducing the chance of high-voltage leaks. This type of system functions the same as other waste spark systems that share one coil for every two spark plugs, as shown in Figure 17–51.

Another coil-on-plug ignition design is a **direct ignition system (DIS),** with individual coils mounted

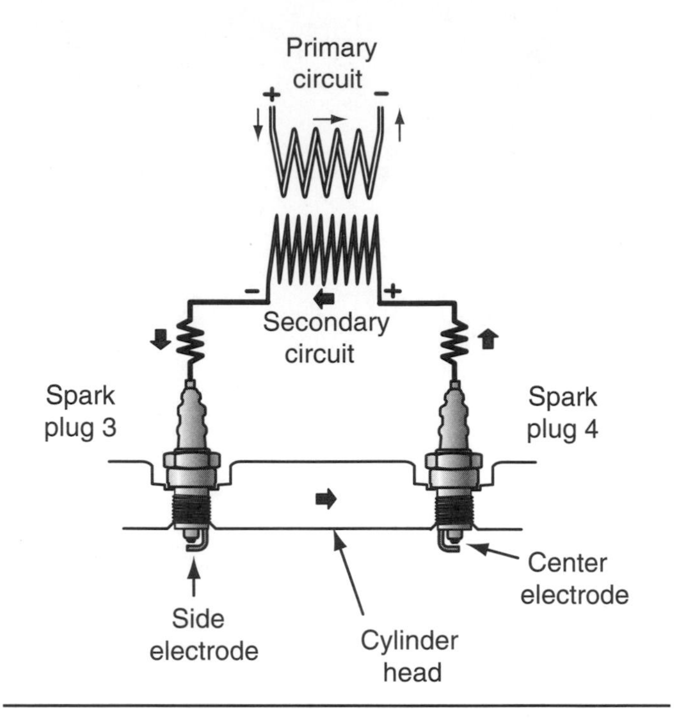

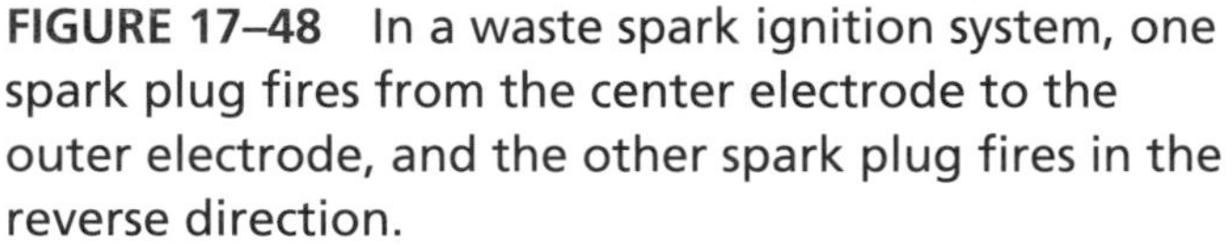

FIGURE 17–48 In a waste spark ignition system, one spark plug fires from the center electrode to the outer electrode, and the other spark plug fires in the reverse direction.

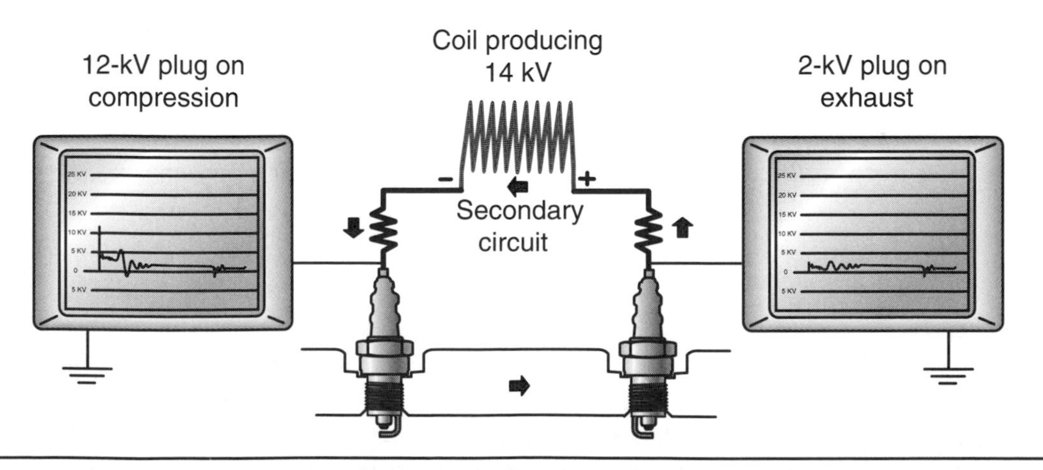

FIGURE 17–49 In a waste spark ignition system, the voltage produced by the coil is divided between the two spark plugs. The cylinder that contains the compression pressure takes more voltage to jump the plug gap.

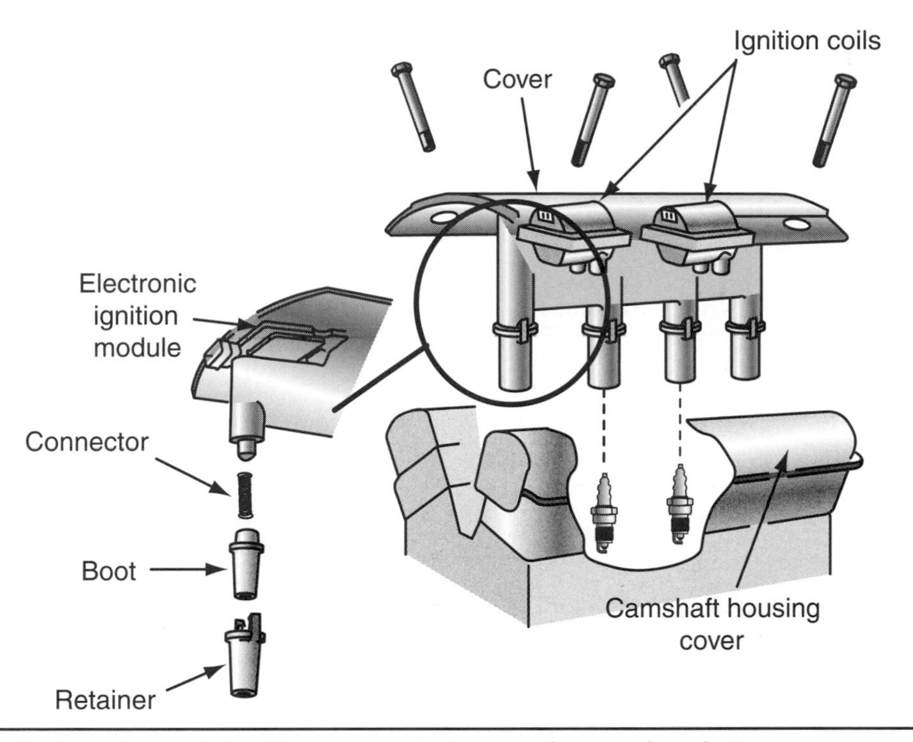

FIGURE 17–50 A typical waste spark ignition system using a coil-over-plug design.

directly to each spark plug on the engine (Figure 17–52). The individual coil assemblies (Figure 17–53) consist of primary and secondary windings, a magnetic core, and a high-voltage diode to rapidly collapse the magnetic field generated in the secondary winding (Figure 17–54). Some DIS designs use a common ignition module to control the individual coils, but some other designs integrate individual ignition modules into each coil-over-plug assembly Figure 17–55). Each coil assembly in Figure 17–55 consists of four wires: battery positive voltage, battery negative voltage, ignition timing control signal from the PCM, and an ignition confirmation signal that the PCM uses to verify the correct operation of the coil assembly. The DIS system has become very popular with vehicle manufacturers because of its better reliability, lower vehicle emissions, and fewer parts (plug wires).

Spark Plugs and Plug Wires

Spark Plugs Every type of ignition system uses spark plugs to ignite the fuel/air mixture in the cylinder. There are three main parts to a spark plug: the steel shell, the ceramic insulator, and a pair of electrodes (Figure 17–56). The steel shell holds the ceramic core and electrodes in a gas-tight assembly and contains the threads and sealing seat.

The center core on most spark plugs contains a resistor between the terminal and the center electrode. The purpose of this resistor is to control **radio frequency interference (RFI)** by suppressing AC voltage spikes while the spark plug is firing. Always use a resistor spark plug when one is specified for the engine. The additional radio frequency interference produced by a nonresistor plug could interfere with the operation of the radio or computer sensor.

Automotive spark plugs are available in either 14- or 18-millimeter diameters. All 18-millimeter plugs feature a tapered seat that seals with a mating seat in the cylinder head. The 14-millimeter plug can have either a tapered seat or a flat metal gasket for sealing (Figure 17–57).

Spark plug reach is an important design characteristic that could cause severe engine damage if it is wrong. The spark plug must be properly placed in the combustion chamber to adequately expose the electrodes to the fuel/air mixture (Figure 17–58). If the spark plug reach is too short, the electrodes may not fire adequately in the fuel/air mixture, and carbon deposits may develop in the

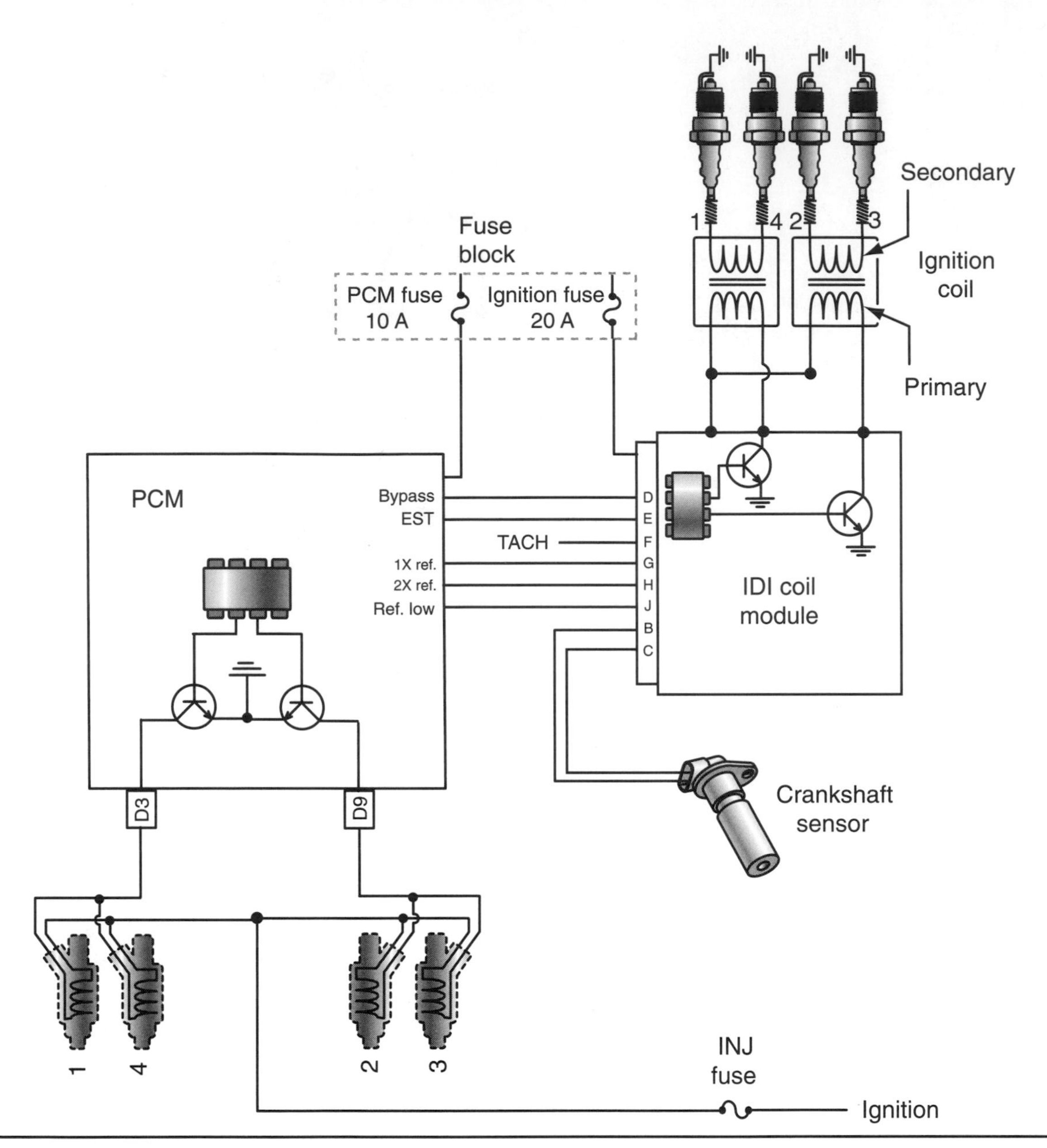

FIGURE 17–51 A wiring diagram for a typical coil-over-plug waste spark ignition system.

formed pocket. If the spark plug reach is too long, the spark plug may interfere with the operation of the piston and/or valves, causing severe engine damage. Exposed plug threads may also get so hot that preignition of the fuel/air mixture may occur, resulting in damage to internal engine components (Figure 17–58).

The spark plug heat range is important for the proper operation and long life of the spark plug (Figure 17–59). The heat range indicates the ability of the spark plug to conduct heat away from the center electrode tip. A cooler plug will transfer away the heat rapidly. A hot plug retains the heat longer. Installing a plug with the correct heat range is important. The spark plug must remain hot enough to burn away deposits and prevent fouling, but not so hot as to melt the plug or cause preignition. Continuous heavy driving conditions require a cooler plug than if the vehicle does a lot of stop-and-go driving.

The gap of the spark plug between the center electrode and the grounding electrode is important for the proper operation of the plug. Figure 17–60 shows an example of a correct spark plug gap being checked by a gap gauge. When the plug is new, the electrode surfaces

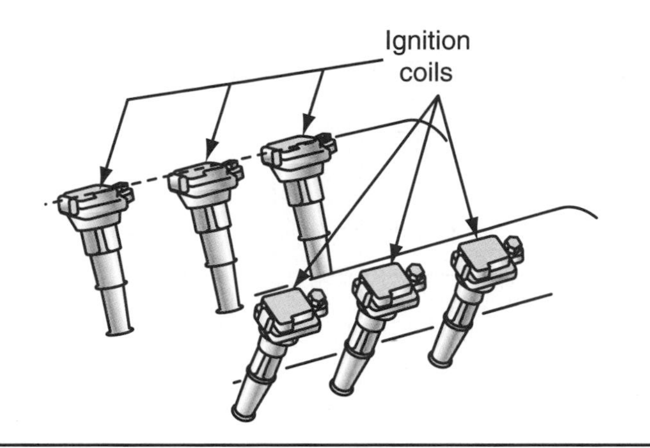

FIGURE 17–52 A direct ignition system (DIC) with individual coils for each spark plug.

FIGURE 17–53 A typical coil-on-plug assembly. (Courtesy of Visteon Corporation)

are squared off and perpendicular to one another. Using a spark plug with an incorrect gap (Figure 17–61) causes the outer electrode to be bent at an angle and shortens the plug's service life.

Spark plugs can provide a lot of information about the internal condition of the engine combustion chamber (Figure 17–62). A normally worn spark plug has minimal deposits on it, and it is light tan or gray in color. The electrode gap on a used plug should have no more than a 0.001-inch change for every 10,000 miles of engine operation, compared to the original gap setting.

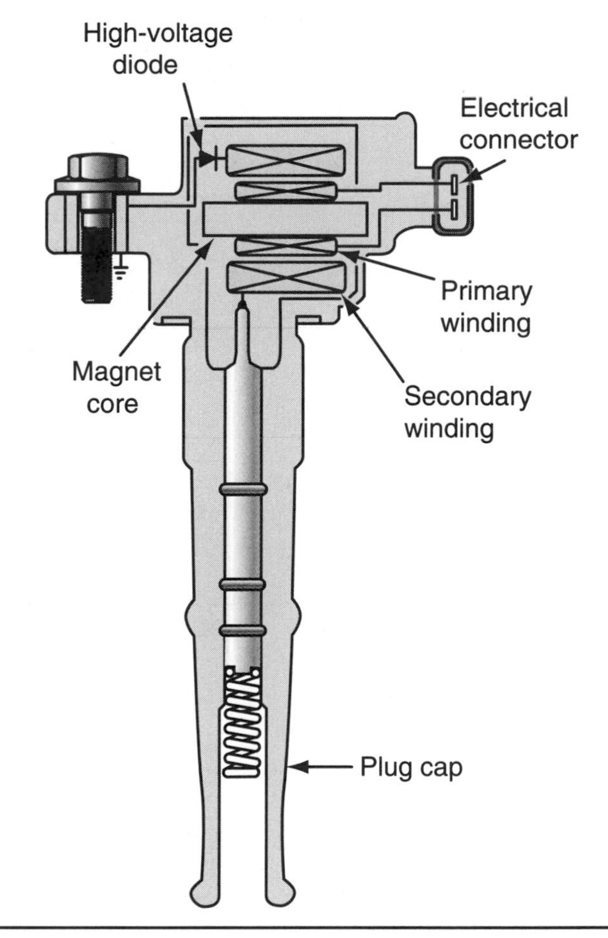

FIGURE 17–54 Parts of an individual coil assembly on a coil-on-plug direct ignition system (DIS).

Many manufacturers recommend the use of antiseize compound on the spark plug threads of engines that use aluminum cylinder heads (Figure 17–63). The antiseize compound should be applied sparingly to the threads to prevent it from contaminating the spark plug electrode. Always refer to the specific manufacturer's recommendations before using antiseize compound.

Plug Wires The ignition cables, or plug wires, deliver the high voltage from the ignition coil to the individual spark plugs. Because this voltage can reach up to 100,000 volts on some of the modern electronic ignition (EI) systems, the insulation surrounding the conductor is just as important as the conductor. Figure 17–64

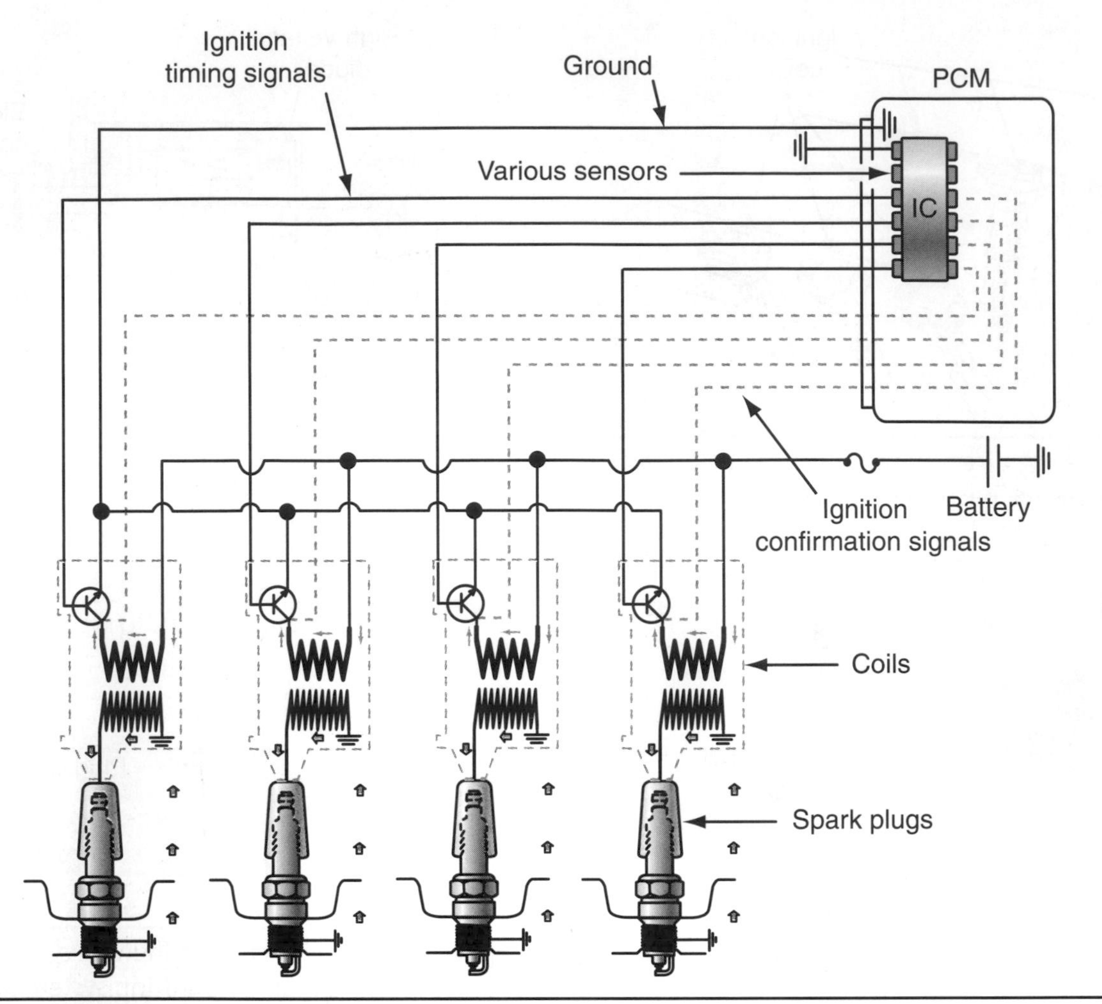

FIGURE 17–55 A direct ignition system (DIS) diagram that incorporates individual ignition modules into each ignition coil assembly.

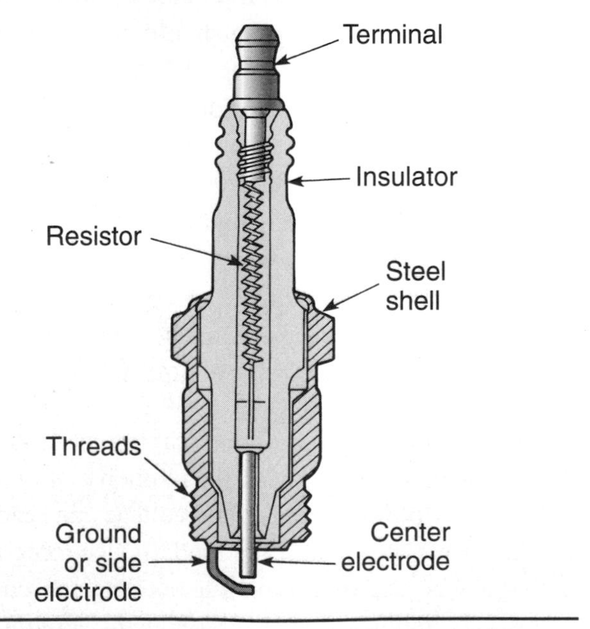

FIGURE 17–56 Components of a basic spark plug.

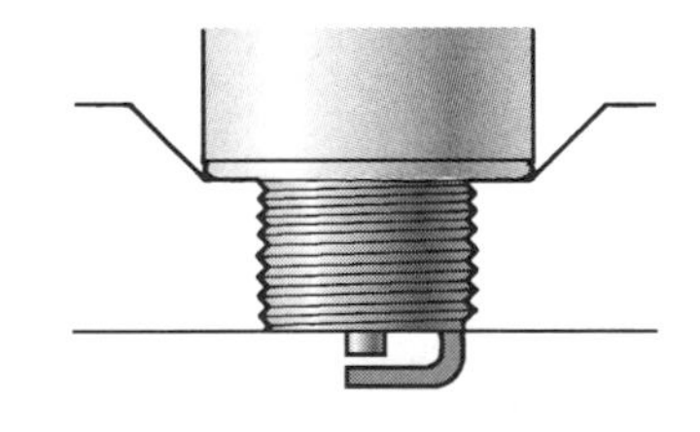

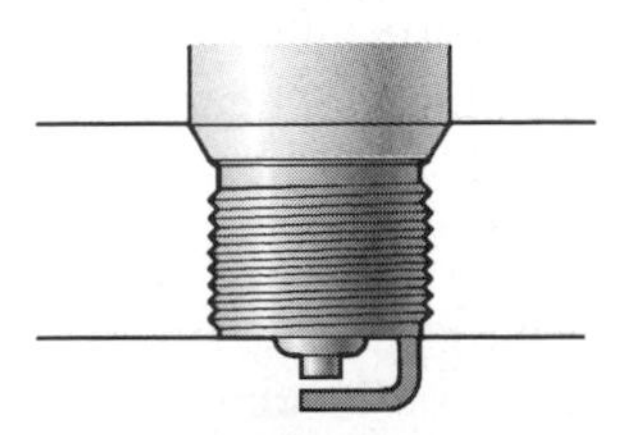

FIGURE 17–57 Spark plug sealing seat designs.

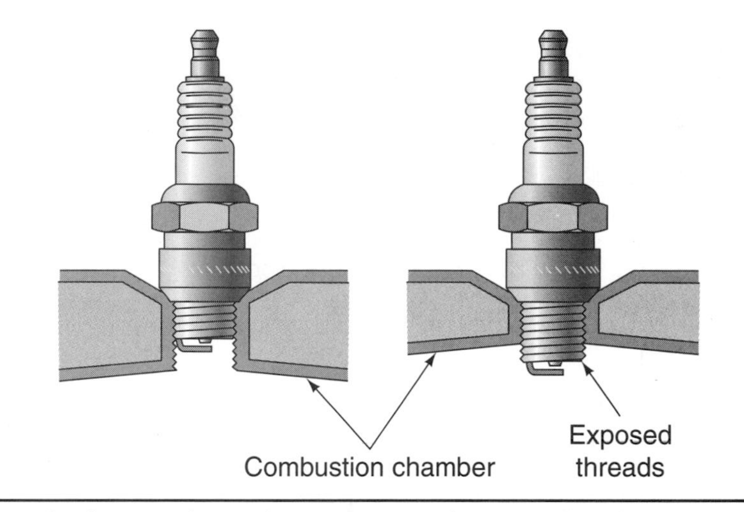

FIGURE 17–58 Incorrect spark plug reach conditions in a combustion chamber.

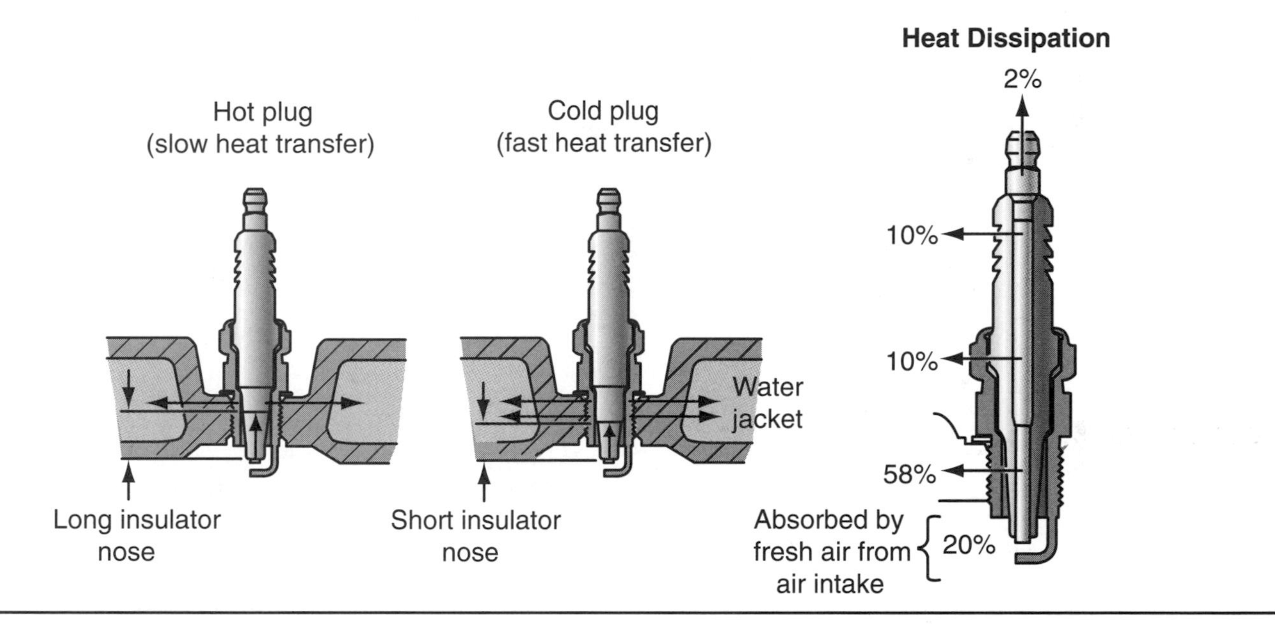

FIGURE 17–59 Heat range transfer of spark plug.

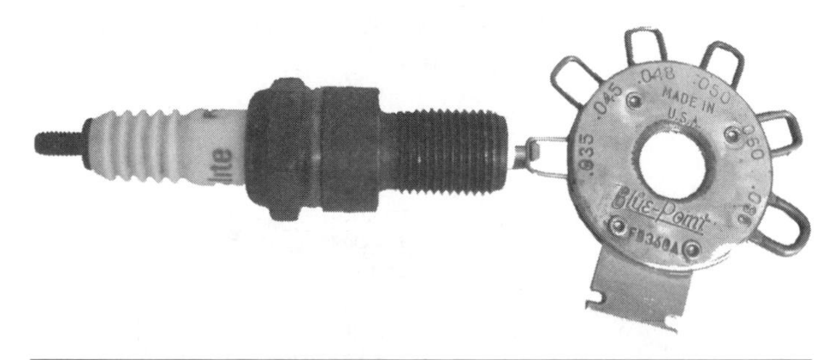

FIGURE 17–60 Checking a spark plug gap with a wire gauge.

(Courtesy of Tim Gilles)

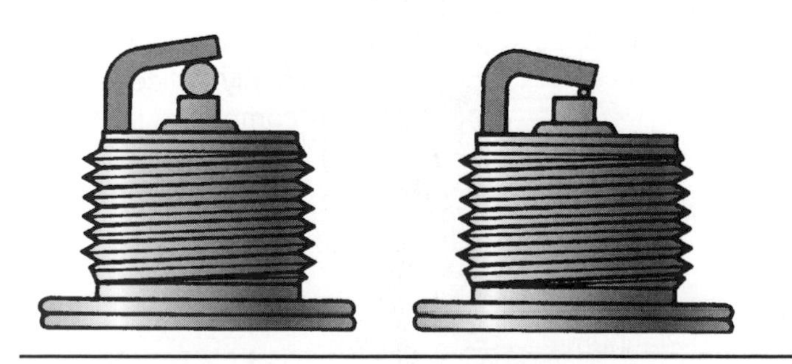

FIGURE 17–61 Incorrect spark plug gaps.

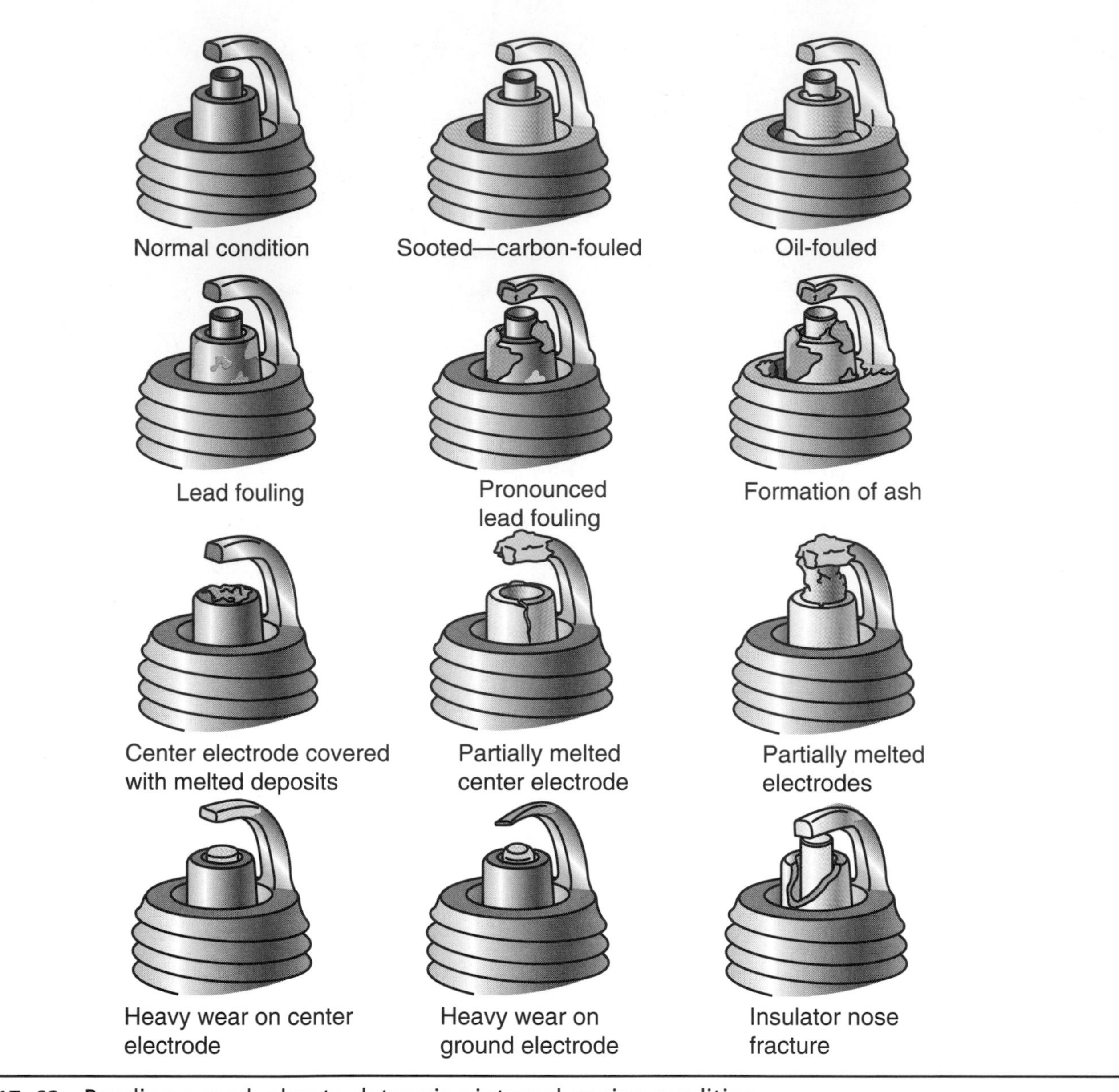

FIGURE 17–62 Reading a spark plug to determine internal engine condition.

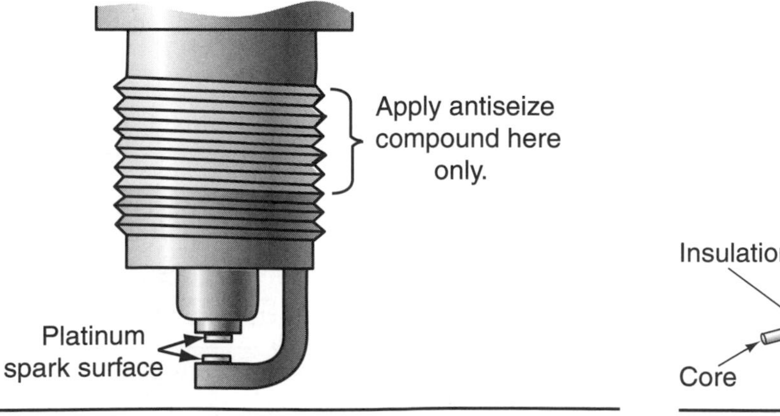

FIGURE 17–63 Using antiseize compound on spark plug threads.

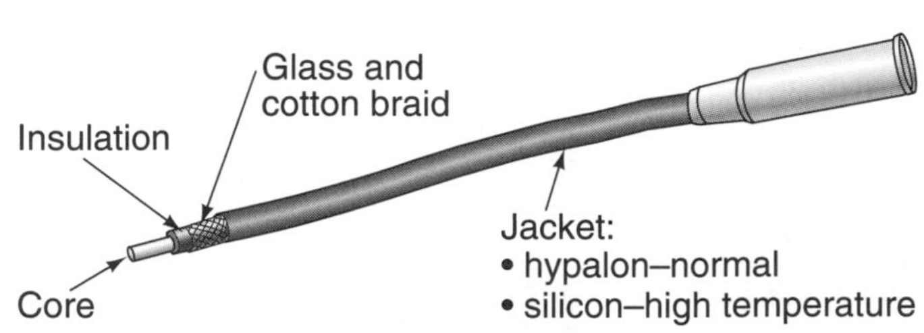

FIGURE 17–64 Typical carbon-impregnated spark plug.

shows an example of a flexible nylon core wire that is impregnated with carbon. The use of carbon in the core suppresses RFI because it offers resistance to the spark and prevents an AC current from being generated as the spark plug fires. An ohmmeter placed at the two ends of this wire can read several thousand ohms per foot, depending on the manufacturer. Always refer to the vehicle service manual for detailed specifications (Figure 17–65).

Another type of spark plug wire that is used to control RFI is **magnetic suppression** wire (Figure 17–66). The wire that is looped around the conductor in the center of the plug wire shields the magnetic field that is produced as current flows through the plug wire, thus preventing RFI from leaving the surface of the plug wire. A magnetic suppression wire may look the same as a typical plug wire, but the resistance specification is different. An ohmmeter placed at the two ends of the plug wire typically reads 0 ohms on a good wire, so refer to the vehicle service manual for the actual specifications of plug wire.

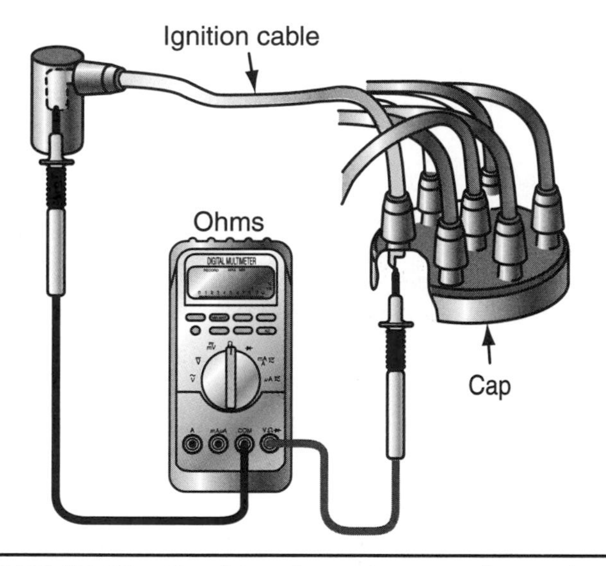

FIGURE 17–65 Checking the resistance of a spark plug wire using an ohmmeter.

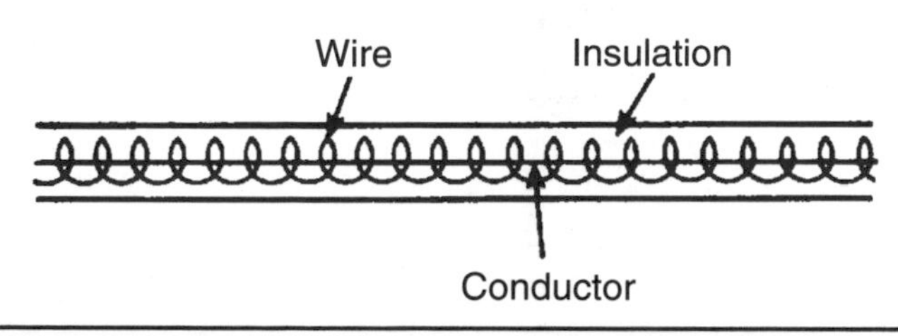

FIGURE 17–66 Typical magnetic suppression secondary cable.

Secondary Voltage Distribution In most older ignition systems, a distributor distributes the secondary voltage by means of a distributor cap and rotor assembly. The distributor cap fits tightly to the top of the distributor housing with hooks, screws, clips, or screws (Figure 17–67).

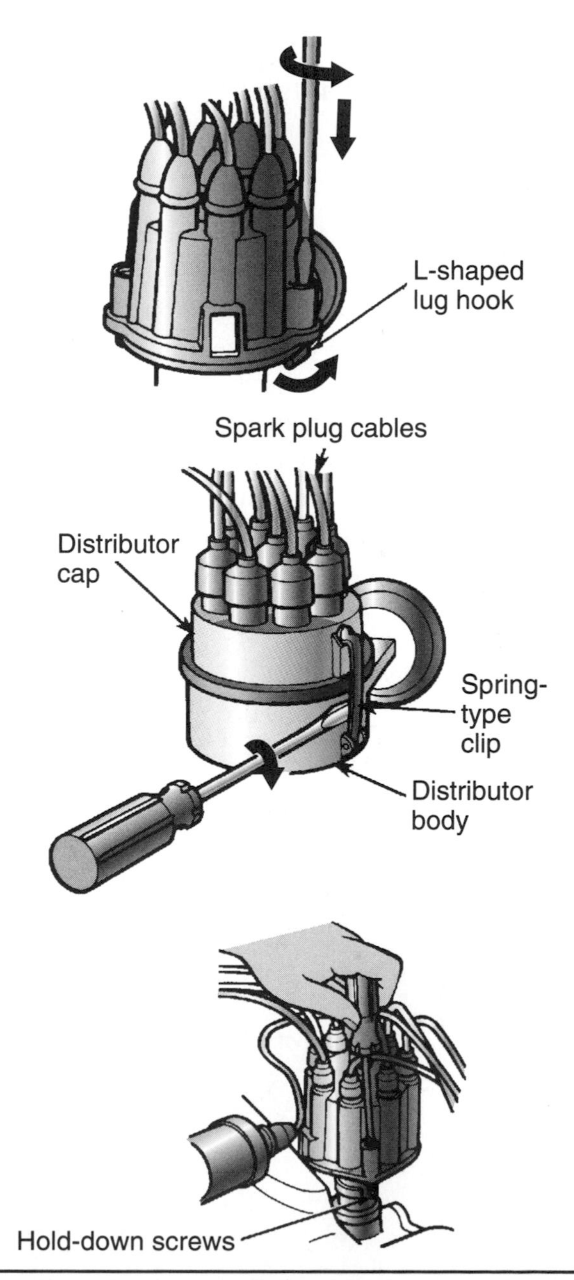

FIGURE 17–67 Typical distributor cap hold-downs. (Courtesy of DaimlerChrysler Corporation)

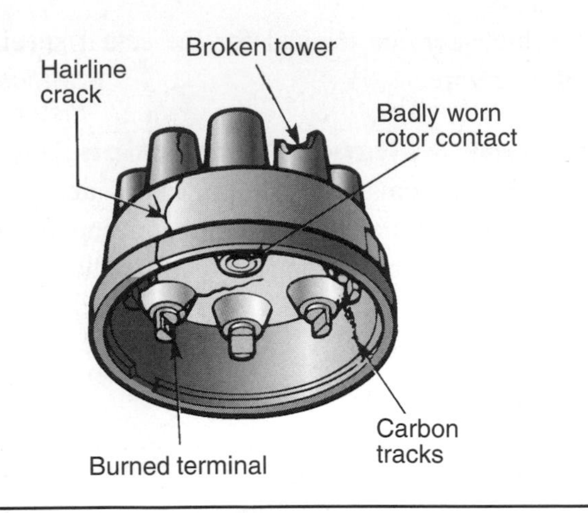

FIGURE 17–68 Typical distributor cap damage to look for.

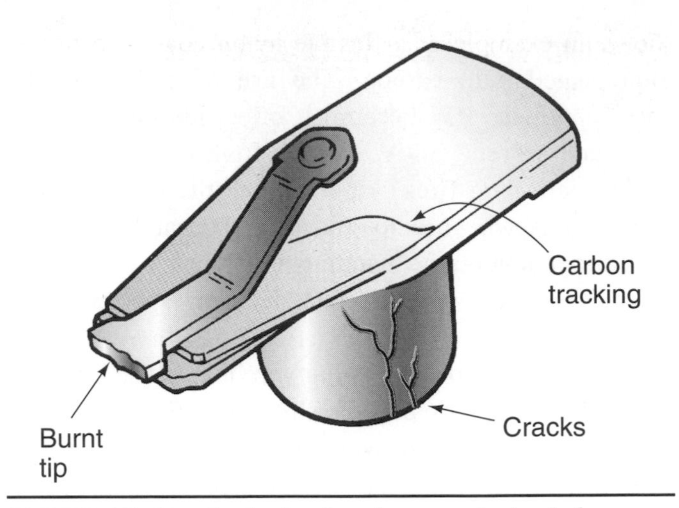

FIGURE 17–69 Typical rotor damage to look for.
(Courtesy of Daimler Chrysler Corporation)

The distributor cap is typically made of a plastic or resin material that can be easily inspected for damage (Figure 17–68). Electrical damage from the secondary voltage can be corroded or burned metal terminals, as well as carbon tracking inside the cap. Carbon tracking is the formation of a carbonized dust line between the high-voltage terminals across the cap surface.

The **rotor** assembly is mounted to the top of the distributor shaft inside the distributor. It is inspected and serviced the same as the distributor (Figure 17–69). As the distributor turns, the rotor directs the high voltage received from the ignition coil to the individual cylinders depending on the engine's **firing order** (Figure 17–70). Spark plug wires from consecutively firing cylinders should cross each cylinder rather than run parallel to one another to prevent a cross-firing condition.

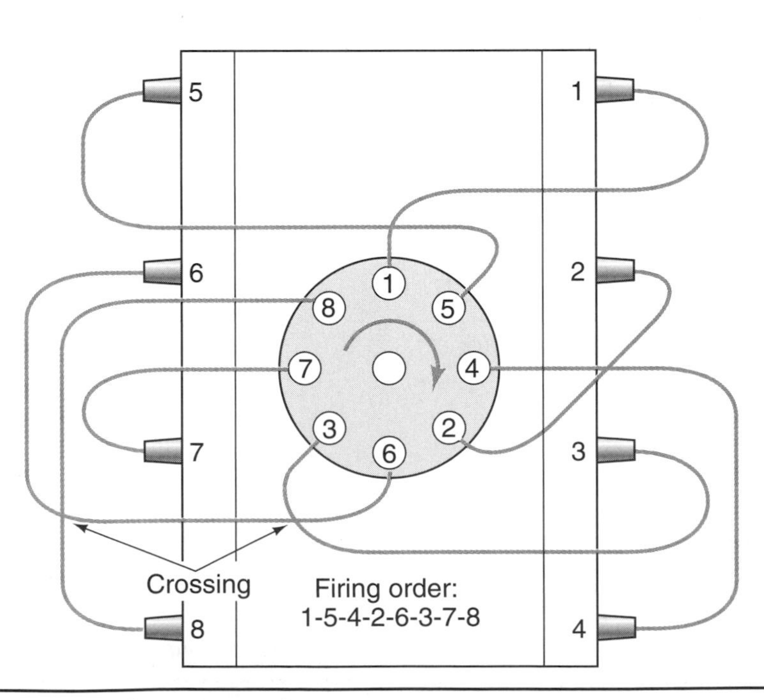

FIGURE 17–70 Typical firing order for a V-8 engine.

Basic Ignition System Testing

Setting Ignition Timing

For an ignition system to operate as designed, it must be timed to the engine. Vehicles with distributors normally operate from a base timing that is set while the engine is at idle. Some vehicles with electronic spark advance need to have the advance process disabled before setting the base timing. Always refer to the vehicle's service manual for exact procedures for setting base ignition timing. Typical timing marks are shown in Figure 17–71. Reading the timing marks requires using a timing light (Figure 17–72) that emits a strobe light to illuminate the timing marks. Adjustment of the ignition timing is normally done by turning the distributor housing in the appropriate direction (Figure 17–73).

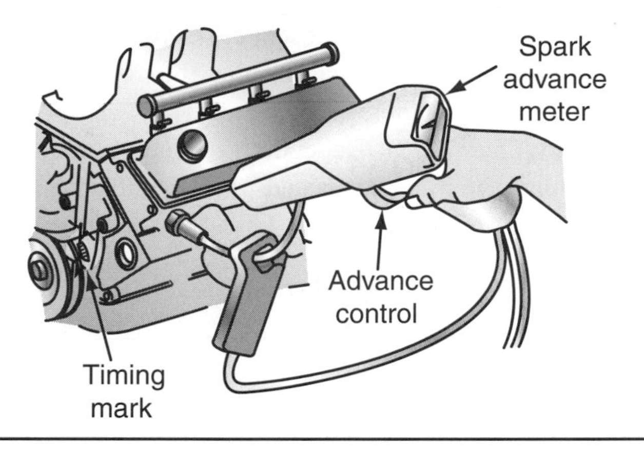

FIGURE 17–72 Using a timing light with advance control features to observe ignition timing marks.

Oscilloscope Testing

An excellent way to check the operation of the ignition system is by means of an oscilloscope. Several types oscilloscopes are on the market, and one of the most popular is a **digital storage oscilloscope (DSO),** which is small and portable (Figure 17–74). The job of the oscilloscope is to convert the electrical signals of the ignition system into a visual image showing voltage changes over a period of time (Figure 17–75).

Oscilloscopes normally have four leads that connect to the ignition system (Figure 17–76). The 1 spark plug is normally the reference cylinder.

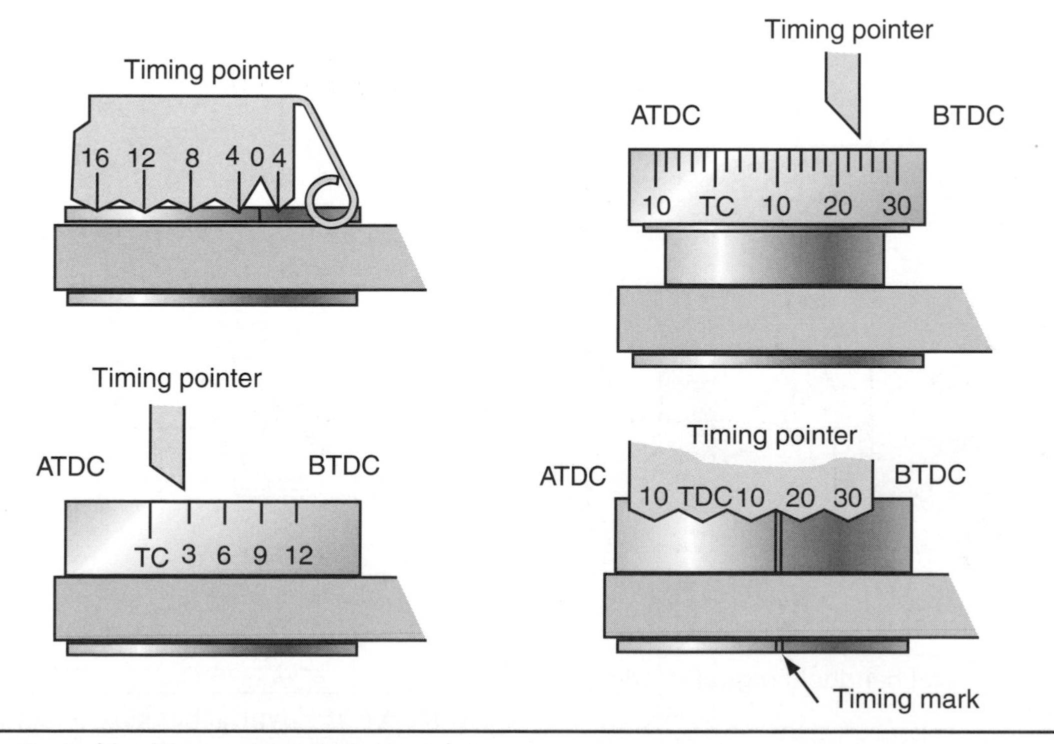

FIGURE 17–71 Typical ignition system timing marks.

Depending on the oscilloscope function selected by the technician, the scope can display the voltage versus time pattern for either the secondary or the primary circuit (Figure 17–77). Depending on the specific type of ignition system being tested, the secondary pattern displays:

- Spark plug firing voltage.
- Spark duration.

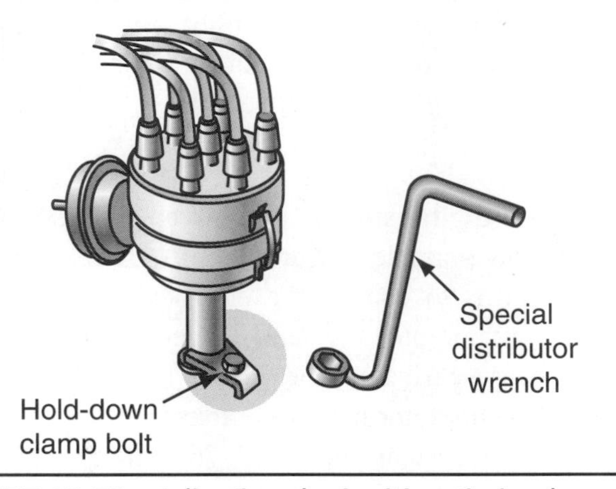

FIGURE 17–73 Adjusting the ignition timing by rotating the distributor housing.

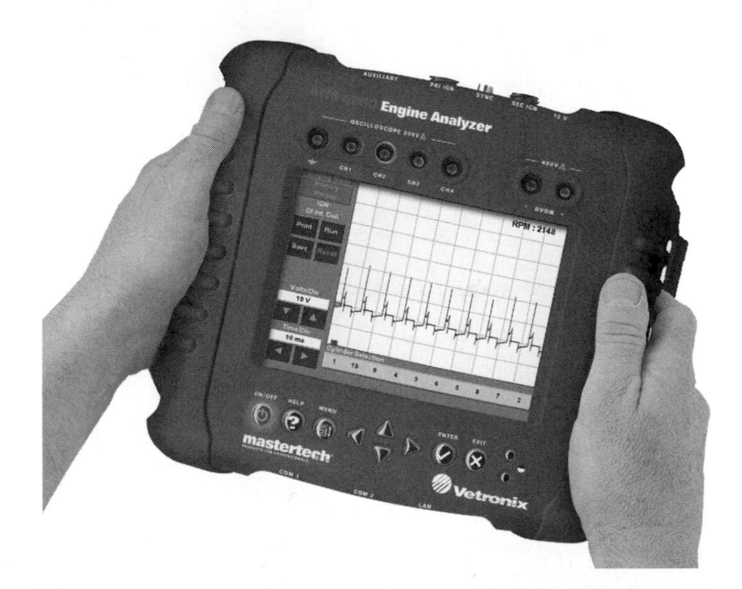

FIGURE 17–74 A typical handheld digital storage oscilloscope (DSO).

(Reprinted with permission from Bosch Diagnostics)

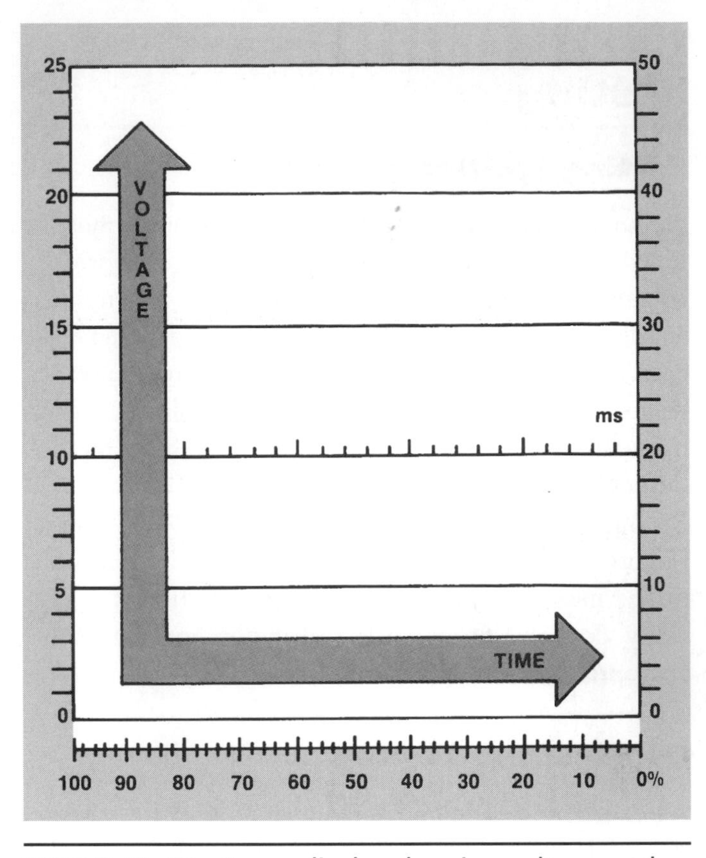

FIGURE 17–75 Scope display showing voltage and time scales.

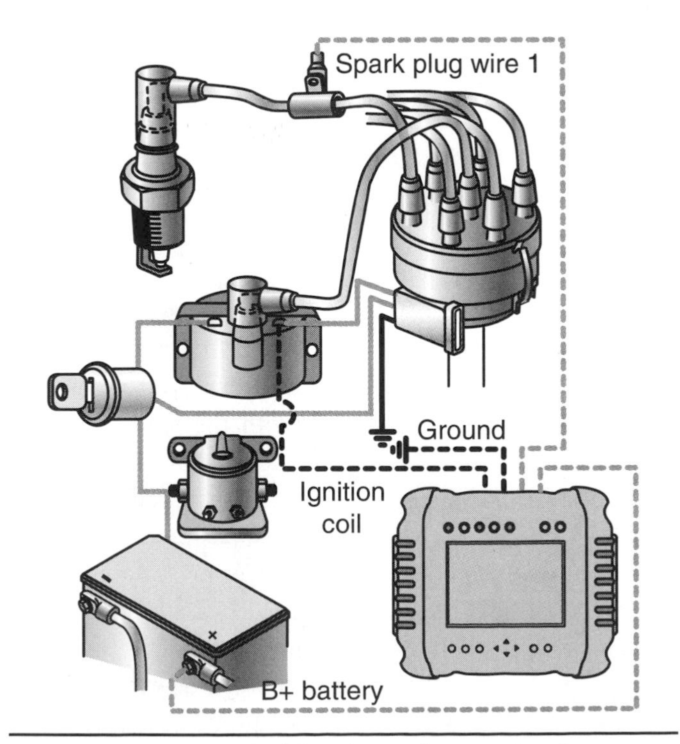

FIGURE 17–76 Typical hookup of scope leads to distributor ignition system.

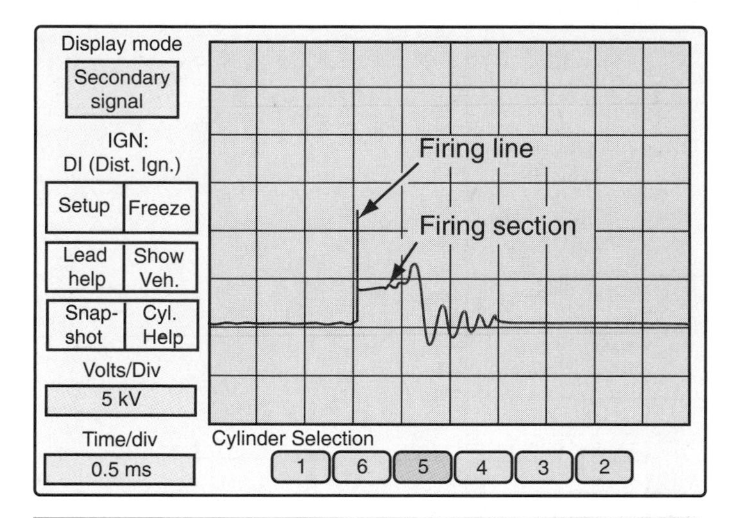

FIGURE 17–77 Typical DSO single-cylinder secondary ignition pattern.

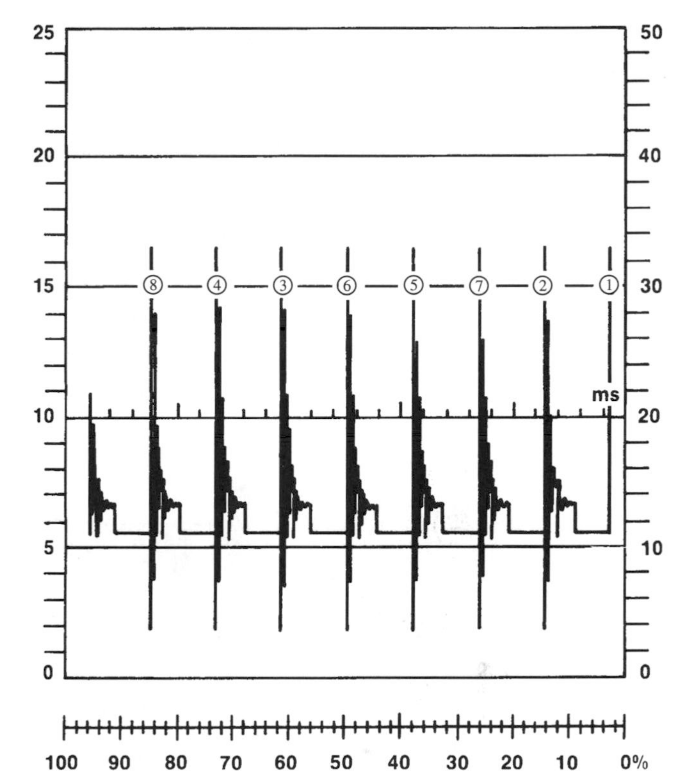

FIGURE 17–78 Typical display pattern

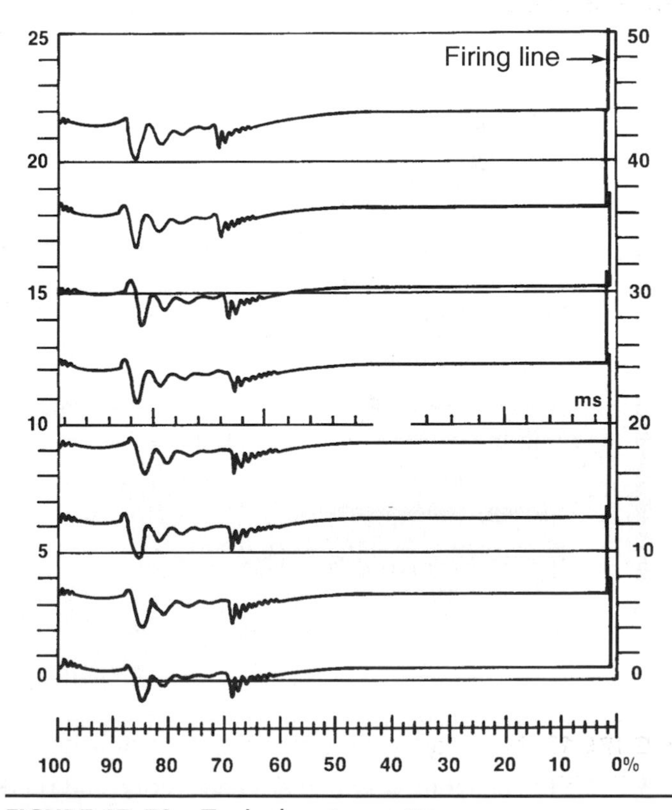

FIGURE 17–79 Typical raster pattern.

- Coil and condenser oscillations.
- Primary switching operation of transistor or breaker points.
- Dwell and cylinder timing accuracy.
- Secondary circuit accuracy.

The oscilloscope has several ways to display the voltage patterns of the primary and secondary circuits. When the display pattern is selected, the scope displays the patterns of all the cylinders in a row, from left to right (Figure 17–78).

Another scope pattern is a raster pattern that stacks the voltage patterns of each cylinder above each other for easy comparison (Figure 17–79). In a raster pattern, the patterns for each cylinder are displayed across the width of the graph, beginning with the spark line and ending with the firing line. This allows for close inspection of the voltage and time trends of the pattern.

A third scope pattern is the superimposed pattern that displays all the patterns on top of each other across the full width of the screen (Figure 17–80). A superimposed pattern allows the technician to compare one cylinder's pattern to the others. A worn or bent distributor shaft is easily seen with this pattern because one or more of the cylinder pattern lines is out of line with the rest.

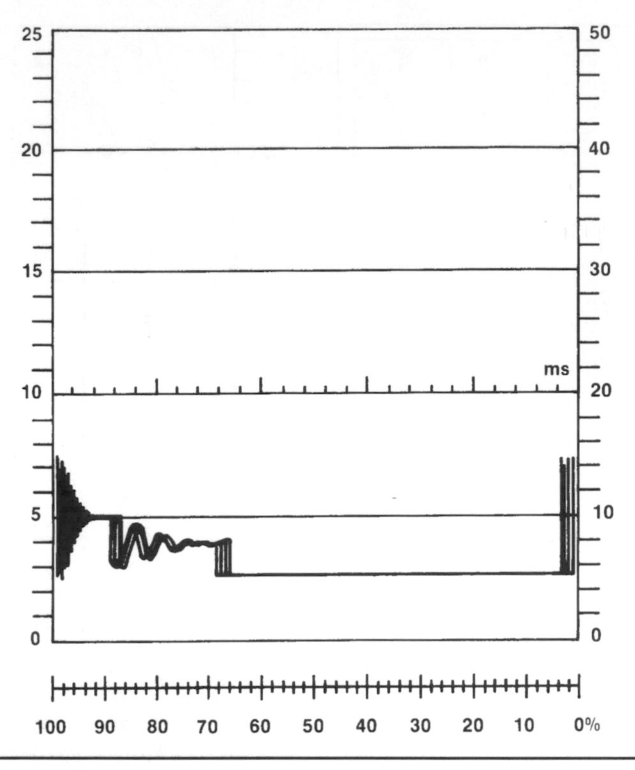

FIGURE 17–80 Typical superimposed pattern.

CASE STUDIES

CASE 1

This case study is typical of an older distributor ignition–related problem that technicians have had to diagnose. Knowledge of how these ignition systems are constructed and of some simple tests procedures will help you perform the diagnosis quickly and efficiently.

Customer Complaint

A customer brings his vehicle into the shop with a possible ignition problem. The vehicle starts while cranking but quits as soon as the ignition key is returned to the run position. The customer changed the ignition coil but the problem remains.

Known Information

- The battery OCV is 12.66 volts.
- The battery voltage while cranking is 11.24 volts.
- The ignition coil has been replaced with a new one.
- The spark plugs and wires are less than one year old.
- The ignition switch has been tested and is OK.
- The simple schematic of ignition system is shown in Figure 17–81.

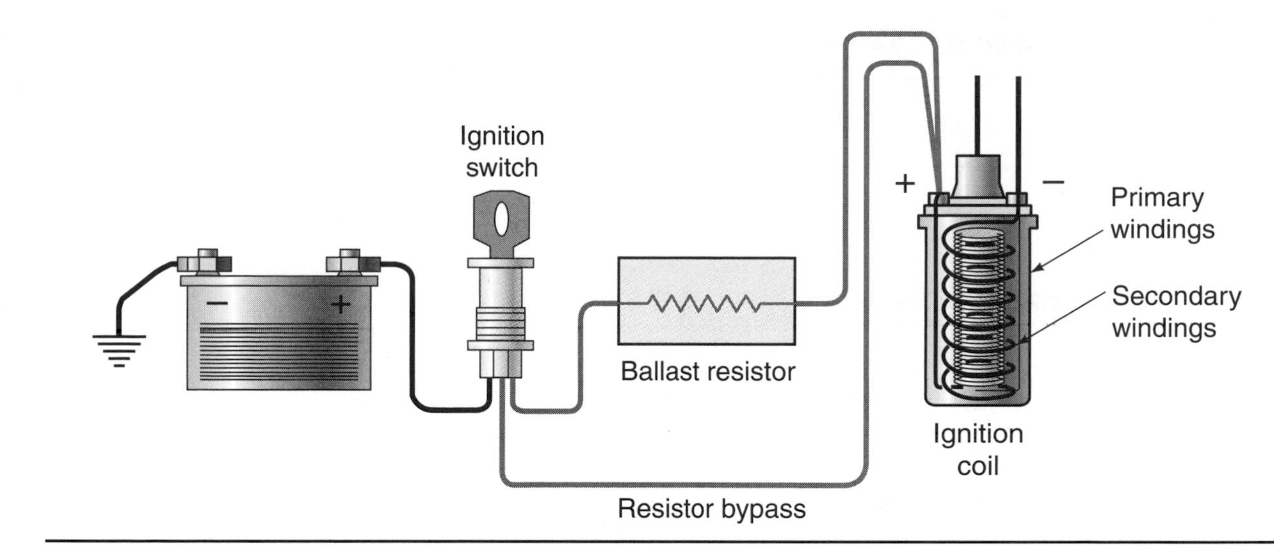

FIGURE 17–81 Simple schematic of ignition system.

Circuit Analysis

Answer the following questions on a *separate* sheet of paper:

1. With the information known, what is the *most* likely cause of the ignition system problem that allows the vehicle to start while it is cranking only? ____________.
2. What preliminary checks should you perform before proceeding with more advanced tests on the ignition system components? ____________ ____________.
3. What is the function of each of the components in Figure 17–81 as it relates to the ignition system?

Component	Function
1.	
2.	
3.	
4.	
5.	

4. What should the voltage be at the positive side of the coil with the key on and the engine off? ____________ Why? ____________
5. What should the voltage be at the positive side of the coil during engine cranking conditions? ____________ Why? ____________
6. What is your analysis of this case study? ____________ ____________.

Hands-On Vehicle Tests

The following six hands-on vehicle checks are included in the NATEF (National Automotive Technicians Education Foundation) Task List. Some tasks are grouped together due to their close association with each other. Complete your answers to the following questions on a *separate* sheet of paper.

Performance Task 1

Task Description
Inspect and test ignition primary circuit wiring and solid-state components; perform necessary action.

Performance Task 2

Task Description
Inspect and test ignition system secondary circuit wiring and components; perform necessary action.

Task Objectives

- ❑ Obtain a vehicle that can be used for these tasks. What model and year of vehicle are you using for these tasks? ____________________.
- ❑ What type of ignition system does this vehicle have? ____________________.
- ❑ Describe the general appearance of the ignition system. ____________________

____________________.
- ❑ Inspect and record the condition of the primary ignition system components.

Component	Condition	Test performed

- ❑ What tools and equipment are useful in servicing the primary ignition system components?

____________________.
- ❑ Inspect and record the condition of the secondary ignition system components.

Component	Condition	Test performed

❑ What tools and equipment are useful in servicing the secondary ignition system components?

____

____.

❑ What special service precautions or service factory service bulletins exist on this ignition system?

____

____.

Task Summary

After performing the preceding NATEF tasks, what can you determine that will be helpful in knowing how to check the ignition primary circuit and components, secondary circuit and components, and ignition wiring harness and connectors? ______________________________

____

____

____.

Performance Task 3

Task Description

Inspect, test, and service a distributor.

Performance Task 4

Task Description

Inspect and test ignition coil(s); perform necessary action.

Performance Task 5

Task Description

Check and adjust ignition system timing and timing advance/retard (where applicable).

Performance Task 6

Task Description

Inspect and test ignition pickup sensor or triggering devices; perform necessary action.

Task Objectives

- ❑ Obtain a vehicle that can be used for these tasks. What model and year of vehicle are you using for these tasks? ______________________.
- ❑ What type of ignition system does this vehicle have? ______________________.
- ❑ Describe the general appearance of the ignition system ______________________

  ______________________.
- ❑ What tests can be performed on the distributor in this vehicle? ______________________

  ______________________.
- ❑ What tools and equipment are being used to inspect and test the distributor? ______________________

  ______________________.
- ❑ List the steps used to check the primary and secondary windings in the ignition coil: ______________________

  ____

  ______________________.
- ❑ What is the resistance reading of the coil primary windings? ______________________.
- ❑ What is the resistance reading of the coil secondary windings? ______________________.
- ❑ List the tools and equipment used to check and adjust the ignition timing: ______________________

  ______________________.
- ❑ What are the recommended factory service manual procedures and precautions for adjusting the ignition timing on this vehicle? ______________________

  ____

  ______________________.
- ❑ What is the current ignition timing setting? ______________________.
- ❑ What is the factory specification for ignition timing? ______________________.
- ❑ What is the ignition timing setting after adjustment? ______________________.
- ❑ What type of primary pickup or triggering mechanism does the ignition system have? ______________________

  ______________________.
- ❑ What are the service manual procedures and specifications for testing the primary triggering switch assembly? ______________________

  ______________________.

- ❑ What is the condition of the primary triggering switch assembly? ____________________

  ____.

After performing the preceding NATEF tasks, what can you determine will be helpful in knowing how to inspect and test the distributor, inspect and test the ignition coil, check and adjust the ignition timing, and inspect and test the primary triggering mechanism? ____________________

____

____

____.

Summary

- ❑ The combustion pressure is greatest when the piston is at 10 to 23 degrees ATDC.
- ❑ Engine detonation occurs when the normal combustion process is too slow, thus causing some unburned fuel/air mixture to ignite and create a colliding flame front.
- ❑ Preignition is caused by an early ignition of the fuel/air mixture by a source other than the ignition system.
- ❑ The primary circuit in an ignition system is the low-voltage circuit.
- ❑ The secondary circuit in an ignition system is the high-voltage circuit.
- ❑ The ignition coil is a pulse transformer that generates a high voltage when a magnetic field moves across a wire.
- ❑ The ignition coil primary windings consist of a couple of hundred turns of a 20- to 24-gauge copper wire coated with an insulating enamel.
- ❑ The ignition coil secondary windings consist of several thousand turns of a very fine copper wire coated with an insulating enamel.
- ❑ The soft iron laminated core in an ignition coil has a low inductive reluctance that freely expands or strengthens the magnetic field around it.
- ❑ Electromagnetism occurs in an ignition coil when current flowing through the primary windings creates magnetic flux lines of force.
- ❑ Self-induction in an ignition coil results when the primary flux lines cross over onto themselves creating a counterelectromotive force (CEMF).
- ❑ Mutual induction in an ignition coil exists when the primary flux lines cross over and create a current in the secondary windings.
- ❑ The capacitor (condenser) in an ignition system absorbs the surges of electrons induced by the collapse of the magnetic fields.
- ❑ The primary coil windings normally have a resistance of 2.0 ohms or less.

- ❑ The secondary windings normally have a resistance of between 5,000 and 20,000 ohms.
- ❑ Ignition coil secondary tracking occurs when the windings short to the coil case.
- ❑ The breaker point ignition system consists of a mechanical set of points in the distributor that open and close to control the ignition primary circuit.
- ❑ Ignition dwell is the amount of time that the primary controlling switch is closed and conducting current.
- ❑ A solid-state ignition system replaces the breaker points with an electronic triggering device and solid-state ignition control module.
- ❑ The magnetic pulse (PM) generator consists of a timing disk and pickup coil assembly that operate on the basis of electromagnetism.
- ❑ The metal detection sensor does not have a permanent magnet, but a weak electromagnet.
- ❑ The Hall effect sensor is based on the principle that if a current is allowed to flow through a thin conducting material, and when that material is exposed to a magnetic field, another voltage is produced.
- ❑ The heart of a Hall effect system is a thin semiconductor gallium arsenate crystal.
- ❑ A Schmitt trigger is a pulse-shaping device that converts the Hall effect signal to a clean square wave digital signal.
- ❑ A photoelectric sensor uses an LED that shines through a rotating slotted disk to a photo cell that converts the intermittent flashes of light into a voltage signal pulse.
- ❑ The distributor centrifugal advance mechanism advances the ignition timing in relation to engine speed.
- ❑ The distributor vacuum advance mechanism advances the ignition timing in relation to the engine vacuum conditions.
- ❑ An electronic ignition (EI) system is a distributorless system consisting of a computer, ignition module, coil pack, position sensors, and associated wiring.
- ❑ A waste spark system fires one of a pair of spark plugs on the exhaust stroke while the mating spark plug fires toward the end of the compression stroke.
- ❑ A coil-over-plug ignition system uses a waste spark system, in which there is one coil for every two spark plugs, and the coil is mounted directly over the spark plugs, eliminating the need for spark plug wires.
- ❑ A direct ignition system (DIS) uses individual coils for each spark plug, mounted directly onto the spark plug.
- ❑ The resistive element in a spark plug controls RFI by suppressing the AC voltage spike produced while the spark plug is firing.
- ❑ The display pattern on a scope screen displays all the cylinder patterns in a row from left to right.
- ❑ The raster pattern on a scope screen displays all the cylinder patterns above each other.
- ❑ The superimposed pattern on a scope screen displays all the cylinder patterns stacked on top of one another.

Key Terms

after-top-dead-center (ATDC)
before-top-dead-center (BTDC)
breaker point ignition
capacitor
centrifugal advance
coil-over-plug
detonation
digital storage oscilloscope (DSO)
direct ignition system (DIS)
distributor ignition (DI)
dwell
E-coil
electronic ignition (EI)
electronic spark timing (EST)
firing order
Hall effect sensor
inductive reluctance
magnetic pulse generator
magnetic suppression
metal detection sensor
mutual induction
photoelectric sensor
pickup coil
PM generator
preignition
primary circuit
pulse transformer
radio frequency interference (RFI)
rotor
saturation
Schmitt trigger
secondary circuit
self-induction
solid-state ignition
vacuum advance
waste spark

Review Questions

Short Answer Essays

1. Describe the difference between engine detonation and preignition.
2. Name the parts of the typical primary ignition system.
3. Name the parts of the typical secondary ignition system.
4. Describe how a coil transforms battery voltage to a voltage high enough to jump a spark plug gap.
5. Describe how a Hall effect sensor works as a triggering mechanism for an ignition system.

Fill in the Blanks

1. In an engine cylinder, the combustion process must begin ________________ on the ________________ stroke.
2. On some ignition systems, to protect the primary switching circuit a ________________ is wired in parallel to the circuit to absorb the surges of ________________ induced by the collapse of the magnetic field.
3. The ________________ in a distributor consists of a set of weights and springs that fly outward as engine revolutions per minute ________________.
4. In a distributor ignition that has no mechanically operated centrifugal or vacuum advance units, the timing advance is controlled totally by the __________ __________ timing function within the PCM.
5. An electronic ignition system that consists of one coil for every two cylinders is called a ________________ system.

ASE-Style Review Questions

1. Technician A says the spark plug wire is part of the secondary ignition system. Technician B says the distributor cap is part of the primary ignition system. Who is correct?
   A. A only
   B. B only
   C. Both A and B
   D. Neither A nor B

2. Technician A says the vacuum advance unit in a distributor may be connected to manifold vacuum. Technician B says the vacuum advance unit in a distributor may be connected to ported vacuum. Who is correct?
   A. A only
   B. B only
   C. Both A and B
   D. Neither A nor B

3. Technician A says that engine detonation occurs when the fuel/air mixture ignites before ignition system spark. Technician B says preignition can occur from hot carbon deposits in the combustion chamber. Who is correct?
   A. A only
   B. B only
   C. Both A and B
   D. Neither A nor B

4. Technician A says the ignition coil is part of the primary system. Technician B says the primary triggering device is normally located after the coil in the circuit. Who is correct?
   A. A only
   B. B only
   C. Both A and B
   D. Neither A nor B

5. Technician A says the coil primary windings consist of a few thousand turns of a 20- to 24-gauge wire. Technician B says the coil secondary windings consist of a few hundred turns of a very fine copper wire. Who is correct?
   A. A only
   B. B only
   C. Both A and B
   D. Neither A nor B

6. Technician A says some coil-on-plug ignition designs use individual coils mounted directly to each spark plug. Technician B says some DIS coils integrate individual ignition modules into each coil assembly. Who is correct?

   A. A only

   B. B only

   C. Both A and B

   D. Neither A nor B

7. Technician A says the capacitor absorbs the surges of electrons in the coil. Technician B says the capacitor conducts electricity anytime the ignition primary system contact points are closed. Who is correct?

   A. A only

   B. B only

   C. Both A and B

   D. Neither A nor B

8. Technician A says the secondary windings in an ignition coil are tested by connecting the ohmmeter leads between the positive and negative terminals of the coil. Technician B says the coil primary resistance should be 2.0 ohms or less. Who is correct?

   A. A only

   B. B only

   C. Both A and B

   D. Neither A nor B

9. Technician A says the amount of time the contact points are closed in a point-style distributor is the ignition dwell time. Technician B says the lower the ignition dwell is, the wider the contact point gap is. Who is correct?

   A. A only

   B. B only

   C. Both A and B

   D. Neither A nor B

10. A distributor with a PM generator is being discussed. Technician A says a permanent magnet is part of the switching assembly. Technician B says the increase and decrease of the voltage in the pick-up coil in relation to zero is what triggers the ignition module to turn the primary circuit on and off. Who is correct?

    A. A only

    B. B only

    C. Both A and B

    D. Neither A nor B

11. Technician A says a Hall effect switch assembly produces an accurate square wave voltage signal. Technician B says the Hall effect switch is a thin carbon and silicon crystal. Who is correct?
    A. A only
    B. B only
    C. Both A and B
    D. Neither A nor B

12. Technician A says that in a waste spark electronic ignition system one ignition coil is used for every two cylinders. Technician B says that in a waste spark electronic ignition system spark plugs also fire on the exhaust stroke. Who is correct?
    A. A only
    B. B only
    C. Both A and B
    D. Neither A nor B

13. Technician A says a Hall effect switch requires its voltage to be amplified, inverted, and shaped into a clean square wave signal. Technician B says a metal detection sensor requires its voltage to be amplified, inverted, and shaped into a clean square wave signal. Who is correct?
    A. A only
    B. B only
    C. Both A and B
    D. Neither A nor B

14. Technician A says a magnetic suppression spark plug wire consists of a heavy copper core that is shielded with a metal jacket. Technician B says a flexible nylon core that is impregnated with carbon can suppress RFI in a spark plug wire. Who is correct?
    A. A only
    B. B only
    C. Both A and B
    D. Neither A nor B

15. Technician A says the firing order in a distributor is controlled by the placement of the secondary wires in the distributor cap. Technician B says the rotor assembly distributes the high voltage from the coil to the plug wires. Who is correct?
    A. A only
    B. B only
    C. Both A and B
    D. Neither A nor B

Appendix A

1. Technician A says the flow of free electrons in a circuit is called current flow. Technician B says that 6.28 billion billion electrons flowing past a point in 1 second is 1 ampere. Who is correct?
   A. A only
   B. B only
   C. Both A and B
   D. Neither A nor B

2. Technician A says the base system of measurement for volts is the kilovolt. Technician B says the milliamp is the base unit of measurement for current. Who is correct?
   A. A only
   B. B only
   C. Both A and B
   D. Neither A nor B

3. Technician A says static electricity can occur whenever the technician slides across a vinyl seat in a car. Technician B says the best way to guard against static electricity is to use a static wrist strap. Who is correct?
   A. A only
   B. B only
   C. Both A and B
   D. Neither A nor B

4. Technician A says the diode acts like a one-way check valve in an electrical circuit. Technician B says a diode allows current to flow in one direction only. Who is correct?
   A. A only
   B. B only
   C. Both A and B
   D. Neither A nor B

5. A circuit has the following four resistors wired in parallel: 10 ohms, 5 ohms, 10 ohms, and 2.5 ohms. What is the resulting total resistance?
   A. 27.5 ohms
   B. 1.25 ohms
   C. 125 ohms
   D. None of the above

6. Technician A says using photovoltaic solar cells provide a way to capture the energy from the sun. Technician B says that a photon is the energy in sunlight that strikes the surface of a photovoltaic cell to create a voltage. Who is correct?
   A. A only
   B. B only
   C. Both A and B
   D. Neither A nor B

7. Technician A says an electrical circuit must be energized to measure the current flow through the circuit. Technician B says an ammeter is connected in series with the component or circuit being measured. Who is correct?
   A. A only
   B. B only
   C. Both A and B
   D. Neither A nor B

8. Two technicians are discussing a multifunction steering column switch. Technician A says the dimmer switch can be part of this style of switch. Technician B says the horn switch can be part of this type of switch. Who is correct?
   A. A only
   B. B only
   C. Both A and B
   D. Neither A nor B

9. Technician A says a diode is forward biased if the positive voltage is connected to the cathode side of the diode. Technician B says a diode is reversed biased if the positive voltage is connected to the anode side of the diode. Who is correct?
   A. A only
   B. B only
   C. Both A and B
   D. Neither A nor B

10. Technician A says a primary cell battery can be recharged to its original condition. Technician B says a secondary cell battery cannot be recharged once its original stored energy has been depleted. Who is correct?
    A. A only
    B. B only
    C. Both A and B
    D. Neither A nor B

11. Two technicians are discussing the results of a starter current draw test on a four-cylinder engine. Technician A says the minimum voltage should not drop below 10.5 volts. Technician B says the amperage draw of 125 amperes is considered satisfactory. Who is correct?
    A. A only
    B. B only
    C. Both A and B
    D. Neither A nor B

12. A vehicle is being tested for a slow-crank condition. A starter current draw test indicates 450 amperes. Technician A says this could be caused by a battery with a CCA rating too large for the system. Technician B says this could be caused by a pinion drive that slips when engaged with the flywheel ring gear. Who is correct?
    A. A only
    B. B only
    C. Both A and B
    D. Neither A nor B

13. Technician A says a delta wound winding consists of three windings connected in a triangular shape with three output leads. Technician B says a delta wound winding provides a high-amperage output for its size. Who is correct?
    A. A only
    B. B only
    C. Both A and B
    D. Neither A nor B

14. Technician A says a thermocouple is installed in the exhaust system of a diesel engine to measure the exhaust temperature. Technician B says a thermocouple operates the same as a mercury glass thermometer. Who is correct?
    A. A only
    B. B only
    C. Both A and B
    D. Neither A nor B

15. Technician A says an IVR can be used on bimetallic gauges. Technician B says the IVR opens if the alternator overcharges the battery. Who is correct?
    A. A only
    B. B only
    C. Both A and B
    D. Neither A nor B

16. Technician A says a halogen headlamp uses a halogen-filled inner bulb sealed in a glass housing. Technician B says in a composite halogen headlamp, the entire headlamp assembly needs to be replaced when it is burned out. Who is correct?
    A. A only
    B. B only
    C. Both A and B
    D. Neither A nor B

17. Technician A says the proper way to find the resistance of a hot bulb parallel circuit is to measure the amperes and volts of the circuit, then calculate the circuit resistance total. Technician B says the hot bulb resistance total of a parallel circuit can be determined by taking the hot bulb resistance of each circuit bulb and finding the reciprocal of the sum of each parallel branch. Who is correct?
    A. A only
    B. B only
    C. Both A and B
    D. Neither A nor B

18. Voltage total of a simple series/parallel circuit is being discussed. There are two series load devices with voltage drops of 6 volts each and four load devices in parallel with a voltage reading across each device of 12 volts. What is the voltage total of this circuit?
    A. 12 volts
    B. 24 volts
    C. 60 volts
    D. None of the above

19. Technician A says a short-to-voltage has an unwanted copper-to-copper connection between two circuits. Technician B says a short-to-voltage may not blow the circuit fuse. Who is correct?
    A. A only
    B. B only
    C. Both A and B
    D. Neither A nor B

20. Technician A says a test light can be connected in series with a cycling circuit breaker to test for a short-to-ground. Technician B says a buzzer can be connected in series with a cycling circuit breaker to test for a short-to-ground. Who is correct?
    A. A only
    B. B only
    C. Both A and B
    D. Neither A nor B

21. Two technicians are discussing how a sneak circuit can affect circuits in an automotive electrical system. Technician A says current can backfeed through a circuit when two circuits share a common power feed. Technician B says current can backfeed through a circuit when two circuits share a common ground path. Who is correct?
    A. A only
    B. B only
    C. Both A and B
    D. Neither A nor B

22. A series circuit has three resistors of 20 ohms each. The voltage drop of each resistor is 4 volts. Technician A says the voltage total is 12 volts. Technician B says the resistance total is 60 ohms. Who is correct ?
    A. A only
    B. B only
    C. Both A and B
    D. Neither A nor B

23. Technician A says a knock sensor produces a small voltage when there is an unusual ignition spark knock. Technician B says the knock sensor is a piezoelectric device that produces a voltage under physical stress or vibration. Who is correct?
    A. A only
    B. B only
    C. Both A and B
    D. Neither A nor B

24. A circuit has four resistors wired in series: 10 ohms, 5 ohms, 4 ohms, and 2.5 ohms. What is the resistance total?
    A. 21.5 ohms
    B. 1.05 ohms
    C. 20.5 ohms
    D. None of the above

25. Technician A says the pitch of a horn is controlled by the voltage going to the horn. Technician B says that in a dual-horn system, the pitch is normally the same for both horns. Who is correct?
    A. A only
    B. B only
    C. Both A and B
    D. Neither A nor B

26. The oil pressure light stays on when the engine is running. Technician A says the engine oil pressure could be low, causing the circuit to operate as designed. Technician B says there could be a ground in the circuit between the indicator light and the pressure switch. Who is correct?
    A. A only
    B. B only
    C. Both A and B
    D. Neither A nor B

27. Two technicians are discussing the operation of NPN transistors. Technician A says the transistor is on when a voltage is applied to the base. Technician B says that, when the transistor is on, current flows from the collector to the emitter. Who is correct?
    A. A only
    B. B only
    C. Both A and B
    D. Neither A nor B

28. Technician A says a potentiometer uses three wires: one movable wiper and two fixed ends. Technician B says a potentiometer can sense a variable voltage drop. Who is correct?
    A. A only
    B. B only
    C. Both A and B
    D. Neither A nor B

29. Technician A says that in a series circuit the voltage of each circuit load device is the same as the total circuit voltage. Technician B says the amperage of each series circuit branch is the same throughout the circuit. Who is correct?
    A. A only
    B. B only
    C. Both A and B
    D. Neither A nor B

30. Technician A says that adding resistive branches to a series circuit increases the total circuit amperage. Technician B says that, if a series circuit is opened, the entire circuit will not operate. Who is correct?
    A. A only
    B. B only
    C. Both A and B
    D. Neither A nor B

31. Technician A says the battery stores the voltage pressure until the electrical system needs it. Technician B says the battery stores the energy in a chemical form. Who is correct?

    A. A only

    B. B only

    C. Both A and B

    D. Neither A nor B

32. Technician A says a CCA rating on a battery determines its ability to provide an amperage load for 30 seconds at 0°F with the terminal voltage dropping no lower than 7.2 volts. Technician B says the CCA rating can be divided be 5.25 to get the approximate A/H rating. Who is correct?

    A. A only

    B. B only

    C. Both A and B

    D. Neither A nor B

33. Technician A says battery temperature above 125°F helps the battery charge faster. Technician B says the battery negative cable should be disconnected before charging, to isolate the vehicle's electrical system from voltage conditions above 15.5 volts. Who is correct?

    A. A only

    B. B only

    C. Both A and B

    D. Neither A nor B

34. Technician A says a wye wound stator winding consists of three output connections and a common neutral junction. Technician B says a wye wound stator winding produces a high-amperage output. Who is correct?

    A. A only

    B. B only

    C. Both A and B

    D. Neither A nor B

35. Technician A says the rule with the wire size AWG standard is that the higher the number is, the larger the wire conductor should be. Technician B says that in the metric system of determining wire size, the smaller the number, the smaller the wire conductor. Who is correct?

    A. A only

    B. B only

    C. Both A and B

    D. Neither A nor B

36. Technician A says a maxi-fuse is a fuse designed as a replacement for a fusible link in a modern vehicle. Technician B says a Pacific fuse element is a fuse designed as a replacement for a fusible link in a modern vehicle. Who is correct?

    A. A only

    B. B only

    C. Both A and B

    D. Neither A nor B

37. Technician A says that analog meters use a pointer that moves across the face of a printed scale. Technician B says that analog meters are not polarity sensitive to the circuit being measured. Who is correct?

    A. A only

    B. B only

    C. Both A and B

    D. Neither A nor B

38. Technician A says the binary code assigned to a low digital signal is 0. Technician B says the binary code assigned to a high digital signal is 2. Who is correct?

    A. A only

    B. B only

    C. Both A and B

    D. Neither A nor B

39. A total of 150 milliamps is flowing through a circuit that consists of three resistors in series: 30 ohms, 40 ohms, and 10 ohms. Technician A says the source voltage of the circuit is 120 volts. Technician B says the voltage is 1.2 volts. Who is correct?

    A. A only

    B. B only

    C. Both A and B

    D. Neither A nor B

40. A vehicle blows a taillight fuse when a trailer is hooked up to the circuit and energized. Technician A says there could be a short in the trailer taillight circuit. Technician B says there could be too much amperage for the vehicle taillight fuse when the trailer is added to the circuit. Who is correct?

    A. A only

    B. B only

    C. Both A and B

    D. Neither A nor B

41. Voltage total of a simple series/parallel circuit is being discussed. There are two series load devices with voltage drops of 6 volts each, and four load devices in parallel with a voltage reading across each device of 12 volts. What is the voltage total of this circuit?

    A. 12 volts

    B. 24 volts

    C. 60 volts

    D. None of the above

42. Technician A says a fused jumper wire can be used to bypass a switch to determine whether a circuit is working. Technician B says a self-powered test light can be a useful tool for checking for circuit continuity. Who is correct?

    A. A only

    B. B only

    C. Both A and B

    D. Neither A nor B

43. Technician A says that backfeed in an electrical circuit occurs when the current in one circuit seeks an alternate path of flow because the original path is interrupted. Technician B says backfeed in an electrical circuit can *only* occur when the insulation breaks down between the wires of two separate circuits. Who is correct?

    A. A only

    B. B only

    C. Both A and B

    D. Neither A nor B

44. Two technicians are discussing the results of a 3-minute charge test. Technician A says a reading of more than 15.5 volts on a conventional battery before 3 minutes time indicates a sulfated battery. Technician B says a load of 40 amperes should be applied to the battery to get an accurate reading. Who is correct?

    A. A only

    B. B only

    C. Both A and B

    D. Neither A nor B

45. Technician A says a negative battery cable is always removed first when removing a battery. Technician B says a negative battery cable should always be connected first when replacing a battery. Who is correct?

    A. A only

    B. B only

    C. Both A and B

    D. Neither A nor B

46. Technician A says the starter solenoid moves a plunger to engage the overrunning clutch. Technician B says the two windings of the starter solenoid are called the pull-in and hold-in windings. Who is correct?

    A. A only

    B. B only

    C. Both A and B

    D. Neither A nor B

47. Technician A says the output winding is rotating in a DC generator. Technician B says the output winding is stationary in a AC generator. Who is correct?

    A. A only

    B. B only

    C. Both A and B

    D. Neither A nor B

48. Technician A says the positive-biased diodes in a rectifier bridge are attached directly to the alternator rear housing. Technician B says the negative-biased diodes in a rectifier bridge are mounted in a heat sink. Who is correct?

    A. A only

    B. B only

    C. Both A and B

    D. Neither A nor B

49. Technician A says that engine detonation occurs when the fuel/air mixture ignites before the ignition system spark. Technician B says preignition can occur from hot carbon deposits in the combustion chamber. Who is correct?

    A. A only

    B. B only

    C. Both A and B

    D. Neither A nor B

50. An ohmmeter is being used to test the windings in an ignition coil. Technician A says the secondary windings in an ignition coil are tested by connecting the ohmmeter leads between the positive and negative terminals of the coil. Technician B says the ignition coil primary resistance should be 2.0 watts or less. Who is correct?

    A. A only

    B. B only

    C. Both A and B

    D. Neither A nor B

Appendix B

TYPICAL AUTOMOTIVE LIGHT BULBS

Trade Number	Design Volts	Design Amperes	Watts: $P = I \times E$
192	13.0	0.33	4.3
194	14.0	0.27	3.8
194E-1	14.0	0.27	3.8
194NA	14.0	0.27	3.8
209	6.5	1.78	11.6
211-2	12.8	0.97	12.4
212-2	13.5	0.74	10.0
214-2	13.5	0.52	7.0
561	12.8	0.97	12.4
562	13.5	0.74	10.0
563	13.5	0.52	7.0
631	14.0	0.63	8.8
880	12.8	2.10	27.0
881	12.8	2.10	27.0
906	13.0	0.69	9.0
912	12.8	1.00	12.8
1003	12.8	0.94	12.0
1004	12.8	0.94	12.0
1034	12.8	1.80/0.59	23.0/7.6
1073	12.8	1.80	23.0
1076	12.8	1.80	23.0
1129	6.4	2.63	16.8
1133	6.2	3.91	24.2
1141	12.8	1.44	18.4
1142	12.8	1.44	18.4
1154	6.4	2.63/.75	16.8/4.5
1156	12.8	2.10	26.9
1157	12.8	2.10/0.59	26.9/7.6
1157A	12.8	2.10/0.59	26.9/7.6
1157NA	12.8	2.10/0.59	26.9/7.6
1176	12.8	1.34/0.59	17.2/7.6
1195	12.5	3.00	37.5
1196	12.5	3.00	37.5
1445	14.4	0.135	1.9
1816	13.0	0.33	4.3
1889	14.0	0.27	3.8
1891	14.0	0.24	3.4
1892	14.4	0.12	1.7
1893	14.0	0.33	4.6

TYPICAL AUTOMOTIVE LIGHT BULBS (*continued*)

Trade Number	Design Volts	Design Amperes	Watts: $P = I \times E$
1895	14.0	0.27	3.8
2057	12.8	2.10/0.48	26.9/6.1
2057NA	12.8	2.10/0.48	26.9/6.1
P25-1	13.5	1.86	25.1
P25-2	13.5	1.86	25.1
R19/5	13.5	0.37	
R19/10	13.5	0.74	
W10/3	13.5	0.25	
37	14.0	0.09	1.3
37E	14.0	0.09	1.3
51	7.0	0.22	1.7
53	14.0	0.12	1.7
55	7.0	0.41	2.9
57	14.0	0.24	3.4
57X	14.0	0.24	3.4
63	7.0	0.63	4.4
67	13.5	0.59	8.0
68	13.5	0.59	8.0
70	14.0	0.15	2.1
73	14.0	0.08	1.1
74	14.0	0.10	1.4
81	6.5	1.02	6.6
88	13.0	0.58	7.5
89	13.0	0.58	7.5
90	13.0	0.58	7.5
93	12.8	1.04	13.3
94	12.8	1.04	13.3
158	14.0	0.24	3.4
161	14.0	0.19	2.7
168	14.0	0.35	4.9

Appendix C

This appendix contains the most common terms and acronyms from the SAE Directive J1930

Acronyms	Name	Acronyms	Name
ABS	Antilock Brake System	EBTCM	Electronic Brake Traction Control Module
A/C	Air Conditioning	EC	Engine Control
ACC	Air Conditioning Clutch	ECM	Engine Control Module
AC	Air Cleaner	ECL	Engine Coolant Level
AdvHEV	Advanced Hybrid	ECT	Engine Coolant Temperature
ADSTWC	Adsorbing Three-Way Catalyst	EEPROM	Electronically Erasable Programmable Read Only Memory
AFC	Air Flow Control	EFE	Early Fuel Evaporation
AFS	Air-Fuel Ratio Sensor	EGR	Exhaust Gas Recirculation
AIR	Secondary Air Injection	EGRT	EGR Temperature
AMT	Automated Manual Transmission	EHOC	Electrically Heated Oxidation Catalyst
AP	Accelerator Pedal	EHTWC	Electrically Heated Three-Way Catalyst
A/T	Automatic Transmission or Transaxle	EM	Engine Modification
B+	Battery Positive Voltage	EPROM	Erasable Programmable Read Only Memory
BARO	Barometric Pressure	EPS	Electric Power Steering
CAC	Charge Air Cooer	ESC	Electronic Spark Control
CAN	Controller Area Network	EST	Electronic Spark Timing
CANP	Canister Purge	EVAP	Evaporative Emission System
CFI	Continuous Fuel Injection	EWP	Electric Water Pump
CL	Closed Loop	FC	Fan Control
CKP	Crankshaft Position Sensor	FEEPROM	Flash Electrically Erasable Programmable Read Only Memory
CKP REF	Crankshaft Reference	FF	Flexible Fuel
CMP	Camshaft Position Sensor	FFS	Flexible Fuel Sensor
CMP REF	Camshaft Reference	FLI	Fuel Level Indicator
CNG	Compressed Natural Gas	FP	Fuel Pump
CO	Carbon Monoxide	FPROM	Flash Erasable Programmable Read Only Memory
CO_2	Carbon Dioxide	FT	Fuel Trim
CPP	Clutch Pedal Position	FTP	Federal Test Procedure
CTOX	Continuous Trap Oxidizer	GCM	Governor Control Module
CTP	Closed Throttle Position	GDI-L	Lean-Burn Gasoline Direct Injection
CVT	Continuously Variable Transmission	GDI-S	Stoichiometric Gasoline Direct Injection
DC	Direct Current	GEN	Generator (Alternator)
DEPS	Digital Engine Position Sensor	GND	Ground
DI	Distributor Ignition System	H_2O	Water
DFI	Direct Fuel Injection	HEV	Hybrid Electric Vehicle
DLC	Data Link Connector	HO_2S	Heated Oxygen Sensor
DTC	Diagnostic Trouble Code	HO_2S1	Upstream Heated Oxygen Sensor
DTM	Diagnostic Test Mode		
DVM	Digital Volt Meter		
DVOM	Digital Volt Ohmmeter		
EBCM	Electronic Brake Control Module		

Acronyms	Name	Acronyms	Name
HO_2S2	Up or Downstream Heated Oxygen Sensor	PNP	Park/Neutral Position Switch
HO_2S3	Downstream Heated Oxygen Sensor	PROM	Program Read Only Memory
HC	Hydrocarbon	PSA	Pressure Switch Assembly
HVS	High Voltage Switch	PSP	Power Steering Pressure
HVAC	Heating Ventilation and Air-Conditioning System	PTOX	Periodic Trap Oxidizer
		PVS	Ported Vacuum Switch
IA	Intake Air	PWM	Pulse-Width Modulated
IAC	Idle Air Control	RAM	Random Access Memory
IAT	Intake Air Temperature	RM	Relay Module
IC	Ignition Control Circuit	ROM	Read Only Memory
ICM	Ignition Control Module	RPM	Revolutions per Minute
IFI	Indirect Fuel Injection	SC	Supercharger
IFS	Inertia Fuel Shutoff	SCB	Supercharger Bypass
IGN	Ignition	SCI	Serial Communications Interface
I/M	Inspection Maintenance	SD	Speed Density
IPC	Instrument Panel Cluster	SDM	Sensing Diagnostic Mode
ISC	Idle Speed Control	SFI	Sequential Fuel Injection
KOEC	Key-On, Engine Cranking	SPL	Smoke Puff Limiter
KOEO	Key-On, Engine Off	SRI	Service Reminder Indicator
KOER	Key On, Engine Running	SRT	System Readiness Test
KS	Knock Sensor	STFT	Short-Term Fuel Trim
KSM	Knock Sensor Module	TAC	Thermostatic Air Cleaner
LPG	Liquified Petroleum Gas	TACH	Tachometer
LTFT	Long-Term Fuel Trim	TB	Throttle Body
MAF	Mass Airflow Sensor	TBI	Throttle Body Injection
MAP	Manifold Absolute Pressure Sensor	TC	Turbocharger
MC	Mixture Control	TCC	Torque Converter Clutch
MDP	Manifold Differential Pressure	TCM	Transmission or Transaxle Control Module
MFI	Multiport Fuel Injection	TFP	Throttle Fluid Pressure
MIL	Malfunction Indicator Lamp	TP	Throttle Position
MPH	Miles per Hour	TPS	Throttle Position Sensor
MST	Manifold Surface Temperature	TR	Transmission Range
MVZ	Manifold Vacuum Zone	TVS	Thermal Vacuum Switch
N_2	Nitrogen	TVV	Thermal Vacuum Valve
N_2O	Nitrous Oxide	TWC	Three-Way Catalyst
N/C	Normally Closed	TWC+OC	Three-Way Catalyst + Oxidation Catalyst
N/O	Normally Open	ULEV	Ultra Low Emission Vehicle
NO_x	Oxides of Nitrogen	V	Volts
NVRAM	Nonvolatile Random Access Memory	VAC	Vacuum
O_2	Oxygen	VAF	Volume Airflow
O_2S	Oxygen Sensor	VCM	Vehicle Control Module
OBD	On-Board Diagnostics	VDC	Variable Displacement Compressor
OBD-I	On-Boards Diagnostics, Generation One	VIN	Vehicle Identification Number
OBD-II	On-Board Diagnostics, Generation Two	VPW	Variable Pulse Width
OC	Oxidation Catalyst	VR	Voltage Regulator
ODM	Output Device Monitor	VS	Vehicle Sensor
OL	Open Loop	VSS	Vehicle Speed Sensor
Ω	Ohms	WU-OC	Warm-Up Catalyst with Oxidation Catalyst
OSC	Oxygen Sensor Storage	WU-TWC	Warm-Up Three-Way Catalytic Converter
PAIR	Pulsed Secondary Air Injection	WOT	Wide-Open Throttle
PCM	Powertrain Control Module	ZEV	Zero Emission Vehicle
PCV	Positive Crankcase Ventilation		

Appendix D

Metric Conversions

	To convert these	To these,	Multiply by:
Temperature	Centigrade degrees	Fahrenheit degrees	1.8 then +32
	Fahrenheit degrees	Centigrade degrees	0.556 after −32
Length	millimeters	inches	0.03937
	inches	millimeters	25.4
	meters	feet	3.28084
	feet	meters	0.3048
	kilometers	miles	0.62137
	miles	kilometers	1.60935
Area	square centimeters	square inches	0.155
	square inches	square centimeters	6.45159
Volume	cubic centimeters	cubic inches	0.06103
	cubic inches	cubic centimeters	16.38703
	cubic centimeters	liters	0.001
	liters	cubic centimeters	1,000
	liters	cubic inches	61.025
	cubic inches	liters	0.01639
	liters	quarts	1.05672
	quarts	liters	0.94633
	liters	pints	2.11344
	pints	liters	0.47317
	liters	ounces	33.81497
	ounces	liters	0.02957
Weight	grams	ounces	0.03527
	ounces	grams	28.34953
	kilograms	pounds	2.20462
	pounds	kilograms	0.45359
Work	centimeter kilograms	inch pounds	0.8676
	inch pounds	centimeter kilograms	1.15262
	meter kilograms	foot pounds	7.23301
	foot pounds	newton meters	1.3558
Pressure	kilograms/sq. cm	pounds/sq. inch	14.22334
	pounds/sq. inch	kilograms/sq. cm	0.07031
	bar	pounds/sq. inch	14.504
	pounds/sq. inch	bar	0.06895

Appendix E

DECIMAL AND METRIC EQUIVALENTS

Fractions	Decimal (in.)	Metric (mm)	Fractions	Decimal (in.)	Metric (mm)
1/64	0.015625	0.397	33/64	0.515625	13.097
1/32	0.03125	0.794	17/32	0.53125	13.494
3/64	0.046875	1.191	35/64	0.546875	13.891
1/16	0.0625	1.588	9/16	0.5625	14.288
5/64	0.078125	1.984	37/64	0.578125	14.684
3/32	0.09375	2.381	19/32	0.59375	15.081
7/64	0.109375	2.778	39/64	0.609375	15.478
1/8	0.125	3.175	5/8	0.625	15.875
9/64	0.140625	3.572	41/64	0.640625	16.272
5/32	0.15625	3.969	21/32	0.65625	16.669
11/64	0.171875	4.366	43/64	0.671875	17.066
3/16	0.1875	4.763	11/16	0.6875	17.463
13/64	0.203125	5.159	45/64	0.703125	17.859
7/32	0.21875	5.556	23/32	0.71875	18.256
15/64	0.234275	5.953	47/64	0.734375	18.653
1/4	0.250	6.35	3/4	0.750	19.05
17/64	0.265625	6.747	49/64	0.765625	19.447
9/32	0.28125	7.144	25/32	0.78125	19.844
19/64	0.296875	7.54	51/64	0.796875	20.241
5/16	0.3125	7.938	13/16	0.8125	20.638
21/64	0.328125	8.334	53/64	0.828125	21.034
11/32	0.34375	8.731	27/32	0.84375	21.431
23/64	0.359375	9.128	55/64	0.859375	21.828
3/8	0.375	9.525	7/8	0.875	22.225
25/64	0.390625	9.922	57/64	0.890625	22.622
13/32	0.40625	10.319	29/32	0.90625	23.019
27/64	0.421875	10.716	59/64	0.921875	23.416
7/16	0.4375	11.113	15/16	0.9375	23.813
29/64	0.453125	11.509	61/64	0.953125	24.209
15/32	0.46875	11.906	31/32	0.96875	24.606
31/64	0.484375	12.303	63/64	0.984375	25.003
1/2	0.500	12.7	1	1.00	25.4

Glossary

A-circuit: A generator regulator curcuit that uses an external grounded field circuit. In the A-circuit, the regulator is on the ground side of the field coil.

alternating current (AC): Electrical current that changes direction from positive to negative.

American wire gauge (AWG): Size standard assigned to wire diameter. The higher the number, the smaller the wire conductor.

ampere: Measurement of the movement of free electrons called current flow. One ampere is equal to 6.28 billion billion free electrons flowing past a point in 1 second.

Ampère, André Marie: French scientist and mathematician, 1775–1836. Established the importance of the relationship between electricity and magnetism.

amp-hour rating (A/H): Battery rating that is the amount of current a fully charged battery can produce over a 20-hour period at 80°F before the terminal voltage reaches a minimum voltage of 10.5 volts.

amplitude: Voltage measurement represented on an oscilloscope screen by the vertical *y* axis.

analog meter: An electrical meter operated by a d'Arsonval movement using a pointer needle that moves across the face of a printed scale.

analog-to-digital (A/D) converter: A device to convert analog signals to digital signals for use by the computer.

analog voltage signal: A signal produced by an input sensor that is continuously variable within a certain voltage range.

anode: The positively charged electrode in a voltage cell.

armature: The rotating component of an electric motor that consists of a conductor wound around a laminated iron core to create a magnetic field.

atom: Smallest part of an element, consisting of protons, neutrons, and electrons.

autofuse: A flat blade type fuse used for circuit protection in modern vehicles.

B-circuit: A generator regulator circuit that is internally grounded. In the B-circuit, the voltage regulator controls the power side of the field circuit.

backfeed: When the current in one circuit seeks an alternate path of flow when the original path is interrupted.

ballast resistor: A resistive element enclosed in a ceramic block. Used to reduce the voltage to the ignition coil.

base: The center layer of a bipolar transistor that activates the switching function of the transistor.

Bendix inertia drive: A starter drive pinion that depends on the inertia movement of the starter armature spinning to mesh the drive pinion forward into the ring gear teeth.

bimetallic strip: Two dissimilar metals, joined together, that, when heated, expand at different rates and cause the strip to bend momentarily until cooled off again.

binary code: A series of numbers represented by 1s and 0s. Any number or word can be translated into a combination of binary 1s and 0s.

bit: A computer term for a binary digit.

bound electron: Inner orbit(s) of electrons rotating around the nucleus of the atom.

byte: A computer term for eight bits of information.

cathode: The negatively charged electrode of a voltage cell.

capacitor: A device used to store electrical charges from a circuit. Also called a condenser.

center high-mounted stop light (CHMSL): A brake light mounted in the high center rear of the vehicle. Required on all vehicles sold in the United States since 1986.

chemical energy: Electricity produced from a battery when two dissimilar metals and an electrolyte are subjected to a completed circuit.

circuit breaker: An electrical circuit load control device. May be designed to be self-resetting or manually reset.

circuit protection device: An in-line fuse or circuit breaker that protects an electrical circuit from excess amperage.

clamping diode: A diode that is connected in parallel with a coil to prevent voltage spikes from the coil from reaching other components in the circuit.

cold cranking ampere rating (CCA): The ability of the battery to provide an amperage load for 30 seconds at 0°F without the terminal voltage dropping below 7.2 volts.

collector: The portion of the bipolar transistor that receives the majority of the current carriers.

commutator: A series of conducting segments located around one end of the armature.

condenser: *See* capacitor.

continuity light: *See* self-powered test light.

conventional battery: A lead-acid battery that uses a lead antimony mixture in the cell plates.

conventional flow theory: An electrical theory stating that current flows from a positive potential to a negative potential.

current flow: The flow of free electrons from positive to negative. Measured in amperes.

daylight running lights (DRL): Headlights that operate automatically during daylight hours at a reduced voltage.

deep cycle: A lead-acid battery that uses approximately 2.75% antimony on the positive plate and a calcium alloy on the negative plate.

delta wound: A stator winding in an AC generator that consists of three connections connected in a triangular shape with three output leads.

depletion zone: *See* PN junction.

depressed park: A wiper circuit that parks the wiper blades off the glass past the end of their normal stroke, usually below the edge of the lower windshield molding.

detonation: A combustion chamber condition that occurs when the normal combustion process is too slow, causing some unburned fuel/air mixture to ignite and create a secondary flame front that collides with the normal flame front.

dielectric: An insulator material.

digital multimeter (DMM): Uses a digital readout display that may also automatically record a measured value. Also known as digital volt ohmmeter (DVOM).

digital storage oscilloscope (DSO): Converts the input voltage signal into digital information for screen display or transfer to other diagnostic equipment. Also known as a scope meter.

digital voltage signal: A signal that produces a square wave from the rapid on/off cycling of the circuit. Also known as a square wave signal.

diode: An electrical one-way check valve that allows current to flow in one direction only.

diode trio: Rectifies current from the stator to create the magnetic field excitation in the rotor field coil.

direct current (DC): Steady-state electrical potential that remains at a specific voltage in one direction.

double-filament bulb: An incandescent bulb with two coiled tungsten filaments that serve more than one electrical function.

duty cycle: The relationship between the amount of on time and off time in a digital signal.

dwell: The amount of time that the primary ignition system points are closed. Measured in degrees.

electrolysis: When the battery electrolyte is converted into hydrogen and oxygen gas. Also called gasing.

electronic erasable programmable read only memory (EEPROM): Computer memory chip that can be programmed only with special test equipment.

electromagnet: A magnet formed when current flows through a coil of wire wrapped around a soft iron core.

electromagnetism: Magnetic field that occurs when a current flows through a conductor.

electromotive force (EMF): *See* voltage

electron: Negatively charged particle of an atom.

electron flow theory: An electrical theory stating that current flows from a negative potential to a positive potential.

electrolyte: A chemical compound of alkali, salt, or acid solution that is capable of conducting electrical energy.

electrostatic discharge (ESD): The process by which charged objects transfer their charge to any neighboring object.

electrostatic field: The field between two oppositely charged plates.

element: A cluster of one type of atom.

emitter: The outer layer of the transistor that supplies the majority of the current carriers.

farad: The standard unit of measurement for a capacitor.

field coils: Heavy copper wire wrapped around an iron core called a pole shoe.

field current: The current going to the field windings of a motor or generator.

fixed resistor: A resistive element normally made of carbon/graphite-based material mixed with a resin bonding agent.

flux density: The concentration of the magnetic lines of force.

forward bias: A positive voltage that is applied to the P-type material or anode of a diode.

free electrons: Valence electrons that become dislodged from their orbit and move to a neighboring atom.

frequency: The number of cycles in a period of time.

full-wave rectification: The conversion of a complete AC voltage signal to a DC voltage.

fusible link: An in-line wire in the circuit that is made of meltable material with a special heat resistant insulation. The link melts when the designed circuit current flow is exceeded.

grid growth: A condition that shorts out the plates in the battery when little metallic fingers grow through the separators to the plates.

grounded circuit: An unwanted path for current to flow directly to ground. Also referred to as a short-to-ground.

half-wave rectification: Rectification of one-half of an AC voltage.

Hall effect switch: A sensor that operates on the principle that if a current is allowed to flow through thin conducting material being exposed to a magnetic field, another voltage is produced.

halogen headlamp: A type of headlamp that uses halogen gas in an inner bulb to allow the tungsten filament to burn brighter.

hard shell connector: An electrical connector that has one to a dozen wires molded separately into the connector.

hydrometer: A testing instrument that measures the specific gravity content of the electrolyte to determine a battery's state-of-charge condition.

incandescent lamp: A device that produces a light as a result of current flowing through a coiled tungsten wire filament inside a glass envelope.

inductive pickup: An adapter available for a meter to measure the strength the magnetic field, generated by the current flow in a circuit.

instrument voltage regulator (IVR): Maintains a constant voltage to the gauge under all battery load and charging conditions.

integrated circuit (IC) chip: A complex circuit consisting of thousands of transistors, diodes, resistors, capacitors, and other electronic devices that are formed on a small silicon chip.

interval wiper: A wiper system that operates in time intervals depending on the delay setting requested by the driver.

joule: Base unit of energy measurement.

keep alive memory (KAM): A volatile RAM memory that retains information as long as it is connected to a battery power source.

Kirchoff's current law: A circuit law for current in a parallel circuit, where the sum of the current in the individual branch circuits equals the total circuit amperage.

Kirchoff's voltage law: A circuit law for voltage in a series circuit, where the sum of the individual voltage drops of each resistive load equals the source voltage.

knock sensor: A piezoelectric device that senses combustion chamber detonation to control spark knock.

lead antimony: A battery grid plate that contains up to a 10% mixture of lead and lead alloys to strengthen the plate grid frame.

lead peroxide: A reddish brown paste material pressed into the battery positive plate.

light-emitting diode (LED): A gallium arsenide diode that converts the energy developed when holes and electrons collide during normal diode operation into light.

logic gates: Electronic circuits that act as gates to output voltage signals depending on different combinations of input signals.

magnetism: A natural invisible force resulting from atoms aligning in certain materials that can then repel or attract other materials.

maintenance-free battery: A lead-acid battery that uses a lead calcium mixture in the cell plates.

matter: Any substance that occupies a space and contains a mass.

maxifuse: A large flat blade-type fuse used as a replacement design for fusible links in electrical circuits.

mercury switch: A vial that contains mercury that conducts electricity when electrical contacts come into contact with the mercury.

metri-pack connector: Similar to a weather-pack connector, except they do not have a seal on the cover half of the connector.

microprocessor: The decision-making chip in the computer that interprets input information, makes calculations, and makes decisions.

minifuse: A small blade-type fuse designed to save space and allow more electrical circuits to have individual fuses.

molecule: A group of elements.

MPMT switch: An electrical switch that contains multiple wipers that work in unison together. Also known as a ganged switch.

multifunction switch: A combination switch on the steering column consisting of a turn signal switch, headlight switch, hazard switch, horn switch, and flash-to-pass switch.

NATEF: The abbreviation for the National Automotive Technician Education Foundation. Dedicated to developing national standards for training automotive technicians.

neutral safety switch: A safety switch that prevents a starter from cranking unless the automatic transmission is in park or neutral or the clutch is pushed in for a manual transmission.

neutrons: Particles of the nucleus of the atom that carry no charge.

nondepressed park: A wiper circuit that parks the wiper blades at the end of their normal stroke, at the edge of the lower windshield molding.

nonpowered test light: Simple test light tool that connects one end to ground and the other end to a voltage source to light the bulb.

normally closed (N/C) switch: A switch denoting that the contacts are normally closed until opened by an outside force.

normally open (N/O) switch: A switch denoting that the contacts are normally open until closed by an outside force.

NPN transistor: A transistor that contains two N-type materials and one P-type material.

nucleus: The core of the atom that contains positively charged protons, and neutrons with no charge.

Ohm, George Simon: German physicist, 1787–1854. Identified the relationship between resistance, current, and voltage in an electrical circuit known as Ohm's law.

Ohm's law: The current in an electrical circuit is inversely proportional to the resistance of the circuit and directly proportional to the voltage.

open circuit: A break in the current flow of a complete circuit.

open circuit voltage: the voltage of a battery in a no-load condition. Normally used on a maintenance-free battery to check the state-of-charge condition.

overrunning clutch drive: A starter drive pinion that wedges spring-loaded rollers between inner and outer housings to engage the pinion teeth into the flywheel teeth.

Pacific fuse element: A quick replacement fuse developed to replace the fusible link.

parallel circuit: A circuit that provides two or more paths for current to flow.

parasitic drain: Unwanted current drain still present in the electrical system after the ignition switch is in the off position.

photons: Pure energy that contains no mass. When light energy strikes the surface of a photovoltaic cell, the energy in the photon is given to a free electron.

piezoelectricity: Electricity produced from pressure being applied to quartz or barium titanate crystals.

PN junction: The area where the anode and cathode material join together in a diode. Also called the depletion zone.

PNP transistor: A transistor that contains two P-type materials and one N-type material.

pole shoe: The soft iron core inside the field coil that helps concentrate and direct the magnetic lines of force in the field coil.

potentiometer: A variable resistor, with three wire connections, that acts as a circuit divider to provide accurate voltage drop readings proportional to movement.

power of ten: Base unit of measurement for the metric system.

preignition: a combustion chamber condition that occurs by an early ignition of the fuel/air mixture, and by a source other than the spark plug and ignition system.

primary wire: Term used for a wiring circuit that carries a low voltage.

primary cell: A cell that cannot be recharged to its original voltage condition.

programmable read only memory (PROM): Memory chip that contains specific data that pertains to the exact vehicle in which a computer is installed.

proton: Positive charged particle of an atom. Protons and neutrons form together to form the nucleus.

pull-to-seat connector: An electrical terminal that is pulled into the connector housing to seat it.

pulse width: The amount of time the voltage is on or off in a digital circuit.

pulse width modulation: The on and off cycling of a component. The pulse width varies, but its frequency remains constant.

push-to-seat connector: An electrical terminal that is pushed into the connector housing to seat it.

random access memory (RAM): Stores temporary information that can be read from or written to by the CPU. RAM can be designated as volatile or nonvolatile.

read only memory: Computer memory chip that stores permanent information used to instruct the computer on what to do in response to input data.

recombination battery: A completely sealed lead-acid battery that uses electrolyte in a gel form.

rectification: The converting of an AC current to a DC current.

relay: A switch that uses low current in a circuit coil to magnetically control the high current in another circuit through a set of mechanical contact points.

reserve capacity (R/C): A battery rating that is the amount of time in minutes a fully charged battery can be discharged at 25 amperes while maintaining a minimum battery voltage of 10.5 volts at 80°F in the event of a charging system failure.

resin core: *See* rosin core.

resistance: Any force or substance that restricts or opposes the flow of current in a circuit.

reverse bias: A positive voltage is applied to the N-type material, or cathode, of a diode.

rheostat: A variable resistor with two terminals, one connected to a fixed end of the resistor, the other end connected to a movable wiper contact.

rosin core: A type of flux cleaning agent that is used with solder to repair electrical connections. Also known as resin core.

rotary switch: An electrical switch that allows for multiple functions to be made from the same control knob. Normally used on an analog meter.

rotor: The component of an AC generator that rotates and creates the magnetic field.

scan tool: A diagnostic tool used to communicate with the vehicle's computer and access stored diagnostic trouble codes.

Schmitt trigger: An electronic circuit used to convert an analog input signal to a clean square wave digital signal.

sealed-beam headlamp: A light assembly consisting of a glass lens fused to a parabolic reflector that is sprayed with vaporized aluminum. The inside of the lamp is filled with argon gas to prevent the filament from oxidizing.

secondary wire: A wiring circuit that carries a high voltage, such as spark plug wires.

secondary cell: A cell that can be recharged once its original stored energy has been depleted.

Seebeak effect: The heating of two dissimilar metals joined at one end produces a voltage at the open ends. Also known as thermoelectricity.

self-powered test light: A simple test light that contains an internal battery that lights the lamp bulb when there is a complete circuit. Also known as a continuity light.

semiconductor: A stable element with exactly four electrons in its structure.

separator: A microporous material used to prevent contact between positive and negative plates.

series circuit: An electrical circuit that provides a single path for current flow from the electrical source through all the circuit's components and back to the source.

series/parallel circuit: An electrical circuit that has some loads in series and some in parallel.

short circuit: An unwanted or accidental bypass of the current flow in a circuit.

short-to-ground: *See* grounded circuit.

short-to-voltage: An unwanted copper-to-copper connection between two separate circuits.

sine wave: A scope pattern that shows the positive and negative voltages generated during one revolution.

single-filament bulb: An incandescent bulb with one coiled tungsten filament.

slip rings: The copper rings at one end of the rotor in an AC generator that provide a current path connection from the carbon brushes to the rotor coil.

solenoid: An electromagnetic device that uses movement of a plunger to exert a pulling or holding force.

specific gravity: The weight of a given volume of liquid in comparison to the weight of the same volume of water with a specific gravity of 1.000.

solar light energy: Electricity generated by using light energy from the sun and capturing it in a photovoltaic solar cell.

solderless connector: Electrical connector that is crimped to the wire with special pliers.

SPDT switch: Single-pole double-throw switch that has one input wire and two output wires.

SPST switch: Single-pole single-throw switch that has one input wire and one output wire. Switch may be on/off controlled, or momentary.

square wave: *See* digital voltage signal.

static electricity: Electricity that is without motion or at rest and that produces an electrical charge whenever two dissimilar nonmetallic materials are rubbed lightly together.

stator: The stationary coil windings in an AC generator in which electricity is produced.

stepped resistor: Typical resistor with two or more fixed resistive values. Normally used to control the current in a heater blower motor circuit.

sulfuric acid: A conductive and reactive mixture of H_2SO_4 consisting of 64% water and 36% sulfuric acid, by weight.

tap splice: A type of solderless connector that joins a new wire to an existing circuit.

thermistor: A variable resistor made from a semiconductor material that changes resistance in relation to temperature changes.

thermocouple: A small device that gives off a voltage when two dissimilar metals are heated. Normally used to monitor exhaust temperatures.

thermoelectricity: Electricity produced when two dissimilar metals are heated to generate an electrical voltage.

three-phase: The resulting overlap of three-phase stator windings produces a three-phase voltage output.

transistor: A three-layer semiconductor used as a very fast switching device.

unintentional open circuit: An unwanted break in the current flow in a electrical circuit.

valence electron: Electrons in the outermost orbit of the atom.

variable resistor: A resistor that provides for an infinite number of resistance values within a set range.

Volta, Alessandro: Italian physicist, 1745–1827. Invented the electrophorus, a device for storing an electrical charge.

voltage: The electrical pressure that causes the free electrons to move through a circuit. Also referred to as EMF or electromotive force.

voltage drop: A resistance in an electrical circuit that reduces the electrical pressure or voltage available after the resistance. The voltage drop can be the result of resistance in the load component, the conductors, connection points, or unwanted resistance.

voltage limiter: Automatically determines the field circuit voltage depending on the required amount of charging.

voltage regulator: Used to control the output voltage of an AC generator based on charging system demand.

voltaic cell: Two unlike metals and an alkali, salt, or acid solution, combined to produce a voltage.

voltaic pile: A series of small round copper and zinc plates separated by cardboard soaked in saltwater.

Watt, James: Scottish inventor, 1736–1819. Developed a comparison to what his steam engines could produce in place of a normal-sized horse, now known as Watt's law.

Watt's law: The amount of work done in 1 second by 1 volt moving a charge of 1 ampere through a resistance of 1 ohm is equal to 1 watt (1 horsepower = 746 watts).

waveform: The picture produced on an oscilloscope screen. Also called a scope trace.

weather-pack connector: Electrical connectors that have a rubber seal on the terminal ends and on the covers of the connector halves. Used on electrical circuits for moisture and corrosion protection.

wye wound: Stator windings that consist of a common junction and three output connections that branch out in a Y pattern.

zener diode: A diode that allows reverse current to flow above a designed-in voltage limit.

Index

A

B

C

D

F

G

P

R

S

T

U

V

W

X

Z